Laser Scanning Applications in Landslide Assessment

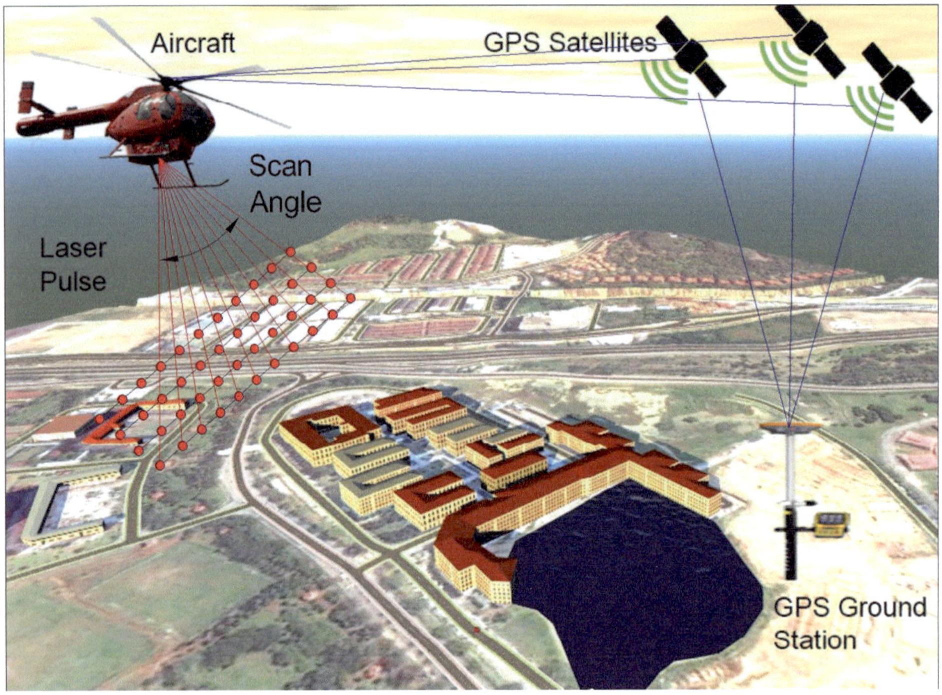

Components of a typical airborne LiDAR system

Biswajeet Pradhan
Editor

Laser Scanning Applications in Landslide Assessment

Springer

Editor
Biswajeet Pradhan
Department of Civil Engineering
University Putra Malaysia
Seri Kembangan
Malaysia

ISBN 978-3-319-55341-2 ISBN 978-3-319-55342-9 (eBook)
DOI 10.1007/978-3-319-55342-9

Library of Congress Control Number: 2017934623

Printed on acid-free paper

This Springer imprint is published by Springer Nature
The registered company is Springer International Publishing AG
The registered company address is: Gewerbestrasse 11, 6330 Cham, Switzerland

To the memory of my loving dad

Purna Chandra Pradhan

Preface

The significant growth of world's population and the rapid expansion of cities generate great challenges for decision makers and put an increasing number of people at hazards. Even though it is impossible for human being to prevent natural disasters, there have been great efforts to create knowledge, design methods, and frameworks to assess, prepare, and mitigate the potential effects of natural hazards. Among natural disasters, landslides pose considerable risks to people's livelihood and to the environment. They cause significant disruption and economic losses by the devastation of major infrastructures such as settlements, transportation, power and communication lines, and other utilities. Several landslide triggering factors such as intense rainfall, earthquakes, volcanic eruptions, hurricanes, and human activities threaten many parts of the world and increase the potential of landslides. Hence, it has been important to put significant efforts to advance landslide studies and design effective and practical tools that could be used by decision makers.

Landslide is defined as "the movement of a mass of rock, debris, or earth down a slope." They result from the failure of hill slope materials driven by the force of gravity. Landslides are also known as slope failure and they are classified into several types according to the type of mass movement. The basic types of landslide movements are fall, topple, slide, flow, and spread. Landslides occur almost worldwide and cause significant disasters with very great impacts to the society. They are studied in many countries, and scientific and engineering fields and a wide variety of innovations have been proposed to enhance our understanding of their mechanisms. In general, landslides occur in a variety of landscapes characterized by the cliff, steep slopes, and unstable geology. There are many other factors contribute to the landslide occurrence such as slope curvature, weathering, water content, sediment availability, climate, vegetation, and anthropogenic inputs. However, most of the time, landslides are triggered by one factor or combination of factors such as heavy rainfall, earthquakes, or glacial erosions. Thus, it is important for scientists to understand the links between these factors and the concept of landslide risk. This has allowed them to accurately predict the distribution of future landslides, estimate and simulate their extents, and quantify their impacts to the human life and property.

Assessment of landslides usually involves several modeling techniques using a wide range of data sources. Overall landslide assessments comprise detection of landslide scarps, prediction of the spatial distribution of potential future landslides, modeling hazards and vulnerability, and estimating landslides risks and their impacts. For the detection of landslide scarps, the key methods are interpretation of aerial photographs, change detection, topographic and geomorphological analysis of laser scanning data. In addition, a wide range of knowledge-based, statistical, and machine learning methods are used for predicting potential future landslides in a given area. For example, the popular methods are analytic hierarchy process, frequency ratio, logistic regression, and support vector machines. On the other hand, techniques for modeling landslide hazards are generally well documented. Their concepts are based on integrating the spatial and temporal variations of triggering factors with potential landslide zones. Furthermore, to model landslide risks, understanding of the elements at risk including exposure information is considered a critical factor. Exposure information is produced by obtaining the best available data, statistics, spatial and attribute data about buildings, demographics, community infrastructure, and agricultural commodities. This information allows to model landslide vulnerability through the use of curves that describe a probable damage severity or economic loss for a particular type of infrastructure when it is subjected to some level of hazard. Finally, modeling of landslide risk is based on statistical information about past events and their estimated impacts. Overall, risk models can be used to perform cost–benefit analysis for various forms of mitigation involving short-term solutions, such as early warning and response, along with long-term solution, such as land use planning and improvements to building codes and infrastructure.

The practical development of landslide risk models requires comprehensive data for each step of modeling. In the recent times, LiDAR (light detection and ranging) is widely used for landslide investigations to create accurate digital elevation models which enable extracting several precise topographic, geomorphological, and hydrological factors used in several steps of overall landslide assessments. The key advantages of using LiDAR for landslide studies are high-resolution landslide contours, which permits identifying landslide scarps and displaced materials and delineating geomorphological features of landslides such as scarps, mobilized material, and foot. Other advantages include automating landslide mapping, penetrating vegetation canopies, and supporting simulations of debris flows and rock falls at small scales due to their high-density points collected over the focused area.

Landslides occur worldwide; however, rainfall-induced slides tend to be much greater in tropical hilly areas. Mountainous terrain and heavy tropical rains put dense populations and infrastructures at risk. Thus, monitoring different types of landslides can be useful for mitigating the effects of these disasters and properly plan for potential future events.

This book at the first describes the fundamental concept of using LiDAR for landslide applications and assessment. A general overview of laser scanning systems in the context of landslide studies is explained to support understating the followed materials of the book. As a preliminary step, landslide and debris flow inventory mapping and characterization is presented with diverse illustrations. This is followed by a detailed landslide susceptibility mapping procedures including optimization of landslide conditioning factors, effects of spatial resolution of DEM, and detailed comparative analysis of a large number of models used in the literature. Besides, identification of debris flow source areas and its assessment using empirical models will be discussed. In addition, landslide risk assessment using multihazard scenarios will be described. Furthermore, LiDAR techniques in rockfall hazard assessments is also investigated and discussed in details.

This book is organized into 17 chapters.

Chapter 1 briefly discusses about the active remote sensing systems, such as light detection and ranging (LiDAR) which are widely used in landslide disaster management and risk mitigation. The main advantage of these technologies is the production of high-resolution digital elevation models (DEMs). Such models allow detailed mapping of terrain and extraction of geomorphological features, which are extremely important in landslide assessment. Therefore, this chapter provides an overview of the use of LiDAR in landslide investigations. First, it introduces the main components of LiDAR systems and the basic concept of laser measurements and then discusses the accuracy and resolution of typical laser scanning systems. Second, it provides information about LiDAR data processing (i.e., point cloud filtering, geometric calibration). Third, it discusses the main products of LiDAR that are useful for landslide investigation and modeling. Finally, it describes and illustrates several landslide applications where LiDAR data are beneficial.

Chapter 2 proposes a semiautomatic supervised approach for the detection of landslides in man-made slopes. Several techniques have been proposed for landslide mapping using remote sensing data in the literature, especially in unstable slope areas. Generally, cut slopes are created to mitigate the risk of land failure for areas that have high probability of failures. This method creates new challenges for landslide mapping in these areas. Five classifiers were evaluated for object-based landslide detection using airborne LiDAR data coupled with orthophotos. This chapter aims to: (1) to prepare spectral, spatial, and texture features, as well as LiDAR-derived parameters for landslide detection using supervised classification schemes; (2) to evaluate five well-known classifiers (i.e., Bayes k-nearest neighbor, support vector machine (SVM), random forest (RF), and decision tree) for landslide detection, thereby determining the best algorithms; and (3) to produce an inventory map of landslides and man-made slopes of the study area using the best classifiers found in the second objective. Results of landslide factor and feature analyses showed that landslides and cut slopes are extremely difficult to separate using only LiDAR-derived parameters. Orthophotos are useful information for the separation of landslides from other features, such as grassland, buildings, and water bodies. Spatial and texture features are also important for landslide detection. A field validation was also applied using a landslide inventory map collected from multiple field investigations. This inventory map shows landslide locations, type, geometry, and direction. Landslide inventory was also used to train the classifier, thus improving sampling accuracy. The result of the analysis showed that SVM and RF achieved relatively high user and producer accuracies, and indicated a good classification of landsides and cut slopes simultaneously. In comparison, SVM performed better than RF for landslide and cut slope classification. Overall assessment indicated that the separation between cut slopes and landslides using LiDAR data and orthophotos in supervised classification is possible and can be improved. The resulting landslide inventories are valuable resources for both the geomorphological investigation of landslide events and hazard assessment and susceptibility analysis in landslide-prone regions.

Chapter 3 discusses about a new approach for detection of different types of landslides such as shallow and deep seated. A good landslide inventory map is a prerequisite for analyzing landslide susceptibility, hazard, and risk as well as for studying the evolution of a landscape affected by landslides. Using traditional methods for landslide detection is challenging because of the presence of dense vegetation in landslide locations and the time-consuming large-scale projects that are concomitant with these methods. Data derived from LiDAR can depict ground surface and provide valuable information on the topographic features of locations hidden under dense vegetation. This study presents an automatic LiDAR-based landslide detection method and discusses its capability to differentiate between shallow and deep-seated landslides as well as its transferability. An existing supervised approach was adopted to optimize segmentation parameters (i.e., scale, shape, and compactness). Subsequently, a correlation-based feature selection technique was used to select relevant attributes for developing the set of rules. The rules were developed using a decision tree algorithm. An object-based approach was applied to identify the locations and characteristics of landslides. To validate the method, the area under the curve was used. The accuracy of landslide detection on the test site was 0.82, and the accuracy of detecting shallow and deep-seated landslides were 0.80 and 0.83, respectively. The intensity derived from the LiDAR data and texture significantly affects the accuracy of differentiating shallow from deep-seated landslides. Therefore, the current study demonstrated that LiDAR data are highly efficient in detecting landslide characteristics in tropical forested areas.

Chapter 4 presents a Taguchi-based Random Forest technique for landslide detection from LiDAR and QuickBird satellite image. Landslide mapping in tropical regions is challenging because of the rapid vegetation growth. Hence, increasing the performance of landslide mapping with remote sensing skills is essential. This chapter proposes an efficient methodology to detect and map the landslide-prone areas located in Bukit Ma'okil, Johor, Malaysia, using an integration of high-resolution LiDAR with high-resolution QuickBird satellite imagery. An object-based classification method was used to distinguish the landslide-prone areas from non-landslide features. The Taguchi technique and Random Forest (RF) methods were employed to optimize the segmentation process and to select important features, respectively. The rule-based technique was also used for object-based classification. The Taguchi optimization applied in the current research allowed the selection of suboptimal segmentation parameters by conducting 25 experiments, each evaluated by kappa coefficient. The application of the RF method significantly contributed in selecting the most relevant features for ruleset development and classification. Landslide and non-landslide locations were detected, and the confusion matrix was used to examine the proficiency and reliability of the results. The overall accuracy was 90%. The current research integrated object-based analysis and optimization method as a pioneering landslide detection application to reduce time for image classification. The successful production of a reliable and accurate landslide inventory map confirmed the efficiency of the methodology. Therefore, the results derived from the proposed method can assist researchers and planners in implementing and expediting landslide inventory mapping.

In contrast to Chap. 4, Chap. 5 presents debris flow detection using LiDAR data in a tropical forested area. Debris flow is one of the most destructive mass-wasting events. Debris flow is also referred as mudslide, lahars, or debris avalanche, which is a rapid mass movement mainly triggered by intense precipitation or rapid snow melt that starts on steep mountain channels. The loosen materials are saturated with water-formed debris flows. Debris flow can be catastrophic because it is associated with the loss of human life and property destruction. Given the rapid population growth, especially in mountainous region, source areas prone to debris flow should be identified. In this study, LiDAR, a high-resolution airborne laser scanning data, was used to obtain debris flow-related parameters. First, a digital elevation model (DEM) was generated from the LiDAR point clouds as a primary source of data. The parameters were constructed in GIS environment, which contains slope, plan curvature and flow accumulation derived from a DEM. The datasets were converted to ASCII grids for importation in Flow-R (Flow path assessment of gravitational hazards at a Regional scale) software. Many softwares were developed to understand debris flow behavior. In this research, Flow-R model was used because it can produce significant results based on the quality of the DEM, thereby obtaining reliable results for identification of debris flow sources. Various DEM resolutions (1, 2, 5, and 10 m) were generated for identification of debris flow source areas and consequent determination of an optimized resolution. Landslide inventory map, which was prepared mostly from field investigation, was used for validation. The landslide inventory map was buffered to 20 and 50 m for each DEM resolution. The results from buffered zones were later used to generate the intersection between the buffered zones and the source area produced from Flow-R. Additionally, high-resolution ortho-images were used as supplementary data to visualize the location of debris flow source areas. The results revealed that DEM with 1-m resolution produced the highest accuracies among all DEM resolutions. According to the sources and landslide inventory data, buffering and intersection were 72% and 93% from 20- and 50-m resolutions, respectively. On the contrary, the DEM of 2-m resolution achieved 45% and 79% of buffering and intersection from 20 and 50 m, respectively. The DEM of 5-m resolution achieved the accuracies of 17% and 31%. Finally, the lowest accuracy was produced by DEM with 10-m resolution at 3% for each 20 and 50 m from buffering and intersection methods. The present findings showed a good compromise between landslide inventory location and modeling source resulting from 1 m DEM resolution. Nevertheless, results obtained from 2 and 5 m still produced significant information about debris flow source areas (but not at an optimum detection), whereas DEM with 10 m produced poor result.

Chapter 6 discusses about the optimization of landslide conditioning factors using LiDAR data. Landslide susceptibility modeling (LSM) is the basic step in overall hazard and risk assessment. This chapter presents the optimization of landslide conditioning factors and an analysis of their effects to improve the accuracy of landslide susceptibility models and provide insights into landslide conditioning factors. A landslide inventory map with 132 landslides was prepared based on multisource remote sensing data. A total of 15 landslide conditioning factors were used, including LiDAR-derived and non-LiDAR-derived factors. First, multicollinearity analysis was conducted to remove highly correlated factors from further analysis. Second, ant colony optimization was used to select significant landslide conditioning factors from the initial 14 factors for further analysis. Data mining techniques, including support vector machine (SVM) and random forest (RF), were used to analyze the effects of the selected landslide conditioning factors on the prediction rate accuracy of the susceptibility models.

Several landslide susceptibility maps were produced for the study area, and the best map was recommended for future land use planning. Results of the multicollinearity analysis showed that the topographic roughness index was highly correlated with the remaining factors, and thus, this factor was removed and not used in LSM. In the factor analysis, 8 underlying factors were extracted from the 15 landslide conditioning factors. All the factors were well represented by the 8 extracted factors because the corresponding communalities (i.e., correlation with the retained factors) were generally high. After multicollinearity and the factor effect were analyzed, 6 experiments classified into 2 main groups were conducted. In the first group, all the 14 factors were examined, whereas the second group included only the LiDAR-derived factors. In the first group, the 3 experiments included 5 factors, 10 factors, and all the 14 factors. In the second group, the 3 experiments involved 3 LiDAR factors, 6 LiDAR factors, and 8 LiDAR factors, which were the total number of LiDAR factors derived from the digital elevation model. These subsets were evaluated using the SVM and RF models. On the one hand, the highest accuracy was achieved using the RF model and 10 factors selected from the 14 initial factors. On the other hand, the lowest accuracy was achieved using the SVM model and only the LiDAR-derived factors. The results showed that LSM should be developed using only significant factors, whereas non-LiDAR factors were important to achieve accurate landslide mapping for a study area.

Chapter 7 discusses about the effect of spatial resolution of DEM in landslide susceptibility mapping. As mentioned previously, landslide susceptibility maps are the main products required for hazard and risk assessments, as well as for land use planning. Spatial data play an essential role in determining the quality of landslide susceptibility maps. Therefore, the spatial resolution of digital elevation models (DEMs) was assessed in this study, and an optimal spatial resolution for landslide susceptibility mapping (LSM) at small-scale catchments was determined. A total of 192 landslide inventories were collected from multisource remote sensing data for the study area. In addition, 13 landslide conditioning factors were derived from a LiDAR-based DEM and existing geodatabases of the study area. Logistic regression was used as the modeling technique to produce landslide susceptibility maps. The accuracy of the susceptibility maps was assessed using several accuracy metrics, namely the area under the curve of a receiver operating characteristic, the kappa coefficient, overall accuracy, and spatial agreement. The spatial agreements were determined using empirical information entropy and average susceptibility values. Results indicated that the importance and multicollinearity of the landslide conditioning factors are sensitive to the spatial resolution and source of the DEM. The optimal spatial resolution was 2 m with a predictive accuracy of 0.963, a kappa coefficient of 0.88, and an overall accuracy that approximates 94.02. The entropy map showed that the produced models generally presented high spatial similarities (entropy value ≤ 0.33), which covered nearly 71% of the study area. Furthermore, the 30-m LiDAR DEM was more capable of predicting future landslides and identifying landslide scarps and flanks than the 30-m DEM based on the Advanced Spaceborne Thermal Emission and Reflection Radiometer. Therefore, a finer spatial resolution does not always guarantee a higher prediction rate. In addition, the selection of DEM spatial resolution and source depends on the objective of the study and the amount of details required in landslide susceptibility maps.

Chapter 8 presents an application of k-nearest neighbor (kNN) and logistic regression (LR) models in landslide susceptibility mapping using LiDAR-derived data. Landslide susceptibility mapping plays an important role in urban planning and disaster management for hilly regions. Such task requires various information on the environmental, geotechnical, and economic aspects of landslides. This paper presents a landslide susceptibility analysis for Bukit Antarabangsa, Ulu Klang, Malaysia, with kNN and LR models. Data on 31 landslide events that occurred in the study area were obtained from different sources. Eleven landslide conditioning factors, including altitude, slope, aspect, curvature, stream power index, topographic wetness index, soil, geology, land use/land cover, distance from rivers, and distance from roads, were considered in landslide susceptibility mapping. The main goal of this study is to examine the efficiency of the kNN algorithm in landslide susceptibility mapping. This algorithm has seldom been adopted in this field of study. Comparative assessment was conducted by applying an LR model to evaluate the reliability of the proposed kNN algorithm. The results of the two models were compared and validated. Same conditioning factors were employed to build both models. The capabilities of kNN and LR methods were evaluated with the area under curve technique. Results show that kNN performs better than the LR model. The success and prediction rates obtained from the testing results of the kNN algorithm are 86.28% and 82.64%, respectively. The success and prediction rates obtained from the validation results of LR are 75.65% and 72.18%, respectively. kNN algorithm can be used in spatial planning and can help in hazard mitigation.

Chapter 9 presents an application of support vector machine (SVM) and its different kernels in landslide susceptibility mapping. The lack of reliable and comprehensive physical approaches for landslide susceptibility mapping (LSM) has motivated the use of statistical and machine learning techniques, such as the frequency ratio, weights of evidence, logistic regression, and artificial neural networks. However, the support vector machine (SVM) has become increasingly popular because of its capability to deal with high-dimensional spaces and perform high-accuracy classification. In SVM, the model is trained on a training dataset with associated input and target output values. This study illustrates the application of a geographical information system-based SVM modeling for mapping landslide susceptibility along Jalan Kota in Bandar Seri Begawan, Brunei, to evaluate the spatial correlation between landslides and each conditioning parameter. These parameters are altitude, slope, aspect, curvature, stream power index, topographic wetness index, topographic roughness index, geology, soil, land use/land cover, rainfall, and distance from rivers, roads, and faults. Furthermore, four kernel types, namely radial basis function (RBF), polynomial, sigmoid, and linear kernels, were applied to examine the performance of SVM kernels. Finally, the efficiency of the output maps was validated using area under curve, which measured the prediction and success rates for each kernel. Among the applied kernel types, RBF performed better than the others, with a success rate of 88.21% and a prediction rate of 82.90%. Results of the validation process proved the reasonable strength and feasibility of SVM (particularly RBF–SVM) in LSM. The proposed model can assist local managers and government officials in Brunei to formulate landslide mitigation strategies.

Chapter 10 discusses about the quality of landslide inventory by using different approaches. Landslide susceptibility modeling (or mapping) has been extensively explored in research. However, its quality is affected by uncertainties in landslide inventory data. The quality of landslide inventory is examined by experts using aerial orthophotos and field investigations, which are time-consuming and costly given several landslide records in the inventory database. Therefore, the current study developed an ensemble method based on the idea of active learning to overcome the landslide inventory data uncertainties. Integrating active learning modeling into landslide susceptibility assessment can improve the accuracy and generalizability of the models as it automatically removes problematic landslide inventories. The specific objective is to evaluate the ensemble disagreement active learning for the spatial prediction of shallow landslides in Cameron Highlands, Malaysia. The study is conducted using LiDAR data (i.e., vertical and horizontal accuracies are 0.15 and 0.3 m, respectively). Nine landslide conditioning factors are prepared and 192 landslide inventories are collected from various sources such as aerial photographs and high-resolution satellite images (i.e., SPOT 5). Results revealed that the active learning approach combined with common models such as support vector machines (SVM) and logistic regression (LR) can improve the performance of the models. The success rates of the SVM and LR models are 0.81 and 0.84, respectively, whereas the prediction rates are 0.75 and 0.84. After the integration of active learning to the models, the success rates increased to 0.88 and 0.89 for the SVM and LR models, respectively. Furthermore, the prediction rates increased by 0.18 and 0.5 accordingly for the SVM and LR models. Therefore, findings indicate that the use of active learning in landslide susceptibility modeling can improve the success and prediction rates of the SVM and

LR models. In addition, this study suggested that the use of active learning can decrease collinearity among the landslide factors, thereby improving the final models.

Landslide susceptibility maps help to understand the spatial distribution of landslide probability, and they also improve landslide risk assessment and land use planning. The advancement in computer hardware and software has improved the accuracy of many landslide susceptibility models. These models are grouped into five categories: expert, bivariate statistical, multivariate statistical, machine learning, and hybrid methods. Each category has several models and possesses respective advantages and disadvantages. The advantage of expert-based models is that they do not require landslide inventory data for model training; however, their disadvantage is the subjectivity of the judgment of the importance of landslide conditioning factors. Bivariate statistical models compute the contribution of landslide conditioning factors for landslide occurrence; however, their main drawback is the assumption of conditional independence. Multivariate statistical models analyze the interaction of all parameters in controlling the occurrence of landslides; their drawback is the collection of data over a large area regarding landslide distribution and factor maps. Machine learning models account for nonlinear relationships and handle uncertainty in landslide inventory data; their drawback is their time-consuming nature and their susceptibility to overfitting of the data. Hybrid models can overcome several of the disadvantages of the individual models, but the complexity of hybrid models is often high. Given the various advantages and disadvantages of the aforementioned methods, today's land use planners face the challenge of selecting the most appropriate model for their needs. Therefore, the main objective of Chap. 11 is to evaluate the performance and sensitivity of 14 models, frequency ratio (FR), statistical index (SI), weights of evidence, logistic regression (LR), partial least squares, discriminant analysis, analytic hierarchy process, fuzzy AHP, support vector machine (SVM), random forest, decision tree, FR–SVM, LR–RF, and SI–LR, to provide clear guidelines for land use planners in selecting the most appropriate model. A test site in Cameron Highlands was selected. The results showed that the best model is the hybrid FR–LR model, with a prediction rate of 0.83. This model could predict over 75% of the landslide inventories in the very high susceptible class. It also demonstrated good spatial agreements with several other models.

In contrary to the previous chapters, Chap. 12 presents a detailed landslide hazard, vulnerability, and risk assessment along a stretch of North–South Expressways in Malaysia. Landslides result in high economic and social losses in Malaysia, especially for highway concessionaries such as the PLUS Expressways Berhad. This study aims to perform landslide vulnerability and risk modeling for cut slopes along the Gua Tempurung area on the North–South Expressway in Malaysia. This area was selected because of the frequent occurrences of landslides along the expressway. Highway concessionaries such as the PLUS Expressways Berhad allot a large portion of their annual budget to ensuring the safety of this expressway and making it resilient against natural hazards. Landslide hazards, vulnerability, and risk zoning maps are considered in the decision-making process involving land use/land cover (LULC) planning and overall road management in prone areas. The accuracy of the results is directly related to the spatial data and the methods for obtaining such data. In the present work, we produced a landslide inventory map depicting the 17 landslide locations identified through a field survey. The landslide inventory data were randomly divided into a training dataset: 60% (10 landslide locations) for training the models and 40% (7 landslide locations) for validation. In the first step, a susceptibility map was constructed using the logistic regression method, in which weights were assigned to each conditioning factor according to its correlation with landslide occurrence. High-resolution LiDAR was used to derive the landslide conditioning factors for the spatial prediction of landslide-prone regions. Eight conditioning factors, namely altitude, slope, aspect, curvature, stream power index (SPI), topographic wetness index (TWI), terrain roughness index (TRI), and distance from river, were used for the weight calculation. The susceptibility mapping results were validated with the area under the curve (AUC). The assessment showed 84.91% and 83.00% success and prediction rates, respectively. In the second stage, a hazard map was calculated using the average of the triggering factor (rainfall) for 2014 because most of the landslides in the inventory took place during this year. Overall landslide susceptibility and hazard maps were prepared for the 5-km corridor of the highway. However, only the cut slopes were considered in the vulnerability and risk analysis because they pose a threat to highway users as a result of their frequent reoccurrence. In the third step, elements at risk, such as risk to road users, relative risk of failure, likely effect on traffic, and likely repair costs, were considered in the vulnerability assessment. Each cut slope was examined under these said elements at risk. Subsequently, a value representing the sensitivity of each slope was assigned and considered as the vulnerability value. Finally, a risk map for each cut slope was produced using the derived vulnerability and hazard information. The map of the risky cut slopes may assist PLUS Expressways Berhad in improving highway management.

Landslide hazard and risk maps are essential for hazard mitigation, risk management, and effective land use planning. Chapter 13 presents a multihazard scenario-based landslide risk maps for the Ringlet area located in Cameron Highlands, Malaysia. The main source of data is a digital elevation model (DEM) produced from a high-resolution LiDAR data. In addition, detailed land use maps, adequate landslide inventory data, and rainfall information were used to implement the proposed method. First, the landslide susceptibility map was produced by the logistic regression (LR) model with 12 landslide conditioning factors: altitude, slope, aspect, curvature, stream power index (SPI), topographic wetness index (TWI), terrain roughness index (TRI), distance from a river, distance from roads, distance from lineament, sediment transport index, and geology. Next, landslide hazard maps were produced using five different scenarios: (1) the average intensity of rainfall in any day in a year, (2) the abnormal intensity of rainfall recorded in a day, (3) 5-year return period, (4) 10-year return period, and (5) 15-year return period with average intensity of rainfall per day. Then, the landslide vulnerability map was produced using an exposure-based method by utilizing the detailed land use map and information from experts and previous works. Finally, five risk maps were produced for the study area using the five hazard scenarios. The results indicated that the LR model could predict the future landslides with an accuracy of 84.87%. The average annual economic risk for landslides was MYR 5,981,379.00 in the study area.

The mapping of debris flow risk areas is an important concern because debris flows could result in social losses in hazardous regions, especially in mountainous areas. However, debris flow risk assessment through procedure-based modeling at a medium scale is complex because of several reasons, such as the complex nature of the phenomenon, the inconsistency of local conditioning factors, and the variability of modeling factors. A wide range of debris flow modeling methods has been explored in literature. An effective modeling approach should provide debris flow susceptibility zonation using only minimum data requirements. In Chap. 14, distributed empirical models are used for medium-scale debris flow susceptibility assessments with a light detection and ranging-derived digital elevation model. For debris flow modeling, Flow-R (Flow path assessment of gravitational hazards at a Regional scale) is applied for path assessment of debris flow at regional and medium scales. The Flow-R model requires minimum data input and is flexible to use because of its simple user interface. The second model, rapid mass movement simulation, is used to simulate the run-out of mass movement on a 3D terrain. Although only a preliminary assessment of debris flow effects is presented, the assessment can be useful for land planners and government agencies in their modeling of debris flows and assessment of further effects. The procedure provided in this work can also be replicated in other areas through detailed analyses based on available input data.

Chapter 15 presents a thorough review on rockfall susceptibility, hazard, and risk assessment using different approaches. Rockfalls occur worldwide and annually cause considerable damage to human life and properties. Therefore, comprehensive research is required to understand the triggering and auxiliary elements of the hazards of rockfalls as well as to assess and identify mitigation processes for these calamities. Such research can be used as a reference for managing future rockfall disasters. Rockfall hazard has recently attracted significant attention and has motivated numerous studies. Moreover, such studies have gained importance in various disciplines with the developments of remote sensing and geographical information system technologies. Current geoinformation techniques have been used to gather information for rockfall analyses. This chapter primarily explains the general principles of and the methodologies for rockfall analyses, including rockfall types, causes, and mechanisms, as well as data sources, modeling approaches, and light detection and ranging techniques for rockfall assessment.

Rockfall magnitude and frequency vary both spatially and temporally. Therefore, eliminating such phenomenon is a challenge. Proper modeling and assessment can aid defining the areas at risk thus remedial the effect of rockfall catastrophe. Chapter 16 describes the location and rockfall characteristics of the study area. The materials used in this study also have been described. Multicriteria method for rockfall source areas identification has been applied in this research. Rockfall trajectories modeling and the velocity associated with them have been explained. Raster modeling using geostatistical method has been applied in this research to represent the spatial distribution of rockfall. Finally, spatial modeling with AHP method has been performed in this study to produce rockfall hazard map. As a result, rockfall trajectories and their characteristics were derived and rockfall hazard map for each scenario was obtained. In addition, barrier location was suggested and its efficiency eliminating rockfall hazard was demonstrated.

In general, this book presents the use of LiDAR in landslide assessments providing useful information and recent findings which will be useful for researchers, graduate and postgraduate students, and decision makers both in government and private agencies. This book describes the main applications of landslides such as supervised/machine learning-based detection and characterization of landslide scarps, spatial prediction of the landslide. It gives a detailed discussion on factor optimization and effects of the spatial resolution of DEM on landslide susceptibility mapping. This book also demonstrates identification of the source of debris flow and its susceptibility assessment by LiDAR data. In addition, this book gives a space for multiscenario landslide hazard assessment using airborne laser scanning data, landslide vulnerability, and risk assessment for multihazard scenarios. Finally, this book describes the LiDAR techniques in rockfall hazard assessment in tropical regions. Many case studies presented in this book help decision makers to follow as guidelines for comprehensive landslide hazard and risk assessment using very high-resolution laser scanning data. This book can be helpful and valuable for new students/researchers to understand the concept and use of LiDAR in many landslide applications. The contribution of each chapter of this book advances the landslide studies, opens new areas, and generates new ideas for better landslide assessment.

I could not have produced this book without the efforts of many people who I would like to thank here. Foremost among them my own research team members at Department of Civil Engineering, Universiti Putra Malaysia, and authors from each of the chapters who worked closely with me to meet the deadlines in developing the scope of each chapter. Thanks to all my coauthors of individual chapters of this book.

The publication of this book would not have been possible without an excellent cooperation from my colleagues at Springer Verlag, Germany. Special thanks to Dr. Nabil Khelifi for motivating and encouraging me to write this book. At Springer Verlag, the efforts from Reyhaneh Majidi are appreciable.

Finally, I would like to thank my family, wife Sheila, my two adorable kids, Krish Pradhan and Darsh Pradhan, for their wonderful support and patience in allowing me to spare time to complete this book.

Kuala Lumpur
April 2017

Biswajeet Pradhan

Contents

Part I
Introduction

Laser Scanning Systems in Landslide Studies

1

Biswajeet Pradhan and Maher Ibrahim Sameen

1.1 Introduction

Remote sensing techniques have undergone rapid and significant improvements in the last few decades. The capability of new and enhanced remote sensing techniques to acquire 3D spatial data and very high-resolution terrain contours allows advanced and effective investigations of landslide phenomena. Data from multi-sensors supplemented with airborne- and ground-based data collection techniques provide useful information for model development, validation, and simulation of natural phenomena (Scaioni et al. 2014). Among these technologies, interferometric synthetic aperture radar and light detection and ranging (LiDAR) are two of the most frequently used methods in landslide studies. Unlike traditional methods, these techniques provide fast and exact mapping of geomorphological elements (Hervás et al. 2003; Ardizzone et al. 2007; Fernández et al. 2008; Guzzetti et al. 2012; Daehne and Corsini 2013; Roering et al. 2013). The emergence of remote sensing (RS) and geographical information systems (GISs) has facilitated the application and extension of various algorithms and methods in landslide studies. New insights into landslide research have been obtained by determining and mitigating failures through these techniques. Without RS and GIS, extensive field work, which requires a considerable budget, is required to identify landslide-prone locations (Van Westen et al. 2006, 2008).

LiDAR is an efficient remote sensing data acquisition technique. Although the first airborne LiDAR system was introduced at the end of the 1990s, its applications are recently increasing, particularly in natural hazards (Lemmens 2011). The main uses of LiDAR in landslide applications are generating high-resolution digital elevation models (DEMs) and investigating detailed geomorphic features that control landslides. A high-resolution DEM can be in raster grids, triangulated irregular networks (TINs), or true 3D point clouds. An accurate DEM allows researchers to derive numerous useful parameters, such as slope, flow direction, curvature, and other terrain and hydrological parameters. These parameters are widely used in landslide investigations. The availability of high-resolution terrain data obtained using LiDAR enables accurate landslide mapping, which can be used in landslide susceptibility mapping as well as in hazard and risk assessments. Recently, the applications of LiDAR in landslide studies have significantly increased because of improved terrain data acquisition over large areas within a short period. Researchers have also determined that LiDAR is more efficient than other landslide mapping techniques because of the following advantages:

- The capability of LiDAR signals to penetrate into vegetation foliage
- The independence of solar radiance, which can generate accurate DEM in forested areas
- Mapping and classifying landslides in tropical regions

LiDAR produces high-resolution contours that allow further extraction of geologic and geomorphic information. Recent and advanced LiDAR data processing tools provide new areas for automated and rapid landslide mapping. Although several researchers claim that traditional landslide mapping techniques, such as field mapping and photograph interpretation, are more accurate than LiDAR-based methods, these techniques require expert knowledge and are relatively time-consuming for large-scale applications (Schulz 2004).

This chapter introduces the basic concepts of LiDAR and its applications in landslide investigations. It presents laser scanning techniques, including system components, measurement theory, and the accuracy and resolution of LiDAR data. Then, it provides information about basic LiDAR data processing, such as filtering and geometric calibration.

B. Pradhan (✉) · M.I. Sameen
Department of Civil Engineering, University Putra Malaysia, Serdang, Malaysia
e-mail: biswajeet24@gmail.com

B. Pradhan (ed.), *Laser Scanning Applications in Landslide Assessment*,
DOI 10.1007/978-3-319-55342-9_1

Subsequently, it discusses the main landslide conditioning factors that can be derived from LiDAR. Afterward, several landslide applications that require high-resolution DEM are described. Finally, this chapter discusses the main implications of this review and summarizes the key points.

1.1.1 Laser Scanning Techniques

LiDAR technologies are well-known and efficient surveying techniques. These technologies have been used in a wide range of applications, and their advantages over photogrammetry are well-documented. To fully present the feasibility of LiDAR systems for landslide studies, this section discusses LiDAR system components, measurement theory, the accuracy and resolution of LiDAR measurements, and the concept of direct georeferencing.

1.1.2 System Components

A laser is a tool that generates and releases a beam or a pulse series of extremely collimated, comprehensible, directional, and in-phase electromagnetic radiation. Laser systems can be utilized to gather huge volumes of 3D information of a terrain at an extremely quick recording rate. The development of laser scanning occurs from two contexts, as guided by the position of a sensor: a ground-based laser scanning system, i.e., terrestrial (TLS) and mobile (MLS), and an airborne-based laser scanning system (ALS). The elementary notions and processing of ALS have been known since the 1990s.

An early application of ALS in Earth science was to assess the topographical alterations of the Greenland ice sheet (Krabill et al. 1995, 1999). This device was an advancement of the electronic distance meter (EDM), which was conceived during the late 1950s and the early 1960s, and the total station, from which initial ALS developments were based on (Dallaire 1974). ALS facilitates an extremely precise measurement of distance. However, 1D measurements are inadequate for gauging the 3D coordinates of single points (X, Y, Z). Moreover, reflectors are frequently required (Bromhead et al. 1988).

A typical LiDAR system consists of several components: the platform; positioning hardware, e.g., global navigation satellite system (GNSS) and inertial measurement unit (IMU); 3D laser scanner(s); photographic/video recording equipment; and computer and data storage (Williams et al. 2013). The platform collects and connects all the components in a single system. Figure 1.1 shows the typical components of ALS systems. The platform should be precisely calibrated for the GNSS, IMU, scanner(s), and video recording equipment to obtain accurate point clouds and orthophotos. Meanwhile, positioning systems provide the best possible position of an aircraft. In periods of poor satellite coverage, the IMU manages the bulk of the positioning workload. In addition to augmenting the Global Positioning System (GPS) in periods of poor satellite coverage, the IMU must continually fill in the gaps between subsequent GPS observations (Williams et al. 2013). Different types of laser scanners are available and suitable for installation onboard an aircraft. A scanner records the point clouds and terrain at certain point spacing. It also records the intensity value, which is a measure of return signal strength that can help distinguish objects of varying reflectivity. In addition, the video recording equipment in LiDAR systems provides colors to individual scan points in the point cloud to represent real-world colors. This process is achieved by mapping red, green, and blue values onto the georeferenced point location.

LiDAR sensors can also be set up on terrestrial and mobile platforms to provide extremely dense point clouds (at present, TLS can capture one million points per second). The components of MLS are similar to those of ALS and include GPS, IMU, and laser scanner; in TLS, however, an IMU is unnecessary (Fig. 1.2). The purpose of IMU in ALS and MLS systems is to measure the accurate position, trajectory, and orientation of a sensor to enable extensive data collection. The accuracy of point clouds collected using MLS and ALS systems generally decreases in mountainous areas because of laser footprint distortion or elongation. By contrast, TLS systems can capture highly accurate and dense point clouds to provide a more appropriate view of steep slopes, where most slope processes typically occur. Other advantages of TLS systems over other systems include their shorter ranges and static locations.

The high accuracy of TLS and MLS point clouds enables simulation of single landslides at fine scales using physically based models. These models typically simulate sliding materials, source areas of landslides, and maximum runout distances. In addition, they provide unique advantages to rockfall modeling. Individual rocks can be modeled because of their highly dense point clouds, and their behavior can be simulated to support susceptibility, hazard, and risk assessments. Furthermore, they provide practical solutions for displacement detection. Active landslides can be effectively monitored in areas where failing mass movements are highly complex and the use of ALS systems is limited.

Although the high accuracy of TLS and MLS makes them more efficient than ALS in many applications, including local scale landslide modeling using physically based approaches, ALS systems are highly effective in regional

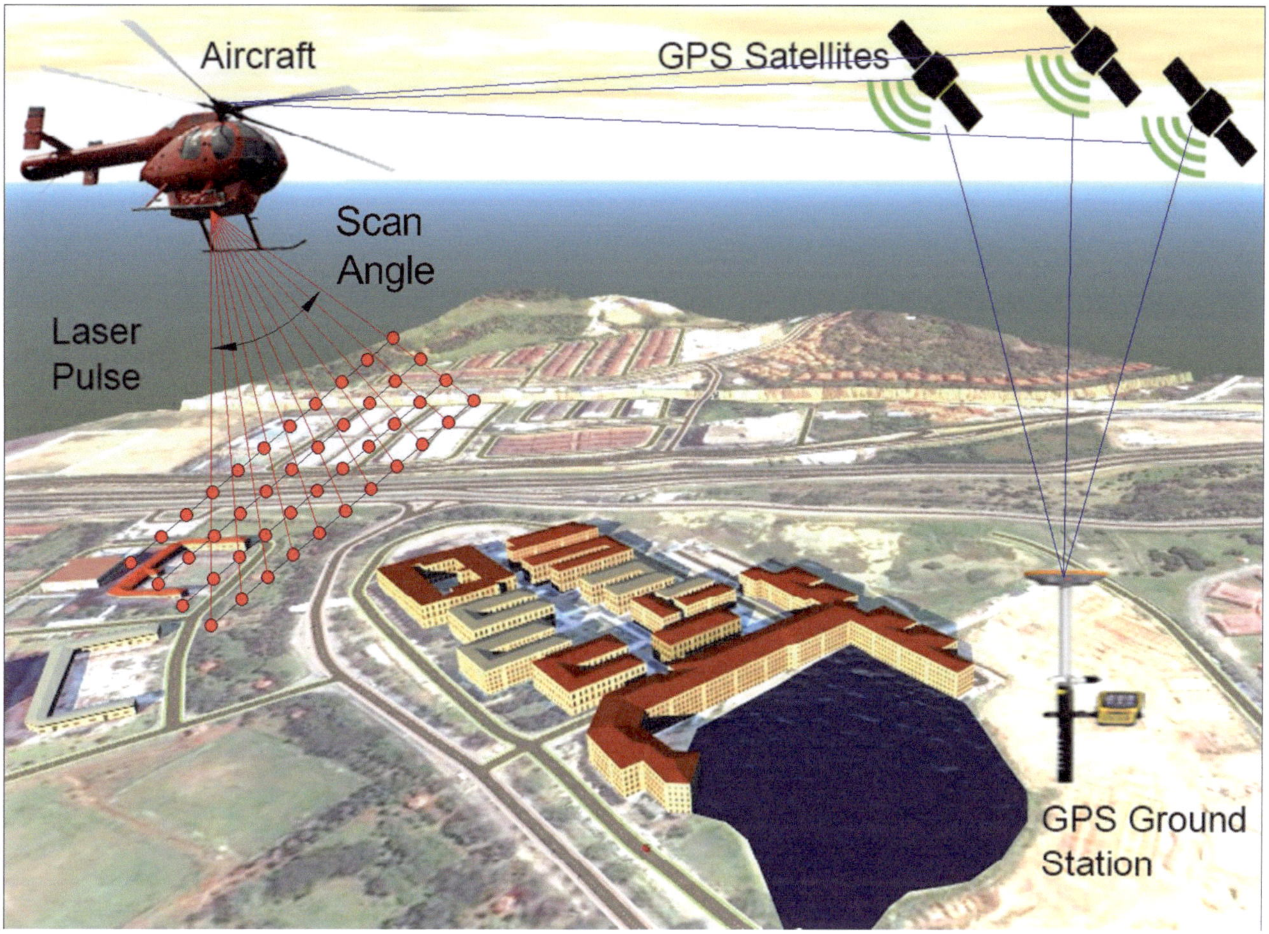

Fig. 1.1 Components of a typical airborne LiDAR system

landslide risk assessments. The large coverage of point clouds captured by ALS systems allows investigation of regional landslides and efficiently manages land uses in a focus area. The applications of ALS in landslides are frequently concentrated on landslide inventory mapping, susceptibility mapping, as well as hazard and risk assessments. By contrast, TLS and MLS are frequently used in small-scale landslide modeling, displacement detection, and to understand the geomorphological processes of landslides in detail. These applications are further discussed and illustrated in Sect. 1.4 of this chapter.

1.1.3 Measurement Theory

A laser scanner comprises a transmitter/receiver of the laser beam and a scanning tool (Hanson 1999). The range can be determined using two different approaches (Wehr and Lohr 1999): the pulse method and the phase method. The phase method can determine the range more precisely than the pulse method but is constrained because of its limited range (Petrie and Toth 2008). The pulse method provides a better range, and thus, is deployed in the majority of TLS and ALS used in Earth surface observations, such as landslide-related research (Baltsavias 1999; Wehr and Lohr 1999).

Ground-based and airborne-based sensors emit laser pulses that are backscattered by different objects, such as vegetation, ground surface, and man-made structures. In addition, sensors log the returning signal. The flight time of a laser pulse is used by pulsed laser scanners to calculate the distance and mechanics, as well as the orientation, of the laser beam in a well-outlined direction. Understanding this line-of-sight direction and the stance (pitch, roll, and yaw) of the instrument aids in ascertaining the position of a reflective surface relative to the device. The range is gauged minus any artificial reflector over a space that is equivalent to the spot dimension (beam width projected on the topography) of the laser, which increases as the distance is extended because of laser beam divergence (Petrie and Toth 2008). The absolute placement of an ALS sensor is outlined using GPS, and its

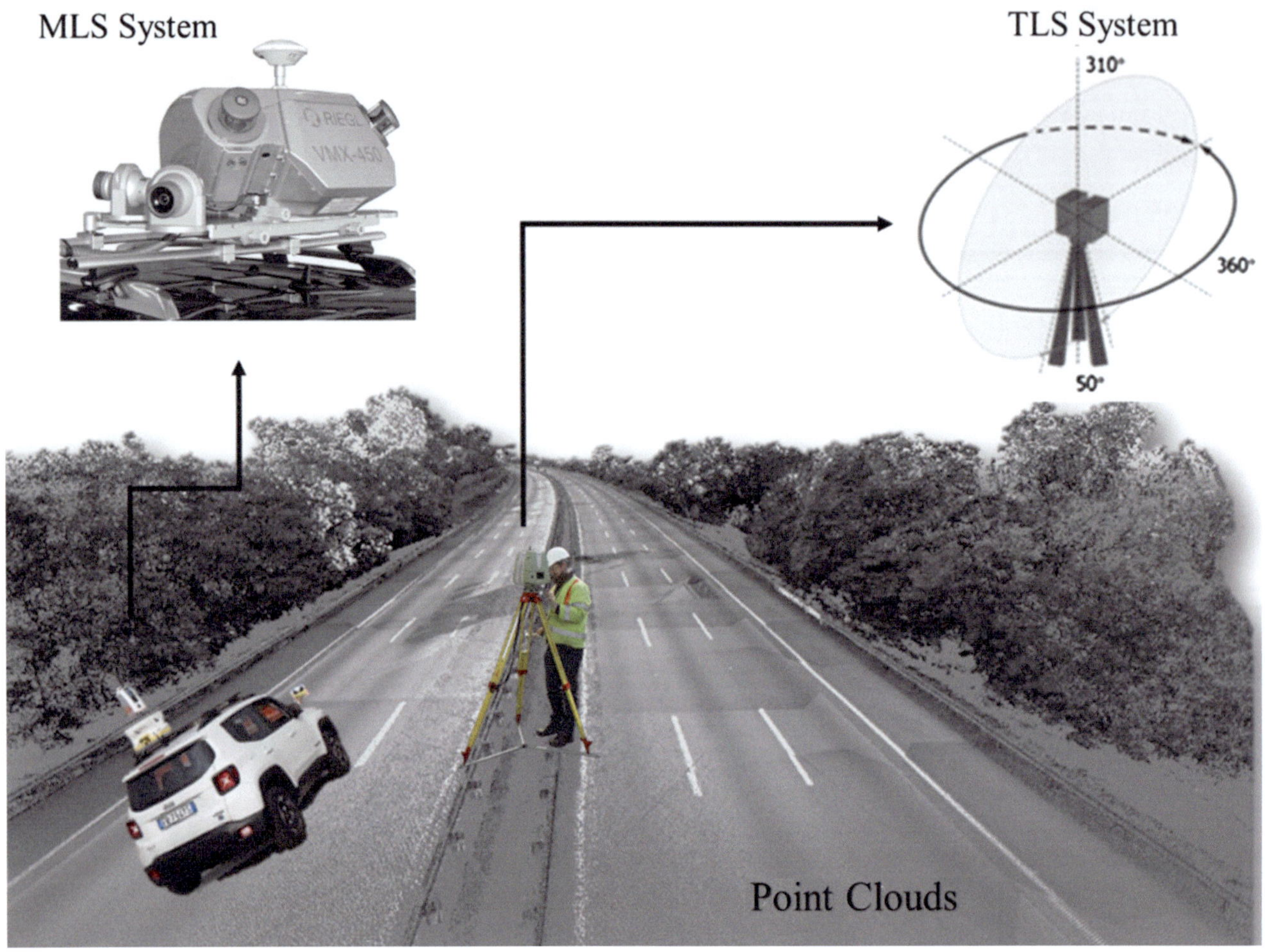

Fig. 1.2 Components of typical mobile and terrestrial LiDAR systems

attitude is logged via an inertial mechanism. Finally, the point cloud coordinates are projected with a precision of approximately 15 cm through ground control points (Habib 2008).

1.1.4 Accuracy and Resolution

The usual precision of a laser instrument is ±1.5 cm within maximum distances of approximately 800–1000 m (Manetti and Steinmann 2007). However, instrumental precision is typically subordinated in practical applications because of adverse conditions, including parallel incident angles, poorly reflecting or extremely uneven surfaces, poor climatic conditions (rain, fog, and hot wind), exceedingly bright ambient settings, and excessive range.

The resolution of a laser scanner is a factor that establishes the extent of the details that can be observed in a point cloud. This resolution is divided into two: angular/spatial and range (Lichti and Jamtsho 2006). Range resolution refers to the capability of the rangefinder to address two objects on the same line of sight. Angular resolution refers to the capability of the rangefinder to address two objects on adjacent lines of sight. This factor is ruled by the following variables: sampling interval or user-defined point spacing and laser beam width, which is contingent on the instrument and the distance.

Apart from the position, the intensity of the reflected signal is obtained. This aspect primarily relies on the type of material (i.e., the roughness and color of the reflective surface), beam wavelength, angle of incidence, and soil moisture (Lichti 2007; Pesci et al. 2008).

The typical point density of ALS and TLS systems ranges from 1 to 100 pts/m^2, whereas that of TLS ranges from 50 to 10,000 pts/m^2 (Jaboyedoff et al. 2012a). However, the point density of laser scanning systems is controlled by several factors, including beam wavelength, type of target materials, soil moisture, and angle of incidence. Consequently, point density can vary from one area to another for the same system.

1.2 LiDAR Data Processing Methods

A LiDAR sensor can generate a huge volume of spatial data with 3D coordinates within a few hours. LiDAR produces irregularly distributed points based on system characteristics and configurations. Raw LiDAR data are unorganized, and thus, a GIS system is necessary to organize such huge amounts of spatial data. After this process, data filtering can be conducted to reconstruct bare-earth ground surfaces and features, such as buildings and trees, and to visualize data. This section describes the two most important LiDAR data processing methods, namely filtering and geometric calibration, to improve understanding of their concept and usage in landslide applications.

1.2.1 LiDAR Digital Surface Model (DSM) Filtering

LiDAR data filtering refers to the process of eliminating non-ground points from LiDAR point clouds to obtain ground points (Vosselman 2000). The concept of filtering is based on the observation that a significant height difference between two nearby points is unlikely to be caused by a steep slope in the terrain. Instead, the higher point is possibly not a ground point (Vosselman 2000). That is, the filtering process is only performed to obtain a digital terrain model (DTM) or a DEM from a DSM. In addition, LiDAR data filtering is necessary to optimize data for analysis and to decrease computational loads.

DSM point cloud filtering in landslide research is based on the close-range photogrammetry technique and is not as common as laser scanning data filtering (Zhan and Lai 2015). LiDAR-based DEM and DSM are widely used in landslide investigations mainly because of their high-resolution elevation data and rapid data acquisition (Bui et al. 2016; Tien Bui et al. 2012; Dou et al. 2015). However, a number of similarities exist between photogrammetry and LiDAR point cloud data filtering methods. A common filtering technique is iterative linear least squares interpolation proposed by Kraus and Pfeifer (1998). This algorithm removes tree measurements from airborne laser scanning data and generates a DTM for a forest area. An adaptive TIN method, which could handle surfaces with discontinuities, was proposed by Axelsson (2000). Vosselman (2000) developed a slope-based filter whose concept relied on the premise that the slope of a natural terrain would be distinctly different from the slopes of non-terrain objects (e.g., trees, buildings). This algorithm was enhanced by Sithole and Vosselman (2001) by using a slope adaptive filter. Lindenberger (1993) proposed a point cloud filtering algorithm based on mathematical morphologic operators. The limitation of this algorithm is its vulnerability to the size of the structural element. Zhang et al. (2003) proposed a method to remove objects using gradually increasing window sizes, which could effectively remove most non-ground points. Recently, a new filtering algorithm based on artificial neural networks (ANNs) was presented by Jahromi et al. (2011) to extract bare-earth points from airborne laser scanning data and efficiently generate a high-quality DTM for an urban area. Zhan and Lai (2015) presented a novel DSM filtering algorithm for landslide monitoring. This algorithm, which was based on multi-constraints, was proposed to solve the problems of vegetation interference and noise points in DSM filtering for landslide monitoring.

1.2.2 Registration

The registration of point clouds is frequently a Euclidean transformation process that combines translation and rotation with respect to a reference point and a coordinate system. Point clouds captured from different stations should be properly registered before data analysis to ensure optimization. The registration of LiDAR point clouds is typically performed using three main methods: target-based, feature-based, and point-to-point methods (Abellán et al. 2014). The first method uses a precise survey from differential GPS (DGPS) or a total station. This method is time-consuming and tedious and requires additional equipment, such as surveying instruments. Nevertheless, researchers have tested this approach on landslide monitoring and obtained satisfactory results (Abellán et al. 2014). In the second method, features, such as power lines and pipelines are usually recognized and used for registration. These methods are frequently utilized in industrial applications and rarely in landslide studies mainly because complex rock slopes seldom present distinctive geometric characteristics. The point-to-point method is generally based on the progressive minimization of the distance between corresponding points in two overlapping point clouds. This method is widely used in geoscience applications; however, a high number of iterations may be required (Abellán et al. 2014). Efficient algorithms should be utilized in this method to produce useful products that can be applied in landslide investigations. Distortions in surface geometry may lead to poor slope data.

1.2.3 Geometric and Radiometric Calibrations

The geometric calibration of LiDAR data aims to remove systematic errors from point clouds. Such errors in LiDAR data are primarily caused by biases in mounting parameters related to system components (e.g., lever arm and boresight angles) and biases in measured ranges and mirror angles

(Zhang et al. 2013). The two main methods of systematic error elimination are system driven (calibration) and data driven (strip adjustment). Calibration methods are based on the physical sensor model that relates the system parameters to the ground coordinates of the LiDAR points. By contrast, the data-driven approach mainly utilizes mathematical models that relate LiDAR strips and the reference frame (Zhang et al. 2015). The effects of systematic errors in system parameters are typically modeled through an arbitrary transformation function between the laser strip and the reference frame coordinate systems. In the study of Habib et al. (2011), significant improvement in horizontal and vertical accuracies was confirmed after removing the effect of estimated biases in the system parameters.

The radiometric correction of the LiDAR data aims to remove the effects of laser energy attenuation caused by atmospheric effects and object surface backscattering (Yan et al. 2012). Such correction can be performed using empirical and physical approaches. An empirical approach does not consider the physical properties of the laser backscattering energy.

LiDAR intensity has been used to study the geomorphology and structure of volcanic surfaces and active landslide bodies (Fornaciai et al. 2010). Fornaciai et al. (2010) reported that vegetation, air fall deposits, epiclastic sediments, and lava flow demonstrated distinctive LiDAR intensity responses. Yan et al. (2012) indicated that the radiometric correction of LiDAR intensity data could significantly improve the accuracy of land cover classification. Land cover data are highly necessary for landform classification, which is an important GIS layer in landslide studies. Wang et al. (2013) found that LiDAR intensity was extremely useful in identifying landslide boundaries. From the aforementioned literature review, radiometrically corrected intensity data are expected to provide more accurate geomorphic features in landslide applications than raw intensity data collected using LiDAR sensors.

1.3 Main LiDAR Data Products Used for Landslide Modeling

The selection and mapping of an appropriate set of conditioning factors associated with landslide events require a priori knowledge of the main contributors to landslides. The most common landslide conditioning factors that can be derived from LiDAR-based DEM are altitude, slope, aspect, profile curvature, plan curvature, topographic wetness index (TWI), topographic roughness index (TRI), stream power index (SPI), and sediment transport index (STI). The following subsections describe each of these conditioning factors derived from LiDAR-based DEM.

1.3.1 Altitude

LiDAR is a technique used to extract very high-resolution DEMs. The current accessibility of DEMs produced using LiDAR sensors enables researchers to improve identification and mapping of slope failures (Derron and Jaboyedoff 2010). LiDAR has the significant advantage of being able to penetrate vegetation canopies and produce valuable information regarding topographic conditions (Slatton et al. 2007). This capability makes LiDAR data distinct compared with other sources, such as aerial photographs, in detecting slope failure in forested regions. From the literature, LiDAR-derived DEM is mostly used for the visual assessment of topographic surfaces (Haneberg et al. 2009) and the semiautomatic identification of landslide characteristics. Ardizzone et al. (2007) reported an improvement in detecting landslide locations using LiDAR-derived DEM compared with analyzing aerial photographs.

1.3.2 Slope

The slope is a measure of change in elevation. It is one of the main landslide conditioning factors used in nearly every landslide susceptibility research. The slope is an important parameter in landslide studies because of its relationship with the driving force of gravitation (Latif et al. 2012). In general, slope angle has a positive linear relationship with landslide occurrence. That is the vertical component of gravity increases with slope degree. The slope can be calculated from 3D grid data acquired using LiDAR or other conventional methods. However, the slope can be accurately estimated using certain mathematical algorithms because LiDAR collects high-resolution elevation data that can be represented in a grid format. Among popular methods, the neighborhood algorithm is one of the techniques proposed to calculate percent slope. This algorithm computes slope for every cell in an elevation grid by analyzing each 3×3 neighborhood. Slope percentage can be converted into slope degree afterward. The result is a grid of slope values that are appropriate for various landslide applications. Figure 1.3 shows the main steps for slope calculation from LiDAR point clouds (Tarboton 1997).

The single triangle shown in Fig. 1.3c is used to calculate slope from an interpolated grid. Slope is represented by the vector (S_1, S_2), where

$$S_1 = \frac{Z_0 - Z_1}{d_1}, \tag{1.1}$$

$$S_2 = \frac{Z_1 - Z_2}{d_2}, \tag{1.2}$$

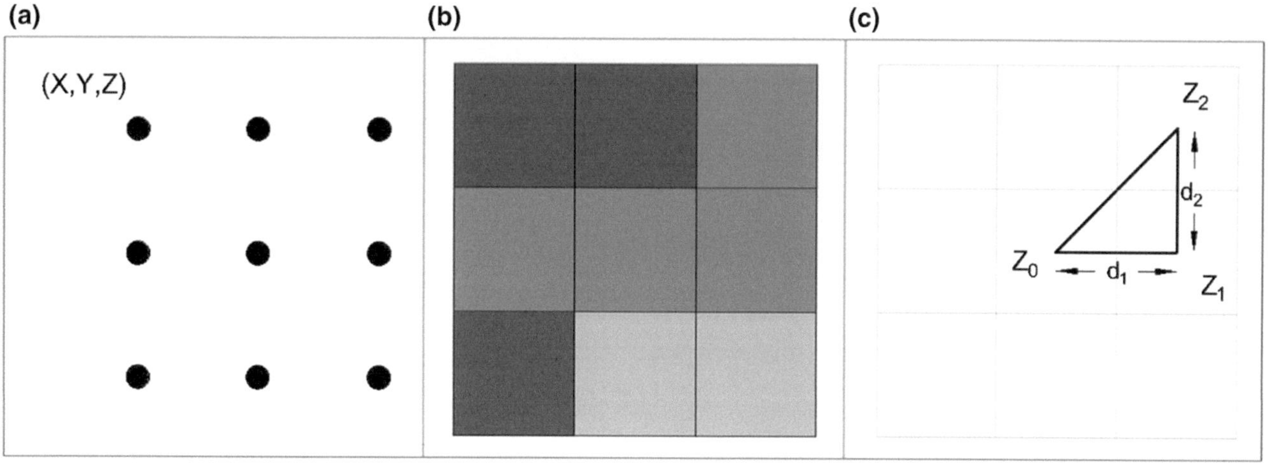

Fig. 1.3 Main steps for slope calculation from LiDAR point clouds. **a** An example of LiDAR point clouds, **b** interpolated grid, and **c** definition of variables for calculating slope on a single facet

where Z_i and d_i denote the elevations and distances between pixels, as labeled in Fig. 1.3c. The magnitude and direction of slope is then computed as follows:

$$S = \sqrt{S_1^2 + S_2^2}, \quad (1.3)$$

$$r = \tan^{-1}\frac{S_2}{S_1}, \quad (1.4)$$

where S is the magnitude of slope, r is the direction of slope, and (S_1, S_2) is a vector that represents the downward slope.

The effect of LiDAR post-spacing and DEM resolution on mean slope estimation was analyzed by Chow and Hodgson (2009). Their analysis indicated that the deviation between mean slope and modeled mean slope decreased with increasing fineness of posting density and DEM spatial resolution. They found that the relationship of the mean slope with varying cell sizes and post-spacing suggested a linear and a logarithmic function, respectively, for all study areas. Moreover, cell size had a more significant effect on the mean slope than LiDAR posting density. In addition, their study also suggested that interpolation methods and their parameters significantly affected DEM generation, and subsequently, other derivatives such as slope, aspect, and curvature.

Chen et al. (2016) presented a study on the relationship between slope angle and landslide size derived from limit equilibrium simulations. Their results indicated that as slope angle increased, sliding mass volume or potential slide size decreased. Conversely, another study based on numerical simulations conducted by Katz et al. (2014) suggested that more material would disintegrate for a given material strength in steeper slopes, and consequently, the produced landslide would be larger. Such contradiction suggests that numerous controls for slope angle influence the size of landslides for a given material strength.

1.3.3 Aspect

Aspect (i.e., slope direction) identifies the downslope direction of the maximum rate of change using eight neighboring pixels (Toutin 2002). Aspect defines slope direction, and subsequently, flow direction. It can affect the physical and biotic features of a slope and can significantly influence its microclimate. In some localities, patterns of soil differences are related to variations in aspect. Slope aspect controls the formation of landslides, such as lineaments, rainfall, wind effects, and exposure to sunlight (Yalcin and Bulut 2007; Pourghasemi et al. 2012). Consequently, aspect indirectly influences landslide and has been used in numerous landslide susceptibility mapping studies worldwide. Aspect maps are generated based on the compass direction that a surface faces at raster pixel locations. It is measured clockwise from 0 (due north) to 360° (due north), thereby coming full circle. Flat areas with no downslope direction are given a value of −1. Figure 1.4b shows an example of an aspect map derived from LiDAR-based DEM. In this example, aspect values continue to increase, thereby indicating that the compass direction calculated using the ESRI algorithm (Burrough et al. 2015). In landslide studies, however, an aspect map is generally classified into nine classes: north, northeast, east, southeast, south, southwest, west, northwest, and flat.

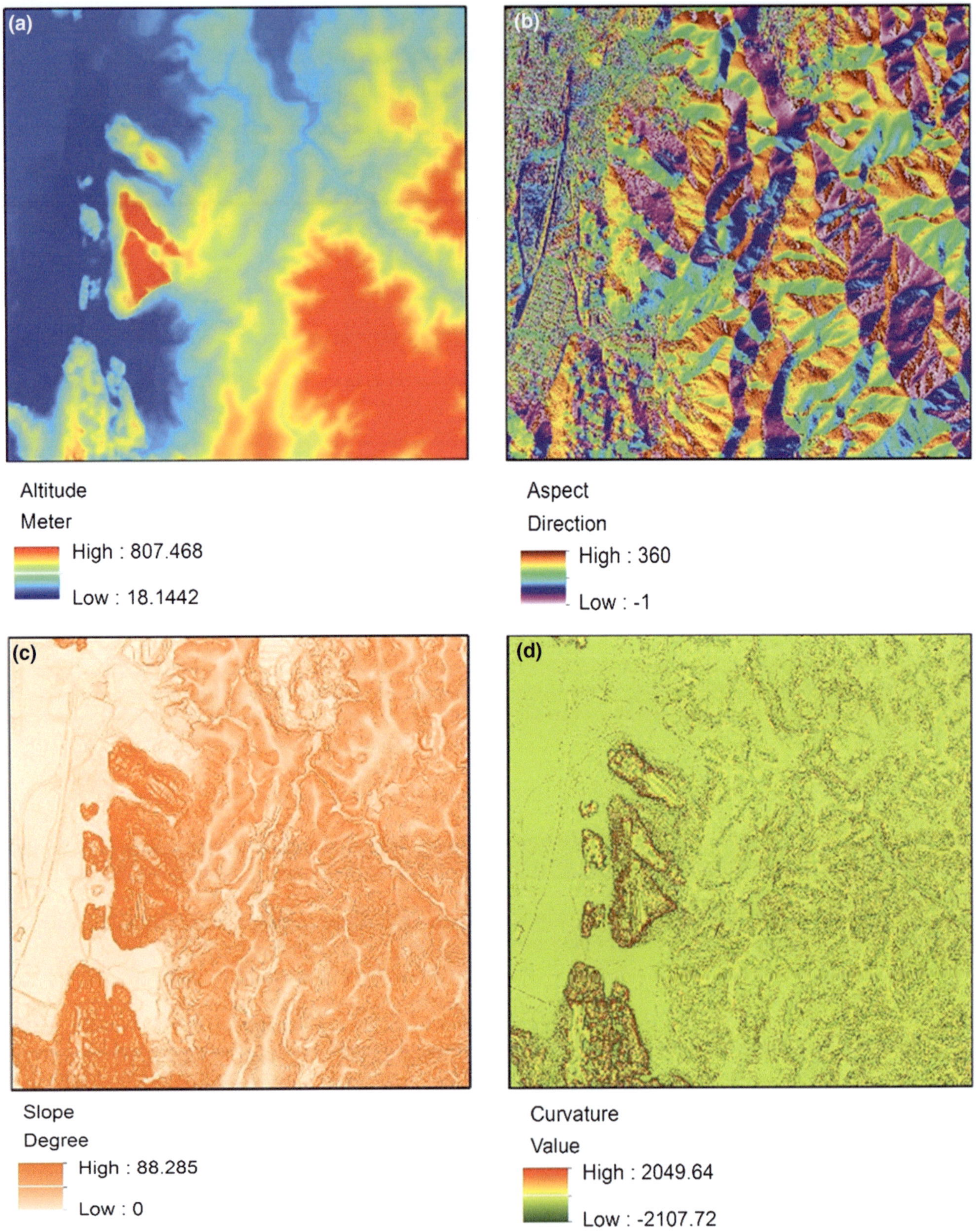

Fig. 1.4 Examples of geomorphic parameters derived from LiDAR data. **a** Altitude, **b** slope aspect, **c** slope angle, and **d** curvature

1.3.4 Curvature

In general, curvature, which is defined as the rate of change in slope degree or aspect, has been reported to affect slope stability. The characterization of slope morphology and flow can be analyzed using a general curvature map (Nefeslioglu et al. 2008). The plan curvature is described as the curvature of a contour line formed by the intersection of a horizontal plane with the surface. It influences the convergence and divergence of flow across a surface. The profile curvature, which is the vertical plane parallel to the slope direction, affects the acceleration and deceleration of downslope flows, and consequently, influences erosion and deposition (Kannan et al. 2013; Kritikos and Davies 2015). In addition, Fernandes et al. (2004) indicated that curvature could be influenced by slope erosion processes, such as the convergence or divergence of water during downhill flow. Furthermore, curvature constitutes one of the landslide-related factors that control landslide occurrence.

1.3.5 Hydrological Factors

In addition to terrain factors, landslides are controlled by several hydrological factors, such as TWI, STI, and SPI. TWI is defined as a steady state wetness index. It is commonly used to quantify topographic control on hydrological processes and is a function of both slope and flow direction. Moreover, TWI describes the effect of topography on the location and size of saturated source areas of runoff generation and is another topographic factor within the runoff model. TWI is expressed as follows:

$$\mathrm{TWI} = \ln\left(\frac{A_s}{\beta}\right), \qquad (1.5)$$

where A_s is the specific catchment area (m^2/m), and β is a sloped angle in degrees. In addition, TWI is an important characteristic of DEM that indicates soil saturation.

STI, which reflects the erosive power of overland flow, is derived by considering transport capacity limiting sediment flux and catchment evolution erosion theories (Devkota et al. 2013; Pradhan and Kim 2014).

$$\mathrm{STI} = \left(\frac{A_s}{22.13}\right)^{0.6}\left(\frac{\sin\beta}{0.0896}\right)^{1.3} \qquad (1.6)$$

SPI, a measure of the erosion power of a stream, is also a factor that contributes to stability within a study area (Conforti et al. 2011; Regmi et al. 2014). SPI is expressed as follows:

$$\mathrm{SPI} = A_s \times \tan\beta. \qquad (1.7)$$

1.4 LiDAR in Landslide Applications

High-resolution DEMs and DSMs derived from most LiDAR systems provide several useful landslide applications. Common applications include landslide scarp identification (Ardizzone et al. 2007), landslide spatial prediction (Ghuffar et al. 2013), landslide hazard and risk assessments, detection of landslide movements, and simulation of debris flow and rockfall (Lan et al. 2010). This section describes the use of LiDAR in the aforementioned applications and then discusses the advantages and disadvantages of LiDAR systems over conventional photogrammetry.

An overview of landslide geometry is necessary to understand the use of LiDAR in various landslide-related applications. Figure 1.5 shows a block diagram of a landslide (right) and a real landslide (left). The failed mass starts from a depletion zone and is deposited in the accumulation zone. The region adjacent to the highest part of the failed materials is called the crown of the landslide. The landslide scarp is the steep rupture surface between the failed body and the terrain. The surface of rupture identifies the interface at the base of the landslide where the materials have slid. The landslide deposit ends with a toe, which is the line (usually bent) between the accumulated material and the untouched terrain. The tip is the point of the landslide deposit that is farthest from the crown.

LiDAR systems collect billions of points that can generate accurate 3D landslide models. These models open new areas for landslide investigations. Accurate LiDAR data facilitate accurate landslide conditioning factor mapping, which allows automatic or semiautomatic detection, spatial prediction, and detailed characterization of landslides. In addition, landslide movements can be detected and analyzed using multi-period LiDAR data. LiDAR data can also be used effectively for debris flow and rockfall simulations. The following subsections describe the use of LiDAR data in various landslide applications.

1.4.1 Landslide Detection (Inventory Mapping)

Landslide inventories report the spatial distribution of existing slope movements, including attribute information about landslide typologies and their activity state (Canuti et al. 2007). The most basic landslide detection method is geomorphological field mapping (Yesilnacar and Topal 2005). This method requires expert knowledge and experience in landslides and the study area. Results can vary significantly depending on the specialists who prepared the map, knowledge of the study area, and the processes

Fig. 1.5 Block diagram of a landslide (*right*) and a real shallow landslide (*left*). *Photo credit* www.uibk.ac.at

involved (van den Eeckhaut 2005). Other methods of landslide detection are mostly based on remote sensing techniques. One of the common approaches for landslide detection is the visual interpretation of aerial orthophotos. Although this approach can produce accurate landslide inventories, it requires expert knowledge and is time-consuming (Pradhan 2013). In recent years, landslide detection is frequently based on LiDAR technology (Jaboyedoff et al. 2012a; Glenn et al. 2006). A DSM can be created by removing vegetation and other objects from a DTM (Schulz 2004). This feature, together with different modifiable angles of shading, enables highly detailed mapping of landslides and other geomorphological features.

Several approaches are available for landslide detection from LiDAR data and orthophotos. Figure 1.6 summarizes the main steps used for different methods of landslide detection. In general, landslide locations are identified from LiDAR using three steps. First, LiDAR data are filtered and interpolated to produce high-resolution DEM. Second, the DEM is used to derive various geomorphological and hydrological factors that are useful for landslide detection and characterization. Finally, machine learning algorithms, object-based image analysis, supervised classification methods or any other feature extraction methods are used to identify landslide locations from the derived factors. Landslide locations can also be identified from orthophotos. In general, two types of approaches are used as follows: visual interpretation and change analysis (Nichol and Wong 2005; Lu et al. 2011).

Eeckhaut et al. (2007) investigated the use of LiDAR in detecting old landslides in forested areas. A landslide inventory map was created by applying the expert knowledge of seven geomorphologists to LiDAR-derived hillshade, slope, and contour line maps in a GIS environment. Their analysis showed that large-scale LiDAR-derived maps could significantly improve field inventories of landslides with a subdued morphology in hilly regions. Schulz (2004) detected landslides from LiDAR data. Landslide deposits and scarps were mapped and classified based on the certainty degree with which they were identified, which depended on the continuity and strength of morphologic features as expressed in LiDAR imagery. Moosavi et al. (2014) evaluated pixel-based and object-oriented approaches optimized using the Taguchi method for landslide inventory mapping. Their study suggested that object-oriented methods outperformed the per-pixel-based methods. Object-oriented approaches provide the shape and geometry of the identified landslides. Recently, Pradhan et al. (2016) proposed a fusion technique based on the wavelet transform and Taguchi methods for automatic landslide detection from LiDAR and very high-resolution satellite images. Wavelet fusion was used to fuse LiDAR elevation data and QuickBird images. Then, Taguchi optimization was applied to select the optimal segmentation parameters of the multi-resolution segmentation algorithm. The results indicated that the proposed method was efficient in identifying landslide locations in an urban environment in a tropical region.

1.4.2 Landslide Susceptibility Modeling

Landslide susceptibility is defined as the proneness of a terrain to generate slope failures (Yalcin 2008). It is typically expressed cartographically and relies on a complex knowledge of slope movements and their controlling factors

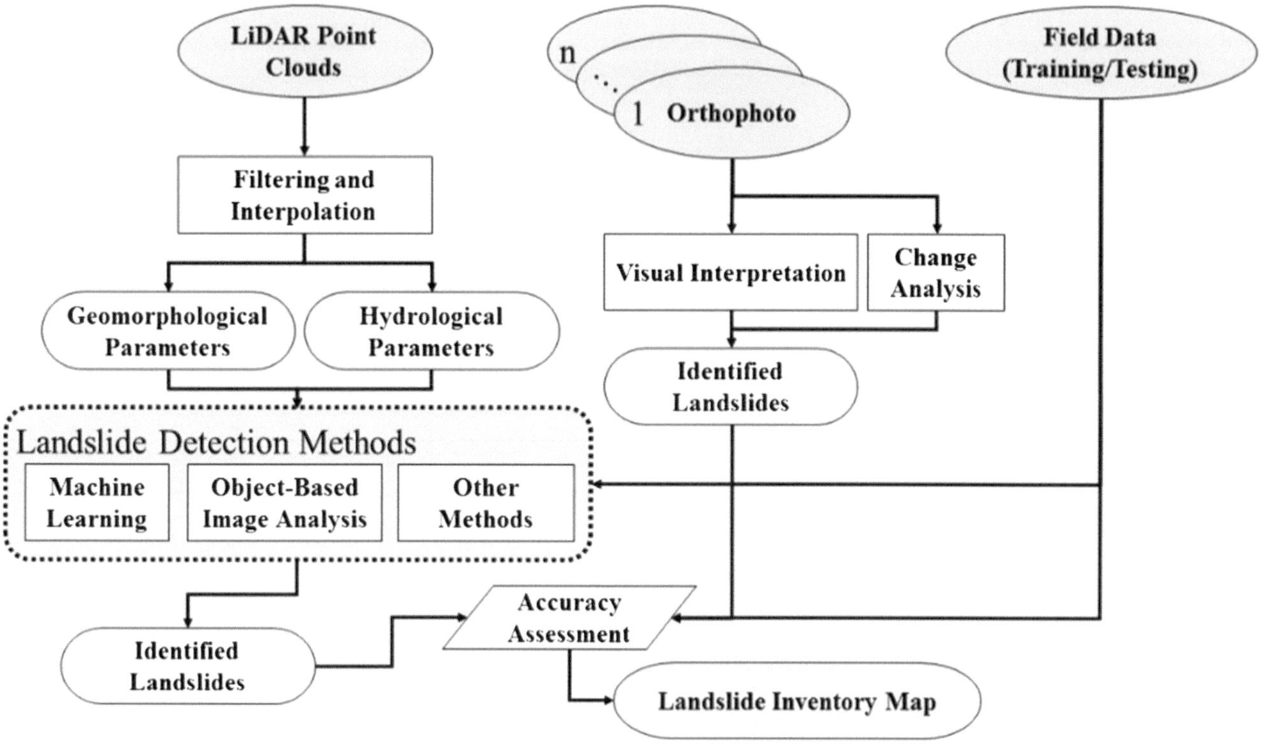

Fig. 1.6 Overview flowchart of landslide detection methods using LiDAR data and orthophotos

(Ayalew et al. 2004). The main causes of slope failure are as follows (de Blasio 2011):

- types of materials involved
- the geometry of materials
- distribution of weight along a slope
- soil moisture and precipitation
- external impulsive forces, such as earthquakes, waves, and volcanic eruptions
- presence/absence of vegetation.

Landslide susceptibility is established by overlaying landslide conditioning factors using the weight of each factor determined via a regression model and landslide inventories (Pourghasemi et al. 2013; Akgun and Erkan 2016). Terrain factors (i.e., altitude, slope, aspect, curvature) and hydrological factors (i.e., TWI, TRI, SPI, STI) are the main landslide conditioning factors that can be derived from LiDAR point clouds. LiDAR data also support the production of accurate landslide inventories. The morphological features of landslides (e.g., scarps, mobilized materials, foot) are easy to delineate based on the hillshades of the produced DEM. Furthermore, high-resolution DEMs and DSMs generated using LiDAR and orthophotos provide images of land use, drainage networks, urban and rural roads, and vegetation structure and density.

Accurate and high-quality landslide susceptibility maps can be produced using LiDAR data and sophisticated algorithms. A high-resolution DEM of LiDAR allows researchers to derive high-resolution landslide conditioning factors, and consequently, high-quality landslide susceptibility maps. Figure 1.7 presents a comparison between LiDAR-based (0.5 m) and ASTER-based (30 m) landslide susceptibility maps produced using the same algorithm and landslide conditioning factors. The LiDAR-based map shows fine details of the landscape, which provide useful information for efficient slope management. By contrast, no detail can be observed in the susceptibility map produced using the ASTER-based DEM. This result indicates the usefulness of LiDAR data for landslide modeling at small catchments.

1.4.3 Detection and Characterization of Landslides Movements

Several methods are available for measuring ground displacements caused by landslide movements. Among the basic methods are theodolite geodetic measurements (Burghaus et al. 2009), GPS (Zhang et al. 2008), and remote sensing-based techniques. The first and second methods are applicable to measuring and monitoring ground deformation at the surface. However, with the recent availability of

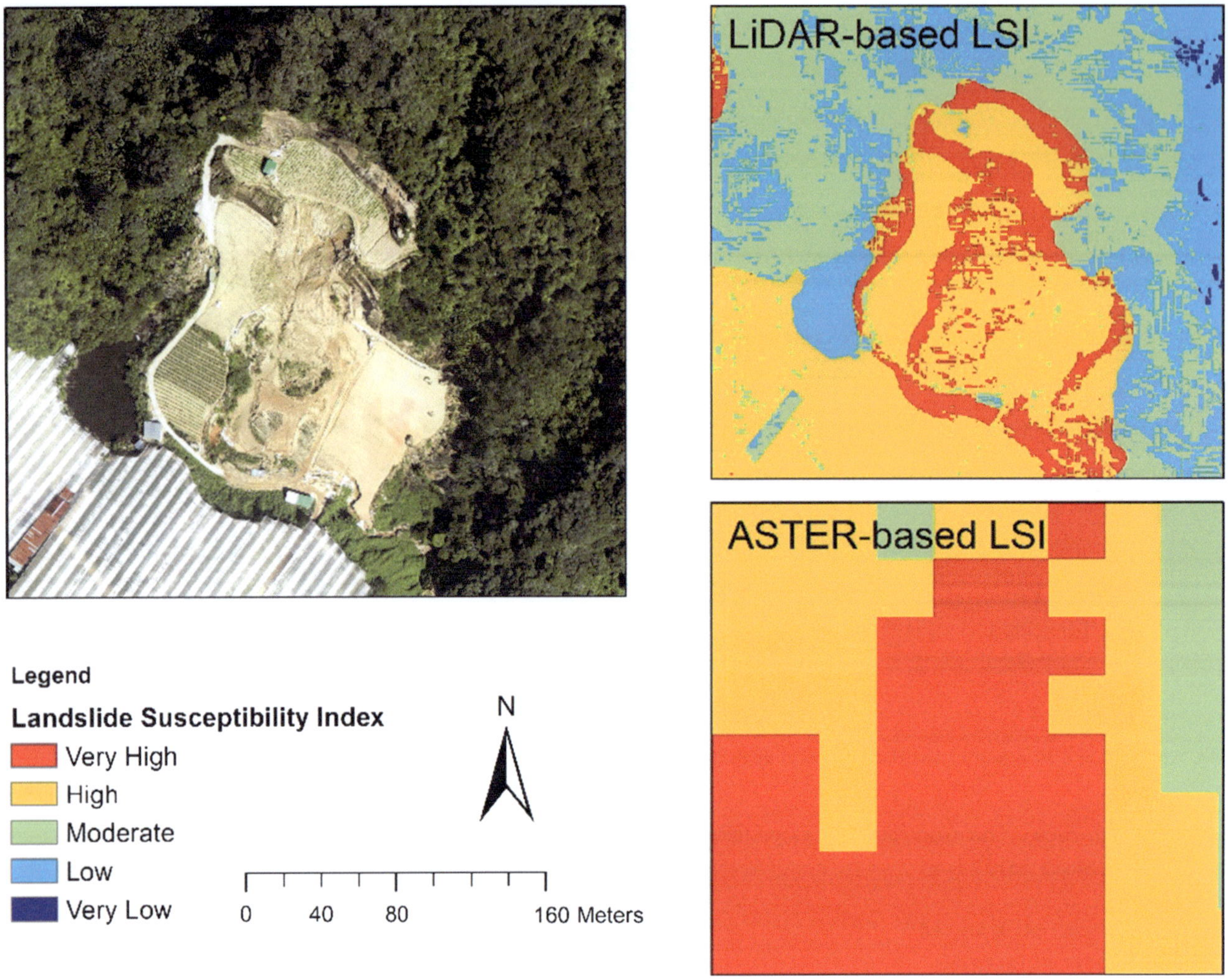

Fig. 1.7 Effects on the spatial resolution of DEM on landslide susceptibility maps

remote sensing techniques, such as LiDAR, performing subsurface measurements has become an easy task, and understanding of landslide behavior can be improved.

Investigation of active landslides can predict future landslide behavior. Therefore, several studies have investigated the movements of active landslides using multi-temporal LiDAR data. A DSM produced from different periods allows a change in detection of a landslide surface. Several techniques can detect changes in geomorphic surface. The commonly extracted parameters are landslide scarp length and affected area. However, LiDAR data provide opportunities to extract the magnitude and direction of complete 3D motion. An efficient algorithm, called range flow, can yield fully automatic dense 3D motion vectors for the entire time series of available LiDAR data (Ghuffar et al. 2013). In addition, adequate time series DSM data provide deeper insight into the dynamics of a landslide. High-resolution digital terrain data allow researchers to derive dense 3D flow fields for tracking landslide deformation. Furthermore, a 3D velocity field over a landslide surface can be simulated using multi-temporal LiDAR data (Schwalbe and Maas 2009). Consequently, a high-resolution DSM derived from dense point clouds provides a qualitative and quantitative method for measuring surface change (geomorphic and anthropogenic changes). These changes can then be directly associated with geomorphic processes and used to infer material dynamics and effects of landslide movement.

One advantage of LiDAR over photogrammetry is that vegetation can be eliminated via laser scanning and data filtering for ground detection (Ghuffar et al. 2013). This process allows producing accurate DTMs with high-quality geometry, which is difficult to achieve using traditional photogrammetry methods. The extraction of an accurate DTM is important for precisely estimating the geomorphological parameters of active landslides. In addition, full-waveform LiDAR can penetrate vegetation that covers landslides, which can be limited in single-return LiDAR.

Another advantage of LiDAR is its capability to set up airborne, mobile, and terrestrial platforms that can enhance landslide investigations in flat urban and hilly/mountainous areas. Terrestrial LiDAR is appropriate for steep cliffs and rock faces because the scanner can be placed in front of it. However, LiDAR may fail to detect shear processes in fast-moving landslides. In such cases, other techniques, such as time domain reflectometry, are suggested (Barendse and Machan 2009).

1.4.4 Simulation of Debris Flow and Rockfall

Debris flows are extremely rapid, surging flows of unsorted, saturated debris in predefined channels (Hungr et al. 2001). Many debris flows form from a debris avalanche that starts on a steep slope and enters a channel. The basic block diagram of debris flow is shown in Fig. 1.8a.

The simulation of debris flow is a complex process because of the variability of controlling factors and the dynamic nature of this phenomenon (Takahashi 2014). Detailed topography and channel materials are the main critical factors. A LiDAR-based DEM provides major input data to debris flow simulation models, including geomorphological parameters, such as slope, curvature, and aspect. LiDAR data are also used to extract the potential source areas of debris flow (Baumann et al. 2011). The identification of debris flow source areas permits the susceptibility analysis of debris flow, which can significantly reduce negative effects on people and infrastructure.

A rock avalanche is the extremely rapid flow-like movement of fragmented rocks from a large rock slide. The initial acceleration is provided by cohesion loss and joint surface roughness reduction. When a rock avalanche mass overrides saturated the surficial soil, a rapid loading, and material entrainment process occur to mobilize long runout movement (Hungr and Evans 2004). The basic block diagram of rockfall is provided in Fig. 1.8b.

1.5 Discussion and Conclusion

Laser scanning systems, including ALS, TLS, and MLS, allow improvements in geological mapping and identification of landslide scarps and displaced materials. They provide automatic landslide detection, mapping, assessment, and modeling. LiDAR products, such as very high-resolution DEMs and slope angles, are extensively used to delineate the geomorphological features of landslides, such as scraps, mobilized materials, and foot. The accuracy of a DEM generated from LiDAR point clouds depends on various factors, including system components, data processing techniques (point clouds and interpolation methods), and signal–target/surface interaction (i.e., backscatter signal strength, laser incidence angle/geometry) (Al-Durgham et al. 2010). In general, the accuracy of an airborne LiDAR-derived DEM is comparable with DGPS-based contours, and an absolute agreement of $\sim$6 cm has been reported in the literature (Al-Durgham et al. 2010; McKean and Roering 2004). In addition, LiDAR systems typically offer the advantage of gathering data over a narrow vertical swath angle (less than 20 off nadir). They also do not suffer from topographic shadowing. Moreover, LiDAR data are considerably easier to process than synthetic aperture radar information. Meanwhile, TLS and MLS systems add finer details to landslide investigations and allow modeling of individual landslides. TLS is highly useful in modeling landslide mechanisms as well as in delineating and estimating volume, whereas MLS provides rapid corridor mapping, thereby allowing roadside slope stability analysis and efficient rockfall simulations.

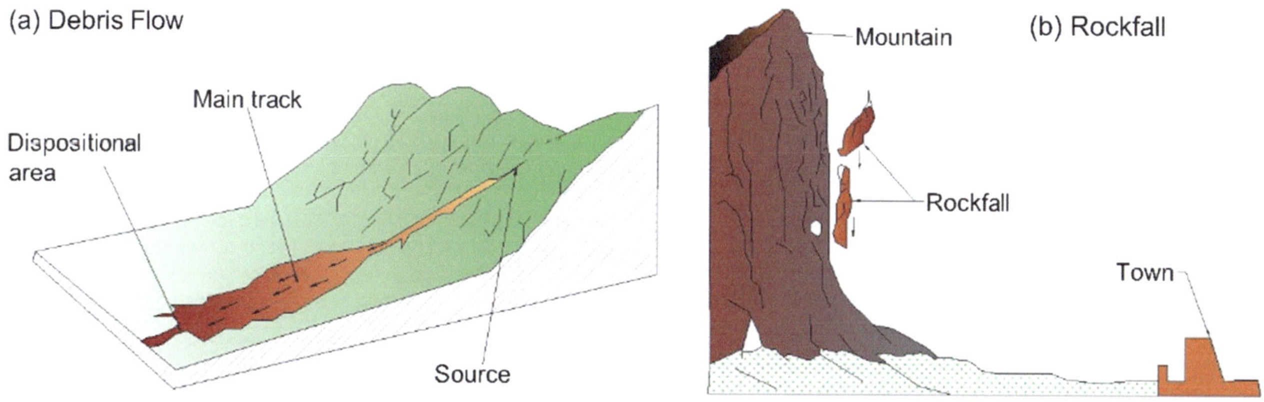

Fig. 1.8 Basic diagrams of debris flow and rockfall

The applications of ALS in landslide mapping are increasing with the advancement of these systems, particularly their vertical and horizontal accuracies. Although the main use of LiDAR point clouds is to generate a very high-resolution DEM, several products can be derived from terrain models, which greatly support landslide mapping and modeling. ALS is highly useful in regional landslide mapping, but its use in rockfall simulations and small-scale modeling is rare. Several studies have shown that LiDAR is effective for landslide inventory mapping, and the process can be automated, which reduces the time spent in the field and the subjectivity of experts. Automatic methods have been applied to produce landslide inventory maps using ALS data. Common methods include unsupervised classification, change detection, and object-based image analysis. The results are generally satisfactory; however, the object-based analysis is preferred over pixel-based methods mainly because per-pixel methods generate outputs that present a salt-and-pepper appearance and are mostly unverifiable on the ground. In addition, object-based methods utilize additional features, such as landslide shape, geometry, and contextual relationships between landslides and other man-made features. Combining LiDAR-derived factors, such as slope angle, slope aspect, drainage, curvature, and geometric features improves landslide inventory mapping and makes automatic landslide mapping possible. In addition, ALS point clouds are highly useful for identifying debris flow sources and producing runout distance susceptibility maps at regional and medium scales. Morphometric features, such as slope, roughness, and curvature, derived from very high-resolution DEMs are essential inputs in debris flow studies. Moreover, a very high accuracy DEM (0.5 m) from LiDAR with sophisticated analysis algorithms and tools can model rockfall propagations. At the regional scale, source areas are typically obtained by applying simple thresholds for the slope angle parameter. Consequently, the structural analysis performed on DEM can be used in rockfall susceptibility and hazard assessments.

TLS and MLS systems are frequently utilized to investigate the detailed characteristics of small-scale landslide modeling. TLS is effective for estimating volumes of mobilized materials and for characterizing rock instabilities. TLS is more efficient than ALS and MLS for such application because rock instabilities are controlled by locally planar structures. Highly dense point clouds captured by TLS and MLS also allow researchers to obtain accurate slope profiles and discontinuity set orientations. TLS enables detailed mapping of channels and section profile evolution for debris flows. In addition, a DEM with fine resolution improves site-specific modeling and monitoring of active landslides. Furthermore, mobile LiDAR is useful in producing detailed 3D models for road corridors and road cuts. It permits detailed structural analysis for rockfall simulation and active area detection.

The usefulness of laser scanning systems increases with their advancement. Improving modern surveying methods, laser scanner accuracy, and hardware capabilities can establish new landslide applications. The current limitations of laser scanning systems should be overcome to improve landslide studies. Handling huge amounts of data, improving data collection methods at shadow areas caused by rugged topography, enhancing data filtering, and optimizing LiDAR data analysis are expected in the next decade. With these improvements in laser scanning systems, landslide scarps can be accurately delineated, a landslide can be monitored at finer details, hazard and risk assessments can be performed at few centimeter scales, and the modeling of debris flow and rockfall can be improved.

References

Abellán, A., Oppikofer, T., Jaboyedoff, M., Rosser, N. J., Lim, M., & Lato, M. J. (2014). Terrestrial laser scanning of rock slope instabilities. *Earth Surface Processes and Landforms, 39*(1), 80–97.

Akgun, A., & Erkan, O. (2016). Landslide susceptibility mapping by geographical information system-based multivariate statistical and deterministic models: In an artificial reservoir area at Northern Turkey. *Arabian Journal of Geosciences, 9*(2), 1–15.

Al-Durgham, M., Fotopoulos, G., & Glennie, C. (2010). On the accuracy of Li-DAR derived digital surface models. *Gravity, geoid and earth observation* (pp. 689–695). Berlin: Springer.

Ardizzone, F., Cardinali, M., Galli, M., Guzzetti, F., & Reichenbach, P. (2007). Identification and mapping of recent rainfall-induced landslides using elevation data collected by airborne LiDAR. *Natural Hazards and Earth System Science, 7*(6), 637–650.

Axelsson, P. (2000). DEM generation from laser scanner data using adaptive TIN models. *International Archives of Photogrammetry and Remote Sensing, 33*(B4/1; PART 4), 111–118.

Ayalew, L., Yamagishi, H., & Ugawa, N. (2004). Landslide susceptibility map-ping using GIS-based weighted linear combination, the case in Tsugawa area of Agano River, Niigata Prefecture, Japan. *Landslides, 1*(1), 73–81.

Baltsavias, E. P. (1999). Airborne laser scanning: Basic relations and formulas. *ISPRS Journal of Photogrammetry and Remote Sensing, 54*(2), 199–214.

Barendse, P. E., Brian, M., & Machan, G. (2009). *In-place microelectromechanical system inclinometer strings: Evaluation of an evolving technology*. In Transportation Research Board 88th Annual Meeting (No. 09-2118).

Baumann, V., Wick, E., Horton, P., & Jaboyedoff, M. (2011, October). Debris flow susceptibility mapping at a regional scale along the National Road N7, Argentina. In *Proceedings of the 14th Pan-American conference on soil mechanics and geotechnical engineering* (pp. 2–6).

Bromhead, E., Curtis, R., & Schofield, W. (1988). Observation and adjustment of a geodetic survey network for measurement of landslide movement. In C. Bonnard & A. A. Balkema (Eds.), *Land-slides-Glissements De Terrains, Proceedings of 5th International Symposium on Landslides* (pp. 383–386). The Netherlands: Rotterdam.

Bui, D. T., Tuan, T. A., Klempe, H., Pradhan, B., & Revhaug, I. (2016). Spatial prediction models for shallow landslide hazards: A comparative assessment of the efficacy of support vector machines, artificial neural networks, kernel logistic regression, and logistic model tree. *Landslides, 13*(2), 361–378.

Burghaus, S., Bell, R., & Kuhlmann, H. (2009). *Improvement of a terrestric network for movement analysis of a complex landslide*. In Presentation at FIG Conference, Eilat, Israel.

Burrough, P. A., McDonnell, R. A., McDonnell, R., & Lloyd, C. D. (2015). *Principles of geographical information systems*. Oxford: Oxford University Press.

Canuti, P., Casagli, N., Catani, F., Falorni, G., & Farina, P. (2007). Integration of remote sensing techniques in different stages of landslide response. *Progress in Landslide Science* (pp. 251–260). Heidelberg: Springer.

Chen, X. L., Liu, C. G., Chang, Z. F., & Zhou, Q. (2016). The relationship between the slope angle and the landslide size derived from limit equilibrium simulations. *Geomorphology, 253,* 547–550.

Chow, T. E., & Hodgson, M. E. (2009). Effects of LiDAR post-spacing and DEM resolution to mean slope estimation. *International Journal of Geographical Information Science, 23*(10), 1277–1295.

Conforti, M., Aucelli, P. P., Robustelli, G., & Scarciglia, F. (2011). Geomorphology and GIS analysis for mapping gully erosion susceptibility in the Turbolo stream catchment (Northern Calabria, Italy). *Natural Hazards, 56*(3), 881–898.

Daehne, A., & Corsini, A. (2013). Kinematics of active earthflows revealed by digital image correlation and DEM subtraction techniques applied to multi-temporal LiDAR data. *Earth Surf Process Landforms, 38,* 640–654.

Dallaire, G. (1974). Electronic distance measuring revolution well underway. *Civil Engineering, 44*(10), 66–71.

De Blasio, F. V. (2011). *Introduction to the physics of landslides: Lecture notes on the dynamics of mass wasting*. Berlin: Springer Science & Business Media.

Derron, M. H., & Jaboyedoff, M. (2010). Preface "LiDAR and DEM techniques for landslides monitoring and characterization". *Natural Hazards and Earth System Sciences, 10,* 1877–1879.

Devkota, K. C., Regmi, A. D., Pourghasemi, H. R., Yoshida, K., Pradhan, B., Ryu, I. C., et al. (2013). Landslide susceptibility mapping using certainty factor, index of entropy and logistic regression models in GIS and their comparison at Mugling–Narayanghat road section in Nepal Himalaya. *Natural Hazards, 65*(1), 135–165.

Dou, J., Yamagishi, H., Pourghasemi, H. R., Yunus, A. P., Song, X., Xu, Y., et al. (2015). An integrated artificial neural network model for the landslide susceptibility assessment of Osado Island, Japan. *Natural Hazards, 78*(3), 1749–1776.

Eeckhaut, M., Poesen, J., Verstraeten, G., Vanacker, V., Nyssen, J., Moeyer-sons, J., et al. (2007). Use of LiDAR-derived images for mapping old landslides under forest. *Earth Surface Processes and Landforms, 32*(5), 754–769.

Fernandes, N. F., Guimarães, R. F., Gomes, R. A., Vieira, B. C., Montgomery, D. R., & Greenberg, H. (2004). Topographic controls of landslides in Rio de Janeiro: Field evidence and modeling. *CATENA, 55*(2), 163–181.

Fernández, T., Jiménez-Perálvarez, J. D., Fernández, P., El Hamdouni, R., Cardenal, F. J., Delgado, J., et al. (2008) Automatic detection of landslide features with remote sensing techniques in the Betic Cordilleras (Granada, Southern Spain). *The International Archives of the Photogrammetry, Remote Sensing and Spatial Information Sciences, XXXVII.* Part B8: 351–356. ISSN 1682–1750.

Fornaciai, A., Bisson, M., Landi, P., Mazzarini, F., & Pareschi, M. T. (2010). A LiDAR survey of Stromboli volcano (Italy): Digital elevation model-based geomorphology and intensity analysis. *International Journal of Remote Sensing, 31*(12), 3177–3194.

Ghuffar, S., Székely, B., Roncat, A., & Pfeifer, N. (2013). Landslide displacement monitoring using 3D range flow on airborne and terrestrial LiDAR data. *Remote Sensing, 5*(6), 2720–2745.

Glenn, N. F., Streutker, D. R., Chadwick, D. J., Thackray, G. D., & Dorsch, S. J. (2006). Analysis of LiDAR derived topographic information for characterizing and differentiating landslide morphology and activity. *Geomorphology, 73,* 131–148.

Guzzetti, F., Mondini, A. C., Cardinali, M., Fiorucci, F., Santangelo, M., & Chang, K. T. (2012). Landslide inventory maps: New tools for an old problem. *Earth-Science Reviews, 112,* 42–66.

Habib, A. (2008). *Accuracy, quality assurance and quality control of LiDAR data, Chap 9. Topographic laser ranging and scanning: Principles and processing* (pp. 269–294). Boca Rotan: CRC Press, Taylor & Francis.

Habib, A. F., Kersting, A. P., Shaker, A., & Yan, W. Y. (2011). Geometric calibration and radiometric correction of LiDAR data and their impact on the quality of derived products. *Sensors, 11*(9), 9069–9097.

Haneberg, W. C., Cole, W. F., & Kasali, G. (2009). High-resolution LiDAR-based landslide hazard mapping and modeling, UCSF Parnassus Campus, San Francisco, USA. *Bulletin of Engineering Geology and the Environment, 68,* 263–276.

Hanson, G. E. (1999). U.S. Patent No. 5,872,354. Washington, DC: U. S. Patent and Trademark Office.

Hervás, J., Barredo, J. I., Rosin, P. L., Pasuto, A., Mantovani, F., & Silvano, S. (2003). Monitoring landslides from optical remotely sensed imagery: The case history of Tessina landslide, Italy. *Geomorphology, 54,* 63–75.

Hungr, O., & Evans, S. G. (2004). The occurrence and classification of massive rock slope failure. *Felsbau, 22*(2), 16–23.

Hungr, O., Evans, S. G., Bovis, M. J., & Hutchinson, J. N. (2001). A review of the classification of landslides of the flow type. *Environmental and Engineering Geoscience, 7*(3), 221–238.

Jaboyedoff, M., Oppikofer, T., Abellán, A., Derron, M.-H., Loye, A., Metzger, R., et al. (2012). Use of LiDAR in landslide investigations: A review. *Natural Hazards, 61,* 5–28.

Jahromi, A. B., Zoej, M. J. V., Mohammadzadeh, A., & Sadeghian, S. (2011). A novel filtering algorithm for bare-earth extraction from airborne laser scanning data using an artificial neural network. *IEEE Journal of Selected Topics in Applied Earth Observations and Remote Sensing, 4*(4), 836–843.

Kannan, M., Saranathan, E., & Anabalagan, R. (2013). Landslide vulnerability mapping using frequency ratio model: A geospatial approach in Bodi-Bodimettu Ghat section. *Arabian Journal of Geosciences, 6*(8), 2901–2913.

Katz, O., Morgan, J. K., Aharonov, E., & Dugan, B. (2014). Controls on the size and geometry of landslides: Insights from discrete element numerical simulations. *Geomorphology, 220,* 104–113.

Krabill, W., Frederick, E., Manizade, S., Martin, C., Sonntag, J., Swift, R., et al. (1999). Rapid thinning of parts of the southern Greenland ice sheet. *Science (New York, N.Y.), 283*(5407), 1522–1524.

Krabill, W., Thomas, R., Martin, C., Swift, R., & Frederick, E. (1995). Accuracy of airborne laser altimetry over the Greenland ice sheet. *International Journal of Remote Sensing, 16*(7), 1211–1222.

Kraus, K., & Pfeifer, N. (1998). Determination of terrain models in wooded areas with airborne laser scanner data. *ISPRS Journal of Photogrammetry and remote Sensing, 53*(4), 193–203.

Kritikos, T., & Davies, T. (2015). Assessment of rainfall-generated shallow landslide/debris-flow susceptibility and runout using a GIS-based approach: Application to western Southern Alps of New Zealand. *Landslides, 12*(6), 1051–1075.

Lan, H., Martin, C. D., Zhou, C., & Lim, C. H. (2010). Rockfall hazard analysis using LiDAR and spatial modeling. *Geomorphology, 118*(1), 213–223.

Latif, Z. A., Aman, S. N. A., & Pradhan, B. (2012, March). Landslide susceptibility mapping using LiDAR derived factors and frequency ratio model: Ulu Klang area, Malaysia. In *IEEE 8th international colloquium on signal processing and its applications (CSPA)*, 2012 (pp. 378–382). IEEE.

Lemmens, M. (2011). *Geo-information: Technologies, applications and the environment* (Vol. 5). Berlin: Springer Science & Business Media.

Lichti, D. D. (2007). Error modeling, calibration and analysis of an AM–CW terrestrial laser scanner system. *ISPRS Journal of Photogrammetry and Remote Sensing, 61*(5), 307–324.

Lichti, D. D., & Jamtsho, S. (2006). Angular resolution of terrestrial laser scanners. *The Photogrammetric Record, 21*(114), 141–160.

Lindenberger, J. (1993). *Laser-Profilmessungen zur Topographischen Ge-landeaufnahme* (Doctoral dissertation, Zugl.: Stuttgart, Univ., Diss., 1992).

Lu, P., Stumpf, A., Kerle, N., & Casagli, N. (2011). Object-oriented change detection for landslide rapid mapping. *IEEE Geoscience and Remote Sensing Letters, 8*(4), 701–705.

Manetti, L., & Steinmann, G. (2007). 3DeMoN ROBOVEC-integration of a new measuring instrument in an existing generic remote monitoring platform. In *7th international symposium on field measurements in geomechanics. FMGM, (LMS-CONF-2008-035)* (pp. 1–12).

McKean, J., & Roering, J. (2004). Objective landslide detection and surface morphology mapping using high-resolution airborne laser altimetry. *Geomorphology, 57,* 331–351.

Moosavi, V., Talebi, A., & Shirmohammadi, B. (2014). Producing a landslide inventory map using pixel-based and object-oriented approaches optimized by Taguchi method. *Geomorphology, 204,* 646–656.

Nefeslioglu, H. A., Duman, T. Y., & Durmaz, S. (2008). Landslide susceptibility mapping for a part of tectonic Kelkit Valley (Eastern Black Sea region of Turkey). *Geomorphology, 94*(3), 401–418.

Nichol, J., & Wong, M. S. (2005). Satellite remote sensing for detailed landslide inventories using change detection and image fusion. *International Journal of Remote Sensing, 26*(9), 1913–1926.

Pesci, A., Teza, G., & Ventura, G. (2008). Remote sensing of volcanic terrains by terrestrial laser scanner: Preliminary reflectance and RGB implications for studying Vesuvius crater (Italy). *Annals of Geophysics, 51*(4), 633–653.

Petrie, G., & Toth, C. K. (2008). Introduction to laser ranging, profiling, and scanning. *Topographic Laser Ranging and Scanning: Principles and Processing*, 1–28.

Pourghasemi, H. R., Pradhan, B., & Gokceoglu, C. (2012). Application of fuzzy logic and analytical hierarchy process (AHP) to landslide susceptibility mapping at Haraz watershed. *Natural hazards, 63*(2), 965–996.

Pourghasemi, H. R., Pradhan, B., Gokceoglu, C., Mohammadi, M., & Moradi, H. R. (2013). Application of weights-of-evidence and certainty factor models and their comparison in landslide susceptibility mapping at Haraz watershed, Iran. *Arabian Journal of Geosciences, 6*(7), 2351–2365.

Pradhan, B. (2013). A comparative study on the predictive ability of the decision tree, support vector machine and neuro-fuzzy models in landslide susceptibility mapping using GIS. *Computers & Geosciences, 51,* 350–365.

Pradhan, B., Jebur, M. N., Shafri, H. Z. M., & Tehrany, M. S. (2016). Data fusion technique using wavelet transform and Taguchi methods for automatic landslide detection from airborne laser scanning data and quickbird satellite imagery. *IEEE Transactions on Geoscience and Remote Sensing, 54*(3), 1610–1622.

Pradhan, A. M. S., & Kim, Y. T. (2014). Relative effect method of landslide susceptibility zonation in weathered granite soil: A case study in Deokjeokri Creek, South Korea. *Natural Hazards, 72*(2), 1189–1217.

Regmi, A. D., Devkota, K. C., Yoshida, K., Pradhan, B., Pourghasemi, H. R., Kumamoto, T., et al. (2014). Application of frequency ratio, statistical index, and weights-of-evidence models and their comparison in landslide susceptibility mapping in Central Nepal Himalaya. *Arabian Journal of Geosciences, 7*(2), 725–742.

Roering, J. J., Mackey, B. H., Marshall, J. A., Sweeney, K. E., Deligne, N. I., Booth, A. M., et al. (2013). Connecting the dots with airborne LiDAR for geomorphic fieldwork. *Geomorphology, 200,* 172–183.

Scaioni, M., Feng, Y., Lu, P., Qiao, G., Tong, X., Li, R., et al. (2014). Close-range photogrammetric techniques for deformation measurement: Applications to landslides. In M. Scaioni (Ed.), *Modern technologies for landslide investigation and prediction* (pp. 13–41). Berlin: Springer.

Schulz, W. H. (2004). Landslides mapped using LiDAR imagery, Seattle, Washington. *US Geological Survey Open-File Report, 1396*(11).

Schwalbe, E., & Maas, H. G. (2009). Motion analysis of fast flowing glaciers from multi-temporal terrestrial laser scanning. *Photogrammetrie-Fernerkundung-Geoinformation, 2009*(1), 91–98.

Sithole, G., & Vosselman, G. (2001). Filtering of laser altimetry data using a slope adaptive filter. *International Archives of Photogrammetry Remote Sensing and Spatial Information Sciences, 34*(3/W4), 203–210.

Slatton, K. C., Carter, W. E., Shrestha, R. L., & Dietrich, W. (2007). Airborne laser swath mapping: Achieving the resolution and accuracy required for geosurficial research. *Geophysical Research Letters, 34*(23).

Takahashi, T. (2014). *Debris flow: Mechanics, prediction and countermeasures*. Boca Raton: CRC Press.

Tarboton, D. G. (1997). A new method for the determination of flow directions and upslope areas in grid digital elevation models. *Water Resources Research, 33*(2), 309–319.

Tien Bui, D., Pradhan, B., Lofman, O., & Revhaug, I. (2012). Landslide susceptibility assessment in Vietnam using support vector machines, decision tree, and Naive Bayes Models. *Mathematical Problems in Engineering*.

Toutin, T. (2002). Impact of terrain slope and aspect on radargrammetric DEM accuracy. *ISPRS Journal of Photogrammetry and Remote Sensing, 57*(3), 228–240.

Van Den Eeckhaut, M., Poesen, J., Verstraeten, G., Vanacker, V., Moeyersons, J., Nyssen, J., et al. (2005). The effectiveness of hillshade maps and expert knowledge in mapping old deep-seated landslides. *Geomorphology, 67*(3), 351–363.

Van Westen, C. J., Castellanos, E., & Kuriakose, S. K. (2008). Spatial data for landslide susceptibility, hazard, and vulnerability assessment: An overview. *Engineering Geology, 102,* 112–131.

van Westen, C. J., van Asch, T. W. J., & Soeters, R. (2006). Landslide hazard and risk zonation—Why is it still so difficult? *Bulletin of Engineering Geology and the Environment, 65,* 167–184.

Vosselman, G. (2000). Slope based filtering of laser altimetry data. *International Archives of Photogrammetry and Remote Sensing, 33* (B3/2; PART 3), 935–942.

Wang, G., Joyce, J., Phillips, D., Shrestha, R., & Carter, W. (2013). Delineating and de-fining the boundaries of an active landslide in the rainforest of Puerto Rico using a combination of airborne and terrestrial LiDAR data. *Landslides, 10*(4), 503–513.

Wehr, A., & Lohr, U. (1999). Airborne laser scanning—An introduction and overview. *ISPRS Journal of Photogrammetry and Remote Sensing, 54*(2), 68–82.

Williams, K., Olsen, M. J., Roe, G. V., & Glennie, C. (2013). Synthesis of transportation applications of mobile LiDAR. *Remote Sensing, 5*(9), 4652–4692.

Yalcin, A. (2008). GIS-based landslide susceptibility mapping using analytical hierarchy process and bivariate statistics in Ardesen (Turkey): comparisons of results and confirmations. *Catena, 72*(1), 1–12.

Yalcin, A., & Bulut, F. (2007). Landslide susceptibility mapping using GIS and digital photogrammetric techniques: A case study from Ardesen (NE-Turkey). *Natural Hazards, 41*(1), 201–226.

Yan, W. Y., Shaker, A., Habib, A., & Kersting, A. P. (2012). Improving classification accuracy of airborne LiDAR intensity data by geometric calibration and radiometric correction. *ISPRS Journal of Photogrammetry and Remote Sensing, 67,* 35–44.

Yesilnacar, E., & Topal, T. (2005). Landslide susceptibility mapping: A com-parison of logistic regression and neural networks methods in a medium scale study, Hendek region (Turkey). *Engineering Geology, 79*(3), 251–266.

Zhan, Z., & Lai, B. (2015). A novel DSM filtering algorithm for landslide monitoring based on multiconstraints. *IEEE Journal of Selected Topics in Applied Earth Observations and Remote Sensing, 8*(1), 324–331.

Zhang, K., Chen, S. C., Whitman, D., Shyu, M. L., Yan, J., & Zhang, C. (2003). A progressive morphological filter for removing nonground measurements from airborne LiDAR data. *IEEE Transactions on Geoscience and Remote Sensing, 41*(4), 872–882.

Zhang, Q., Wang, L., Zhang, X. Y., Huang, G. W., Ding, X. L., Dai, W. J., & Yang, W. T. (2008). Application of multi-antenna GPS technique in the stability monitoring of roadside slopes. In *Landslides and engineered slopes: From the past to the future. Proceedings of the tenth international symposium on landslides and engineered slopes* (pp. 1367–1372). Xi'an: Taylor & Francis.

Zhang, Y., Xiong, X., & Hu, X. (2013). Rigorous LiDAR strip adjustment with tri-angulated aerial imagery. *ISPRS Annals of Photogrammetry, Remote Sensing and Spatial Information Sciences, 5*(W2), 361–366.

Zhang, Y., Xiong, X., Zheng, M., & Huang, X. (2015). LiDAR strip adjustment using multifeatures matched with aerial images. *IEEE Transactions on Geoscience and Remote Sensing, 53*(2), 976–987.

Part II

Landslide Detection Techniques

2 A Supervised Object-Based Detection of Landslides and Man-Made Slopes Using Airborne Laser Scanning Data

Biswajeet Pradhan and Ali Alsaleh

2.1 Introduction

In recent years, airborne-derived products from light detection and ranging (LiDAR) measurements, such as high-resolution digital elevation models (DEMs), slope, curvature, shaded relief, and maps of landslides obtained from beneath dense vegetation, are becoming increasingly important for producing a detailed landslide inventory map (Eeckhaut et al. 2007). LiDAR applications include the construction of DEMs and shaded relief maps, detection of historical landslides under forested area, creation of topographic contours, tracking of multitemporal digital terrain model (DTM) of landslides, hydrological modeling, landform and/or soil classification, and understanding fine-scale landslide patterns (McKean and Roering 2004). The rapidness of LiDAR technology in landslide mapping of terrains through quantitative or visual analysis provides several advantages.

Landslide inventory maps provide baseline information of landslide types, location, distribution, and boundaries in landslide-prone areas. In addition, landslide inventory provides information on displacement and slope measurements that affect a failure (Galli et al. 2008). Moreover, landslide inventories are significantly useful for various purposes, such as recording of landslide magnitude, implementing the initial stage for landslide susceptibility, and hazard and risk assessments.

Object-based image analysis (OBIA) is a well-known technique resulting from the recent advances in computer vision and machine intelligence, with the main purpose of automatically extracting both man-made and natural objects from remote sensing images (Akcay and Aksoy 2008). OBIA, in which the information content of an object is used to classify a landscape, is a step toward replicating human interpretation process (Navulur 2006). In addition, OBIA can detect landslides accurately and meaningfully by integrating contextual information to image analysis (Martha 2011), which reduces the time and cost for producing a decent landslide inventory map, especially in large areas. Several techniques have been proposed for landslide mapping, such as field observation and aerial photointerpretation; however, these techniques have some limitations, such as lack of proper resolution for aerial photographs required for the mapping of small-scale landslides caused by morphologic feature obscuration by thick vegetation cover, time-consuming, and difficulties in field mapping (Gorum et al. 2011). Remote sensing data and methods have been proven efficient in landslide mapping because of their wide area coverage, relatively cheap cost, and remarkably high-resolution data for landslide mapping, in which even minor landslides can be mapped easily. Landslide inventory maps resulting from the application of OBIA techniques can be easily converted to GIS data, which is considered as an initial stage for a more advance analysis, such as susceptibility, and hazard and risk analysis.

Landslides can be triggered by various factors. These factors can be man-made (such as mineral mining, road cutting, and urbanization) or natural (e.g., extreme rainfall events and earthquakes (Zêzere et al. 1999). However, in tropical areas, a rapid and accurate method for landslide mapping is required because of the rapid growth of vegetation that covers the land surface characteristics in those areas. Furthermore, several cut slopes are generally created to mitigate the risk of land failure for areas that have high probability to fail because of the frequent occurrence of landslides. These requirements create new challenges for landslide identification and mapping in these areas. Thus, new methods should be developed for automatic landslide detection to produce high-quality landslide inventory map.

B. Pradhan (✉) · A. Alsaleh
Department of Civil Engineering, University Putra Malaysia, Serdang, Malaysia
e-mail: biswajeet24@gmail.com

B. Pradhan (ed.), *Laser Scanning Applications in Landslide Assessment*,
DOI 10.1007/978-3-319-55342-9_2

2.2 Study Area and Data

2.2.1 Location of Study Area

Taman Ringlet is the first town along the Cameron Highlands stretched from Tapah, Cameron Highlands, Peninsular Malaysia. This town is a small hill that is well known for tea and strawberry farming. The town is located at an altitude of 1140 m above the sea level. Geographically, Taman Ringlet is located at latitude 04° 24′ 45″N and longitude 101° 23′ 30″E.

Three different subsets were selected for analysis, as shown in Fig. 2.1. One of the subsets was used to develop the methodology of landslide detection, whereas the other two were used to test the proposed method in other areas. The process of selecting subset locations was carefully implemented, with each subset having the same land cover classes but with different coverage percentage. Figure 2.1 shows that the training site and Testing Site 1 have more urban coverage compared to subset Testing Site 2, where thick vegetation covers almost the entire area.

2.2.2 LiDAR Data

Study and data collection was implemented over Ringlet and nearby surrounding area, which covered a total area of 25 km^2. The LiDAR data were recorded for the entire 25 km^2 with a flight height of 1510 m. Data were obtained on January 15, 2015. Data capturing performed well with eight points per square m and gave a pulse rate frequency of 25,000 Hz. Furthermore, the captured data were within the root-mean-square of 0.15 and 0.3 in the vertical and horizontal axes, respectively; thus, the accuracy of the captured data was reasonable. Along with LiDAR point clouds, orthophotos were also collected by the same system, as shown in Fig. 2.2.

2.2.3 Geological Characteristics of the Study Area

Cameron Highlands District is located in the eastern part of the main range, which is composed of granites (Bignel and

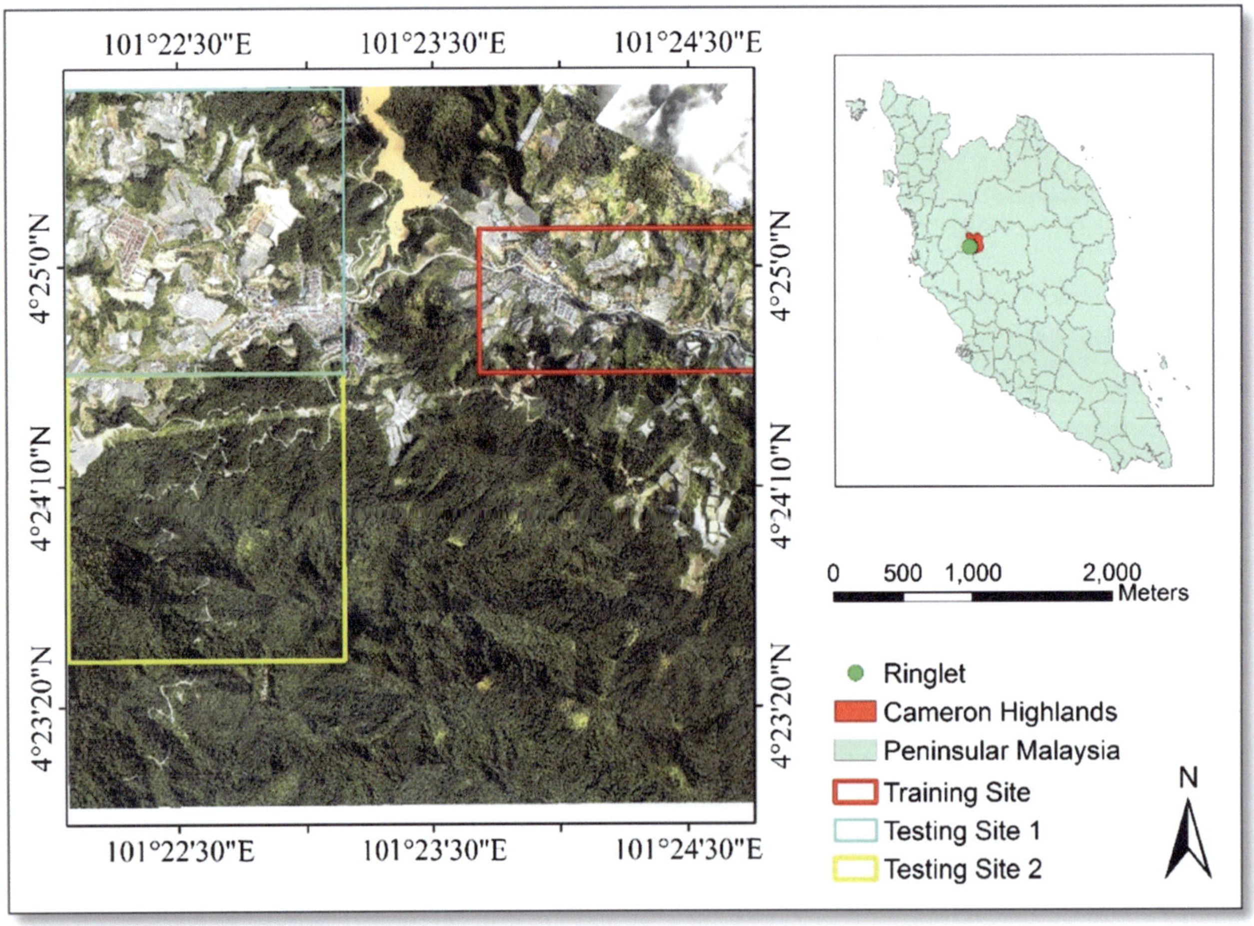

Fig. 2.1 Geographic location of the study area

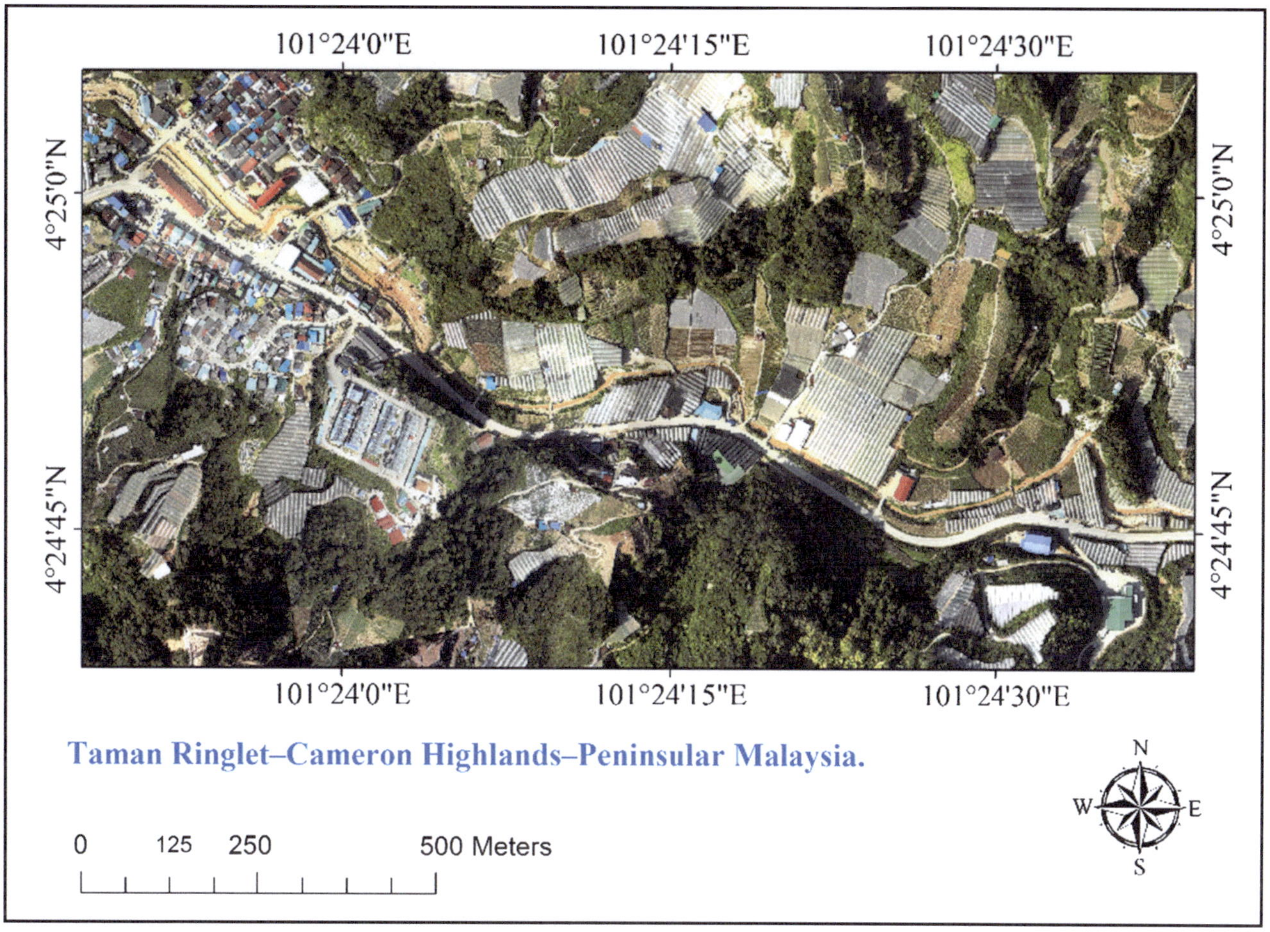

Fig. 2.2 Aerial photograph of the training site

Snelling 1977). However, scattered outliers (roof pendants) of metasediments can be seen clearly in Fig. 2.3. Bignel and Snelling (1977) classified the granites of Cameron Highlands District as megacrystic biotite granite. Cobbing et al. (1992) mentioned that the associated microgranite and some of the granites may be mineralized and may contain muscovite. Furthermore, Chow et al. (2003) stated that metasediments consist of phyllite, schist, limestone, and slate. They also stated that minor intercalations of volcanic rocks and sandstones were found. Figure 2.3 shows the geological map of the study area and its surrounding areas. Post-Triassic–Mesozoic granites comprise most of the granite rocks, whereas a few are patches of metamorphic rocks that are mostly composed of Silurian–Ordovician schist, phyllite, limestone, and sandstone. As for the soil type, steep land soil covers the entire scene.

2.3 Methodology

Figure 2.4 shows the overall flowchart of the methodology implemented in this study. High-resolution LiDAR data with 1-m spatial resolution were used as a main data along with aerial photographs covering the Ringlet and its surrounding regions. For the ancillary data, a landslide inventory map showing the location of historical landslides was used. The overall methodology comprises three main phases: The first phase is the pre-processing and preparation of data; the second phase is image segmentation and object creation; and the final phase is image classification, and the detection of landslides and man-made slopes. The third step also includes result validation using a landslide inventory map created from field investigations based on site visits.

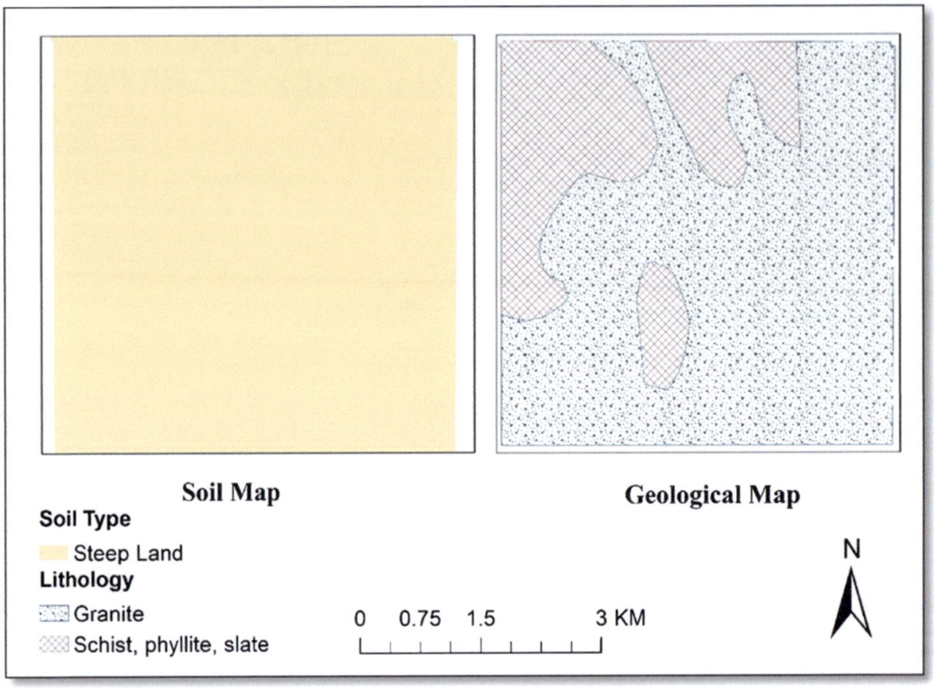

Fig. 2.3 Geological characteristic maps of the entire study area

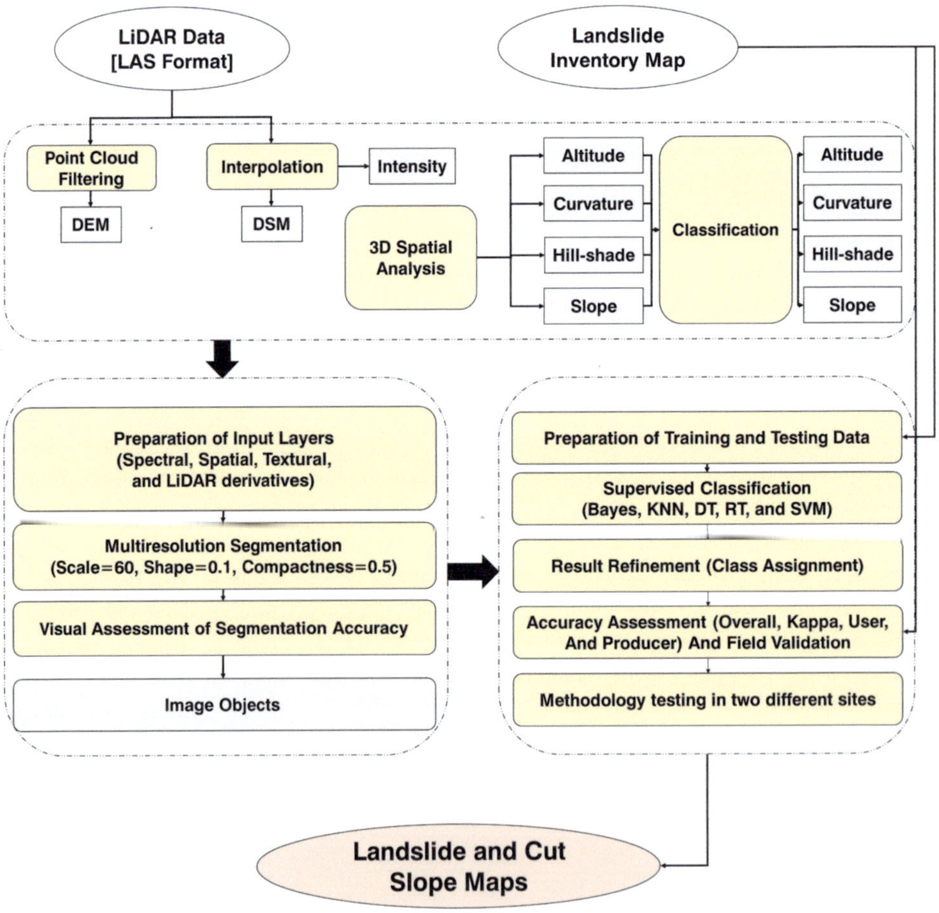

Fig. 2.4 Overall methodology flowchart

2.3.1 Data Pre-processing and Preparation of Landslide Factors

Calibrated raw LiDAR data are typically processed into flight lines of 3D points and saved in LASer (LAS) format. The first step of LiDAR data pre-processing is a visual review of the flight lines to inspect the breaks or gaps between or within flight lines (Chen et al. 2004). Afterward, LiDAR data were validated with several control points. These data were obtained from the field to ensure accurate collection. Typically, these known features are in open areas free of vegetation, such as airport runways, roads, and targeted survey points. Some systematic errors can be corrected by additional bore sighting and calibration; however, if the digital data are inaccurate or corrupted, the mission may have to be repeated (Chen et al. 2004). Basically, the two former steps should be done before the acquisition team leaves the field. If errors are discovered at this stage, the system can be redeployed for another mission. After the LiDAR data were reviewed and a few reference points were validated, noise points were then filtered out; these noises could be of an extremely high or low value with unrealistic elevation values or with unexpected values in the project area (Fang and Huang 2004). After noise removal, layer extraction was done by attributing all the remaining points into layers using the multiple return system (Hodgson et al. 2005).

Subsequently, landslide conditioning factors (i.e., digital surface model (DSM), DTM, curvature, slope, hill shade, and altitude) were prepared. This process starts with the conversion of the LAS format to raster using ArcMap software by applying the Triangular Irregular Network (TIN)-based interpolation to create the DSM layer (Fig. 2.5a) with a spatial resolution of 1 m. Next, DEM layer (Fig. 2.5b) was generated by filtering out the point cloud into ground and non-ground points using the former interpolation technique. The 3D spatial analysis tool in ArcMap software was used to extract slope, curvature, hill shade, and altitude layers (Fig. 2.5c–f, respectively) from the DSM and DEM layers. Evidently, slope is the principal factor affecting landslide occurrences (Pradhan and Lee 2010). Slope is considered as an important factor for land stability because of its direct impact on landslide phenomenology (Martha et al. 2011). That is, a steeper slope means higher risk of landslide caused by gravity-induced high shear (Long 2008). The hill-shade map shows a good image and movement of the terrain, which supports landslide mapping (Olaya 2009). Curvature layer defines the convex/concave character of the surface. Curvature values are calculated as positive, negative, and zero values, which refers to concavity, convexity, and flattening ground surface, respectively (Pradhan and Lee 2010). Plan curvature is considered important, because it reliably indicates convergence and divergence of slope surfaces in depletion (concave forms of the landslide crowns, tension cracks and depressions, and zones of local water accumulation) and accumulation zones (convex forms of the landslide foot and toe; (Ohlmacher 2007).

2.3.2 Image Segmentation

Before classifying a feature of interest, such as landslides, delineating image objects that separately or aggregately discriminate a specific feature (i.e., trees, buildings, and parcels of land) is important. This process is called image segmentation, which divides an image into objects or regions based on the homogeneity of pixel values (Martha 2011). The precision and quality of segmentation have a direct impact on the accuracy of the generated classification map (Laliberte et al. 2004).

This research analysis was conducted in eCognition software. Several types of algorithms can be found for the purpose of image segmentation, with multiresolution, quad tree, and chessboard being the most efficient ones (Definiens 2007). These algorithms provide an effective application for segmentation and perform good accuracy results.

Multiresolution algorithm, which belongs to the region-based algorithm category, was utilized in this study for segmentation (Möller et al. 2007). This algorithm performs various steps, which is initiated with one pixel and continues until all the criteria specified by the user are covered (Benz et al. 2004). Multiresolution segmentation algorithm uses three parameters: scale, shape, and compactness. Selecting the value of these three parameters should be carefully implemented to achieve meaningful classification results (Gibril et al. 2016). In this study, a trial-and-error approach was used to select the parameters, and evaluation was based on visual interpretation.

2.3.3 Classification

Image segmentation was also examined visually. The software calculated different parameters for each object, including the mean of slope, curvature, DEM, brightness, and density, as well as geometrical parameters, such as shape index, texture, length/width, area, and compactness. Each of these parameters was later used to classify an image object into several classes using supervised object-based classification approach.

2.3.3.1 Classifier

Classifier algorithm allows analysts to apply machine-learning functions in a two-step process. First, a classifier is trained using the classified domain objects as training samples. The trained classifier was stored as a string variable in the configuration settings. Second, the trained classifier was applied to the domain, classifying the image objects according to the trained parameters. Classification accuracy ensures the proper selection of the sampling method (Chen et al. 2014). In this study, training samples were selected randomly and distributed fairly over the entire study area, in which 60% of the samples were used for training purpose and 40% were used for testing the result.

Generally, five different algorithms [i.e., Bayes, *k*-nearest neighbor (*k*-NN), decision tree (DT), random forest (RF), and support vector machine (SVM)] can be applied to the classifier algorithm. Each of the aforementioned algorithms is best suited for a specific purpose. In this study, these algorithms were tested to identify the optimum algorithm for landslide and cut slope detection.

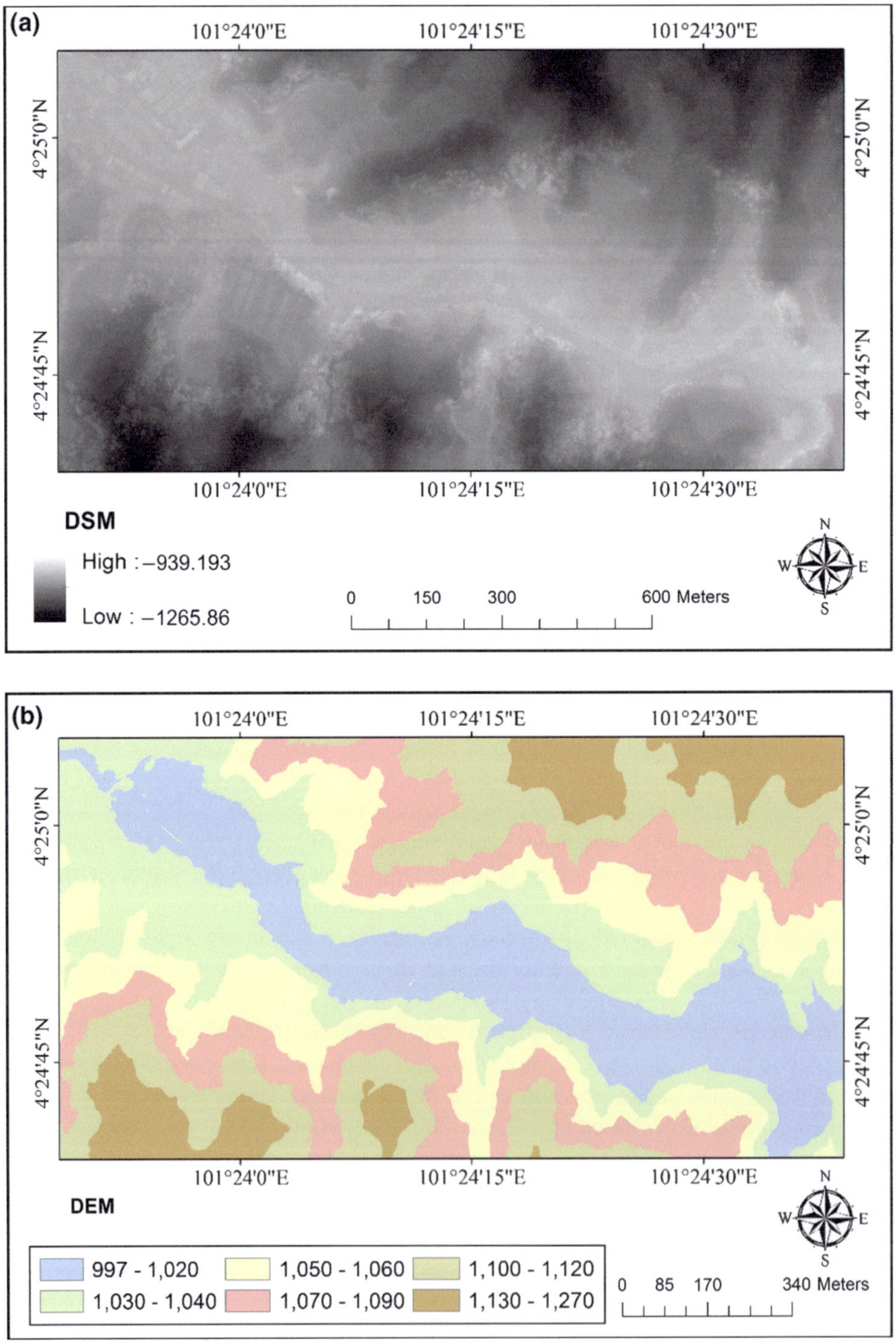

Fig. 2.5 Landslide conditioning factors: **a** DSM; **b** DEM; **c** slope; **d** curvature; **e** altitude; and **f** hill shade

Fig. 2.5 (continued)

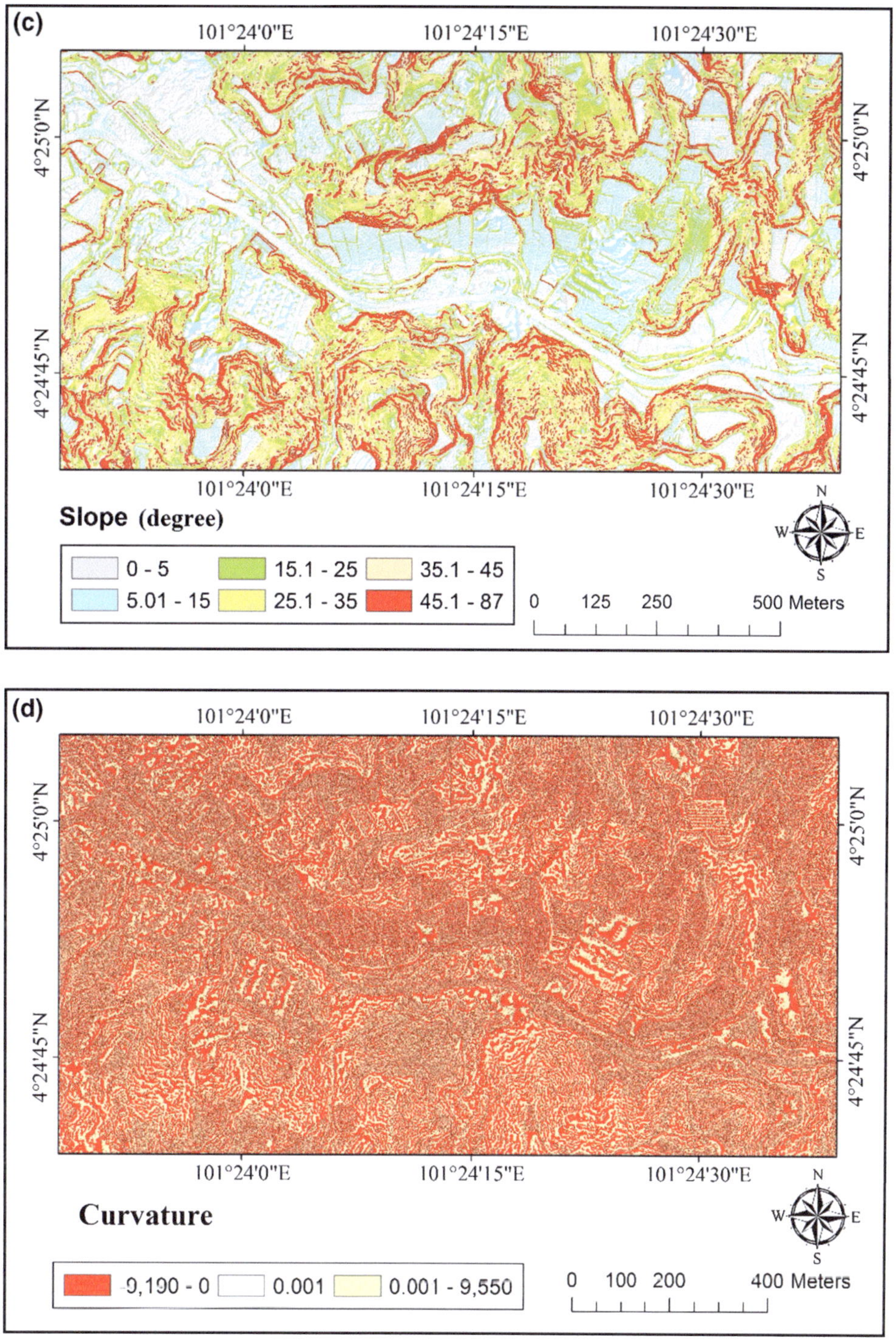

2.3.3.2 Bayes

Conditional independence assumption is a machine-learning classification system derived from Bayes' theorem, which strongly supports the assumption of the independence between features (Soria et al. 2011). One advantage of this classifier is the simplicity of its construction, which does not need any complicated estimation schemes of iterative parameters (Wu et al. 2008). In addition, Naive Bayes (NB) classifier is unaffected by noise or irrelevant attributes. Numerous successful experiments and studies of this classifier have been conducted in the literature (Xie et al. 2005).

Given an observation consisting of k-attributes x_i, $i = 1, 2\ldots, K$ (x_i is a landslide conditioning factor), and y_j, j = landslide, man-made slopes are the output class. NB

Fig. 2.5 (continued)

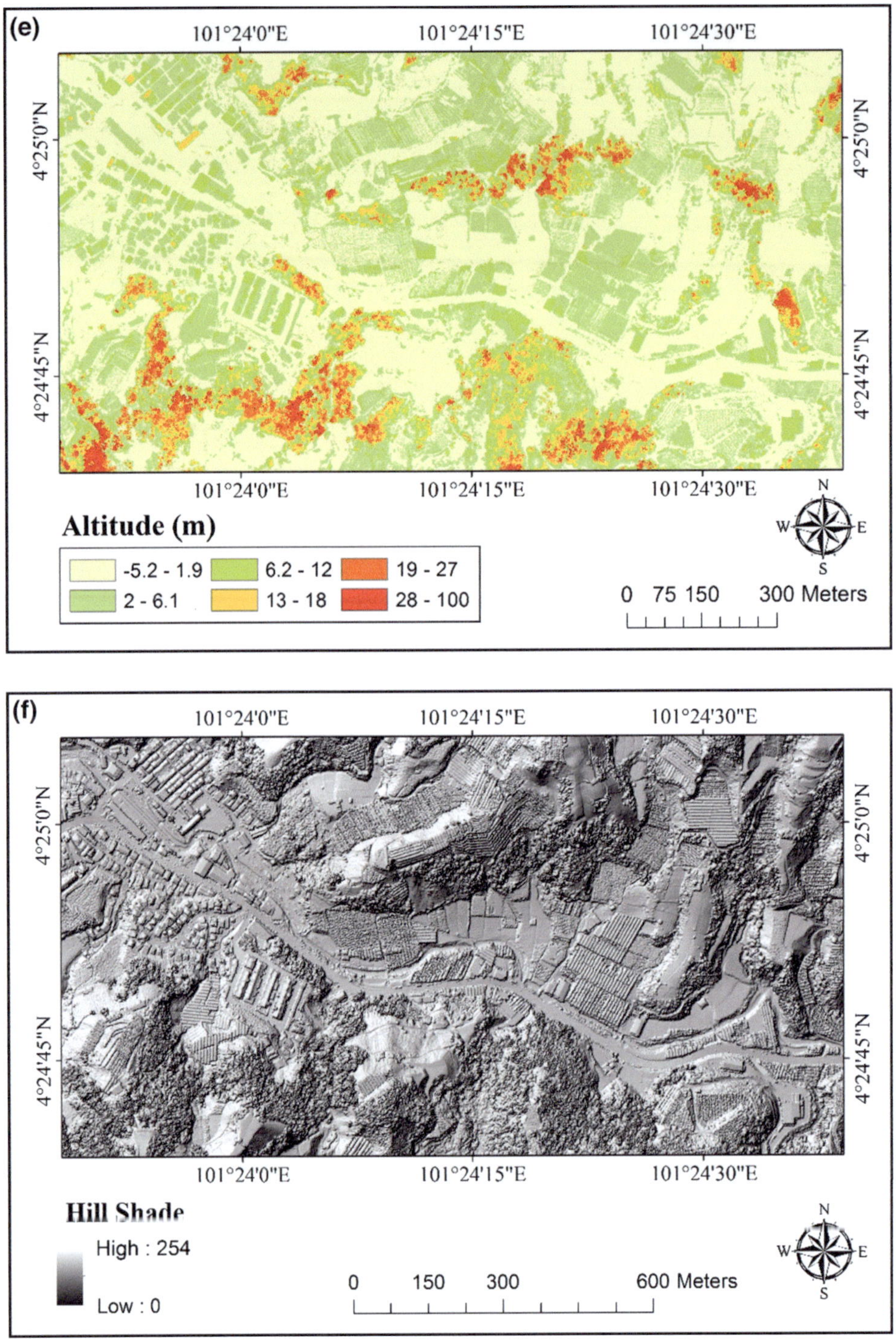

estimates the probability $P(y_j/x_i)$ for all possible output class. The class can be predicted depending on the largest posterior probability using Eq. (2.1).

$$y_{NB} = \underset{y_j \in [\text{landslide, non-landslide}]}{\arg\max} P(y_j) \prod_{i=1}^{n} P(x_i/y_j) \quad (2.1)$$

The prior probability $P(y_j)$ can be estimated using the proportion of the observations with output class y_j in the training dataset. Conditional probability is calculated using Eq. (2.2).

$$p\left(\frac{x_i}{y_j}\right) = \frac{1}{\sqrt{2\pi}\delta} e^{-(x_i-\mu)^2/2\delta^2}, \quad (2.2)$$

where μ is mean, and δ is standard deviation of x_i.

The Bayes classifier has a simple design and assumptions and was applied successfully in many practical situations.

The basic assumption of this classifier (conditional independence) is rarely true in real-world applications (Zhang 2004). Caruana and Niculescu-Mizil (2006) applied a comprehensive comparison with other classification methods, which showed a better performance compared with other approaches (e.g., boosted tree). Friedman et al. (1997) stated that Bayes classifier requires only a small number of training data to evaluate the necessary classification parameters, which is considered an advantage.

2.3.3.3 *k*-NN

k-NN is one of the simplest algorithms (Mitchell 1997); *k*-NN classifies pixel instance x containing x_i coordinates (including an n-dimensional input space $x = (x_1, x_2, \ldots, x_n)| x \in R^n$, where dimensions represent the values of the conditioning factors related to that particular pixel) by class values c_j of the k-closest neighboring pixels x_r surrounding x (c_j is previously assigned in the training set by a practitioner as $fc(x_r)$). The nearest neighbors are defined in terms of Euclidean distance $d(x, x_r)$. Thus, the classifier initially calculates the distances to k-neighbors for each x instance in the training set. Subsequently, a simple voting system assigns c_j class value (landslide class) to that particular pixel by class, which predominates the neighboring instances (Eq. 2.3), or alternatively assigns its mean value to the pixel if the data are ordinal numerical [Eq. (2.4); Fig. 2.6].

$$f_c' \leftarrow \arg\max \sum_{i=1}^{k} f(d(x, x_r), f_c(x_r)), \quad \forall (x \wedge x_r) \in \text{nominal data type} \tag{2.3}$$

$$f_c' \leftarrow \arg\max \frac{\sum_{i=1}^{k} f_{c(x_r)}}{k}; \quad \forall (x \wedge x_r) \in \text{ordinal data type} \tag{2.4}$$

Typically, *k*-NN does need conventional training/testing procedures; f_c' is simply calculated based on the remaining (testing) part of the dataset, which is similar with the training mode (Varmuza and Filzmoser 2016). The number of neighbors is necessarily an odd number (k = 1, 3, 5, 7…) to avoid even votes. Closer neighbors tend to have a greater impact; thus, it is further desirable to ponder the proximity of each neighbor, thereby upgrading to weighted *k*-NN (Mitchell 1997). Thus, the algorithm becomes global (Sheppard's method) but requires sorting and weighting of distances per each pixel element (and each conditioning factor is assigned to it) in the training set, resulting in a hardware-demanding and time-consuming procedure.

Therefore, *k*-NN classifier can be biased if all, relevant and irrelevant, conditioning factors are fed together to the algorithm because it builds a weighted or regular *k*-NN relation per each conditioning factor, thus misleading the classification. In other words, *k*-NN is extremely sensitive to

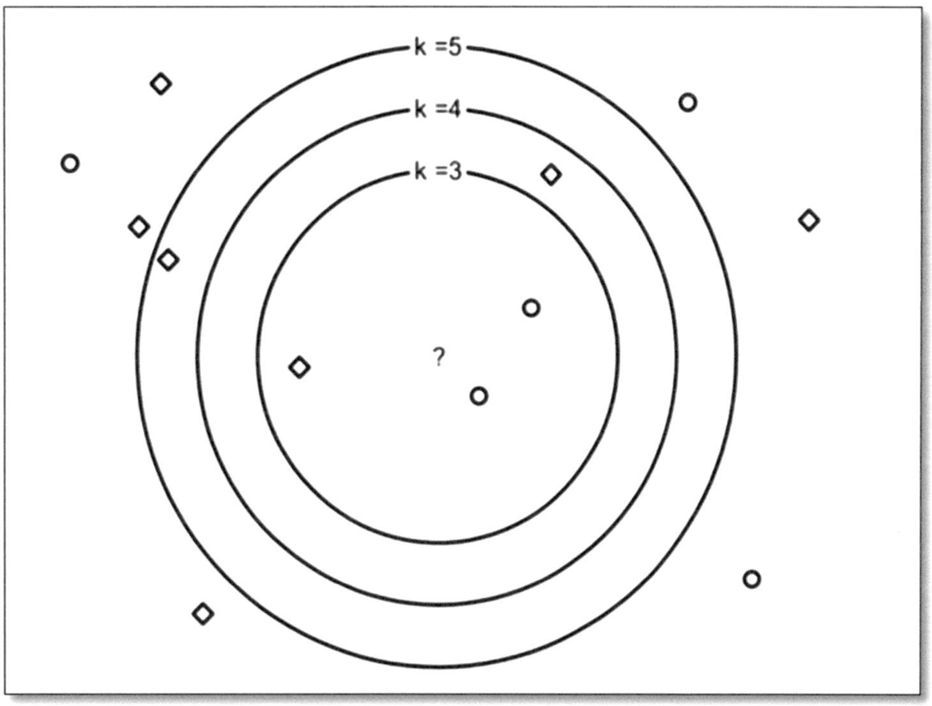

Fig. 2.6 *k*-NN classification principle. Unclassified instance (?) is classified by the majority of neighbors into landslide (*circle*) or non-landslide (*square*) instance. Note that for $k = 3$, the instance is classified as landslide; for $k = 4$, the instance remains unclassified (2:2 even votes); and for $k = 5$, the instance is classified as non-landslide

the relevance of the conditioning factor with landslide occurrence; thus, a strict attribute selection should be performed prior to the analysis. Alternatively, Euclidean distance axis can be stretched in the case of weighted k-NN, so different conditioning factors will have different weights according to their relevance. Nonetheless, this process does not solve the computational demands of this algorithm, especially when mixed data types are present, which results in a double procedure because of varied distance calculations.

Nevertheless, the distances are the classification criteria; thus, k-NN algorithm is straightforward and does not require a true black-box model. Furthermore, the algorithm can originate from a remarkably sparse data that are randomly sampled throughout the training set, which are sometimes convenient but are of little relevance to the concept of landslide assessment and prediction of the spatial landslide distribution. k-NN classifier is also convenient for experimenting because it only needs one parameter, which is the number of k-neighbors k, to be optimized.

2.3.3.4 SVM

Support vector machine (SVM) is a supervised learning method that analyzes data and recognizes patterns. In other words, given a labeled training data (supervised learning), the algorithm outputs an optimal hyperplane that categorizes new samples (Vapnik and Vapnik 1998). SVM classifies the original entry space into a more detailed feature space using training samples. Thereafter, the ideal hyperplane within this feature space is assigned by doubling the class boundary margins (Abe 2005). The nearest training samples to the ideal hyperplane are called support vectors. After determining the decision surface, it will be used to classify new data. Consider a training dataset of instance-labeled pairs (x_i, y_i) with $x_i \in R^n$, $y_i \in (1, -1)$, and $i = 1 \ldots m$. In this study of landslide and man-made cut slope detection, x is a vector of entry space, which includes slope, curvature, hill shade, soil type, distance to road, and altitude.

The two classes (1, −1) stand for the pixels of landslide and man-made slope, respectively. Finding the ideal hyperplane separation that discriminates the two classes from the set of training data is the aim of SVM classification. In case of linear data separation, a separating hyperplane can be defined as follows:

$$y_i(w \cdot x_i + b) \geq 1 - \delta_i, \qquad (2.5)$$

where w is a coefficient vector that determines the orientation of the hyperplane in the feature space, b is the offset of the hyperplane from the origin, and δ_i is the positive slack variables (Cortes and Vapnik 1995).

Determining an optimal hyperplane leads to solving the following optimization problem using Lagrangian multipliers (Samui 2008):

$$\text{minimize} \quad \sum_{i=1}^{n} a_i - \frac{1}{2}\sum_{i=1}^{n}\sum_{j=1}^{n} a_i a_j y_i y_j (x_i x_j) \qquad (2.6)$$

$$\text{subjected to} \quad \sum_{i=1}^{n} a_i y_i = 0, \quad 0 \leq a_j \leq C, \qquad (2.7)$$

where a_i are Lagrange multipliers, C is the penalty, and the slack variables δ_i allow the violation of penalized constraint.

The decision function, which is used to classify new data, can then be written as

$$g(x) = \text{sign}\left(\sum_{i=1}^{n} y_i a_i x_i + b\right). \qquad (2.8)$$

In some cases, where determining the separating hyperplane is impossible through the linear kernel function, data entry can be transferred to a high-dimensional feature space using a few nonlinear kernel equations. The classification decision equation is then written as

$$g(x) = \text{sign}\left(\sum_{i=1}^{n} y_i a_i k(x_i x_j) + b\right), \qquad (2.9)$$

where $k(x_i x_j)$ is the kernel function.

2.3.3.5 DT

Decision tree (DT) is a nonparametric supervised learning method that is usually used for data mining. In this method, a series of decisions are made to segment the data into homogeneous subgroups. DT model is more likely to look like a tree with several branches. In some cases, DT can be remarkably complex with the involvement of a large number of splits and nodes. DT aims to build a model that can estimate the value of a target variable depending on several input variables regarded as training samples. The tree model can be learned by breaking the main set into subsets depending on an attribute value test. Thereafter, this operation is repeated for each derived subset in a repetitive manner called recursive partitioning (Last et al. 2002). Once the subset of all nodes has the same value as the target variable, or when the breaking operation does not add any more value to the predictions, the recursion step is considered complete. The main objective of using a tree-building algorithm is to determine a set of if-then logical or split conditions.

Important DT Parameters

The minimum number of samples needed per node that are defined by the parameter is called Min sample count. Finding the optimum-sized tree can be challenging, because the prediction of a tree with a few splits may be inaccurate. Conversely, a tree with a multitude splits will add unnecessary complications to the analysis operation. Cross-validation can be performed to address this issue by setting cross-validation folds using eCognition parameters. In this process, computing for the classification tree is done by learning the samples and then evaluating the prediction accuracy by testing these samples. Cross-validation gives a poor indication in cases where the test sample cost is more than the learning sample cost and a good indication in instances with a different-sized tree.

2.3.3.6 RF

Random forest (RF) is a machine-learning algorithm used for the purpose of classification and regression, as proposed by (Breiman 2001). This supervised method was successfully applied in several areas and domains. Remote sensing field is one of the major domains and has been applied in landslide detection (Chen et al. 2014), urban trees (Puissant et al. 2014), agricultural soil mapping (Grimm et al. 2008), and biomass estimation (Mutanga et al. 2012). RF is a multiple DT classifier based on classification and regression tree [CART; (Breiman et al. 1984)]. This method implements a bootstrap sampling for each DT, which enables the estimate calculation of errors to be based on the remaining instances, which is known as "out-of-bag" (OOB). RF applies a different process to determine the best split threshold, in comparison with CART. RF is considered as a random subset of the original set of the feature, whereas CART considers all variables at each node. Users can estimate the variables per the number of node by using the square root of the total variable number. Two mechanisms, sampling and the use of random variables for each node, generate significantly different uncorrelated trees. Furthermore, having a relatively large number of trees is necessary to obtain the full variability of the training data, which gives good classification performance with high accuracy. The final step is assigning a feature into a class by considering the votes of all the trees in the forest. The class will then be assigned based on majority voting. The RF package (Liaw and Wiener 2002) for the open-source statistical language R (R Development Core Team 2013) was used for all experiments in this study.

Random forest (RF) has several advantages. First, RF is a nonparametric method; thus, the values of variables are not required to follow a particular statistical distribution. Second, it is insensitive to overfitting and noise. Furthermore, RF is relatively fast compared with other techniques, such as the boosted method (Breiman 2001). The calculation time for training RF is defined by Eq. (2.10).

$$cT\sqrt{MN}\log N, \tag{2.10}$$

where c is a constant dependent of data complexity (i.e., small or large dataset), T is the number of tree, M is the number of variables, and N is the number of instances (Breiman 2003).

When RF and SVM, whose complexities vary between N^2 (when c is small) and N^3 (when c is large), are compared (Bottou and Lin 2007), RF will give a better adaption for larger datasets. Also, RF requires less tuning (Rodriguez-Galiano et al. 2012) and can implement the actual measures of variables, which can be estimated by alternating the value of variables on the OOB sample and calculating the difference in OOB errors before and after the alternation process. Those measures are used to analyze and interpret the classification (Rodriguez-Galiano et al. 2012) and define the type of sensor (Guo et al. 2011). Otherwise, defining the scale of segmentation (Duro et al. 2012) is more suitable for identifying a particular geographic object.

2.3.3.7 Landslide and Cut Slope Detection

The supervised landslide detection and cut slope detection were done in two successive steps. The first step was training the classifier with an adequate number of samples. The samples were selected randomly based on landslide inventory data. In this study, 60% of the samples were selected for training the classifier for the classes: landslide, cut slope, and non-landslide. These samples were examined based on the aerial photographs, slope, and hill-shade layer of the study area to ensure that each sample was selected accurately. The classifiers were then trained using these samples. In the second step, in each classifier method, several user-defined parameters should be carefully selected. In this study, the user-defined parameters were selected based on a trial-and-error approach. Table 2.1 shows the classifiers with their user-defined parameters that were selected for supervised landslide detection.

2.3.3.8 Validation

The efficiency and quality of the presented methodology for each study and research must be properly examined and tested, which can be achieved by a proper validation technique. In this study, the validation was done in three steps: The first step is to examine the classification results visually;

Table 2.1 Selected value of each parameter for each classification algorithm used

Classifier	Parameter	Selected value
Bayes	NA	NA
k-NN	*K*	1
SVM	Kernel type	Linear
	C	1
DT	Depth	0
	Min sample count	0
	Use of surrogates	Yes
	Max categories	16
	Cross-validation folds	3
	Use of 1 standard error (SE) rule	No
	Truncate pruned tree	Yes
RF	Depth	0
	Min sample count	0
	Use of surrogates	Yes
	Max categories	16
	Active variables	0
	Max tree number	50
	Forest accuracy	0.01
	Termination criteria type	Both

the second step is to transfer the methodology to a different subset of the study area and examine the ability of the method to detect landslides and cut slopes; and the final step is field validation, which is a site visit to the field, and is necessary to confirm the location and boundary of few landslides detected by the methodology presented in this study.

Visual Interpretation

In the first validation step, where the results of each classifier are examined visually, some classifiers (e.g., Bayes) produced results with high level of uncertainty and misclassification, thereby making visual interpretation useful for the rejection of the result of such classifier. In addition, a few classifiers have numerous user-defined parameters (e.g., DT and RF), which need to be fine-tuned; visual interpretation is considerably useful for this purpose.

Transferring to Testing Subset

The efficiency and quality of the presented methodology should be properly examined and tested. In this research, the study area was divided into three different subsets, and the same methodology was replicated on the testing sites to examine its validity and accuracy. Dividing the study area into three different subsets was implemented carefully. The first subset (Fig. 2.7a) has the smallest land area (2 km^2) and was used to develop the method. The first subset was easy to process and interpret because of its small size. Moreover, the training subset has various types of land-use classes (e.g., landslides, cut slopes, vegetation, and urbanization) that are fairly distributed over the entire study area, thereby making it a challenging task during the development of the methodology. The other two subsets, which have a larger land area of 4 km^2, were used to test the proposed method. The first testing subset (Fig. 2.7b) is considerably similar to the training subset but larger in size. Conversely, the second subset (Fig. 2.7c) has different distribution of land-use classes; the vegetation covers almost 85% of the entire area.

Field Validation

Multiple field visits were conducted using Global Positioning System (GPS) devices to examine the location and the boundary of detected landslides. Documenting these landslides in the field was challenging, because most of the landslides are in private farms, and other landslides are within thick-vegetated forests. Only landslides parallel to the road or in open areas were well documented through multiple field inspections. Most of the landslides were covered by vegetation and became invisible because of the rapid growth of vegetation in tropical areas, thereby posing a new

Fig. 2.7 Subsets of the study area

challenge for documentation. Figure 2.8 shows some examples of landslides from the study area of Ringlet.

Accuracy Assessment

Accuracy assessment is based on the comparison of supervised object-based classification result with an actual landslide inventory map. Actual landslide inventory data are generally derived from ground truth data, ground reference data, or other dependable datasets. Performing accuracy assessment of the features detected through remote sensing images is highly pertinent (Lillesand et al. 2004).

Ground truth information (or reference data source) consists of gathered observation about features and phenomena that are captured by data. In terms of validation, ground truth information is used for the accuracy assessment of landslide inventory map. Land cover maps from remotely sensed data have minor practical value without accuracy assessment. Accuracy assessment aims to evaluate the pattern classification landslide location map. In addition, this assessment reports the importance of classification schemes, so other researchers can easily interpret and apply them. Accuracy assessment involves two steps:

Fig. 2.8 Photographs taken during field validation

1. Collection of ground truth data. Ground truth data are independent from the training data that have been used in the process of image classification. Based on the methodology, ground truth data are collected in specific locations to be found in remotely sensed data. Sources of ground truth data consist of high-spatial resolution remotely sensed data, such as aerial photography, a high-resolution satellite imagery (including Ikonos, Quick-bird, and Worldview2), or a field survey measurement using GPS.

2. Comparison. Ground truth data are compared with data that are determined from different classes to classify the imagery.

A confusion matrix is a cross-tabulation of the classified and actual class labels for the study area (Foody 2004). This matrix is a square array of dimension $r \times r$, where r is the number of categories. Confusion matrix represents the correlation between two samples of measurement from the classified region. The overall, user, and producer accuracies, and the kappa coefficient can be measured using confusion matrix. The overall accuracy is attained by dividing the aggregate of the main diagonal entries of the confusion matrix by the entire number of samples. The kappa coefficient (K) was measured using Eq. (2.11).

$$K = \frac{\theta 1 - \theta 2}{1 - \theta 2} \tag{2.11}$$

2.4 Results

2.4.1 Landslide Detection Results

Several supervised classification methods were applied for landslide detection, including Bayes, DT, RF, k-NN, and SVM. The results of classification using k-NN, DT, and Bayes algorithms showed poor accuracy results, because most of the landslides were not detected correctly. Furthermore, landslides were misclassified as man-made cut slopes and bare lands in some cases. SVM and RF algorithms performed better compared with the previous three; many landslides were correctly detected, positioned, and delineated. Two testing sites were used to evaluate the consistency of the used classifiers for landslide detection. This section presents the results of landslide detection in the two sites. The first testing site contained several landslides, while few were detected in the second testing site.

2.4.2 Results of Landslide Detection in the Training Site

2.4.2.1 RF

In this study, RF classifier was also used for landslide detection. Results of RF landslide detection are shown in Fig. 2.9. Initial observations for the map indicate that this method performed better than k-NN, DT, and, Bayes algorithms. Evidently, most of the landslide inventories were detected accurately. Few cut slopes were misclassified as landslides, as shown in the northeastern part of the study area. RF detected 30 out of the 40 landslide inventories found in the study area. However, some landslides were undetected despite being visible in the middle part of the study area. The challenge with RF algorithm is that it requires the fine-tuning of several parameters. The current study optimized the parameters by trial-and-error approach. However, best results were not achieved. Using more robust algorithms for fine-tuning the parameters of RF could improve the landslide detection results.

2.4.2.2 SVM

Support vector machine (SVM) has been widely used for landslide susceptibility mapping, and its advantages are well established in several landslide studies. In this study, SVM was used for landslide detection. Figure 2.10 shows the results of SVM landslide detection. Results indicate that SVM is the best among the other four methods; most of the landslides were detected, few cut slopes were misclassified as landslides, and few landslides were undetected. SVM works on the concept of optimization and error reduction; therefore, it performs well for landslide detection. Having accurate landslide inventories is difficult; thus, a methodology that can detect and reduce errors is significantly important. SVM utilizes this concept; thus, it detected landslides accurately, leaving only few undetected.

Figure 2.10c shows a landslide photograph taken during the field visit. The capturing angle does not show the entire boundary of the landslide; thus, proper documentation of the landslide was challenging.

2.4.3 Results of Landslide Detection in Testing Site 1

2.4.3.1 RF

The result of RF landslide detection for Testing Site 1 is shown in Fig. 2.11. The first examination of the map shows that several landslides were accurately detected, and few cut slopes were misclassified. This shows the main difference between the result of RF and those of other methods presented previously. The RF algorithm tends to separate landslides from cut slopes better than Bayes, DT, and k-NN techniques. Although RF requires several user-defined parameters for fine-tuning, its results are better than the other techniques, using several combinations of the parameters.

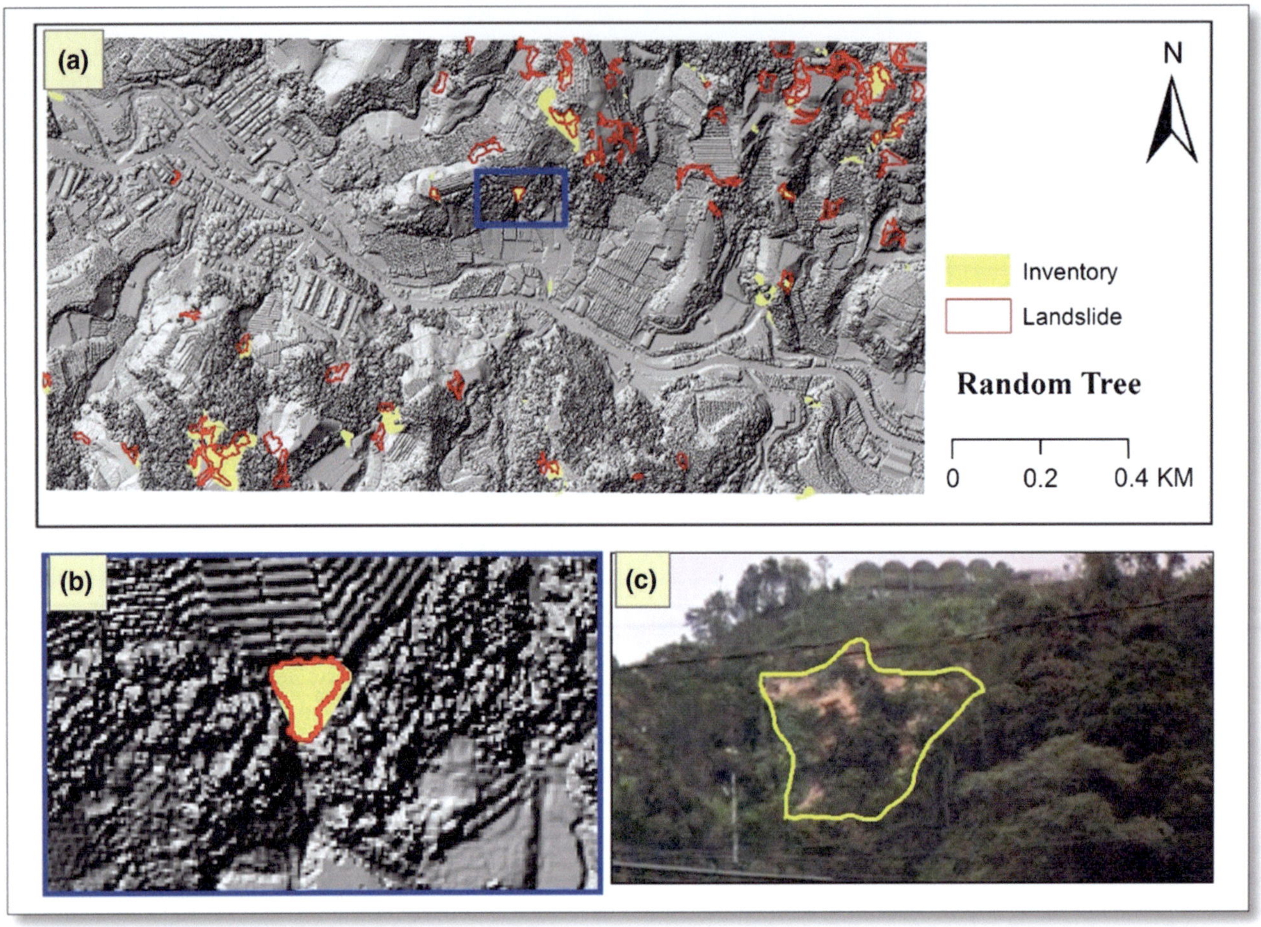

Fig. 2.9 Detected landslides using RF classification algorithm

2.4.3.2 SVM

A landslide inventory map for Testing Site 1 was produced by SVM method, as shown in Fig. 2.12. This method is shown to be suitable for the training site. When the results were examined, the algorithm produced a good landslide inventory map; the landslides were accurately detected, and few cut slopes were misclassified.

2.4.4 Results of Landslide Detection in Testing Site 2

2.4.4.1 RF

Figure 2.13 shows the results of landslides detected using RF for Testing Site 2. Results are far from excellent, as many landslides were undetected. However, this method remains better than Bayes, DT, and *k*-NN, because few cut slopes were misclassified. This result shows that RF is a good classifier for landslide detection in the presence of man-made slopes.

2.4.4.2 SVM

Figure 2.14 shows the result of SVM landside detection in the presence of man-made slopes for Testing Site 2. SVM produced an accurate landslide inventory map with few undetected landslides. In addition, results show that SVM is better than RF through visual examination. Several landslides in the upper left part of the study area were detected by SVM but not by RF. However, both the SVM and RF performed well in landslide detection when man-made slopes are present in the study area.

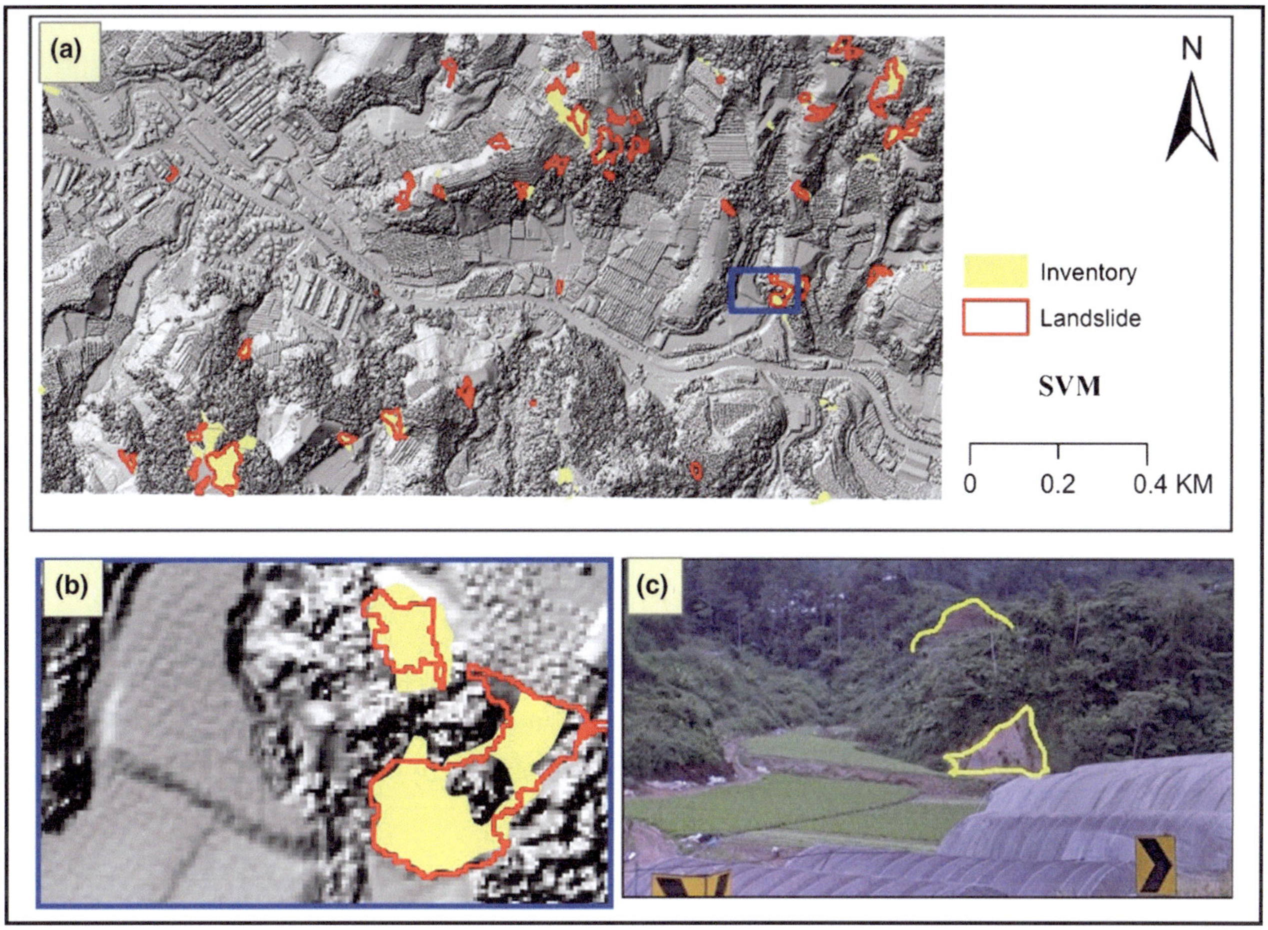

Fig. 2.10 Detected landslides using SVM classification algorithm

2.4.5 Cut Slope and Landslide Detection Results

This section presents the results obtained from landslide and cut slope mapping of the training site and two testing sites. Figure 2.15 shows the detected landslides and cut slopes by SVM and RF models. These two algorithms were considered as the most effective among others because of their high accuracy. The map shows the detected landslides in dark blue, the cut slopes in light blue, and the other features in pink. Both landslides and cut slopes are randomly distributed in the study. However, the study area exhibited a clustered pattern of landslides in the upper right corner, characterized by having a high slope, concave curvature, and is mostly barren.

Figures 2.16 and 2.17 show the landslides and man-made slopes of the two testing sites. The landslides are shown in dark blue, whereas man-made slopes are highlighted in light blue. The landslides and cut slopes are randomly distributed in the study area. The area has large and small landslides and cut slopes. Landslides may also vary in types in this study such as landslides and debris flows. Figure 2.17 illustrates few landslides and cut slopes in Testing Site 2. Most of the landslides and man-made slopes are located in the north part of the area; the middle and south parts are mostly forested area. Some landslides may have occurred in forest area, which could not be detected because LiDAR point clouds only have one return. Multiple LiDAR data returns are important to detect landslides in forested areas. Overall, 123

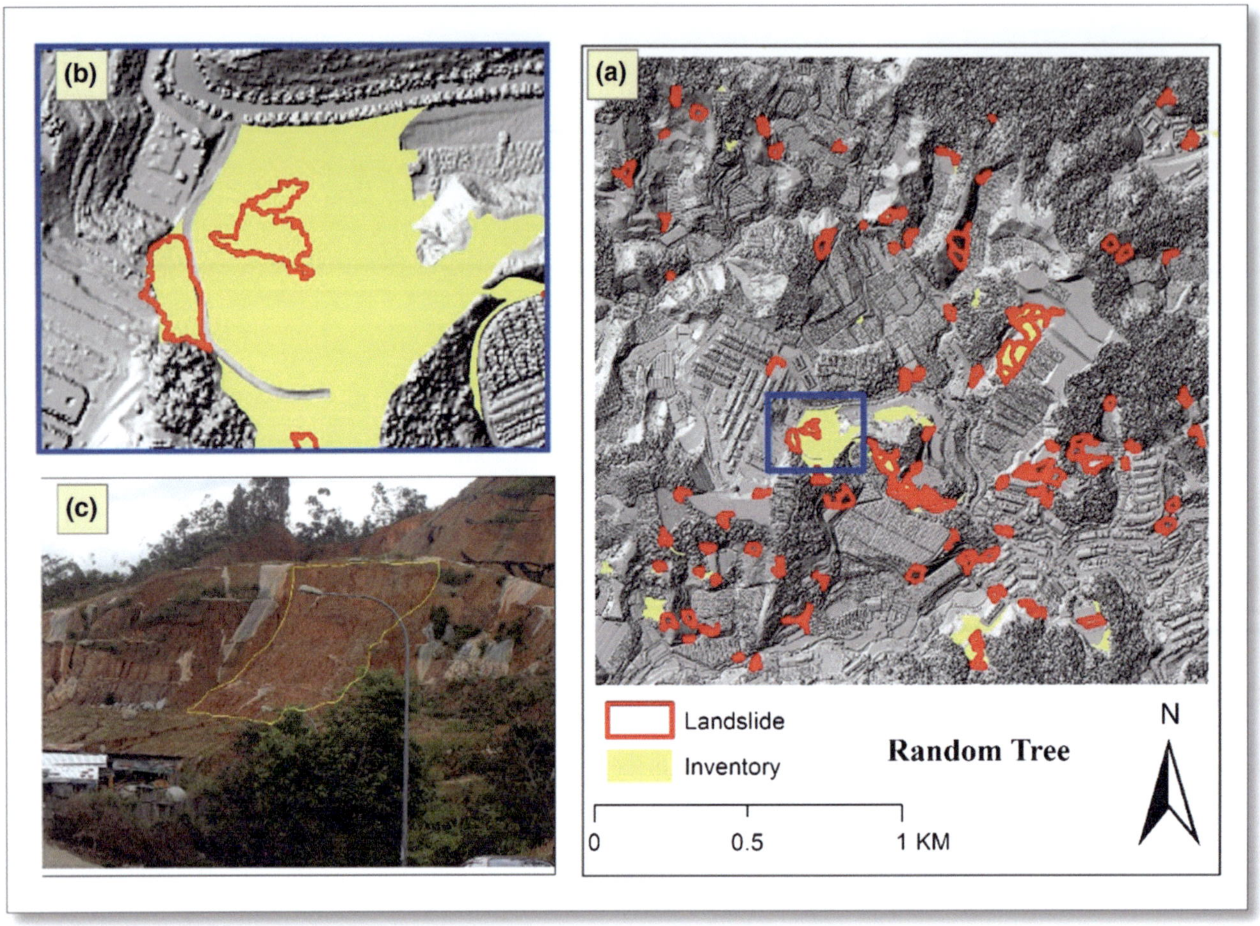

Fig. 2.11 Detected landslides using RF classification algorithm

landslides and 205 cut slopes were detected in the first testing site. Similarly, 18 landslides and 51 man-made slopes were detected in the second testing site.

2.4.6 Results of Image Segmentation

Once the input layers were prepared, spectral and LiDAR-based features were combined in one raster dataset for segmentation. Then, a multiresolution segmentation algorithm was utilized for segmentation. The parameters of segmentation were set as scale (60), shape (0.1), and compactness (0.5). These values were selected based on trial-and-error experiments in eCognition software. Segmentation result of the training site is shown in Fig. 2.18. Landslide features are accurately delineated. Accurate segmentation is important for efficient landslide detection by various features. For example, in Fig. 2.18a, the segments show that the landslide scarp is accurately defined, whereas in Fig. 2.18b, the landslide scarp is only partially defined. Moreover, in Fig. 2.18c, the landslide scarp is defined inaccurately.

Some landslides are defined accurately because of fewer variations in slope, curvature, and altitude values. Landslide scarps are defined inaccurately, because the slope, curvature, and height values vary significantly within the landslide objects. Therefore, one landslide scarp may be segmented as two or more landslide objects, thereby reducing the accuracy of landslide detection as several spatial features can be useless.

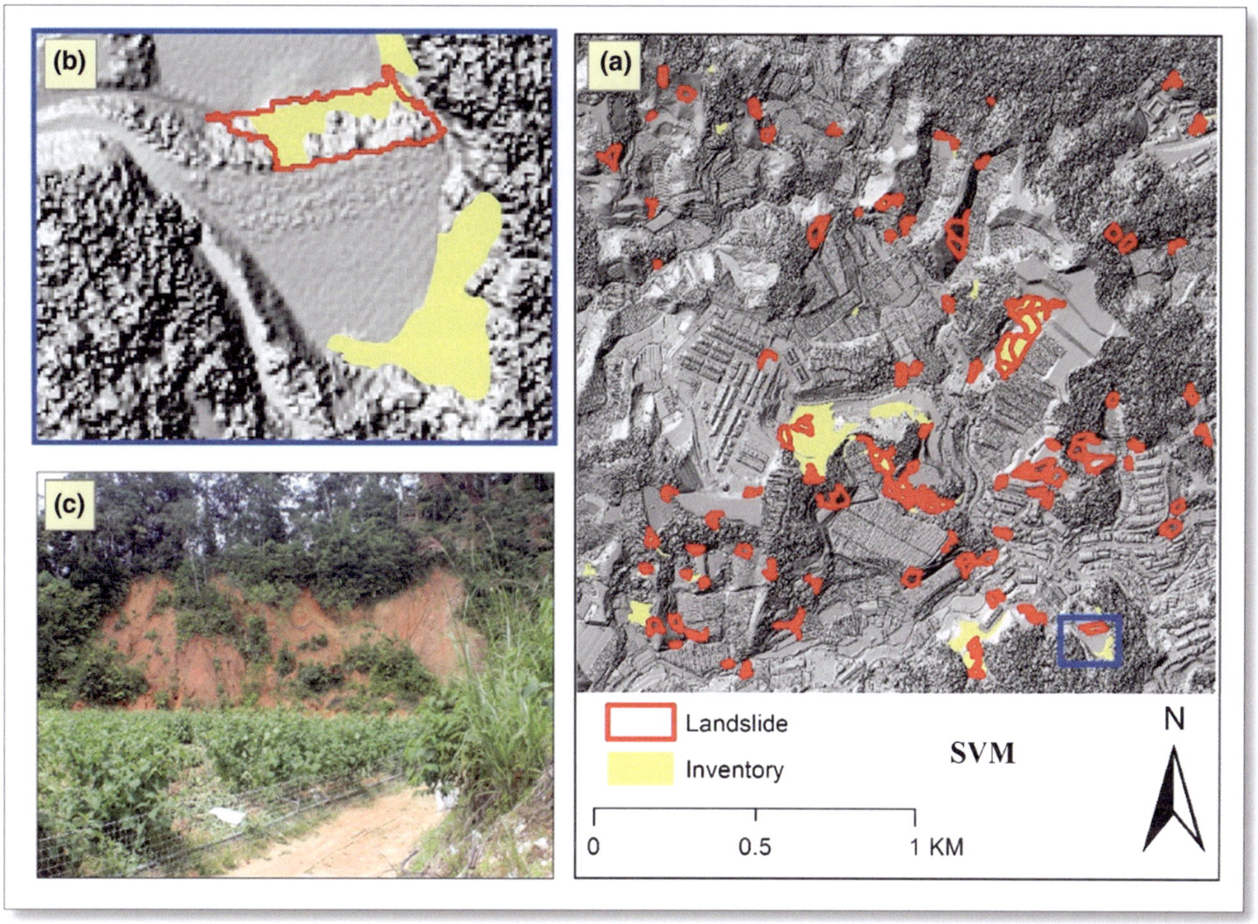

Fig. 2.12 Detected landslides using SVM classification algorithm

2.4.7 Results of Accuracy Assessment

Table 2.2 shows the overall accuracies and kappa indices of landslide detection using several classifiers and datasets. For the training site, the SVM method had the highest overall accuracy (0.90) and kappa index (0.83), in contrast to the DT algorithm that had the lowest overall accuracy and kappa index of 0.61 and 0.37, respectively. In general, the accuracy assessment shows that SVM and RF performed well in landslide detection compared to other methods. For Testing Site 1, the highest and lowest overall accuracies were 0.80 and 0.61 for SVM and Bayes methods, respectively. Moreover, the highest and lowest kappa indices were 0.74 and 0.33 for the same classifiers, respectively. Thus, RF is considered as a good classifier for landslide detection. Results confirmed that SVM and RF are the best methods for landslide detection. For Testing Site 2, RF had the highest overall accuracy of 0.91, followed by SVM with 0.90. The lowest overall accuracy was achieved by Bayes algorithm with 0.65. The kappa indices indicate that SVM is better than RF and other methods. The kappa index of SVM and RF is 0.85 and 0.80, respectively. Quantitative assessments

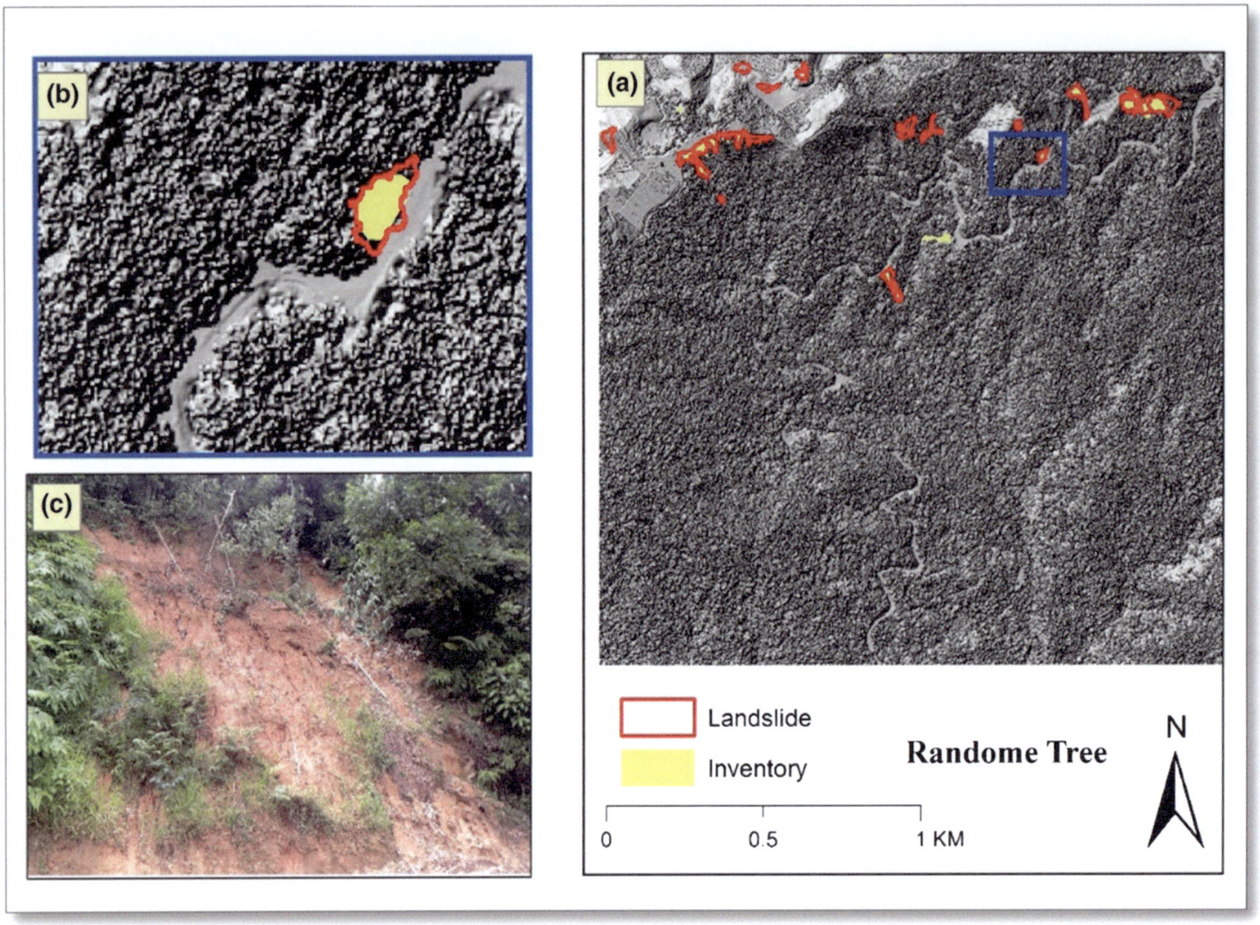

Fig. 2.13 Detected landslides using RF classification algorithm

show that SVM and RF can be good classifiers for supervised landslide detection in LiDAR data and orthophotos.

The study aimed to detect landslides in the presence of man-made slopes and produce an accurate inventory map. Man-made or cut slopes create challenges in landslide detection because their geometry and geomorphology are relatively similar. This study attempted to separate these classes. Table 2.3 shows the user and producer accuracies obtained from various classifiers for cut slope and landslide classes. Although Bayes method achieved the highest user accuracy, evaluation of simultaneous user and producer accuracies is important. This evaluation ensures that the detected landslides are accurate and that only few landslides will be undetected. The highest user and producer accuracies for landslide class were achieved by Bayes and SVM methods, whereas the highest user and producer accuracies for cut slope were observed for Bayes (1) and RF methods (0.90). However, SVM and RF achieved relatively high user and producer accuracies simultaneously, indicating a good classification of landsides and cut slopes. SVM performed better than RF for landslide and cut slope classification.

For Testing Site 1, the highest user and producer accuracies for landslide class were achieved by RF and Bayes algorithms, whereas the highest user and producer accuracies for cut slope were found for SVM and RF algorithms. Kappa indices showed that SVM and RF are best for landslide and cut slope separation. The highest kappa index was achieved by SVM (0.78, 0.78) and RF (0.67, 0.83) for landslide and cut slope classes, respectively. In addition, the user and producer accuracies and kappa indices for Testing Site 2

Fig. 2.14 Detected landslides using SVM classification algorithm

confirmed that SVM and RF classifiers are the best algorithms for landslide detection. Overall assessment shows that the separation between cut slopes and landslides using LiDAR data and orthophotos through supervised classification is possible and can be improved. The current study achieved satisfactory results of landslide detection and separation between landslides and man-made slopes; however, further research is needed to detect the type of and improve the accuracy in cut slopes. The proposed supervised framework provides a rapid and efficient guideline for landslide mapping, which is valuable for landslide susceptibility mapping, and hazard and risk assessments.

2.5 Discussion

Several methods of determining segmentation parameters, such as supervised and Taguchi approaches (Gibril et al. 2016), were reviewed. In supervised approaches, segmentation is usually optimized based on multiple features found in the image of the study area. Conversely, in Taguchi approaches, segmentation parameters are optimized for a single feature only. Because several types of landslides are present in the study, optimizing the parameters for only one feature without considering the type of landslide created a huge challenge for the Taguchi approach. Supervised method

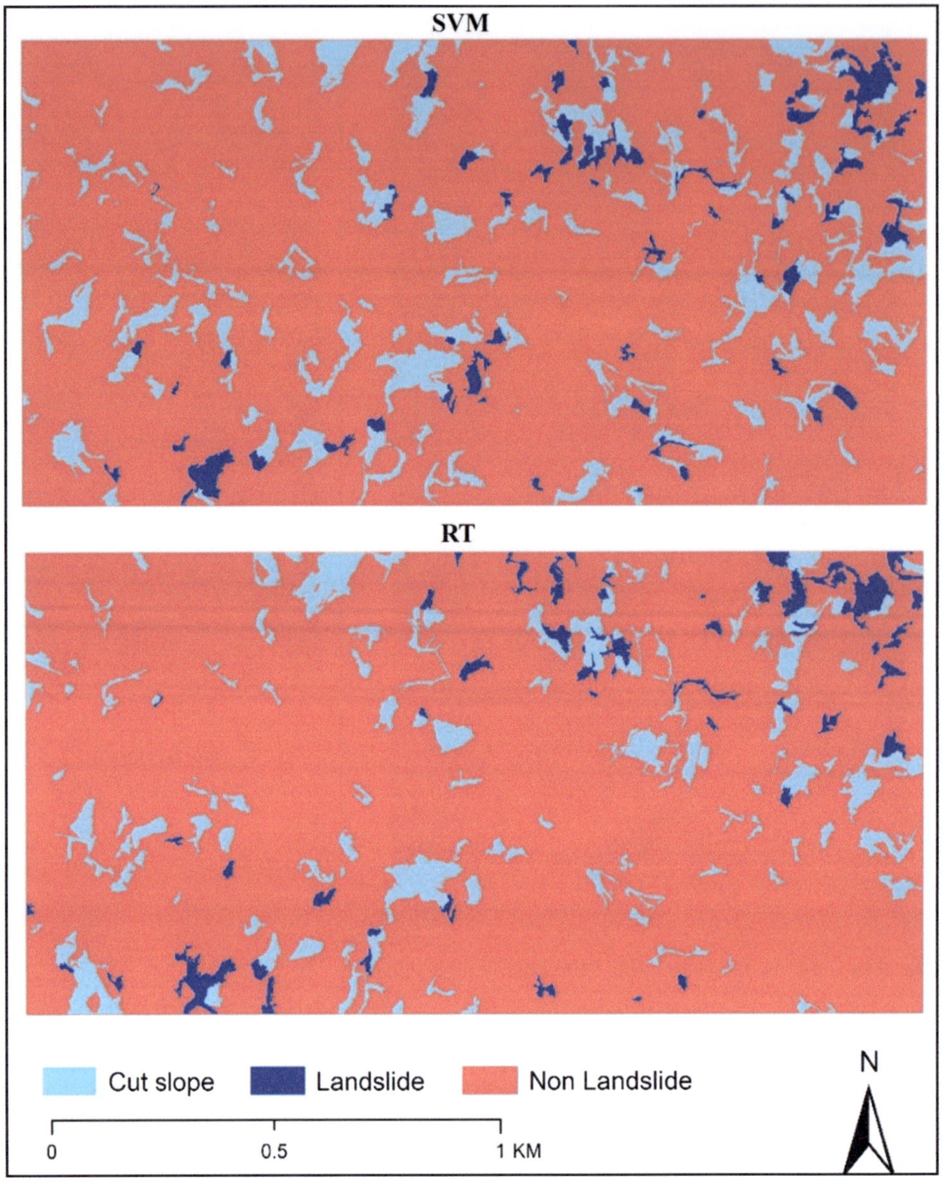

Fig. 2.15 Landslide and cut slope mapping (Training Site)

is a good option after the Taguchi method. Although supervised approaches are user independent and require less time than trial-and-error method, they also depend on the selected subsegments that aimed to be merged into a target segment.

Preparing the input layers, selecting a classifier, and the fine-tuning of user-defined parameters of the classifier are important in supervised landslide detection (W. Chen et al. 2014). The current study analyzed several input layers derived from LiDAR-based DEM and DSM for improved landslide detection. Significant layers were selected based on their importance for landslide detection using the training site and analyzed through trial-and-error approach. Overall, 13 features, including spectral, LiDAR, spatial, and texture, were used. Furthermore, several classifiers were analyzed by measuring their accuracies for landslide detection. The user-defined parameters of the classifiers were also fine-tuned by trial-and-error method.

From LiDAR point clouds, six features were produced: DEM, DSM, height, slope, curvature, and hill shade. Training site elevation ranged from 997 to 1270 m. Conversely, the height feature showed that the height of objects in the study areas varies from 0 m (flat objects) to 100 m (hilly lands). In addition, slope of the study area includes flat and hilly lands. The slope ranged from 0 to 87°. As in

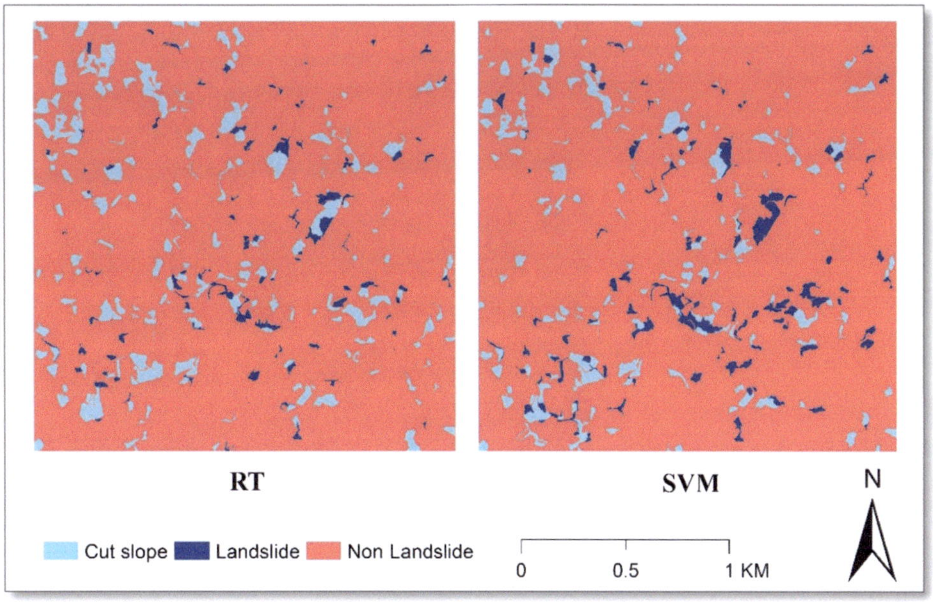

Fig. 2.16 Landslide and cut slope mapping (Testing Site 1)

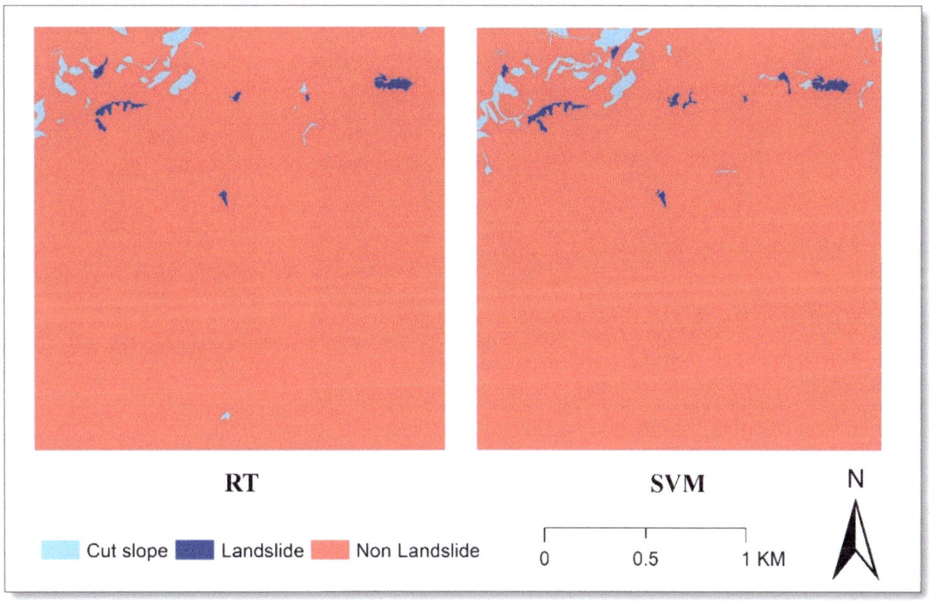

Fig. 2.17 Landslide and cut slope mapping (Testing Site 2)

segmentation results which is shown in Fig. 2.18, the study area has flat, concave, and convex type of curvature surfaces.

Figure 2.19 shows that landslides and cut slopes are difficult to separate using only LiDAR-derived parameters because both landslides and cut slopes have relatively similar characteristics. Therefore, investigating other parameters and orthophotos is important.

Three spectral bands (R, G, and B) of the orthophotos were used for landslide detection. Orthophotos are useful information to separate landslides from other features, such as grassland, buildings, and water bodies. Analysis of typical values of spectral bands for landslide and non-landslides is presented in Fig. 2.20. The chart presents the minimum, maximum, and mean values of RGB

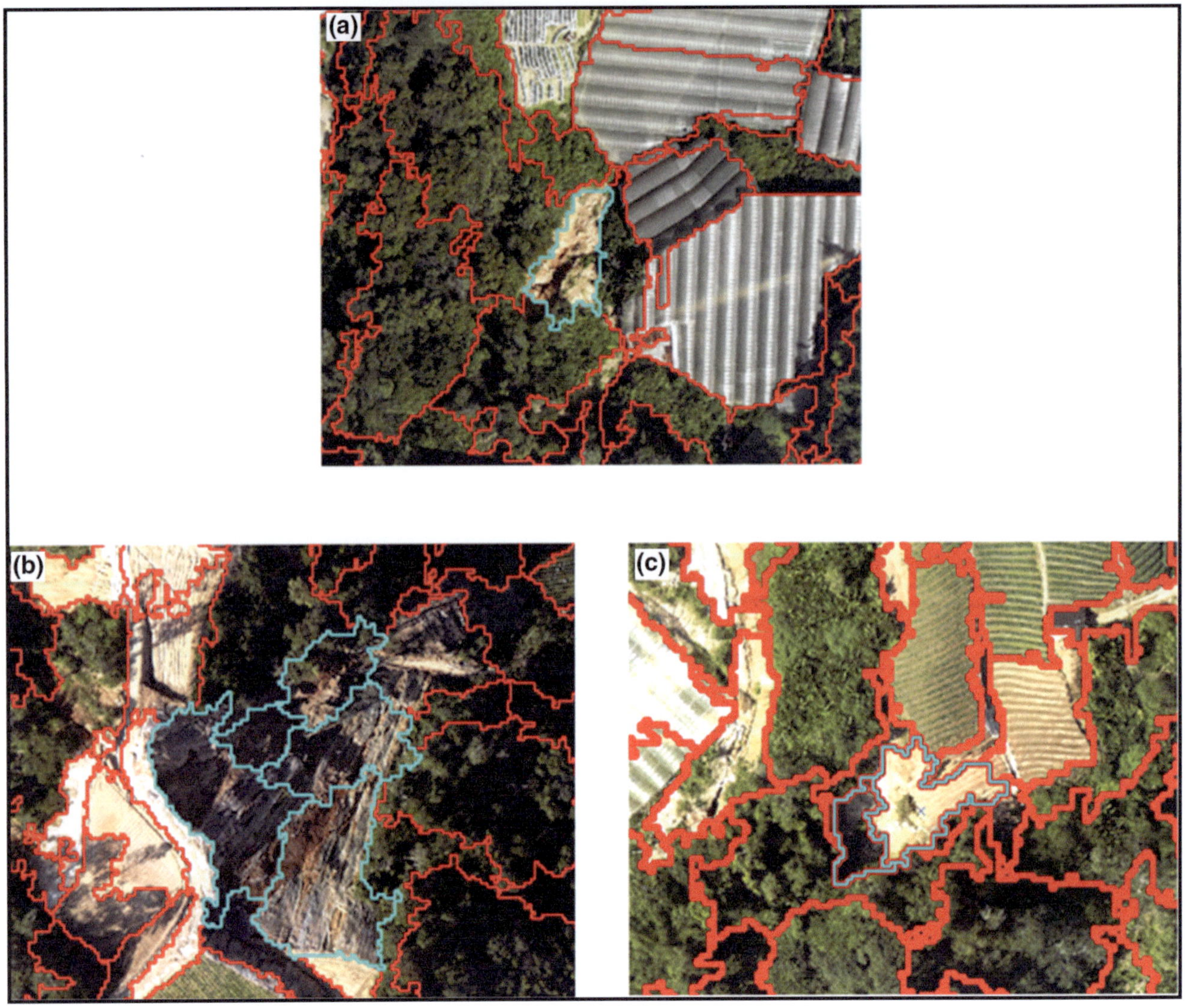

Fig. 2.18 Segmentation results

Table 2.2 Overall accuracies and kappa indices of landslide detection

Dataset	Classifier	Overall accuracy	Kappa index
Training Site	Bayes	0.75	0.46
	k-NN	0.65	0.4
	DT	0.61	0.37
	RF	0.82	0.7
	SVM	0.9	0.83
Testing Site 1	Bayes	0.61	0.33
	k-NN	0.72	0.51
	DT	0.72	0.53
	RF	0.78	0.64
	SVM	0.86	0.74
Testing Site 3	Bayes	0.65	0.37
	k-NN	0.71	0.43
	DT	0.78	0.56
	RF	0.91	0.8
	SVM	0.9	0.85

Table 2.3 User and producer accuracies obtained from various classifiers for cut slope and landslide classes

Dataset	Classifier	Class	User	Producer	KIA
Training Site	Bayes	Landslide	1	0.5	0.44
		Cut slope	1	0.26	0.22
		Others	0.7	1	1
	k-NN	Landslide	0.37	0.45	0.27
		Cut slope	0.54	0.65	0.53
		Others	0.82	0.71	0.4
	DT	Landslide	0.58	0.35	0.25
		Cut slope	0.42	0.75	0.59
		Others	0.77	0.67	0.36
	RF	Landslide	0.92	0.6	0.53
		Cut slope	0.69	0.9	0.86
		Others	0.86	0.88	0.72
	SVM	Landslide	0.82	0.95	0.93
		Cut slope	0.8	0.8	0.74
		Others	0.98	0.92	0.83
Testing Site 1	Bayes	Landslide	0.85	0.2	0.15
		Cut slope	0.66	0.4	0.27
		Others	0.58	0.96	0.84
	k-NN	Landslide	0.77	0.45	0.37
		Cut slope	0.51	0.75	0.63
		Others	0.82	0.8	0.55
	DT	Landslide	0.6	0.45	0.34
		Cut slope	0.52	0.73	0.62
		Others	0.86	0.82	0.62
	RF	Landslide	0.94	0.58	0.52
		Cut slope	0.56	0.91	0.86
		Others	0.91	0.8	0.6
	SVM	Landslide	0.79	0.76	0.71
		Cut slope	0.69	0.71	0.65
		Others	0.93	0.93	0.82
Testing Site 2	Bayes	Landslide	1	0.33	0.26
		Cut slope	1	0.33	0.26
		Others	0.58	1	1
	k-NN	Landslide	0.4	0.66	0.54
		Cut slope	0.75	0.5	0.44
		Others	0.83	0.76	0.37
	DT	Landslide	0.66	0.57	0.5
		Cut slope	0.8	0.44	0.36
		Others	0.8	0.96	0.85
	RF	Landslide	1	0.71	0.67
		Cut slope	0.85	0.85	0.83
		Others	0.9	0.96	0.87
	SVM	Landslide	1	0.83	0.78
		Cut slope	1	0.83	0.78
		Others	0.83	1	1

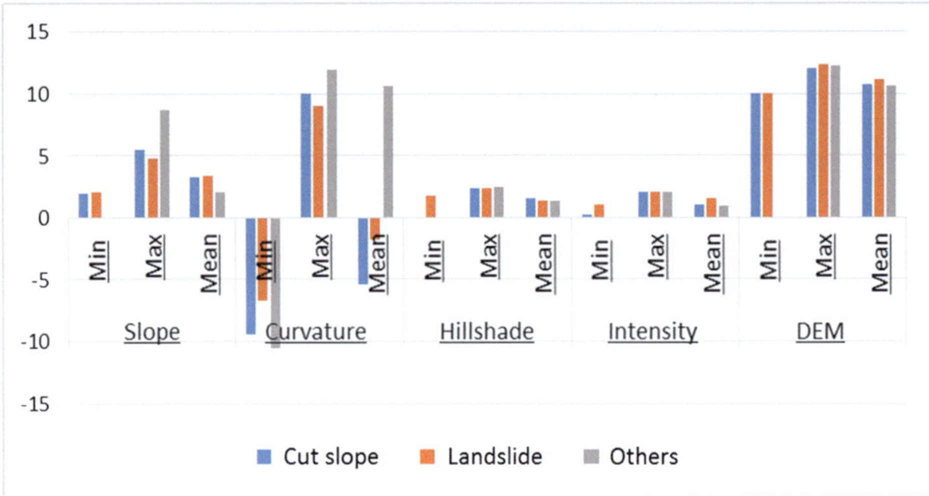

Fig. 2.19 Typical values of LiDAR-based features for landslide and non-landslide features. *Note* (1) slope ×10; (2) hill shade, intensity, and DEM ×102

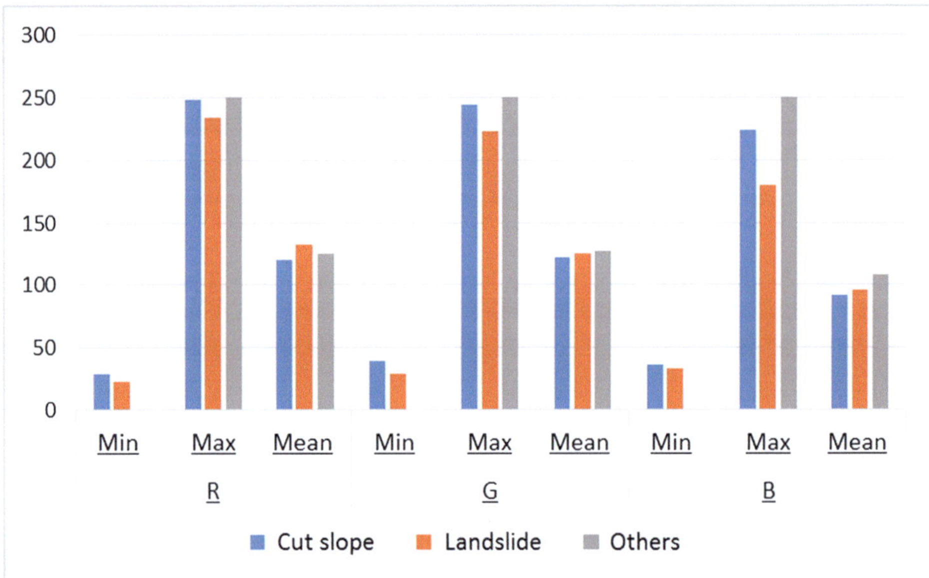

Fig. 2.20 Typical values of spectral bands for landslide and non-landslide features

bands for landslide, cut slope, and other features in the study area. The three classes have approximately similar average, minimum, and maximum values of RGB bands. However, slight differences in these bands can also separate the three classes at some extent depending on the classifier used. When these parameters are combined with other features, they can work more significantly in landslide and cut slope detection.

Spatial and texture features are also important for landslide detection. Area, length/width, compactness, shape index, rectangular fit, and Gray-Level Co-Occurrence Matrix (GLCM) mean are the spatial and texture features used in this study. Figure 2.21 shows the extracted values of these features for landslide and non-landslide features.

The chart of spatial parameters reveals that landslide and cut slope can be well separated using these parameters than other previously discussed parameters. The values of most of the parameters for landslide, cut slope, and other features are different. Minimum, maximum, and average values vary for the three classes. Area, length/width, and texture are the most important parameters for the separation of landslide and cut slope features in the study area.

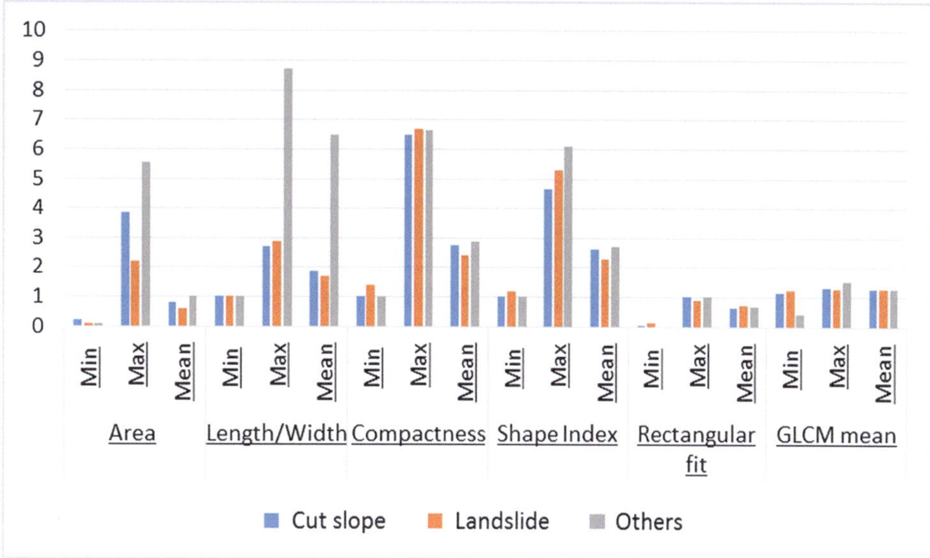

Fig. 2.21 Typical values of landslide and non-landslide features. *Note* (1) area ×103; (2) GLCM mean ×102

2.6 Conclusion

Landslides are one of the most destructive natural disasters that mountainous areas, such as Malaysia, suffer from and are known to threaten human lives and properties. Landslide inventory maps are valuable sources of information and are essential for various studies and investigations, such as landslide susceptibility, and hazard and risk assessments, as well as various decision-making processes and polices. Providing an efficient method in detecting and distinguishing landslides and man-made slopes is a challenging task; most methods concentrate on landslide detection only, and these methods require much time and are costly.

This study aims to provide a rapid and accurate method that can create a landslide and man-made slope inventory map semiautomatically. To achieve this goal, few steps were applied: (1) the preparation and analysis of several spatial, spectral, and texture features, and LiDAR-derived parameters; and (2) the evaluation of five well-known classifiers to determine the best algorithm for landslide and man-made slope detection.

In general, five classifiers, i.e., Bayes, DT, *k*-NN, RF, and SVM, were evaluated to propose a semiautomatic supervised landslide and man-made slope detection approach using airborne LiDAR data coupled with orthophotos.

The research findings provided an effective solution for supervised and semiautomatic landslide and man-made mapping in tropical areas. Analysis showed that RF and SVM are the suitable classifiers for object classification using LiDAR data. The accuracies of these models were consistent in the three subsets of data that were used for validations. Overall evaluations of the studied classifiers showed that using supervised classification at object level, separation between cut slopes and landslides using LiDAR data and orthophotos is possible and can be improved.

References

Abe, S. (2005). *Support vector machines for pattern classification* (Vol. 53). Berlin: Springer.

Akcay, H. G., & Aksoy, S. (2008). Automatic detection of geospatial objects using multiple hierarchical segmentations. *IEEE Transactions on Geoscience and Remote Sensing, 46*(7), 2097–2111.

Benz, U. C., Hofmann, P., Willhauck, G., Lingenfelder, I., & Heynen, M. (2004). Multi-resolution, object-oriented fuzzy analysis of remote sensing data for GIS-ready information. *ISPRS Journal of Photogrammetry and Remote Sensing, 58*(3), 239–258.

Bignel, F., & Snelling, G. (1977). The geochronology of the main range Batholith: Cameron highlands road and Gunong Bujang Melaka. *Overseas Geol Miner Resour, 47,* 3–35.

Bottou, L., & Lin, C.-J. (2007). Support vector machine solvers. *Large Scale Kernel Machines*, 301–320.

Breiman, L. (2001). Random forests. *Machine Learning, 45*(1), 5–32.

Breiman, L. (2003). RF/tools: A class of two-eyed algorithms. *Paper presented at the SIAM workshop.*

Breiman, L., Friedman, J., Stone, C. J., & Olshen, R. A. (1984). *Classification and regression trees*. Boca Raton: CRC Press.

Caruana, R., & Niculescu-Mizil, A. (2006). An empirical comparison of supervised learning algorithms. *Paper presented at the proceedings of the 23rd international conference on machine learning.*

Chen, L. C., Teo, T.-A., Shao, Y.-C., Lai, Y.-C., & Rau, J.-Y. (2004). Fusion of LiDAR data and optical imagery for building modeling. *International Archives of Photogrammetry and Remote Sensing, 35* (B4), 732–737.

Chen, W., Li, X., Wang, Y., Chen, G., & Liu, S. (2014). Forested landslide detection using LiDAR data and the random forest algorithm: A case study of the Three Gorges, China. *Remote Sensing of Environment, 152,* 291–301.

Chow, W., Zakaria, M., Ferdaus, A., & Nurzaidi, A. (2003). Geological terrain mapping. JMG unpublished report. JMG. SWP. GS, 16, 1–42.

Cobbing, E., Pitfield, P., Darbyshire, D., & Mallick, D. (1992). The granites of the SE Asian tin belt. British Geological Survey, Overseas Memoir No. 10: HMSO, London.

Cortes, C., & Vapnik, V. (1995). Support-vector networks. *Machine Learning, 20*(3), 273–297.

Definiens, A. (2007). *Definiens developer 7 reference book* (pp. 21–24). München: Definiens AG.

Duro, D. C., Franklin, S. E., & Dubé, M. G. (2012). Multi-scale object-based image analysis and feature selection of multi-sensor earth observation imagery using random forests. *International Journal of Remote Sensing, 33*(14), 4502–4526.

Eeckhaut, M., Poesen, J., Verstraeten, G., Vanacker, V., Nyssen, J., Moeyersons, J., et al. (2007). Use of LiDAR-derived images for mapping old landslides under forest. *Earth Surface Processes and Landforms, 32*(5), 754–769.

Fang, H.-T., & Huang, D.-S. (2004). Noise reduction in LiDAR signal based on discrete wavelet transform. *Optics Communications, 233* (1), 67–76.

Foody, G. M. (2004). Thematic map comparison. *Photogrammetric Engineering & Remote Sensing, 70*(5), 627–633.

Friedman, N., Geiger, D., & Goldszmidt, M. (1997). Bayesian network classifiers. *Machine Learning, 29*(2–3), 131–163.

Galli, M., Ardizzone, F., Cardinali, M., Guzzetti, F., & Reichenbach, P. (2008). Comparing landslide inventory maps. *Geomorphology, 94*(3), 268–289.

Gibril, M. B. A., Bakar, S. A., Yao, K., Idrees, M. O., & Pradhan, B. (2016). Fusion of RADARSAT-2 and multispectral optical remote sensing data for LULC extraction in a tropical agricultural area. *Geocarto International*, 1–14.

Gorum, T., Fan, X., van Westen, C. J., Huang, R. Q., Xu, Q., Tang, C., et al. (2011). Distribution pattern of earthquake-induced landslides triggered by the 12 May 2008 Wenchuan earthquake. *Geomorphology, 133*(3), 152–167.

Grimm, R., Behrens, T., Märker, M., & Elsenbeer, H. (2008). Soil organic carbon concentrations and stocks on Barro Colorado Island —digital soil mapping using Random Forests analysis. *Geoderma, 146*(1), 102–113.

Guo, L., Chehata, N., Mallet, C., & Boukir, S. (2011). Relevance of airborne lidar and multispectral image data for urban scene classification using Random Forests. *ISPRS Journal of Photogrammetry and Remote Sensing, 66*(1), 56–66.

Hodgson, M. E., Jensen, J., Raber, G., Tullis, J., Davis, B. A., Thompson, G., et al. (2005). An evaluation of lidar-derived elevation and terrain slope in leaf-off conditions. *Photogrammetric Engineering & Remote Sensing, 71*(7), 817–823.

Laliberte, A. S., Rango, A., Havstad, K. M., Paris, J. F., Beck, R. F., McNeely, R., et al. (2004). Object-oriented image analysis for mapping shrub encroachment from 1937 to 2003 in southern New Mexico. *Remote Sensing of Environment, 93*(1), 198–210.

Last, M., Maimon, O., & Minkov, E. (2002). Improving stability of decision trees. *International Journal of Pattern Recognition and Artificial Intelligence, 16*(02), 145–159.

Liaw, A., & Wiener, M. (2002). Classification and regression by random forest. *R News, 2*(3), 18–22.

Lillesand, T. M., Kiefer, R. W., & Chipman, J. (2004). *Remote sensing and image interpretation*. New York: Wiley.

Long, N. T. (2008). *Landslide susceptibility mapping of the mountainous area in A Luoi district, Thua Thien Hue province*, Vietnam. Faculty of Engineering, Department of Hydrology and Hydraulic Engineering, Vrije Universiteit Brussel, Belgium.

Martha, T. R. (2011). *Detection of landslides by object oriented image analysis. University of Twente, Faculty of Geo-Information Science and Earth Observation*. Enschede, The Netherlands: ITC Printing Department.

Martha, T. R., Kerle, N., Van Westen, C. J., Jetten, V., & Kumar, K. V. (2011). Segment optimization and data-driven thresholding for knowledge-based landslide detection by object-based image analysis. *IEEE Transactions on Geoscience and Remote Sensing, 49* (12), 4928–4943.

McKean, J., & Roering, J. (2004). Objective landslide detection and surface morphology mapping using high-resolution airborne laser altimetry. *Geomorphology, 57*(3), 331–351.

Mitchell, T. M. (1997). *Machine learning* (Vol. 45, p. 37). Burr Ridge, IL: McGraw Hill.

Möller, M., Lymburner, L., & Volk, M. (2007). The comparison index: A tool for assessing the accuracy of image segmentation. *International Journal of Applied Earth Observation and Geoinformation, 9* (3), 311–321.

Mutanga, O., Adam, E., & Cho, M. A. (2012). High density biomass estimation for wetland vegetation using WorldView-2 imagery and random forest regression algorithm. *International Journal of Applied Earth Observation and Geoinformation, 18,* 399–406.

Navulur, K. (2006). *Multispectral image analysis using the object-oriented paradigm*. Boca Rotan: CRC Press.

Ohlmacher, G. C. (2007). Plan curvature and landslide probability in regions dominated by earth flows and earth slides. *Engineering Geology, 91*(2), 117–134.

Olaya, V. (2009). Basic land-surface parameters. *Developments in Soil Science, 33,* 141–169.

Pradhan, B., & Lee, S. (2010). Landslide susceptibility assessment and factor effect analysis: Backpropagation artificial neural networks and their comparison with frequency ratio and bivariate logistic regression modelling. *Environmental Modelling and Software, 25* (6), 747–759.

Puissant, A., Rougier, S., & Stumpf, A. (2014). Object-oriented mapping of urban trees using Random Forest classifiers. *International Journal of Applied Earth Observation and Geoinformation, 26,* 235–245.

Rodriguez-Galiano, V. F., Ghimire, B., Rogan, J., Chica-Olmo, M., & Rigol-Sanchez, J. P. (2012). An assessment of the effectiveness of a random forest classifier for land-cover classification. *ISPRS Journal of Photogrammetry and Remote Sensing, 67,* 93–104.

Samui, P. (2008). Slope stability analysis: A support vector machine approach. *Environmental Geology, 56*(2), 255–267.

Soria, D., Garibaldi, J. M., Ambrogi, F., Biganzoli, E. M., & Ellis, I. O. (2011). A 'non-parametric' version of the naive Bayes classifier. *Knowledge-Based Systems, 24*(6), 775–784.

Team, R. C. (2013). R: *A language and environment for statistical computing*.

Vapnik, V. N., & Vapnik, V. (1998). *Statistical learning theory* (Vol. 1). New York: Wiley.

Varmuza, K., & Filzmoser, P. (2016). *Introduction to multivariate statistical analysis in chemometrics*. Boca Rotan: CRC Press.

Wu, X., Kumar, V., Quinlan, J. R., Ghosh, J., Yang, Q., Motoda, H., et al. (2008). Top 10 algorithms in data mining. *Knowledge and Information Systems, 14*(1), 1–37.

Xie, Z., Zhang, Q., Hsu, W., & Lee, M. L. (2005). Enhancing SNNB with local accuracy estimation and ensemble techniques. *Paper presented at the international conference on database systems for advanced applications*.

Zêzere s, J. L., de Brum Ferreira, A., & Rodrigues, M Ls. (1999). The role of conditioning and triggering factors in the occurrence of landslides: A case study in the area north of Lisbon (Portugal). *Geomorphology, 30*(1), 133–146.

Zhang, H. (2004). The optimality of naive Bayes. *AA, 1*(2), 3.

Optimized Rule Sets for Automatic Landslide Characteristic Detection in a Highly Vegetated Forests

3

Biswajeet Pradhan and Mustafa Ridha Mezaal

3.1 Introduction

The rapid expansion of cities and the continuously increasing population in urban areas lead to the establishment of settlements in mountainous areas. This phenomenon has increased the impact of natural disasters, particularly landslides, in these mountainous areas. Landslides result in severe property losses, human casualties, and environmental damage. (2) Data interpretation is frequently based on the expert knowledge and experience of an analyst, as well as his or her familiarity with the area (Chen et al. 2014; Malamud et al. 2004). (3) Finally, additional errors can be introduced while translating image interpretation results into thematic maps (Malamud et al. 2004). High-resolution LiDAR-derived DEMs can depict ground surfaces and provide valuable information on the topographic features of possible landslide-affected areas that are covered by dense vegetation (McKean and Roering 2004). In addition, high-resolution DEMs can be utilized to identify landforms with a scale of a meter to a few meters and provide useful information about rocky and densely vegetated areas (Tarolli 2014; Van Westen et al. 2008). Minimal changes in terrain information can be easily detected using LiDAR data (Chen et al. 2015). In general, LiDAR data have a definite advantage because of their capability to penetrate vegetation canopies and provide valuable information on topographic conditions. This advantage makes LiDAR data different from other data sources, such as aerial photographs, in terms of detecting slope failure under dense vegetation (Pradhan et al. 2016). LiDAR data and their derivatives, such as hillshade, surface roughness, slope, and contour maps, provide significant and valuable information about active geological processes, such as landslides, which reshape the topography of an area. Overall, LiDAR data can serve as a promising tool for enhancing landslide inventory maps (Kasai et al. 2009). However, distinguishing different types of landslides is important for studying the geomorphological development of hillsides and the mitigation of landslide hazards (Dou et al. 2015; Lin et al. 2013).

The remainder of this paper is organized as follows: The study area and data set are described in Sect. 2. The research methodology and the types of landslides, namely shallow and deep-seated landslides, are explained in detail in Sect. 3. The results are presented and discussed in Sect. 4. Finally, a brief summary, including the main findings and future directions, is presented in Sect. 5.

3.2 Types of Landslides

A landslide is the motion of the mass of debris, rocks, or a portion of the earth down a slope under the impact of gravity (Cruden and Varnes 1996; Guzzetti et al. 2012). Landslides are classified as either shallow or deep-seated, depending on the movement characteristics and landslide volume (Brunetti et al. 2009; Guzzetti et al. 2012). Shallow and deep-seated landsides differ in terms of size, volume, and damage influence (Zêzere et al. 2005). However, landslide mass volume is difficult to evaluate (Brunetti et al. 2009). Shallow landslides are typically associated with short high-intensity rainfall, whereas most large-scale deep-seated landslides result from the interaction between long-term rainfall and natural denudation processes (Zêzere et al. 2005). A landslide can demonstrate a sliding, flowing, falling, or toppling movement, but numerous landslides also exhibit a combination of two or more types of these movements either at the same time or during their lifetime (Cruden and Varnes 1996). A high-resolution DEM is necessary to study the characteristics of different types of landslides (Dou et al. 2015). Several techniques with various image sources have been utilized to distinguish landslide (Chang et al. 2012; Chen et al. 2014; Cruden and Varnes 1996; Dou et al. 2015; Korup 2006; Ma et al. 2016; Rau et al. 2012; Yu et al. 2015; Zêzere et al. 2005). Figure 3.1 shows a simple diagram

B. Pradhan (✉) · M.R. Mezaal
Department of Civil Engineering, University Putra Malaysia, Serdang, Malaysia
e-mail: biswajeet24@gmail.com

B. Pradhan (ed.), *Laser Scanning Applications in Landslide Assessment*,
DOI 10.1007/978-3-319-55342-9_3

illustrating the different landslide types. Meanwhile, Fig. 3.2 shows the general sketch of shallow and deep-seated landslides which is overlaid on contour map.

3.2.1 Deep-Seated Landslide

Deep-seated landslides are characterized by slope movements occurring on high-relief-energy hillsides; the span of this landslide type is comparable with the entire slope, but the displacements are relatively small compared with the slope itself (Goudie 2004; Kellerer-Pirklbauer et al. 2010). This type of landslides can be reactivated during intense events, and they can evolve into destructive failures. Deep-seated landslides are generally difficult to recognize in the field, especially when they happen in densely forested areas where quick revegetation occurs, such as in tropical forests. Vegetation indices are particularly helpful in recognizing deep-seated landslides (Vohora and Donoghue 2004). In general, optical images are unsuitable for analyzing this type of slope failures because they are usually covered by vegetation. To identify and map deep-seated landslides, the automation of recognition procedures and the integration of optical sensors with other remote sensing techniques have been proven to be highly effective (Delgado et al. 2011; Moine et al. 2009), and LiDAR data have been found useful (Agliardi et al. 2009). Automatic algorithms have been applied to achieve this task, such as supervised classification methods (Kasai et al. 2009) and standard signal processing techniques. Chen et al. (2015) showed that LiDAR data with 1-m resolution are sufficient for mapping the geomorphology of forested areas and identifying deep-seated landslides.

3.2.2 Shallow Landslide

A shallow landslide is the movement of a mixture of water, soil, and debris; this landslide type starts on steep slopes during periods of intense rainfall (Bugnion et al. 2009). Thus, this landslide type is a threat to infrastructure, buildings, roads, and railways. The use of remote sensing images for shallow landslide detection is common; however, field data are necessary for verification. This concern is primarily ascribed to the swiftness and the spatial diffusion of these slope failures. Optical sensors, which can extract geomorphic features and cover a wide area, are suitable for studying shallow landslides (Gao and Maro 2010). Visual interpretation has also been proven useful for mapping this landslide type because the boundaries are frequently distinct after a shallow landslide event. Furthermore, automatic and semi-automatic classification algorithms have been tested in recent years (Heleno et al. 2015; Ma et al. 2016; Stumpf et al. 2014; Wiegand et al. 2013). The use of very-high-resolution (VHR) satellite images (e.g., GeoEye and WorldView) can provide high accuracy for detecting shallow landslides and creating databases of susceptible areas (Zizioli et al. 2014). The automated analysis of airborne laser scanning data has been used for post-event analysis, such as for mapping earthquake- or typhoon-triggered shallow landslides (Rau et al. 2012) and rainfall-induced landslides (Bai et al. 2012). Visual interpretations, such as shaded relief maps, slope

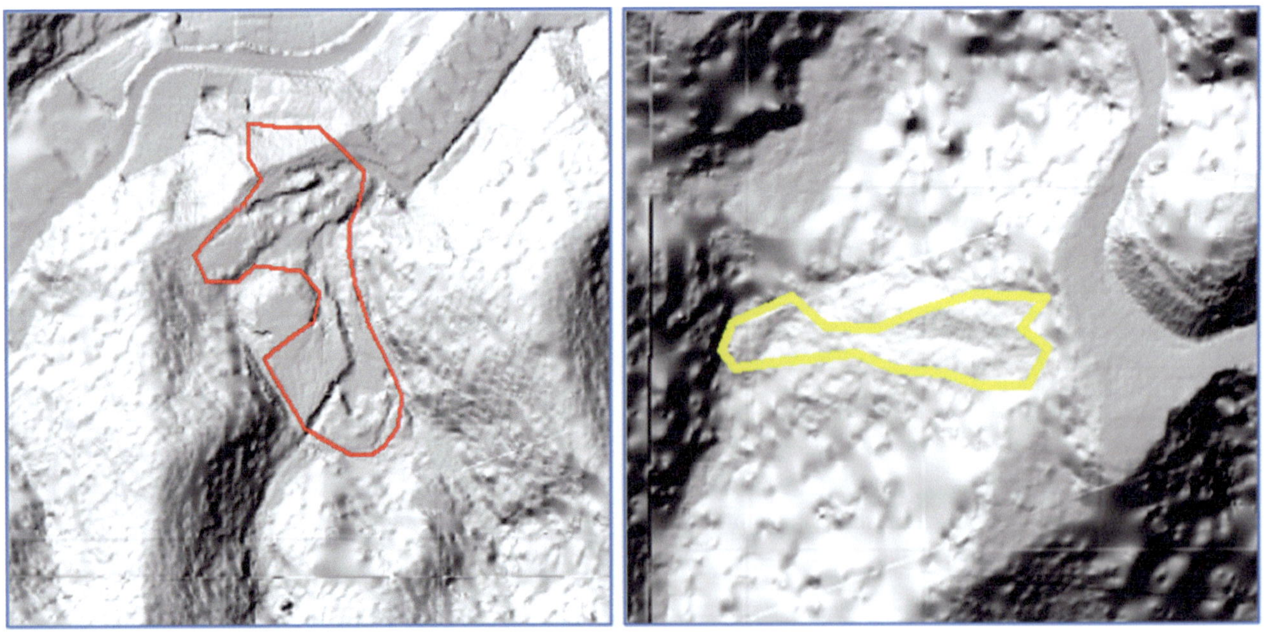

Fig. 3.1 Shallow landslide (*red polygon*) and deep-seated landslide (*yellow polygon*) in the study area

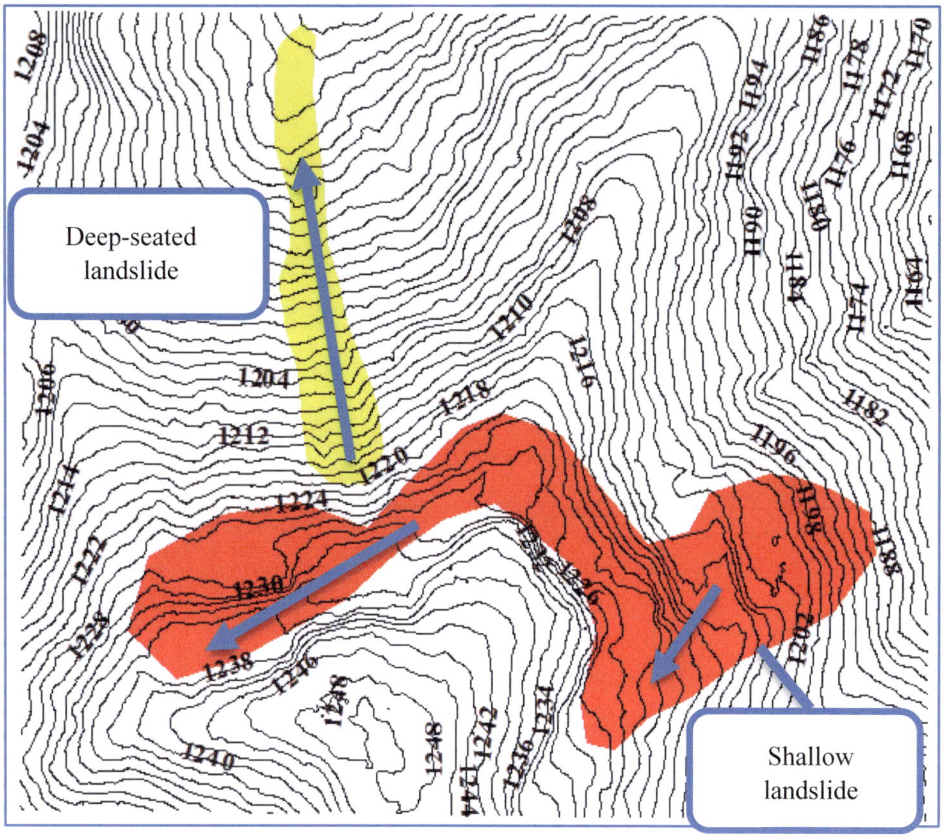

Fig. 3.2 General sketch of types of landslide

maps, and contour maps, derived from LiDAR-based DEMs were used by (Đomlija et al. 2014) to create a catalog of shallow landslides. In addition, rainfall-induced shallow landslides have been predicted using empirical rainfall thresholds or spatially distributed, physically based numerical models (Hong et al. 2015; Vennari et al. 2014). Attempts have been made to forecast this type of landslide using numerical models with both predisposing factors, e.g., landform curvature (van Asselen and Seijmonsbergen 2006) and the main triggering factor, i.e., rainfall (Segoni et al. 2009).

3.3 Study Area

The Cameron Highlands in Malaysia is a tropical rain forest area located in the latitude range 4°22′52″–4°25′48″ N and the longitude range 101°22′30″–101°23′30″ E. The Cameron Highlands is an active landslide site, and the selected subset has an aerial coverage of approximately 24.38 km^2. The average annual rainfall in the area is 2660 mm. Its average daytime temperature is moderate (~24 °C), whereas its nighttime temperature is 14 °C. A large portion of the area (80%) is forested, and the land slopes range from 80° to flat areas (0) Fig. 3.3.

3.4 Material and Method

The multistep overall workflow of the current research is presented in Fig. 3.4. First, remote sensing data and landslide inventories were preprocessed to remove the noise and outliers from the data and prepare them for the subsequent steps. Second, a high-resolution digital elevation model (DEM) (0.5 m) was derived from the LiDAR point clouds and used to generate other LiDAR-derived products and landslide conditioning factors, such as slope, aspect, height (nDSM), hillshade, and intensity. Third, the LiDAR-derived products were combined by correcting their geometric distortions and incorporating them into a coordinate system. Then, the combined products were prepared for feature extraction using a geographic information system (GIS). Fourth, a multiresolution algorithm was used to segment the data and create image objects. In this step, the fuzzy logic supervised approach proposed by Zhang et al. (2010) was used to select the best combination of segmentation parameters (i.e., scale, shape, and compactness). Fifth, relevant attributes were selected using the correlation-based feature selection method (CFS), which ranks the attributes from the most important to the least important. Rule sets were developed by applying a decision tree (DT) algorithm using the training set of the landslide inventories and the

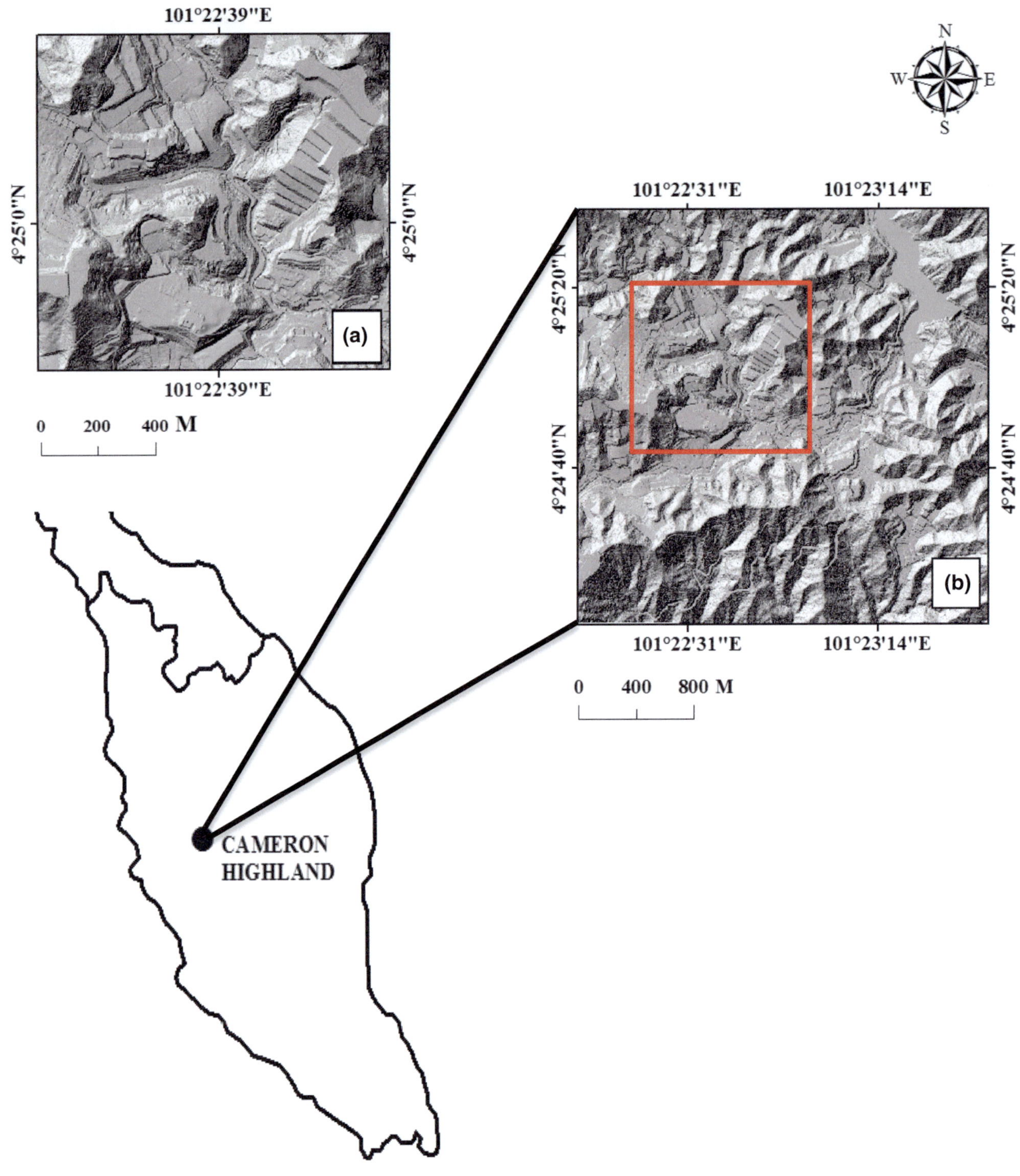

Fig. 3.3 Study area showing **a** analysis area; and **b** test area

selected relevant attributes. Rule sets were developed for extracting five land types that were enumerated as follows: (1) bare soil, (2) cut slope, (3) building, (4) vegetation, and (5) landslide. The validation of the rule sets was based on the receiver operating characteristic curves and the area under the curve (AUC) values. Finally, the results were exported into GIS to detect the precise location and direction of the landslides by overlaying the classified segments and the hillshade on the slope and aspect data to visualize other characteristics of landslide (i.e., runoff, volume, and width).

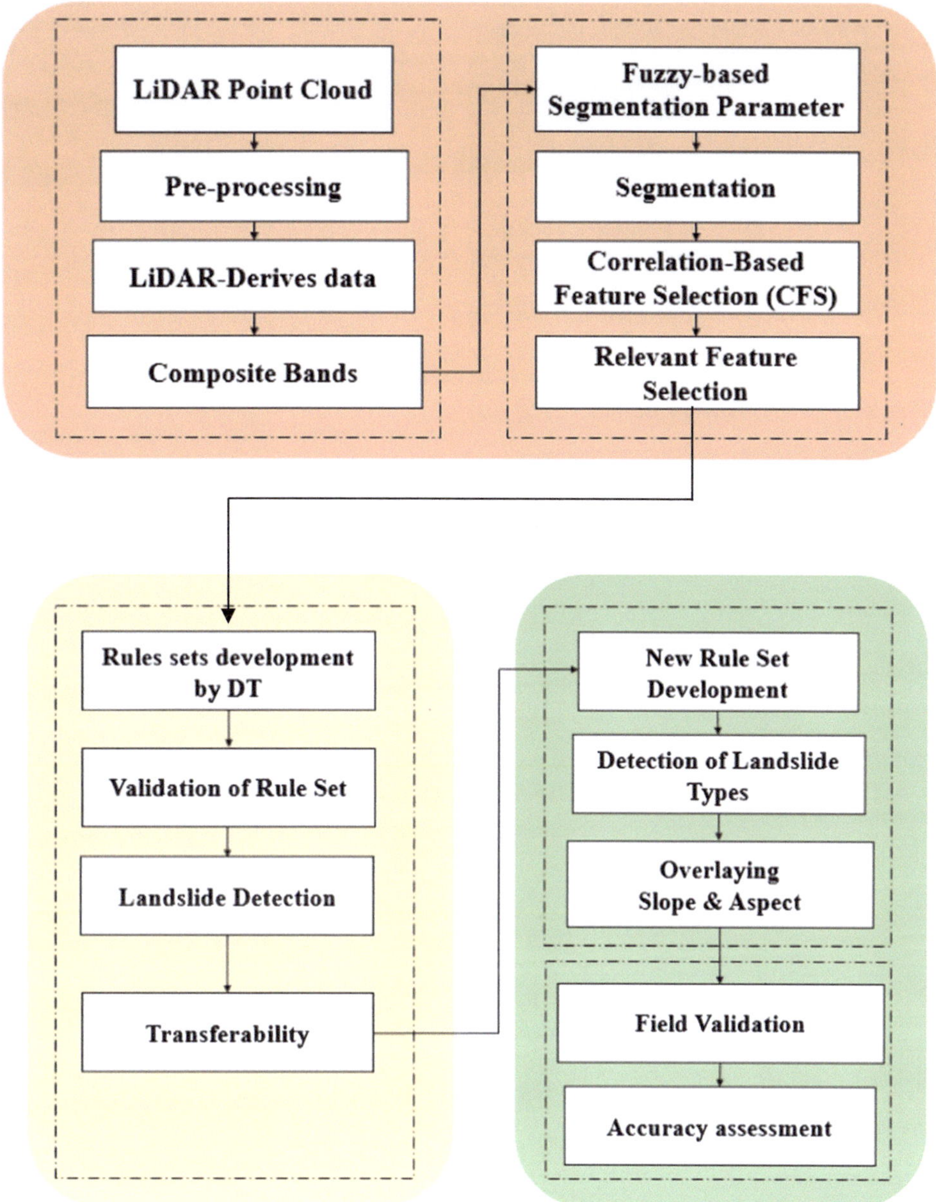

Fig. 3.4 An overview of the method adopted in the present study

3.4.1 LiDAR Data

The LiDAR point cloud was acquired over 25 km^2 of the ringlet and surrounding area in Cameron Highlands at a flying height about 1510 m. The data were captured on January 15, 2015. The point density is closely 8 points per square meter with a 25,000 Hz pulse rate frequency. The absolute accuracy of the LiDAR data should meet the root-mean-square errors (RMSE) of 0.15 m in the vertical axis and 0.3 in the horizontal axis. Orthophotos were also acquired by the same system for the study area to support landslide identification and characterization.

A digital elevation model at 0.5-m spatial resolution was interpolated from LiDAR point clouds after removing the non-ground points, with the spatial reference of GDM2000_Peninsula_RSO. The derived DEM helps in generating a number of derived layers that support landslide identification and characterization . In the current study, hillside, intensity, height (nDSM), slope, and aspect were derived from the LiDAR-based DEM Fig. 3.2, (Fig. 3.5).

3.4.2 Object-Based Image Analysis (OBIA)

An object-based approach considers the spatial, spectral, and texture attributes of the features in the classification process; thus, it is different from other classification approaches such

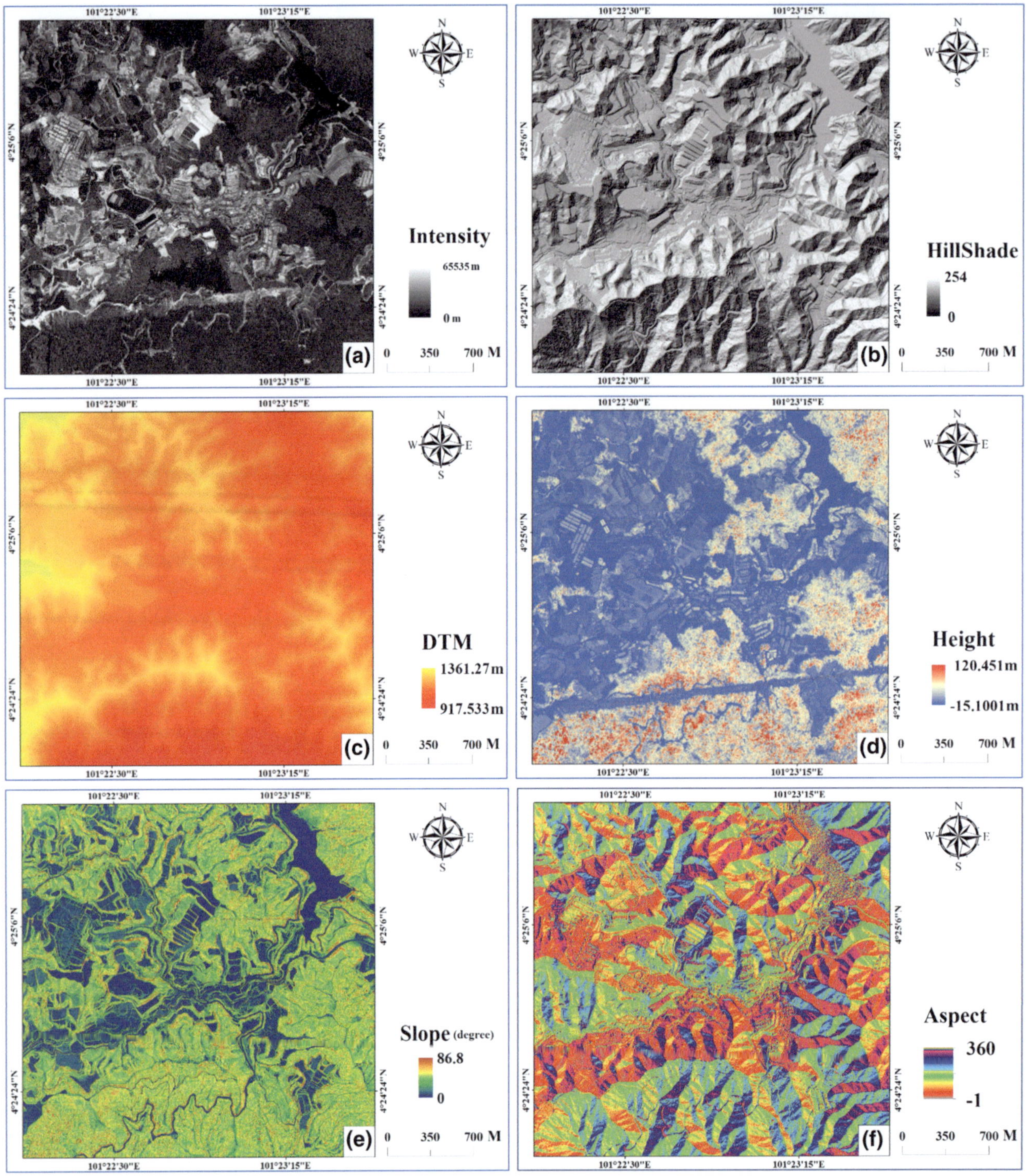

Fig. 3.5 LiDAR-derived data used for identification the location and direction of landslide include **a** intensity; **b** hillshade; **c** DTM; **d** height; **e** slope; and **f** aspect

as traditional pixel-based (Pradhan et al. 2016). The results of an object-based approach are more accurate (Rau et al. 2014) than pixel-based classification (Tehrany et al. 2014). As one of the most significant classification schemes, OBIA can provide valuable information for landslide inventory mapping (Guzzetti et al. 2012). It is considered suitable for landslide inventory mapping because its image segmentation and classification processes resemble human knowledge, and existing knowledge can be transferred into a machine algorithm in the form of rule sets, which have been implemented in landslide mapping (Barlow et al. 2003; Pradhan et al. 2016; Stumpf and Kerle 2011).

Image segmentation is the first step of OBIA (Pradhan et al. 2016). In this step, homogeneous pixels are grouped into non-overlapping regions based on spectral and geometric characteristics (Pal and Pal 1993). Multiresolution segmentation is a common algorithm that has been utilized in various earth science studies (Blaschke 2010). Numerous features of the objects can be used in multiresolution segmentation. In the current study, three parameters, namely scale, shape, and compactness, are selected for the analysis (Pradhan et al. 2016). Determining these parameters using the traditional trial-and-error method is a time-consuming and work-intensive procedure (Pradhan et al. 2016). Therefore, automatic and semiautomatic methods have been used to identify the optimal parameters in various studies (Anders et al. 2011; Belgiu and Drăguţ 2014; Drăguţ et al. 2010). Among the advanced methods for automatically selecting segmentation parameters are the Taguchi optimization technique proposed by and the fuzzy logic supervised approach presented by (Zhang et al. 2010). Therefore, automatic methods that utilize optimization algorithms (Pradhan et al. 2016) can best reduce the time required for selecting the segmentation parameters.

The selection of a small (possibly the minimum) feature set yields the best possible classification results, which is desirable for practical reasons (Kursa and Rudnicki 2010). Therefore, significant attributes should be selected to enhance the results of landslide detection in a certain area (Kursa and Rudnicki 2010). The selection of important (or relevant) attributes can help differentiate between landslide and non-landslide areas as well as better classify landslides according to their types (Van Westen et al. 2008). Several studies have used multiple attribute integration for landslide detection (Borghuis et al. 2007; Danneels et al. 2007; Hervás and Rosin 1996; Tapas R Martha, Kerle, Jetten, van Westen, & Kumar, 2010; Tapas Ranjan Martha et al. 2011; Moine et al. 2009; Stumpf and Kerle 2011). Chen et al. (2014) used 10 attributes for landslide identification; their results showed that the selected relevant attributes provided valuable information for landslide identification. The current study primarily aims to optimize the parameters of segmentation and the attributes for developing transferable rule sets for landslide detection and their characteristics by using high-resolution LiDAR data.

3.4.3 Image Segmentation

Image segmentation, which is a fundamental step of OBIA, is realized using both spatial and spectral information (Darwish et al. 2003). Segmenting the presented object by delineating their boundaries directly affects the quality and performance of the classification process. Various segmentation algorithms have been previously elaborated and applied to remote sensing data sets (Dey et al. 2010). The purpose of these algorithms is to determine relatively homogenous and powerful segments. In the current study, the image segmentation process was implemented by first identifying the three main parameters. These parameters control the segmentation results and affect the classification process (Möller et al. 2007; Tian and Chen 2007). A supervised fuzzy logic approach was used to select the optimal parameters because the traditional method for identifying the aforementioned parameters is time-consuming.

3.4.4 Correlation-Based Feature Selection

Selecting only the relevant attributes enhances the quality of landslide identification and classification in a particular area (Kursa and Rudnicki 2010). Working with a large number of attributes generates several problems. First, the algorithms are slowed down because numerous resources have to be considered (Kursa and Rudnicki 2010). Second, a higher number of attributes than the number of observations result in low accuracy (Kohavi and John 1997). Third, irrelevant input features may lead to overfitting (Chen et al. 2014). Therefore, important attributes should be selected to improve the accuracy of the feature extraction results. In the current study, CFS was performed using Weka 3.7 software to select the relevant attributes. The method established by (Li et al. 2016) was adopted in this study. The CFS algorithm was applied to all the LiDAR-derived data and the additional texture and geometric features. CFS was performed to determine the feature subsets to be used for developing the rules for landslide identification and characterization. The CFS method has two basic steps: ranking the initial attributes and eliminating the least important attributes through an iterative process.

3.4.5 DT Classifiers

DT methods are data mining techniques that generate a graphical illustration of the feature classification process (Daniel 2014). In OBIA, the most important phase is the construction of the image interpretation model (knowledge) for the segmented objects. However, executing OBIA with other classifiers that are considered "black boxes" may be difficult; by contrast, DT classifiers are like "white boxes"; that is, users can easily interpret the links between the variables of different classes and the explanatory variables of remote sensing data (Li et al. 2016). The model generated via a DT method can be either a predictive or a descriptive model. Basically, establishing DT classifiers does not require the elaborate setting of the domain knowledge or attribute; consequently, it has become popular for exploratory knowledge discovery. This type of method can provide a rapid and powerful mode of showing the structures of a data set and handle a high-dimensional data set. In the current study, a DT algorithm was used to generate the rule sets for landslide identification and characterization using CFS-derived attributes.

3.4.6 Landslide Mapping

An object-based classification method using a DT algorithm was used to produce a landslide inventory map for the study area. In this method, training data were required. Landslide training samples were collected from the orthophotos of the locations where landslides were identified via visual interpretation. The image objects were then classified using the DT algorithm, and rule sets were generated for landslides, non-landslide features, and the two types of landslides (i.e., shallow and deep-seated). The important features, namely mean slope, area, mean intensity, and gray level co-occurrence matrix (GLCM) homogeneity, were selected using the CFS method in Weka 3.7 software. Furthermore, classification was executed using the J48 algorithm in Weka 3.7. Consequently, rule sets were developed to differentiate landslides from non-landslide features, such as vegetation, building, bare soil, and man-made slopes. In addition, other rule sets were developed to distinguish the two types of landslides. The classification results were validated using the ROC method and through field investigations.

3.5 Results

3.5.1 Segmentation Parameters Selected Using a Fuzzy Logic Supervised Approach

The optimal segmentation parameters (i.e., scale, shape, and compactness) were selected using the fuzzy logic supervised approach developed by (Zhang et al. 2010). The best values of the segmentation parameters were determined based on an adequate number of training samples, which included shallow and deep-seated landslides. Table 3.1 shows the selected values for the three parameters. For example, the initial segmentation parameter values inputted into the fuzzy-based optimization tool were 20, 0.3, and 0.1 for scale, shape, and compactness, respectively, as shown in (Fig. 3.6a). From these initial values and after 100 iterations, the best values for scale, shape, and compactness derived from the optimization tool were 46.37, 0.37, and 0.401, respectively.

3.5.2 Attributes Selected Using the CFS Approach

The use of a large number of attributes can decrease the accuracy of landslide detection because of the presence of irrelevant attributes. Therefore, significant attributes were selected using the CFS method. Table 3.2 shows the 10 attributes selected out of the initial 35 attributes, including spatial, texture, geometric, and LiDAR derivatives. As shown in Table 3.2, the most important attribute is the mean intensity of LiDAR data, followed by the mean digital terrain model DTM or altitude. Texture attributes, such as the gray level difference vector contrast, GLCM homogeneity, and GLCM StdDev, were also significant in detecting landslides and determining their types using LiDAR data. In addition, area, height, and GLCM standard were the last three important attributes.

3.5.3 Rule Sets Developed for Landslide Detection and Characterization

Rules sets were developed using the DT algorithm, and the optimized attributes were selected via CFS. In general, six sets of rules were established for extracting various features or land cover types, including vegetation, bare soil, cut slope, building, and landslide. The rules developed for landslide identification included mean slope, area, mean intensity, and GLCM homogeneity. The thresholds for these attributes are provided in Table 3.3. Figure 3.7 illustrated the results of rule set based on analysis area. For example, the selected threshold for the mean slope attribute is 28°. This result indicates that most of the landslides in the study area fall from a relatively highly sloped area, and the chance of landslides occurring in flat terrain areas is minimal. The rules developed for identifying cut slopes included attributes such as mean height, GLCM StdDev, and area of objects. On the basis of the rules developed for landslide and cut slope

Table 3.1 Segmentation parameters

Initial parameters				Iteration (optimal parameters)		
No.	Scale	Shape	Compactness	Scale	Shape	Compactness
1	20	0.3	0.1	46.37	0.37	0.401
2	50	0.1	0.1	73.52	0.52	0.5
3	80	0.1	0.1	100.33	0.65	0.65

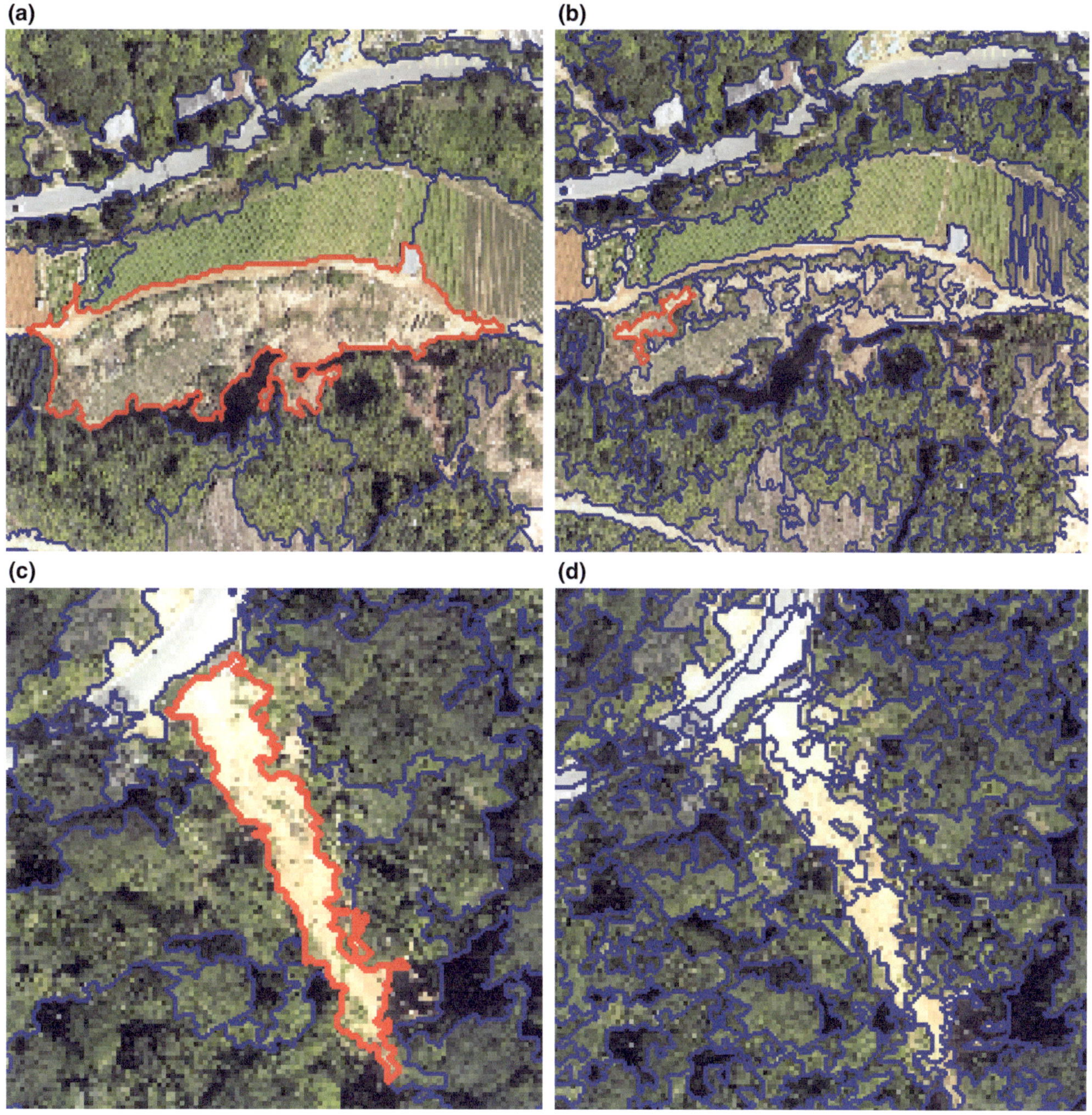

Fig. 3.6 **a** Initial segmentation for training site; **b** optimal segmentation for training site; **c** initial segmentation for test site; and **d** optimal segmentation for test site

detection, intensity and texture attributes (i.e., GLCM StdDev and GLCM homogeneity) are the most important attributes to differentiate the two land cover types.

The deep-seated landslides (10) in the study area are less than the shallow landslides (29). The detection accuracy for deep-seated landslides is higher than that for shallow landslides. (Lin et al. 2013) demonstrated that LiDAR data could significantly aid in identifying deep-seated landslides, particularly for densely vegetated areas. This current study showed that texture and LiDAR intensity attributes were more significant for distinguishing between the two landslide types than other attributes. Table 3.4 shows that GLCM homogeneity and mean intensity attributes can effectively differentiate shallow from deep-seated landslides.

3.5.4 Model Transferability

In hilly and densely vegetated terrains, such as the Cameron Highlands, discriminating between landslides and man-made

Table 3.2 Outcomes of the attributes selection for landslide detection

Attribute	No. times	Rank
Mean intensity	20	1
Mean DTM	20	2
GLDV contrast	20	3
GLCM homogeneity	20	4
GLCM StdDev	20	6
Mean slope	20	7
Area	20	8
Height	20	9
GLCM standard	20	10

Table 3.3 Rules defined for landslide identification from non-landslide

Features	Rules define	Attributes
Vegetation	5.1 < Mean slope > 1.3 Area > 2837	LiDAR derivatives Spatial
Bare soil	GLCM standard <= 26.6 Area <= 4905	Spatial Spatial
Cut slope	Mean height <= 113.8 GLCM StdDev > 30.3 Area <= 2837	LiDAR derivatives Spatial Spatial
Building	Mean slope > 2.3 Mean height > 113.8	LiDAR derivatives LiDAR derivatives
Landslide	Mean slope > 28.0748 Area > 4905 Mean intensity <= 255.49 GLCM homogenous > = 0.0496	LiDAR derivatives Spatial LiDAR derivatives Spatial

slopes and distinguishing between the two types of landslides are challenging. Consequently, the results of transferability of the developed rule sets form analysis area to the entire area were tested as shown in Fig. 3.8. Two types of landslides were differentiated, namely shallow and deep-seated, by developed new rule set see Table 3.4. The results are shown in Fig. 3.9. Stumpf et al. (2011) said that the overall accuracy of landslide detection applied to other areas even if the same method was used for model development would frequently decrease. This decline in accuracy is ascribed to various reasons, including differences in landslide characteristics and environmental conditions. Differences in the sensors used, spatial resolutions of images, and illumination conditions are among the other challenges mentioned in a recent study (Rau et al. 2014).

3.6 Discussion

The identification of landslides and their types in densely vegetated areas, such as the Cameron Highlands, is challenging because of several reasons, including the presence of man-made slopes, dense vegetation, and hilly areas. This study presented a method for automatically detecting landslides and their types by using high-resolution LiDAR data. The quantitative results of landslide identification demonstrated the robustness of the method. In addition, this study showed that optimizing the segmentation parameters, namely scale, shape, and compactness, by using the fuzzy logic supervised approach was satisfactory for differentiating non-landslide (i.e., vegetation and cut slope) from landslide features and between the two landslide types. Creating accurate objects through the optimized segmentation process allowed the use of spatial, texture, and geometric attributes for feature identification. Accurate segmentation is necessary to distinguish deep-seated from shallow landslides because landslides can be classified according to their geometric and texture attributes (Table 3.4).

The selection of the optimal attributes, which are relevant to a landslide, mainly relies on the experience of the analysts. Thus, establishing an attribute selection method is imperative to detect landslides and their characteristics. The relevant attributes selected using the CFS algorithm simplify the rule sets used to detect landslides and their types. In addition, the rules developed with less optimized attributes can be transferable to the entire study area. The optimized attributes for detecting landslides and determining their types included LiDAR-derived data (i.e., slope, height, and intensity), texture attributes (i.e., GLCM StdDev and GLCM homogeneity), and geometric attributes. Computation time and reliance

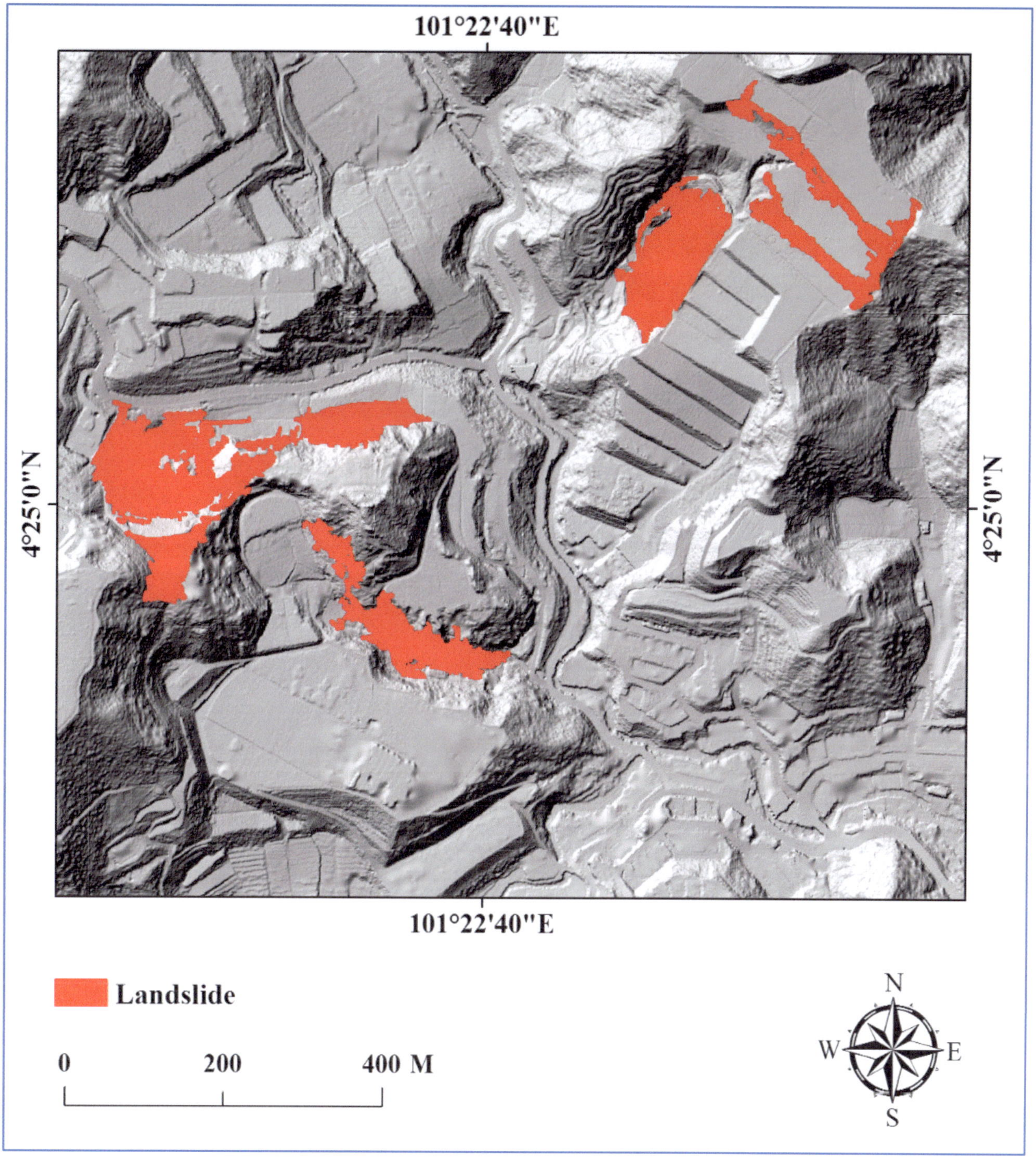

Fig. 3.7 Analysis area for detecting the location of landslide

Table 3.4 Rules defined to differentiate between types of landslide

Feature types	Rules define	Attributes
Deep-seated	GLCM homogenous <= 0.05 GLCM homogenous (Direction = 0°) > 0.038	Spatial Spatial
Shallow	GLCM homogenous > 0.05 Mean intensity <= 414.3 GLCM homogenous (Direction = 0°) <= 0.038	Spatial LiDAR derivatives Spatial

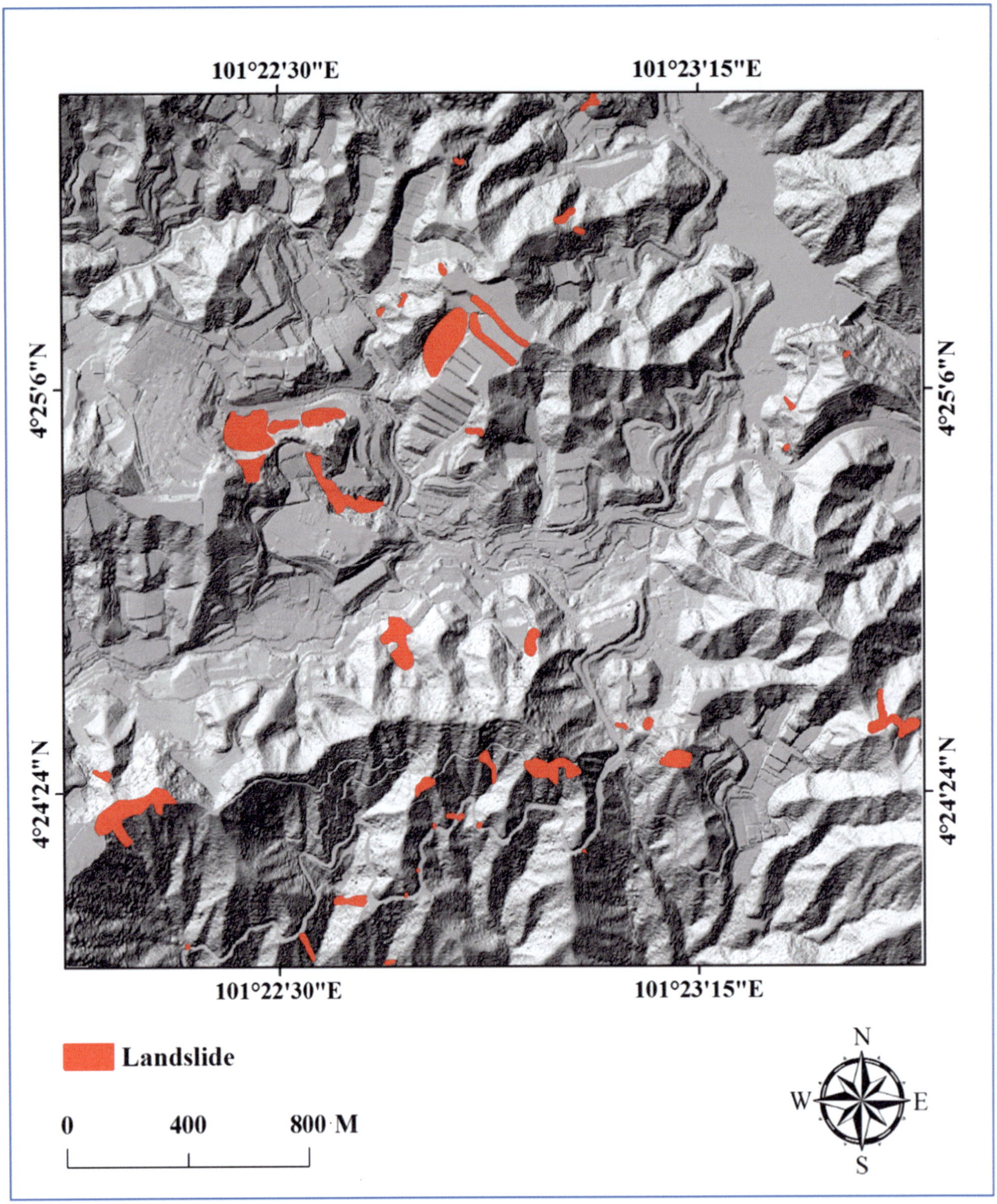

Fig. 3.8 Test area showing the locations of landslide in whole study area after transferability

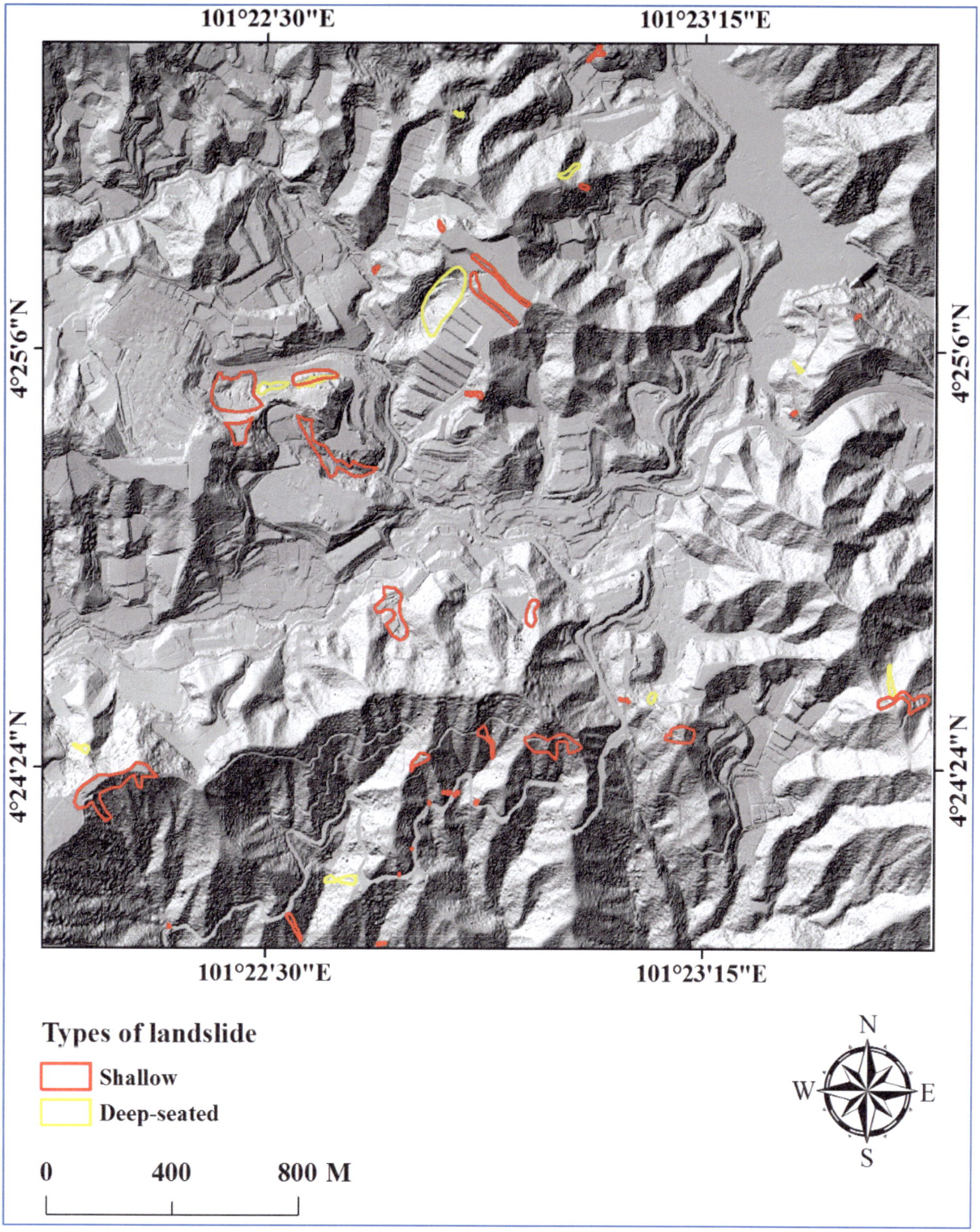

Fig. 3.9 Location of the shallow and deep-seated landslides in the study area

on the expert knowledge of the analyst were reduced to a greater extent with the proposed rule sets than with existing complex rule sets based on image classification.

Field investigation is challenging and tedious for mapping landslide inventories (Dou et al. 2015). Although the visual interpretation of remote sensing data using this method is reliable, this method is both costly and time-consuming. A new landslide detection method, which utilizes LiDAR-derived data, is effective for mapping geomorphic features and landslides, particularly in densely vegetated areas. The natural features of the earth surface are difficult to recognize using traditional methods (Chen et al. 2015). By contrast, a LiDAR-derived DEM provides additional terrain data, such as curvature, slope, and hillshade, which can help better describe the landscape of an area and identify landslides and their types. The current study used a LiDAR-derived DEM, as well as texture and geometric features, for landslide detection and characterization. The results show that LiDAR data can be effectively used to analyze and visualize terrains that are difficult to explore because of the presence of dense vegetation. Furthermore, the locations of landslides were identified using the proposed method, and their directions were visualized by overlaying the slope and aspect factors from the LiDAR-derived data. The landslides were also rendered in 3D to visualize their other characteristics, such as width, length, runoff distance, and depth as shown in Fig. 3.10.

3.6.1 Accuracy Assessment

Various methods and accuracy metrics, such as kappa indices, overall accuracy, user's accuracy, and producer's accuracy, have been applied to estimate the accuracy of remote sensing products (Dou et al. 2015). The concept of a confusion matrix is frequently adopted to simplify the calculation of these accuracy metrics (Radoux and Bogaert 2014). In recent studies, however, the use of kappa indices has been criticized by several researchers (Pontius and Millones 2011). Mondini et al. (2011) utilized ROC plots to

Fig. 3.10 3D perspective of detected landslides for the study area

evaluate remote sensing outcomes by plotting the true positive value against the positive predictive value (PPV). ROC plotting is one of the common methods used in engineering and signal processing; therefore, the use of this method to evaluate the quality of deterministic and probabilistic detection and forecast systems is extremely helpful. In the current study, the landslide detection results were assessed using the ROC plotting method and further verified through field investigations. The accuracy assessment is performed as follows: The locations of the landslides were first collected from the study area using a handheld Global Positioning System (GPS) device with an accuracy of 5 m. Subsequently, the ROC curves for shallow and deep-seated landslides were generated, and the AUCs were calculated. Figure 3.10 shows the ROC curves for landslide detection and their types as detected by the method proposed in the current study. The estimated AUC of landslide detection was 0.82, and the accuracy rates after transferability for shallow and deep-seated landslides were 0.80 and 0.83, respectively. The estimated accuracy rates indicated the effectiveness of the proposed approach for detecting landslides and distinguishing their types (Fig. 3.11).

3.6.2 Field Investigation

A field investigation was conducted as an additional assessment method to examine the reliability of the proposed method. A handheld GPS device (GeoExplorer 6000) was used to identify the locations of the landslides, as shown in Fig. 3.12. The information acquired from field measurements allowed for the assessment of the precision and reliability of the produced landslide inventory map. In addition, the field investigation demonstrated that the landslides detected using the proposed methodology were accurate. Thus, the current methodology can identify landslide locations, distinguish landslide types, and produce a reasonable acceptable landslide inventory map for the Cameron Highlands.

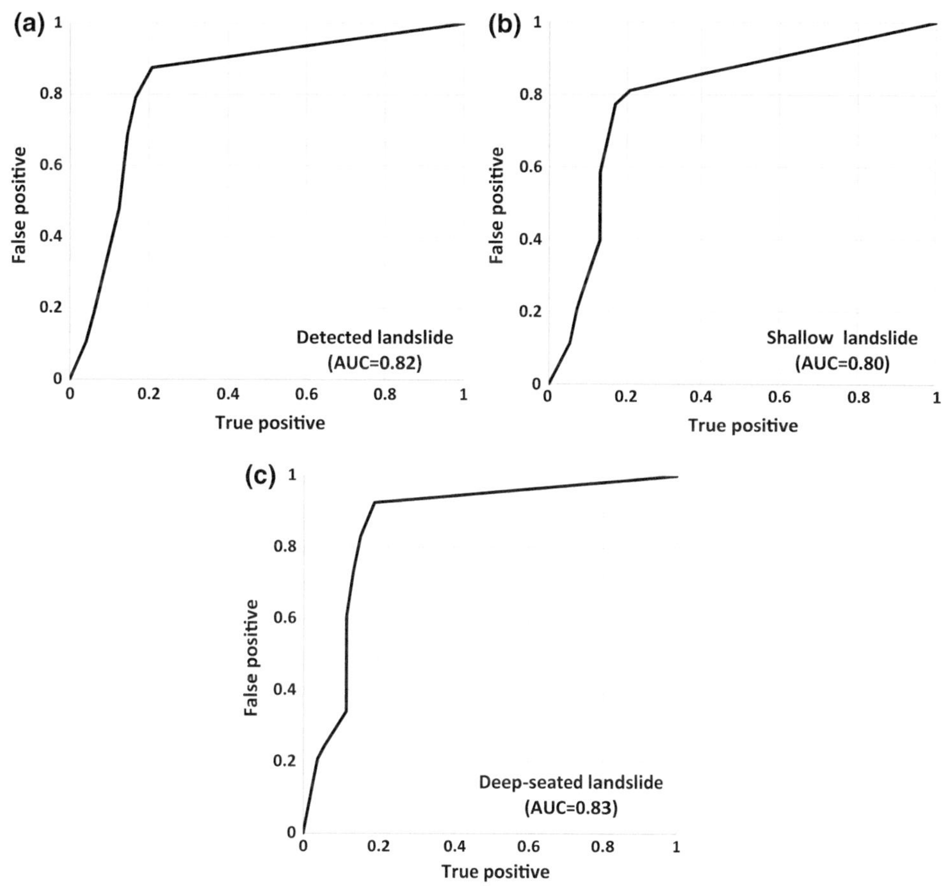

Fig. 3.11 ROC curve for **a** detected landslides and their characteristics; **b** shallow landslide; and **c** deep-seated

Fig. 3.12 Field photographs taken in the study area during field investigation

3.7 Conclusion

Identifying landslides and their types (i.e., shallow and deep-seated) in tropical regions is a difficult task. In this study, an optimized object-oriented rule set was developed to detect landslides in the Cameron Highlands and differentiate between their types. The main data sources were high-resolution airborne LiDAR point clouds. Optimized segmentation was performed using an existing fuzzy logic supervised approach, and the important attribute subset was selected using the CFS algorithm. The overall accuracy of landslide detection using the proposed method was 0.82, and the prediction accuracy rates for shallow and deep-seated landslide detection were 0.80 and 0.83, respectively. The optimization of the segmentation parameters and attributes improved the computational efficiency of the workflow and enhanced the transferability of the rule sets into different spatial subsets within the Cameron Highlands. LiDAR data were effective and useful for identifying landslides and distinguishing their types. In addition, the use of LiDAR data allowed for the identification of other characteristics of landslides, such as orientation and runoff distance, by overlaying LiDAR-derived slope and aspect factors on detected landslide scarps. The optimization of segmentation parameters and the selection of attributes could also improve the computational efficiency of the workflow and enhance the transferability of the rule sets into different spatial subsets within the Cameron Highlands in Malaysia. This study suggests that developing rule sets based on optimized techniques and using VHR airborne LiDAR-derived data and spatial attributes are effective in identifying the locations of landslides and distinguishing their types in tropical regions. This proposed automatic landslide detection method can be an important geospatial solution for managing landslide hazards and conducting landslide risk assessments.

References

Agliardi, F., Crosta, G. B., Zanchi, A., & Ravazzi, C. (2009). Onset and timing of deep-seated gravitational slope deformations in the eastern Alps, Italy. *Geomorphology, 103*(1), 113–129.

Anders, N. S., Seijmonsbergen, A. C., & Bouten, W. (2011). Segmentation optimization and stratified object-based analysis for semi-automated geomorphological mapping. *Remote Sensing of Environment, 115*(12), 2976–2985.

Bai, S., Wang, J., Zhang, Z., & Cheng, C. (2012). Combined landslide susceptibility mapping after Wenchuan earthquake at the Zhouqu segment in the Bailongjiang Basin, China. *Catena, 99,* 18–25.

Barlow, J., Martin, Y., & Franklin, S. (2003). Detecting translational landslide scars using segmentation of Landsat ETM + and DEM data in the northern Cascade Mountains, British Columbia. *Canadian Journal of Remote Sensing, 29*(4), 510–517.

Belgiu, M., & Drăguţ, L. (2014). Comparing supervised and unsupervised multiresolution segmentation approaches for extracting buildings from very high resolution imagery. *ISPRS Journal of Photogrammetry and Remote Sensing, 96,* 67–75.

Blaschke, T. (2010). Object based image analysis for remote sensing. *ISPRS Journal of Photogrammetry and Remote Sensing, 65*(1), 2–16.

Borghuis, A., Chang, K., & Lee, H. (2007). Comparison between automated and manual mapping of typhoon-triggered landslides from SPOT-5 imagery. *International Journal of Remote Sensing, 28*(8), 1843–1856.

Brunetti, M., Guzzetti, F., & Rossi, M. (2009). Probability distributions of landslide volumes. *Nonlinear Processes in Geophysics, 16*(2), 179–188.

Bugnion, L., Volkwein, A., & Denk, M. (2009). *Artificial full scale shallow landslides.* Paper presented at the EGU General Assembly Conference Abstracts.

Chang, K.-T., Liu, J.-K., & Wang, C.-I. (2012). An object-oriented analysis for characterizing the rainfall-induced shallow landslide. *Journal of Marine Science and Technology, 20*(6), 647–656.

Chen, R.-F., Lin, C.-W., Chen, Y.-H., He, T.-C., & Fei, L.-Y. (2015). Detecting and characterizing active thrust fault and deep-seated landslides in dense forest areas of Southern Taiwan Using Airborne LiDAR DEM. *Remote Sensing, 7*(11), 15443–15466.

Chen, W., Li, X., Wang, Y., Chen, G., & Liu, S. (2014). Forested landslide detection using LiDAR data and the random forest algorithm: A case study of the Three Gorges, China. *Remote Sensing of Environment, 152,* 291–301.

Cruden, D. M., & Varnes, D. J. (1996). Landslides: investigation and mitigation. Chapter 3-Landslide types and processes. *Transportation research board special report* (247).

Daniel, S. (2014). *Predictive modeling of trust to Social Media content.*

Danneels, G., Pirard, E., & Havenith, H.-B. (2007). *Automatic landslide detection from remote sensing images using supervised classification methods.* Paper presented at the 2007 IEEE International Geoscience and Remote Sensing Symposium.

Darwish, A., Leukert, K., & Reinhardt, W. (2003). *Image segmentation for the purpose of object-based classification.* Paper presented at the International Geoscience and Remote Sensing Symposium.

Delgado, J., Vicente, F., García-Tortosa, F., Alfaro, P., Estévez, A., Lopez-Sanchez, J., et al. (2011). A deep seated compound rotational rock slide and rock spread in SE Spain: Structural control and DInSAR monitoring. *Geomorphology, 129*(3), 252–262.

Dey, V., Zhang, Y., & Zhong, M. (2010). *A review on image segmentation techniques with remote sensing perspective*: na.

cĐomlija, P., Bernat, S., Mihalić, S. A., & Benac, Č. (2014). Landslide inventory in the area of Dubračina River Basin (Croatia) *Landslide science for a safer geoenvironment* (pp. 837–842). Berlin: Springer.

Dou, J., Chang, K.-T., Chen, S., Yunus, A. P., Liu, J.-K., Xia, H., et al. (2015a). Automatic case-based reasoning approach for landslide detection: Integration of object-oriented image analysis and a genetic algorithm. *Remote Sensing, 7*(4), 4318–4342.

Dou, J., Paudel, U., Oguchi, T., Uchiyama, S., & Hayakavva, Y. S. (2015). Shallow and deep-seated landslide differentiation using support vector machines: A case study of the Chuetsu Area, Japan. *Terrestrial, Atmospheric & Oceanic Sciences, 26*(2).

Drăguţ, L., Tiede, D., & Levick, S. R. (2010). ESP: A tool to estimate scale parameter for multiresolution image segmentation of remotely sensed data. *International Journal of Geographical Information Science, 24*(6), 859–871.

Gao, J., & Maro, J. (2010). Topographic controls on evolution of shallow landslides in pastoral Wairarapa, New Zealand, 1979–2003. *Geomorphology, 114*(3), 373–381.

Goudie, A. (2004). *Encyclopedia of geomorphology* (Vol. 2). UK: Psychology Press.

Guzzetti, F., Mondini, A. C., Cardinali, M., Fiorucci, F., Santangelo, M., & Chang, K.-T. (2012). Landslide inventory maps: New tools for an old problem. *Earth-Science Reviews, 112*(1), 42–66.

Heleno, S., Matias, M., Pina, P., & Sousa, A. (2015). Automated object-based classification of rain-induced landslides with VHR multispectral images in Madeira Island. *Natural Hazards & Earth System Sciences Discussions, 3*(9).

Hervás, J., & Rosin, P. L. (1996). *Landslide mapping by textural analysis of ATM data.* Paper presented at the Proceedings of the Thematic Conference on Geologic Remote Sensing.

Hong, Y., He, X., Cerato, A., Zhang, K., Hong, Z., & Liao, Z. (2015). Predictability of a physically based model for rainfall-induced shallow landslides: Model development and case studies. In *Modern technologies for landslide monitoring and prediction* (pp. 165–178). New York: Springer.

Kasai, M., Ikeda, M., Asahina, T., & Fujisawa, K. (2009). LiDAR-derived DEM evaluation of deep-seated landslides in a steep and rocky region of Japan. *Geomorphology, 113*(1), 57–69.

Kellerer-Pirklbauer, A., Proske, H., & Strasser, V. (2010). Paraglacial slope adjustment since the end of the Last Glacial Maximum and its long-lasting effects on secondary mass wasting processes: Hauser Kaibling, Austria. *Geomorphology, 120*(1), 65–76.

Kohavi, R., & John, G. H. (1997). Wrappers for feature subset selection. *Artificial Intelligence, 97*(1), 273–324.

Korup, O. (2006). Effects of large deep-seated landslides on hillslope morphology, western Southern Alps, New Zealand. *Journal of Geophysical Research: Earth Surface, 111*(F1).

Kursa, M. B., & Rudnicki, W. R. (2010). Feature selection with the Boruta package: Journal.

Li, M., Ma, L., Blaschke, T., Cheng, L., & Tiede, D. (2016). A systematic comparison of different object-based classification techniques using high spatial resolution imagery in agricultural environments. *International Journal of Applied Earth Observation and Geoinformation, 49,* 87–98.

Lin, C.-W., Tseng, C.-M., Tseng, Y.-H., Fei, L.-Y., Hsieh, Y.-C., & Tarolli, P. (2013). Recognition of large scale deep-seated landslides in forest areas of Taiwan using high resolution topography. *Journal of Asian Earth Sciences, 62,* 389–400.

Ma, H.-R., Cheng, X., Chen, L., Zhang, H., & Xiong, H. (2016). Automatic identification of shallow landslides based on Worldview2 remote sensing images. *Journal of Applied Remote Sensing, 10*(1), 016008.

Malamud, B. D., Turcotte, D. L., Guzzetti, F., & Reichenbach, P. (2004). Landslide inventories and their statistical properties. *Earth Surface Processes and Landforms, 29*(6), 687–711.

Martha, T. R., Kerle, N., Jetten, V., van Westen, C. J., & Kumar, K. V. (2010). Characterising spectral, spatial and morphometric properties of landslides for semi-automatic detection using object-oriented methods. *Geomorphology, 116*(1), 24–36.

Martha, T. R., Kerle, N., van Westen, C. J., Jetten, V., & Kumar, K. V. (2011). Segment optimization and data-driven thresholding for knowledge-based landslide detection by object-based image analysis. *IEEE Transactions on Geoscience and Remote Sensing, 49* (12), 4928–4943.

McKean, J., & Roering, J. (2004). Objective landslide detection and surface morphology mapping using high-resolution airborne laser altimetry. *Geomorphology, 57*(3), 331–351.

Moine, M., Puissant, A., & Malet, J.-P. (2009). *Detection of landslides from aerial and satellite images with a semi-automatic method.* Application to the Barcelonnette basin (Alpes-de-Hautes-Provence, France): Paper presented at the Landslide processes-from geomorphologic mapping to dynamic modelling.

Möller, M., Lymburner, L., & Volk, M. (2007). The comparison index: A tool for assessing the accuracy of image segmentation. *International Journal of Applied Earth Observation and Geoinformation, 9* (3), 311–321.

Mondini, A., Guzzetti, F., Reichenbach, P., Rossi, M., Cardinali, M., & Ardizzone, F. (2011). Semi-automatic recognition and mapping of rainfall induced shallow landslides using optical satellite images. *Remote Sensing of Environment, 115*(7), 1743–1757.

Pal, N. R., & Pal, S. K. (1993). A review on image segmentation techniques. *Pattern Recognition, 26*(9), 1277–1294.

Pontius, R. G., Jr., & Millones, M. (2011). Death to Kappa: Birth of quantity disagreement and allocation disagreement for accuracy assessment. *International Journal of Remote Sensing, 32*(15), 4407–4429.

Pradhan, B., Jebur, M. N., Shafri, H. Z. M., & Tehrany, M. S. (2016). Data fusion technique using wavelet transform and taguchi methods for automatic landslide detection from airborne laser scanning data and quickbird satellite imagery. *IEEE Transactions on Geoscience and Remote Sensing, 54*(3), 1610–1622.

Radoux, J., & Bogaert, P. (2014). Accounting for the area of polygon sampling units for the prediction of primary accuracy assessment indices. *Remote Sensing of Environment, 142,* 9–19.

Rau, J.-Y., Chang, K.-T., Shao, Y.-C., & Lau, C.-C. (2012). Semi-automatic shallow landslide detection by the integration of airborne imagery and laser scanning data. *Natural Hazards, 61*(2), 469–480.

Rau, J.-Y., Jhan, J.-P., & Rau, R.-J. (2014). Semiautomatic object-oriented landslide recognition scheme from multisensor optical imagery and DEM. *IEEE Transactions on Geoscience and Remote Sensing, 52*(2), 1336–1349.

Segoni, S., Leoni, L., Benedetti, A., Catani, F., Righini, G., Falorni, G., et al. (2009). Towards a definition of a real-time forecasting network for rainfall induced shallow landslides. *Natural Hazards and Earth System Sciences, 9*(6), 2119–2133.

Stumpf, A., & Kerle, N. (2011). Object-oriented mapping of landslides using Random Forests. *Remote Sensing of Environment, 115*(10), 2564–2577.

Stumpf, A., Lachiche, N., Malet, J.-P., Kerle, N., & Puissant, A. (2014). Active learning in the spatial domain for remote sensing image classification. *IEEE Transactions on Geoscience and Remote Sensing, 52*(5), 2492–2507.

Tarolli, P. (2014). High-resolution topography for understanding Earth surface processes: Opportunities and challenges. *Geomorphology, 216,* 295–312.

Tehrany, M. S., Pradhan, B., & Jebuv, M. N. (2014). A comparative assessment between object and pixel-based classification approaches for land use/land cover mapping using SPOT 5 imagery. *Geocarto International, 29*(4), 351–369.

Tian, J., & Chen, D. M. (2007). Optimization in multi-scale segmentation of high-resolution satellite images for artificial feature recognition. *International Journal of Remote Sensing, 28*(20), 4625–4644.

van Asselen, S., & Seijmonsbergen, A. (2006). Expert-driven semi-automated geomorphological mapping for a mountainous area using a laser DTM. *Geomorphology, 78*(3), 309–320.

Van Westen, C. J., Castellanos, E., & Kuriakose, S. L. (2008). Spatial data for landslide susceptibility, hazard, and vulnerability assessment: An overview. *Engineering Geology, 102*(3), 112–131.

Vennari, C., Gariano, S., Antronico, L., Brunetti, M., Iovine, G., Peruccacci, S., et al. (2014). Rainfall thresholds for shallow landslide occurrence in Calabria, southern Italy. *Natural Hazards and Earth System Sciences, 14*(2), 317–330.

Vohora, V., & Donoghue, S. (2004). *Application of remote sensing data to landslide mapping in Hong Kong*. Remote Sensing and Spatial Information Sciences: International Archives of Photogrammetry.

Wiegand, C., Rutzinger, M., Heinrich, K., & Geitner, C. (2013). Automated extraction of shallow erosion areas based on multi-temporal ortho-imagery. *Remote Sensing, 5*(5), 2292–2307.

Yu, T.-T., Wang, T.-S., & Cheng, Y.-S. (2015). Analysis of factors triggering shallow failure and deep-seated landslides induced by single rainfall events. *Journal of Disaster Research, 10*(5), 966–972.

Zêzere, J. L., Trigo, R. M., & Trigo, I. F. (2005). Shallow and deep landslides induced by rainfall in the Lisbon region (Portugal): Assessment of relationships with the North Atlantic Oscillation. *Natural Hazards and Earth System Science, 5*(3), 331–344.

Zhang, Y., Maxwell, T., Tong, H., & Dey, V. (2010). *Development of a supervised software tool for automated determination of optimal segmentation parameters for ecognition*: na.

Zizioli, D., Meisina, C., Bordoni, M., & Zucca, F. (2014). Rainfall-triggered shallow landslides mapping through Pleiades images In *Landslide science for a safer geoenvironment* (pp. 325–329). New York: Springer.

4 Integration of LiDAR and QuickBird Data for Automatic Landslide Detection Using Object-Based Analysis and Random Forests

Biswajeet Pradhan, Maher Ibrahim Seeni and Haleh Nampak

4.1 Introduction

Landslide inventories are indispensable in producing landslide susceptibility, hazard, and risk maps. Landslide inventory maps are produced by detecting landslide locations or scarps. The greatest target of inventory map is to supply the criterion information on the distribution and movement triggered by one or multiple events and to evaluate parameters that affect the slope failure (Galli et al. 2008). The conventional methods used to create landslide inventory include aerial photograph interpretation, field observations, and historical records of landslides (Nandi and Shakoor 2010). However, these conventional methods are limited by inadequate aerial photograph resolution desired to detect small landslides, morphologic structures covered by vegetation, and laborious and time-consuming nature of detailed field mapping in rough terrains.

Recently, high-resolution digital elevation models (DEMs) and landslide maps obtained from airborne products are becoming increasingly popular for generating accurate landslide inventories (Eeckhaut et al. 2007). Light detection and ranging (LiDAR), high-resolution satellite images (i.e., IKONOS, QuickBird, and SPOT), and interferometric synthetic aperture radar (InSAR) are salient examples of using advanced technologies (Calabro et al. 2010; Murillo-García et al. 2015; Singh et al. 2005; Zhang and Gruen 2006). Several advantages of the techniques are in their rapid cost-effective data acquisition over large geographic areas; fusion of various spatial, spectral, and multitemporal data products that generate topographic information and analytical interpretation; and enhanced viewing capabilities beneath dense vegetation. Based on the comparison of traditional and advanced techniques for landslide recognition conducted by (Guzzetti et al. 2012), they found that the accuracy of landslide maps can be enhanced with modern techniques. Furthermore, they can positively affect all derived outcomes and investigations, such as landslide modeling, susceptibility, and risk assessments.

As mentioned, the use of very high-resolution DEMs is one of the common methods in analyzing surface morphology. The high-resolution DEM derived by LiDAR sensors provides opportunities for researchers to identify and map slope failures (Derron and Jaboyedoff 2010). Optical satellite images have several advantages over aerial photographs, panchromatic images, and LiDAR data as they record multispectral information of the terrain in specific portions. The main privilege of using LiDAR data is its capability to penetrate vegetated locations and to attain valuable and useful information on topographic features (Slatton et al. 2007). The superiority of LiDAR data makes it preferable compared with other data sources such as aerial photographs for landslide detection in forested areas (Brardinoni et al. 2003).

However, very high-resolution sensors have caused more challenges considering the lower spectral bands and higher sensitivity to co-registration errors. Hence, object-based image analysis (OBIA) has become a widely used technique in various geoscience fields to extract geometric and contextual information of multisource data (Blaschke 2010). The OBIA techniques are used in natural disaster monitoring and assessment, as well as in other environmental studies (Blaschke 2010; Castillejo-González et al. 2009; Van Den Eeckhaut and Hervás 2012). In these techniques, the image is segmented into homogeneous regions composed of similar pixels and then, objects are classified into sets of features related to spectral, spatial, and contextual properties.

Feature selection in high-dimensional datasets is also an important element to increase the accuracy of the algorithm for classification. Several feature selection methods are used in conjunction with OBIA (Laliberte et al. 2007). The most widely used examples are artificial neural networks (Van Coillie et al. 2011), support vector machines, genetic programming (Van Coillie et al. 2007), RF (Pal 2005), *K*-nearest neighbor algorithm, kernel-based feature extraction

B. Pradhan (✉) · M.I. Seeni · H. Nampak
Department of Civil Engineering, University Putra Malaysia, Serdang, Malaysia
e-mail: biswajeet24@gmail.com

B. Pradhan (ed.), *Laser Scanning Applications in Landslide Assessment*,
DOI 10.1007/978-3-319-55342-9_4

(Arenas-Garcia and Camps-Valls 2008), and cellular automata (Wang et al. 2011).

An accurate landslide inventory map can be constructed using a set of criteria and a detailed analysis. Therefore, the availability of very high-resolution satellite images in conjunction with OBIA techniques can improve the feature extraction and effectively integrate data from multiple sources. Currently, the integration of QuickBird imagery with a LiDAR-derived DEM was conducted to map the landslide locations in the study area. Moreover, advanced methods, such as Taguchi technique, RF method, and rule-based classification, were used instead of traditional classification techniques for segmentation, feature selection, and object classification.

4.2 Study Area

Bukit Ma'okil, Malaysia, was selected as the study area to implement landslide detection considering its proneness to landslides. This area is located at the zone of 3°11N–3°12N latitude and 101°45E–101°46E longitude. Area coverage is 24 km^2, covering a small part of Johor Bahru state (Fig. 4.1). From April to June, the highest temperature is between 29 and 32 °C. Annual rainfall is very high averaging between 2500 and 3000 mm annually. The two pronounced wet seasons are from September to December and February to May. The rainfall peaks are between November and December and March to May. The geomorphology of the area consists of undulating plateau and hilly terrain. The geology of the areas consists of quaternary and Devonian granite (Yusof et al. 2015). In recent years, several landslides were recorded along highways, main roads, and stream scouring the sides of the streams.

4.3 Data Used

Two data sources, namely LiDAR derived DEM and QuickBird imagery, were utilized to create the landslide inventory. Airborne LiDAR is a popular remote sensing method used for the digital presentation of the topographic surface of regions with small to large coverage (Rau et al. 2014). This method uses a laser sensor positioned on an airplane to record the distance from the sensor to various points on earth. A total of 100 points are recorded in each square based on a number of conditions such as elevation, speed, and type of sensor (Saeidi et al. 2014). Moreover, terrain condition is an important factor. LiDAR vector point data were gathered over −24 km^2 of Ma'okil on August 3, 2013. The data were recorded approximately 112,732,461 data points. QuickBird imagery using high-resolution satellites has become popular for mapping. The satellite contains panchromatic (61–72 cm spatial resolution) and multispectral sensors (2.44–2.88 m spatial resolution) captured on February 8, 2014. The sensor covered 16.5–19 km in the across-track direction. The significant advantage of these data over conventional aerial photography is the provision of multitemporal landslide maps with a revisiting rate of one day (Cheng et al. 2004).

4.4 Methodology

The overall workflow of the automatic landslide detection proposed in the current study is presented in Fig. 4.2. The process consists of three main stages. First is the data preprocessing and preparation. Second is the Taguchi optimization used to select the best combination of segmentation parameters. Once the input data was segmented, RF algorithm was adopted to select relevant spectral, spatial, and texture features for landslide detection. In addition, relevant features were used to develop the rulesets for the landslide detection using knowledge-based approach. Afterward, the developed rulesets were applied on the identified processed data and landslide prone areas. Finally, the proposed workflow was validated using landslide inventory data collected by various methods, such as interpretation of aerial photographs, field survey, and previous studies.

4.4.1 Data Preprocessing and Preparation

The original LiDAR data was in LAS format. First, the LAS data was converted into raster format in GIS with (1 m) spatial resolution using a TIN-based interpolation (Fig. 4.3a) creating the DSM. Afterward, the point clouds were filtered into ground and non-ground points to generate DEM using the same interpolation technique (Fig. 4.3b). Two new DEM and DSM datasets were derived from these processes used for landslide detection. The slope map was also generated from LiDAR derived from DEM with a resolution of 1 m using 3D spatial analysis tool of ArcGIS (Fig. 4.3c). In addition, slope is one of the most significant factors causing slope instability given that the shear stress in soil or in other unconsolidated material increases with progressive inclination. Normalized difference vegetation index (NDVI) is one of the most important data layers for landslide detection. This index represents the greenness of the vegetated earth surface, and it is an essential parameter for class separation of vegetation and non-vegetation in case of landslide. Currently, this layer was generated from the NIR and red bands of QuickBird image using Eq. 4.1 as follows:

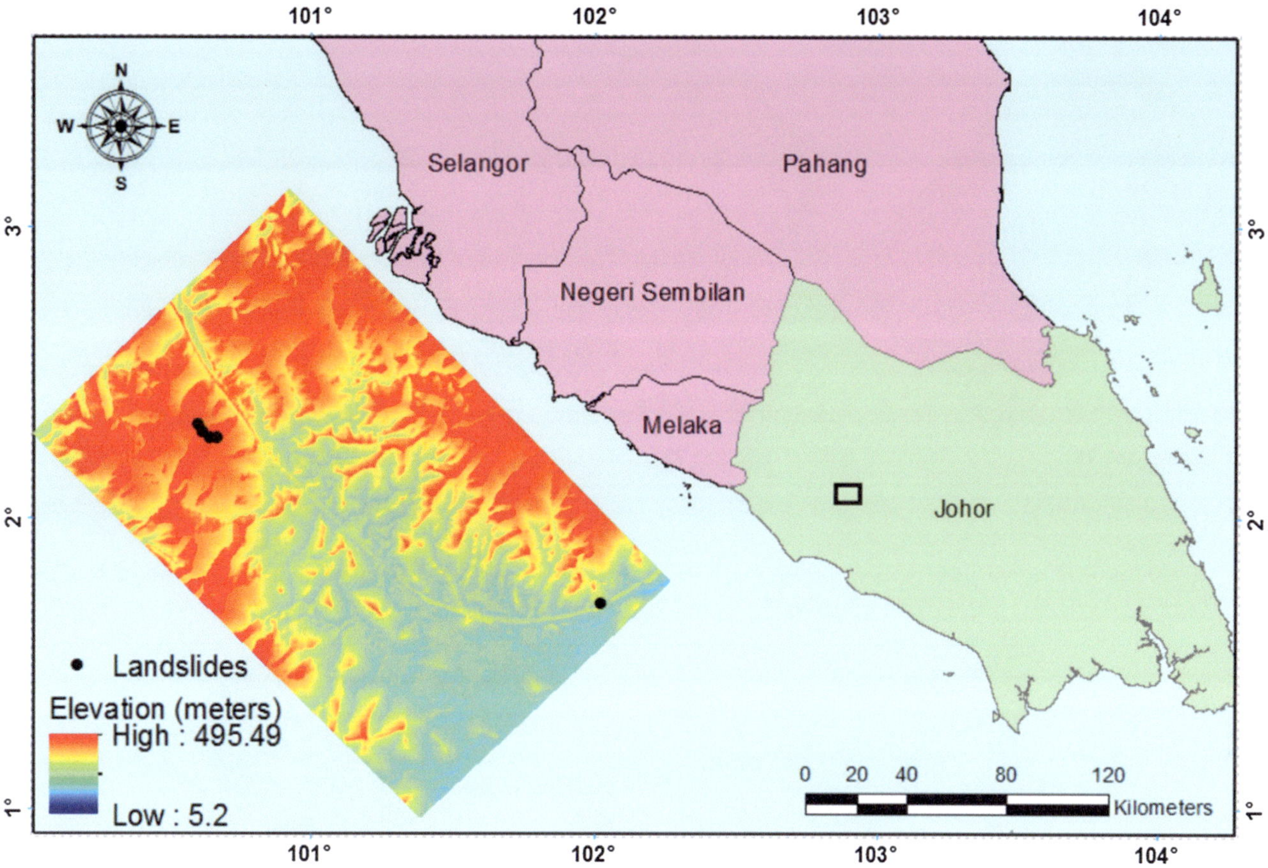

Fig. 4.1 Location of the study area used in the current research to test a developed methodology for detecting landslides from integrated data

Fig. 4.2 Overall method used to identify landslide prone areas from LiDAR and QuickBird data

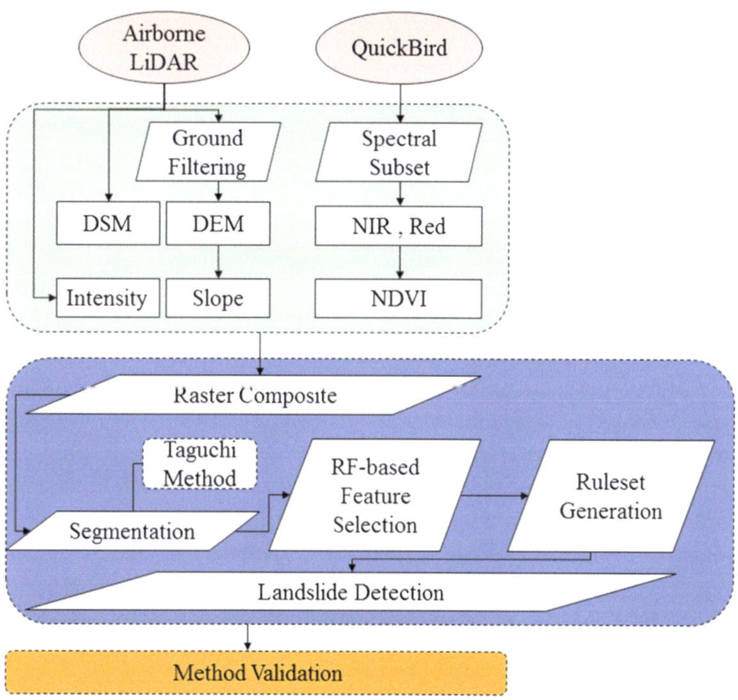

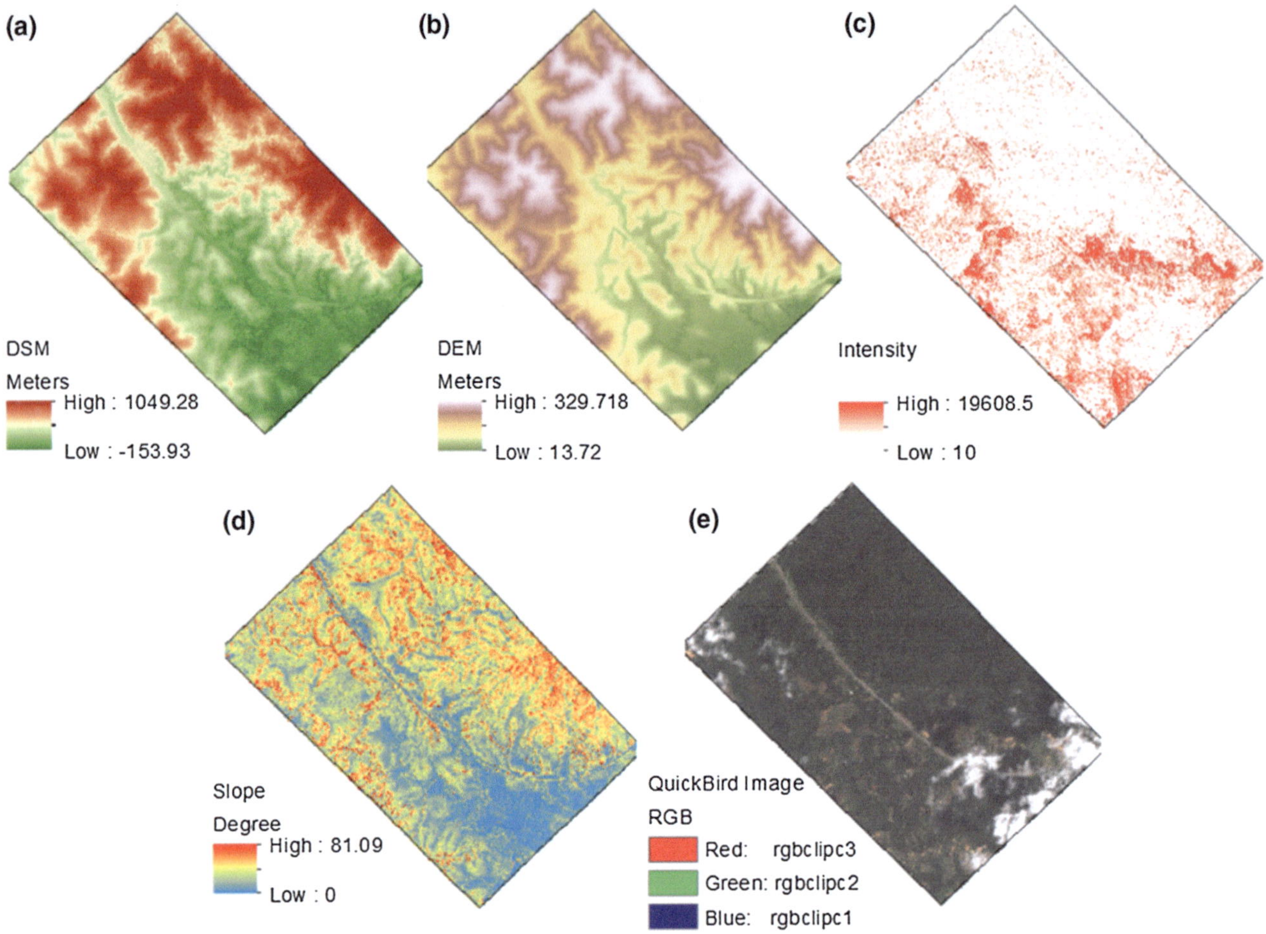

Fig. 4.3 Input parameters used in the object—based analysis for landslide prone area identification: **a** LiDAR-derived DSM; **b** LiDAR-derived DEM; **c** LiDAR intensity; **d** slope; and **e** QuickBird image

$$\mathrm{NDVI} = \frac{\mathrm{RED} - \mathrm{NIR}}{\mathrm{RED} + \mathrm{NIR}} \quad 1 < \mathrm{NDVI} > -1 \tag{4.1}$$

4.4.2 Segmentation Using Taguchi Method

Image segmentation is an essential step for the object based automatic feature extraction process; it is a spatial clustering technique that completely subdivides the image into non-overlapping areas or segments (Chen et al. 2009). The two main segmentation categories are edge-based and region-based methods. In the edge-based method, image edges are distinguished by thresholding the image gradient attained from a differentiation filter, whereas in the region-based method, a group of pixels is distinguished either by growing a "seed" pixel until the homogeneity criteria (scale) is fulfilled or by splitting the entire image initially and then merging the homogeneous regions (Möller et al. 2007). These methods require several user-defined parameters to guide the segmentation process. The Taguchi tables facilitate easy, stable experiment designs (Moosavi et al. 2014). Therefore, only 25 experiments were selected for assessment by the Taguchi method in terms of the three segmentation parameters (i.e., scale, shape, and compactness). Moreover, the plateau objective function (POF) was measured for each test to evaluate segmentation accuracy in each of the 25 experiments. POF is a combination of a spatial autocorrelation index and a variance indicator.

The following general steps were performed to apply the Taguchi method: (a) The process objective was determined in the beginning, and the possible values of a particular parameter for the process were defined. (b) The parameters (scale, compactness, and shape) that can influence the process were then defined, and the variable values that can affect the performance were presented; thus, the level was defined by the user depending on the effect of the parameter. For instance, a shape value can vary from 0 to 1.0. When the level increases, the number of experiments conducted increases as well. (c) An orthogonal array was created to design the condition and to determine the number of experiments.

Selecting the orthogonal array depends on the number of levels and parameters. (d) The experiments were then applied once the appropriate array was selected. The effect of each parameter on the performance was then measured.

The loss function can be calculated as follows (Borghuis et al. 2007):

$$l(y) = k_c(y - T)^2 \qquad (4.2)$$

where T is target value of y, and the measured value, y, as a loss function. k_c is the constant in the loss function that can be calculated by considering the acceptable interval as follows:

$$k_c = \frac{C}{\Delta^2} \qquad (4.3)$$

where C is loss associated with sp limit and Δ is deviation of specification from target value. When the three parameters that affect the segmentation process were defined, the level of each parameter should also be defined. The level refers to the probable value of each parameter in terms of maximum, minimum, and current values. In case of a large gap between the minimum and maximum values of a specific parameter, more levels were added to the parameter. The proper array was selected after the number of parameters and levels were defined. A constant array was found for the Taguchi method. Each array is selected depending on the parameters and levels. In the case of three parameters and two levels, L25 array was selected (Table 4.1). The array assumes that the number of levels equals each parameter. Otherwise, the assumption is based on the highest value.

Furthermore, the POF was measured for each test to examine segmentation quality using each testing combination and to determine the optimum segmentation parameters. POF is the combination of a spatial autocorrelation index and a variance indicator. Signal-to-noise (S/N) ratio was measured to assess the testing segmentation outcomes. A higher S/N ratio represents higher segmentation accuracy. The S/N ratio is calculated as follows:

$$\mathrm{SNR} = -10\,\mathrm{Log}_{10}\left(\frac{1}{n}\sum\frac{1}{y_i^2}\right) \qquad (4.4)$$

where n is the number of repetitions under similar test situations (n = 1 in this study) and y are the POF values obtained from each segmentation test. The S/N ratio table was then achieved, and the optimum condition was determined.

4.4.3 Feature Selection Using RF Method

Random forest is a classification algorithm that directly provides measures of variable importance (Genuer et al. 2015). These outstanding features render it appropriate to classify remote sensing data, such as multispectral (Pal 2005) or multisource data (Gislason et al. 2006). Variable importance measure was implemented based on the estimation of permutation importance, which was regarded as a reliable indicator (Strobl et al. 2007). When the training set for a particular tree was drawn by sampling with replacement, nearly one-third of the cases were excluded. The out-of-bag (OOB) data can be used to estimate the test accuracy and the permutation importance measure.

The importance of variable m can be estimated by randomly permuting all the values of the mth variable in the OOB samples for each tree. The common measure for variable importance (Breiman 2001) is prediction accuracy (i.e., the number of observations classified correctly) before and after permuting variable m averaged over all trees. The high decrease in prediction accuracy indicated the importance of the variable. Therefore, to select the most relevant features, RFs were iteratively fit to the data. A fraction of the features (the least important ones) was eliminated at each iteration, and a new forest was built. The fraction was fixed at 0.2 by default. This allows a relatively rapid operation and increases the resolution as the number of chosen features decreases. After fitting all forests, the selected set of features was the OOB error rate, falling within the u = 1 standard error of the minimum error rate of all forests.

4.4.4 Rule Definition for Classification

Rule-based classification considers the spectral, spatial, textural, and contextual characteristics of each segment. Once objects were detected and segmented using the multiresolution algorithm optimized by the Taguchi method, eCognition calculates various parameters such as size, compactness, and shape of each object. These parameters are applied as class discriminators in the object-oriented classification. These object attributes are called object features. In this research, a procedure proposed by (Díaz-Uriarte and De Andres 2006) was adopted, and the numbers of the first and second RFs were set as 2000 and 500, respectively, to specify the feature subsets used in building the final RF classifiers. An initial feature rank was constructed using the

Table 4.1 Factors and their levels used for image segmentation optimization

Factor	Description	Level 1	Level 2	Level 3	Level 4	Level 5
A	Scale	5	10	20	30	40
B	Shape	0.1	0.3	0.5	0.8	1
C	Compactness	0.2	0.3	0.4	0.5	0.8

first RF, and then the least significant at 20% of features were excluded using the second RF by an iterative process. All landslide pixels and an equal number of non-landslide pixels were chosen randomly as the sets for feature selection. Therefore, setting the rules was the final step in classifying the image. The rules were created based on the importance of feature subsets and their values, as well as the general knowledge of the author on landslide mechanisms and characteristics.

4.4.5 Validation

The quality of a landslide inventory map depends on its accuracy and on the type and certainty of the information represented in the map. The accuracy of the generated map is described according to the completeness of the map and the thematic precision of the information on the map (Galli et al. 2008). The confusion matrix was used to examine the efficiency of the classification map and the detected landslide locations. The confusion matrix is a cross-tabulation of the classified and actual class labels for the study area (Foody and Mathur 2004). This matrix is a square array of dimension $r \times r$, where r is the number of categories. The confusion matrix represents the correlation among two samples of measurements taken from the classified region. Overall accuracy, recall, and precision are the common indicators in measuring the accuracy of extracted landslides. The overall accuracy is determined by dividing the aggregate of the main diagonal entries of the confusion matrix by the entire number of samples. However, the recall or true positive rate (TP) is the proportion of correctly identified positive cases, and precision (P) is the proportion of the corrected predicted positive cases.

4.5 Results

For the application of the Taguchi method, the choices for the three parameters, namely scale, compactness, and shape, were defined and listed in five levels (Table 4.1). Thus, 243 combinations can be created for segmentation, and the analysis of these combinations requires a significant amount of time. However, the use of the Taguchi optimization method reduced the number of combinations to 25 experiments. Every 25 segmentation prototypes were also tested based on the Taguchi orthogonal array, and the estimated POF and SNR of each available combination were computed (Table 4.2).

The segmentation quality and the optimum combination parameters were selected based on the S/N ratio and the mean effect plot for POF (Table 4.3 and Fig. 4.4). The boldface figure refers to the maximum value of the S/N ratios of a certain factor which represents the optimum conditions for segmentation. The precision of this segmentation was assessed visually as the boundaries of most objects were detected accurately (Fig. 4.5).

Another stage in classifying the image segments is selecting the most significant variables through the application of the RF method. The RF method procedure involves the creation of new RFs iteratively ignored the least important 20% of the variables in each iteration. This procedure is repeated until only the most significant subset of variables are left. For each RF, the OOB error is computed as a criterion to choose the model with the smallest OOB error. This variable selection simplifies the model, thereby helping enhance the classification accuracy. Table 4.4 represents the importance of various features based on their potential to distinguish landslide and non-landslide elements. Table 4.5 shows that the values derived from mean DEM, mean DSM, and standard deviation DSM dominated the variable importance ranking.

Figure 4.6 shows the created rules and their categories for the data used in this research. For instance, the mean of DEM and the standard deviation (stdev) of DSM values were used as important indices in differentiating landslide and non-landslide locations. The gray level co-occurrence matrix (GLCM) and mean NIR were also considered as other significant indices to determine the classified landslide inventory map. Various rules were examined to obtain the best results by applying the final rules shown in Fig. 4.6. The application of the created rules classified the image segments and detected the homogeneous groups. The entire image was then classified based on the defined rules, and the classification map was produced. Figure 4.7 shows the classified map with the detected landslide location in the study area. Table 4.5 shows the results of the accuracy assessment based on the confusion matrix for landslide inventory analysis. The achieved accuracy for the classification of both landslide and non-landslide locations was 90% (overall). This result revealed that the locations were accurately detected as inventory. In addition, the location of the identified landslides was also plotted in 3D to better understand their potential effects on the highway in the study area (Fig. 4.8). Some statistics were also calculated to clearly interpret the results.

Furthermore, the detected landslides were further validated by a field visit to the study area to quantitatively assess the landslide identification. The landslide inventory used in the accuracy assessment was collected from various sources, thus the field visit was important to validate the proposed method. Figure 4.9 presents parts from the study area showing one of the detected landslides. The landslide location is highly susceptible for landslide considering the hilly terrain as steep slopes (Fig. 4.9a). Identifying landslides can prevent more landslides in the future by implementing small and low-cost projects. The identification of landslide prone areas and the

Table 4.2 L25 orthogonal array, Moran's Index, intra-segment variance, and POF values

Scale	Shape	Compactness	Moran's index (I)	Intra-segment variance	Normalized POF
1	1	1	0.1194	82.19	1.343
1	2	2	−0.0202	82.33	1.417
1	3	3	−0.0970	121.39	1.446
1	4	4	0.4592	247.25	1.114
1	5	5	−0.2178	3428.21	0.522
2	1	2	−0.1300	194.72	1.441
2	2	3	−0.2255	196.58	1.492
2	3	4	−0.3378	309.06	1.517
2	4	5	−0.3591	583.53	1.447
2	5	1	−0.4321	602.26	1.480
3	1	3	−0.6605	442.43	1.649
3	2	4	−0.9714	497.08	1.797
3	3	5	−1.0910	503.68	1.858
3	4	1	−1.0245	534.71	1.814
3	5	2	−0.8979	542.88	1.744
4	1	4	−0.3476	321.45	1.519
4	2	5	−0.4498	402.31	1.549
4	3	1	−1.0251	502.65	1.824
4	4	2	−0.9704	468.11	1.805
4	5	3	−0.3762	511.23	1.477
5	1	5	−1.0680	488.97	1.851
5	2	1	−0.8767	435.60	1.765
5	3	2	−1.1204	545.87	**1.861**
5	4	3	−0.2669	477.89	1.429
5	5	4	0.7665	509.66	0.872

The bold text represents the important/significant values.

Table 4.3 Signal-to-Noise Ratios (Larger is better)

Level	A	B	C
1	0.8167	3.8117	4.259
2	3.3766	4.0659	**4.3112**
3	**4.964**	**4.5639**	3.5023
4	4.234	3.51	2.418
5	3.5178	0.9575	2.4186
Delta	4.1473	3.6064	1.8932
Rank	1	2	3

The bold text represents the important/significant values.

implementation of supported projects can reduce potential risks to the community and near highways (Fig. 4.9b).

4.6 Discussions

The detection of landslides using knowledge-based methods without training data is difficult because various types of landslides, data sources, and landslides are hard to separate from slope bare soil and man-made cut slopes. Currently, the integration of RF and object-based analysis simplified the process by selecting only relevant features with a certain degree of importance for landslide detection. Therefore, knowledge-based rules are easily defined, and these rules are most probably transferable to other study areas. Features unique for landslide identification were used in the rule-based classification to accurately detect and separate landslides from other features with slightly similar

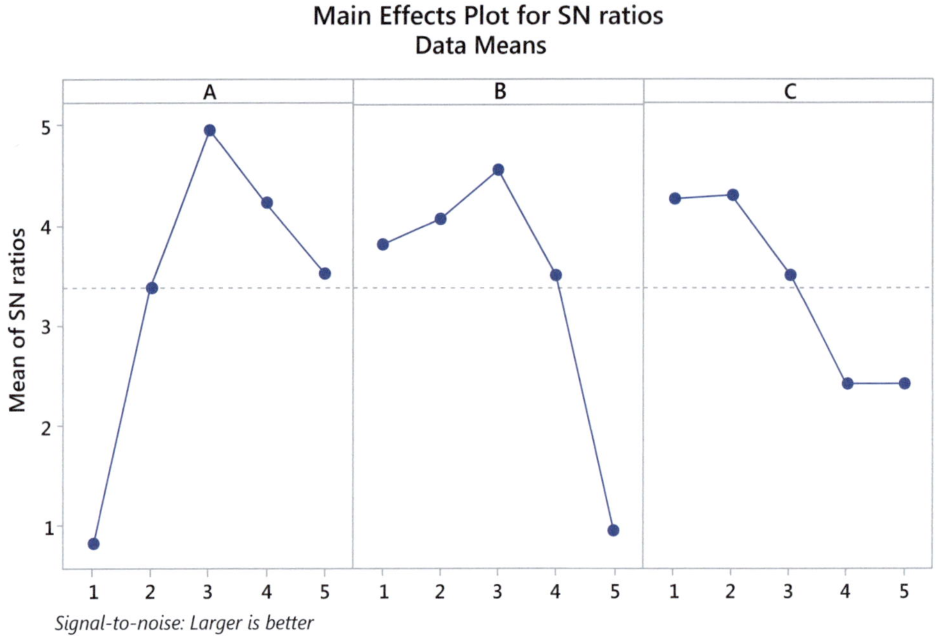

Fig. 4.4 Graph shows the SNR values calculated by Taguchi used to determine best combination of segmentation parameters (larger is better)

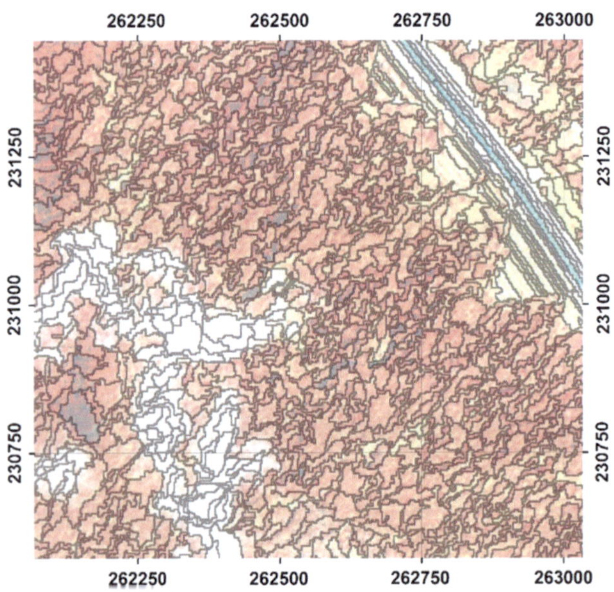

Fig. 4.5 Results of image segmentation using the optimum selected parameters

characteristics such as cut slopes. The separation of landslides from cut slopes and slope bare soil can only be achieved by integrating spectral, spatial, and texture rules. However, using these attributes to accurately detect landslides and separate them from other features requires the identification of the accurate boundary of the landslide objects. This issue was considered in the current research by using an optimized segmentation process. The Taguchi method permitted the selection of optimal combination of segmentation parameters. Thus, accurate landslide objects were defined to help precisely identify landslide prone areas using the spatial and texture information of the objects. If the landslide objects were not defined well, then the use of spatial and texture information is not useful given that the unique characteristics of landslides are undefined. Therefore, the accurate segmentation of landslide objects created using Taguchi and multiresolution algorithms was an essential step for landslide detection.

Even though the quantitative accuracy assessment showed the robustness of the proposed method, adding contextual-based rules to the rulesets proposed can improve the accuracy of the landslide detection. This was beyond the scope of this study considering the difficultly in addressing the conjunction of the current proposed workflow. An in-depth understanding of landslide features and their correlation with other features are necessary to construct such contextual-based rules for object-based analysis. Therefore, this limitation should be improved in future research. Other possible spaces for improvement include considering the type of landslides in the classification process and the selection of segmentation parameters not only based on the accuracy of created objects, but also based on the transferability to other study areas.

This study improved landslide detection by integrating LiDAR-derived DEM and very high-resolution satellite imagery. The study revealed that this integration is important as the feature selection analysis by RF showed the importance of NIR band in landslide identification. In addition, the selection of LiDAR-derived DSM as an important feature for landslide identification confirmed

Table 4.4 Importance scores of spatial/spectral features selected by random forest algorithm

<table>
<tr><th>Variable</th><th>Standard score</th><th>Relative score</th><th>Graphical score</th></tr>
<tr><td>Mean DEM</td><td>0.02724</td><td>100</td><td>||||||||||||||||||||||||||||||||||||||||||||||||</td></tr>
<tr><td>Mean DSM</td><td>0.01968</td><td>72.25505</td><td>||||||||||||||||||||||||||||||||||</td></tr>
<tr><td>Std DSM</td><td>0.01777</td><td>65.25907</td><td>|||||||||||||||||||||||||||||||</td></tr>
<tr><td>Mean red</td><td>0.01353</td><td>49.68443</td><td>|||||||||||||||||||||||</td></tr>
<tr><td>Mean blue</td><td>0.01174</td><td>43.09243</td><td>||||||||||||||||||||</td></tr>
<tr><td>GLCM contrast</td><td>0.01017</td><td>37.32635</td><td>|||||||||||||||||</td></tr>
<tr><td>GLCM homogeneity</td><td>0.00899</td><td>33.02039</td><td>|||||||||||||||</td></tr>
<tr><td>NDVI</td><td>0.00825</td><td>30.30717</td><td>||||||||||||||</td></tr>
<tr><td>GLCM dissimilarity</td><td>0.00771</td><td>28.31526</td><td>|||||||||||||</td></tr>
<tr><td>Mean green</td><td>0.00641</td><td>23.55235</td><td>|||||||||||</td></tr>
<tr><td>Max. difference</td><td>0.00611</td><td>22.41656</td><td>||||||||||</td></tr>
<tr><td>Std green</td><td>0.00585</td><td>21.46415</td><td>||||||||||</td></tr>
<tr><td>Std intensity</td><td>0.00548</td><td>20.13314</td><td>|||||||||</td></tr>
<tr><td>Std slope</td><td>0.00476</td><td>17.48251</td><td>||||||||</td></tr>
<tr><td>Std NIR</td><td>0.00423</td><td>15.51774</td><td>|||||||</td></tr>
<tr><td>GLCM StdDe</td><td>0.00377</td><td>13.84709</td><td>||||||</td></tr>
<tr><td>Skewness DSM</td><td>0.00289</td><td>10.59902</td><td>|||||</td></tr>
<tr><td>Mean NIR</td><td>0.00278</td><td>10.19153</td><td>||||</td></tr>
<tr><td>Mean Slope</td><td>0.00234</td><td>8.60536</td><td>||||</td></tr>
<tr><td>GLCM correlation</td><td>0.00131</td><td>4.79398</td><td>||</td></tr>
<tr><td>Brightness</td><td>0.00115</td><td>4.20923</td><td>||</td></tr>
<tr><td>Length/width</td><td>0.00112</td><td>4.09631</td><td>|</td></tr>
<tr><td>GLCM entropy</td><td>0.00107</td><td>3.93436</td><td>|</td></tr>
<tr><td>Std blue</td><td>0.0007</td><td>2.57535</td><td>|</td></tr>
<tr><td>Asymmetry</td><td>0</td><td>0</td><td></td></tr>
<tr><td>Compactness</td><td>0</td><td>0</td><td></td></tr>
<tr><td>GLCM Angular_2</td><td>0</td><td>0</td><td></td></tr>
<tr><td>GLCM mean</td><td>0</td><td>0</td><td></td></tr>
<tr><td>MEAN intensity</td><td>0</td><td>0</td><td></td></tr>
<tr><td>Shape index</td><td>0</td><td>0</td><td></td></tr>
<tr><td>Skewness blue</td><td>0</td><td>0</td><td></td></tr>
<tr><td>Skewness DEM</td><td>0</td><td>0</td><td></td></tr>
<tr><td>Skewness green</td><td>0</td><td>0</td><td></td></tr>
<tr><td>Skewness intensity</td><td>0</td><td>0</td><td></td></tr>
<tr><td>Skewness NIR</td><td>0</td><td>0</td><td></td></tr>
<tr><td>Skewness red</td><td>0</td><td>0</td><td></td></tr>
<tr><td>Skewness slope</td><td>0</td><td>0</td><td></td></tr>
<tr><td>Std red</td><td>0</td><td>0</td><td></td></tr>
<tr><td>Std DEM</td><td>0</td><td>0</td><td></td></tr>
</table>

Table 4.5 Detailed accuracy by class

		Predicted	
		Landslide	Non-landslide
Actual	Landslide	4	1
	Non-landslide	0	5

Overall accuracy = 90%
Recall (landslide) = 0.80
Recall (non-landslide) = 1
Precision (landslide) = 1
Precision (non-landslide) = 0.83

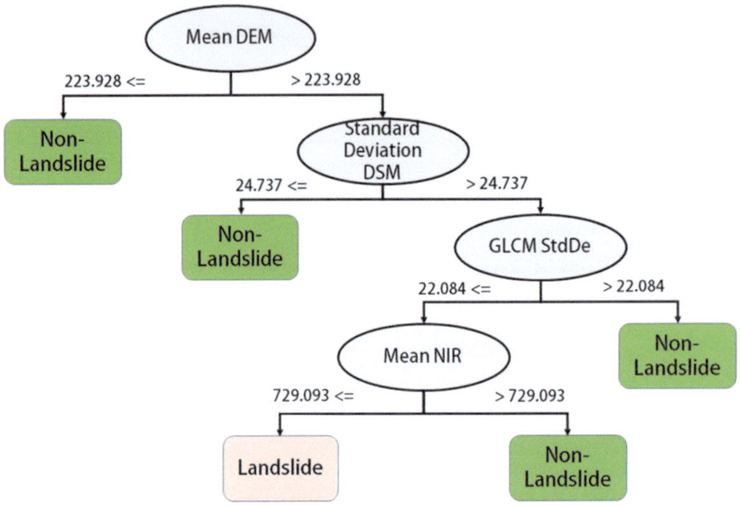

Fig. 4.6 Rules developed for landslide identification using in object-based analysis

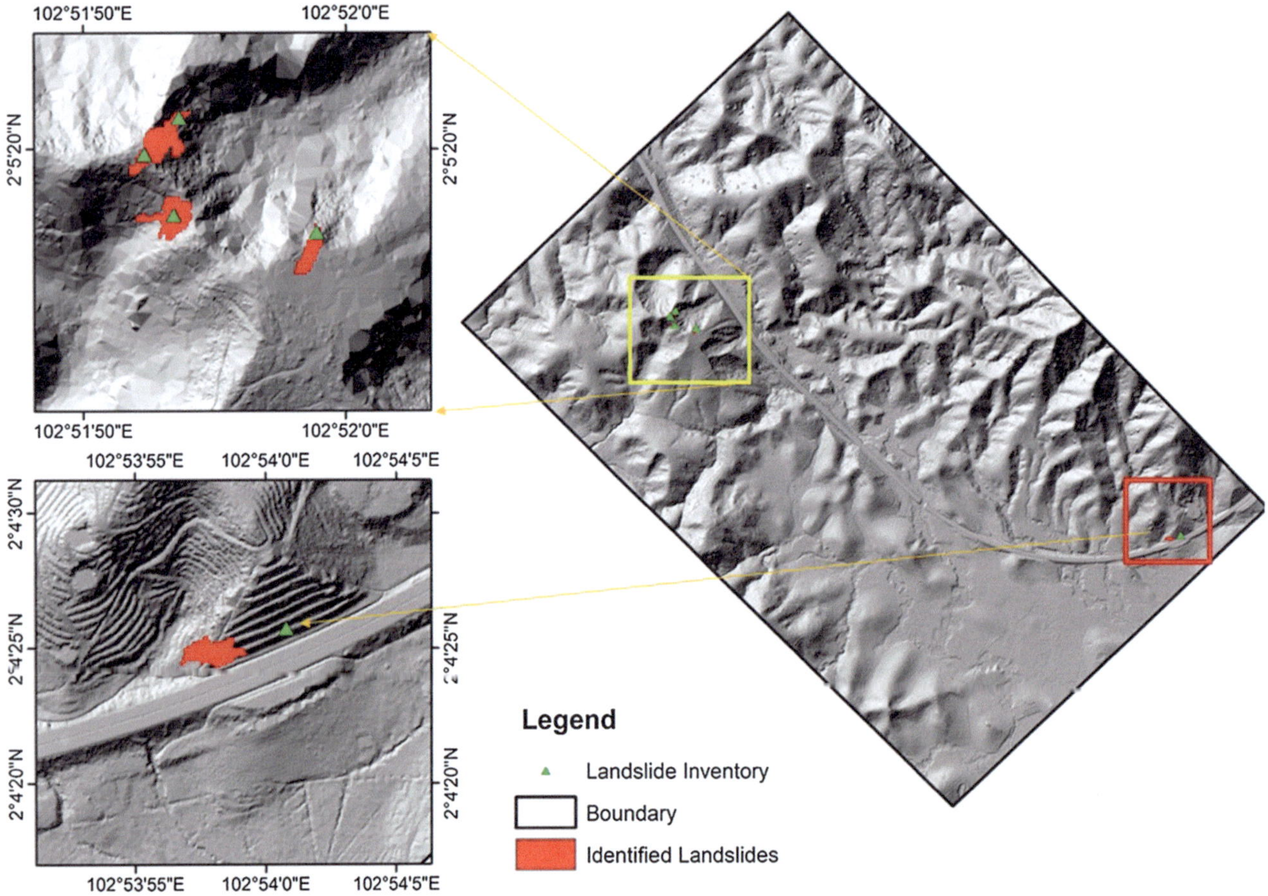

Fig. 4.7 Results of landslide detection using object-based method. Results overlapped on landslide inventory show that 4 landslides are correctly identified (*upper left figure*), and only one landslide was not detected accurately (*lower left figure*)

that the integration was critical for landslide detection. The rapid growth of vegetation in tropics, a major problem in landslide detection, was addressed by using NIR band in the rule-based classification. The use of NIR permits differentiation bare soils from newly growth vegetation. The methodology of detecting landslides in tropics with only the use of LiDAR data should be improved possibly through the use of contextual

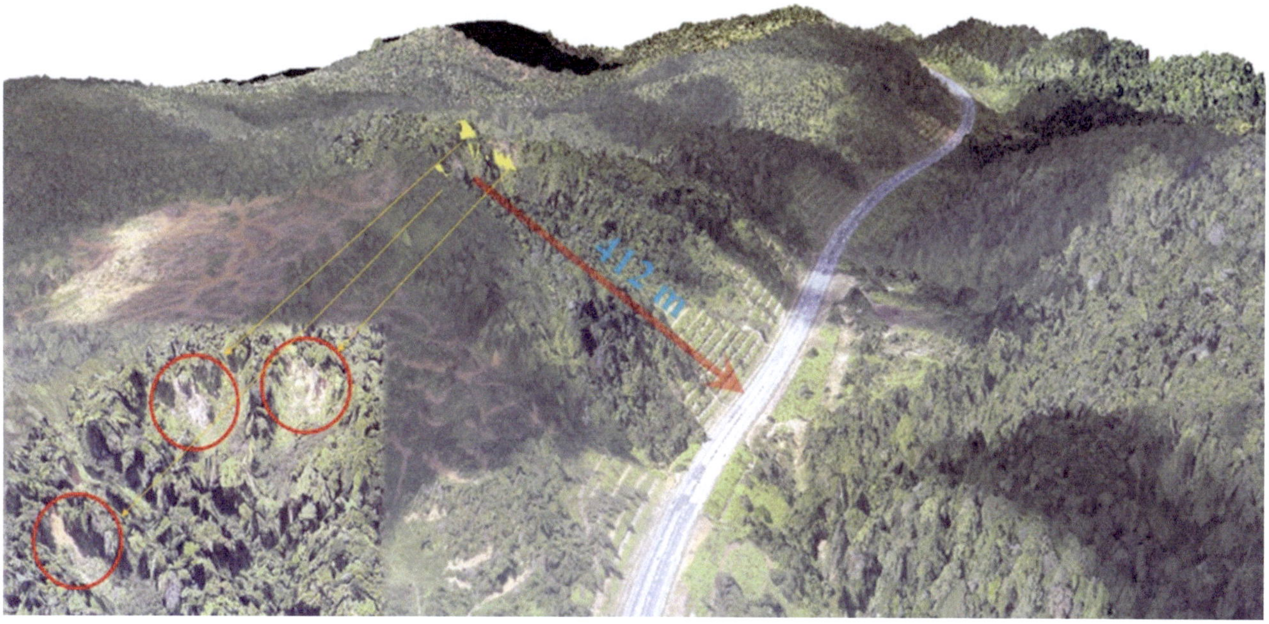

Fig. 4.8 Location of the identified landslides in 3D view perceptive showing its potential risk on highway

Fig. 4.9 Field verifications of detected landslide

information from the scene as stated earlier or by incorporating additional information on roads, landuse, and streams in the classification process.

4.7 Conclusion

The current research mainly aimed to automatically produce landslide inventory map with the integration of efficient methods for object-oriented image classification, such as the Taguchi technique for optimization of image segmentation, RF method for selection of most relative feature, and rule-based technique for differentiating between landslide and non-landslide features. Two sets of data sources, including LiDAR and very high-resolution QuickBird imagery were used to extract the required information for landslide mapping. The classification performance was considerably enhanced and simplified using detailed information and efficient techniques. Application of object-based classification methods compared with the pixel-based techniques required more time. However, they dominated the shortcomings of pixel-based techniques. The spatial, spectral, and textural features of each object were considered in the object-based classification procedure. Rule-based object-oriented classification was chosen to discriminate between classes. The first stage of object-oriented classification was segmentation performed through multiresolution technique. Optimized object segmentation and selection of the most relevant features are required to create rules for an accurate and rapid object-based classification of landslide locations.

The optimum combination of the segmentation of parameter was determined through the Taguchi method which has been proven as an efficient method for such applications. The orthogonal arrays generated from the Taguchi method simplified the ideal combination for segmentation by employing fewer examinations. A feature section method based on the RF method was also used to estimate the useful features in the image. A total of 23 features were selected as the members of the feature subset, and some of those features had the highest relative scores, such as mean DEM, mean DSM, std DSM, mean Red, mean Blue, GLCM Contrast, GLCM Homogeneity, GLCM Dissimilarity, and mean Green. The outcomes obtained from the feature selection confirmed that the proposed features can deliver operative information for landslide detection. Therefore, the applicable rules were defined and applied to the image to generate the classified landslide map. In addition, the validation was also performed using the confusion matrix method. The overall accuracy of the classification map showed that it can be used as a valid inventory map, which is highly valuable in planning and disaster management.

References

Arenas-García, J., & Camps-Valls, G. (2008). Efficient kernel orthonormalized PLS for remote sensing applications. *IEEE Transactions on Geoscience and Remote Sensing, 46*(10), 2872–2881.

Blaschke, T. (2010). Object based image analysis for remote sensing. *ISPRS Journal of Photogrammetry and Remote Sensing, 65*(1), 2–16.

Borghuis, AM., Chang, K., Lee, HY. (2007). Comparision between automated and manual mapping of typhoon-trigerred landslides from SPOT-5 imagery. *International Journal of Remote Sensing, 7* (8), 1843–1856.

Brardinoni, F., Slaymaker, O., & Hassan, M. A. (2003). Landslide inventory in a rugged forested watershed: A comparison between air-photo and field survey data. *Geomorphology, 54*(3), 179–196.

Breiman, L. (2001). Random forests. *Machine Learning, 45*(1), 5–32.

Calabro, M., Schmidt, D., & Roering, J. (2010). An examination of seasonal deformation at the Portuguese Bend landslide, southern California, using radar interferometry. *Journal of Geophysical Research: Earth Surface, 115*(F2).

Castillejo-González, I. L., López-Granados, F., García-Ferrer, A., Peña-Barragán, J. M., Jurado-Expósito, M., de la Orden, M. S., et al. (2009). Object-and pixel-based analysis for mapping crops and their agro-environmental associated measures using QuickBird imagery. *Computers and Electronics in Agriculture, 68*(2), 207–215.

Chen, Y., Su, W., Li, J., & Sun, Z. (2009). Hierarchical object oriented classification using very high resolution imagery and LIDAR data over urban areas. *Advances in Space Research, 43*(7), 1101–1110.

Cheng, K., Wei, C., & Chang, S. (2004). Locating landslides using multi-temporal satellite images. *Advances in Space Research, 33*(3), 296–301.

Derron, M.-H., & Jaboyedoff, M. (2010). Preface "LIDAR and DEM techniques for landslides monitoring and characterization". *Natural Hazards and Earth System Science, 10*(9), 1877–1879.

Díaz-Uriarte, R., & De Andres, S. A. (2006). Gene selection and classification of microarray data using random forest. *BMC Bioinformatics, 7*(1), 1.

Eeckhaut, M., Poesen, J., Verstraeten, G., Vanacker, V., Nyssen, J., Moeyersons, J., et al. (2007). Use of LIDAR-derived images for mapping old landslides under forest. *Earth Surface Processes and Landforms, 32*(5), 754–769.

Foody, G. M., & Mathur, A. (2004). Toward intelligent training of supervised image classifications: Directing training data acquisition for SVM classification. *Remote Sensing of Environment, 93*(1), 107–117.

Galli, M., Ardizzone, F., Cardinali, M., Guzzetti, F., & Reichenbach, P. (2008). Comparing landslide inventory maps. *Geomorphology, 94*(3), 268–289.

Genuer, R., Poggi, J.-M., & Tuleau-Malot, C. (2015). VSURF: An R package for variable selection using random forests. *The R Journal, 7*(2), 19–33.

Gislason, P. O., Benediktsson, J. A., & Sveinsson, J. R. (2006). Random forests for land cover classification. *Pattern Recognition Letters, 27*(4), 294–300.

Guzzetti, F., Mondini, A. C., Cardinali, M., Fiorucci, F., Santangelo, M., & Chang, K.-T. (2012). Landslide inventory maps: New tools for an old problem. *Earth-Science Reviews, 112*(1), 42–66.

Laliberte, A., Rango, A., Herrick, J., Fredrickson, E. L., & Burkett, L. (2007). An object-based image analysis approach for determining fractional cover of senescent and green vegetation with digital plot photography. *Journal of Arid Environments, 69*(1), 1–14.

Möller, M., Lymburner, L., & Volk, M. (2007). The comparison index: A tool for assessing the accuracy of image segmentation. *International Journal of Applied Earth Observation and Geoinformation, 9*(3), 311–321.

Moosavi, V., Talebi, A., & Shirmohammadi, B. (2014). Producing a landslide inventory map using pixel-based and object-oriented approaches optimized by Taguchi method. *Geomorphology, 204*, 646–656.

Murillo-García, F. G., Alcántara-Ayala, I., Ardizzone, F., Cardinali, M., Fiourucci, F., & Guzzetti, F. (2015). Satellite stereoscopic pair images of very high resolution: A step forward for the development of landslide inventories. *Landslides, 12*(2), 277–291.

Nandi, A., & Shakoor, A. (2010). A GIS-based landslide susceptibility evaluation using bivariate and multivariate statistical analyses. *Engineering Geology, 110*(1), 11–20.

Pal, M. (2005). Random forest classifier for remote sensing classification. *International Journal of Remote Sensing, 26*(1), 217–222.

Rau, J.-Y., Jhan, J.-P., & Rau, R.-J. (2014). Semiautomatic object-oriented landslide recognition scheme from multisensor optical imagery and DEM. *IEEE Transactions on Geoscience and Remote Sensing, 52*(2), 1336–1349.

Saeidi, V., Pradhan, B., Idrees, M. O., & Latif, Z. A. (2014). Fusion of airborne LiDAR with multispectral SPOT 5 image for enhancement of feature extraction using Dempster-Shafer theory. *IEEE Transactions on Geoscience and Remote Sensing, 52*(10), 6017–6025.

Singh, L. P., Van Westen, C., Ray, P. C., & Pasquali, P. (2005). Accuracy assessment of InSAR derived input maps for landslide susceptibility analysis: A case study from the Swiss Alps. *Landslides, 2*(3), 221–228.

Slatton, K. C., Carter, W. E., Shrestha, R. L., & Dietrich, W. (2007). Airborne laser swath mapping: Achieving the resolution and accuracy required for geosurficial research. *Geophysical Research Letters, 34*(23).

Strobl, C., Boulesteix, A.-L., Zeileis, A., & Hothorn, T. (2007). Bias in random forest variable importance measures: Illustrations, sources and a solution. *BMC Bioinformatics, 8*(1), 1.

Van Coillie, F., Lievens, H., Joos, I., Pizurica, A., Verbeke, L., De Wulf, R., et al. (2011). Training neural networks on artificially generated data: A novel approach to SAR speckle removal. *International Journal of Remote Sensing, 32*(12), 3405–3425.

Van Coillie, F. M., Verbeke, L. P., & De Wulf, R. R. (2007). Feature selection by genetic algorithms in object-based classification of IKONOS imagery for forest mapping in Flanders, Belgium. *Remote Sensing of Environment, 110*(4), 476–487.

Van Den Eeckhaut, M., & Hervás, J. (2012). State of the art of national landslide databases in Europe and their potential for assessing landslide susceptibility, hazard and risk. *Geomorphology, 139,* 545–558.

Wang, F., Hasbani, J.-G., Wang, X., & Marceau, D. J. (2011). Identifying dominant factors for the calibration of a land-use cellular automata model using rough set theory. *Computers, Environment and Urban Systems, 35*(2), 116–125.

Yusof, N. M., Pradhan, B., Shafri, H. Z. M., Jebur, M. N., & Yusoff, Z. (2015). Spatial landslide hazard assessment along the Jelapang Corridor of the North-South Expressway in Malaysia using high resolution airborne LiDAR data. *Arabian Journal of Geosciences, 8* (11), 9789–9800.

Zhang, L., & Gruen, A. (2006). Multi-image matching for DSM generation from IKONOS imagery. *ISPRS Journal of Photogrammetry and Remote Sensing, 60*(3), 195–211.

Part III

Debris Flow Source Identification

5 Debris Flow Source Identification in Tropical Dense Forest Using Airborne Laser Scanning Data and Flow-R Model

Biswajeet Pradhan and Suzana Binti Abu Bakar

5.1 Introduction

Debris flow and related landslide processes can cause significant hazard to human kind and economic loss annually. Debris flow is a type of mass movement or landslide (Kuriakose 2006; U.S Geological Survey 2004). Varnes (1978) defined debris flow as a sudden mass movement, in which a combination of loose soil, rock, organic matter and water moves as a flowing slurry. In an earlier paper, Hutchinson (1988) defined debris flow as a mixture of sand, silt, clay and coarse materials, such as gravel, cobbles and boulders, with variable amounts of water that travels down under the influence of gravity in high density. Youssef and Pradhan (2013) specified that moving downward the slope causes debris flows when poorly sorted sediments or loose overburden materials are saturated with water. Several terms related to mass movement include debris floods, lahars, debris torrents or debris slides (Varnes 1978; Johnson 1984; Pierson and Costa 1987; Pradhan and Lee 2009, 2010a; Youssef and Pradhan 2013).

Debris flow can initiate from shallow landslides (Gabet and Mudd 2006; Park et al. 2016). According to Rahman (2014), the most common types of landslides in Malaysia are shallow slides where the slide surface is usually less than 4 m deep; this type occurs during or immediately after an intense rainfall. Shallow landslide commonly occurs in mountainous terrain and is triggered by earthquakes or intense rainfall. Pimiento (2010a, b) studied shallow landslide, which is also known as slope failure where movement involves earth or debris from superficial deposits that are mainly soil and colluvium. Cruden and Varnes (1996) described landslide classification based on the material (rock, debris and earth) and movement (fall, topple, slide, spread and flow).

Monitoring debris flow is considerably complex because it occurs in rugged and inaccessible terrain where field work can be challenging, dangerous and time-consuming (Elkadiri et al. 2014). In a recent study by Rickenmann (2016), the mechanical description of debris flow depends on the material composition and solid–fluid mixture. Three main factors, namely slope gradient, sediment availability and water input, are identified as critical conditions for debris flows to occur (Takahashi 2007).

New sensors and techniques, such as airborne laser scanning (ALS) by light detection and ranging (LiDAR) and terrestrial laser scanning (TLS), provide high-resolution topographic data contributing to improved representation of land surface (Martinaszabova and Stanislavhroncek 2015). Availability of high-resolution digital elevation models (DEM) from ALS and TLS facilitates detailed hazard mapping related to mass movement and flood in mountainous areas (Jaboyedoff et al. 2012). Many landslide-related researches have been conducted in Malaysia. Most of these published works are related to landslide susceptibility, hazard and risk assessment. Most recently, LiDAR data were used in landslide-related research in Malaysia. Abdulwahid and Pradhan (2016) used a high-resolution airborne LiDAR data to obtain landslide conditioning factors for spatial prediction of landslide hazard areas. Jebur et al. (2014) used high-resolution ALS data to optimise landslide conditioning factors in a tropical forest area of Malaysia. Yusof et al. (2015) used high-resolution airborne LiDAR data for spatial landslide hazard assessment along the Jelapang Corridor of the North–South Expressway in Malaysia.

Huat and Ali (2012) mentioned that ALS coupled with ortho-rectified photographs can be remarkably useful for detailed geomorphological mapping, especially in tropical forests. LiDAR data can also be significant for identifying potential debris flow areas because of the ability of laser pulse to penetrate forest canopy, thereby revealing hidden features, such as scarp areas. However, research concerning debris flow in Malaysia has not been fully developed and is limited. The present study mainly aims to identify an optimum DEM resolution for identification of debris flow areas in a tropical environment where thick vegetation hinders the identification process.

B. Pradhan (✉) · S.B.A. Bakar
Department of Civil Engineering, University Putra Malaysia, Serdang, Malaysia
e-mail: biswajeet24@gmail.com

B. Pradhan (ed.), *Laser Scanning Applications in Landslide Assessment*,
DOI 10.1007/978-3-319-55342-9_5

5.2 Formation of Debris Flow

Nettleton et al. (2005) characterised debris flow based on topographic and geological characteristics of locations. Debris flow can be formed into two types, namely hill slope (open-slope) and channelled debris flows (Evans 1982; Chen et al. 2012). Figure 5.1 shows the illustration of hill slope and channelled debris flows.

5.2.1 Hill Slope (Open-Slope) Debris Flow

Cruden and Varnes (1996) stated that hill slope debris flow forms its own path down the valley slopes as a track before it flows down into the deposition area. This type of flow often initiates during rainfall with a slide-detached materials, such as rock and upland debris slide. Materials accumulate on a lower area, that is, the deposition area, where the slope gradient is low. Channels and levees can also be found in a deposition area (Chen et al. 2012).

5.2.2 Channelled Debris Flow

Nettleton et al. (2005) defined this type of debris flow formed in an existing channel, such as valleys, gullies, depressions or hollows; in these areas, the materials consist of 80% of solid weight, and the flow presents high density as that of wet concrete. Channel bed failure is the main cause of channelled debris flow (Chen et al. (2012).

5.3 Previous Study and Flow-R Software

Flow-R (flow path assessment of gravitational hazards at a regional scale) is developed under MATLAB specifically to identify potential source areas and propagation extent of debris flow (Horton et al. 2013). Flow-R has been used in many countries for identification debris flow source area and production of susceptibility maps. Such studies include those of Canton de Vaud (Horton et al. 2008) and Val de Bagnes (Jaboyedoff et al. 2012) in Switzerland. In addition, local institutions, universities and government agencies have benefited from this software because it is freely downloadable from www.flow-r.org (Blais-Stevens and Behnia 2016). Numerous studies on debris flow sources and propagation areas using Flow-R software have been found in the literature. Research done by Fischer et al. (2012) in Norway considered using DEM of varying resolutions, i.e. 5, 10 and 25 m, to evaluate the effect of DEM on source area identification using Flow-R software. They used lithology map, plan curvature, slope angle and flow accumulation as parameters. The results revealed that the modelling result of source area presents a good correlation with field observations. Blais-Stevens and Behnia (2016) investigated the sources of debris flow by using qualitative heuristic method and Flow-R along the Yukon Alaska Highway Corridor, Canada. They used several parameters, such as geology, curvature, proximity to drainage, slope aspect, slope angle and flow accumulation, which were obtained from 10-m DEM resolution. Their results revealed that 80–90% of debris flow source is identified using the heuristic method and Flow-R. They also identified the

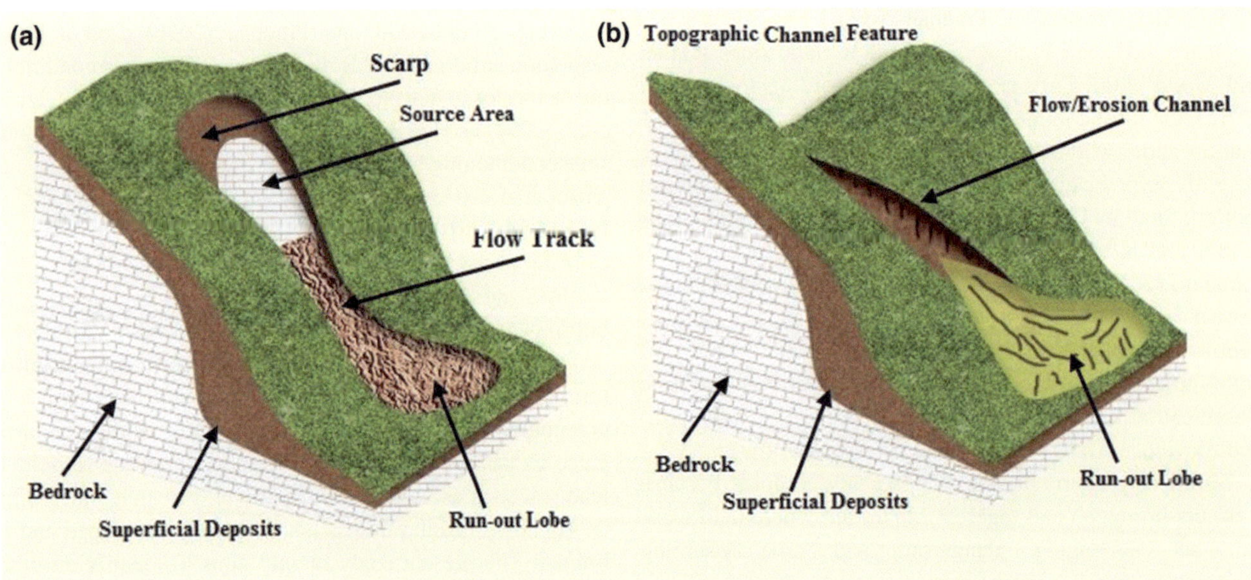

Fig. 5.1 Formation of debris flow: **a** hill slope and **b** channelled debris flows (modified after Nettleton et al. (2005)

potential debris flow areas. Realino et al. (2015) identified debris flow sources using interferometric synthetic aperture radar from a 5-m DTM resolution. Their research identified six potential source areas, which were also verified in the field. Elkadiri et al. (2014) assessed a remote sensing-based approach for debris flow susceptibility assessment using artificial neural network (ANN) and logistic regression (LR) modelling. The parameters used for source detection are slope, aspect, topographic position index, flow accumulation, normalised difference vegetation index, stream power index, topographic wetness index and distance-to-drainage line. Their results revealed that ANN and LR produce 96.1 and 96.3% prediction accuracies, respectively. Kasim et al. (2016) and Jamaludin et al. (2014) studied the triggering mechanisms and characteristic of debris flow at eight debris flow locations in Peninsular Malaysia. They revealed that triggering mechanisms depend on hydrological, morphometric, geological and geotechnical factors, as well as the duration and intensity of rainfall.

This paper demonstrates the identification of potential source areas using Flow-R model at Ringlet, Cameron Highlands, Malaysia. The algorithms and models and the details of their utilisation were explained. Data from a high-resolution ALS (LiDAR) were used to derive terrain information and debris flow-related parameters, such as altitude, slope angle, flow accumulation and plan curvature. Subsequently, high and low resolutions of DEM (1, 2, 5 and 10 m) were extracted from LiDAR data. The specific objectives of this study are as follows: (1) to identify the effect of DEM resolutions on the detection of debris flow sources using ALS (LiDAR) data and (2) produce debris flow source map within the study area.

5.4 Study Area and Data

5.4.1 Study Area

This study was conducted in Ringlet, Cameron Highlands, Malaysia. The area is commonly prone to landslide activities (such as translational, rock falls, rotational and debris flows) and gully erosions due to running water. According to the local newspaper reports (The Star, 2015) dated 9 November, debris flow that occurred along the road causes a traffic standstill for about 6 h when trees, soil and rocks fell on the road. Ringlet is located between longitude 101° 22′ 59.99″E and latitude 4° 24′ 59.99″N with an altitude of 1200 m above mean sea level (Fig. 5.2). The area is situated on the main range known as Banjaran Titiwangsa. The geological setting of Ringlet is mainly composed of granite. Nevertheless, scattered outliers (roof pendants) of metasediments are also present. The metasediments consist of schist, phyllite, slate and limestone (Chow et al. 2003). The geology map of the study area is Post-Triassic–Mesozoic granite comprising granite rocks and a few patches of metamorphic rocks comprising Silurian–Ordovician schist, phyllite, limestone and sandstone (Pradhan and Lee 2010). The average annual rainfall in Ringlet area is around 2660 mm, and rainfall intensity is a triggering mechanism that affects the hill slope and causes gully erosion (Abdulwahid and Pradhan 2016).

5.4.2 Data Used

Airborne Laser Scanning data (LiDAR) data were used. These data were acquired on 15 January 2015 using an aircraft flying at an altitude of approximately 1500 m above mean sea level and resulted in nearly eight points per square metre with a 25,000-Hz pulse rate frequency. The LiDAR vector point data were recorded over 25 km^2 of the Ringlet area. The absolute accuracy of the LiDAR data satisfied the root-mean-square errors of 0.15 m in the vertical axis and 0.3 m in the horizontal axis. Acquiring a reliable inventory map is a crucial task in landslide susceptibility assessment (Jebur et al. 2014). In the present study, the landslide inventory map was prepared by multiple field campaigns. The existing landslide inventory for the past 15 years was used as a reference; in this inventory, historical records of landslides were obtained using remote sensing methods, such as archived 1:10,000–1:50,000 aerial photograph, SPOT 5 panchromatic satellite image, IKONOS and QuickBird satellite images, for visual detection of landslide occurrences in Ringlet area. The locations of the landslide event were drawn on 1:25,000 maps; they were also plotted closely based on site description, aerial photo-interpretation and archived database and verified using field observations (Pradhan and Lee 2010). A total of 29 landslides were identified in the study area (Fig. 5.4). The orthophoto was acquired during the LiDAR data acquisition campaign with 0.09 m spatial resolution and georeferenced to Geocentric Datum of Malaysia (GDM2000). Visual interpretation was performed by using high-resolution orthophoto, in which analyst could detect the fresh landslide scars or geomorphological features, such as initiation zone, debris deposit zone and flow track. The lithology map was prepared from a 1:63,300-scale geological map and collected from the Department of Geology and Mineral Sciences, Malaysia. In this study area, different types of lithology, such as granite and schist, phyllite and slate, were present (Fig. 5.3). The drainage networks and contours of 10 m were generated from the Global Mapper Software. The land use/land cover map of Ringlet area for 2015 was obtained from the Jabatan Perancangan Bandar dan Desa Pahang (Figs. 5.4, 5.5).

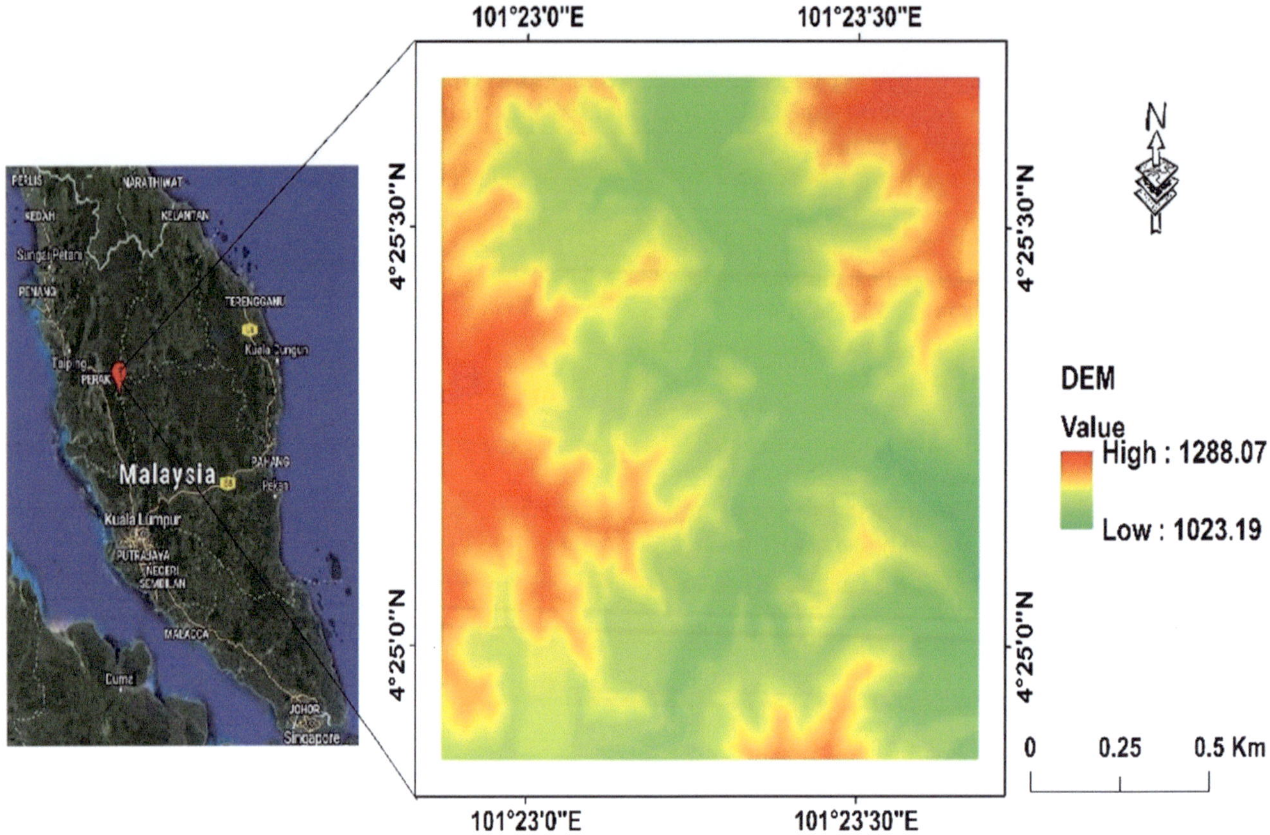

Fig. 5.2 Digital elevation model (DEM) map of Ringlet, Cameron Highlands, Malaysia

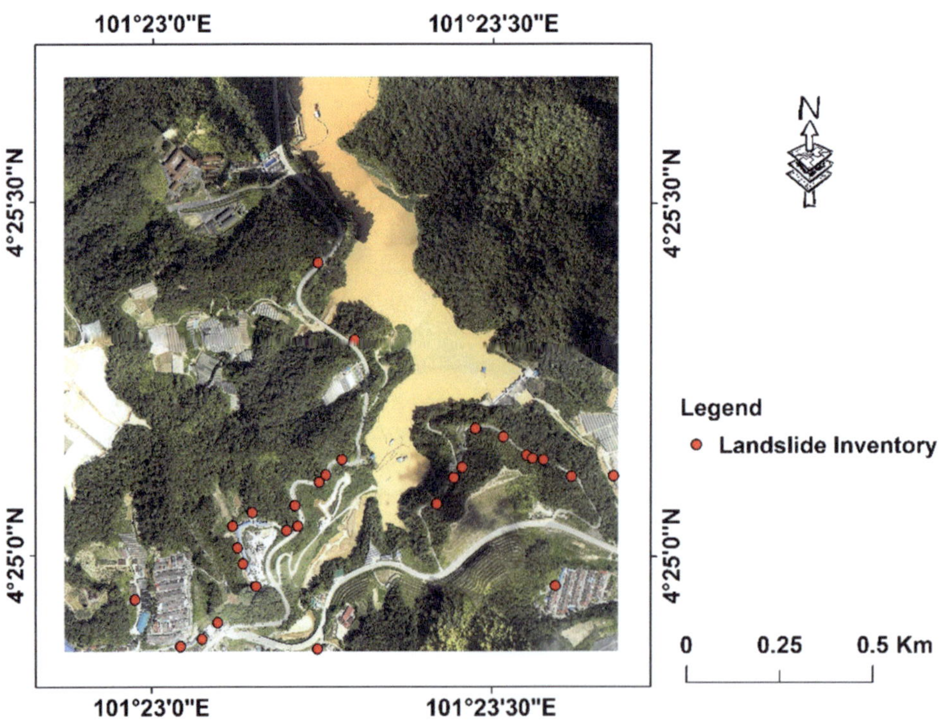

Fig. 5.3 Landslide inventory map shown on the orthophoto

Fig. 5.4 Lithology map

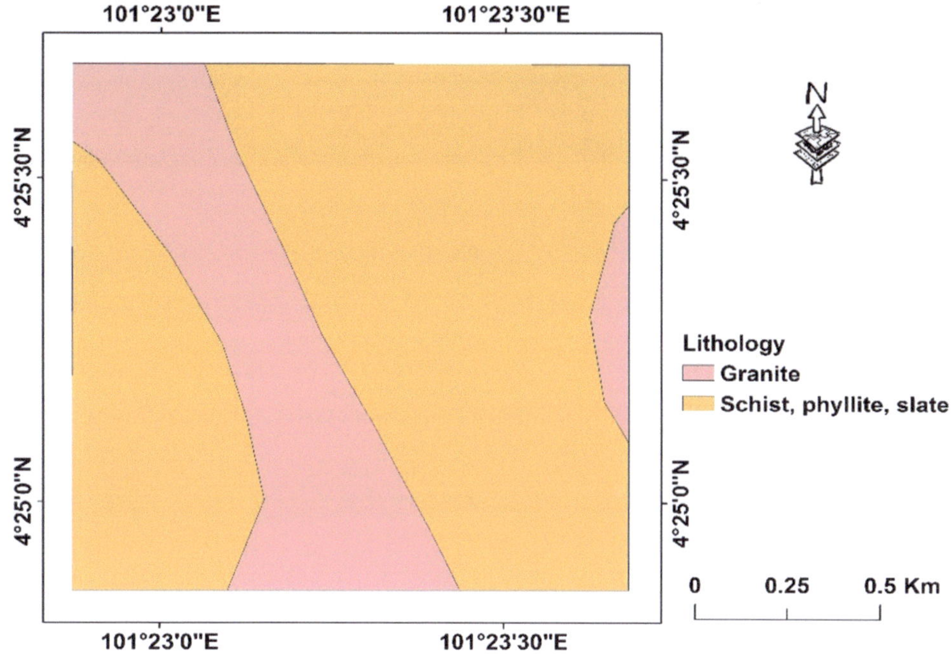

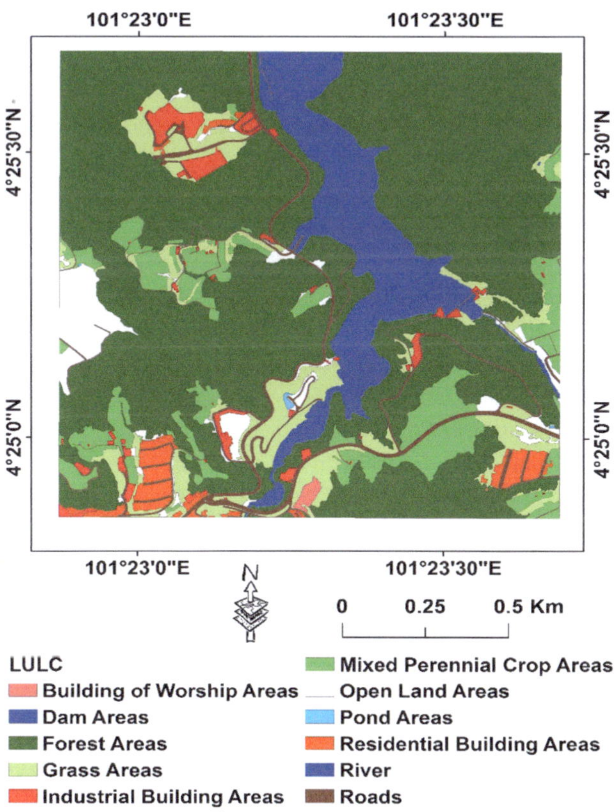

Fig. 5.5 Land use/land cover map

5.5 Methodology

Figure 5.6 illustrates the data processing workflow adopted in this research. The process started with LiDAR data processing, followed by DEM generation, parameters of source area, validation and finally results and discussion.

5.5.1 LiDAR Data Processing

Data processing was performed using Global Mapper 17, ArcGIS 10.2 and Flow-R software. Figure 5.6 shows the data processing workflow for source area assessment using Flow-R software. The DTM 0.5 grid derived from the *x*, *y* and *z* point clouds was processed using the Global Mapper 17 software. The DTM consisted of 17 tiles with 36 million point clouds covering the entire Ringlet, Cameron Highlands. To avoid complex computation, the area was then subset to a small area of interest that consists of four tiles with 9 million point clouds only. The DTM represented the bare ground surface without any objects. Afterwards, the DEM was generated from DTM using binning (average value) grid method because it is efficient for large point datasets and suitable for complex mapping. The DEM was then resampled to 1-, 2-, 5- and 10-m spatial resolutions. Various resolutions should be created to identify the source

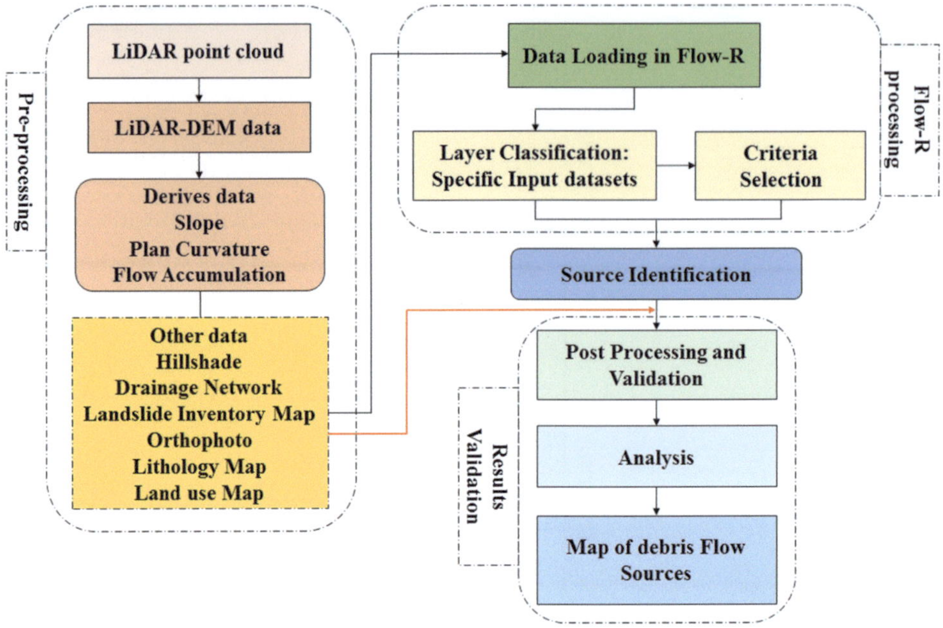

Fig. 5.6 Data processing workflow for source area assessment

areas of debris flow because different resolutions produce difference results, depending on the DEM quality and accuracy (Horton et al. 2013). When the DEM resolution is low, some information on significant areas might be lacking due to the roughness of the DEM. When the DEM resolution is considerably high, the extent of the debris flow sources might be enlarged and time-consuming.

5.5.2 Source Area Delineation

The primary dataset in this research was grid-based DEM, which was fed into Flow-R software. Blais-Stevens and Behnia (2016) revealed that result accuracy is based on the quality of the DEM. In the present study, 0.5-m DTM was used to generate DEM resolutions of 10, 5, 2 and 1 m to illustrate the effect of data resolution on the assessment of debris flow source area. In Flow-R software, the index-based approach is used; in this approach, the grid cells of input data are classified based on favourable option, subsequently excluded and ignored by the software (Horton et al. 2013). Their work revealed that the favourable option is activated when the user specifies the criteria of initiation of debris flow on the Flow-R software. The software later excludes the data when the initiation is unlikely to happen and ignores them when no decision can be made on the parameter. All the input data are combined in ASCII format generated from ArcGIS software before loading into Flow-R software according to their specific rule, and the results are subsequently generated. Figure 5.7 illustrates the main frame of Flow-R software.

5.5.3 Source Parameters

Three important parameters are required to identify possible locations of debris flow. Takahashi (1981) and Rickenmann and Zimmermann (1993) demonstrated that debris flow initiation is based on slope angle, plan curvature and flow accumulation parameters. Delmonaco et al. (2003) pointed out that sediment availability, water input and slope gradient are the three main factors affecting the possible initiation of debris flow directly or indirectly. DEM can be used to obtain the morphological data using GIS-based approach, where water input, slope gradient and curvature can be generated directly from DEM (Horton et al. 2011).

5.5.3.1 Slope Angle

Slope angle plays a vital role in the initiation of debris flow. According to Takahashi (1981), the slope angle is the main criterion for debris flow susceptibility. The slope angle is different for initiation, transportation and deposition zones. Most debris flow occurs when the slope angle is higher than 15° (Rickenmann and Zimmermann 1993; Takahashi 1981). Ortigao and Kanji (2004) specified that debris flow starts when the slope angle is above 20°–25°. In the present research, we considered the low slope angle as the initiation

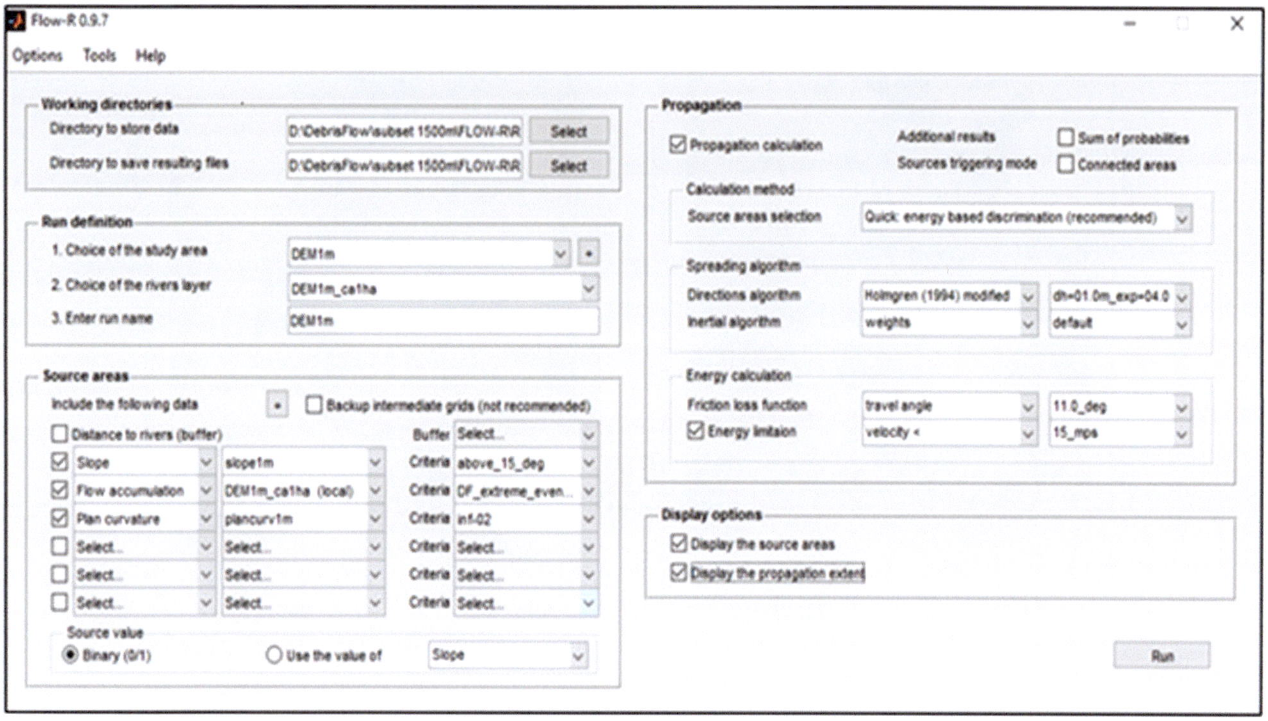

Fig. 5.7 Flow-R software user interface

of debris flow. Therefore, the selection of slope angle with less than 15° threshold in the Flow-R software was excluded from the analysis. Figure 5.8a, b, c and d show the slope with 1-, 2-, 5- and 10-m resolutions, respectively. Blais-Stevens and Behnia (2016) specified that the number of debris flow sources increases with increasing slope angle. However, their research stated that at an angle range of 34°–37°, debris flow is unlikely to occur due to the existence of rock scarps. At a slope angle higher than 40°, the slope comprises weathered rock, which made the rock hard and strong. Consequently, this situation is unlikely to initiate debris flow (Dai and Lee 2001).

5.5.3.2 Flow Accumulation (Water Input)

Water input can be obtained from DEM information. This flow accumulation, also known as upslope contributing area, is widely used in distributed hydrological models (Erskine et al. 2006). Flow accumulation presents information regarding the amount of watershed flowing through the cell. Flow accumulation aims to identify any active stream, drainage network, gully or hidden flow in the study area. The presence of the active stream causes the channelled debris flow with the surface water run-off (transportation zone) (Nettleton et al. 2005). According to Elkadiri et al. (2014), flow accumulation indicates the process of water flow, convergence and infiltration of fluid into pores or solids. Horton et al. (2013) developed the following empirical formula of debris flow initiation zone threshold between slope angle and contributing area based on the studies of Rickenmann and Zimmermann (1993) and Heinimann (1998).

Equation 5.1 shows the threshold for rare event as follows:

$$\begin{aligned} \tan\beta\,\text{thres} &= 0.32\,\text{Suca}^{-0.2}, \text{ if Suca} < 2.5\ \text{km}^2 \\ \tan\beta\,\text{thres} &= 0.26, \text{ if Suca} \geq 2.5\ \text{km}^2 \end{aligned} \tag{5.1}$$

Equation 5.2 shows the threshold for extreme event as follows:

$$\begin{aligned} \tan\beta\text{thres} &= 0.31\,\text{Suca}^{-0.15}, \text{ if Suca} < 2.5\ \text{km}^2 \\ \tan\beta\text{thres} &= 0.26, \text{ if Suca} \geq 2.5\ \text{km}^2 \end{aligned} \tag{5.2}$$

where tan βthres is the slope gradient, and Suca is the surface of the upslope contributing area. Horton et al. (2013) specified that the criterion of flow accumulation in Flow-R software is 1-ha threshold with the slope relationship of extreme event. Extreme event was used in this research to avoid any lacking or possibility of source areas. Horton et al. (2008) mentioned that when the user uses rare event, the sources might be limited, and a small area can still contribute to the initiation of debris flow. Flow direction was generated firstly in ArcGIS, and the obtained product was used to generate flow accumulation. Figure 5.9a, b, c and d show the products of flow accumulation for 1-, 2-, 5- and 10-m resolutions, respectively.

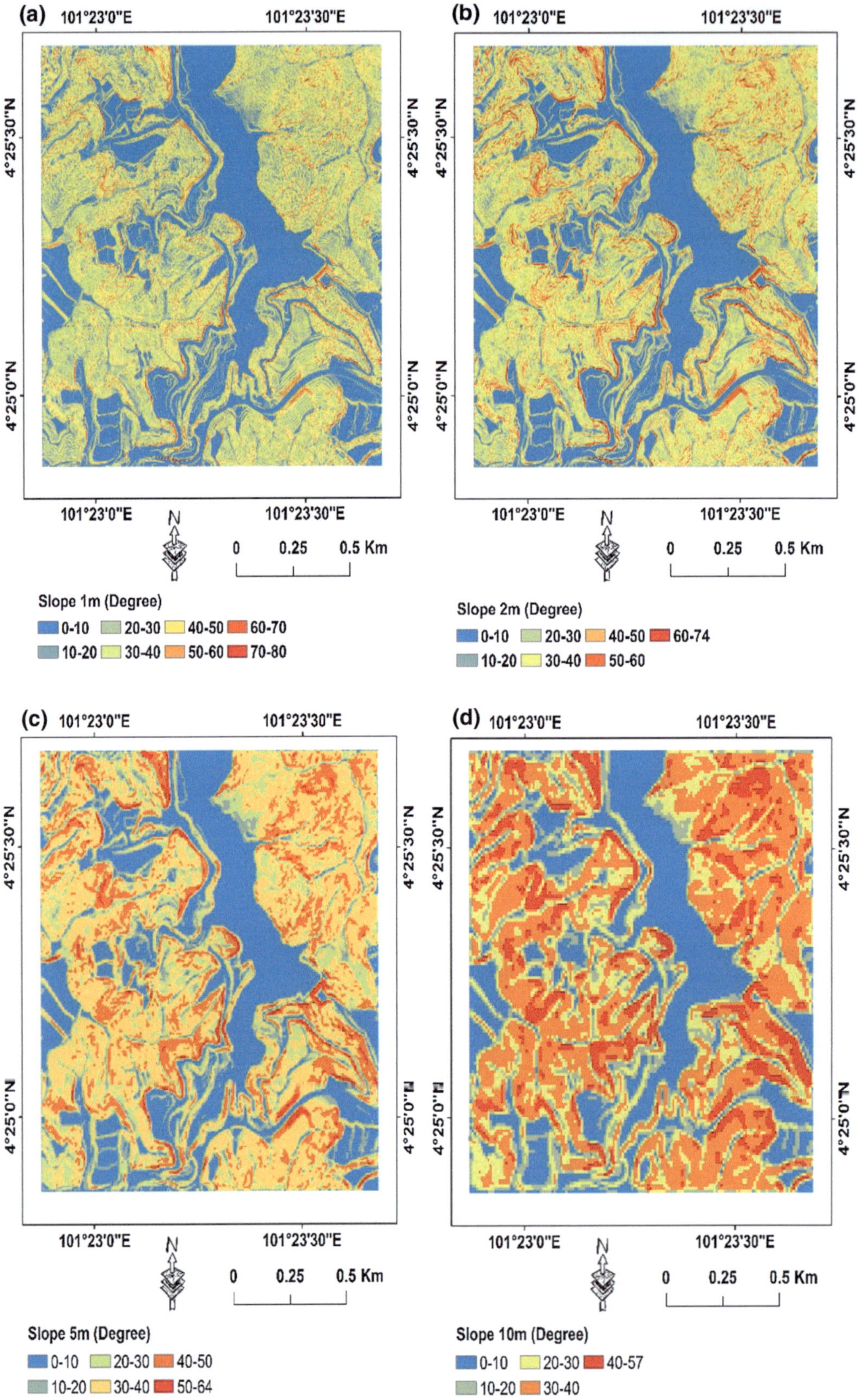

Fig. 5.8 Slope angle sourced at different DEM resolutions: **a** 1 m, **b** 2 m, **c** 5 m and **d** 10 m

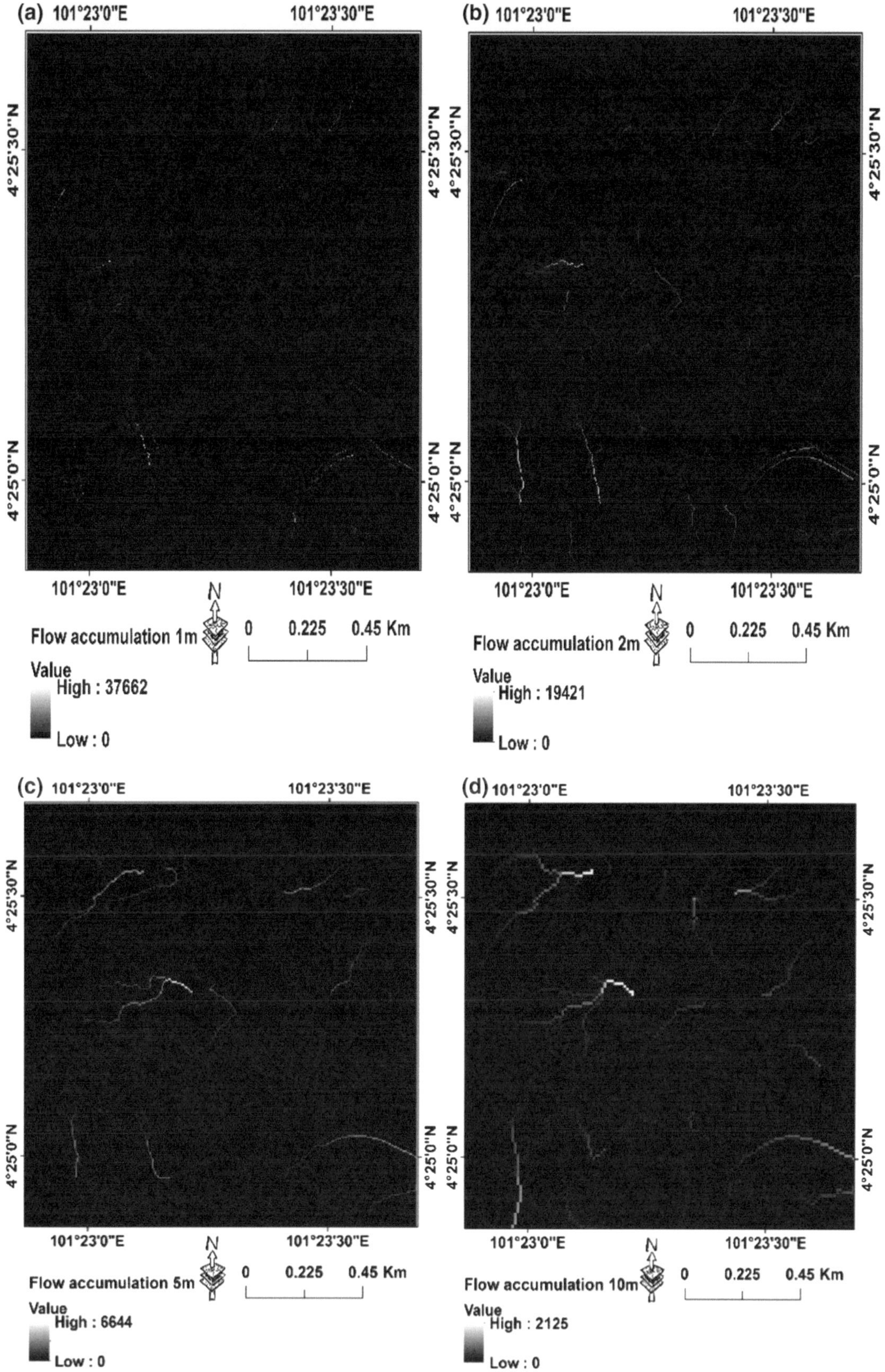

Fig. 5.9 Flow accumulation obtained from different DEM resolutions; **a** 1 m; **b** 2 m; **c** 5 m; and **d** 10 m

5.5.3.3 Plan Curvature (Sediment Availability)

Plan curvature is the third main parameter used to identify the sources of debris flow. Plan curvature consists of three types, namely flat, convex and concave. According to Delmonaco et al. (2003), debris flow tends to occur when the curvature is concave. Blais-Stevens and Behnia (2016) also pointed out that debris flow occurs in gullies where the plan curvature is concave. No standard threshold values exist for plan curvature. Horton et al. (2013) revealed that curvature threshold values may vary with location. In a study conducted in Norway by Fischer et al. (2012), they used the value $-1.5/100\ m^{-1}$ for mapping debris flow susceptibility map. In Korea, plan curvature values ranging from −2/100 to $-1/100\ m^{-1}$ were used (Park et al. 2013). Another study by Horton et al. (2013) in Switzerland used the curvature value of $-2/100\ m^{-1}$. These studies confirmed that threshold values vary with location. In the present study, the values tested were between −1/100 and $-2/100\ m^{-1}$. Our plan curvature of $-2/100\ m^{-1}$ provided an improved representation of gullies for the source areas of debris flow. According to the results, the plan curvature of $-2/100\ m^{-1}$ was used in 1-, 2-, 5- and 1-m resolutions in Flow-R software. The plan curvature results generated in ArcGIS are shown in Fig. 5.10a, b, c and d, where positive value indicates convex, negative value indicates concave, and zero value indicates flat curvature.

5.5.4 Landslide Inventory and Other Datasets

In this research, hillshade, orthophoto and landslide inventory map consisting of 29 landslide locations were used to validate the source areas of debris flow. Drainage network was generated directly from Global Mapper 17 software, and postprocessing was subsequently performed. Lithology map was used to obtain knowledge regarding the geological setting of the area. However, the lithology map was not included in Flow-R software because it was only characterised by two types: (a) granite and (b) metasediment. According to the research conducted by Kasim et al. (2016) and Jamaludin et al. (2014) on debris flow in Peninsular Malaysia, these two types of lithology are prone to debris flow, and no mapped geology type can be excluded as debris flow source areas.

5.5.5 Modelling of Source Areas Using Flow-R

Index-based approach was used to identify the possible source areas of debris flow. This approach considered three topographic characteristics, namely slope angle, plan curvature and flow accumulation, for the initiation of debris flow. These parameters were extracted from DEM using GIS-based approach. When the parameters' threshold satisfied the criteria, Flow-R calculated the start cells of the source areas. On the basis of thorough literature and trial-and-error technique, the criteria selected for the parameters were as follows:

- Slope angle: above 15°
- Flow accumulation: 1 ha with slope relationship extreme event,
- Plan curvature: $-2/100\ m^{-1}$.

For the run-out and spread modelling, the following criteria were generally used:

- Direction algorithm for propagation: Modified Holmgren's with an exponent of 4 and
- Energy calculation: the friction loss function with a travel angle of 11° and velocity threshold of 15 mps.

The modelling of source areas based on the probabilistic and energetic algorithms was performed on the basis of the defined source areas and DEM (Fischer et al. 2012). Starting from the start cells of the source areas, the probability of the debris flow path was calculated (Fig. 5.11).

5.6 Results and Discussion

This section presents the results of debris flow source area identification.

5.6.1 Effect of DEM Resolution on Source Detection

Figure 5.11a, b, c and d illustrate the effect of resolutions by using modified Holmgren's algorithm on DEM resolutions (1, 2, 5 and 10 m) produced by Flow-R software. The obtained result indicated the source areas of debris flow. Figure 5.12a, b, c and d show the inset map (of Fig. 5.11) displaying an enlargement of the calculated source areas of different DEM resolutions produced in Flow-R software. Figure 5.13 shows the combination of different DEM resolutions (1, 2, 5 and 10 m) of the sources at the same area.

Figures 5.11, 5.12 and 5.13 show that the effect of DEM resolutions varies among 1-, 2-, 5- and 10-m resolutions. Results obtained from DEM resolutions of 5 and 10 m (Fig. 5.14) lacked a large number of observed source areas due to the roughness of the DEM. The DEM 2-m resolution produced a significantly improved result on the detection of the major source areas compared with DEM 5- and 10-m resolutions. However, the DEM of 1 m enlarged the coverage of the source area compared with those of other DEM

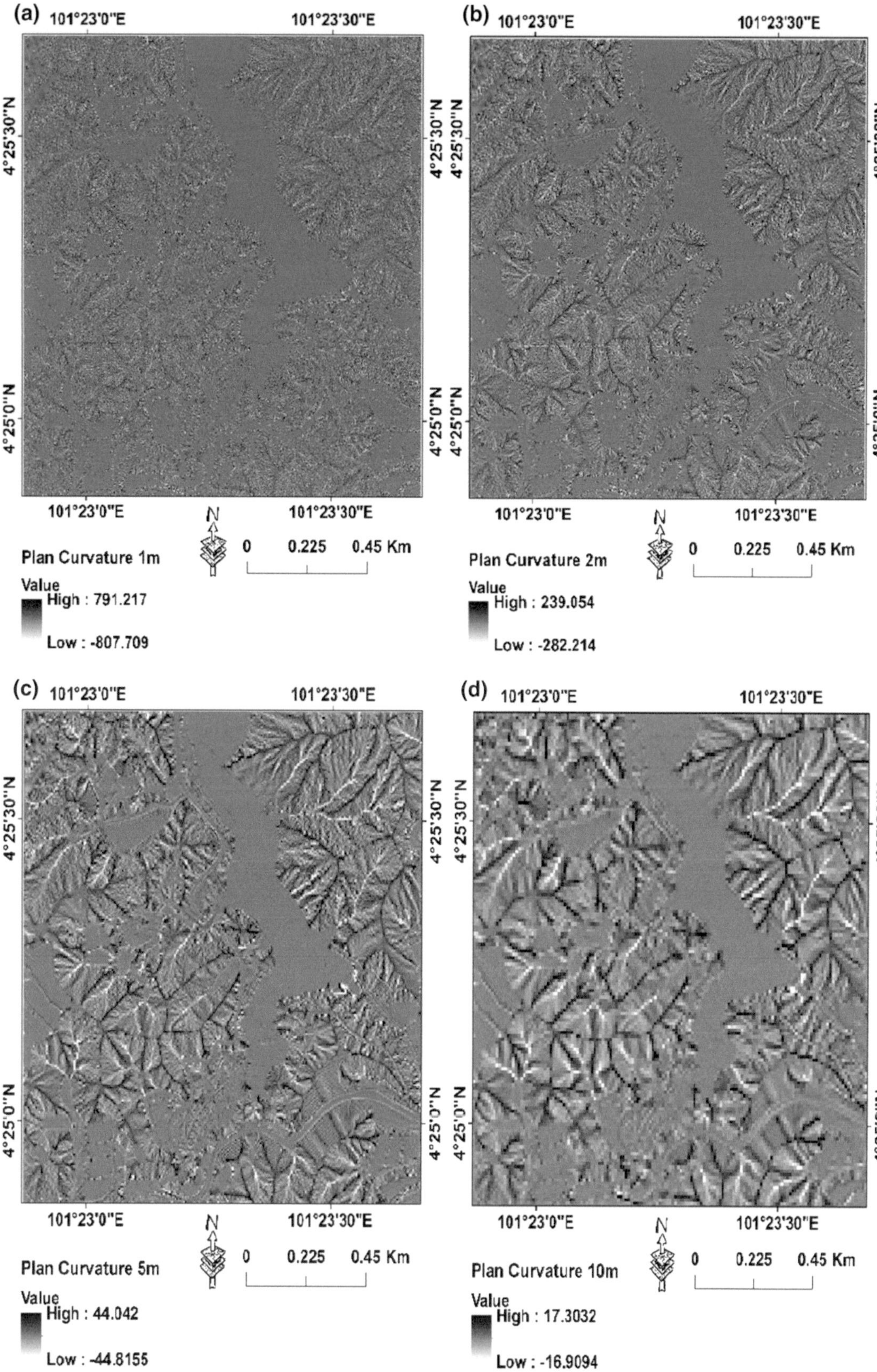

Fig. 5.10 Plan curvature values obtained from different DEM resolutions; **a** 1 m; **b** 2 m; **c** 5 m; and **d** 10 m

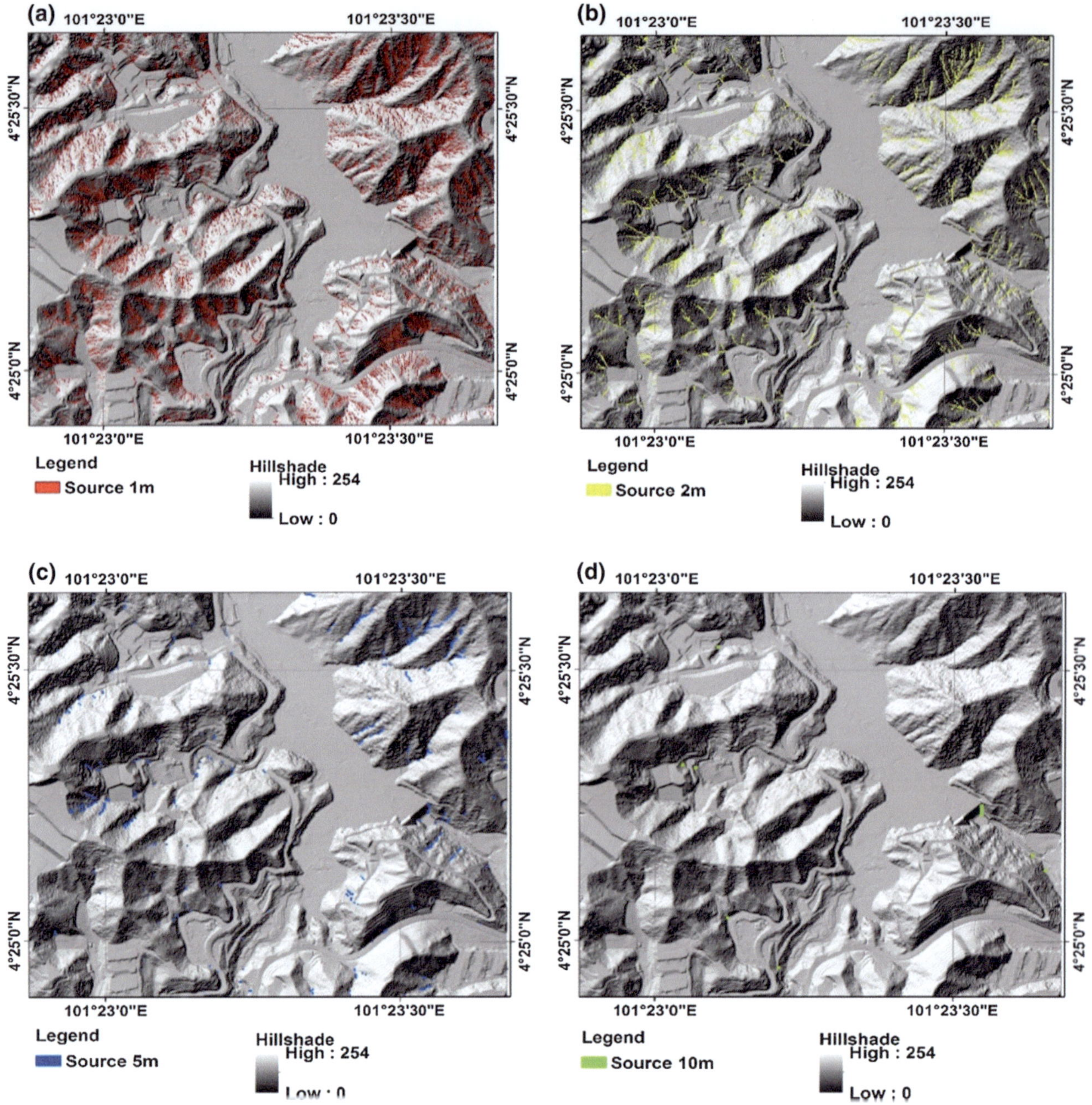

Fig. 5.11 Effect of different DEM resolutions; **a** 1 m; **b** 2 m; **c** 5 m; and **d** 10 m

resolutions. In general, the results obtained from DEM 1 m showed a good match between the modelled source areas and cross-sectional profile. Furthermore, the curvature was concave, which showed the presence of gullies at the top detected by DEM 1 m. The result was consistent with that of previous studies conducted by Zhang and Montgomery (1994) who revealed that the low resolution of DEM shows no significant information due to lacking information. Quinn et al. (1995) pointed out that the resolution and data volume should be increased to simulate the hydrological and geomorphological processes. Such increase is due to that high resolution can detect the presence of gullies and torrents better than fine DEM resolution, as shown in DEM 1-m resolution. Results showed that 1-m resolution produced a significant result to capture the maximum topographic variability of the natural gullies and channels existing in that area compared with that of DEM 2 m. By contrast, DEM 5 and 10 m presented insignificant effect.

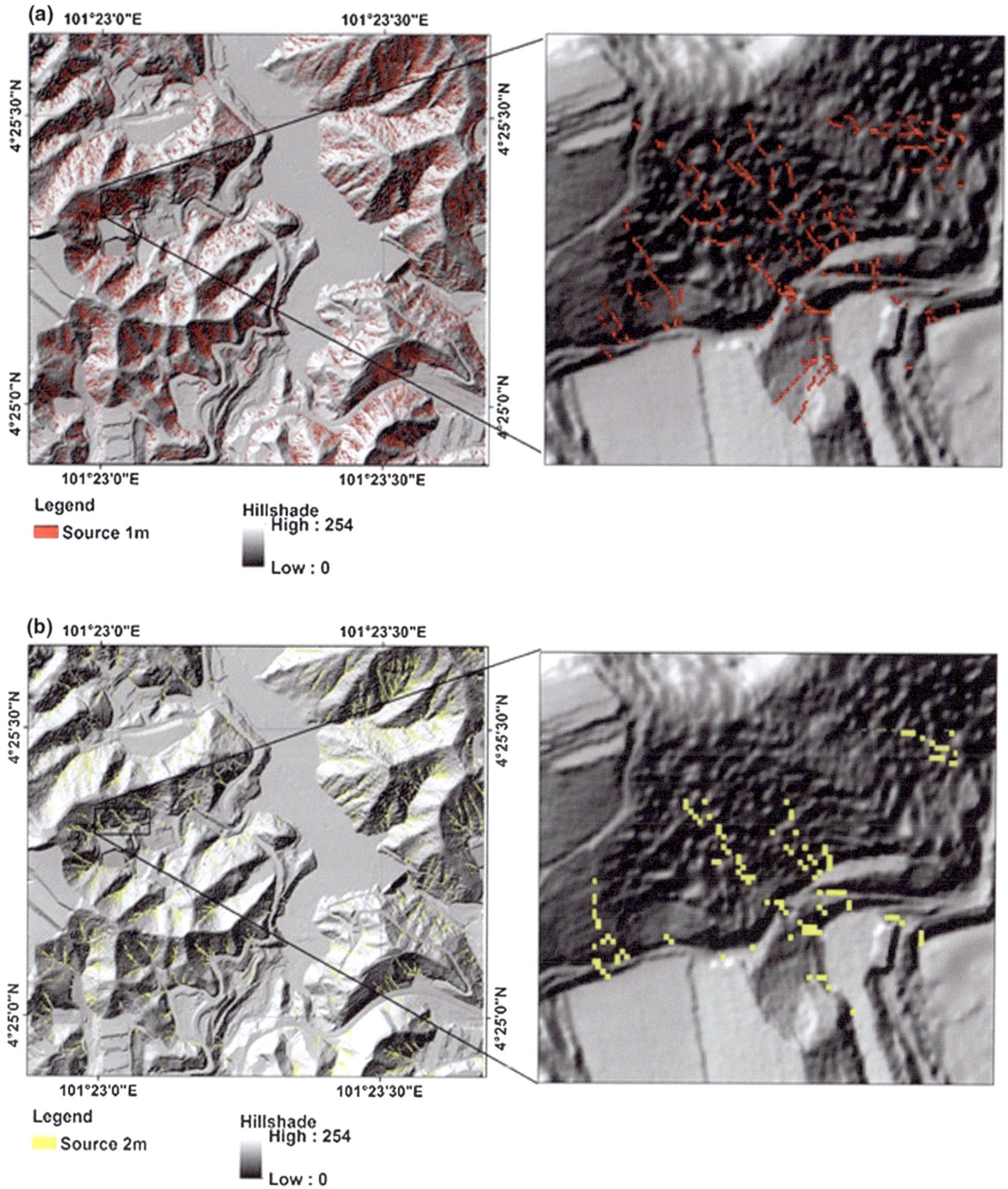

Fig. 5.12 Inset map displaying an enlarged calculated source areas by different DEM resolutions; **a** 1 m; **b** 2 m; **c** 5 m; and **d** 10 m

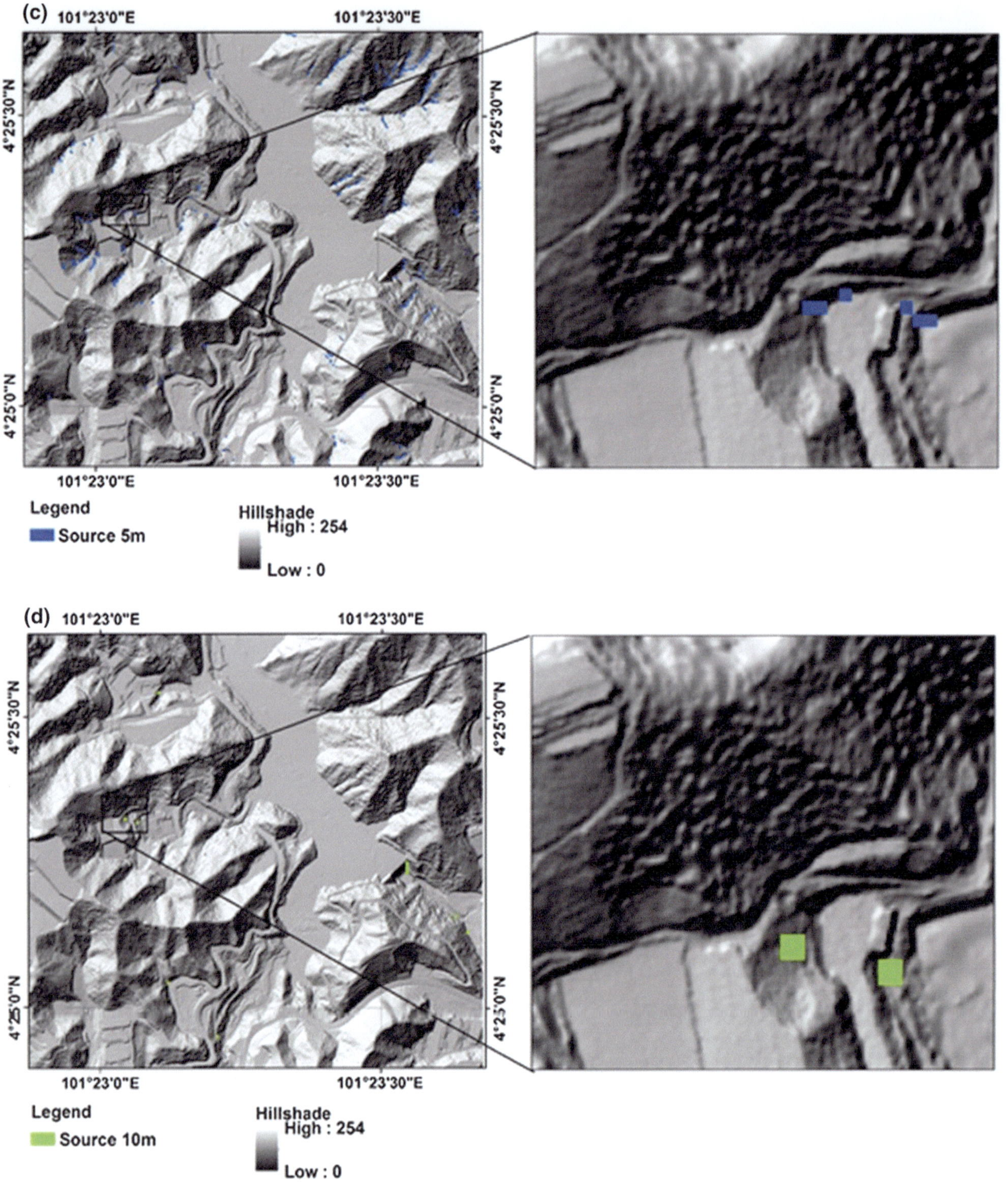

Fig. 5.12 (continued)

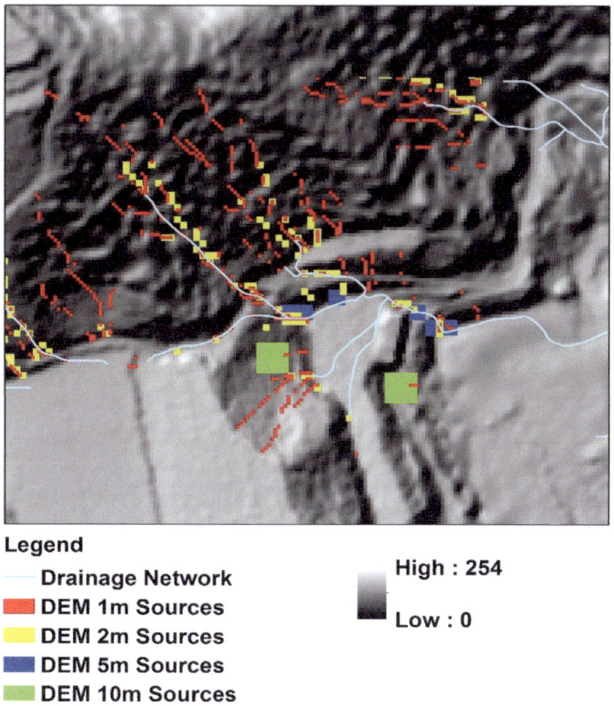

Fig. 5.13 Combination of DEM resolution sources of the same area

5.6.2 Density of Cell Directed from Different DEM Resolutions

The density of cells detected from 1-, 2-, 5- and 10-m resolutions varied. Table 5.1 shows the statistics calculated for each subset of DEMs. For DEM 1 m, the density of cells was 331 with an area of 618 m^2, where the mean and standard deviations were 1.867 and 1.610, respectively. For DEM 2 m, the total number of cells detected for the sources was 78, with an area of 450 m^2. The mean and standard deviations calculated were 5.769 and 4.157, respectively. The DEM of 5 and 10 m detected only a small amount at 4 and 2 cells, respectively. The mean and standard deviations calculated for DEM 5 and 10 m were 37.5 and 12.5 and 100 and 0, respectively.

5.6.3 Cross Section and Longitudinal Sections

To understand the characteristics of debris flow process, three main concepts, namely initiation, transition and deposition zones, were used. The main focus of this research was the source of debris flow. Thus, initiation zone was primarily studied. Debris flow commonly occurs when the

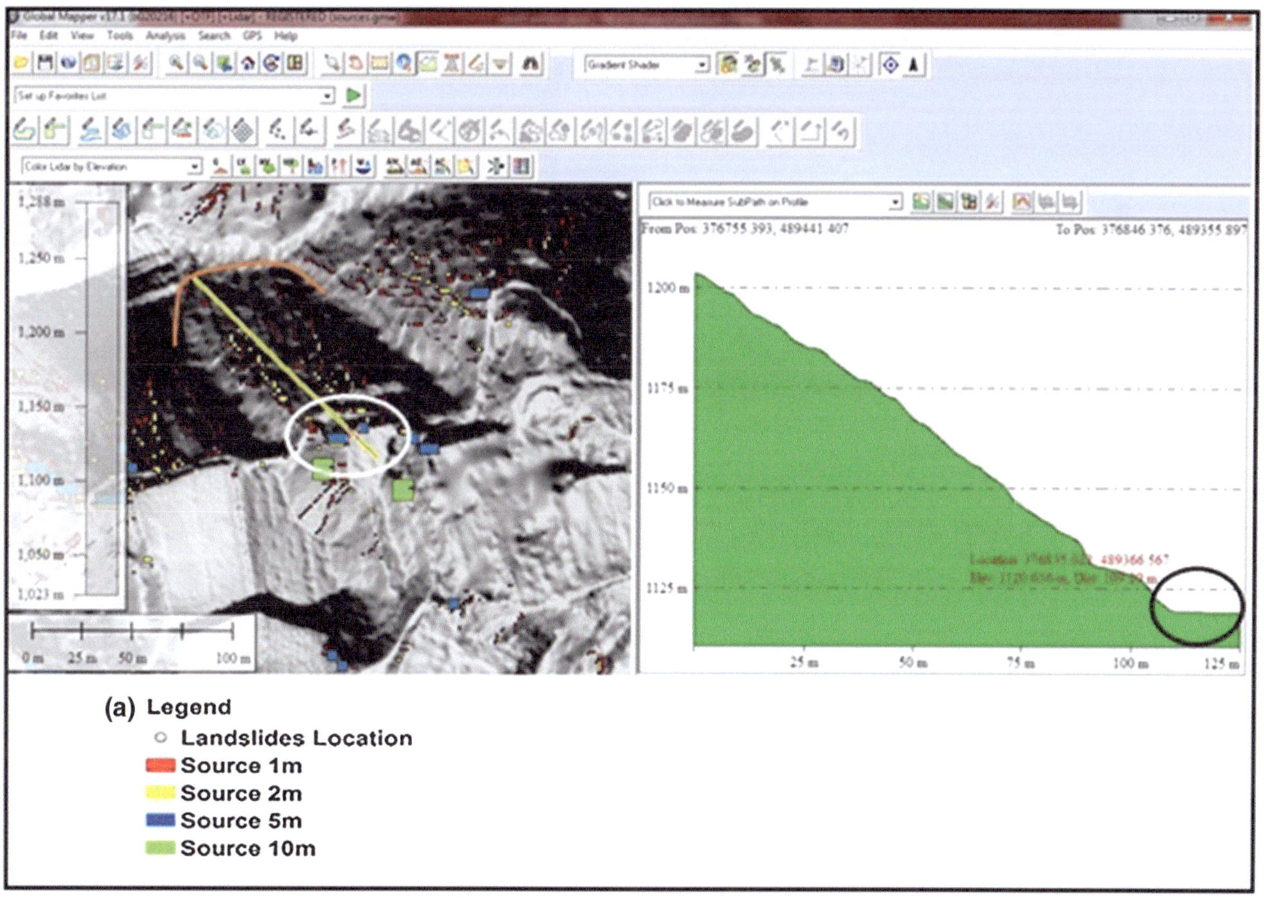

Fig. 5.14 Subset area 1; **a** longitudinal section; and **b** cross section

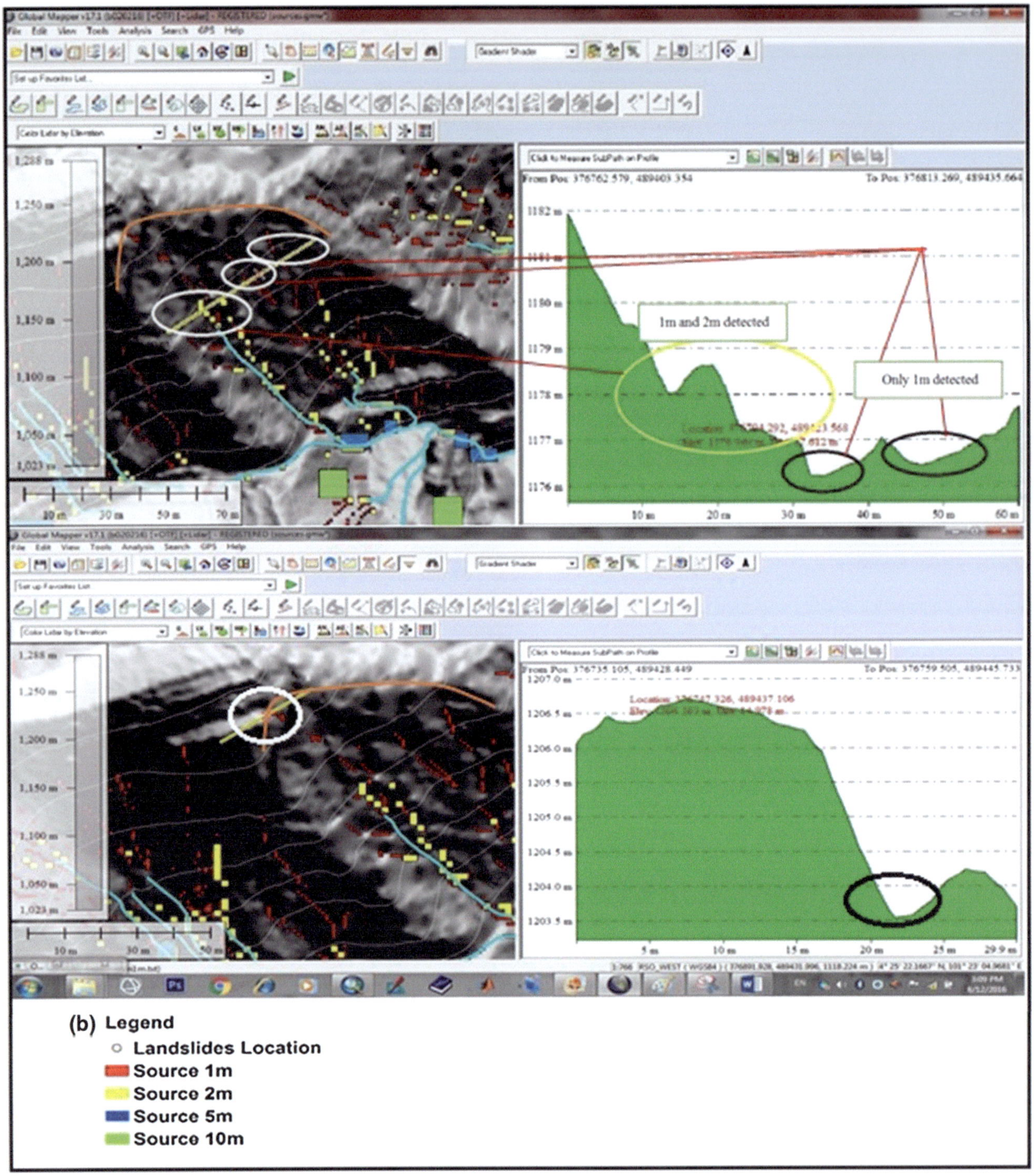

Fig. 5.14 (continued)

plan curvature is concave. Consequently, at the initiation zone, such flow usually contains steep slope failure in the headwall or side slope of a gully or stream channel. At the initiation zone, the source areas form a V or U shape. In this zone, the vegetation is sparse and contains the landslide scarps from previous landslide occurrences. Harris (2008) specified that the plan curvature for debris flow sources can be either linear or hyperbolic concave or a combination of the two, thereby increasing the velocity of debris flow when moving downward. The information that can be extracted from the longitudinal section was the deposition area whose curvature was convex. Hence, the cross and longitudinal

Table 5.1 Statistical results for DEM at 1-m resolution

Statistics	DEM 1 m	DEM 2 m	DEM 5 m	DEM 10 m
Cell count	331	78	4	2
Minimum	1	0	25	100
Maximum	14	28	50	100
Sum (area m^2)	618	450	150	200
Mean	1.610	5.769	37.5	100
Standard deviation	1.867	4.157	12.5	0

sections provided user with an understanding of the source area and allowed its differentiation from the deposition area.

5.6.3.1 Subset Area 1

Plan curvature is the third main parameter used to identify the sources of debris flow. Plan curvature consists of three types, namely flat, convex and concave. According to Delmonaco et al. (2003), debris flow tends to occur when the curvature is concave. Blais-Stevens and Behnia (2016) also pointed out that debris flow occurrences. Figure 5.14a, b illustrate the cross and longitudinal sections of the subset area 1. Visual interpretation on orthophoto and hillshade images showed the presence of semicircular or crown-shaped shallow landslide scarp hidden in the thick forest at the initiation zone (orange colour), where the DEM 1 m detected the source area. Figure 5.14b shows that DEM 1 m detected the presence of flow paths/channels or gullies at the steep angle with a V shape below the scarp. DEM 2 m detected the source area at low angle. Both 1- and 2-m resolutions detected the presence of gullies and channels. Nevertheless, 2-m resolution lacked the natural topography on two other places highlighted by 1 m (Fig. 5.14b). Moreover, the cross section showed the presence of gullies/channels that will transport the sediment. DEM 5- and 10-m resolutions provided no significant information regarding the source area flowing from the steep terrain, but they indicated the deposition area of the debris flow (Fig. 5.14a, longitudinal section) where the curvature is flat. Therefore, the results from DEM 5 and 10 m are negligible.

5.6.3.2 Subset Area 2

Figure 5.15 shows the subset of the second area where the white colour points indicate the location of landslide from the inventory. Landslides were located along the road. Debris flow can be initiated from previous landslide occurrence. The cross section in Fig. 5.15b illustrated that DEM resolution of 1 m can detect natural topography of the present channels or gullies, whereas DEM 2 m can only detect extent of the low area. The cells detected on both DEM of 5 and 10 m indicated the deposition area (Fig. 5.15a, longitudinal section) where the curvature is flat.

5.6.3.3 Subset Area 3

Figure 5.16 illustrates the subset of the third area of debris flow sources that could not be detected by the landslide inventory location. The cross section (Fig. 5.16a) showed that DEM 1 m detected the sources at the steep terrain below the scarp. The source areas of DEM 1 m formed a dendritic shape moving downward along the drainage network. The cross section also illustrated that DEM 1 m can considerably detect the channels than that of DEM 2 m, where some information was lacking. DEM 5-m source cells were detected along the channel moving downward to the deposition area, whereas DEM 10 m could not detect the source at the area. The orthophoto showed the area of the location. The vegetation around the surrounding was sparse, and the landslide scar was hidden in the thick forest.

5.6.3.4 Subset Area 4

According to the landslide inventory, seven locations of debris flow were identified. Figure 5.17 shows the debris flow location mapped in the inventory (white colour points), where the sources can only be detected by DEM 1 and 2 m. The cross section (Fig. 5.17a) illustrated that DEM 1 m can provide a better extent of the source areas compared with that of DEM 2 m. The presence of scarp in the thick forest and multiple channels indicated that DEM 1 m produced a significant result for source area detection. The orthophoto showed the area of the location below.

5.6.4 Visual Interpretation from Orthophoto and Hillshade (Potential Areas)

Figure 5.18a, b, c, d, e and f illustrate the visual observation from the orthophotos and hillshade images whose locations can be described as new potential source areas for debris flow. The landslide inventory map showed that these locations have not been mapped or located. The visual interpretation from orthophotos and hillshade images illustrated the semicircular or crown shape of shallow landslide scarp, where the scarp/scar was active or fresh. The results obtained from DEM 1 m showed that the shallow failure scars were the initiation points of the sources of debris flow (Fig. 5.18). The scarp commonly exists on steep slope where the source of debris flow is initiated. The shallow failure of the scarps in the lithology map was identified as granite and schist, phyllite and slate. Figure 5.18 shows the presence of flow paths or gullies where the curvature is concave to transport the sediments from the source areas moving downward into the channels. The deposition area usually presented convex

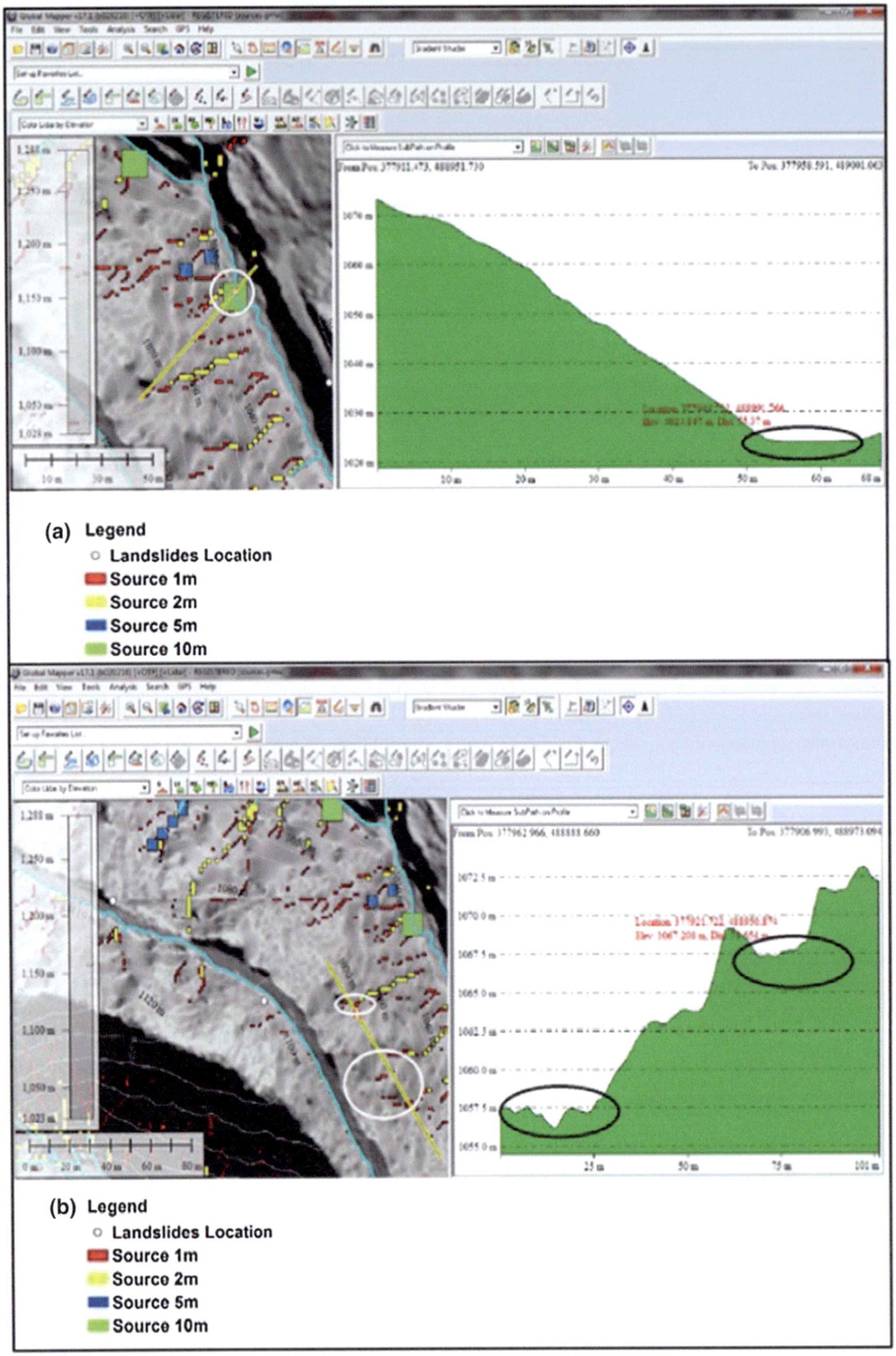

Fig. 5.15 Subset area 2; **a** longitudinal section; and **b** cross section

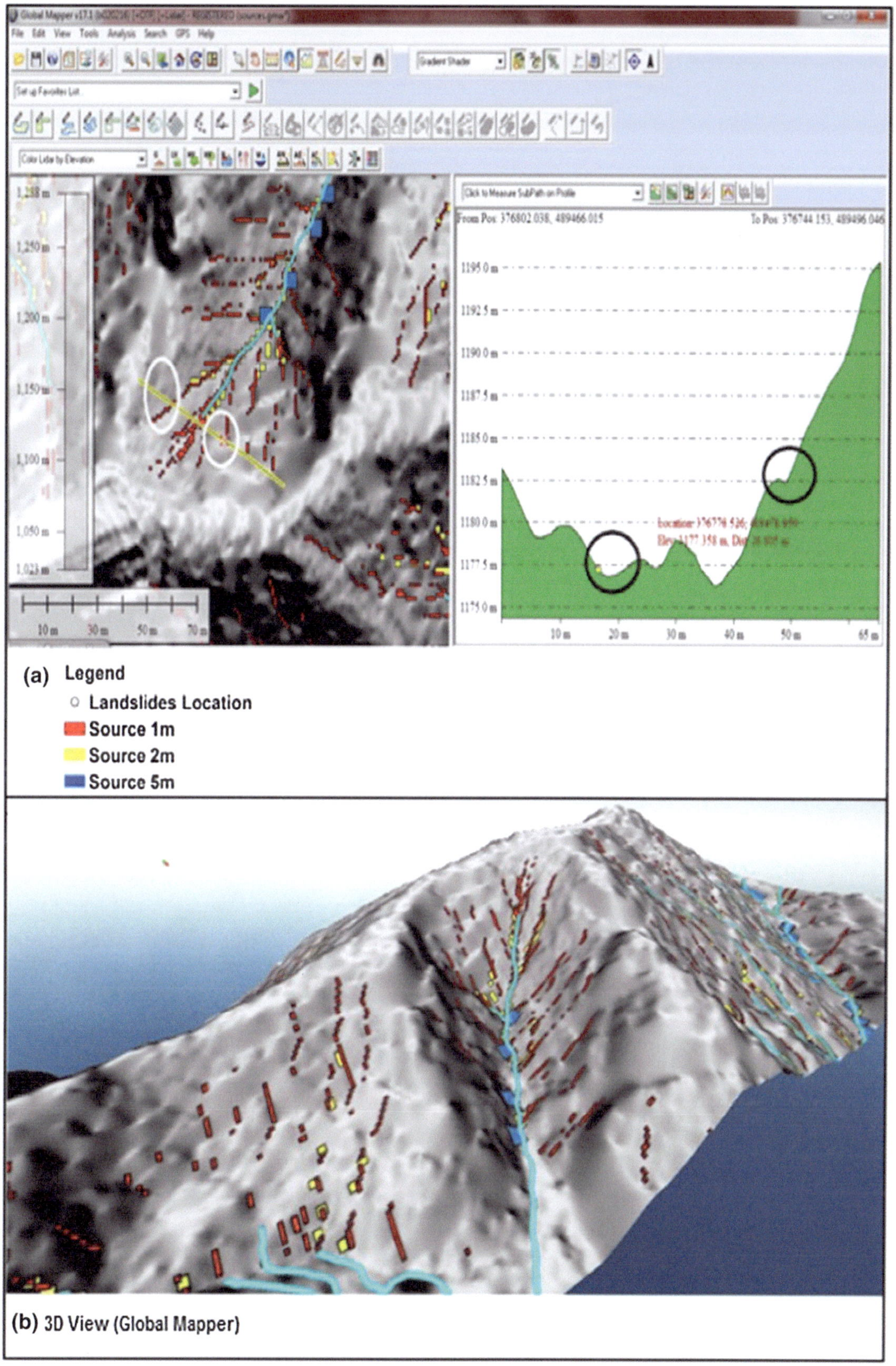

Fig. 5.16 Subset area 3; **a** cross section; and **b** 3D view from global mapper

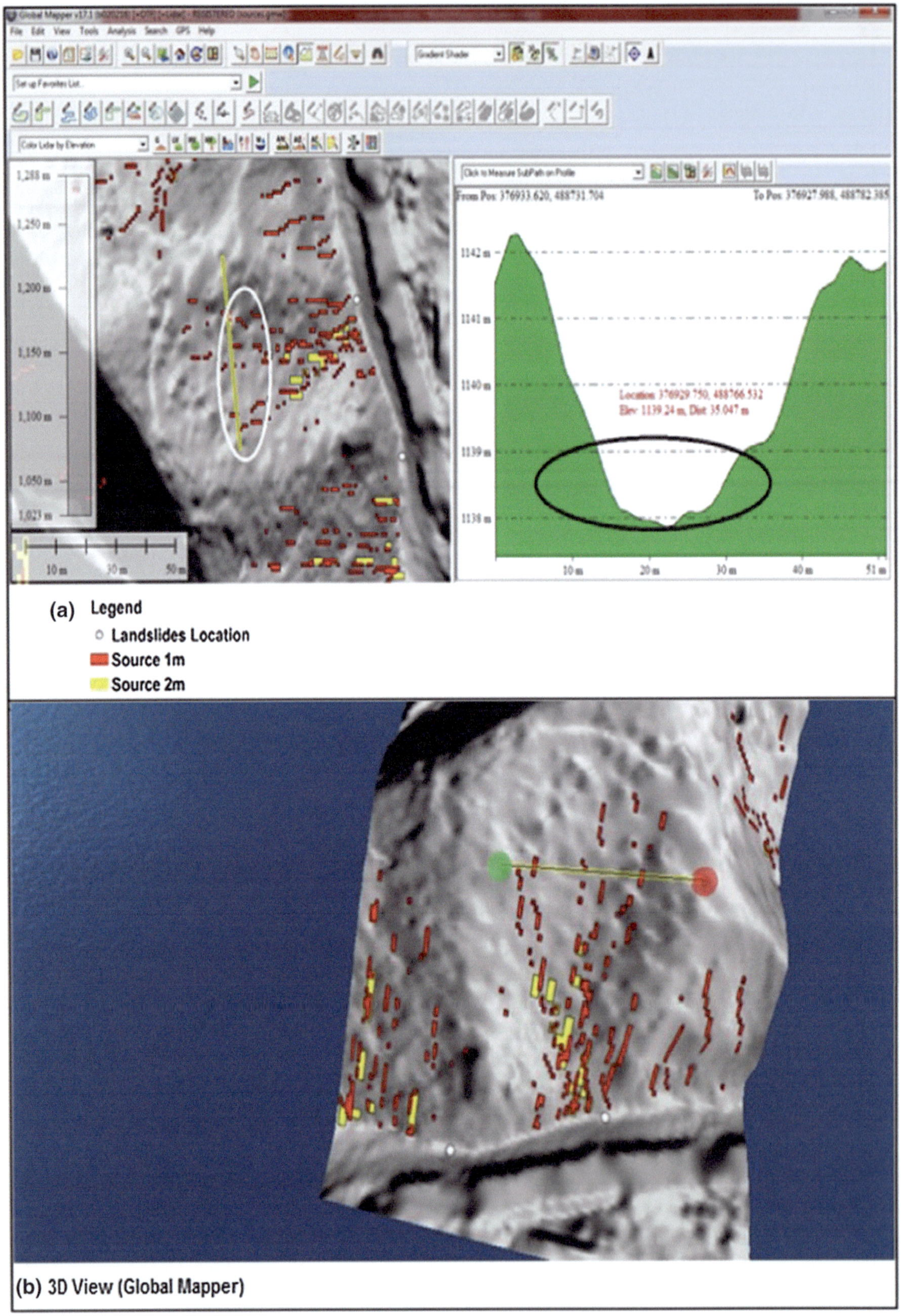

Fig. 5.17 Subset area 4; **a** cross section; and **b** 3D view from global mapper

or flat curvature. These new potential areas were located in a thick forest with steep terrain, where the scarp was hidden. Fortunately, the high-resolution of orthophoto images and airborne LiDAR data presents potential to address these inadequacies. The landslide scarps are shown in orange line in the image.

5.6.5 Slope Coverage

According to the results produced from different DEM resolutions, the slope gradient for the source areas detected by DEM 1 m started from the slope angle range of 50°–70° within an altitude of 1259 m and above mean sea level. This

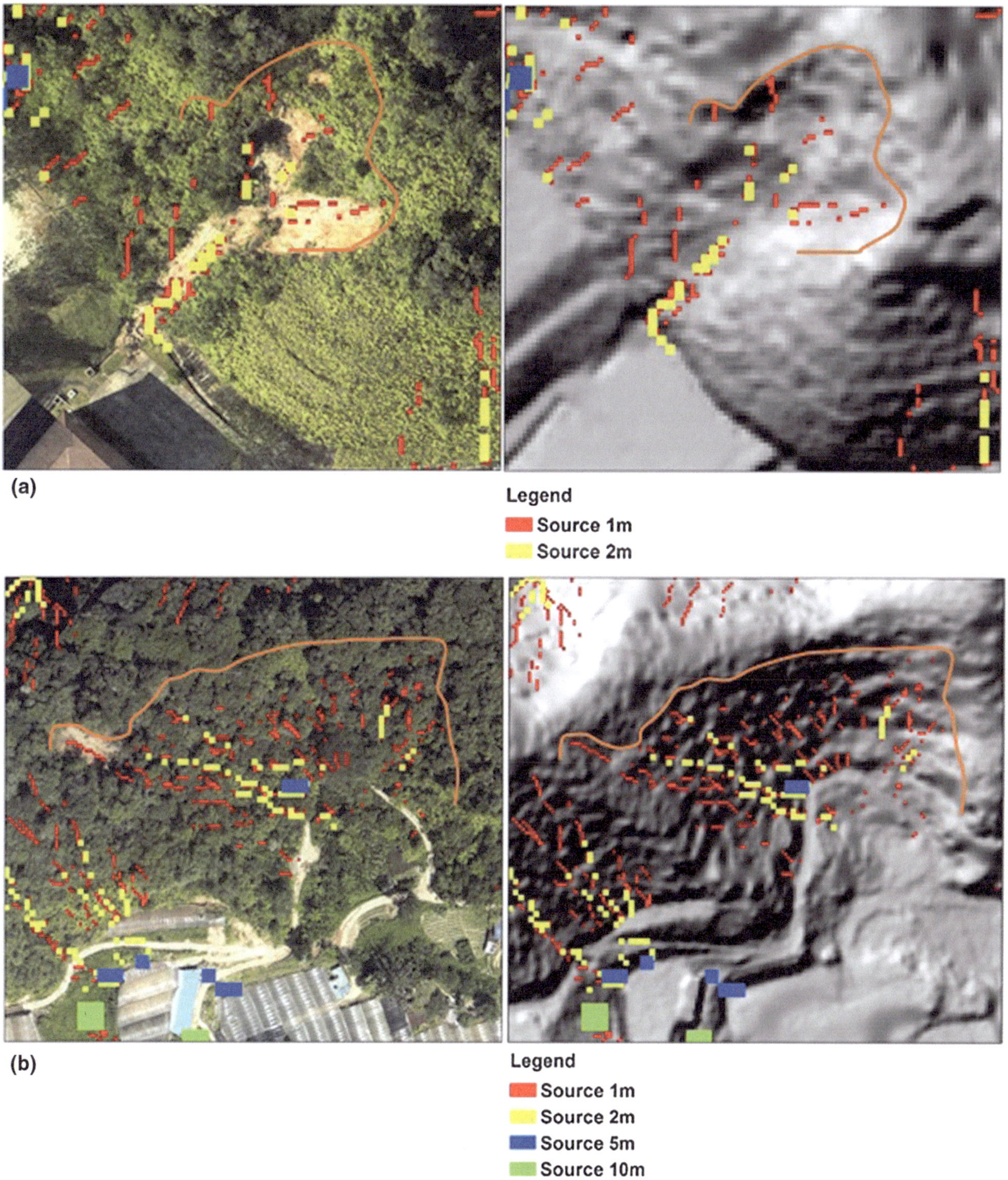

Fig. 5.18 Visual observation of fresh scar/scarp (*orange line*) from orthophoto and hillshade images

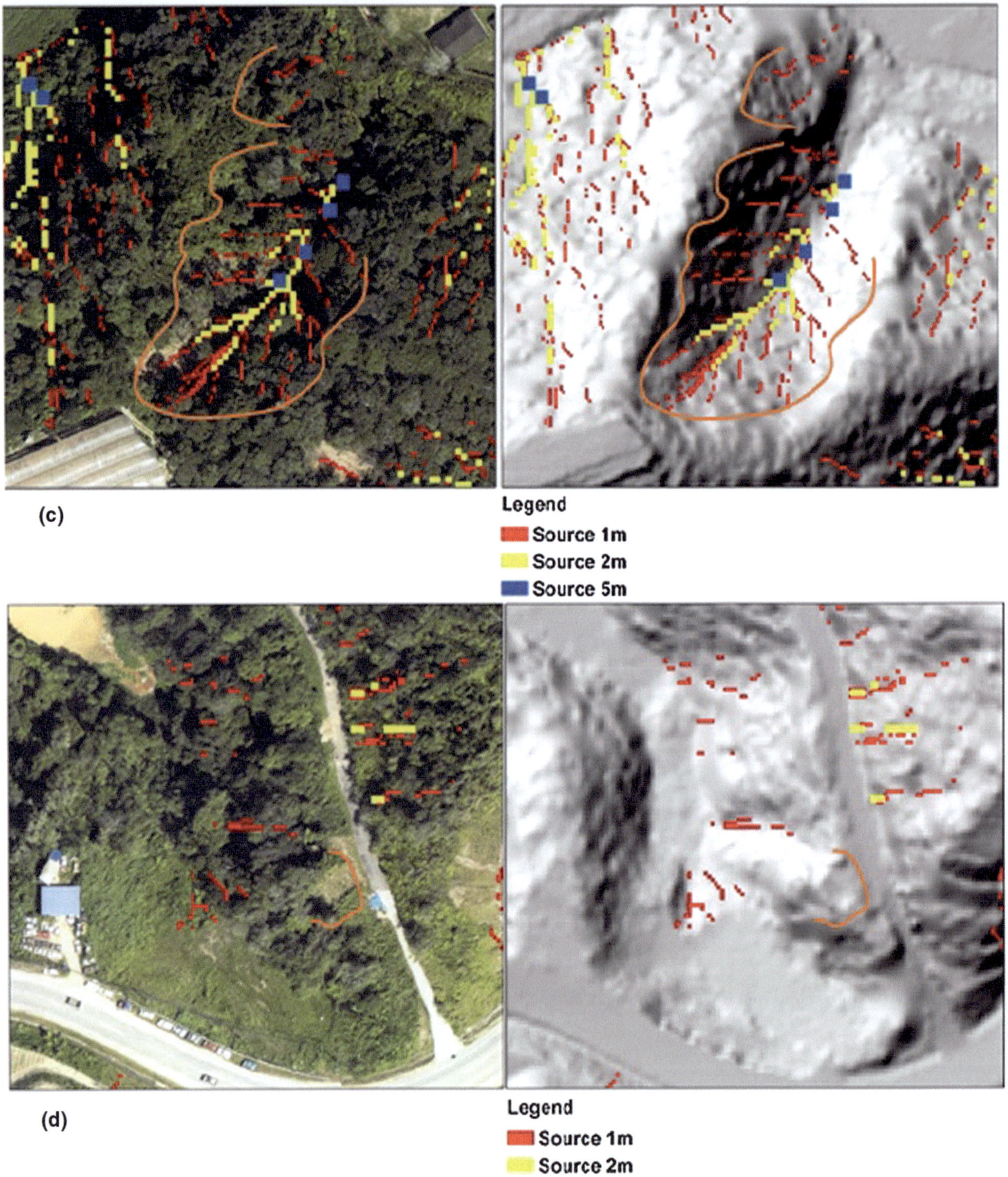

Fig. 5.18 (continued)

result was consistent with that of previous study by Puglisi et al. (2011) who clarified that the slope is a high-angle phenomena, and the soil is moderately thick. For DEM 2 m, the slope angle started at 40°–50° for source areas at 1178 m altitude. The slope angle for the source areas detected by DEM 5 m varied from 30° to 40° at 1134-m altitude. The slope angles varying between 25° and 55° were described as low-angle phenomena, where the soil lied on the smooth

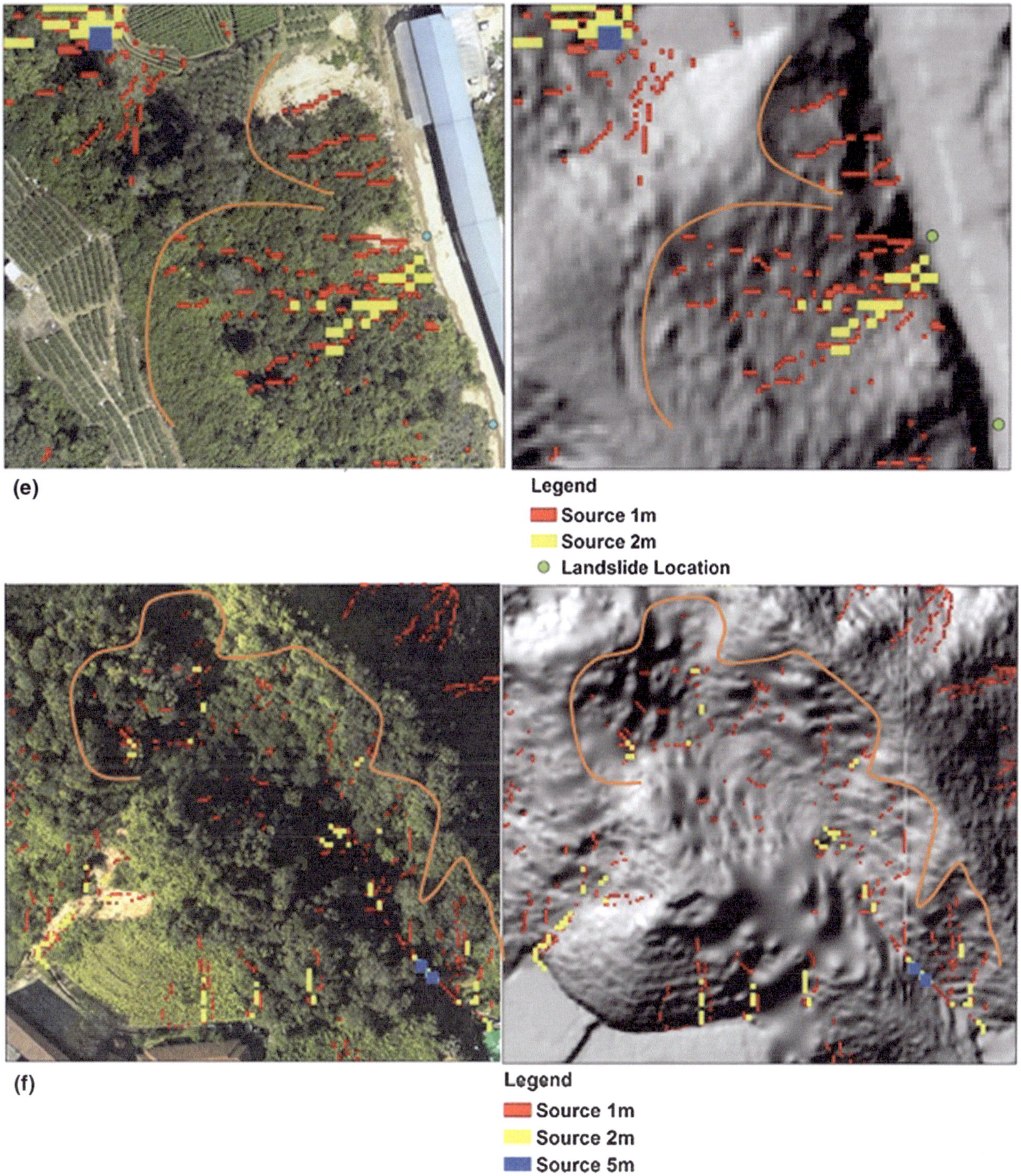

Fig. 5.18 (continued)

slopes (Puglisi et al. 2011). The source areas detected by DEM 10 m started at an altitude of 1117 m and slope angles within 20°–30°. Furthermore, according to Horton et al. (2013), some of the areas are detected at the minimum slope of 5°–10°, and debris flow deposit starts with the slope at 5°.

The source areas followed dendritic drainage networks, as shown by DEM resolutions 1 and 2 m and some in 5 m. DEM 10 m lacked significant information. In addition, the slope gradient indicated a debris flow deposit. The source areas detected from DEMs 1, 2 and 5 m exhibited a concave

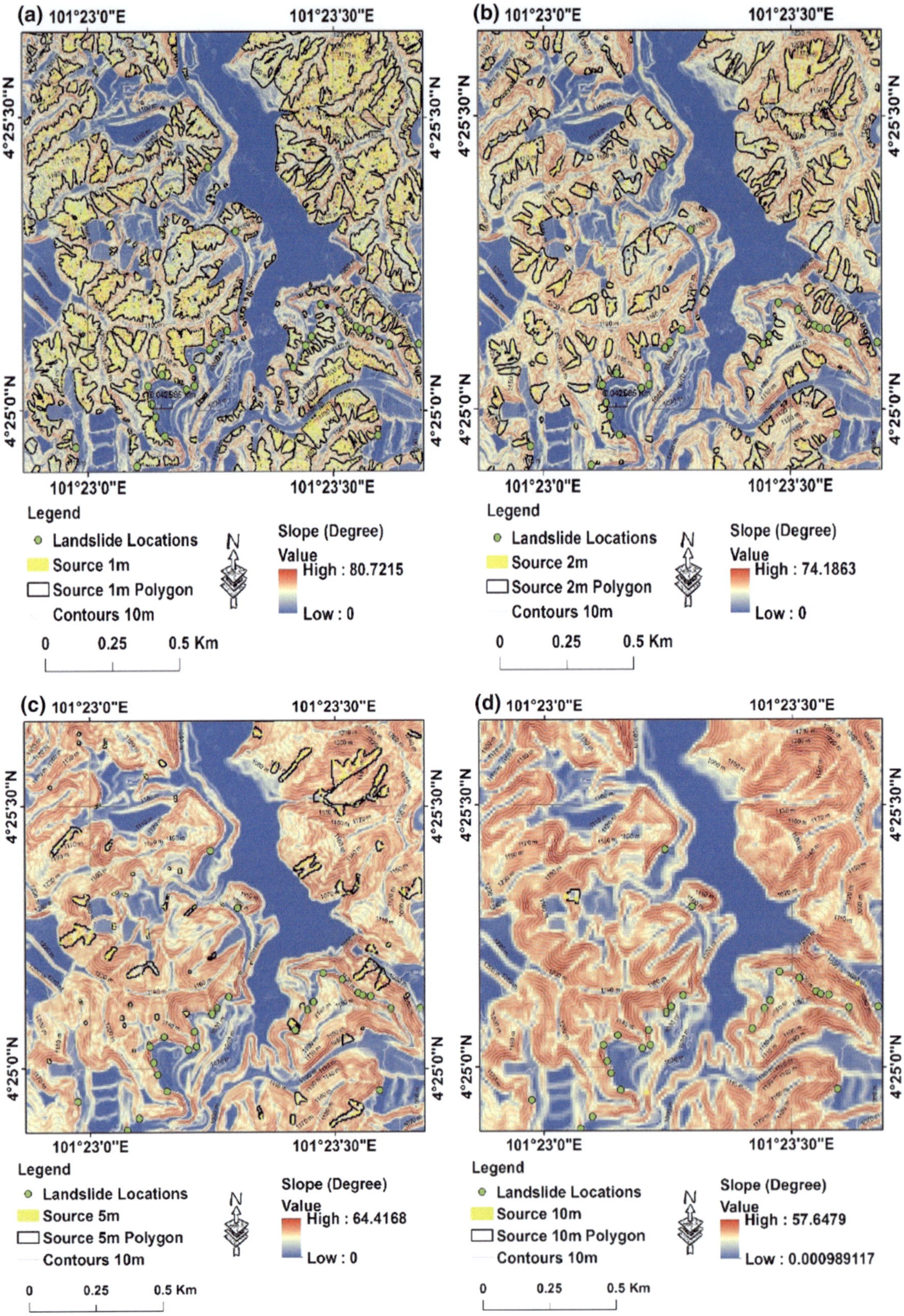

Fig. 5.19 Slope coverage for **a** 1 m; **b** 2 m; **c** 5 m; and **d** 10 m

Table 5.2 Probability method according to the landslide inventory location

Methods	DEM 1 m (%)	DEM 2 m (%)	DEM 5 m (%)	DEM 10 m (%)
Buffering and intersection, 20 m	72	45	17	3
Buffering and intersection, 50 m	93	79	31	3

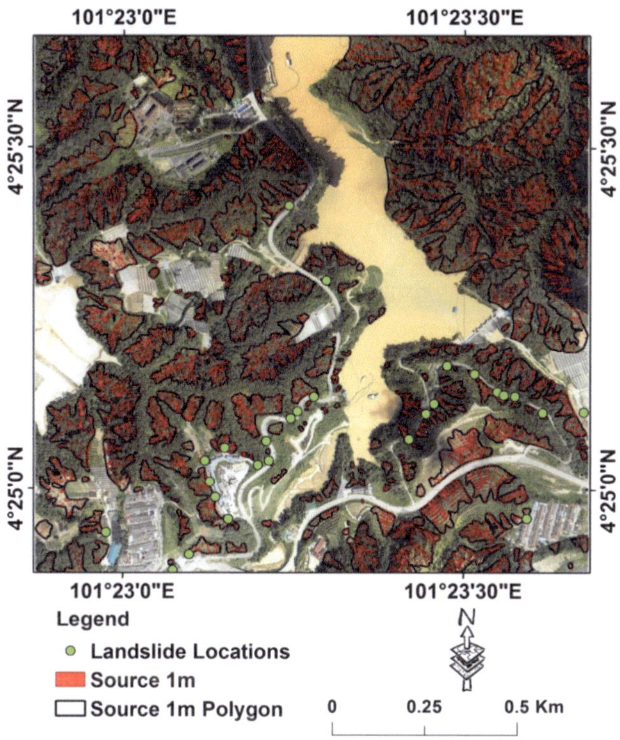

Fig. 5.20 Map of the source areas of debris flow

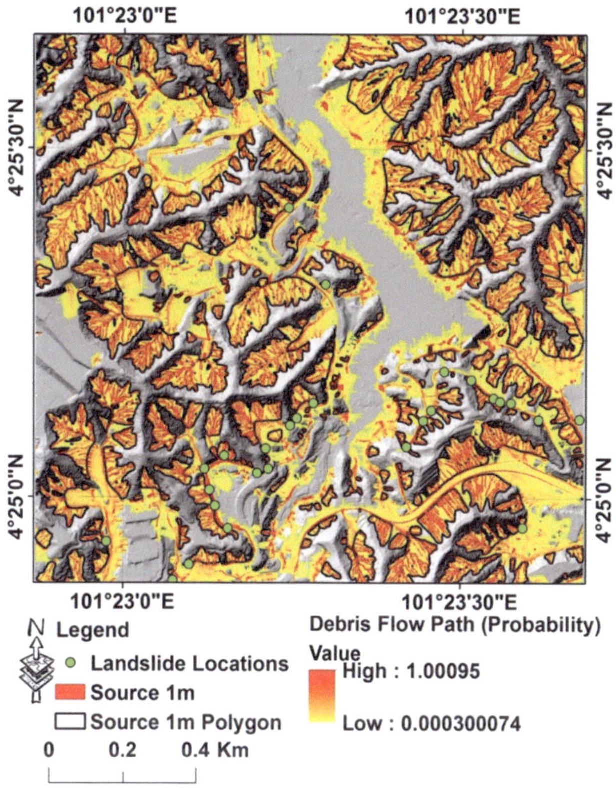

Fig. 5.21 Map of the source areas of debris flow with the debris flow probability value

curvature, where gullies are present for the possibility of sediment transportation along the channel. However, DEM 1 m provided an extended coverage of the source areas. In these areas, the scarps were found on high-steep slopes, where the debris flow was initiated. Figure 5.20a, b, c and d show the slope coverage for DEM resolutions of 1, 2, 5 and 10 m (Fig. 5.19).

5.6.6 Geological Setting up the Study Area

According to the geological map provided by the Department of Geology and Mineral Sciences, Malaysia, and depicted in Fig. 5.2, the geological setting of Ringlet is mainly composed of granite. Kasim et al. (2016) studied the triggering mechanism and characteristic of debris flow in Peninsular Malaysia and revealed that the occurrence of debris flow is mostly located in the main range of granite. In addition, Jamaludin et al. (2014) pointed out that areas covered by granite, volcanic, metamorphic and sedimentary rocks also initiate the debris flow in Peninsular Malaysia, where the slope angle ranges from 50° to 70°. Hence, these two types of lithology existing in Ringlets and comprising granite rocks and a few patches of metamorphic rocks (schist, phyllite and slate) are prone to debris flow initiation. Most of these types are formed in upstream and downstream areas.

5.6.7 Validation

To validate the accuracy of the results produced by Flow-R software, the probability of the source extent produced by Flow-R and past event recorded from the landslides inventory map were obtained by using buffering and intersection methods of 20 and 50 m. Horton et al. (2011) and Blais-Stevens and Behnia (2016) pointed out that these methods are necessary to rate the zone based on its proximity (recorded events) to the source areas. Moreover, the buffer and intersection methods were performed to observe the worst scenario of the distance from the debris flow source areas detected by Flow-R software to the landslide inventory locations. Table 5.2 shows the validation results of the probability detection method from source areas by Flow-R software and the landslide inventory map. DEM 1-m resolution achieved the highest accuracies with 72 and 93% of buffering and intersection from 20 and 50 m from the sources and landslide inventory data, respectively. DEM 2-m resolution achieved 45 and 79% of buffering and intersection from 20 and 50 m, respectively. DEM 5-m resolution achieved the accuracies of 17 and 31%. Finally, the lowest accuracies were produced by DEM 10-m resolution at 3% for both 20 and 50 m from buffering and intersection methods.

The DEM validation results also revealed that DEM 1 m from 20-m buffering and intersection from source area can detect 21 out of 29 from landslide inventory data, and 5 out of 7 were detected from recorded debris flow. DEM 2 m detected 13 out of 29 from landslide inventory and 2 out of 7 from recorded debris flow. DEM 5 m only detected 5 out of 29, and DEM 10 m detected 1 out of 29. Both DEM resolutions 5 and 10 m could not detect the recorded debris flow event from the inventory. However, from the 50-m buffering and intersection of source areas, DEM 1 m achieved the highest detection performance with 27 out of 29 and DEM 2 m with 23 out of 29; both of these resolutions also detected 7 out of 7 of past debris flow event from the inventory. DEM 5 m detected 9 out of 29, and DEM 10 m detected only 1 out of 29 with zero detection of past debris flow event.

5.6.8 Map of the Source Areas of Debris Flow at Ringlet

Figure 5.20 illustrates the final map of the source areas of debris flow at Ringlet, Cameron Highlands, Malaysia. The map was generated from Flow-R software, and DEM 1-m resolution was chosen according to the validation result produced. Data indicated a slightly better extent ot source area detection of DEM 1 compared with those of other DEMs. Figure 5.21 shows the map of the source areas with debris flow probability path indicating the source area (red colour) and deposition area (yellow colour).

5.7 Conclusion

The quantitative method using Flow- R software was used to assess the source areas of debris flow in Ringlet, Cameron Highlands, Malaysia. Results showed that the steep channels are possible zones of the initiation of debris flow, with the presence of flow paths or gullies to transport the sediment. Validation was carried out on the basis of the landslide inventory locations for the past 15 years and the source area generated by Flow-R. Trial-and-error technique was performed to establish the threshold value for plan curvature. In this case, the suitable threshold of $-2/100\ m^{-1}$ was used after the threshold value of $-1/100\ m^{-1}$ failed to detect gullies and hidden flows. High DEM quality and resolution can be used to remove inaccurate source area. However, using high resolution will result in complex computation. In this case, DEM resolutions 1, 2, 5 and 10 m were generated within 8, 2, 1 h and less than 30 min, respectively. Cautions should also be observed for high-resolution data because improper threshold can be misleading. Moreover, the knowledge of the study area is needed. Jaboyedoff et al. (2012) revealed that the identified areas are often larger than the observed events on the field. Their result indicated that the map should always represent the worst-case scenario, and field observation should be performed. Flow-R software was specifically developed by Horton et al. (2013) to detect debris flow source area. This software yields useful results with minimum dataset. The use of modified algorithm of Holmgren's in this research was consistent with the observation made by Horton et al. (2013) who pointed out that the modified version improves the source extent and makes it less sensitive to small features and less dependent on DEM resolution. The approach also presented some limitations. Some of the source areas were in thick forest, and no historical landslide inventory was mapped on such area (along the road). Data require further verification. Nevertheless, all the landslide inventory locations mapped along the road produced significant result regarding the source areas generated in Flow-R. In addition, potential areas were identified, despite that the landslide inventory map cannot detect (sources generated from Flow-R) and verify with visual interpretation from high-resolution orthophoto. Further research should be carried out on the field to improve the result and remove inaccurate data.

References

Abdulwahid, W. M., & Pradhan, B. (2016). *Landslide vulnerability and risk assessment for multi-hazard scenarios using airborne laser scanning data (LiDAR)*. Berlin: Springer-Verlag.

Blais-Stevens, A., & Behnia, P. (2016). Debris flow susceptibility mapping using a qualitative.heuristic method and flow-R along the Yukon Alaska highway corridor, Canada. *Natural Hazards Earth System Sciences, 16*, 449–462.

Chen, H. X., Zhang, L. M., Chang, D. S., & Zhang, S. (2012). Mechanisms and runout characteristics of the rainfall-triggered debris flow in Xiaojiagou in Sichuan Province, China. *Natural Hazards, 62*, 1037–1057.

Chow, W. S., Zakaria, M., Ferdaus, A., & Nurzaidi, A. (2003). *Geological terrain mapping*. JMG unpublished report, JMG.SWP. GS 16/2003, pp. 1–42.

Cruden, D., & Varnes, D. J. (1996). Landslide types and processes. In A. Turner & R. Shuster (Eds.), *Landslides: Investigation and mitigation* (pp. 36–75). Washington, DC: National Academy Press.

Dai, F. C., & Lee, C. F. (2001). Terrain-based mapping of landslide susceptibility using a geographical information system: A case study. *Canadian Geotechnical Journal, 38,* 911–923. doi:10.1139/cgi-38-5-911

Delmonaco, G., Leoni, G., Margottini, C., Puglisi, C., & Spizzichino, D. (2003). Large scale debris-flow hazard assessment: A geotechnical ap-proach and GIS modelling. *Natural Hazard Earth Systems, 3,* 443–455.

Elkadiri, R., & Sultan, M., Youssef, A. M., Elbayoumi, T., Chase, R., Ali, B., Al-katheeri, M. M., & Survey, S. G. (2014). A remote-sensing-based approach for debris-flow susceptibility assessment using artificial neural networks and logistic regression modeling. *IEEE Journals of Selected Topics in Applied Earth Observations and Remote Sensing, 7,* 4818–4835.

Erskine, R., Green, T., Ramirez, J., & MacDonald, L. (2006). Comparison of grid-based algorithms for computing upslope contributing area. *Water Resources Research*, 42.

Evans, S. G. (1982). Landslides and surficial deposits in urban areas of British Columbia: A review. *Canadian Geotechnical Journal, 19,* 269–288.

Fischer, L., Rubensdotter, L., Sletten, K., Stalsberg, K., Melchiorre, C., Horton, P., & Jaboyedoff, M. (2012). *Debris flow modeling for susceptibility mapping at regional to national scale in Norway*. UK: Taylor Francis Group, pp. 723–729.

Gabet, E. J., & Mudd, S. M. (2006). The mobilization of debris flows from shallow landslides. *Geomorphology, 74,* 207–218. Available from: www.elsevier.com/locate/geomorph

Harris, R. (2008). Geospatial assessment of debris flow hazards for Cedar Valley, Utah [Internet]. University of Redlands. Available from: http://inspire.redlands.edu/gis_gradproj

Heinimann, H. R., Hollenstein, K., Kienholz, H., Krummenacher, B., & Mani, P. (1998). *Methoden zur analyse und Bewertung von Naturgefahren*. Wald und Landschaft (BUWAL), Bern: Bundesamt für Umwelt.

Horton, P., Jaboyedoff, M., & Bardou, E. (2008). Debris flow susceptibility mapping at a regional scale. In J. Locat, D. Perret, D. Turmel, D. Demers & S. Leroueil (Eds.), *Proceedings of the 4th Canadian conference on geohazards*, Qu´ebec, Canada, May 20–24, 2008, pp. 339–406.

Horton, P., Jaboyedoff, M., Rudaz, B., & Zimmermann, M. (2013). Flow-R, a model for susceptibility mapping of debris flows and other gravitational hazards at a regional scale. *Natural Hazards and Earth Systems Sciences, 13,* 869–885.

Horton, P., Jaboyedoff, M., Zimmermann, M., Mazotti, B., & Longchamp, C. (2011). Flow-R, a model for debris flow susceptibility mapping at a regional scale—some case studies. In *Proceedings of 5th international conference debris-flow hazards Mitig Mech Predict Assess*, pp. 875 884.

Huat, L. T., & Ali, F. (2012). Application of air borne laser scanning and ortho-rectified photograph geomorphological mapping works. *Electronic Journal of Geotechnical Engineering, 17 H,* 1015–1023.

Hutchinson, J. N. (1988). General report: Morphological and geotechnical parameters of landslides in relation to geology and hydrogeology. In C. Bonnard (Ed.), *Proceedings of fifth international symposium on landslides* (Vol. 1, pp. 3–35). Rotterdam: Balkema.

Jaboyedoff, M., Choffet, C. H., Derron, M.-H., Horton, P., Loye, A., Longchamp, C., Mazotti, B., Michoud, C., & Pedrazzini, A. (2012). Preliminary slope mass movements susceptibility mapping using DEM and LiDAR DEM. In B. Pradhan & M. Buchroithner (Eds.), *Terrigenous mass movements: Detection, modelling, early warning and mitigation using geoinformation technology* (Vol. 5, pp. 109–170). Berlin Heidelberg, Germany, Berlin: Springer-Verlag. doi:10.1007/978-3-642-25495-6

Jamaludin, S., Abdullah, C. H., & Kasim, N. (2014). *Rainfall intensity and duration for debris flow triggering in Peninsular Malaysia. Landslide Sci a safer geoenvironment*. Switzerland: Springer International Publishing. 1.

Jebur, M. N., Pradhan, B., & Tehrany, M. S. (2014). Optimization of landslide conditioning factors using very high-resolution airborne laser scanning (LiDAR) data at catchment scale. *Remote Sensing Environment [Internet], 152,* 150–165. Available from: doi:10.1016/j.rse.2014.05.013

Johnson, A. M. (1984). Debris flow. In: D. Brunsden, D. B. Prior (Eds.), *Slope instability* (pp. 257–361). New York: Wiley.

Kasim, N., Taib, K. A., Mukhlisin, M., & Kasa, A. (2016). Triggering mechanism and characteristic of debris flow in Peninsular Malaysia. *American Journal of Engineering and Research*, 112–119.

Kuriakose, S. L. (2006). *Effect of vegetation on debris flow initiation*. The Netherlands: International Institute For Geo-Information Science And Earth Observation Enschede.

Martinaszabova & Stanislavhroncek. (2015). Investigating geohazards lidar reveals the turbulent life of mountain slopes. *LiDAR Magazine, 5*(6), 5–7. Retrieved from www.lidarnews.com

Nettleton, I. M., Martin, S., Hencher, S., & Moore, R. (2005). 4 4.1 Debris flow types and mechanisms. Earth, (August). Retrieved from www.scotland.gov.uk

Ortigao, J. A. R., & Kanji, M. A. (2004). Landslide classification and risk management. In J. A. R. Oritgao & A. S. F. J. Sayao (Eds.), *Handbook of slope stabilization*. Heidelberg: Springer Verlag.

Park, D., Lee, S., Nikhil, N. V., Kang, S., & Park, J. (2013). Debris flow hazard zonation by probabilistic analysis (Mt. Woomyeon, Seoul, Korea). *International Journal of Innovative Research in Science, Engineering Technology [Internet], 2,* 2381–2390. Available from: www.ijirset.com

Park, D. W., Lee, S. R., Vasu, N. N., Kang, S. H., & Park, J. Y. (2016). Coupled model for simulation of landslides and debris flows at local scale. *Natural Hazards [Internet], 81,* 1653–1682. Available from: doi:10.1007/s11069-016-2150-2

Pierson, T. C., & Costa, J. E. (1987). A rheologic classification of subaerial sediment-water flows. In J. E. Costa & G. F. Wieczorek (Eds.), *Debris flows/avalanches: Process, recognition and mitigation. Reviews in Engineering Geology*. Geol Soc Am VII:1–12.

Pimiento, E. (2010a). *Shallow landslide susceptibility: Modelling and validation. [place unknown]*. Sweden: Lund University.

Pimiento, E. (2010b). *Shallow landslide susceptibility; Modeling and validation*. Master Thesis, Lund University.

Pradhan, B., & Lee, S. (2009). Landslide risk analysis using artificial neural network model focusing on different training sites. *International Journal of Physical Sciences, 3*(11), 1–15.

Pradhan, B., & Lee, S. (2010a). Regional landslide susceptibility analysis using back- propagation neural network model at Cameron Highland, Malaysia. *Landslides, 7*(1), 13–30.

Pradhan, B., & Lee, S. (2010). *Regional landslide susceptibility analysis using back-propagation neural network model at Cameron [Internet],* pp. 13–30. Available from: http://link.springer.com/article/10.1007/s10346-009-0183-2#page-1

Puglisi, C., Falconi, L., Lentini, A., Leoni, G., & Prada, C. R. (2011). Debris flow risk assessment in the Aguas Calientes Village (Cusco, Perù) Debris flow risk assessment in the Aguas Calientes. In *Proceedings of second world landslide forum [Internet]. [place unknown]*, pp. 1–6. Available from: https://www.researchgate.net/publication/266688811

Quinn, P., Beven, K., & Lamb, R. (1995). The in(a/tanβ) index: How to calculate it and howto use it within the topmodel framework. *Hydrological Processes, 9,* 161–182. doi:10.1002/hyp.3360090204.

Rahman, H. A. (2014). Brief review an overview of environmental disaster in Malaysia and Preparedness Strategies. *Iranian Journal of Public Health [Internet], 43*:17–24. Available from: http://ijph.tums.ac.ir

Realino, V., Llanes, F., Resma, M. D., Obrique, J., Quina, C., Gacusan, R., et al. (2015). *Debris flow numerical modelling using high resolution digital terrain models of Ilocos Sur, Philippines.* Nationwide Oper Assess Hazards [Internet]. 4:1–7. Available from: http://blog.noah.dost.gov.ph/2015/07/21/debris-flow-numerical-modelling-using-high-resolution-digital-terrain-models-of-ilocos-sur-philippines/

Rickenmann, D. (2016). Debris-flow hazard assessment and methods applied in engineering practice. *International Journal of Erosion Control Engineering, 9,* 80–90.

Rickenmann, D., & Zimmermann, M. (1993). The 1987 debris flows inSwitzerland: documentation and analysis. *Geomorphology, 8,* 175–189. doi:10.1016/0169-555X(93)90036-2.

Takahashi, T. (1981). Debris flow. *Annual Review of Fluid Mechanics, 13*(1), 57–77.

Takahashi, T. (2007). Estimation of potential debris flows and their hazardous zones: Soft countermeasures for a disaster. *Journal of Natural Disaster Science, 3,* 57–89.

U.S Geological Survey. (2004). *Landslide types and processes.* Online http://pubs.usgs.gov/fs/2004/3072/fs-2004-3072.html

Varnes DJ. 1978. Slope movement types and processes. In: Schuster RL, Krizek RJ, editors. Landslides analysis and control, transportation. Special Report 176. Washington (DC): Transport Research Board, National Research Council; p. 1133.

Youssef, A. M., & Pradhan, B. (2013). *Debris flow impact assessment caused by 14 April 2012 rainfall along the Al-Hada Highway.* Saudi Soc Geosci: Kingdom of Saudi Arabia using high-resolution satellite imagery.

Yusof, N. M., Pradhan, B., Shafri, H. Z. M., Jebur, M. N., & Yusoff, Z. (2015). Spatial landslide hazard assessment along the Jelapang Corridor of the North-South Expressway in Malaysia using high resolution airborne LiDAR data. *Arab J Geosci., 8,* 9789–9800.

Zhang, W., & Montgomery, D. (1994). Digital elevation model grid size, landscape representation, and hydrologic simulations. *Water Re- sour. Res., 30,* 1019–1028. doi:10.1029/93WR03553

Part IV

Landslide Susceptibility Modelling

6 Landslide Susceptibility Modeling: Optimization and Factor Effect Analysis

Biswajeet Pradhan and Maher Ibrahim Sameen

6.1 Introduction

Landslides are considered devastating natural geohazards worldwide; they pose significant threats to human life and result in socioeconomic losses in many countries (Mahalingam et al. 2016). A literature search shows that considerable efforts have been exerted to develop new ideas and tools that can improve the mitigation of landslide effects. One field that is attracting the attention of an increasing number of researchers worldwide is landslide susceptibility modeling (LSM). LSM is the basic information required for hazard and risk assessments; it is also a critical component in disaster management and mitigation (Pradhan and Lee 2009; Bui et al. 2015; Gaprindashvili and van Westen 2016). Significant studies on landslide susceptibility mapping were conducted in the last decades, thereby creating new ideas and research directions for future studies. The optimization of landslide conditioning factors (Jebur et al. 2014), the study of the effects of landslide sampling procedures (Hussin et al. 2016), the development of novel and hybrid models (Moosavi and Niazi 2015), and the analysis of the effects of landslide factors (Guo and Hamada 2013) are among recent and significant research directions in landslide susceptibility studies.

Landslides are triggered by several factors that create challenges for researchers in analyzing and predicting different types of landslides. In general, geomorphological, topographical, geological, and hydrological factors are among the factors that are widely studied and considered in LSM (Pradhan 2013; Pereira et al. 2013). However, landslide conditioning factors, such as slope, aspect, land use, distance to road, and vegetation density are not consistent among studies. In addition, the quality and quantity of data can also vary, thereby affect the accuracy of LSM. Therefore, a detailed analysis and comprehensive investigation of the input data before LSM is performed are important to increase the accuracy of landslide susceptibility models. In addition, recent advances in light detection and ranging (LiDAR) technology enable landslide researchers to collect high-quality data (Kasai et al. 2009). Nevertheless, challenges remain because of the variability in topography and other conditions of different study areas.

Several studies have attempted to provide insights into landslide conditioning factors and have investigated these factors for LSM. Mahalingam et al. (2016) evaluated landslide susceptibility mapping techniques using LiDAR-derived factors in Oregon City. The results of their study showed that only a few factors were necessary to produce satisfactory maps with a high predictive capability (area under the curve >0.7). Qin et al. (2013) investigated uncertainties caused by digital elevation map (DEM) error in LSM. The uncertainty assessment showed that modeling techniques could have varying sensitivities to DEM errors. Mahalingam and Olsen (2015) assessed the influences of the source and spatial resolution of DEMs on derivative products used in landslide mapping. Their study showed that a fine resolution would not necessarily guarantee high predictive accuracy in landslide mapping, and the source of the datasets would be an important consideration in LSM. The effects of landslide conditioning factor combinations on the accuracy of LSM were explored by Meten et al. (2015). In their study, the accuracy of LSM was improved by removing certain landslide conditioning factors based on their correlations with other factors. Kayastha (2015) conducted a study on factor effect analysis using the frequency ratio (FR) model in Nepal. The results indicated that using all nine causative factors produced the best success rate accuracy of over 80%. However, in the study of Vasu and Lee (2016), an LSM with 13 relevant factors selected from the initial 23 factors presented a success rate of 85% and a prediction rate of 89.45%. Hussin et al. (2016) evaluated the effects of different landslide sampling procedures on a statistical susceptibility model. The study demonstrated that the highest success rates were obtained when sampling shallow

B. Pradhan (✉) · M.I. Sameen
Department of Civil Engineering, University Putra Malaysia, Serdang, Malaysia
e-mail: biswajeet24@gmail.com

B. Pradhan (ed.), *Laser Scanning Applications in Landslide Assessment*,
DOI 10.1007/978-3-319-55342-9_6

landslides as 50 m grid points and debris flow scarps as polygons. The highest prediction rates were achieved when the entire scarp polygon method was used for both landslide types. The sample size test using the landslide centroids showed that a sample of 104 debris flow scarps was sufficient to predict the remaining 941 debris flows, whereas 161 shallow landslides were the minimum number required to predict the remaining 1451 scarps.

The current study used 15 landslide conditioning factors and an adequate number of landslide inventories to investigate the optimization of landslide conditioning factors and conduct a factor effect analysis for developing landslide susceptibility models in the Cameron Highlands, western Malaysia. After multicollinearity and factor effect analyses were performed, Ant colony optimization (ACO) was utilized to select significant landslide conditioning factors among the initial 14 factors for further analysis. Data mining techniques, including support vector machine (SVM) and random forest (RF), were used to analyze the effects of the selected landslide conditioning factors on the prediction rate accuracy of the susceptibility models. Details and discussions on the obtained results are presented in the remainder of this chapter.

6.2 Study Area and Landslide Inventory Data

The Cameron Highlands is a tropical rain forest district located in western Malaysia at the northwestern tip of Pahang. It is approximately 200 km from Kuala Lumpur. Previous studies have reported several landslides in this region, which have caused significant damages to properties (Khan 2010). The lithology of the Cameron Highlands mainly consists of Quaternary and Devonian granite and schist (Pradhan and Lee 2010). The granite in the Cameron Highlands is classified as megacrysts biotite granite (Pradhan and Lee 2010). A subset that occupies a surface area of approximately 25 km^2 was selected for the current study because of the frequent occurrence of landslides in this area (Fig. 6.1). The lowest and highest altitudes are 889.61 and 1539.49 m, respectively.

Multisource remote sensing images and geographic information system (GIS) data were used to collect and prepare a landslide inventory database for LSM. Remote sensing data, including archived 1: 10,000–1: 50,000 aerial photographs, SPOT 5 panchromatic satellite images, and high-resolution LiDAR-based orthophotos, were used to

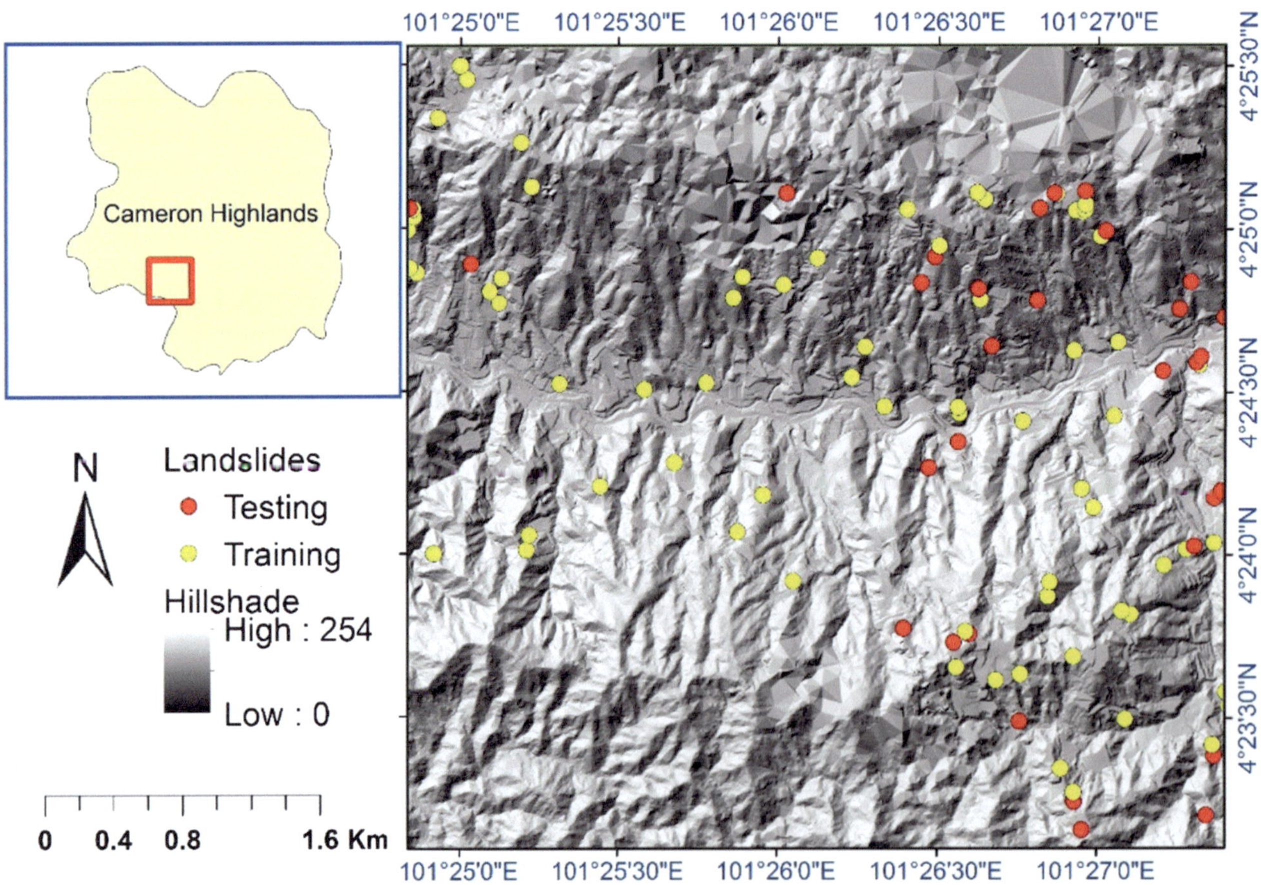

Fig. 6.1 Geographic location of the study area and the landslide inventory map created by using multisource remote sensing data

visually detect landslide occurrences in the study area. In addition, all historical landslide reports, newspaper records, and archived data for the period under examination were collected. The locations of the individual landslides were drawn on 1:25,000 maps based on the site description, archived database, and aerial photograph interpretation. Field observations were performed to confirm fresh landslide scarps. In the aerial photographs and SPOT 5 satellite images, historical landslides could be observed as breaks in the forest canopy, bare soil, or geomorphological features, such as head and side scarps, flow tracks, and soil and debris deposits below a scarp. These landslides were then classified and sorted based on their modes of occurrence. Most of the landslides are shallow rotational, whereas a few are translational. A few landslides that occurred in flat areas were not considered, and thus eliminated from the analysis. To create a database for assessing the surface area and number of landslides in the study area, landslides were mapped within an area of 25 km^2. The landslide inventory map is shown in Fig. 6.1.

6.2.1 Preparation of Landslide Conditioning Factors

A geospatial database that contained 15 landslide conditioning factors was prepared for susceptibility analysis in GIS. Some factors were derived from a LiDAR-based DEM and Advanced Spaceborne Thermal Emission and Reflection Radiometer (ASTER) images, whereas others were digitized from GIS layers collected from government agencies. First, a DEM at 0.5 m spatial resolution was created from LiDAR point clouds using a multiscale curvature algorithm and inverse distance weighted (IDW) interpolation techniques implemented in ArcGIS 10.3. Subsequently, slope, aspect, profile, and plan curvature were derived from the generated DEM at 0.5 m spatial resolution using the spatial analysis tools of GIS. In the case of curvature, negative curvatures represent concave surfaces, zero curvatures represent flat surfaces, and positive curvatures represent convex surfaces. In addition, four hydrological factors, namely the topographic wetness index (TWI), the topographic roughness index (TRI), the stream power index (SPI), and the sediment transport index (STI), were derived from the slope and flow accumulation layers. The land cover map was prepared from SPOT 5 satellite images (10 m spatial resolution) using a supervised classification method. The map was verified via field survey. Then, 10 classes of land cover types were identified, including water bodies, transportation, agriculture, residential, and bare land. The normalized difference vegetation index (NDVI) map was generated from SPOT 5 satellite images (10 m spatial resolution). The NDVI value was calculated using the formula NDVI = (IR − R)/(IR + R), where IR and R denote the energy reflected in the infrared and red portions, respectively, of the electromagnetic spectrum. Finally, distance to road, distance to river, and distance to lineament were calculated based on the Euclidean distance method using the GIS layers.

Several studies have explained the contributing factors of a landslide. The significance of a particular factor depends on site-specific conditions. In the current study, soil and lithology were not used because the study area consists of only one type of soil and lithology. However, 15 factors were used, namely altitude, slope, aspect, profile curvature, plan curvature, land use, TWI, TRI, SPI, STI, NDVI, vegetation density, distance to road, distance to river, and distance to the fault. The succeeding paragraphs briefly describe these factors.

Altitude is controlled by several geological and geomorphological processes. Landslides typically occur at intermediate elevations because slopes tend to be covered by a layer of thin colluvium, which is prone to landslides. In this study, the lowest and highest altitudes were 889.61 and 1539.49 m, respectively. The altitude layer was reclassified into six classes using the quantile classification method, as shown in Fig. 6.2d.

The slope is a measure of the rate of change in elevation in the direction of the steepest descent and is considered the main cause of landslides. The slope gradient map of the study area was divided into six slope angle classes. The study area has flat regions. The highest slope was observed at 80° (Fig. 6.2e).

Aspect is defined as the slope direction measured (in degrees) from the north in a clockwise direction. It ranges from 0° to 360°. Parameters, such as exposure to sunlight, rainfall, and dry winds control the concentration of soil moisture, which in turn, determines landslide occurrence (Fig. 6.2f).

Plan curvature is described as the curvature of a contour line formed by the intersection of a horizontal plane with the surface. It influences the convergence and divergence of flow across a surface. Profile curvature, in which the vertical plane is parallel to the slope direction, affects the acceleration and deceleration of downslope flows and, consequently, influences erosion and deposition. Plan and profile curvature maps were reclassified into three classes, namely convex, flat, and concave lands, with negative, zero, and positive values, respectively (Figs. 6.2g and h).

In addition to the topographical factors, land use, NDVI, and vegetation density are key conditioning factors that contribute to the occurrence of landslides. Sparsely vegetated areas are more prone to erosion and increased instability than forests. Vegetation strengthens the soil through an interlocking network of roots that forms erosion-resistant mats that stabilize slopes. Evapotranspiration controls the wetness of slopes. NDVI is frequently considered a

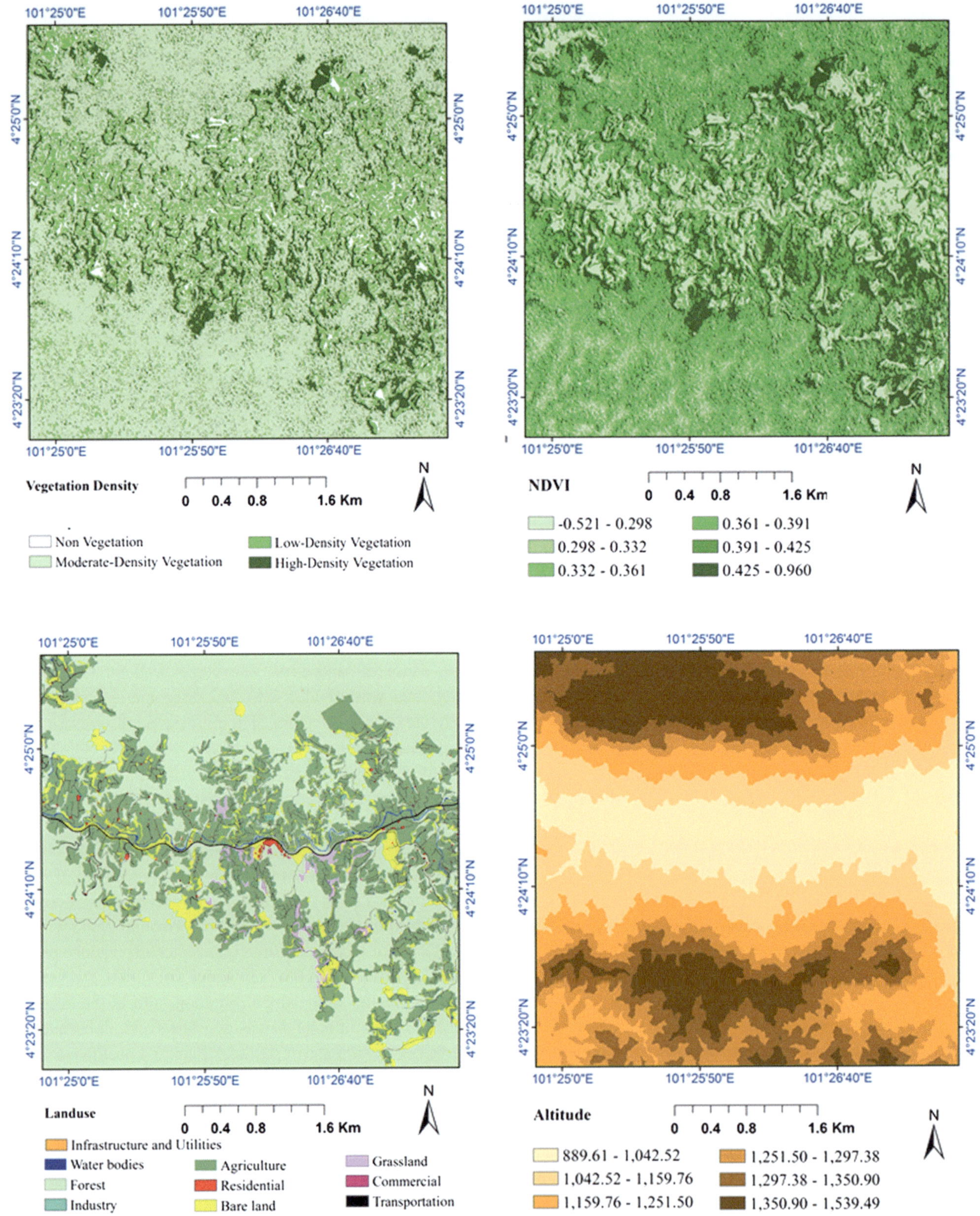

Fig. 6.2 Landslide conditioning factor used in the current study

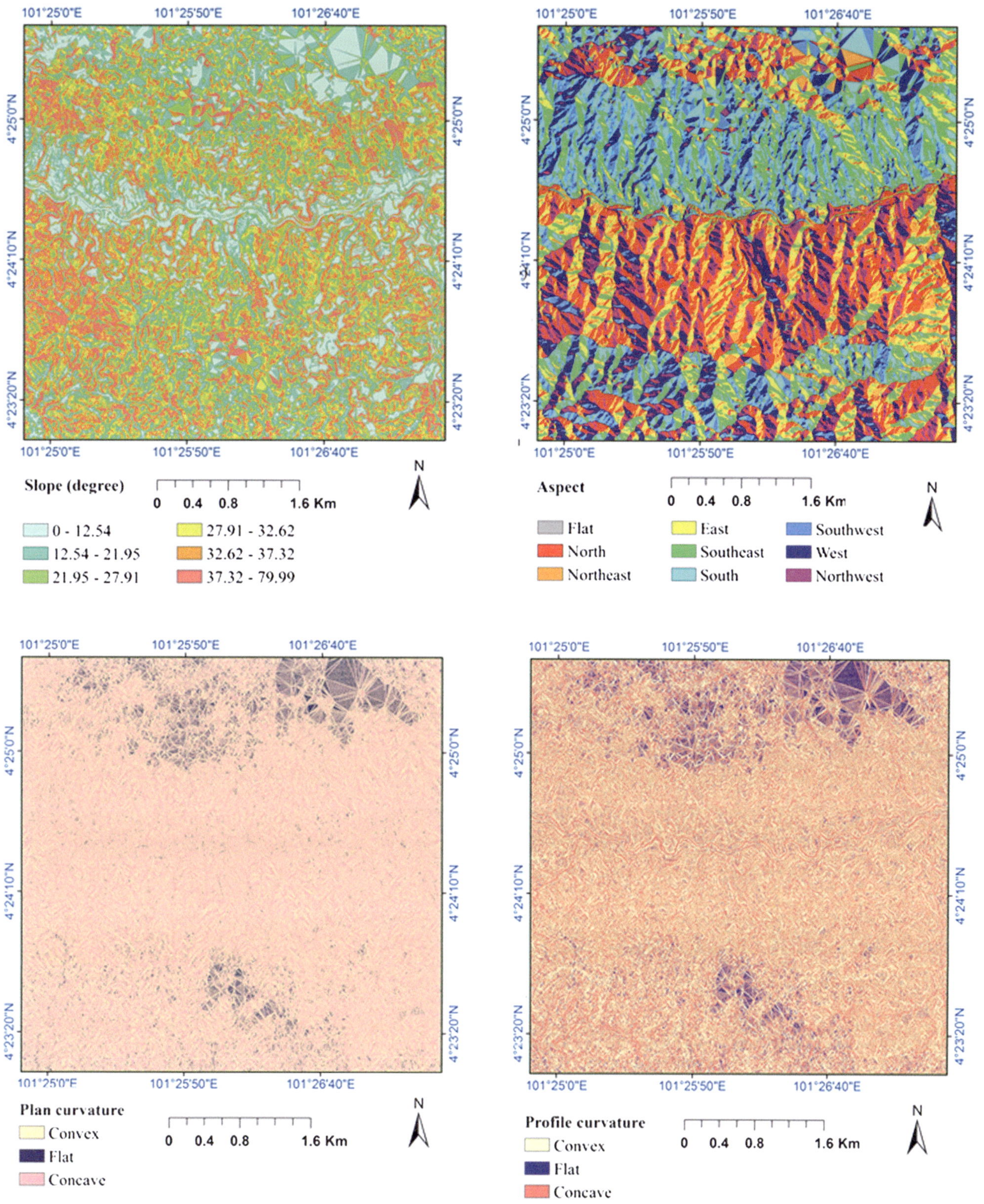

Fig. 6.2 (continued)

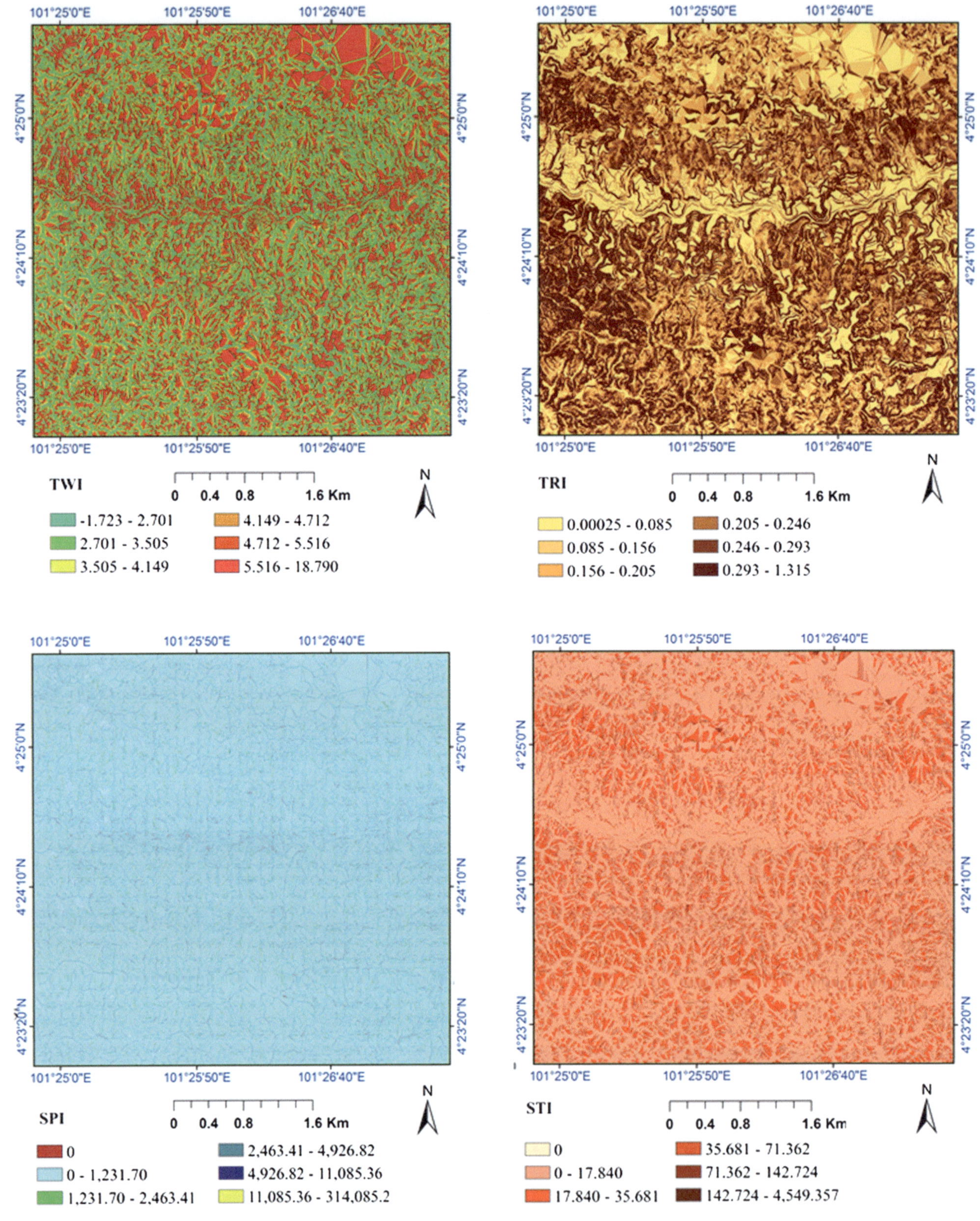

Fig. 6.2 (continued)

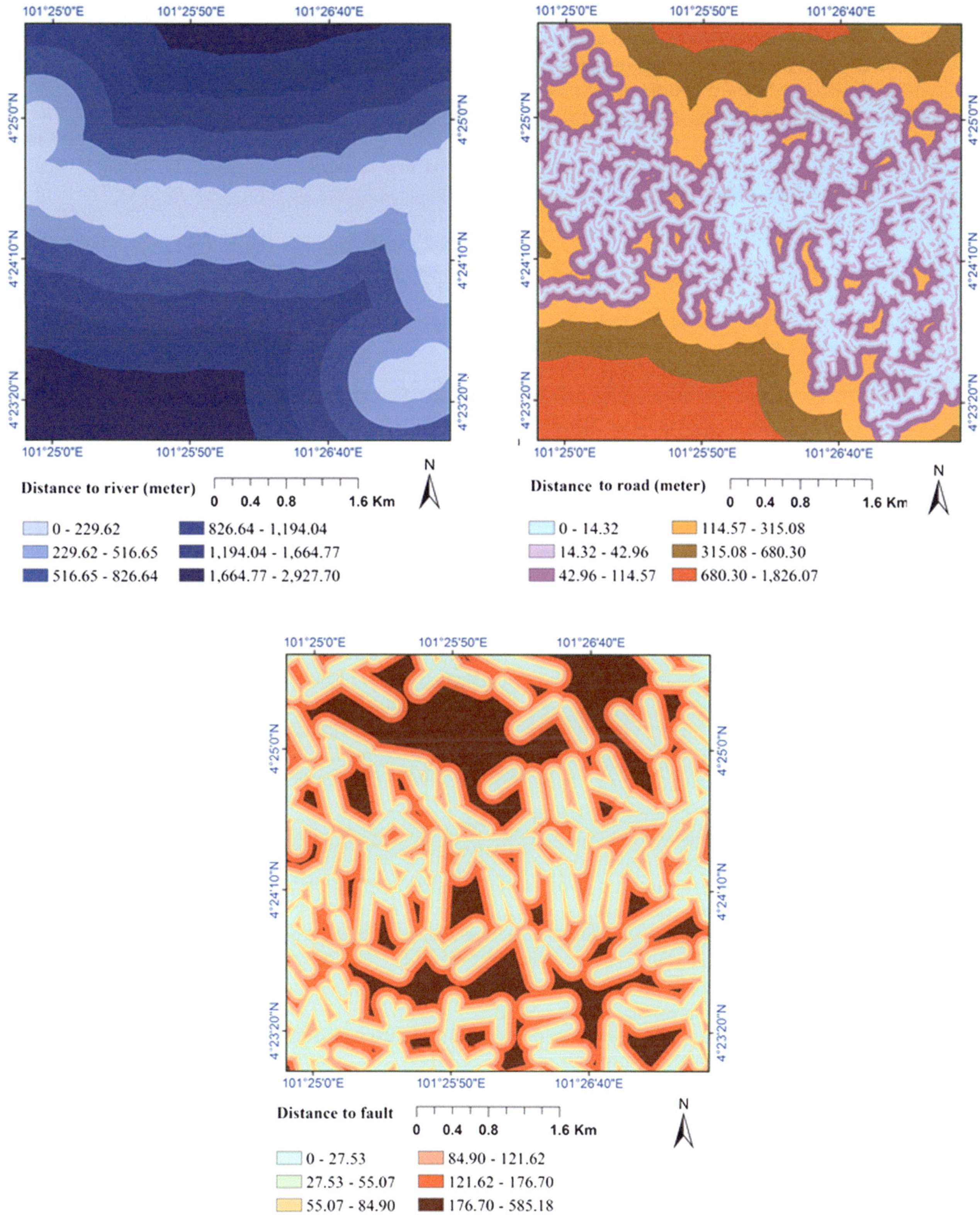

Fig. 6.2 (continued)

controlling factor in landslide susceptibility mapping. In general, when the value of NDVI is high, the area covered by vegetation is large. Furthermore, a relatively low vegetation coverage can easily lead to a landslide incident. In this study, a land use layer that consisted of 10 classes was used for LSM. Vegetation density was reclassified into four classes, namely non-vegetation, low vegetation, moderate vegetation, and dense vegetation (Fig. 6.2a). NDVI was reclassified into six classes starting from the lowest value of −0.521 to 0.96 (Fig. 6.2b).

Four hydrological factors were also used for LSM in the current study. TWI describes the effects of topography on the location and size of saturated source areas of runoff generation. This index is calculated using $\mathrm{Ln}[AS/\tan(\beta)]$, where AS is the specific catchment area of each cell, and β represents the slope gradient (in degrees) of the topographic heights. SPI, which is a measure of the erosion power of a stream, is also considered a factor that contributes to the stability of the study area. This index is expressed as $SPI = AS \times \tan(\beta)$, where AS is the area of a specific catchment, and β is the local slope gradient measured in degrees. STI, which reflects the erosive power of overland flow, is derived by considering transport capacity limiting sediment flux and catchment evolution erosion theories. TRI is another important factor that affects landslide susceptibility. These hydrological factors were reclassified into six classes using the quantile method and then applied in LSM.

Anthropogenic factors, such as distance to roads, distance to rivers, and distance to faults, have been considered important factors that influence landslides. Extensive excavations, application of external loads, and vegetation removal are some of the most common actions that occur along road network slopes during their construction. The intermittent flow regime of a hydrological network and gullies encompasses erosive and saturation processes, thereby increasing pore water pressure and leading to landslides in areas adjacent to drainage channels. In addition, geological faults are important triggering factors of landslides. The fracturing and shearing degree plays an important role in determining slope instability. Proximity (buffers) to these structures increases the likelihood of landslides given that selective erosion and the movement of water along fault planes promote these phenomena. The aforementioned layers were reclassified into six classes using the quantile method.

6.3 Methodology

6.3.1 Overall Research Flow

This study encompasses four methodological steps. The first step is the multicollinearity and factor effect analyses. In the second step, relevant factors among the initial 15 landslide conditioning factors are selected using ACO. The third step involves the application of the susceptibility models using several experiments that aim to analyze the effects of relevant factors. In the last step, susceptibility models are validated using receiver operator characteristic (ROC) curves. The overall workflow of this study is shown in Fig. 6.3.

6.3.2 Selection of Relevant Factors Using ACO

ACO is a metaheuristic optimization technique whose applications have developed significantly. The advantages of ACO include a probabilistic decision in terms of artificial pheromone trails and local heuristic information. These advantages enable the exploration of a larger number of solutions compared with that of greedy heuristics (Gottlieb et al. 2003). The overall workflow of the ACO-based landslide factor selection is presented in Fig. 6.4. First, ants were generated and then placed randomly on a graph, i.e., each ant starts with one random landslide factor. The number of ants placed on the graph may be set to be equal to the number of factors of the data; each ant initiates a path construction at a different factor. The ants traverse nodes probabilistically from their initial positions until a traversal stopping criterion is satisfied. The resulting subsets are gathered and evaluated. When an optimal subset has been found or when the algorithm has been executed a certain number of times, the process stops and the best encountered factor subset is outputted. If none of these conditions hold, then the pheromone is updated, a new set of ants are created, and the process is reiterated.

6.3.3 Susceptibility Models

In this study, susceptibility maps were produced using two data mining approaches: SVM and RF. These algorithms were used to determine whether the results were consistent or the performance of the susceptibility models with

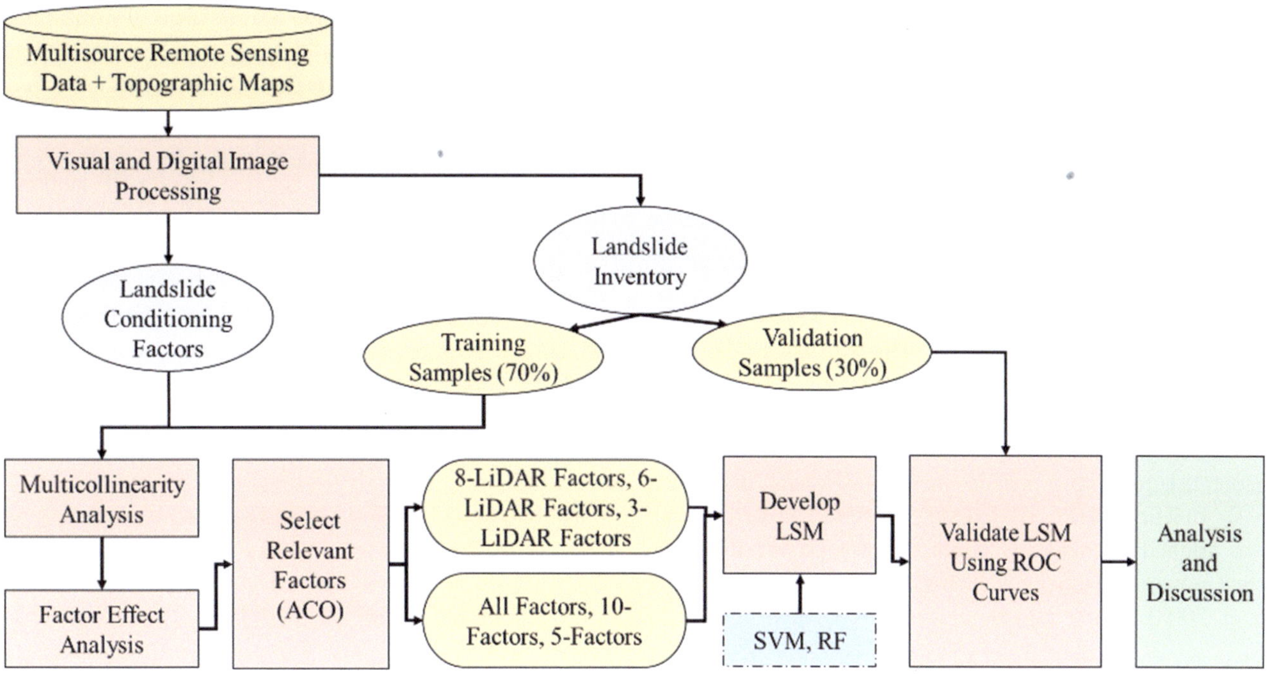

Fig. 6.3 Overall research activities used to optimize landslide conditioning factors, conduct factor effect analysis, and develop improved susceptibility models

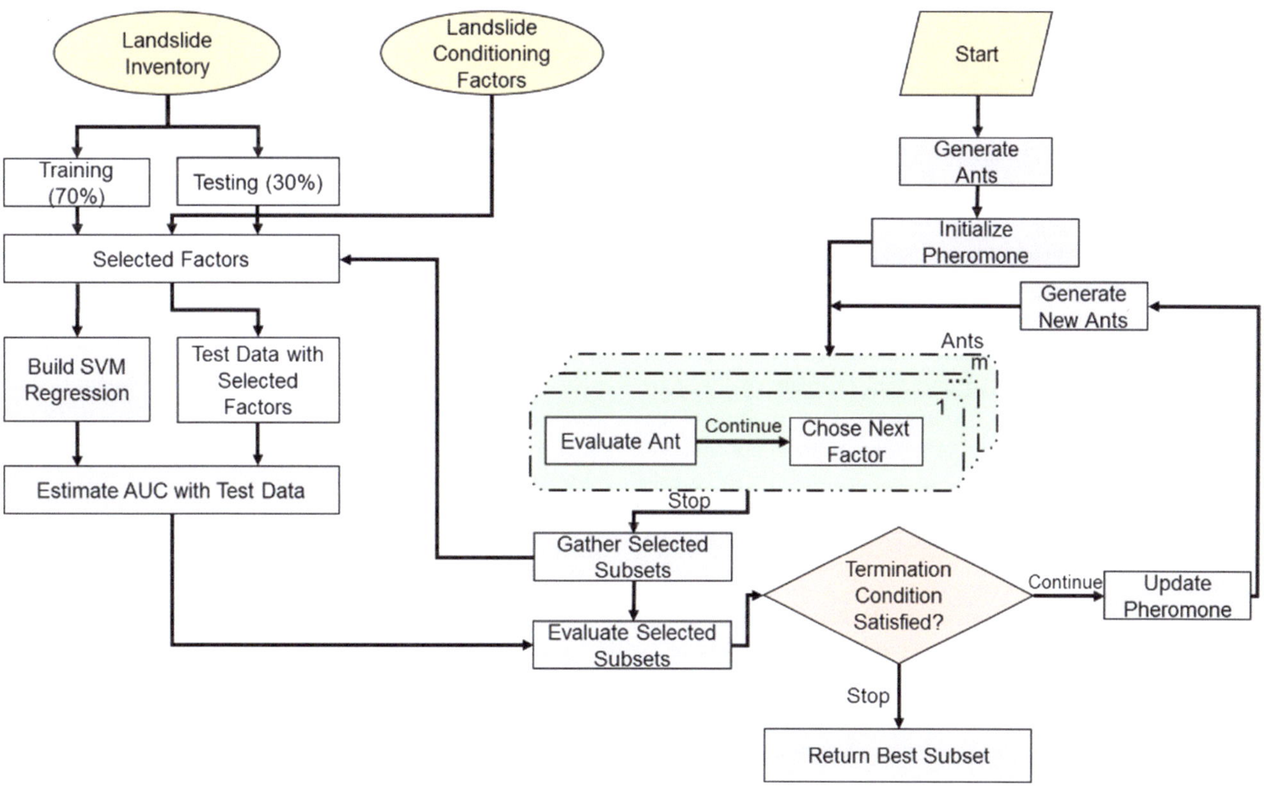

Fig. 6.4 Overall workflow of factor subset selection by ACO method

significant factors varied from one model to another. The subsequent sections briefly describe the basic concept of the algorithms.

6.3.3.1 SVM

SVM was originally developed by Vladimir and Vapnik (1995) as a more recent machine learning method than artificial neural networks. SVM uses the training data to convert the original input space implicitly into high-dimensional feature space based on kernel functions (Brenning 2005). Subsequently, the optimal hyperplane in the feature space is determined by maximizing the margins of class boundaries (Abe 2005). Therefore, SVM training is modeled by constraining the duality optimal solution. In general, kernel types include linear, polynomial, and radial basis function (RBF) or Gaussian kernels. The RBF kernel was applied in this study because it was proven to be the most powerful kernel for addressing nonlinear cases (Yao et al. 2008).

6.3.3.2 RF

RF is an ensemble machine learning method that generates numerous classification trees that are combined to compute a classification (Breiman et al. 1984; Breiman 2001). Hansen and Salamon (1990) indicated that a necessary and sufficient condition for an ensemble of classification trees to be more accurate than any of its individual member was that the members of the ensemble must perform better than random members and should be diverse. RF increases diversity among classification trees by resampling the data with replacement and randomly changing the predictive variable sets over different tree induction processes. The RF algorithm involves two main user-defined parameters that require appropriate specifications: the number of trees (k) and the number of predictive variables. A predictive variable may be numerical or categorical, and translation into the design variables is unnecessary. An unbiased estimate of the generalization error is obtained during the construction of an RF. The proportion of misclassifications (%) overall out-of-bag (OOB) elements is called the OOB error. The OOB error is an unbiased estimate of the generalization error. Breiman (2001) proved that RF produces a limiting value of the generalization error. As the number of trees increases, the generalization error always converges. The value of k must be set sufficiently high to allow this convergence. The RF algorithm estimates the importance of a predictive variable by examining the OOB errors. An increase in the OOB error is relative to predictive variable importance.

The advantages of RF include resistance to overtraining and the capability to grow a large number of RF trees without creating a risk of overfitting. RF algorithm data do not need to be rescaled, transformed, or modified; they are also resistant to outliers in predictors. In this study, the number of trees in an RF was fixed at 500 for RF modeling after a primary analysis, and m sampled at each node was set at 3 to analyze the combined contributions of subsets of features while maintaining fast convergence during iterations. No calibration set is required to regulate the parameters (Micheletti et al. 2014). The importance and standardized rank of each landslide variable were calculated. The ranks were then used to overlay landslide factors and generate the susceptibility maps.

6.4 Results

6.4.1 Multicollinearity Analysis

Multicollinearity analysis is an important step in LSM. The existence of a near-linear relationship among factors can create a division-by-zero problem during regression calculations. This problem can cause the calculations to be aborted and the relationship to be inexact; division by an extremely small quantity still distorts the results. Therefore, analyzing landslide conditioning factors before LSM is important. In multicollinearity analysis, collinear (dependent) factors are identified by examining a correlation matrix constructed by calculating R2. Various quantitative methods for detecting multicollinearities, such as pairwise scatter plots, estimation of the variance inflation factor (VIF), and investigation of eigenvalues in a correlation matrix, are available. In this study, multicollinearity was detected by calculating the VIF values of each landslide conditioning factor. In addition, communalities similar to R2 were calculated for each factor (Costello 2009). Communality shows how well a variable is predicted by the retained factors. Table 6.1 presents the estimated communalities and VIF values for each landslide conditioning factor. The second column of Table 6.1 indicates that some factors, such as land use, distance to road, distance to river, slope, STI, TWI, and TRI, exhibit strong linear relationships with other factors. These factors may negatively affect the regression analysis. However, VIF values are quantitative measures that are typically used to conclude whether a factor has a problem. In some studies, a VIF greater than two was considered problematic, whereas in other studies, a VIF greater than 10 was considered problematic (Garrosa et al. 2010). To solve the

Table 6.1 Calculated communalities and VIF values for each landslide conditioning factor

Factors	Communality	VIF
Aspect	0.053	1.14
Land use	0.566	3.15
Vegetation density	0.044	2.9
NDVI	0.069	2.93
Distance to lineament	0.001	1.25
Distance to road	0.576	3.74
Distance to river	0.626	4.15
Altitude	0.35	2.47
Slope	0.608	9.02
Profile curvature	0.015	1.11
Plan curvature	0.1	1.25
SPI	0.311	1.57
STI	0.684	2.77
TWI	0.638	2.46
TRI	0.589	39.79

multicollinearity problem, factors can be excluded from further analysis or other sampling techniques should be examined. In this study, factors with VIF values greater than 10 (e.g., TRI) were removed from further analysis.

6.4.2 Factor Analysis

The previous section shows that multicollinearity analysis identifies landslide factors that exhibit the problem of having a strong correlation with other remaining factors. To determine underlying factors that are responsible for correlations in data, factor analysis was conducted in the current study. Factor analysis is an investigative method that is applied to a set of observed variables; it aims to identify underlying factors from which observed variables are generated (Roscoe et al. 1982). Factor analysis using the principal component extraction method was applied in this study to determine the factors that underlay the data. Figure 6.5 shows the graph of the underlying factors versus the eigenvalues calculated based on the correlation matrix. The graph provides information about the factors. It was used to determine how well the selected number of components fit the data. The graph indicated that the first eight factors accounted for the majority of the total variability in the data (given by the eigenvalues). The remaining factors accounted for a minimum amount of the variability (nearly zero) and were likely insignificant.

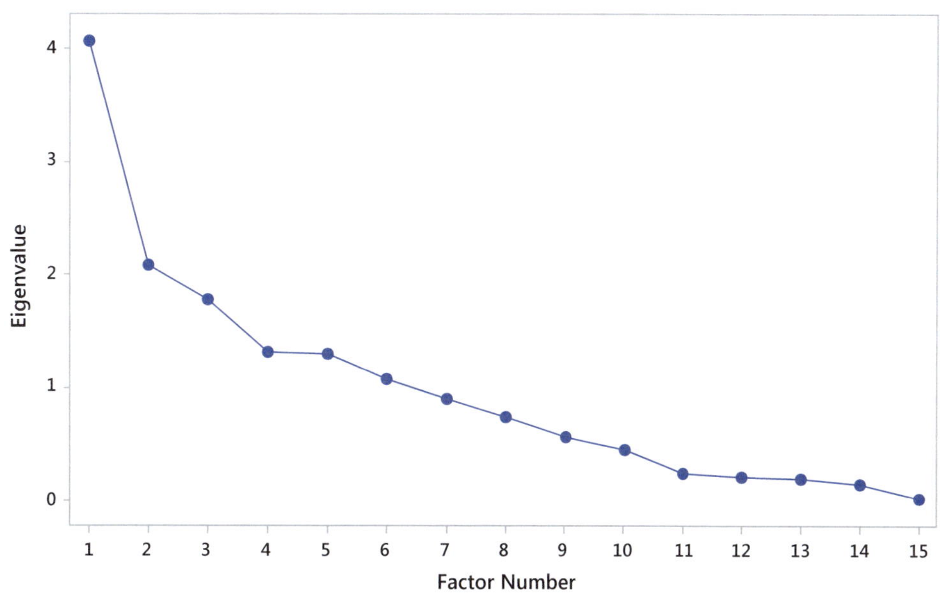

Fig. 6.5 Graph of factors versus the corresponding eigenvalues calculated based on the correlation matrix

Table 6.2 presents the sorted unrotated factor loadings and communalities resulting from the factor analysis. Communalities describe the proportion of variability of each variable that is explained by the factors. When a communality is closer to 1, the variable is better explained by the factors. Variance demonstrates the variability in the data explained by each factor (i.e., the variance is equal to the eigenvalue). Meanwhile, %Var shows the proportion of variability in the data explained by each factor.

In the factor analysis, 8 factors were extracted from the 15 variables. All the variables were well-represented by the 8 selected factors given that the corresponding communalities were generally high. For example, 0.974 or 97.4% of the variability in aspect and profile curvature was explained by the 8 factors. In addition, the 8 selected factors explained most of the total data variation (0.881 or 88.1%, Table 6.2). Furthermore, Table 6.2 shows the variable loading on each factor. For example, distance to river (−0.823), distance to road (−0.796), land use (0.795), slope (0.779), TRI (0.77), altitude (−0.656), TWI (−0.322), and NDVI (0.324) have large absolute loadings on factor 1. This result indicates that this subset of variables can be reduced into fewer variables. By contrast, STI (0.84), TWI (0.782), SPI (0.718), and plan curvature (−0.402) have large absolute loadings on factor 2. This finding suggests that these factors can be combined and reduced into fewer theoretical factors. In addition, land use, NDVI, and vegetation density have large absolute loadings on factor 3, thereby suggesting that a theoretical factor can combine these three interrelated factors. Furthermore, several factors, including slope, aspect, and profile curvature, have large loadings on factor 4. LiDAR-derived factors and distance to the road have large absolute loadings on factor 5. SPI, distance to lineament, and both curvature layers have a few underlying factors. Aspect and profile curvature have large positive loadings on factor 7. Plan and profile curvatures have large absolute loadings on factors 5, 6, and 8. This finding indicates that these two variables can be combined into one variable. This resulting variable can be the total curvature, which has not been used in the current study.

6.4.3 ACO-Based Factor Selection

Table 6.3 shows the landslide conditioning factors and their corresponding codes used in the subsequent tables. This section describes the six experiments conducted in this study to analyze the effects of landslide conditioning factors on LSM.

The six experiments were classified into two main groups. The first group included all the 14 factors (Table 6.4), whereas the second group contained only the LiDAR-derived factors. In the first group, the three experiments included 5 factors, 10 factors, and the produced susceptibility models that used all the 14 factors. In the second group, the three experiments involved 3 LiDAR factors, 6 LiDAR factors, and 8 LiDAR factors, which were the total number of LiDAR factors derived from the DEM. These subsets were evaluated using the SVM and RF models. The selected factors and the prediction accuracy rate of both models are presented in Table 6.4. The results showed that using all the conditioning factors did not necessarily guarantee the highest accuracy. In the case of the first group, the highest accuracy was achieved with either 10 or 14 factors when the RF model was used. In the case of the SVM model, using all the 14 factors produced the highest accuracy. In the three experiments in the first group, the RF model performed better than the SVM model. However, no significant difference was found between using all the 14 factors and using only 10 factors in the susceptibility analysis for both the SVM and RF models. In the experiments in the second group, accuracy decreased by approximately 0.16 on average. This result indicated that some factors, such as land use, vegetation density, and NDVI, were important for predicting landslides in the study area. The highest accuracy was achieved using the RF model with 8 LiDAR factors. The RF model with only 3 factors selected via ACO performed better than the SVM model with 8 LiDAR factors. In the SVM model, the findings indicated that using only 3 LiDAR factors yielded better results than using 6 factors mainly because the selected individual factors in the subset with 3 factors were more important than those selected in the subset with 6 factors. Consequently, including additional factors to LiDAR-derived factors was necessary for accurate LSM in the study area. The RF model performed better than the SVM model even with fewer factors. The second subset of the first group, which had 10 factors that included LiDAR-derived and non-LiDAR-derived factors, was recommended to produce landslide susceptibility maps in the study area for land use planning.

6.4.4 Landslide Susceptibility Models

In the current study, four landslide susceptibility maps were produced for the study area (Fig. 6.6). These maps were generated using the SVM and RF models with the best subsets of the two groups as described in the previous section. The first examination of the maps showed no spatial agreement among the susceptibility classes of the four models. For example, the maps produced using a combination of LiDAR and non-LiDAR factors were different from those produced using only LiDAR factors. In addition, the two maps produced using the SVM and RF models with the significant factors selected among the 14 factors were different. The apparent difference was mainly observed in the middle part of the study area. The map produced using the

Table 6.2 Sorted unrotated variable loadings on extracted factors resulted from factor effect analysis

Variable	Factor1	Factor2	Factor3	Factor4	Factor5	Factor6	Factor7	Factor8	Communality
Distance to river	−0.823	−0.025	−0.194	−0.04	−0.345	0.131	0.12	−0.021	0.868
Distance to Road	−0.796	−0.04	−0.289	−0.176	−0.252	−0.019	0.083	0.166	0.848
Land use	0.795	−0.003	0.304	0.182	0.105	0.074	−0.025	−0.206	0.817
Slope	0.779	−0.14	−0.175	−0.303	−0.413	0.21	0.016	0.081	0.97
TRI	0.77	−0.138	−0.174	−0.295	−0.434	0.216	0.025	0.084	0.971
Altitude	−0.656	−0.045	−0.125	0.188	−0.427	0.29	−0.042	−0.257	0.818
STI	0.236	0.84	0.015	−0.196	−0.112	0.092	0.134	0.189	0.875
TWI	−0.322	0.782	0.048	0.033	0.121	−0.175	0.13	0.089	0.789
SPI	0.104	0.718	0.051	−0.022	−0.009	0.403	−0.114	0.037	0.707
Vegetation Density	0.265	0.102	−0.851	0.278	0.14	−0.061	−0.002	−0.005	0.905
NDVI	0.324	0.109	−0.835	0.216	0.173	−0.043	−0.033	−0.07	0.899
Distance to Lineament	−0.05	−0.005	0.164	0.691	−0.002	0.586	−0.124	0.084	0.872
Aspect	0.274	−0.119	0.112	0.404	−0.153	−0.073	0.817	0.11	0.974
Plan Curvature	−0.213	−0.402	−0.09	−0.182	0.503	0.374	0.056	0.534	0.93
Profile Curvature	−0.157	0.003	−0.112	−0.442	0.448	0.424	0.369	−0.474	0.974
Variance	4.0649	2.0748	1.7765	1.3129	1.2884	1.0738	0.8964	0.7292	13.217
% Var	0.271	0.138	0.118	0.088	0.086	0.072	0.06	0.049	0.881

Table 6.3 Assigned code of each landslide conditioning factor

Factor	Code				
Aspect	1	Distance to Road	6	Plan curvature	11
Land use	2	Distance to river	7	SPI	12
Vegetation density	3	Altitude	8	STI	13
NDVI	4	Slope	9	TWI	14
Distance to lineament	5	Profile curvature	10		

RF model exhibited nearly moderate and very high susceptibility in the middle part of the study area, whereas the map produced using the SVM model exhibited high and very high susceptibility in the same area. The southeastern part of the study area had very low and low susceptibility based on the RF model, whereas its susceptibility was moderate and high based on the SVM model. Consequently, no exact spatial agreement was found on the susceptibility classes in most parts of the study area based on the two models. The susceptibility maps produced using only LiDAR-derived factors are different from those produced using the significant factors selected among the 14 factors. However, spatial agreements were found among the susceptible zones in the northern, middle, and southern parts of the study area when the RF- and SVM-generated maps were compared.

6.4.5 Validation

The ROC curve is a graph with a false positive rate plotted on the x-axis and a true positive rate plotted on the y-axis. It uses a visual comparison of the performance of the methods. The area under the ROC curve (AUC) shows the global accuracy statistics for each model. If the AUC (which varies from 0.5 to 1) increases, then the prediction performance of the method increases (Erener and Düzgün 2010). Figure 6.7 shows the plotted ROC curves and the estimated AUC values for the four susceptibility maps described in previous section. On the one hand, the highest accuracy was achieved using the RF model with 10 factors selected among the 14 initial factors. On the other hand, the lowest accuracy was achieved using the SVM model with only LiDAR-derived factors.

6.5 Discussion and Conclusion

In this study, we optimized landslide conditioning factors and conducted a factor effect analysis to provide useful information about landslide susceptibility analysis in the Cameron Highlands, Malaysia. This study first identified problematic factors by calculating VIF values during multicollinearity analysis. As mentioned earlier, problematic factors can disrupt or distort the regression results. Therefore, removing these factors is an essential step in LSM. The communality of each variable was calculated from the correlation matrix. The communalities indicated that land use (0.566), distance to road (0.576), distance to river (0.626), altitude (0.35), slope (0.608), SPI (0.311), STI (0.684), TWI (0.638), and TRI (0.589) demonstrated relatively strong correlations with other factors. However, only TRI was problematic (given by the VIF) based on the selected threshold (VIF > 10 was considered problematic), and thus, it was excluded from LSM. In addition, slope had a relatively high VIF of approximately 10. However, slope is the most important factor for LSM, and thus, it has been retained. In future studies, this problem could be solved by using different sampling procedures, such as landslide polygons instead of the centroid of landslides, which was adopted in the current study. The use of different sampling procedures or the removal of inaccurate landslide inventories may solve the problem of collinear factors.

Factor analysis was conducted to identify underlying factors. The eigenvalues showed that the first 8 factors accounted for the majority of the total variability in the data. The remaining factors accounted for a minimal amount of the variability (approximately 0) and were likely insignificant. Therefore, 8 factors were extracted from the 15 landslide conditioning factors. The corresponding communalities were generally high, and thus, the landslide-related variables were well-represented by the 8 factors. The highest percentage of over 97% of the variability in aspect and profile curvature was explained through these 8 extracted factors. In general, the factor effect analysis suggested reducing the number of landslide conditioning factors by combining some of the factors into fewer theoretical factors. For example, plan and profile curvature were highly recommended to be combined (Table 6.2). To achieve such combination, a comprehensive analysis of landslide conditioning factors is required. In addition, distance to river (−0.823), distance to road (−0.796), land use (0.795), slope (0.779), TRI (0.77), altitude (−0.656), TWI (−0.322), and NDVI (0.324) were found to have large absolute loadings on factor 1. This result indicated that this subset of variables could be reduced into fewer theoretical factors.

Thereafter, ACO was used to select significant variable subsets from the available variables. The SVM and RF classification models were adopted to evaluate the selected

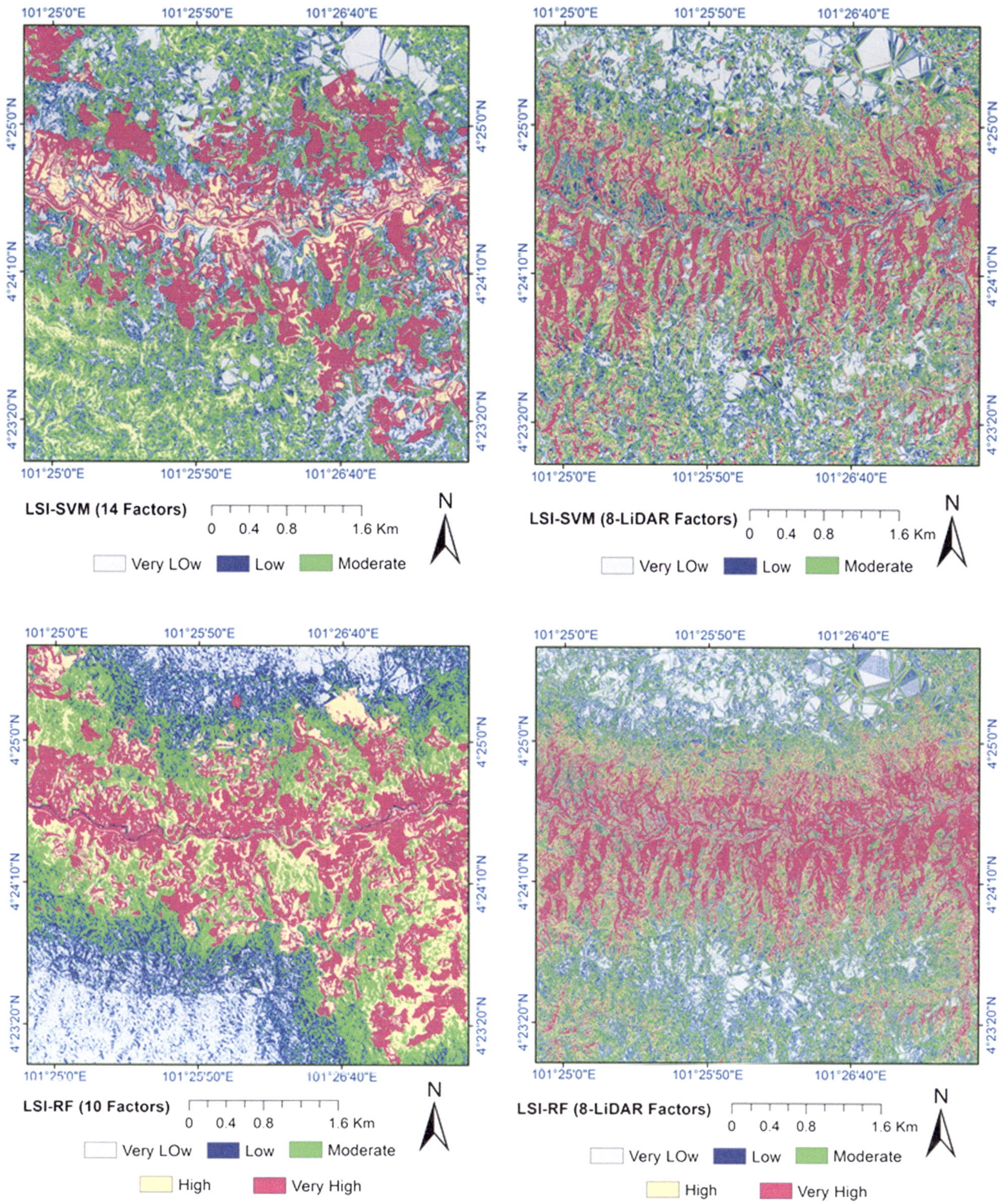

Fig. 6.6 Landslide susceptibility maps

subsets. A total of six experiments were conducted in the study to analyze the effects of landslide conditioning factors on LSM. These experiments were as follows: 5 factors, 10 factors, all the 14 factors, 3 LiDAR factors, 6 LiDAR factors, and 8 LiDAR factors. The evaluation of the six experiments showed that the RF model with 10 landslide factors selected from among the 14 factors produced the best result (AUC = 0.95). In addition, a significant decrease in

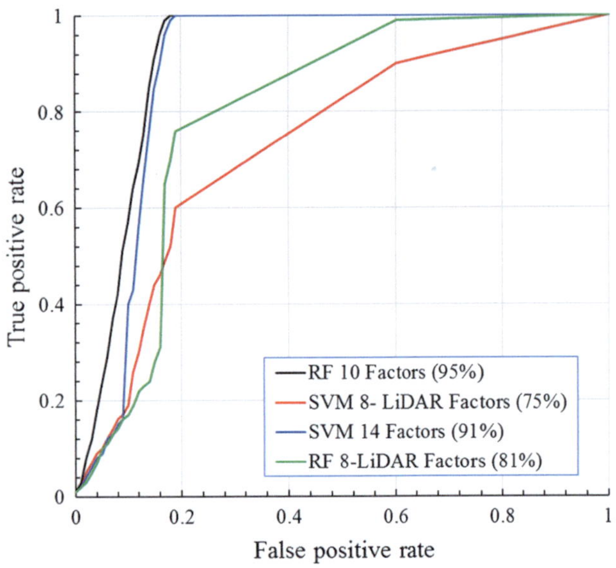

Fig. 6.7 ROC curves of the produced susceptibility map

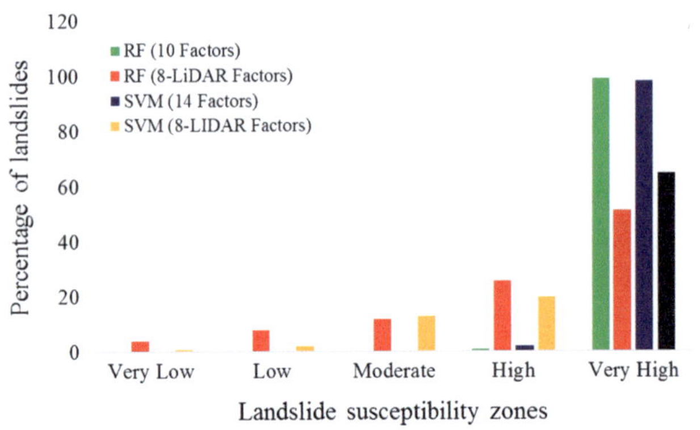

Fig. 6.8 Percentages of landslide inventories in each susceptibility zone

accuracy was observed when only the LiDAR-derived factors were used. Factors, such as land use, vegetation density, and NDVI were found to be important for predicting landslides in the study area.

In this study, 4 landslide susceptibility maps were produced for the study area. The susceptibility maps produced using only LiDAR-derived factors were different from those produced using significant factors selected from all the 14 factors. This study showed that spatial agreement on susceptibility zones decreased by adding non-LiDAR factors in the analysis. A visual interpretation of the susceptibility maps indicated spatial agreements on susceptible zones in the northern, middle, and southern parts of the study area when LiDAR-based factors were used. Therefore, statistical validation methods, such as ROC curves and spatial agreement analysis should be considered to decide whether a map can be used for land use planning. In addition, Fig. 6.8 shows the percentages of landslides in each susceptibility class. The graph shows that most of the landslides are located in high and very high susceptibility zones.

In general, the RF model performed better than the SVM algorithm regardless of the combination of factors used for modeling. Although the parameters of the SVM algorithm were fine-tuned in the current study, concluding that RF should be used for LSM in the Cameron Highlands would be difficult. This study suggests that significant attention should be directed toward analyzing input landslide factors. Moreover, problematic factors and observations should be removed. Several factors are typically derived from a LiDAR DEM, and thus, collinearity can be found among these factors. Therefore, additional factors, including non-LiDAR factors, should always be used in LSM. Sometimes, factors such as distance to the road have a strong correlation with land use. The careful design of classification schemes when producing land use maps is recommended.

Table 6.4 Results of factor subset selection of ACO-based experiments

Dataset	Experiment	Total number of factors	Selected factors	AUC	
				SVM	RF
All data	5-Factors	14	[7 8 6 5 9]	0.83	0.89
	10-Factors	14	[2 10 4 8 3 1 12 6 14 7]	0.89	0.95
	14-Factors	14	[9 10 4 3 5 12 11 1 7 8 2 6 14 13]	0.91	0.95
Only LiDAR	3-Factors	8	[3 4 2]	0.72	0.77
	6-Factors	8	[6 1 5 4 8 7]	0.69	0.70
	8-Factors	8	[4 5 2 8 7 6 3 1]	0.75	0.81

For example, roads can be classified into different classes based on road type or width. Such classification can reduce the correlation among landslide factors, and thus improve LSM.

This study examined the optimization of landslide conditioning factors and conducted a factor effect analysis to improve understanding of susceptibility models. However, several issues should be considered in future studies. First, the effects of landslide sampling procedures and the spatial resolution of DEMs should be investigated in detail. Attention should also be directed toward developing new theoretical factors in future studies. LiDAR-derived factors can be reduced into fewer factors, which can decrease collinearity among factors. Quantitative accuracy indicators, such as AUC, may be insufficient when deciding which algorithm or LSM approach should be used. Therefore, new indicators that consider spatial agreements on susceptible classes should be developed. In summary, comprehensive analysis on landslide conditioning factors should be conducted to improve understanding of LSM in the future.

References

Abe, S. (2005). *Support vector machines for pattern classification* (Vol. 2). London: Springer.

Breiman, L. (2001). Random forests. *Machine Learning, 45*(1), 5–32.

Breiman, L., Friedman, J., Stone, C. J., & Olshen, R. A. (1984). *Classification and regression trees*. Boca Raton: CRC press.

Brenning, A. (2005). Spatial prediction models for landslide hazards: review, comparison and evaluation. *Natural Hazards and Earth System Science, 5*(6), 853–862.

Bui, D. T., Tuan, T. A., Klempe, H., Pradhan, B., & Revhaug, I. (2015). Spatial prediction models for shallow landslide hazards: A comparative assessment of the efficacy of support vector machines, artificial neural networks, kernel logistic regression, and logistic model tree. *Landslides*, 1–18.

Costello, A. B. (2009). Getting the most from your analysis. *Pan, 12*(2), 131–146.

Erener, A., & Düzgün, H. S. B. (2010). Improvement of statistical landslide susceptibility mapping by using spatial and global regression methods in the case of More and Romsdal (Norway). *Landslides, 7*(1), 55–68.

Gaprindashvili, G., & Van Westen, C. J. (2016). Generation of a national landslide hazard and risk map for the country of Georgia. *Natural Hazards, 80*(1), 69–101.

Garrosa, E., Rainho, C., Moreno-Jimenez, B., & Monteiro, M. J. (2010). The relationship be-tween job stressors, hardy personality, coping resources and burnout in a sample of nurs-es: A correlational study at two time points. *International Journal of Nursing Studies, 47*(2), 205–215.

Gottlieb, J., Puchta, M., & Solnon, C. (2003). A study of greedy, local search, and ant colony optimization approaches for car sequencing problems. In *Applications of evolutionary computing* (pp. 246–257). Berlin Heidelberg: Springer.

Guo, D., & Hamada, M. (2013). Qualitative and quantitative analysis on landslide influential factors during Wenchuan earthquake: A case study in Wenchuan County. *Engineering Geology, 152*(1), 202–209.

Hansen, L. K., & Salamon, P. (1990). Neural network ensembles. *IEEE Transactions on Pattern Analysis and Machine Intelligence, 12*(10), 993–1001.

Hussin, H. Y., Zumpano, V., Reichenbach, P., Sterlacchini, S., Micu, M., van Westen, C., & Bălteanu, D. (2016). Different landslide sampling strategies in a grid-based bi-variate statistical susceptibility model. *Geomorphology, 253*, 508–523.

Jebur, M. N., Pradhan, B., & Tehrany, M. S. (2014). Optimization of landslide conditioning factors using very high-resolution airborne laser scanning (LiDAR) data at catchment scale. *Remote Sensing of Environment, 152*, 150–165.

Kasai, M., Ikeda, M., Asahina, T., & Fujisawa, K. (2009). LiDAR-derived DEM evaluation of deep-seated landslides in a steep and rocky region of Japan. *Geomorphology, 113*(1), 57–69.

Kayastha, P. (2015). Landslide susceptibility mapping and factor effect analysis using frequency ratio in a catchment scale: A case study from Garuwa sub-basin, East Nepal. *Arabian Journal of Geosciences, 8*(10), 8601–8613.

Khan, Y. A. (2010). Monitoring of hill-slope movement due to rainfall at Gunung Pass of Cameron Highland district of Peninsular Malaysia. *International Journal of Earth Sciences and Engineering, 3*, 06–12.

Mahalingam, R., & Olsen, M. J. (2015). Evaluation of the influence of source and spatial reso-lution of DEMs on derivative products used in landslide mapping. *Geomatics, Natural Hazards and Risk*, 1–21.

Mahalingam, R., Olsen, M. J., & O'Banion, M. S. (2016). Evaluation of landslide susceptibility mapping techniques using lidar-derived conditioning factors (Oregon case study). *Geomatics, Natural Hazards and Risk*, 1–24.

Meten, M., PrakashBhandary, N., & Yatabe, R. (2015). Effect of landslide factor combinations on the prediction accuracy of landslide susceptibility maps in the Blue Nile Gorge of Central Ethiopia. *Geoenvironmental Disasters, 2*(1), 1–17.

Micheletti, N., Foresti, L., Robert, S., Leuenberger, M., Pedrazzini, A., Jaboyedoff, M., & Kanevski, M. (2014). Machine learning feature selection methods for landslide susceptibility mapping. *Mathematical Geosciences, 46*(1), 33–57.

Moosavi, V., & Niazi, Y. (2015). Development of hybrid wavelet packet-statistical models (WP-SM) for landslide susceptibility mapping. *Landslides*, 1–18.

Pereira, S. D. S., Zêzere, J. L. G. M. D., & Bateira, C. (2013). Technical note: Assessing predictive capacity and conditional independence of landslide predisposing factors for shallow landslide susceptibility models. *Natural Hazards and Earth System Sciences, n. 12* (2012), 979–988.

Pradhan, B. (2013). A comparative study on the predictive ability of the decision tree, support vector machine and neuro-fuzzy models in landslide susceptibility mapping using GIS. *Computers and Geosciences, 51*, 350–365.

Pradhan, B., & Lee, S. (2009). Landslide risk analysis using artificial neural network model focusing on different training sites. *International Journal of Physical Sciences, 3*(11), 1–15.

Pradhan, B., & Lee, S. (2010). Regional landslide susceptibility analysis using back-propagation neural network model at Cameron Highland, Malaysia. *Landslides, 7*(1), 13–30.

Qin, C. Z., Bao, L. L., Zhu, A. X., Wang, R. X., & Hu, X. M. (2013). Uncertainty due to DEM error in landslide susceptibility mapping. *International Journal of Geographical Information Science, 27*(7), 1364–1380.

Roscoe, B. A., Hopke, P. K., Dattner, S. L., & Jenks, J. M. (1982). The use of principal component factor analysis to interpret particulate compositional data sets. *Journal of the Air Pollution Control Association, 32*(6), 637–642.

Vasu, N. N., & Lee, S. R. (2016). A hybrid feature selection algorithm integrating an extreme learning machine for landslide susceptibility modeling of Mt. Woomyeon, South Korea. *Geomorphology, 263*, 50–70.

Vladimir, V. N., & Vapnik, V. (1995). The nature of statistical learning theory.

Yao, X., Tham, L. G., & Dai, F. C. (2008). Landslide susceptibility mapping based on support vector machine: a case study on natural slopes of Hong Kong, China. *Geomorphology, 101*(4), 572–582.

Effects of the Spatial Resolution of Digital Elevation Models and Their Products on Landslide Susceptibility Mapping

7

Biswajeet Pradhan and Maher Ibrahim Sameen

7.1 Introduction

Landslides are among the destructive natural disasters that cause significant damage to human life and properties worldwide. Numerous researchers have attempted to provide an understanding of landslide causes and related problems. An important and simple analysis method that has been used in landslide studies is landslide susceptibility mapping /modeling (LSM). LSM is fundamental to hazard and risk assessments, and it is widely used by governments for planning land use and strategic projects. LSM requires landslide conditioning factors and landslide inventories, which can be acquired using remote sensing and field surveying techniques. The output of LSM is a map that shows the degree of landslide susceptibility of an area.

Digital elevation models (DEMs) are the main data commonly used in LSM. Various types of landslide conditioning factors can be derived from DEMs, including slope, aspect, curvature, topographic wetness index (TWI), topographic roughness index (TRI), stream power index (SPI), and sediment transport index (STI). The spatial resolution of a DEM and its derivatives directly affect the accuracy and quality of landslide susceptibility maps. Few studies have investigated the effects of the spatial resolution and source of a DEM on LSM (Lee et al. 2004; Chang et al. 2016).

The recent literature indicates that a number of efforts have been made to develop an understanding of landslide conditioning factors and their effects on LSM (Althuwaynee et al. 2016). Various landslide conditioning factors have been used in LSM, including those derived from DEMs and at different spatial resolutions depending on the sources of DEMs. The most common DEM-derived factors used are slope, aspect, curvature, TWI, TRI, STI, and SPI (Chalkias et al. 2014; Huang et al. 2015; Pradhan 2010; Pradhan and Kim 2016; Pradhan et al. 2016). However, many researchers have also used other conditioning factors, such as stream sinuosity, local relief, and slope angle (Raman and Punia 2012; Quinn 2014). Various factor-selection techniques to optimize these factors have been explored in many studies. Park (2015) conducted a factor contribution analysis for LSM using maximum entropy modeling. The analysis indicated that the distances from lineaments and slope layers were the most influential factors. Guri and Patel (2015) analyzed 15 landslide conditioning factors using 2 different landslide inventory datasets. The results showed that 76.5% of all the landslides were predicted in 24% of the total landslide-susceptible area using the best combination of 8 parameters. Meinhardt et al. (2015) optimized landslide conditioning factors using a new method called "omit error". This statistical method involves omitting a parameter to obtain the weightings, which describe how strong each parameter improves or reduces the objective function. After this optimization, the 9 remaining input parameters were weighted using the omit error method to produce the best susceptibility map with a success rate of 92.9% and a prediction rate of 92.3%, which were 4.4 and 4.2% higher than the success rate and prediction rate obtained using the basic statistical index method with 13 input parameters, respectively.

By contrast, few studies have explored the effects of the spatial resolution of landslide conditioning factors on LSM. Oh et al. (2012) extracted landslide conditioning factors from a DEM based on the Advanced Spaceborne Thermal Emission and Reflection Radiometer (ASTER) and examined their applications in LSM. The validation results for the landslide susceptibility map produced using these factors showed 84.78% frequency ratio and 84.20% logistic regression (LR) prediction accuracy. Mahalingam et al. (2016) analyzed the influence of the spatial resolution and source of a DEM on LSM using ASTER, National Elevation Dataset (NED), and LiDAR data. The results showed that the 10 m spatial-resolution DEM derived from LiDAR data yielded higher predictive accuracy in several modeling

B. Pradhan (✉) · M.I. Sameen
Department of Civil Engineering, University Putra Malaysia, Serdang, Malaysia
e-mail: biswajeet24@gmail.com

B. Pradhan (ed.), *Laser Scanning Applications in Landslide Assessment*,
DOI 10.1007/978-3-319-55342-9_7

approaches for producing landslide susceptibility maps. However, at a resolution of 10 m, the output maps based on NED and ASTER had higher misclassification rates than the LiDAR-based output. Moreover, the 30 m LiDAR output exhibited improved results over the 10 m NED and 10 m ASTER outputs, thereby indicating that a finer resolution does not necessarily result in a higher predictive accuracy in LSM. Overall, the study suggested that DEM source is an important consideration, and it can significantly influence the accuracy of a landslide susceptibility analysis. Qin et al. (2013) simulated DEM error fields at a 25 m resolution with different magnitudes and spatial autocorrelation levels. The study indicated that the overall uncertainty in LSM could be sensitive to the spatial resolution and quality of DEMs.

The research on the effects of the spatial resolution of DEMs on the resultant landslide susceptibility maps is limited, and thus, this paper presents a detailed analysis and discussion of this topic. First, a geodatabase consisting of 13 landslide conditioning factors was constructed, and a 0.5 m LiDAR-based DEM was resampled to various spatial resolutions (1, 2, 3, 5, 10, 20, and 30 m). In this study, 30 m LiDAR-based and ASTER-based DEMs were compared to analyze the effects of resampling and DEM source. In addition, the effects of DEM spatial resolution on multicollinearity and factor optimization were also investigated and discussed. LR was used as the LSM technique. Evaluations were conducted using several accuracy metrics, such as the area under the curve of a receiver operating characteristic (AUC of ROC), overall accuracy, the kappa coefficient, and spatial agreement.

7.2 Study Area

The study area is the Cameron Highlands, which is located in the west of Malaysia. It is a tropical rain forest situated at the northwestern tip of Pahang. The Cameron Highlands is approximately 200 km from Kuala Lumpur, the capital of Malaysia. Many landslides were reported and documented in the study area in previous studies, which indicated that most of the landslides caused significant damage to properties. The lithology in the Cameron Highlands mainly consists of Quaternary and Devonian granite and schist (Pradhan and Lee 2010). The granite in the Cameron Highlands is classified as megacrystic biotite granite (Pradhan and Lee 2010). A subset, which occupies a surface area of ~ 25 km^2, was selected for the current study because landslides frequently occur in this area (Figs. 7.1 and 7.2). The lowest and highest elevations are 1153 and 1765 m, respectively.

7.3 Data

Data from various remote sensing and field sources were acquired and used. Landslide inventories were collected from aerial photographs, satellite images, and high-resolution orthophotos. Landslide conditioning factors were derived from LiDAR point clouds, an ASTER DEM, and geospatial databases of the Cameron Highlands.

7.3.1 Landslide Inventory

Adequate landslide inventories are required for LSM. Thus, remote sensing data and geographic information system (GIS) functions were used to collect and prepare landslide inventories. The remote sensing data, including archived 1: 10,000–1: 50,000 aerial photographs, SPOT 5 panchromatic satellite images, and high-resolution orthophotos, were used to visually recognize landslide scarps in the study area. In addition, historical landslide reports, newspaper records, and archived data were collected during the study period. Field observations were used to confirm the landslide scarps. In the aerial photographs and SPOT 5 satellite images, historical landslides could be observed as breaks in the forest canopy, bare soil, or geomorphological features, such as head and side scarps, flow tracks, and soil and debris deposits below a scarp. A geodatabase was constructed to assess the surface area and number of landslides in the study area. A total of 192 landslides were mapped in an area of 25 km^2.

7.3.2 Landslide Conditioning Factors

A total of 13 landslide conditioning factors, namely slope, aspect, altitude, TWI, TRI, normalized difference vegetation index (NDVI), vegetation density, land use, distance to road, distance to river, distance to fault, plan curvature, and profile curvature, were analyzed.

A DEM for the study area was generated from LiDAR point clouds at a scale of 0.5 m and then resampled into other spatial resolutions for analysis. In addition, another DEM was generated from the ASTER Global DEM (http://gdem.ersdac.jspacesystems.or.jp/) at a scale of 30 m for comparison. The landslide conditioning factors were extracted using these DEMs and ArcGIS 10.3.

The altitude in the study area ranges from 1152 to 1765 m (Fig. 7.3a). The slope of the study area extends from 0° to 82.41° (Fig. 7.3b). Slope angle is an important factor in landslide development because of its relationship to the

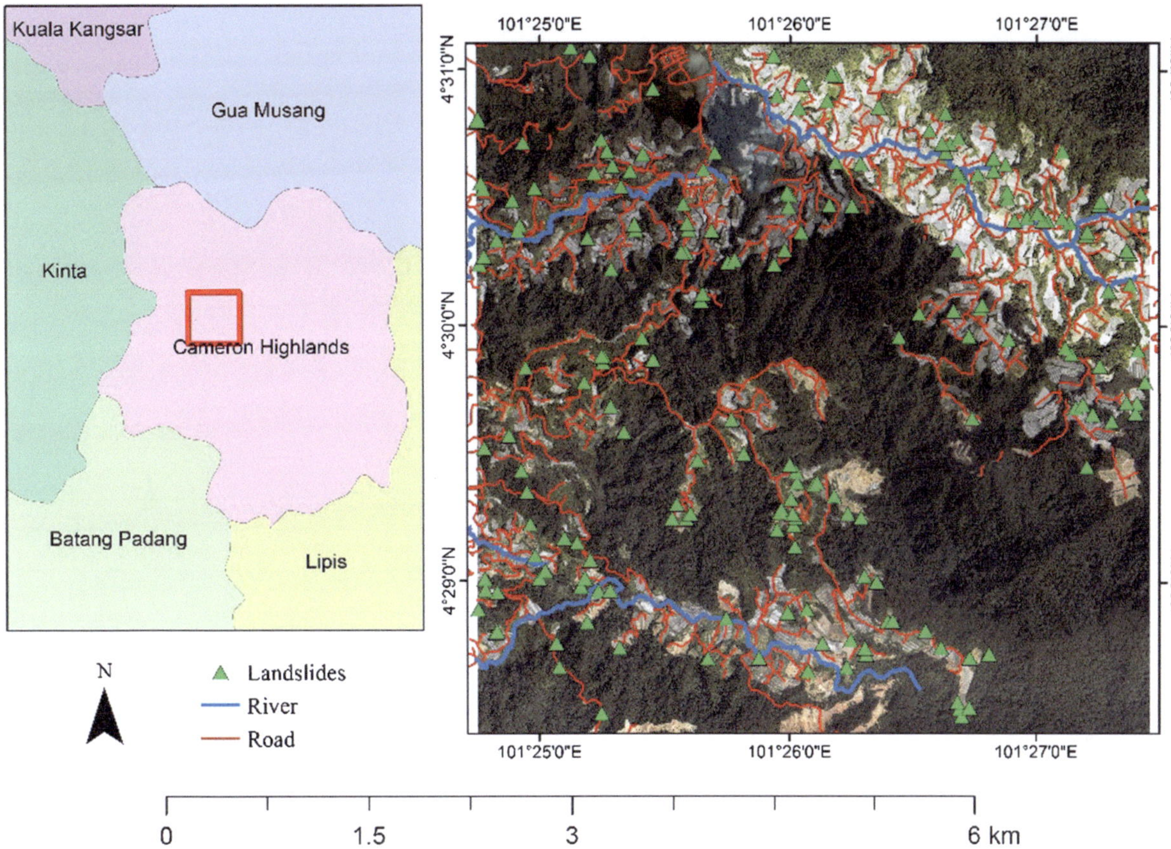

Fig. 7.1 Study area map showing the landslide locations

driving force of gravitation. In general, and under certain favorable conditions, the likelihood of a landslide increases as slope angle increases. The constructed aspect map (Fig. 7.3c) had nine classes: flat (−1°), north (337.5°–360°, 0°–22.5°), northeast (22.5°–67.5°), east (67.5°–112.5°), southeast (112.5°–157.5°), south (157.5°–202.5°), southwest (202.5°–247.5°), west (247.5°–292.5°), and northwest (292.5°–337.5°). Aspect can affect the physical and biotic features of a slope and significantly influence its microclimate. In some localities, the variation in soil patterns is related to differences in aspect. Therefore, slope aspect indirectly affects landslide occurrence, and it has been applied in many studies on landslide susceptibility assessment.

TWI is defined as a steady state wetness index. It is commonly used to quantify topographic control on hydrological processes. This index is a function of both slope and flow direction. The formula for TWI is as follows:

$$\mathrm{TWI} = \ln\left(\frac{A_s}{\beta}\right), \tag{7.1}$$

where A_s is the specific catchment area (m^2/m) and β is the slope angle in degrees. The TWI in this study ranges from

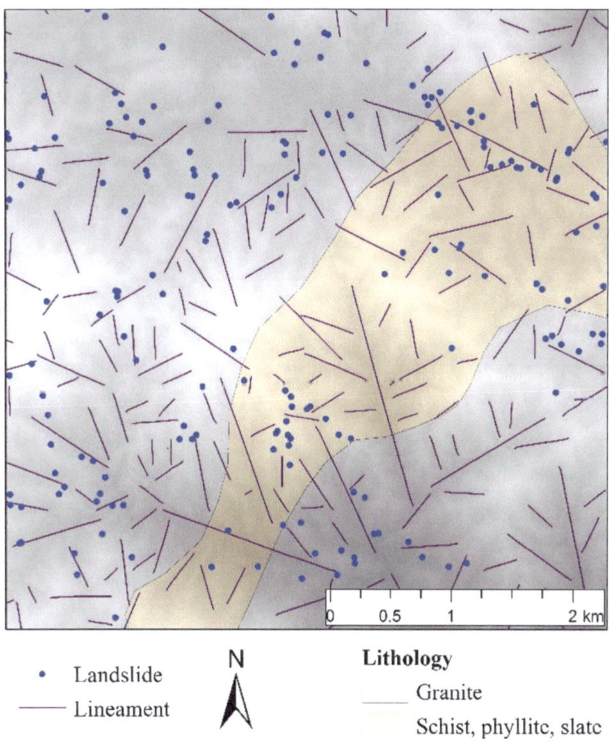

Fig. 7.2 Lithological and structural map of study area

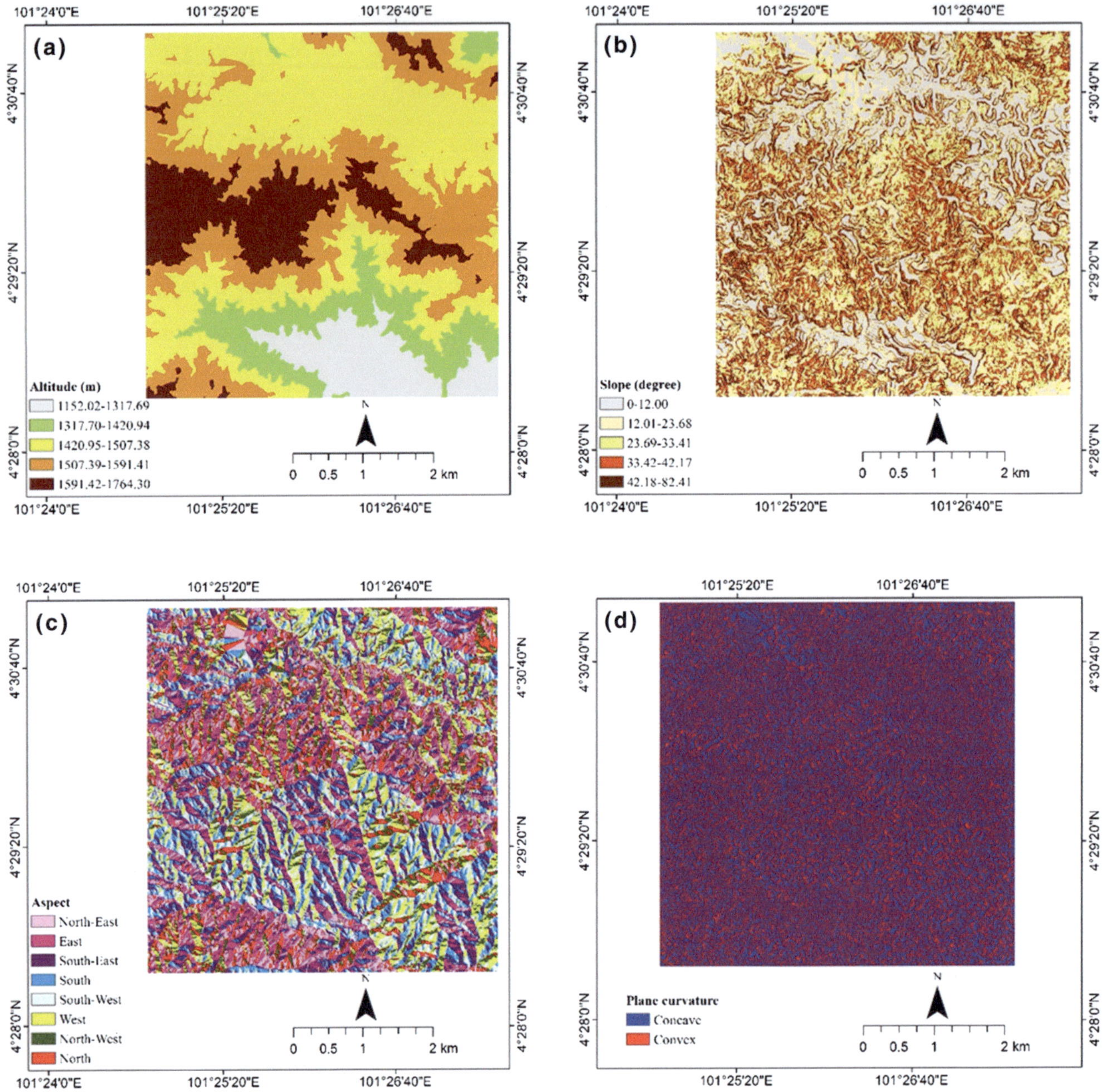

Fig. 7.3 Landslide conditioning factors used in the current study, **a** altitude, **b** slope, **c** aspect, **d** plane curvature, **e** profile curvature, **f** TRI, **g** TWI, **h** distance to fault, **i** distance to road, **j** distance to river, **k** land use, **l** NDVI, and **m** vegetation density

−5 to 20 (Fig. 7.3g). In addition, TRI is an important characteristic of DEM because this factor indicates soil saturation. It is used in hydrological simulation and in estimating runoff, soil moisture, and groundwater depth (Mukherjee et al. 2013). In this study, the TRI ranges from 0 to 221 (Fig. 7.3f).

NDVI is a spectral index used to analyze remote sensing measurements, typically but not necessarily from a space platform, and to assess whether the target being observed contains live green vegetation. The NDVI ranges from −0.818 to 0.96 (Fig. 7.3l) in this study.

NDVI is calculated using the following common formula:

$$\mathrm{NDVI} = \frac{\mathrm{NIR} - R}{\mathrm{NIR} + R}, \tag{7.2}$$

where NIR and R denote the near-infrared and red bands, respectively. Thus, the NDVI varies between −1.0 and +1.0. Vegetation density is another important factor in landslide susceptibility assessment. In this study, vegetation density was classified into non-vegetation, low-density vegetation, moderate-density vegetation, and high-density vegetation (Fig. 7.3m).

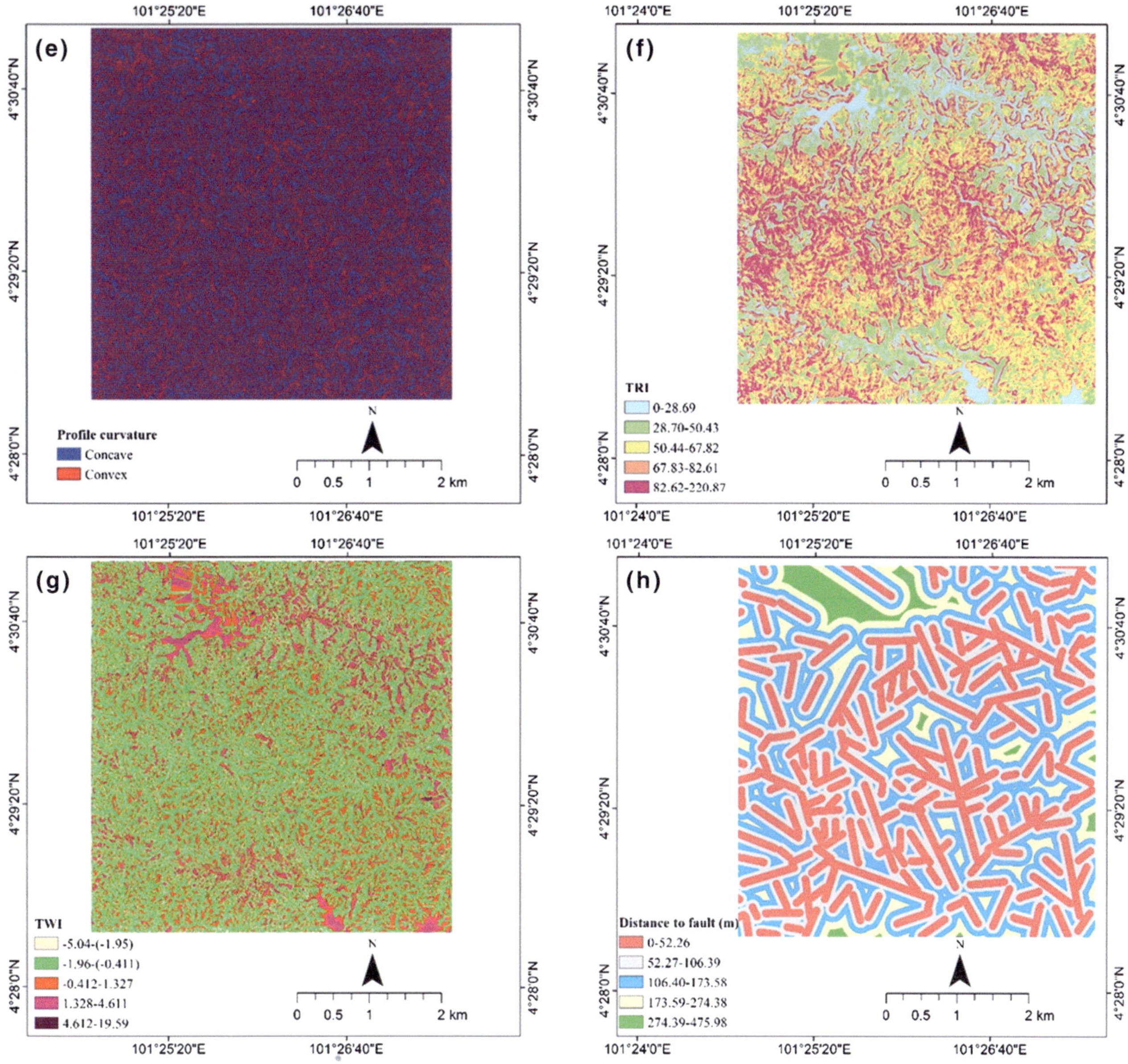

Fig. 7.3 (continued)

Land use, which reflects human activities, also has a major effect on natural resources, soil, and plants. The land use map was produced using 2.5 m pan-sharpened SPOT 5 satellite images, and 10 land use classes were identified (Fig. 7.3k). Furthermore, road and river networks were extracted from the topographic map at a scale of 1: 20,000 and used to construct the distance-to-road maps (Fig. 7.3i) and the distance-to-river maps (Fig. 7.3j), respectively. The fault lines were extracted from the geological map at a scale of 1:63,300 and used to construct the distance-to-fault map, with values ranging from 0 to 476 m (Fig. 7.3h).

7.4 Methods

7.4.1 LSM Via LR

LR is a statistical approach used to establish a univariate regression model by creating a nonlinear relationship between a dependent variable and several independent variables (Eker et al. 2015). The relationship between landslide occurrence and landslide conditioning factors can be quantitatively defined as follows:

$$P = \frac{1}{1 + e^{-(\beta_0 + \beta_1 X_1 + \beta_2 X_2 \ldots \beta_n X_n)}}, \quad (7.3)$$

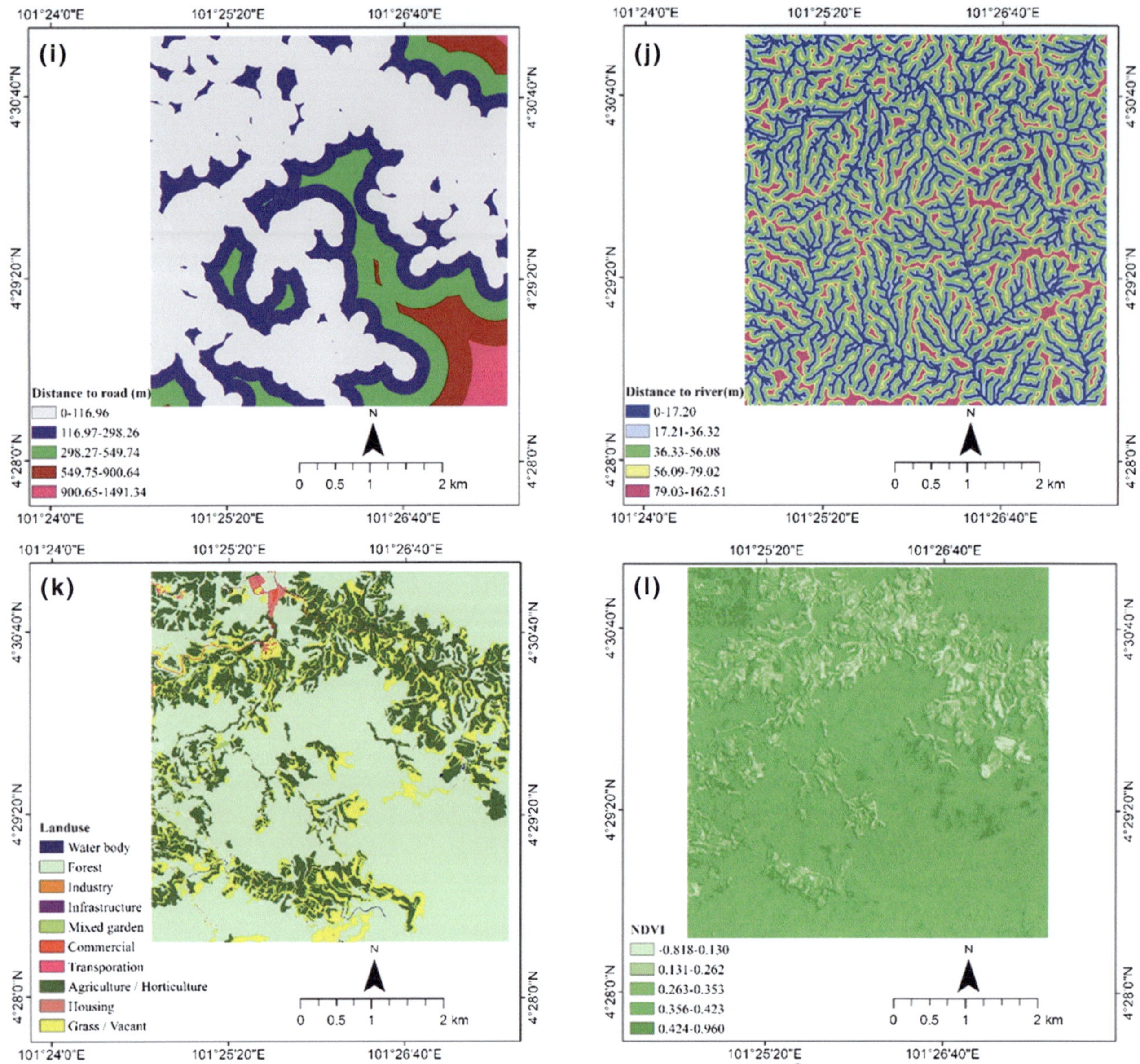

Fig. 7.3 (continued)

where P is the estimated probability of occurrence, β_0 is the intercept of the model, $\beta_i\,(i = 0, 1, 2, \ldots, n)$ are the slope coefficients of the LR model, and $X_i\,(i = 0, 1, 2, \ldots, n)$ are the landslide conditioning factors.

7.4.2 Multicollinearity Analysis

Multicollinearity analysis was performed to detect the existence of any collinear (dependent) variables in the dataset. In such analysis, tolerance is the percentage of the variance in a given landslide conditioning factor; this variance cannot be explained using other factors (Norusis 2006). When tolerance is approximate to 0, the multicollinearity is high, and the standard error of the regression coefficients will be inflated. A variance inflation factor (VIF) greater than ~ 4 is generally considered problematic (Norusis 2006). The multicollinearity analysis was conducted by excluding the redundant landslide conditioning factors to reduce collinearity. The remaining dependent and independent variables would subsequently form the multiple regression models, which could be used to produce the landslide susceptibility maps. The redundant landslide conditioning factors can affect the precision of the model and lead to unreliable predicted values as a result of the multicollinearity phenomenon (Alin 2010). Consequently, a multicollinearity

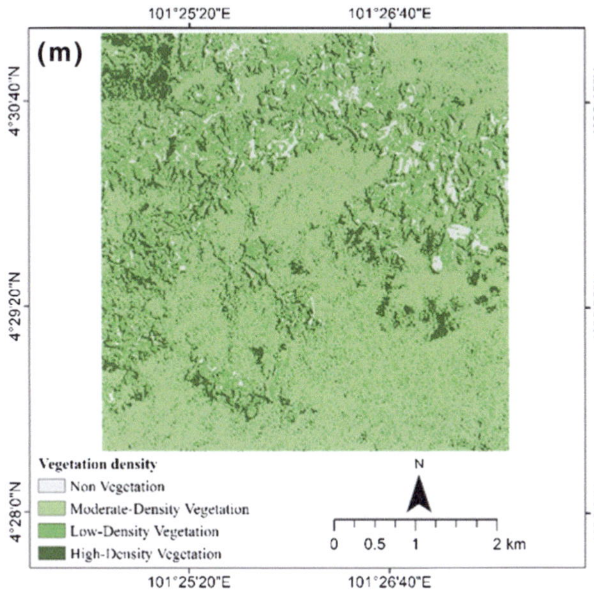

analysis is essential to show the relationship among independent variables.

7.4.3 Validation of Landslide Susceptibility Models

The validation of landslide susceptibility models is an essential step to compare the different models and determine the reliability of the modeling process. It also provides a meaningful interpretation of results (Pourghasemi et al. 2013). Success rate and prediction rate are frequently used in landslide susceptibility analyses. The success rate indicates how well the resultant maps classify the areas of existing landslides, and the prediction rate shows how well the models and causal factors predict future landslides by using a validation dataset (typically 30% of the original dataset is selected randomly). In this study, the AUC of ROC, the kappa coefficient, overall accuracy, spatial agreement, and landslide density graphs were used to validate the landslide susceptibility models.

7.4.3.1 ROC Method

The ROC method works by creating specific rate curves (i.e., success rate and prediction rate curves), which explain the percentage of known landslides that fall under each defined level of susceptibility and are represented in a cumulative frequency diagram (Chung and Fabbri 2003; Intarawichian and Dasananda 2011). In this method, the success rate curve is derived by comparing the susceptibility map with the landslides used in modeling (i.e., the training set), and the prediction rate curve can be created by validating the landslide inventory (Pradhan and Kim 2014). In the rate curve, the *y*-axis normally represents the cumulative percentage of the observed landslide occurrences in different susceptibility classes, whereas the *x*-axis corresponds to the cumulative percentage of the area of the susceptibility classes. The total area under the rate curve can be used to qualitatively determine the prediction accuracy of the susceptibility map; a larger area indicates that higher accuracy is achieved (Lee 2005; Mathew et al. 2009; Intarawichian and Dasananda 2011; Pourghasemi et al. 2013).

7.4.3.2 Kappa Coefficient and Overall Accuracy

A confusion matrix can be constructed by establishing a cross tabulation of the true positive and true negative values. Sensitivity can be defined as the probability that a test result will be positive when the event is present, whereas specificity is the probability that a test result will be negative when the event is absent. A higher sensitivity indicates that the model can better identify unstable zones from the inventory, whereas a higher specificity suggests that the model can better identify stable zones from the inventory (Guzzetti et al. 2006). In addition, the kappa coefficient is a quantitative statistical value of agreement on the presence or absence of specific events. This coefficient is a measure of the difference between the observed agreement and the expected agreement standardized to a scale of −1 to 1. A value of 1 represents perfect agreement, 0 represents an agreement by chance, and negative values signify disagreement.

7.4.3.3 Landslide Density Graphs

A landslide density graph can be constructed by plotting landslide density, which is the ratio of pixels with landslide occurrence to the ratio of pixels without landslide occurrence per classified susceptible zone in a diagram. This graph provides information about the landslide distributions in different susceptible classes in an output of LSM.

7.4.3.4 Evaluation of Spatial Agreement Between Susceptibility Maps

Statistical accuracy metrics, such as the AUC of ROC and the kappa coefficient, do not provide information about the susceptibility classes in which the susceptibility maps spatially agree or disagree. Therefore, the maps were further evaluated via empirical information entropy (EIE), and the average susceptibility of the produced susceptibility maps were calculated using DEMs with different spatial resolutions. This analysis enabled the production of a map that could be used to determine the similarities and dissimilarities among the assigned susceptibility ranks. EIE is expressed as follows:

$$H = -\sum_{i=1}^{n} P(i)\log(P(i)), \quad (7.4)$$

where $P(i)$ is the likelihood of the occurrence of susceptibility rank (class) i, which ranges from 1 to 4 in this study (with 1 being the lowest susceptibility class), and n is the number of susceptibility classes, such that $P(i)$ becomes the probability of the occurrence of class i when the ranks (classes) of different susceptibility maps are given in a grid. H is at its minimum when all the maps agree on the same susceptibility rank, whereas H is at its maximum when the spatial agreement among the susceptibility maps is minimal.

In addition, another map is produced by calculating the average values of the susceptibility classes in all the maps produced in this study to show the spatial agreements and disagreements among the produced maps.

7.5 Results

7.5.1 Effects of the Spatial Resolution of DEM on LSM

The average point spacing of the LiDAR point clouds is ~0.7 m. Therefore, the first DEM was produced at a spatial resolution of 0.5 m. In addition, seven DEMs with various spatial resolutions were produced by resampling the 0.5 m DEM. Figure 7.4 shows the first DEM derived from the LiDAR point clouds and the resampled DEMs with different spatial resolutions. Figure 7.4 shows the extremely small subset of the study area; thus, the differences among the DEMs could be easily detected. Figure 7.4 shows a considerable difference between the first DEM (0.5 m) and the 30 m DEM. The 0.5 m DEM presents a smooth boundary of lands with different elevations, whereas the 30 m DEM is nearly pixelated and exhibits detail loss.

Figure 7.5 shows the difference between the LiDAR-derived DEM and the ASTER DEM both at a spatial resolution of 30 m. Although the two DEMs have the same spatial resolution, the LiDAR-derived DEM contains more details, particularly at the boundaries of features. Quantitatively, the difference between the lowest altitudes of these DEMs is 5 m, whereas the difference between their highest altitudes is 75 m. The reasons for the differences are attributed to the data acquisition from the sensors and to the interpolation techniques used for resampling. The LiDAR data were primarily acquired at a spatial resolution of 0.7 m and further resampled into 30 m. The potential information stored in every cell of a 0.7 m grid is diluted when it is resampled into 30 m. Overall, however, both DEMs exhibit a similar Earth surface topography. High altitudes are located in the middle part of the study area, whereas low altitudes are mostly located in the southern part of the study area.

Figure 7.6 shows a graph that compares the estimated altitudes from the LiDAR-derived DEM and the ASTER DEM for some landslide locations in the study area. The graph shows that both datasets generally agree in terms of the altitude of landslide locations. For certain landslide locations, such as 4, 16, and 19, the altitudes differ by ~15 m on average.

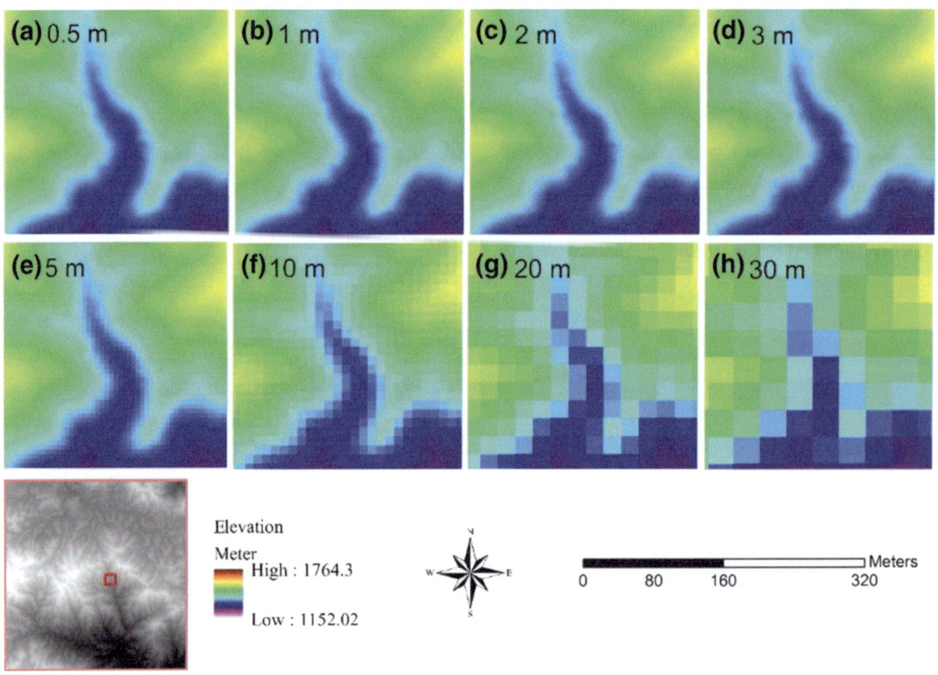

Fig. 7.4 DEMs derived from LiDAR point clouds and iteratively resampled from the 0.5 m DEM

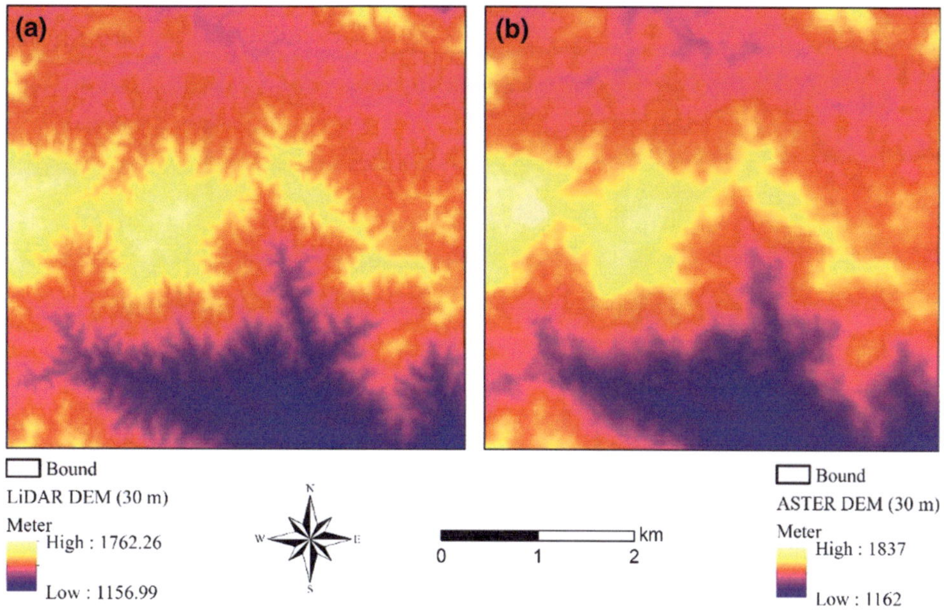

Fig. 7.5 Comparison between ASTER DEM and LiDAR-based DEM, **a** LiDAR DEM (30 m), **b** ASTER DEM (30 m)

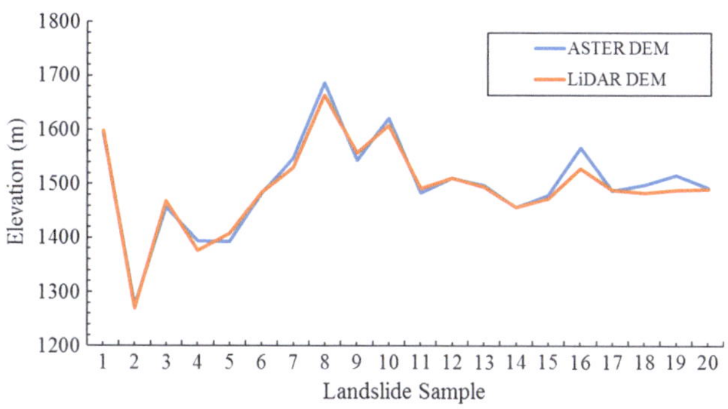

Fig. 7.6 Estimated altitudes from ASTER and LiDAR-based DEMs of randomly selected samples

7.5.2 Results of the Multicollinearity Analysis

Prior to the application of LR modeling, a multicollinearity analysis was conducted to exclude the highly correlated landslide conditioning factors from further analysis. Figure 7.7 shows the calculated respective VIFs of the landslide conditioning factors for every DEM product. Table 7.1 shows the detailed calculations from the multicollinearity analysis of the landslide conditioning factors. The VIF, R2 with the remaining factors, and tolerance are also shown for each factor. Reducing the spatial resolution of the DEM layer directly affects multicollinearity among landslide conditioning factors. Slope and TRI are the most affected factors. Their VIFs decrease from 4.2 to ~2 by reducing DEM spatial resolution from 0.5 to 30 m. Distance to lineament, STI, SPI, aspect, and altitude are the factors whose VIFs are least affected by the reduction in DEM spatial resolution. The VIFs of the remaining factors are also affected by the change in DEM spatial resolution. Overall, no marked VIF variation pattern was detected during the analysis. The VIFs of TWI, plan curvature, profile curvature, and distance to river increase with decreasing DEM spatial resolution. By contrast, the VIFs of vegetation density, NDVI, distance to road, TRI, land use, and slope decrease with decreasing DEM spatial resolution. Furthermore, the VIFs of the remaining factors change without any apparent pattern.

7.5.3 Results of the Sensitivity Analysis

A sensitivity analysis is typically conducted to determine the contribution of each landslide conditioning factor. In this study, a chi-square method was applied to evaluate the contributions of landslide conditioning factors. In this method, the *p*-values of each factor were calculated to determine its importance in predicting landslides in the study area. The analysis showed that the importance of a landslide conditioning factor depends on the spatial resolution and source of the DEM. For example, Table 7.2 indicates that the p-value of the factor slope is 0.15 (insignificant at the 95% confidence level) when the spatial resolution of the DEM is 0.5 m. By contrast, slope is statistically significant at the 95% confidence level at DEM spatial resolutions of 1, 2, 3, 5, 10, 20, and 30 m. In addition, the analysis showed that slope is insignificant at the 95% confidence level when the ASTER DEM is used as the data source. Thus, DEM source should be considered in LSM. The importance of landslide conditioning factors is sensitive to DEM spatial

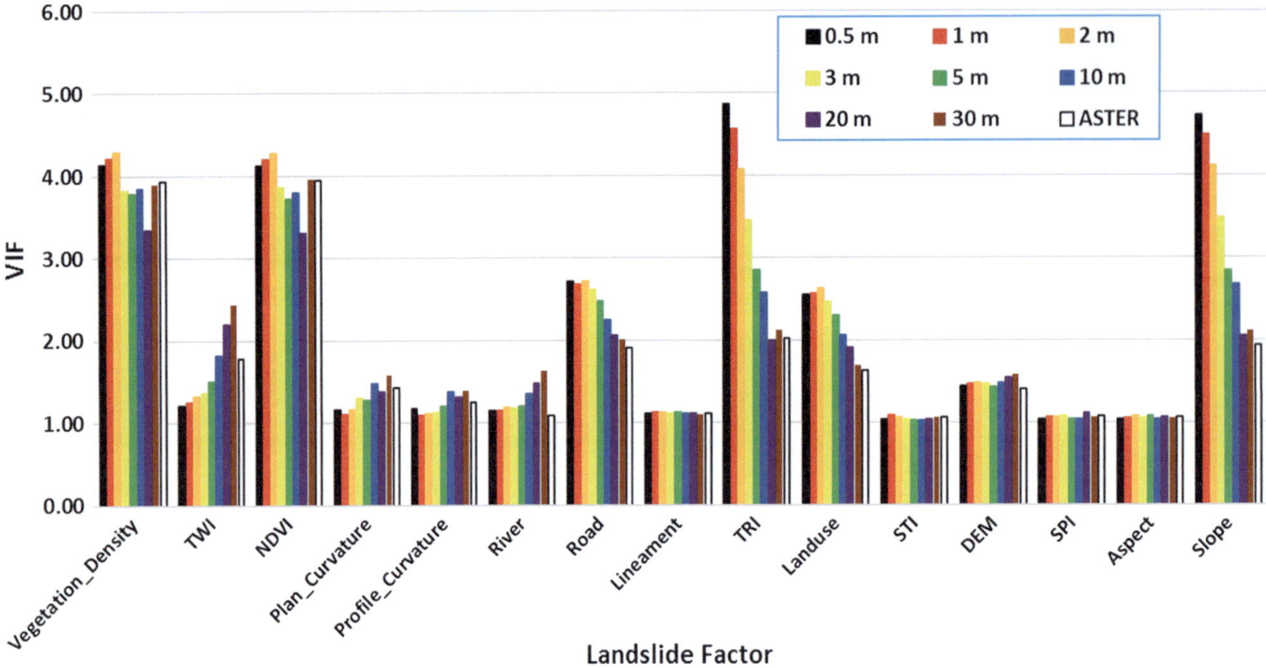

Fig. 7.7 Estimated VIF of each landslide conditioning factor

resolution, and thus, no factor has been removed in the analysis to achieve an objective comparison among the landslide susceptibility models (Fig. 7.8).

7.5.4 Results of the Landslide Susceptibility Mapping

Overall, nine landslide susceptibility maps were produced for the study area using the DEMs with different spatial resolutions (Fig. 7.9). The maps were created using the LR modeling approach and reclassified using the natural break method in GIS. The susceptibility classes are low, moderate, high, and very high. The very high and high susceptibility classes are mostly distributed in the northern and western parts of the study area. The percentages of the very high and high susceptibility classes in the landslide susceptibility map derived from the 0.5 m DEM are lower compared with those in the map derived from DEMs with higher spatial resolutions. The high susceptibility class significantly increased in the map produced using the 3 m DEM. The difference among the landslide susceptibility maps produced using the ASTER DEM and the 30 m LiDAR-derived DEM is the slight increase in the very high susceptibility class in the ASTER-derived map.

7.5.5 Validation of the Landslide Susceptibility Models

The landslide susceptibility models were validated using several methods. Table 7.3 shows the validation of LSM via AUC of ROC, the kappa coefficient, and overall accuracy. The validation shows that the highest success rate (0.969) and prediction rate (0.968) were obtained by the 3 m DEM. By contrast, the lowest success rate (0.890) and prediction rate (0.879) were obtained by the ASTER DEM. In addition, no significant difference was apparent among the success rates and prediction rates for the spatial resolutions less than 10 m. However, the success rates and prediction rates significantly decreased for the spatial resolutions greater than 10 m. The highest kappa coefficient was 0.88 for the LSM produced using the 2 m DEM, whereas the lowest kappa coefficient was 0.645 for the LSM derived from the ASTER DEM. Furthermore, the lowest (82.29%) and highest (94.02%) overall accuracies were obtained by the LSM produced from the ASTER DEM and the 2 m LiDAR-derived DEM, respectively.

In addition, the success rates and prediction rates of the LSMs produced using the ASTER DEM and the 30 m LiDAR-derived DEM have no significant difference.

Table 7.1 Estimated VIF for landslide conditioning factors

Landslide factor	0.5 m			1 m			2 m			3 m			5 m		
	VIF	R^2	Tolerance	VIF	R^2	Tolerance	VIF	R^2	Tolerance	VIF	R^2	Tolerance	VIF	R^2	Tolerance
Vegetation density	4.14	0.76	0.24	4.21	0.76	0.24	4.29	0.77	0.23	3.82	0.74	0.26	3.78	0.74	0.26
TWI	1.22	0.18	0.82	1.25	0.20	0.80	1.33	0.25	0.75	1.36	0.27	0.73	1.50	0.34	0.66
NDVI	4.13	0.76	0.24	4.20	0.76	0.24	4.28	0.77	0.23	3.86	0.74	0.26	3.73	0.73	0.27
Plan curvature	1.16	0.14	0.86	1.11	0.10	0.90	1.16	0.14	0.86	1.31	0.24	0.76	1.28	0.22	0.78
Profile curvature	1.17	0.15	0.85	1.09	0.08	0.92	1.11	0.10	0.90	1.13	0.12	0.88	1.21	0.17	0.83
River	1.15	0.13	0.87	1.15	0.13	0.87	1.18	0.16	0.84	1.18	0.15	0.85	1.20	0.17	0.83
Road	2.71	0.63	0.37	2.68	0.63	0.37	2.72	0.63	0.37	2.61	0.62	0.38	2.47	0.60	0.40
Lineament	1.12	0.11	0.89	1.12	0.11	0.89	1.12	0.11	0.89	1.12	0.10	0.90	1.12	0.11	0.89
TRI	4.87	0.79	0.21	4.57	0.78	0.22	4.08	0.75	0.25	3.45	0.71	0.29	2.86	0.65	0.35
Land use	2.54	0.61	0.39	2.56	0.61	0.39	2.63	0.62	0.38	2.46	0.59	0.41	2.29	0.56	0.44
STI	1.05	0.04	0.96	1.09	0.08	0.92	1.06	0.06	0.94	1.04	0.04	0.96	1.02	0.02	0.98
DEM	1.44	0.30	0.70	1.47	0.32	0.68	1.48	0.32	0.68	1.47	0.32	0.68	1.43	0.30	0.70
SPI	1.04	0.04	0.96	1.07	0.06	0.94	1.07	0.06	0.94	1.07	0.07	0.93	1.04	0.04	0.96
Aspect	1.04	0.04	0.96	1.05	0.05	0.95	1.07	0.07	0.93	1.06	0.05	0.95	1.07	0.07	0.93

Landslide factor	10 m			20 m			30 m			ASTER		
	VIF	R^2	Tolerance	VIF	R^2	Tolerance	VIF	R^2	Tolerance	VIF	R^2	Tolerance
Vegetation density	3.85	0.74	0.26	3.35	0.70	0.30	3.89	0.74	0.26	3.93	0.75	0.25
TWI	1.82	0.45	0.55	2.20	0.55	0.45	2.43	0.59	0.41	1.78	0.44	0.56
NDVI	3.79	0.74	0.26	3.31	0.70	0.30	3.96	0.75	0.25	3.95	0.75	0.25
Plan curvature	1.48	0.32	0.68	1.39	0.28	0.72	1.58	0.37	0.63	1.43	0.30	0.70
Profile curvature	1.38	0.28	0.72	1.32	0.24	0.76	1.39	0.28	0.72	1.25	0.20	0.80
River	1.35	0.26	0.74	1.48	0.33	0.68	1.63	0.39	0.61	1.09	0.08	0.92
Road	2.24	0.55	0.45	2.06	0.52	0.49	2.01	0.50	0.50	1.91	0.48	0.52
Lineament	1.12	0.11	0.89	1.11	0.10	0.90	1.10	0.09	0.91	1.12	0.10	0.90
TRI	2.57	0.61	0.39	2.00	0.50	0.50	2.12	0.53	0.47	2.02	0.51	0.50
Land use	2.06	0.51	0.49	1.91	0.48	0.52	1.70	0.41	0.59	1.64	0.39	0.61
STI	1.03	0.03	0.97	1.04	0.04	0.96	1.06	0.06	0.94	1.06	0.06	0.94
DEM	1.48	0.33	0.67	1.55	0.35	0.65	1.58	0.37	0.63	1.40	0.28	0.72
SPI	1.05	0.04	0.96	1.11	0.10	0.90	1.07	0.07	0.93	1.08	0.08	0.92
Aspect	1.05	0.04	0.96	1.06	0.06	0.94	1.05	0.05	0.95	1.06	0.06	0.94
Slope	2.68	0.63	0.37	2.04	0.51	0.49	2.11	0.53	0.47	1.94	0.48	0.52

Table 7.2 Landslide conditioning factor optimization by Chi-square method

Factor Omitted	DF	0.5 m		1 m		2 m		3 m		5 m	
		Deviance	*P*-value	Deviance	*P*-value	Deviance	*P*-value	Deviance	*P*-value	Deviance	*P*-value
All	15	532.34	0.00	532.34	0.00	532.34	0.00	532.34	0.00	532.34	0.00
Vegetation Density	1	115.44	0.27	115.34	0.66	123.44	0.23	130.96	0.89	155.24	0.35
TWI	1	114.65	0.51	115.19	0.84	122.75	0.38	138.29	0.01	155.11	0.39
NDVI	1	115.31	0.30	115.28	0.72	124.17	0.14	130.98	0.85	154.70	0.56
Plan Curvature	1	114.34	0.73	116.54	0.24	122.30	0.58	136.15	0.02	163.79	0.00
Profile Curvature	1	117.91	0.05	116.43	0.26	122.56	0.45	131.60	0.42	157.24	0.09
River	1	114.73	0.48	116.43	0.26	122.81	0.36	133.27	0.13	155.73	0.24
Road	1	123.08	0.00	120.46	0.02	127.84	0.02	149.86	0.00	175.67	0.00
Lineament	1	119.26	0.02	120.54	0.02	126.09	0.04	133.48	0.11	155.94	0.21
TRI	1	114.63	0.53	115.28	0.72	121.99	0.98	131.98	0.31	154.56	0.65
Landuse	1	234.29	0.00	228.76	0.00	232.36	0.00	225.91	0.00	242.14	0.00
STI	1	119.13	0.03	115.17	0.89	122.08	0.76	132.10	0.28	155.13	0.38
DEM	1	115.12	0.34	115.47	0.57	122.42	0.51	132.02	0.30	155.79	0.23
SPI	1	114.23	0.93	118.44	0.07	121.99	0.97	139.41	0.00	154.47	0.73
Aspect	1	114.73	0.48	115.17	0.88	121.99	0.93	131.04	0.75	154.49	0.72
Slope	1	116.34	0.15	121.84	0.01	125.14	0.08	141.86	0.00	160.67	0.01
None(Model)	15	114.22		115.15		121.99		130.94		154.36	

Factor Omitted	DF	10 m		20 m		30 m		ASTER	
		Deviance	*P*-value	Deviance	*P*-value	Deviance	*P*-value	Deviance	*P*-value
All	15	532.34	0.00	532.34	0.00	532.34	0.00	532.34	0.00
Vegetation Density	1	188.13	0.18	248.18	0.80	245.08	0.28	264.36	0.36
TWI	1	187.99	0.19	250.69	0.11	244.87	0.33	263.53	0.99
NDVI	1	186.63	0.57	248.15	0.86	244.45	0.46	264.39	0.35
Plan Curvature	1	190.90	0.03	248.48	0.55	244.81	0.34	264.06	0.47
Profile Curvature	1	193.57	0.01	248.16	0.85	245.19	0.26	264.00	0.49
River	1	187.58	0.26	249.04	0.34	245.31	0.24	272.74	0.00
Road	1	236.50	0.00	323.72	0.00	324.23	0.00	354.70	0.00
Lineament	1	187.11	0.37	248.13	0.95	244.08	0.67	263.98	0.50
TRI	1	186.71	0.52	248.85	0.40	247.89	0.05	264.63	0.29
Landuse	1	253.14	0.00	277.76	0.00	273.45	0.00	286.95	0.00
STI	1	189.25	0.09	248.26	0.71	243.95	0.83	264.11	0.45

(continued)

Table 7.2 (continued)

Factor Omitted	DF	10 m		20 m		30 m		ASTER	
		Deviance	*P*-value	Deviance	*P*-value	Deviance	*P*-value	Deviance	*P*-value
DEM	1	191.64	0.02	254.03	0.02	248.26	0.04	270.29	0.01
SPI	1	187.52	0.27	248.44	0.57	244.34	0.51	263.64	0.74
Aspect	1	186.34	0.84	248.16	0.85	245.11	0.27	263.53	0.97
Slope	1	190.60	0.04	252.39	0.04	252.80	0.00	263.55	0.90
None(Model)	15	186.30		248.12		243.90		263.53	

However, the kappa coefficients and overall accuracies of these LSMs considerably differ. These results indicate that the LSM produced using the LiDAR dataset is more capable of predicting future landslides probably because the LiDAR DEM contains more information given that it has been resampled from the 0.5 m DEM.

The spatial agreement evaluation was performed for every grid via EIE and based on the average susceptibility values of all the models. This analysis allowed for the production of two maps, which were used to assess the spatial agreements among the produced models (Fig. 7.10). The entropy map showed that the models produced in this study generally presented high spatial similarities because the lowest two entropy values (≤ 0.33) covered nearly 71% of the study area. The maximum similarities were apparent in most parts of the study area (Fig. 7.10a). Although rarely observed, the highest dissimilarities generally occurred in the southwestern and northwestern parts of the study area. In addition, dissimilarities generally occurred in high-slope areas. This outcome could be attributed to the differences in the elevations estimated from the DEMs.

7.6 Discussion

7.6.1 Effects of DEM Spatial Resolution on LSM

The capability of the datasets to delineate triggering zones and deposits is illustrated in Fig. 7.9. The fine spatial resolution of LiDAR has helped extract details such as landslide probability in hilly and flat areas (Fig. 7.9a). However, these details are compromised in Figs. 7.8f–i, in which the spatial resolutions are ≥ 10 m. The difference between the LiDAR and ASTER datasets is demonstrated by the difference in the hilly areas, where the details are delineated by the LiDAR sensor. By contrast, the details have been compromised in the ASTER dataset.

In fine-resolution maps, susceptibility zones can be accurately extracted at the parcel level to allow for efficient land use planning and slope management. However, this task becomes challenging in low-spatial-resolution maps because details and susceptibility zones are difficult to identify. Landslides can be delineated from 10 m spatial-resolution susceptibility maps; however, a finer spatial resolution is more suitable for LSM. In addition, identification of landslide scarps and flanks also requires a finer spatial resolution. Thus, the spatial resolution should be selected based on the information source.

Statistically, finer spatial resolutions do not always guarantee higher prediction rates. This study shows that a 3 m spatial resolution has a higher predictive accuracy rate than a 0.5 m spatial resolution. However, based on

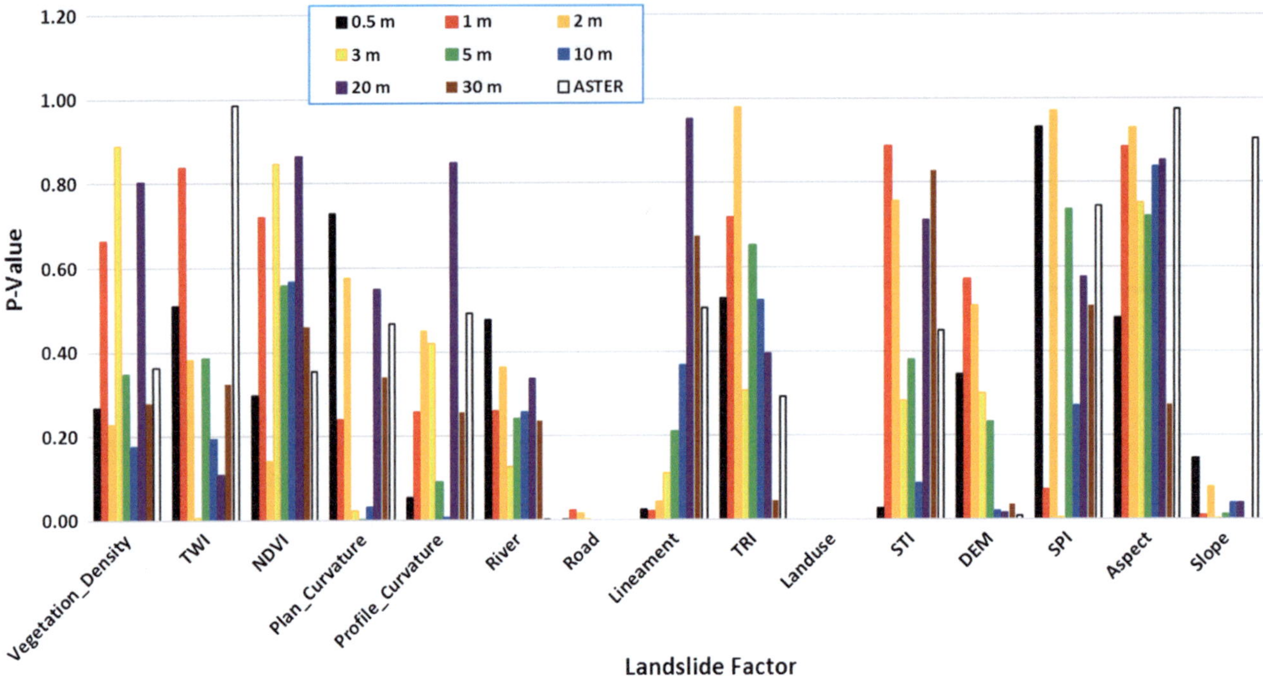

Fig. 7.8 Estimated p-values of each landslide conditioning factor

Table 7.3 Accuracy assessment of landslide susceptibility models produced at different spatial resolutions

DEM resolution	ROC area	PRC area	Kappa	Overall accuracy
0.5 m	0.961	0.967	0.854	92.70
1 m	0.962	0.953	0.859	92.96
2 m	0.957	0.963	0.88	94.02
3 m	0.969	0.968	0.838	91.92
5 m	0.953	0.945	0.828	91.40
10 m	0.944	0.932	0.817	90.88
20 m	0.907	0.890	0.682	84.11
30 m	0.909	0.889	0.713	85.67
ASTER (30 m)	0.890	0.879	0.645	82.29

other statistical metrics, such as the kappa coefficient and overall accuracy, 2 m spatial resolution is the most powerful for LSM. Compared with the information from the ASTER sensor, the information that was stored using the LiDAR sensor at a fine resolution allows for the production of relatively satisfactory landslide susceptibility maps. Therefore, storing valuable ground information for LSM is imperative. The ASTER data compromise the ground information, thereby introducing challenges in delineating detailed information about the landslides in the study area.

7.6.2 Relationship Between Landslide Density and Susceptibility Zones

The estimated landslide density in each susceptibility zone indicates how well the established LR model predicts future

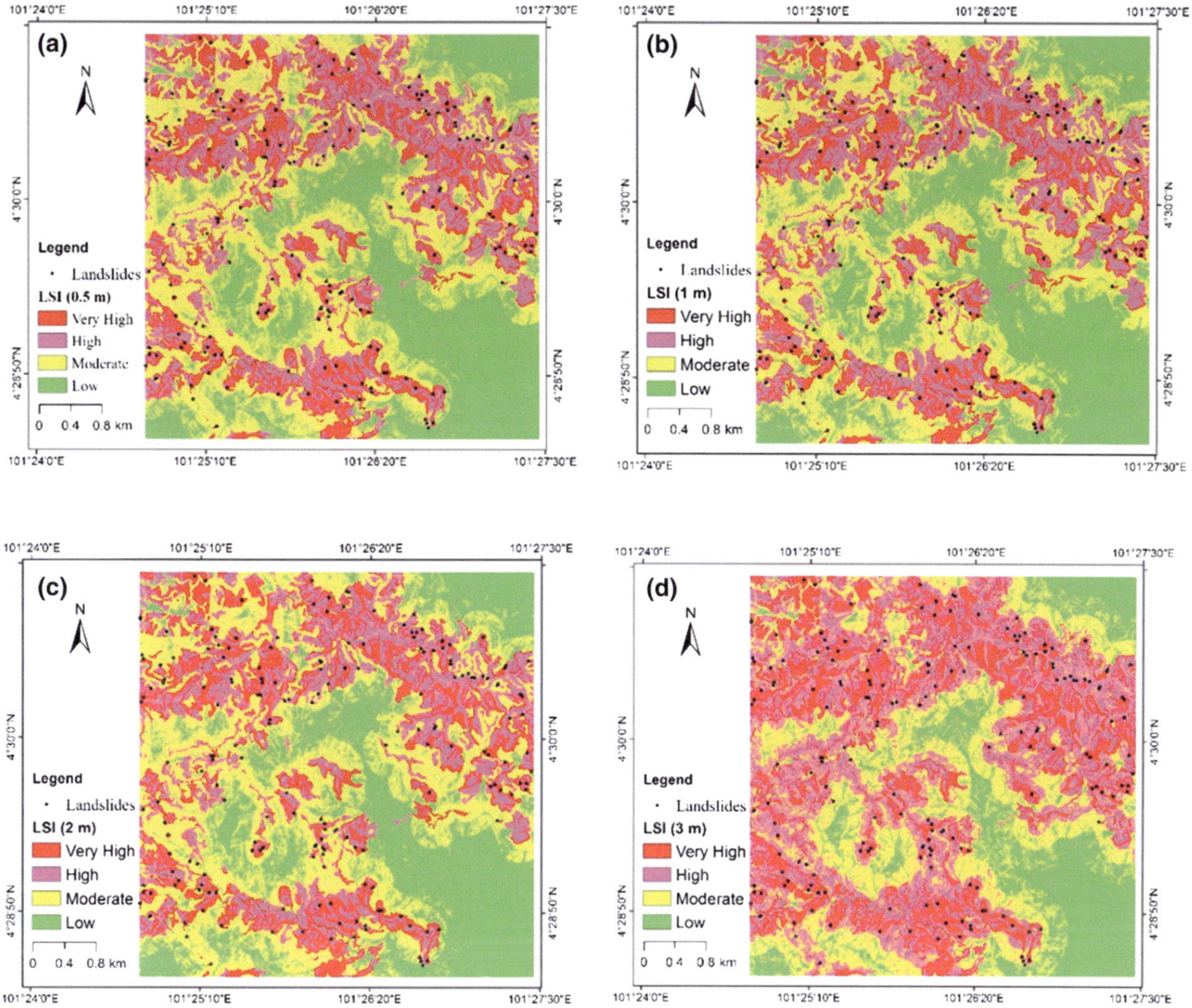

Fig. 7.9 Landslide susceptibility maps using various spatial resolution **a** 0.5 m DEM, **b** 1 m DEM, **c** 2 m DEM, **d** 3 m DEM, **e** 5 m DEM, **f** 10 m DEM, **g** 20 m DEM, **h** 30 m DEM, and **i** ASTER DEM

landslides in the study area. Therefore, zonal statistical functions were used to calculate the number of landslides in each susceptibility zone, and the results were plotted on a bar chart (Fig. 7.11). The analysis showed that the average number of landslides in the very high and high susceptibility zones were 160 and 25 landslides, respectively, thereby indicating that the established LR model is capable of predicting landslide-susceptible areas.

In addition, the estimated number of landslides in each susceptibility zone shows that prediction accuracy for

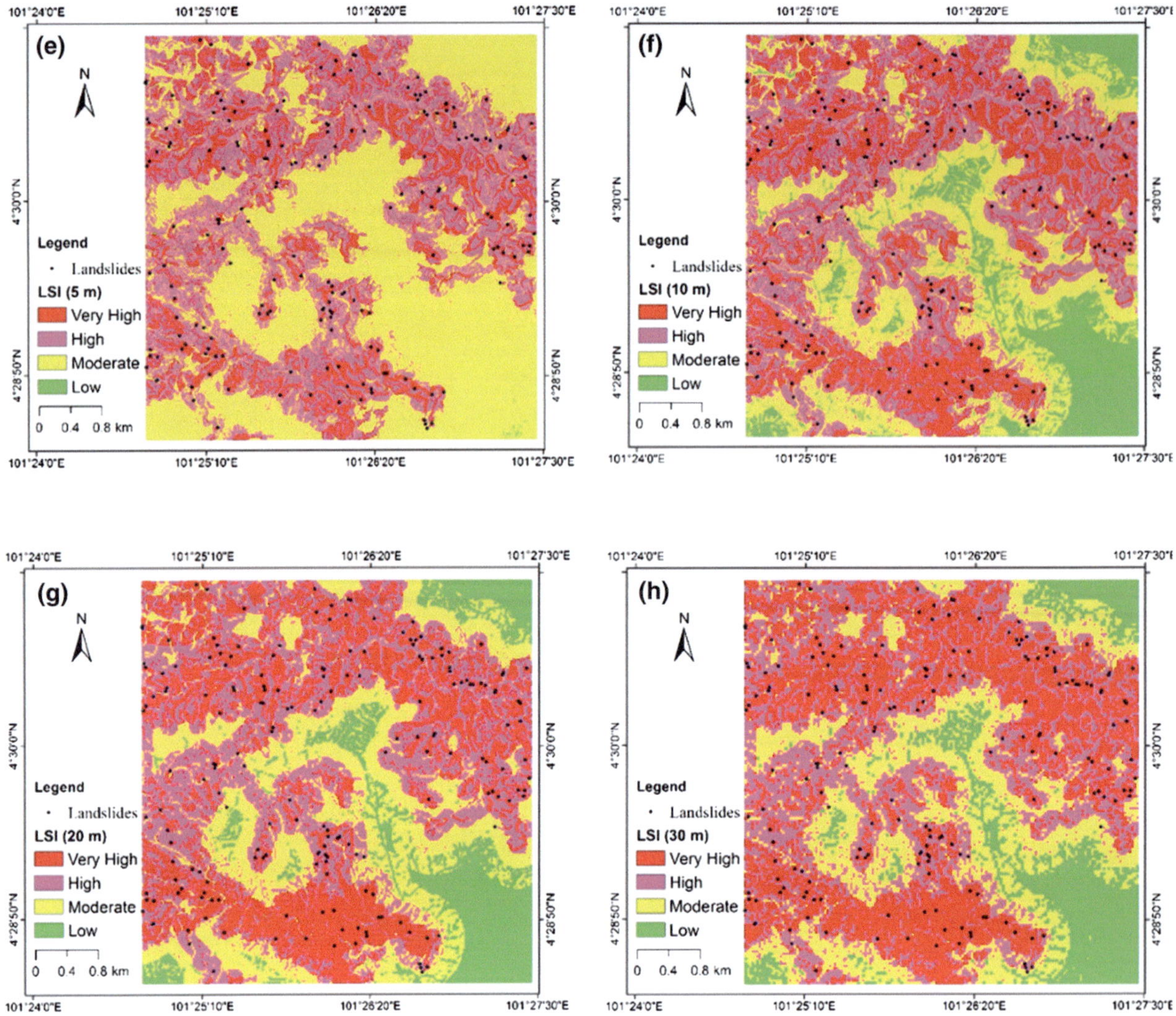

Fig. 7.9 (continued)

future landslides is sensitive to the spatial resolution of landslide conditioning factors. As shown in Fig. 7.11, the estimated number of landslides decreased with decreasing DEM spatial resolution in the very high susceptibility zone. By contrast, in the high susceptibility zone, the estimated number of landslides increased with decreasing DEM spatial resolution.

Furthermore, the analysis indicated that the ASTER DEM and the 30 m LiDAR DEM differed in terms of the estimated number of landslides in the very high susceptibility zone. The number of landslides in the very high susceptibility zone in the landslide susceptibility map produced using the ASTER dataset was 135, whereas that obtained using the 30 m LiDAR dataset was 143. This difference suggests that the DEM source also affects the accuracy and quality of landslide susceptibility maps.

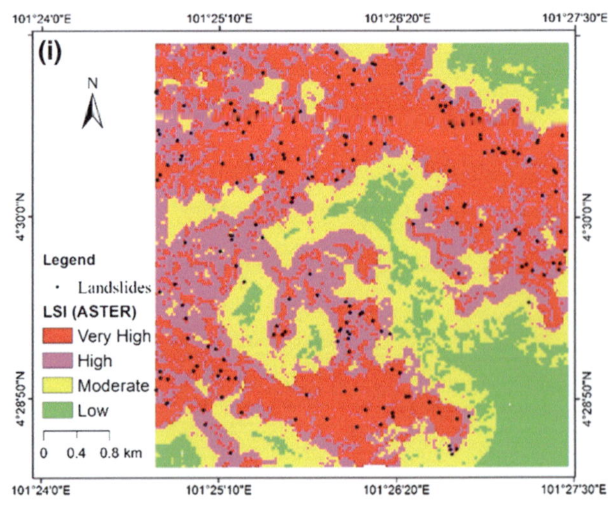

Fig. 7.9 (continued)

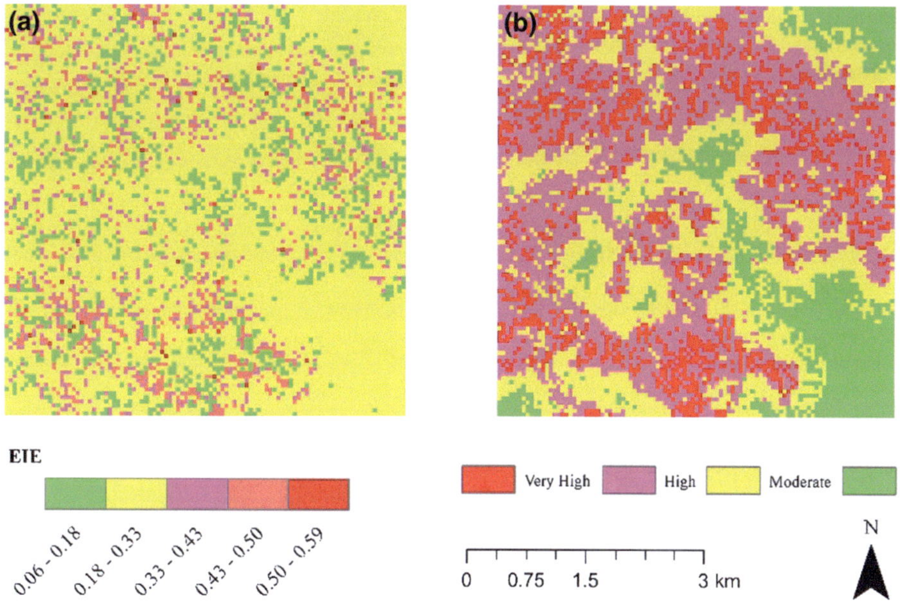

Fig. 7.10 Evaluation of spatial agreements of landslide susceptibility maps, **a** empirical information entropy (H), and **b** average landslide susceptibility values

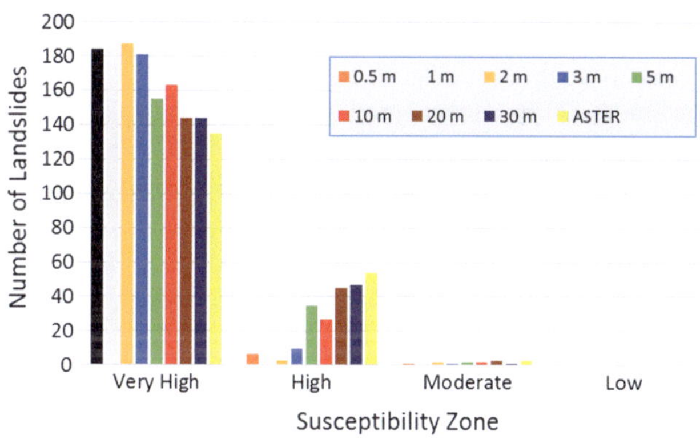

Fig. 7.11 Number of landslides within landslide susceptibility zones

7.7 Conclusion

The main objective of this study was to assess the effects of DEM spatial resolution on LSM in tropical areas. Two datasets were used, namely, LiDAR point clouds with an average point spacing of 0.7 m and an ASTER DEM with 30 m spatial resolution. Consequently, thirteen landslide conditioning factors were derived, and the 0.5 m LiDAR DEM was resampled into various spatial resolutions (1, 2, 3, 5, 10, 20, and 30 m). For each dataset of a different spatial resolution, a landslide susceptibility index was developed using the LR model and evaluated with various accuracy metrics. The AUC of ROC, the kappa coefficient, overall accuracy, and spatial agreement by EIE were used to validate the landslide susceptibility models.

The main finding of this study is that the multicollinearity and importance of a landslide conditioning factor are sensitive to the spatial resolution of the DEM. In addition, the analysis indicated that a finer spatial resolution does not always guarantee a higher prediction rate. The optimal spatial resolution is 2 m based on the accuracy metrics. However, DEM spatial resolution has a weaker effect on the spatial distribution of landslide-prone areas in LSM. The entropy analysis showed that the models spatially agreed in 71% of the study area in terms of landslide susceptibility class. Furthermore, the comparison of the LiDAR sensor with the ASTER sensor showed that the collection of detailed ground information using the LiDAR sensor provided better LSM.

Several limitations were observed in the current study. First, the resampling/interpolation methods were not evaluated, although these methods might have an effect on LSM. Therefore, this concern should be addressed in future studies. In addition, the landslide sampling procedure and the number of landslide inventories might also affect LSM; thus, both factors should be carefully studied to improve landslide susceptibility assessment.

References

Alin, A. (2010). Multicollinearity. Wiley Interdisciplinary Reviews: Computational Statistics, 2(3), 370–374.

Althuwaynee, O. F., Pradhan, B., & Lee, S. (2016). A novel integrated model for assessing land-slide susceptibility mapping using CHAID and AHP pair-wise comparison. *International Journal of Remote Sensing, 37*(5), 1190–1209.

Chalkias, C., Kalogirou, S., & Ferentinou, M. (2014). Landslide susceptibility, Peloponnese Peninsula in South Greece. *Journal of Maps, 10*(2), 211–222.

Chang, K. T., Dou, J., Chang, Y., Kuo, C. P., Xu, K. M., & Liu, J. K. (2016). Spatial resolution effects of digital terrain models on landslide susceptibility analysis. *ISPRS-International Ar-chives of the Photogrammetry, Remote Sensing and Spatial Information Sciences*, XLI-B8, 33–36.

Chung, C. J. F., & Fabbri, A. G. (2003). Validation of spatial prediction models for landslide hazard mapping. *Natural Hazards, 30*(3), 451–472.

Eker, A. M., Dikmen, M., Cambazoğlu, S., Düzgün, Ş. H., & Akgün, H. (2015). Evaluation and comparison of landslide susceptibility mapping methods: A case study for the Ulus district, Bartın, northern Turkey. *International Journal of Geographical Information Science, 29*(1), 132–158.

Guri, P. K., & Patel, R. C. (2015). Spatial prediction of landslide susceptibility in parts of Garhwal Himalaya, India, using the weight of evidence modelling. *Environmental monitoring and assessment, 187*(6), 1–25.

Guzzetti, F., Reichenbach, P., Ardizzone, F., Cardinali, M., & Galli, M. (2006). Estimating the quality of landslide susceptibility models. *Geomorphology, 81*(1), 166–184.

Huang, J., Zhou, Q., & Wang, F. (2015). Mapping the landslide susceptibility in Lantau Island, Hong Kong, by frequency ratio and logistic regression model. *Annals of GIS, 21*(3), 191–208.

Intarawichian, N., & Dasananda, S. (2011). Frequency ratio model based landslide susceptibility mapping in lower Mae Chaem watershed, Northern Thailand. *Environmental Earth Sciences, 64* (8), 2271–2285.

Lee, S. (2005). Application and cross-validation of spatial logistic multiple regression for land-slide susceptibility analysis. Geosciences Journal, 9(1), 63–71.

Lee, S., Choi, J., & Woo, I. (2004). The effect of spatial resolution on the accuracy of landslide susceptibility mapping: A case study in Boun, Korea. Geosciences Journal, 8(1), 51–60.

Mahalingam, R., Olsen, M. J., & O'Banion, M. S. (2016). Evaluation of landslide susceptibility mapping techniques using lidar-derived conditioning factors (Oregon case study). *Geomatics, Natural Hazards and Risk, 7*(6), 1–24.

Mathew, J., Jha, V. K., & Rawat, G. S. (2009). Landslide susceptibility zonation mapping and its validation in part of Garhwal Lesser Himalaya, India, using binary logistic regression analysis and receiver operating characteristic curve method. *Landslides, 6*(1), 17–26.

Meinhardt, M., Fink, M., & Tünschel, H. (2015). Landslide susceptibility analysis in central Vietnam based on an incomplete landslide inventory: Comparison of a new method to calculate weighting factors by means of bivariate statistics. *Geomorphology, 234*, 80–97.

Mukherjee, S., Mukherjee, S., Garg, R. D., Bhardwaj, A., & Raju, P. L. N. (2013). Evaluation of topographic index in relation to terrain roughness and DEM grid spacing. *Journal of Earth System Science, 122*(3), 869–886.

Norusis, M. J. (2006). *SPSS 15.0 guide to data analysis*. Upper Saddle River, NJ: Prentice Hall.

Oh, H. J., Park, N. W., Lee, S. S., & Lee, S. (2012). Extraction of landslide-related factors from ASTER imagery and its application to landslide susceptibility mapping. *International Journal of Remote Sensing, 33*(10), 3211–3231.

Park, N. W. (2015). Using maximum entropy modeling for landslide susceptibility mapping with multiple geoenvironmental data sets. *Environmental Earth Sciences, 73*(3), 937–949.

Pourghasemi, H. R., Jirandeh, A. G., Pradhan, B., Xu, C., & Gokceoglu, C. (2013). Landslide susceptibility mapping using support vector machine and GIS at the Golestan Province, Iran. *Journal of Earth System Science, 122*(2), 349–369.

Pradhan, A. M. S., & Kim, Y. T. (2014). Relative effect method of landslide susceptibility zonation in weathered granite soil: A case study in Deokjeokri Creek, South Korea. *Natural hazards, 72*(2), 1189–1217.

Pradhan, A. M. S., & Kim, Y. T. (2016). Evaluation of a combined spatial multi-criteria evaluation model and deterministic model for landslide susceptibility mapping. *Catena, 140*, 125–139.

Pradhan, A. M. S., Kang, H. S., Lee, S., & Kim, Y. T. (2016). Spatial model integration for shallow landslide susceptibility and its runout using a GIS-based approach in Yongin, Korea. *Geocarto International*, 1–22.

Pradhan, B. (2010). Application of an advanced fuzzy logic model for landslide susceptibility analysis. *International Journal of Computational Intelligence Systems, 3*(3), 370–381.

Pradhan, B., Lee, S. (2010). Regional landslide susceptibility analysis using back-propagation neural network model at Cameron Highland. Malaysia. *Landslides, 7*(1), 13–30.

Qin, C. Z., Bao, L. L., Zhu, A. X., Wang, R. X., & Hu, X. M. (2013). Uncertainty due to DEM error in landslide susceptibility mapping. International *Journal of Geographical Information Science, 27*(7), 1364–1380.

Quinn, P. E. (2014). Landslide susceptibility in sensitive clay in eastern Canada: Some practical considerations and results in development of an improved model. *International Journal of Image and Data Fusion, 5*(1), 70–96.

Raman, R., & Punia, M. (2012). The application of GIS-based bivariate statistical methods for landslide hazards assessment in the upper Tons river valley, Western Himalaya, India. *Georisk: Assessment and Management of Risk for Engineered Systems and Geohazards, 6*(3), 145–161.

8 Spatial Prediction of Landslide-Prone Areas Through *k*-Nearest Neighbor Algorithm and Logistic Regression Model Using High Resolution Airborne Laser Scanning Data

Biswajeet Pradhan and Mustafa Neamah Jebur

8.1 Introduction

Rapid urban growth and climate change in recent years have resulted in many environmental problems and increased risks due to natural disasters. Landslides, floods, earthquakes, and tsunamis are natural hazards that must concern governments worldwide because of the incalculable and irrecoverable damages they can cause. Landslides are one of the most destructive natural hazards in the world; they exert numerous negative effects on many lives and properties (Antronico et al. 2015; Dehnavi et al. 2015; Fiorucci et al. 2015; Jebur et al. 2014a; Vranken et al. 2013). Landslides are expected to continue to occur because of the significant expansion of urban areas, deforestation actions, etc. (Jebur et al. 2014b; Rahman et al. 2014; Runyan and D'odorico 2014; Ueno et al. 2015). Hence, areas that are susceptible to landslides must be identified and mapped to assist planners in establishing safe urbanizations. Landslide susceptibility maps can be used in other researches such as hazards, risks, and vulnerability studies. Decision makers and planners need to identify landslide-prone locations to plan subsequent actions in consideration of the damage induced by landslides. Landslide susceptibility mapping relies on the methods used and the quality and scale of conditioning factors. Thus, a spatial landslide database must be established before implementing an analysis. Various types of datasets have been utilized in different studies. Selection of factors may be accomplished based on the information extracted from field investigations and related literature (Nandi and Shakoor 2010). The accuracy of landslide susceptibility maps relies on the quality of the selected landslide occurrence distribution, the conditioning factors and the adopted methodology. Therefore, data quality is one of the factors that control the quality of landslide susceptibility maps. Selecting the most suitable method is vital; hence, various methods have been developed and examined by many researchers (Akgun 2012; Ayalew and Yamagishi 2005; Saito et al. 2009; Tien Bui et al. 2012a).

The emergence of remote sensing (RS) and geographical information systems (GISs) has facilitated the application and extension of various algorithms and methods in landslide studies (Xu et al. 2012). New insights into landslide research have been obtained by determining and mitigating failures through these techniques. Without RS and GIS, extensive fieldwork, which requires considerable budget, is required to identify landslide-prone locations.

Landslide susceptibility mapping methods can be generally divided into two categories, namely qualitative and quantitative (Guzzetti et al. 1999). Most qualitative methods rely on previous landslide occurrences to determine locations that are prone to such a disaster in the future. Two of the popular qualitative approaches are weighted linear combination and analytic hierarchy process (AHP) (Ayalew and Yamagishi 2005; Ayalew et al. 2004; Yalcin et al. 2011). Ayalew and Yamagishi (2005) observed that these methods are appropriate for regional studies because different outputs are affected by the changing views of experts. Different conditions may be involved in the decision of an expert and may thus positively or negatively affect the prediction. However, the opinions of experts may vary. Chen et al. (2011) noted that a high value of subjectivity is involved in qualitative methods and could lead to questionable outcomes. Several researchers, such as Chen et al. 2011, stated that the disadvantage of AHP is that it requires the opinion of various experts from different regions. Moreover, human errors, which increase the possibility of uncertainty, are involved.

Statistical methods, which are considered quantitative approaches, are favored over qualitative approaches (Van Beek and Van Asch 2004). The reason for such preference is that expert knowledge is always involved in qualitative approaches and thus causes uncertainty in the results. Therefore, statistical methods have become popular. These methods rely on numerical expressions of the relationship

B. Pradhan (✉) · M.N. Jebur
Department of Civil Engineering, University Putra Malaysia, Serdang, Malaysia
e-mail: biswajeet24@gmail.com

B. Pradhan (ed.), *Laser Scanning Applications in Landslide Assessment*,
DOI 10.1007/978-3-319-55342-9_8

between conditioning factors and landslide occurrence. Various statistical techniques, including frequency ratio (FR) and logistic regression (LR), exist and have been employed in landslide modeling. Most statistical approaches require the establishment of strict assumptions prior to the study. LR as a multivariate statistical analysis (MSA) can overcome this difficulty by offering an efficient method of statistical analysis and also allowing for the incorporation of bivariate statistical analysis (BSA). Many researchers have employed LR in landslide susceptibility mapping (Akgun 2012; Felicísimo et al. 2013; Jebur et al. 2014c). In LR analysis, the correlation between landslide occurrence and conditioning factors is considered (Kavzoglu et al. 2014). FR can also be performed with BSA, in which the weight of each class of each factor is assigned by calculating its effect on landslide occurrence. The determination of each weight can be accomplished by analyzing the correlation between landslide conditioning factors and landslide inventory. The assigned weight is correlated to the landslide density in each class. The FR approach considers the impact of the classes of each factor on landslide occurrence. However, this approach disregards the relation between landslide occurrence and conditioning factors.

Machine learning methods are popular in natural hazard studies which can handle data from different measurement scales. For instance, artificial neural network (ANN), which is one of the machine learning methods, has been widely utilized in landslide susceptibility mapping in various case studies (Choi et al. 2010; Pradhan et al. 2010). The drawback of ANN is that it requires a large amount of data for processing (Pradhan and Lee 2010a). Furthermore, weak prediction results can be obtained when the testing data have values beyond the range of training data. Yilmaz (2009) indicated that the main disadvantage of ANN is that the input and output analyses and training operations are time-consuming. Moreover, ANN requires the data to be converted into another format, such as ASCII, to be utilized in the process. Such procedure may be difficult because a large amount of data is involved. Other machine learning techniques, such as support vector machine (SVM) (Tien Bui et al. 2012a) and decision tree (DT) (Saito et al. 2009), are examples of machine learning algorithms. Machine learning methods are efficient approaches for landslide modeling because of their capability to process large amounts of data with reasonably higher precision than statistical methods (Nefeslioglu et al. 2010; Pradhan 2013b). Moonjun (2007) established the distinction among SVM, DT, LR, and AHP in landslide modeling and found that although these approaches are all acceptable, SVM outperforms the other algorithms in most areas.

The disadvantage of machine learning algorithms is that their analysis requires a large amount of time to some extent (Moonjun 2007). Considering that each method has disadvantages, researchers have developed advanced ensemble methods to address the drawbacks of individual ones. Jang (1993) employed the Takagi–Sugeno rule format to introduce the adaptive network-based fuzzy inference system (ANFIS), which is widely utilized in complicated system modeling (Tien Bui et al. 2012b). ANFIS is a combination of ANN and the fuzzy interface system (FIS) and results in high performance and minimum input. Pradhan (2013b) compared the performance of SVM, ANFIS, and DT in mapping landslide susceptibility in Malaysia and found that ANFIS provides results that are more accurate than those of DT and SVM; however, ANFIS is time-consuming, which limits its application (Chau et al. 2005). Tien Bui et al. (2012b) demonstrated that ANFIS requires more factors than the LR model in landslide mapping.

As demonstrated by previous studies, several algorithms have limitations in landslide modeling. Hence, the objective of the current research is to obtain accurate results through the use of ensemble methods. These methods have been proven effective in the related literature and can increase the accuracy and overcome the disadvantages of stand-alone algorithms (Wan et al. 2012). For example, ANFIS, which is an ensemble of ANN and FIS, has been proven to be more accurate than stand-alone ANN and FIS. Another example of model combination is the hybrid integration of ANN with the genetic algorithm (ANN–GA). ANN–GA utilizes the advantages of both models. Chau et al. (2005) evaluated the strength and validity of ANN–GA, ANFIS, and LR by comparing their performance. LR has the highest uncertainty among the three; ANN–GA and ANFIS perform better than LR because of their nonlinear nature.

According to the aforementioned studies, the information in the spatial database, its quality and appropriateness, the type and performance of each model affect the prediction capability of landslide susceptibility maps. *k*-nearest neighbor (kNN) is the most widely utilized algorithm in non-parametric density estimation (Brahim-Belhouari et al. 2005). Despite its simplicity, the approach performs very well. The non-parametric kNN algorithm performs landslide inventory prediction, where all conditioning factors are acquired simultaneously using similar implied field data (Ohmann and Gregory 2002). Marjanović et al. (2009) used SVM and kNN algorithms, to map the landslide susceptible areas in Fruška Gora Mountain, in vicinity of Novi Sad, NW Serbia. Another study was done by de Souza and Ebecken (2012) which utilized kNN method to perform landslide analysis in Rio de Janeiro city. Cheng et al. (2013) used kNN classifier to classify the sub-images into landslide areas and non-landslide areas based on the object distribution. kNN is an important standard algorithm. This study examined the efficiency of the kNN algorithm in landslide susceptibility mapping. Although machine learning approaches have some advantages over statistical approaches, the required processing time in machine learning methods limits the applicability of such methods in natural hazard studies, particularly when various types of data, such as nominal, scale, and ordinal, are employed in the analysis. The goal of this study is to determine if the required time can be reduced with the kNN algorithm. In order to determine the efficiency of kNN, LR method was used to be compared with kNN. Therefore, another aim of this study is to compare prediction rates from both kNN and LR applications and their usefulness in landslide susceptibility mapping. kNN and LR (MSA) were applied to estimate the correlation between conditioning factors and landslide occurrence.

8.2 Study Area and Data

The case study is a landslide-prone area in Ulu Klang, Bukit Antarabangsa. The location was selected with different spatial datasets. The approximate area boundary is at a longitude range of 101° 44′ 21.657″E–101° 47′ 11.058″E and latitude of 3° 9′ 57.492″N to 3° 14′ 5.062″N. The area of coverage is 26 km^2 (Fig. 8.1). The study area has a tropical climate similar to that of the entire Malaysia, and the average monthly temperature is between 29 and 32 °C. Monthly rainfall ranges from 85 to 240 mm. The lack of appropriate planning and design for hillside expansion has exposed Bukit Antarabangsa to landslide risks. Five lives were lost in the extensive landslide that occurred in Bukit Antarabangsa at 3:30 am on December 6, 2008. This disaster caused thousands of people to leave their homes. This area was selected because of the frequent catastrophic landslides that have occurred in recent years. The dominant land uses include urban and vegetation. Loam and clay are the predominant types of soil; these types are known to have very weak structures (Althuwaynee et al. 2012). The geologic structure of the study area is reasonably homogeneous and contains three rock types: vein quartz, acid intrusives, and schist. The landslides that occurred in the area were generally triggered by heavy monsoon rainfall.

8.2.1 Landslide Inventory Map

One of the most important factors in predicting landslides is the inventory map, which can represent single or numerous occurrences of landslides in a specific area (Tien Bui et al. 2012a). Various resources are utilized to establish an inventory map, and historical landslide data, such as aerial photos, field surveys, QuickBird satellite imagery, and government agency records, are collected (Althuwaynee et al. 2012; Hassaballa et al. 2013). In this study, a landslide inventory map of the area was prepared with reports, satellite and aerial photos, and GPS surveying. A total of 31 landslide incidents were employed in the analysis (Fig. 8.1). To produce a susceptibility map, the landslide occurrences were divided into two classes, namely testing and training. Various researchers have recommended that roughly 70% of landslide incidents can be utilized for training and the remaining 30% for validation (Tehrany et al. 2013; Xu et al. 2013). Using the above-mentioned method, 22 cases of landslide occurrences (70%) were randomly selected and utilized for the training of the algorithms (kNN and LR). The remaining nine incidents (30%) were employed for validation (Fig. 8.1). The dependent layer, which represents landslide, was constructed with landslide cases with a value of 1. Similarly, an equal number of cases were assigned to the areas with no possibility of landslide occurrence (i.e., flat areas) with a value of 0.

Various factors, such as unsaturated hydraulic conductivity, water pressure, and soil moisture condition, contribute to the increase in trapped water and pore water pressure. One out of three slip surfaces results in a shallow landslide, as in the case of most landslides in Bukit Antarabangsa. The locations of landslides are represented by demarcated marks at the middle of the head scarps. The presented points cannot represent the entire area covered by a particular landslide event. However, the above-mentioned limitation does not have an effect on the result because 85% of the landslides in the analyzed area (Bukit Antarabangsa) have a width of less than 30 m.

8.2.2 Landslide Conditioning Factors

The other main step in susceptibility analysis is to establish the conditioning factors that have a relationship with landslide occurrence. Prior to constructing a suitable set of conditioning factors that have a correlation with landslide occurrence, the primary contributor to landslides must be initially determined (Guzzetti et al. 1999). Conditioning factors contain topological information, such as altitude, slope, aspect, and curvature, and hydrological parameters, such as stream power index (SPI) and topographic wetness index (TWI). Other conditioning factors, such as soil, geology, land use/land cover (LULC), distance from rivers, and distance from roads, were also considered in this study (Yalcin 2008). In various regions, topographical factors are considered significant elements. However, in other areas, other environmental or geological parameters are deemed substantial. The availability of thematic layers varies comprehensively and relies on the method, scale, and type of collected data. Therefore, the conditioning factors utilized in this study were selected depending on the parameters most widely used by researchers. In other words, the selected parameters are those utilized by many other researchers;

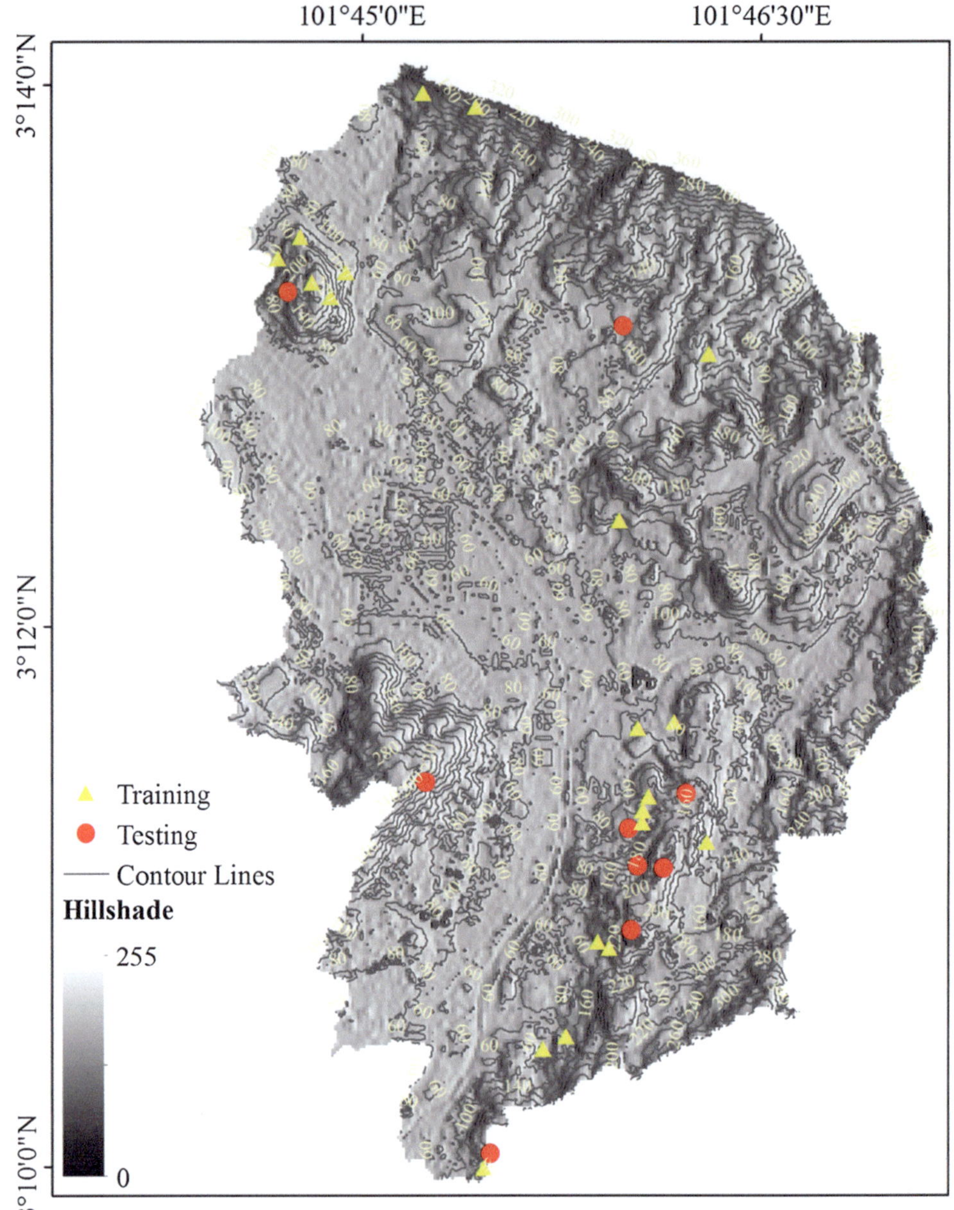

Fig. 8.1 Training and testing landslide events mapped on Bukit Antarabangsa, Ulu Klang, Malaysia hill-shaded map

these parameters have a significant impact on the potential of the terrain.

Related conditioning factors that have a spatial correlation with landslide occurrence were prepared to construct the probabilistic model. The data were resampled to 20 m. Topographic factors were acquired from a local Malaysian (Kertau-RSO) coordinate system with a grid size of 20 m. Altitude, slope, curvature, SPI, TWI, distance from rivers, and distance from roads were defined as scalar data. Geology, soil, and LULC were defined as nominal data. All conditioning factors are with a grid size of 20 m. Two of the important factors that play significant roles in landslide occurrence are altitude and slope degree. These factors are widely utilized in landslide susceptibility analysis (Moonjun 2007; Pradhan and Lee 2010a). Thus, slope and altitude maps constructed from a digital elevation model were included in the analysis (Figs. 8.2a and 8.2b). One of the other main conditioning factors is aspect, which has been employed in various studies (Fig. 8.2c) (23-IEEE). The impact of curvature on the process of landslide occurrence during downhill flow is the divergence or convergence of water (33-IEEE). Therefore, this parameter is related to landslide occurrence (Fig. 8.2d). Hydrology-related parameters, such as SPI and TWI, were calculated with Eqs. 8.1 and 8.2 as follows:

$$\mathrm{SPI} = A\ \tan\beta/b, \tag{8.1}$$

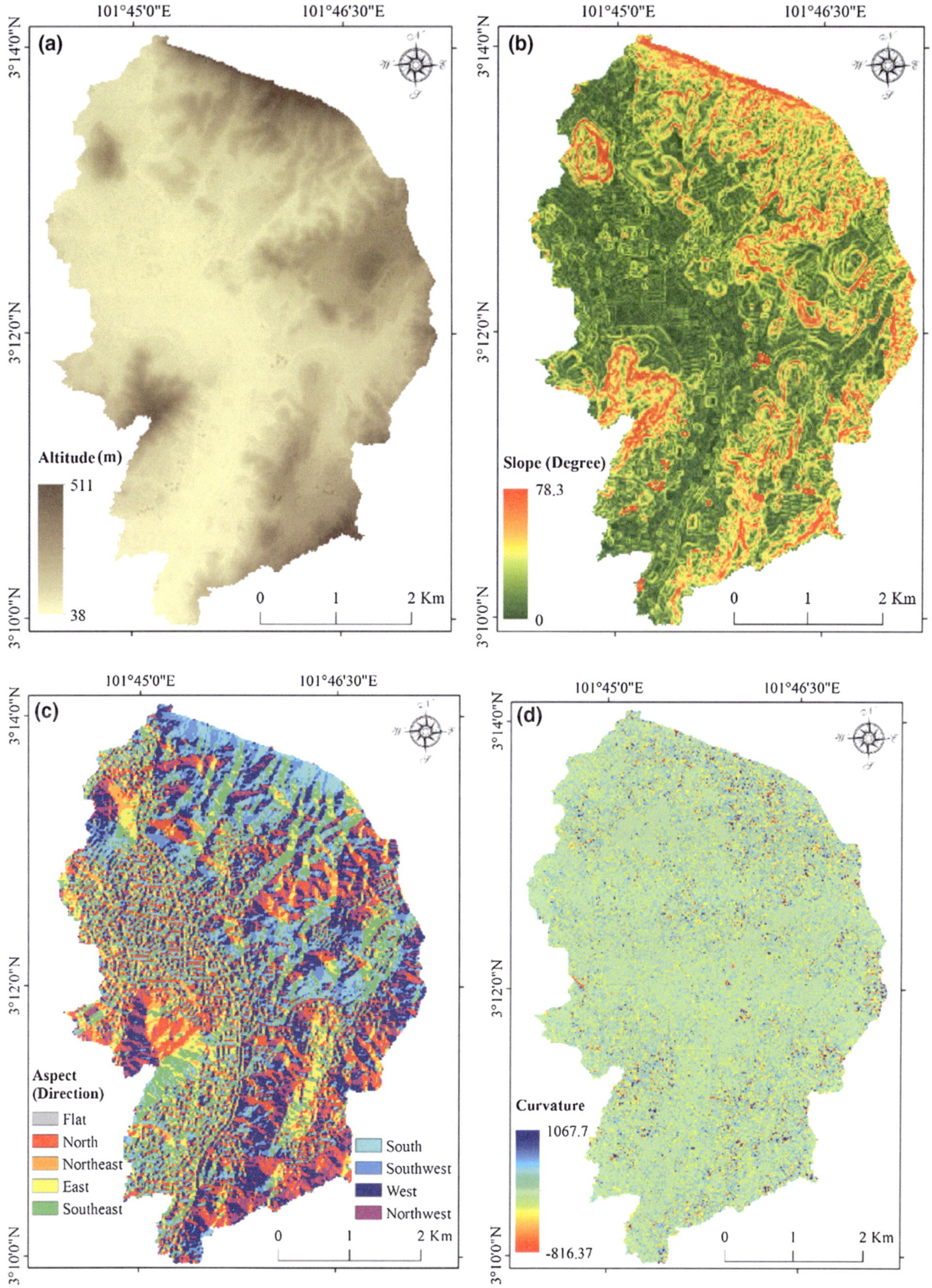

Fig. 8.2 Input thematic layers: **a** altitude, **b** slope, **c** aspect, **d** curvature, **e** SPI, **f** TWI, **g** Soil, **h** geology, **i** LULC, **j** distance from river, and **k** distance from road

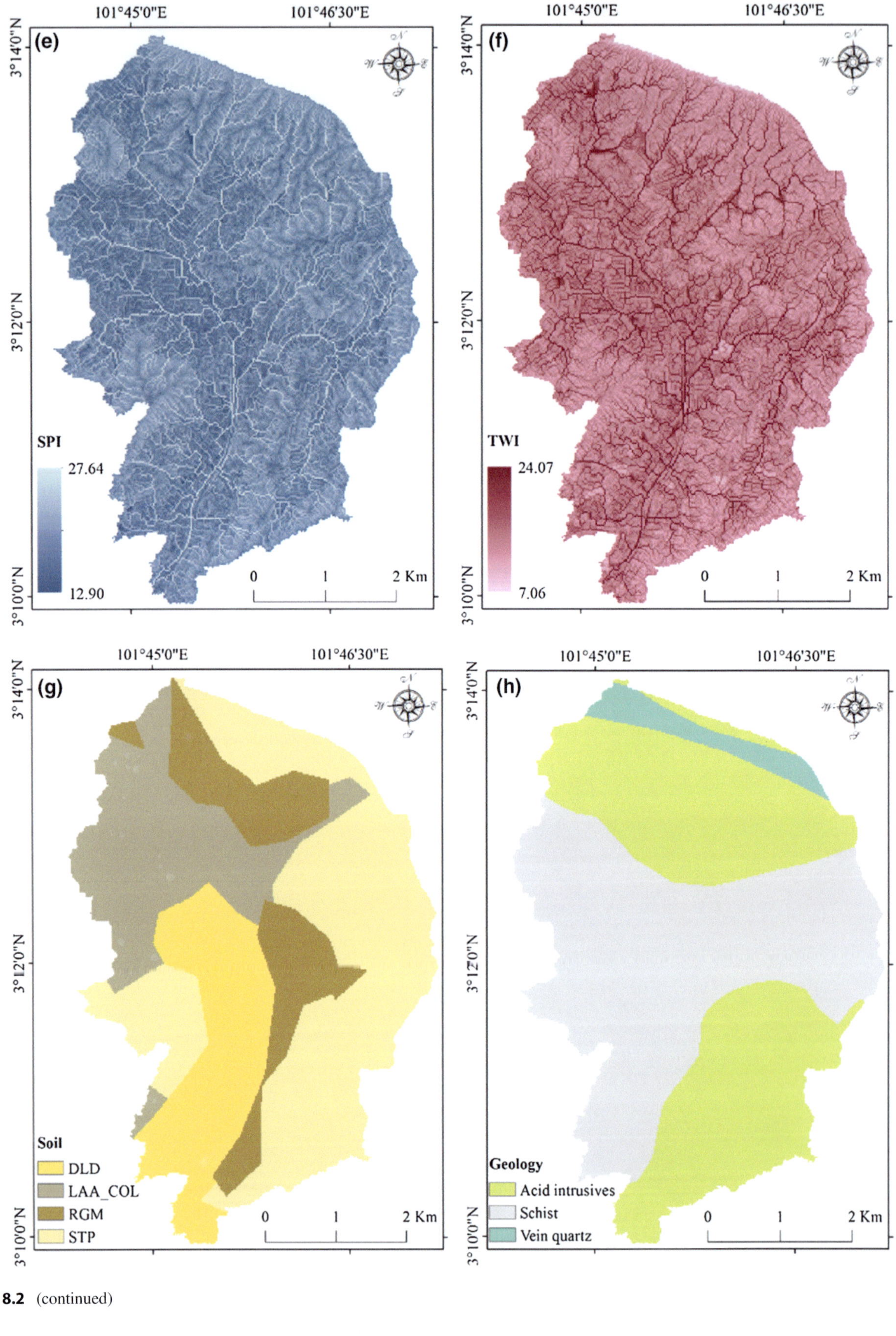

Fig. 8.2 (continued)

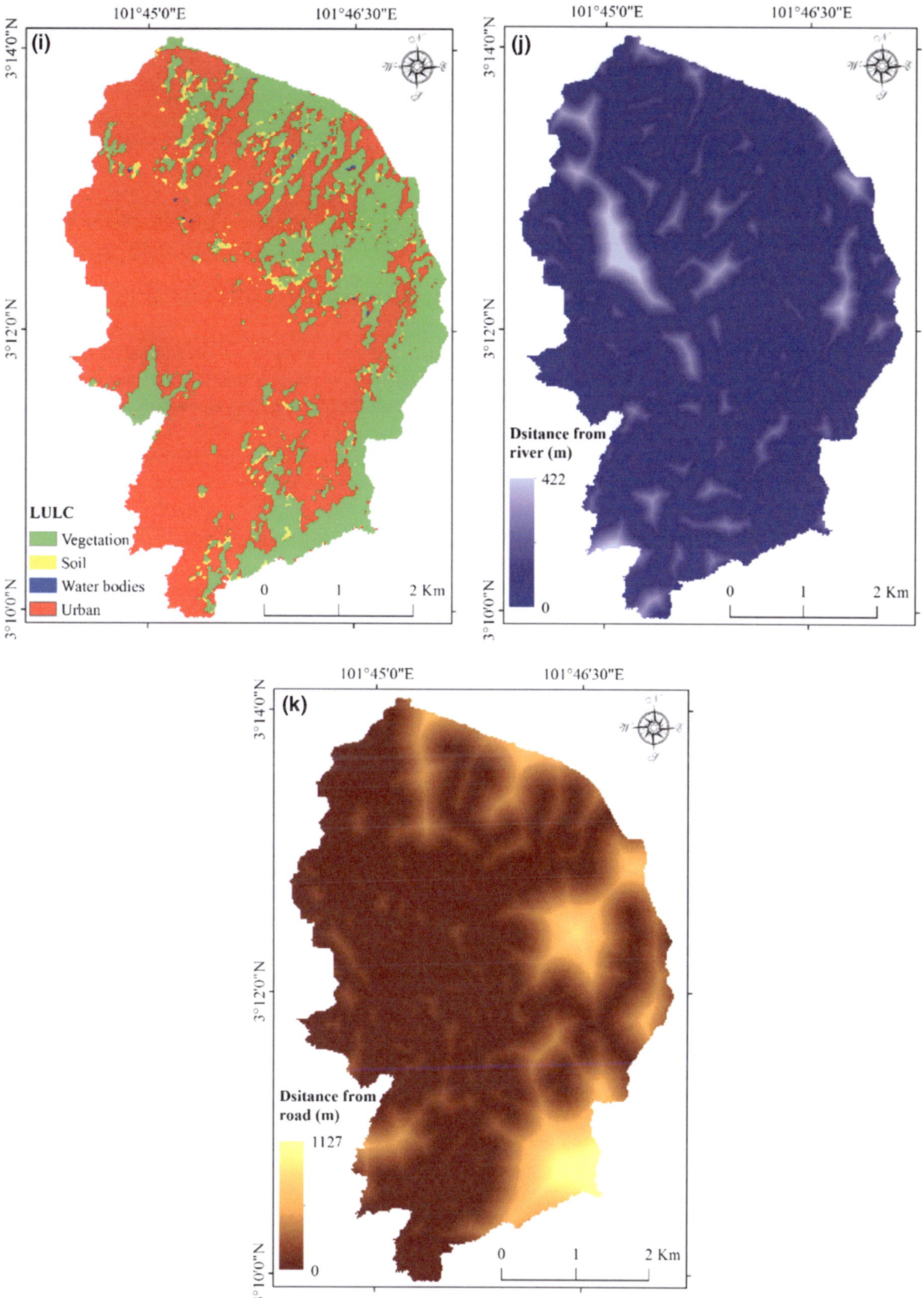

Fig. 8.2 (continued)

$$\mathrm{TWI} = \log_e(A/b \tan\beta), \quad (8.2)$$

where b (m) is the width of the cell through which the water flows, A (m^2) is the flow accumulation, and β (radian) is the slope (Regmi et al. 2010). In a particular catchment area, SPI is considered the amount of erosive power of flowing water depending on the presumption that discharge is proportional to the catchment. Net erosion can be calculated with SPI in the profile region and tangential convexity area (flow acceleration and convergence zones); net deposition can be calculated in the areas with profile concavity (zones of decreasing flow velocity) (Fig. 8.2e). TWI is the degree of accumulated water in a particular region (Fig. 8.2f).

8.3 Methodology

The types of soil employed in the analysis include disturbed land (DLD), local alluvium colluvium association (LAA-COL), rengam (RGM), and steep land (STP) (Fig. 8.2g). One of the important factors in various studies on landslide apportionment is geology. The current study area contains three types of lithology, namely vein quartz, acid intrusives, and schist (Fig. 8.2h). The main LULC types include urbanized areas followed by agricultural areas (particularly for oil palm, rubber, and paddies). A significant change has occurred in the urban areas because of deforestation. The distances from rivers and roads were calculated with the Euclidean distance tool in ArcGIS. The results were employed in the subsequent analysis (Figs. 8.2j, k and 8.3).

8.3.1 kNN

Landslide applications mostly require appropriate factors for all landslide variables (Brahim-Belhouari et al. 2005). Traditional landslide modeling algorithms have their drawbacks and limitations in this aspect. The non-parametric kNN algorithm performs landslide inventory prediction, where all conditioning factors are acquired simultaneously using similar implied field data (Ohmann and Gregory 2002). Through such technique, the weight of a conditioning factor

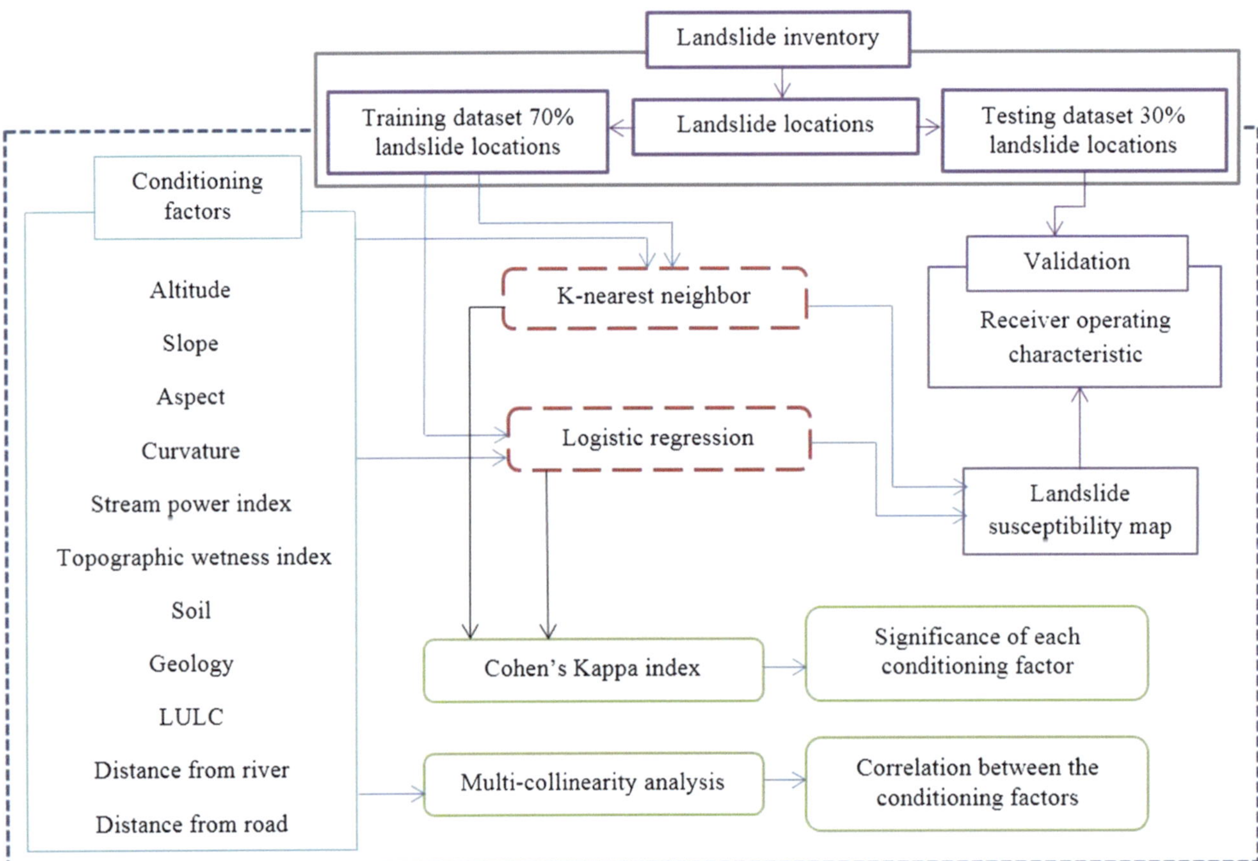

Fig. 8.3 Overall flowchart used in this study

is estimated as the sum of each factor over landslide events. In the kNN algorithm, prediction m_p for particular area P for scale factor M is determined as

$$m_p = \sum_{i=1}^{k} w_{i,p}\, m_i, \tag{8.3}$$

where m_i is the M variable value in reference plot i. The variables $w_{i,p}$, which represent the pixel weights, are estimated as

$$W_{i,p} = \begin{cases} d_{p_i,p}^{-t} / \sum_{j \in \{i_1(p),\ldots,j_k(p)\}} d_{p_j,p}^{-t}, & \text{if and only if } i \in \{i_1(p),\ldots,i_k(p)\}, \\ 0 \end{cases} \tag{8.4}$$

where the representation of the field plot and target set pixel is i and p, respectively. The corresponding area with the landslide spots is p_j, and $\{i_1(p),\ldots, i_k(p)\}$ is the set of k pixels that are nearest to area p. Katila and Tomppo (2001) and Franco-Lopez et al. (2001) demonstrated the impact of various t-values on prediction errors. In the current study, the nearest metric (shown in Eq. 8.4) was employed for the nominal factor (i.e., soil or geology); however, the following equation was utilized to predict the impact of the scale factor (i.e., altitude or slope).

$$d_{p_j p}^2 = \sum_{l=1}^{n_f} \omega_{l,f}^2 (f_{l,p_j} - f_{l,p})^2 \tag{8.5}$$

In the case of categorical (nominal) factors, instead of using the weighted average as in the case of the scale factor, the prediction median classes for the nearest neighbors can be utilized. The predicted landslide location has the highest sum value of weights w_i and p. Theoretically, when the weights are utilized, equal sums are rarely obtained. The probability is null if a neighbor is considered. If a similarity exists in the sum of two or more classes, then the selection of one class will be randomly processed through those with the highest value. The categorical factors included in this study are soil, geology, and LULC, which were predicted through this approach.

8.3.2 LR

One of the popular MSA algorithms is LR, which considers several parameters that have a possible influence on landslide occurrence. In LR method, data distribution does not necessarily need to be normal. Data distribution can be nominal, continuous, or a combination of both (Lee and Pradhan 2007). With LR technique, analyzing the correlation between continuous or nominal data (i.e., conditioning factors) and the binary dependent variable is possible (Shirzadi et al. 2012). In this study, the landslide events were utilized as dependent and binary variables. The absence and presence of landslides were assigned values of 0 and 1, respectively. The weight of each conditioning factor was extracted through LR and was then employed in GIS to establish the final susceptibility map. Multivariate analysis, where the coefficient values were calculated, was performed with SPSS software. In the LR algorithm, the impact of each conditioning factor can be deduced from the coefficient value (Ayalew and Yamagishi 2005). The landslide susceptibility map was produced with the following formula.

$$P = \frac{1}{1 + e^{-z}}, \tag{8.6}$$

where P represents the probability of landslides on an s-shaped curve ranging between 0 and 1. Furthermore, z represents the linear assembly, where LR involves fitting an equation of the following form to the data.

$$z = b_0 + b_1x_1 + b_2x_2 + b_3x_3 + b_nx_n, \tag{8.7}$$

where b_0 is the constant value of the algorithm, b_i (i = 0, 1, 2,. …, n) is the coefficients of the LR algorithm, and x_i (i = 0, 1, 2, …, n) shows the conditioning factors (Lee and Pradhan 2007). Finally, the probability map was divided into various categories to produce the susceptibility map (Ohlmacher and Davis 2003). Different categorization methods, such as equal interval (Kavzoglu et al. 2014), standard deviation (Günther et al. 2013), natural break (Ayalew et al. 2004), and quantile (Ayalew et al. 2004), can be used in the GIS environment for classification. In this research, ideal optimization was achieved with the quantile technique. Quantile method classifies the landslide susceptibility index into classes with equal area. When other methods were employed, the landslide-prone areas were exaggerated, and most of the areas in Bukit Antarabangsa were classified as highly susceptible. Tehrany et al. (2014b) mentioned that although the quantile technique classifies highly different discrete into the same class, the outcome is reasonably acceptable and good landslide susceptibility mapping can be achieved. Various researchers have utilized the quantile technique, and its efficiency in susceptibility mapping has been proven (Ayalew and Yamagishi 2005; Tehrany et al. 2013).

8.4 Results

8.4.1 kNN

The conditioning factors with the training layer were utilized for analysis to measure the probability map of landslides. By using Eqs. 8.4 and 8.5 for the conditioning factors, the probability map was produced and ranged from 0 to 1. In the processing of landslide susceptibility, 11 of both categorical and continuous parameters were used to produce two probability maps with values from 0 to 1. An area is considered likely to be prone to landslides when the value is close to 1. The scale outcome is the probability index, and further analysis was conducted to produce the final susceptibility map. No framework or rule exists to classify the probability index, which makes the task difficult (Ayalew et al. 2004). The problem of reformatting the probability map into various classes brings uncertainty in landslide susceptibility modeling. This phenomenon is due to human error given that susceptible zones are usually defined based on expert knowledge and opinion. A few classification methods exist. Four of the most popular classification methods, namely quantile, natural break, standard deviation, and equal interval methods, were tested in this study (Tehrany et al. 2014a). Each method has its own procedure of classifying values; thus, each method may produce different results. For example, standard deviation recognizes the mean of input data and consequently partitions the data into classes using the standard deviation of the mean. The number of categories in this algorithm is constant, which is inappropriate for susceptibility research because a target number of classes is required. Equal interval method is unsuitable because it emphasizes one categorical susceptible class. Natural break method defines the boundary of each category depending on the inherent nature of the input data whenever a significant change in value occurs. In this study, a probability histogram was derived and examined; no significant change in value was detected. Thus, the natural break scheme is inappropriate for the current study. When the quantile method was employed, the derived classes resulted in an equal area format, where each category has a similar amount of pixels. Hence, the quantile algorithm was used because it is the best fit for the current research.

8.4.2 LR

The LR algorithm, a popular algorithm in various landslides studies, was applied to determine the relationship between landslide occurrence and landslide influence parameters (Kavzoglu et al. 2014; Regmi et al. 2014). The coefficients calculated with LR are listed in Table 8.1. A positive coefficient value for a particular factor means that the influence of this factor in the selected catchment increases the possibility of landslide occurrence. By contrast, a negative coefficient value means that the existence of a particular factor decreases the probability of landslide occurrence (Chauhan et al. 2010).

Significant probability (Sig) was computed through LR processing. The significant influence on landslides can be determined with the Sig factor (Papadopoulou-Vrynioti et al. 2013). A conditioning factor considered statistically has a significant effect on landslides if its Sig value is <0.05. The results show that slope, aspect, SPI, TWI, and distance from roads are the most important independent parameters with values of (0.02), (0.03), (0.03), (0.03), and (0.04), respectively. Other independent parameters, such as altitude, soil, geology, curvature, LULC, and distance from rivers, with Sig values higher than 0.05 are not as effective as the others. To obtain the landslide probability index, the regression coefficients for each factor were defined with Eq. 8.4.

$$\begin{aligned} Z =& (0.92\text{Slope}) - (0.079\text{Altitude}) + (0.038\text{Aspect}) \\ & + (0.202\text{Curvature}) + (25.318\text{SPI}) - (20.513\text{TWI}) \\ & + \text{Soil} + \text{Geology} + \text{LULC} - (0.073\text{River}) \\ & - (0.085\text{Road}) - 255.931 \end{aligned} \tag{8.8}$$

Next, the probability map was measured with Eq. 8.8. The values of the probability map range from 0 to 1. The

Table 8.1 Multicollinearity diagnostics of the conditioning factors

Layer	Altitude	Slope	Aspect	Curvature	SPI	TWI	Soil	Geology	LULC	River	Road
	VIF										
Altitude	1	1.295	1.027	1.093	2.149	2.400	1.245	1.099	1.352	1.071	1.291
Slope	2.415	1	1.029	1.097	2.080	2.247	1.263	1.098	1.503	1.088	1.791
Aspect	2.420	1.300	1	1.097	2.279	2.647	1.265	1.100	1.478	1.089	1.792
Curvature	2.416	1.300	1.029	1	2.232	2.636	1.266	1.101	1.502	1.088	1.791
SPI	2.286	1.186	1.029	1.074	1	1.298	1.253	1.089	1.494	1.089	1.780
TWI	2.198	1.103	1.029	1.092	1.118	1	1.254	1.093	1.498	1.084	1.772
Soil	2.384	1.296	1.028	1.097	2.255	2.622	1	1.080	1.493	1.048	1.783
Geology	2.420	1.295	1.029	1.097	2.254	2.268	1.242	1	1.502	1.087	1.778
LULC	2.181	1.299	1.012	1.096	2.265	2.639	1.258	1.100	1	1.087	1.777
River	2.384	1.298	1.029	1.096	2.280	2.634	1.219	1.099	1.500	1	1.769
Road	1.746	1.299	1.029	1.096	2.263	2.618	1.260	1.092	1.490	1.075	1

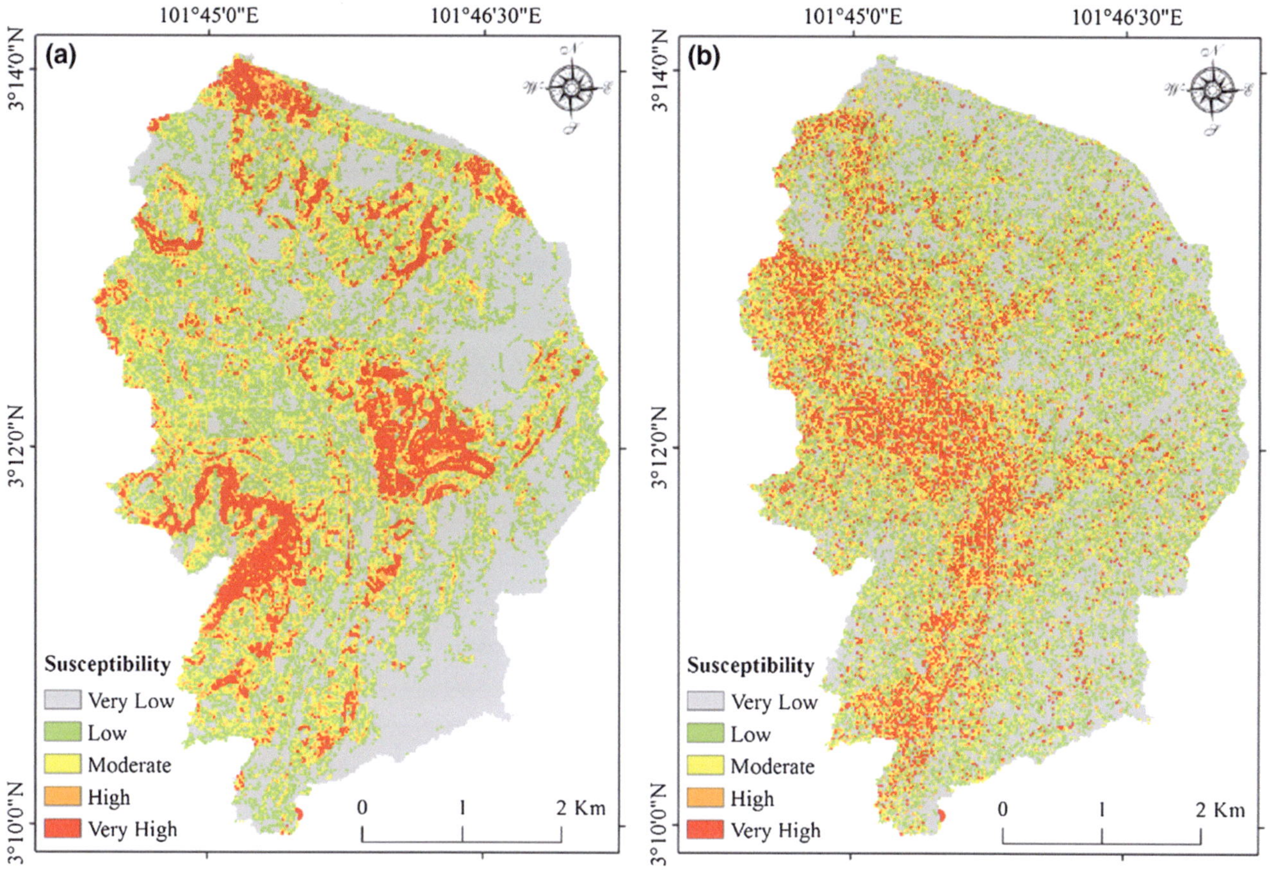

Fig.8.4 Landslide susceptibility maps derived from **a** kNN and **b** LR

probability map represents the prediction of landslides for each pixel obtained from the used independent variables. Finally, quantile method was implemented; the landslide susceptibility map was acquired and divided into five classes of landslide susceptibility as shown in Fig. 8.4.

8.5 Discussions

As mentioned previously, the landslide probability maps for both algorithms (kNN and LR) were divided into five classes, namely high, very high, moderate, low, and very low (Fig. 8.4) (Ayalew and Yamagishi 2005; Pradhan 2013a; Pradhan and Lee 2010b). The resulting maps show the probability of landslide occurrence in Bukit Antarabangsa, Ulu Klang, Malaysia. Huabin et al. (2005) reported that two factors need to be considered to obtain an effective landslide susceptibility map. The first factor is the overlapping of the conditioning factors in the inventory map to identify landslide-prone areas. The second factor is that the coverage area of landslide-prone areas should be small. The susceptible maps generated with the two algorithms are shown in Fig. 8.4.

The result obtained from kNN is visually different from the one derived from LR. kNN detected specific zones as highest susceptible areas, while LR distributed these regions. In both classified maps, the highest landslide-prone areas are located in the urban areas. On the other hand, vegetation areas are classified as less susceptible regions. It is due to the impact of vegetation on controlling the soil movement. To determine the most precise method of determining landslide-prone regions, the obtained results must be compared. Through the area under curve (AUC) technique, the success and prediction rates of the two susceptibility maps were calculated based on data on previous landslide events. AUC is considered as one of the most popular methods to assess the efficiency of the generated landslide maps which produces both success and prediction rates. Prediction and success rates should be evaluated as an essential outcome of every program. The validation process is implemented by comparing the existing landslide events with the acquired landslide probability map. As it has been mentioned in data used section, landslide events should be divided into two datasets of training and testing. If the AUC is calculated using training dataset, it produces success rate; if AUC is calculated using testing dataset, it gives perdition rate.

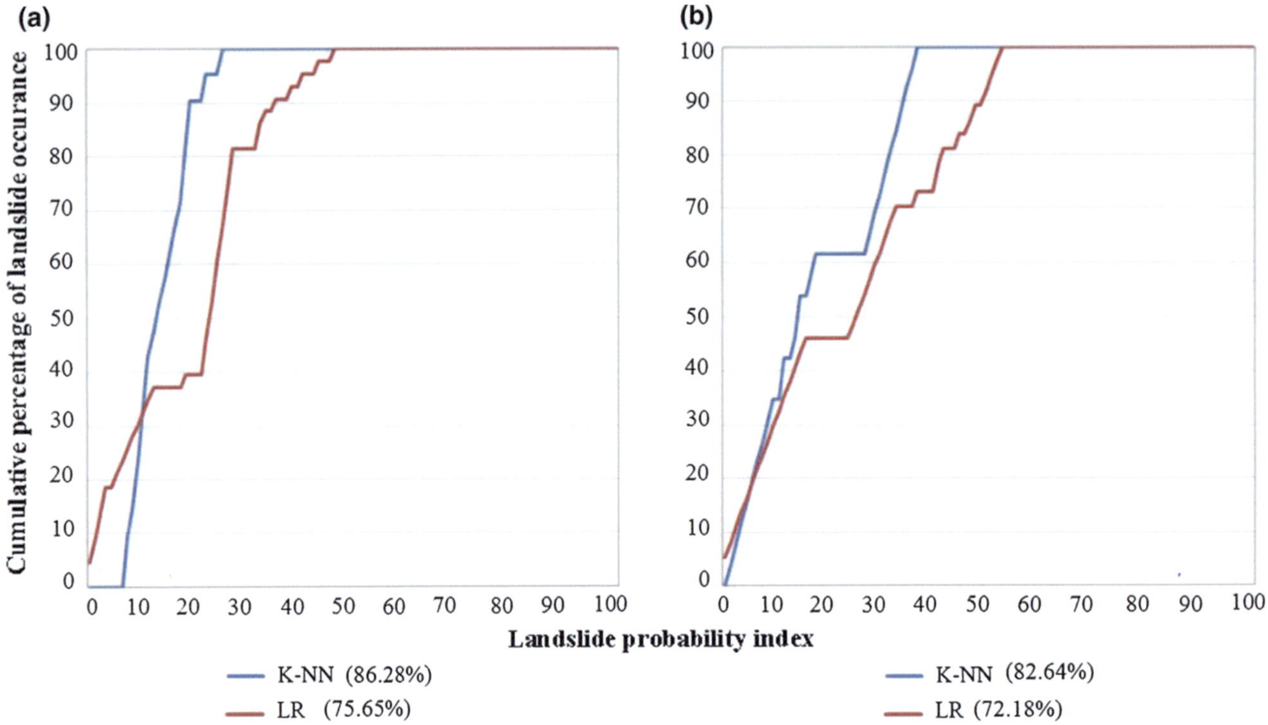

Fig. 8.5 Graphic representation of the cumulative frequency diagram presenting the cumulative landslide occurrence (%; *y*-axis) in landslide probability index rank (%; *x*-axis): **a** success rate and **b** prediction rate

Success rate shows how well the model worked with training dataset, and the prediction rate shows how well the model can predict the landslide-prone areas. The representation of AUC for LR and kNN is shown in Fig. 8.5.

The success rate should be analyzed to identify the algorithm that is suitable for a selected study. Success rate shows the efficiency of the algorithms in handling the case study. A low success rate indicates that the method utilized is unsuitable for a particular study. In the current study, both LR and kNN exhibited reasonable and acceptable success rates of 75.65 and 86.28%, respectively. These results indicate that the kNN algorithm is useful and can be utilized in the subsequent analysis. The actual efficiency of the obtained result is expressed by the prediction rate. The prediction rates obtained for LR and kNN were 72.18 and 82.64%, respectively. In terms of prediction rates, the kNN algorithm exhibited higher efficiency and reasonable results.

As discussed in a previous study, a difference of approximately 14% in the prediction accuracy of two algorithms means that several of the utilized conditioning factors generated some noise in the analysis (Chang et al. 2007). Therefore, multicollinearity could exist (Zhu and Huang 2006). When perfect correlation exists among the input factors, the prediction of the regression model would be imperfect. In other words, collinearity means that two parameters have a linear relation. Multicollinearity is applicable to situations where more than two parameters are involved. Regression models, such as LR, are very sensitive to collinearity (Ozdemir 2011). Therefore, the use of parameters that are not significant in the regression analysis will reduce the prediction accuracy of the model. Hence, a multicollinearity analysis was performed. The variance inflation factor (VIF) is shown in Table 8.1. A VIF value greater than 10 denotes the existence of multicollinearity (Ozdemir 2011). No VIF value greater than 10 was obtained. Therefore, no actual multicollinearity exists among the input parameters.

In both algorithms, the highly prone areas have sharp slopes and high altitudes. Altitude and slope thus have significant effects on landslide occurrence. Areas with weak rock types, such as vein quartz, are considered susceptible. As speculated, low areas with sharp slopes and close distance to a river have low landslide susceptibility. Another investigation was performed to determine the significance of each particular conditioning factor. Each factor was excluded, and the model was implemented to calculate AUC. This technique was applied to both kNN and LR algorithms, and the result is shown in Table 8.2.

In the case of LR, a significant reduction was observed when the slope factor was eliminated from the analysis. This phenomenon shows that the slope layer has a large impact on the performance of LR. In the case of kNN, the most significant parameter was the altitude layer. The measured accuracy was reduced when altitude layer was

Table 8.2 The relative important of landslide conditioning factors for the two models

Layer	AUC (prediction)	
	LR	kNN
Without altitude	66.93	70.09
Without slope	62.65	77.45
Without aspect	67.77	80.03
Without curvature	66.74	76.72
Without SPI	66.27	74.91
Without TWI	67.03	73.82
Without soil	67.26	75.61
Without geology	69.12	76.69
Without LULC	67.05	79.2
Without river	66.05	78.3
Without road	67.17	78.83

eliminated from the analysis. Meanwhile, removing other factors produced a slight difference in the accuracy of the algorithm.

8.6 Conclusion

Various algorithms and methods have been applied in the literature to construct landslide susceptibility maps. Several of these techniques have disadvantages, including difficulties in data management or time-consuming analysis. A few well-defined algorithms have not been utilized in landslide modeling and need to be tested. The use of a new algorithm involves different aspects, such as time and accuracy. This research assessed the potential application of a kNN algorithm in landslide susceptibility mapping for Bukit Antarabangsa, Ulu Klang, Malaysia. LR, which is a well-established technique in landslide research, was also employed. Its efficiency was compared with that of the kNN model.

Landslide occurrence is correlated with different conditioning factors. Eleven landslide conditioning factors, namely altitude, slope, aspect, curvature, SPI, TWI, soil, geology, LULC, distance from rivers, and distance from roads were acquired from various sources and used in this study. Similar conditioning factors were used in both kNN and LR to examine the effectiveness of the kNN method in landslide mapping. The probability of landslide occurrence for LR and kNN was derived, and the impact of each factor on landslide occurrence was assessed. kNN was implemented, and its result was compared with the output of LR method. AUC charts were utilized to validate each method by calculating both success and prediction rates. kNN and LR acquired prediction values of 82.64 and 72.18%, respectively. The kNN model has better prediction capability than the LR method. However, one disadvantage of kNN is that extensive analysis is required to estimate the susceptibility output because all the training data need to be stored. The proposed model has the potential to perform better than other popular methods and slightly increased the precision of the processing for the study area. The obtained probability maps for each method were classified through quantile technique. Two landslide susceptibility outputs were mapped. The obtained susceptibility maps provide prediction information on the spatial component of landslides without knowing "when" and "how frequent" the next landslide would occur. The validation results indicated that the landslide susceptibility maps produced in this study are of good quality. Therefore, the findings of this research may help governments and researchers predict and manage future landslides. In addition, the proposed algorithm can be utilized because it is simple, scalable, and has an acceptable and reasonable prediction capability. The proposed method can be useful for complicated problems (i.e., landslides analysis), although it requires large amounts of analysis compared with other algorithms. This study can be enhanced in the future by considering more conditioning factors. However, adding other parameters may increase the amount of analysis time required by the algorithm. Research should be performed on other case studies to examine the kNN algorithm. Testing other unused or unexamined algorithms is also recommended to help determine the most suitable technique for landslide susceptibility mapping. The results of this study verify the superiority of the kNN algorithm over other methods and can be used as a reference for future studies.

References

Akgun, A. (2012). A comparison of landslide susceptibility maps produced by logistic regression, multi-criteria decision, and likelihood ratio methods: A case study at İzmir, Turkey. *Landslides, 9,* 93–106.

Althuwaynee, O. F., Pradhan, B., & Lee, S. (2012). Application of an evidential belief function model in landslide susceptibility mapping. *Computers & Geosciences, 44,* 120–135.

Antronico, L., Borrelli, L., Coscarelli, R., & Gullà, G. (2015). Time evolution of landslide damages to buildings: The case study of Lungro (Calabria, southern Italy). *Bulletin of Engineering Geology and the Environment, 74,* 47–59.

Ayalew, L., & Yamagishi, H. (2005). The application of GIS-based logistic regression for landslide susceptibility mapping in the Kakuda-Yahiko mountains, Central Japan. *Geomorphology, 65,* 15–31.

Ayalew, L., Yamagishi, H., & Ugawa, N. (2004). Landslide susceptibility mapping using GIS-based weighted linear combination, the case in Tsugawa area of Agano river, Niigata Prefecture, Japan. *Landslides, 1,* 73–81.

Brahim-Belhouari, S., Bermak, A., Shi, M., & Chan, P. C. (2005). Fast and robust gas identification system using an integrated gas sensor technology and gaussian mixture models. *IEEE Sensors Journal, 5,* 1433–1444.

Chang, K.-T., Chiang, S.-H., & Hsu, M.-L. (2007). Modeling typhoon-and earthquake-induced landslides in a mountainous watershed using logistic regression. *Geomorphology, 89,* 335–347.

Chau, K., Wu, C., & Li, Y. (2005). Comparison of several flood forecasting models in Yangtze River. *Journal of Hydrologic Engineering, 10,* 485–491.

Chauhan, S., Sharma, M., & Arora, M. K. (2010). Landslide susceptibility zonation of the Chamoli region, Garhwal Himalayas, using logistic regression model. *Landslides, 7,* 411–423.

Chen, Y.-R., Yeh, C.-H., & Yu, B. (2011). Integrated application of the analytic hierarchy process and the geographic information system for flood risk assessment and flood plain management in Taiwan. *Natural Hazards, 59,* 1261–1276.

Cheng, G., Guo, L., Zhao, T., Han, J., Li, H., & Fang, J. (2013). Automatic landslide detection from remote-sensing imagery using a scene classification method based on BoVW and pLSA. *International Journal of Remote Sensing, 34,* 45–59.

Choi, J., Oh, H.-J., Won, J.-S., & Lee, S. (2010). Validation of an artificial neural network model for landslide susceptibility mapping. *Environmental Earth Sciences, 60,* 473–483.

de Souza, F. T., & Ebecken, N. F. (2012). A data based model to predict landslide induced by rainfall in Rio de Janeiro city. *Geotechnical and Geological Engineering, 30,* 83–94.

Dehnavi, A., Aghdam, I. N., Pradhan, B., & Varzandeh, M. H. M. (2015). A new hybrid model using step-wise weight assessment ratio analysis (SWARA) technique and adaptive neuro-fuzzy inference system (ANFIS) for regional landslide hazard assessment in Iran. *Catena, 135,* 122–148.

Felicísimo, Á. M., Cuartero, A., Remondo, J., & Quirós, E. (2013). Mapping landslide susceptibility with logistic regression, multiple adaptive regression splines, classification and regression trees, and maximum entropy methods: A comparative study. *Landslides, 10,* 175–189.

Fiorucci, F., Reichenbach, P., Rossi, M., Cardinali, M., & Guzzetti, F. (2015). Integration of landslide susceptibility products in the environmental plans. In: *EGU General Assembly Conference Abstracts* (Vol. 17, pp 12404).

Franco-Lopez, H., Ek, A. R., & Bauer, M. E. (2001). Estimation and mapping of forest stand density, volume, and cover type using the k-nearest neighbors method. *Remote Sensing of Environment, 77,* 251–274.

Günther, A., Reichenbach, P., Malet, J.-P., Van Den Eeckhaut, M., Hervás, J., Dashwood, C., et al. (2013). Tier-based approaches for landslide susceptibility assessment in Europe. *Landslides, 10,* 529–546.

Guzzetti, F., Carrara, A., Cardinali, M., & Reichenbach, P. (1999). Landslide hazard evaluation: A review of current techniques and their application in a multi-scale study, Central Italy. *Geomorphology, 31,* 181–216.

Hassaballa, A. A., Althuwaynee, O. F., & Pradhan, B. (2013). Extraction of soil moisture from RADARSAT-1 and its role in the formation of the 6 December 2008 landslide at Bukit Antarabangsa, Kuala Lumpur. *Arabian Journal of Geoscience, 7,* 1–10.

Huabin, W., Gangjun, L., Weiya, X., & Gonghui, W. (2005). GIS-based landslide hazard assessment: An overview. *Progress in Physical Geography, 29,* 548–567.

Jang, J.-S. (1993). ANFIS: Adaptive-network-based fuzzy inference system. *IEEE Transaction on Systems, Man, and Cybernetics, 23,* 665–685.

Jebur, M., Pradhan, B., & Tehrany, M. (2014a). Manifestation of LiDAR-derived parameters in the spatial prediction of landslides using novel ensemble evidential belief functions and support vector machine models in GIS. *IEEE Journal of Selected Topics in Applied Earth Observations and Remote Sensing, 8*(2), 674–690.

Jebur, M. N., Pradhan, B., & Tehrany, M. S. (2014b). Detection of vertical slope movement in highly vegetated tropical area of Gunung pass landslide, Malaysia, using L-band InSAR technique. *Geosciences Journal, 18,* 61–68.

Jebur, M. N., Pradhan, B., & Tehrany, M. S. (2014c). Optimization of landslide conditioning factors using very high-resolution airborne laser scanning (LiDAR) data at catchment scale. *Remote Sensing of Environment, 152,* 150–165.

Katila, M., & Tomppo, E. (2001). Selecting estimation parameters for the Finnish multisource national forest inventory. *Remote Sens Environment, 76,* 16–32.

Kavzoglu, T., Sahin, E. K., & Colkesen, I. (2014). Landslide susceptibility mapping using GIS-based multi-criteria decision analysis, support vector machines, and logistic regression. *Landslides, 11,* 425–439.

Lee, S., & Pradhan, B. (2007). Landslide hazard mapping at Selangor, Malaysia using frequency ratio and logistic regression models. *Landslides, 4,* 33–41.

Marjanović, M., Bajat, B., Kovačević, M. (2009). Landslide susceptibility assessment with machine learning algorithms. In: *Intelligent Networking and Collaborative Systems, 2009. INCOS'09. International Conference on* (pp 273–278).

Moonjun, R. (2007). Application of artificial neural network and decision tree in a GIS based predictive soil mapping for landslide vulnerability study: A case study of Hoi Num Rin sub–Watershed, Thailand (104pp). The Netherlands: ITC, Enschede.

Nandi, A., & Shakoor, A. (2010). A GIS-based landslide susceptibility evaluation using bivariate and multivariate statistical analyses. *Engineering Geology, 110,* 11–20.

Nefeslioglu, H., Sezer, E., Gokceoglu, C., Bozkir, A., & Duman, T. (2010). Assessment of landslide susceptibility by decision trees in the metropolitan area of Istanbul, Turkey. *Mathematical Problems in Engineering, 2010*(901095), 15.

Ohlmacher, G. C., & Davis, J. C. (2003). Using multiple logistic regression and GIS technology to predict landslide hazard in northeast Kansas, USA. *Engineering Geology, 69,* 331–343.

Ohmann, J. L., & Gregory, M. J. (2002). Predictive mapping of forest composition and structure with direct gradient analysis and nearest-neighbor imputation in coastal Oregon, USA. *Canadian Journal of Forest Research, 32,* 725–741.

Ozdemir, A. (2011). Using a binary logistic regression method and GIS for evaluating and mapping the groundwater spring potential in the Sultan mountains (Aksehir, Turkey). *Journal of Hydrology, 405,* 123–136.
Papadopoulou-Vrynioti, K., Bathrellos, G. D., Skilodimou, H. D., Kaviris, G., & Makropoulos, K. (2013). Karst collapse susceptibility mapping considering peak ground acceleration in a rapidly growing urban area. *Engineering Geology, 158,* 77–88.
Pradhan, B. (2013a). A comparative study on the predictive ability of the decision tree, support vector machine and neuro-fuzzy models in landslide susceptibility mapping using GIS. *Computers & Geosciences, 51,* 350–365.
Pradhan, B. (2013b). A comparative study on the predictive ability of the decision tree, support vector machine and neuro-fuzzy models in landslide susceptibility mapping using GIS. *Computers & Geosciences, 51,* 350–365.
Pradhan, B., & Lee, S. (2010a). Delineation of landslide hazard areas on Penang Island, Malaysia, by using frequency ratio, logistic regression, and artificial neural network models. *Environmental Earth Sciences, 60,* 1037–1054.
Pradhan, B., & Lee, S. (2010b). Landslide susceptibility assessment and factor effect analysis: Backpropagation artificial neural networks and their comparison with frequency ratio and bivariate logistic regression modelling. *Environmental Modelling and Software, 25,* 747–759.
Pradhan, B., Lee, S., & Buchroithner, M. F. (2010). A GIS-based back-propagation neural network model and its cross-application and validation for landslide susceptibility analyses. *Computers, Environment and Urban Systems, 34,* 216–235.
Rahman, A.-U., Khan, A. N., & Collins, A. E. (2014). Analysis of landslide causes and associated damages in the Kashmir Himalayas of Pakistan. *Natural Hazards, 71,* 803–821.
Regmi, N. R., Giardino, J. R., McDonald, E. V., & Vitek, J. D. (2014). A comparison of logistic regression-based models of susceptibility to landslides in western Colorado, USA. *Landslides, 11,* 247–262.
Regmi, N. R., Giardino, J. R., & Vitek, J. D. (2010). Modeling susceptibility to landslides using the weight of evidence approach: Western Colorado, USA. *Geomorphology, 115,* 172–187.
Runyan, C. W., D'odorico, P. (2014). Bistable dynamics between forest removal and landslide occurrence. *Water Resources Research, 50,* 1112–1130.
Saito, H., Nakayama, D., & Matsuyama, H. (2009). Comparison of landslide susceptibility based on a decision-tree model and actual landslide occurrence: The Akaishi Mountains, Japan. *Geomorphology, 109,* 108–121.
Shirzadi, A., Saro, L., Joo, O. H., & Chapi, K. (2012). A GIS-based logistic regression model in rock-fall susceptibility mapping along a mountainous road: Salavat Abad case study, Kurdistan, Iran. *Natural Hazards, 64,* 1639–1656.
Tehrany, M. S., Lee, M.-J, Pradhan, B., Jebur, M. N., & Lee, S. (2014a). Flood susceptibility mapping using integrated bivariate and multivariate statistical models. *Environmental Earth Science, 72* (10), 1–s15.
Tehrany, M. S., Pradhan, B., & Jebur, M. N. (2013). Spatial prediction of flood susceptible areas using rule based decision tree (DT) and a novel ensemble bivariate and multivariate statistical models in GIS. *Journal of Hydrology, 504,* 69–79.
Tehrany, M. S., Pradhan, B., & Jebur, M. N. (2014b). Flood susceptibility mapping using a novel ensemble weights-of-evidence and support vector machine models in GIS. *Journal of Hydrology, 512,* 332–343.
Tien Bui, D., Pradhan, B., Lofman, O., & Revhaug, I. (2012a). Landslide susceptibility assessment in Vietnam using support vector machines, decision tree, and Naive Bayes Models. *Mathematical Problems in Engineering, 2012.*
Tien Bui, D., Pradhan, B., Lofman, O., Revhaug, I., & Dick, O. B. (2012b). Landslide susceptibility mapping at Hoa Binh province (Vietnam) using an adaptive neuro-fuzzy inference system and GIS. *Computers & Geosciences, 45,* 199–211.
Ueno, K., Kurobe, K., Imaizumi, F., & Nishii, R. (2015). Effects of deforestation and weather on diurnal frost heave processes on the steep mountain slopes in south central Japan. *Earth Surface Processes and Landforms,* 40(15), 2013–2025.
Van Beek, L., & Van Asch, T. W. (2004). Regional assessment of the effects of land-use change on landslide hazard by means of physically based modelling. *Natural Hazards, 31,* 289–304.
Vranken, L., Van Turnhout, P., Van Den Eeckhaut, M., Vandekerckhove, L., & Poesen, J. (2013). Economic valuation of landslide damage in hilly regions: A case study from Flanders, Belgium. *Science of the Total Environment, 447,* 323–336.
Wan, S., Lei, T.-C., & Chou, T.-Y. (2012). A landslide expert system: Image classification through integration of data mining approaches for multi-category analysis. *International Journal of Geographical Information Science, 26,* 747–770.
Xu, C., Dai, F., Xu, X., & Lee, Y. H. (2012). GIS-based support vector machine modeling of earthquake-triggered landslide susceptibility in the Jianjiang River watershed, China. *Geomorphology, 145,* 70–80.
Xu, C., Xu, X., Dai, F., Wu, Z., He, H., Shi, F., et al. (2013). Application of an incomplete landslide inventory, logistic regression model and its validation for landslide susceptibility mapping related to the May 12, 2008 Wenchuan earthquake of China. *Natural Hazards, 68,* 883–900.
Yalcin, A. (2008). GIS-based landslide susceptibility mapping using analytical hierarchy process and bivariate statistics in Ardesen (Turkey): Comparisons of results and confirmations. *Catena, 72,* 1–12.
Yalcin, A., Reis, S., Aydinoglu, A., & Yomralioglu, T. (2011). A GIS-based comparative study of frequency ratio, analytical hierarchy process, bivariate statistics and logistics regression methods for landslide susceptibility mapping in Trabzon, NE Turkey. *Catena, 85,* 274–287.
Yilmaz, I. (2009). A case study from Koyulhisar (Sivas-Turkey) for landslide susceptibility mapping by artificial neural networks. *Bulletin of Engineering Geology and the Environment, 68,* 297–306.
Zhu, L., & Huang, J.-F. (2006). GIS-based logistic regression method for landslide susceptibility mapping in regional scale. *J Zhejiang univ-Sc A, 7,* 2007–2017.

Spatial Prediction of Landslides Along Jalan Kota in Bandar Seri Begawan (Brunei) Using Airborne LiDAR Data and Support Vector Machine

9

Biswajeet Pradhan, Mustafa Neamah Jebur and Saleh Abdullahi

9.1 Introduction

A landslide is one of the most dangerous natural hazards that can cause considerable damage to human life and properties (Yin et al. 2009; Pradhan and Lee 2010; Jebur et al. 2014). The effective mapping of landslide-prone areas can help reduce and even avoid the tragic loss of lives and economic devastation. Recently, rapid urban growth in rural environments, particularly in environmentally sensitive areas, has generated considerable interest in landslide assessment studies and, subsequently, has increased the number of researchers in this field (Gokceoglu and Sezer 2009). In addition, the availability of spatial data in costly and time-effective digital format has enabled users to apply sophisticated data-driven approaches in landslide-related analysis (Marjanović et al. 2011). Landslide susceptibility mapping (LSM) is one of these analyses; it identifies landslide-prone areas, thereby providing valuable information about the spatial probability of landslide occurrence. This information can assist planners and decision makers in land management and infrastructure planning (Xu et al. 2012; Regmi et al. 2014). The process of LSM has been presented using various techniques based on site specification and the strength of approaches in different studies (Bonham-Carter 1994; Chacón et al. 2006). The performance of these techniques is based on running different aggregation models to investigate the interaction of several site-specific triggering parameters that cause landslides. Addressing these parameters individually and considering their mathematical relationships are essential. However, understanding these relationships to predict landslide susceptibility is difficult (Cui et al. 2010). One of the main objectives of LSM is to investigate these relationships according to background knowledge and then to identify areas prone to future mass movements. This objective can be attained by evaluating the spatial distribution of previous landslides with respect to the physical properties (e.g., geological and geomorphological properties and land use categories) of a site to create a susceptibility map (Ballabio and Sterlacchini 2012). In recent decades, numerous studies have been conducted to develop geographic information system (GIS)-based LSM by focusing on conditioning factors (Lee and Pradhan 2007; Pradhan and Lee 2010; Pradhan 2011; Bui et al. 2013; Pourghasemi et al. 2013). These studies report the inefficiencies of physical models resulting from the complexity, nonlinearity, and ambiguity of the relations between the involved factors and landslide occurrence; these inefficiencies have encouraged the current researchers to apply more sophisticated statistical approaches (Micheletti et al. 2011). In general, the most common approaches for LSM are the deterministic approach (Westen and Terlien 1996), the heuristic approach (Barredo et al. 2000; van Westen 2000), the statistical approach (Ramani et al. 2011; Youssef et al. 2015), the support vector machine (SVM) (Wan and Lei 2009; Pourghasemi et al. 2013), the neuro-fuzzy-based method (Oh and Pradhan 2011; Akgun et al. 2012), fuzzy logic (Pradhan 2011), and artificial neural networks (ANNs) (Chi et al. 2002; Wang et al. 2015). To develop stronger and applicable LSM, these methods should be integrated into remote sensing (RS) and GIS technologies. Several benefits that can be gained from RS and GIS technologies include high data quality with regular updating and a wide coverage, a variety of data sources, and rapid and accurate analysis. Numerous RS- and GIS-integrated approaches have been proposed to evaluate landslide-prone areas (Sarkar et al. 2008; Konadu and Fosu 2009; Pradhan and Lee 2010).

Although the aforementioned approaches have certain capabilities to extract susceptible areas and create landslide susceptibility maps, these approaches have their respective advantages and disadvantages; the latter reduces the efficiency of the predictive models. For example, the dependency of the analytic hierarchy process (AHP) on expert knowledge is the main source of its uncertainty; this method

B. Pradhan (✉) · M.N. Jebur · S. Abdullahi
Department of Civil Engineering, University Putra Malaysia, Serdang, Malaysia
e-mail: biswajeet24@gmail.com

B. Pradhan (ed.), *Laser Scanning Applications in Landslide Assessment*, DOI 10.1007/978-3-319-55342-9_9 is mainly suitable for regional studies (Subramanian and Ramanathan 2012). The lack of spatial relationship among variables is the main drawback of the frequency ratio (FR) technique. Logistic regression (LR), as a multivariate statistical analysis, overcomes this problem. Although LR has long been used in natural hazard assessment, it cannot evaluate the classes of each contributing factor (Tehrany et al. 2014). An ANN works like a "black box"; thus, the esoteric performance of this method and the relationship among its variables are difficult to understand (Pradhan and Buchroithner 2010). In addition, an ANN is not a time-effective model for a large number of variables (Ghalkhani et al. 2013). By contrast, the structure of fuzzy logic is more understandable than that of an ANN. However, similar to AHP, fuzzy logic also results in uncertainty because it is an expert knowledge-based process (Tilmant et al. 2002). Despite the successful application of the aforementioned approaches, machine learning models have recently received more attention in LSM applications. Concepts and details about machine learning and its application in geospatial data processing are presented by Pozdnoukhov (2009).

Recent works include comparative studies of various approaches to evaluate their advantages and disadvantages. Pradhan (2013) compared the performance of the decision tree (DT) algorithm, SVM, and an adaptive neuro-fuzzy inference system for the LSM of Penang Hill, Malaysia. The analysis reported the difficulty in defining the rule for DT and in selecting parameters for SVM modeling. In addition, all the three approaches require advanced computer systems to run an analysis (Chau et al. 2005). Marjanović et al. (2011) compared three machine learning algorithms (SVM, DT, and LR) and then compared the selected algorithm with AHP. From the aforementioned comparisons, SVM outperformed the other approaches in all the evaluation metrics. Ballabio and Sterlacchini (2012) applied and tested all the procedural steps for LSM using SVM modeling. An SVM susceptibility map was compared with LR, linear discriminant analysis, and naïve Bayes classifier. The results demonstrated the feasibility and capability of SVM to outperform other models in terms of accuracy and generalization capacity. Other researchers, including Brenning (2005), Lifeng and Youshui (2006), and Yilmaz (2009), have compared SVM with other models for LSM. All these studies strongly agree that the performance of SVM is better than those of the other approaches. This finding may be attributed to the training phase of SVM modeling, with associated input and target output values (Xu et al. 2012). Micheletti et al. (2011) applied SVM to map landslide susceptibility in the Canton of Vaud (Switzerland) by using efficient feature selection methods, which enhanced model interpretability. Xu et al. (2012) evaluated the mapping power of SVM modeling in earthquake-triggered LSM using GIS software. They applied three groups of positive and negative training samples with four different kernel functions: radial basis function (RBF), polynomial (PL), sigmoid (SIG), and linear (LN). Among the 12 results, SIG was the least skillful when used with the centroid data of all the studied landslides as positive training samples. Pourghasemi et al. (2013) implemented more kernel classifiers (i.e., LN, PL degree of 2, PL degree of 3, PL degree of 4, RBF, and SIG) for LSM in Golestan Province, Iran. Among these classifiers, RBF provided better results based on the evaluation of all six output maps using success and prediction rate methods. Yao et al. (2008) first reviewed one-class and two-class SVM methods and then presented their applications to LSM. The results of both methods were compared with the LR model, and the two-class SVM performed better than LR and one-class SVM in terms of prediction accuracy. The current study introduces SVM modeling approaches for the LSM of a landslide-prone area in Jalan Kota, Brunei. Four kernel types of SVM (RBF, PL, SIG, and LN) are included in the process to evaluate their performance in the entire modeling procedure, classification, and output maps.

9.2 Data and Methodology

9.2.1 Landslide Inventory Map

The proposed model was applied to a landslide-prone area in Jalan Kota, which is located in the northeast of Brunei (Bandar Seri Begawan) and spreads over approximately 26.84 km^2 of a hilly landscape (Fig. 9.1). Most landslides in Jalan Kota occur in unstable ground with a high degree of slope angle. In addition, heavy rainfall, along with less dense and less compact soil properties, increases the probability of landslide occurrence. In general, landslide susceptibility maps are produced based on various inventory maps. Such maps can be used to assess and mitigate landslide hazard or risk on a regional level. Furthermore, these inventory maps can have significant effects on investigating the relationship between the spatial distribution of landslide occurrence and landslide triggering factors. Extensive site surveys and observations were performed to create landslide inventory maps of the study area. Approximately 90 landslide locations were identified in Jalan Kota using aerial photos (spatial resolution: 0.61 m) taken by the QuickBird satellite on April 1, 2011. From the total number of 90 landslides in the inventory map, 70% were randomly selected for the training data set, and 30% were selected for the testing data set. Subsequently, the dependent layer was created using the training data set. This layer comprised three classes: 1, −1, and 0. The value of 1 was assigned to areas with landslide occurrence. The value of −1, which was equal in number to the training locations, was assigned to areas where landslide did not occur. The value of 0 was assigned to areas that

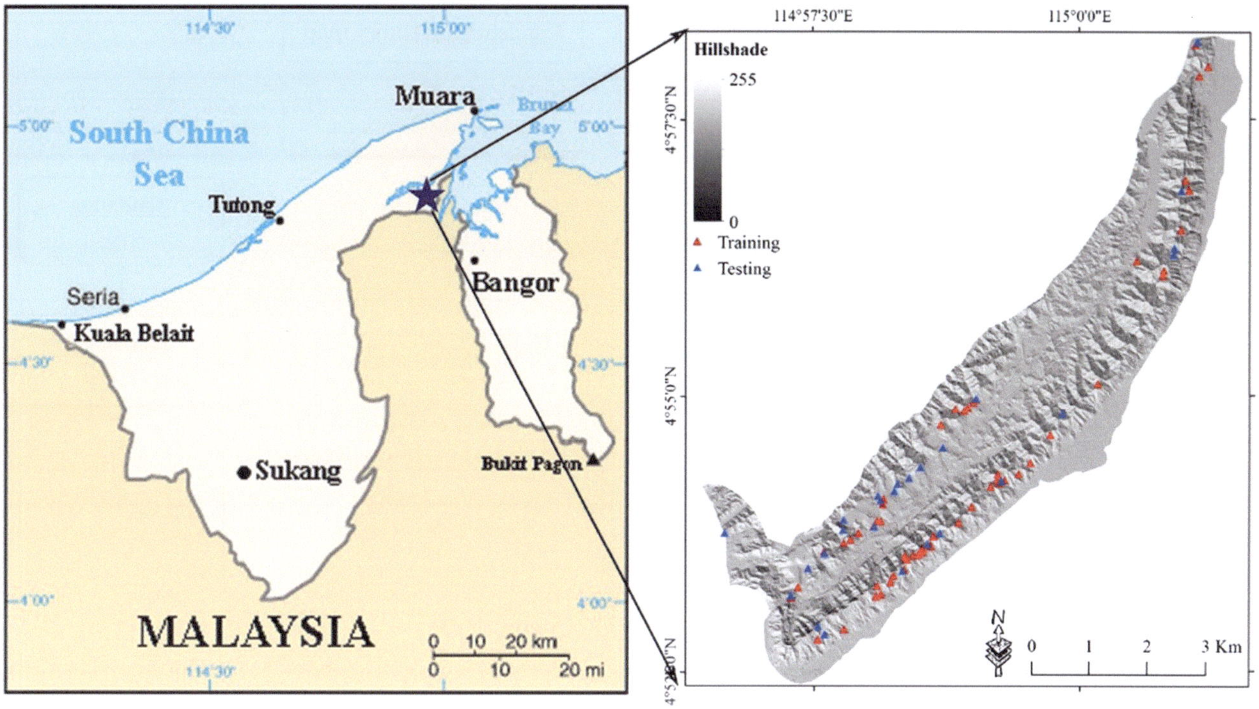

Fig. 9.1 Landslides location map with hill-shaded map of Jalan Kota, Brunei

should be predicted. The remaining landslide events were used as testing data. The methodology flowchart of the research is provided in Fig. 9.2.

9.2.2 Landslide Conditioning Factors

As explained earlier, landslide occurrence is caused by several factors, which should be considered and evaluated in susceptibility mapping (Liu and de Smedt 2005). Through knowledge from the literature and field observation, several factors (in nominal, ordinal, or scale format) were selected to train the model and predict the potential distribution of landslides (Park et al. 2013). Altitude, slope angle, aspect, curvature, stream power index (SPI), topographic wetness index (TWI), topographic roughness index (TRI), geology, soil, land use/land cover (LULC), rainfall, and distance from rivers, roads, and faults were used as the conditioning factors to prepare the susceptibility maps (Table 9.1). Each factor was resized to fit a 5 m × 5 m grid. The grid of the study area was constructed with 1889 columns and 2168 rows (1,073,753 pixels; 26.84 km^2).

Altitude is one of the significant parameters because it is controlled by various geological and geomorphological processes (Gritzner et al. 2001; Dai and Lee 2002; Ayalew and Yamagishi 2005). Slope variation and instability depend on elevation, which directly affects landslide occurrence. In high slope angle degree, the level of gravity-induced shear stress increases, thereby increasing the probability of landslide occurrence. The slope angle of the study area was produced from the altitude map. Slope aspect is also considered important because of site characteristics, such as exposure to sunlight and dry wind (Baeza and Corominas 2001). These characteristics control soil moisture concentration, which directly affects landslide occurrence (Magliulo et al. 2008). Therefore, an aspect map was divided into 10 classes (flat, north, northeast, east, southeast, south, southwest, west, northwest, and north) as shown in Fig. 9.3c. Continuous variables (such as slope angle and elevation), which were represented by real number vectors, were entered as scale input for the modeling (Pourghasemi et al. 2012). Curvature was divided into three classes (convex, flat, and concave).

The spatial distribution of landslides is correlated with distance to existing faults. In general, landslide occurrence is high along faults and is decreased as distance from faults increases (Xu et al. 2012). The Euclidean distance analysis for existing faults of the study area is presented in Fig. 9.3j. Proximity to rivers also affects landslide occurrence because of the reduced slope stability of eroded riverbanks (Xu et al. 2012; Umar et al. 2014). Similarly, roadsides or slopes affected by roads have high landslide probability occurrence

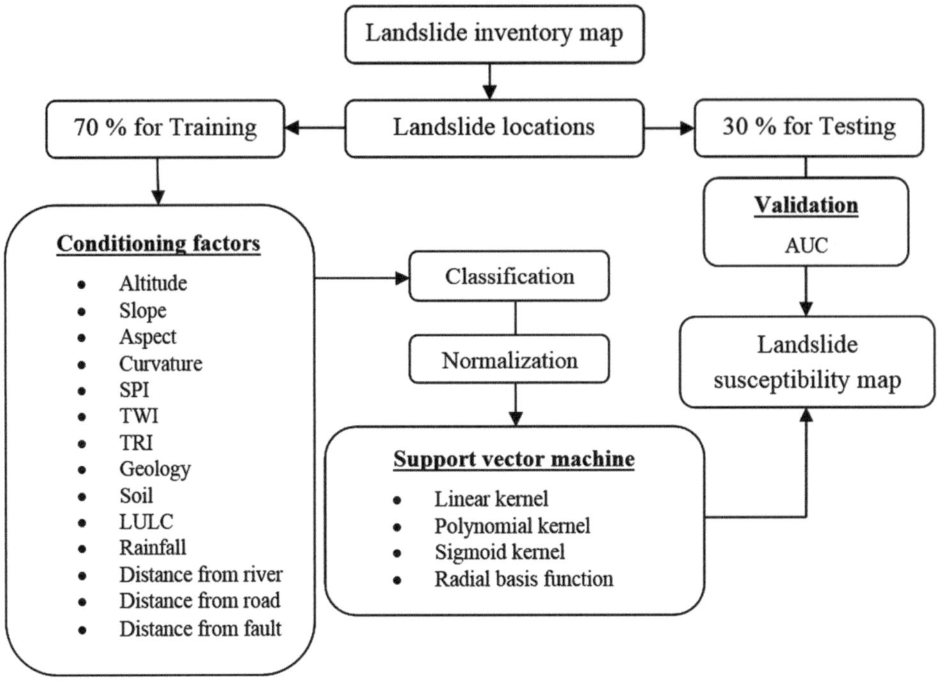

Fig. 9.2 Overall methodology flowchart

Table 9.1 Controlling factors and their classes

Controlling factors	No. of classes	Classes
Altitude		Scale
Slope		Scale
Aspect	10	Flat, N, NE, E, SE, S, SW, W, NW, N
Curvature	3	Flat, concave, convex
SPI		Scale
TWI		Scale
TRI		Scale
Geology	3	Belait, Pleistocene, Recent
Soil	5	ME, NY, PD, RG, TA
LULC	5	Barren area, building, road, vegetation, water bodies
Rainfall		Scale
Distance from river		Scale
Distance from road		Scale
Distance from fault		Scale

(Pourghasemi et al. 2013). Therefore, the Euclidean distance process was run for river and road layers to create proximity maps of these linear features (Figs. 9.3h, n).

Rainfall data were collected from meteorological stations around the study area. However, the application of geostatistical interpolation to calculate rainfall value for unknown points was required. Among various geostatistical interpolation approaches, inverse distance weighting (IDW) was utilized via line density operation in ArcGIS 10. The main logic behind this technique is that the value of an unsampled point is the weighted average of the known value points of the surrounding. Therefore, this computation was used to extract the unknown value of rainfall from adjacent known samples (Chen and Liu 2012). Other water-related factors, such as SPI and TWI, were computed using the following equations:

$$\mathrm{TWI} = \ln\left(\frac{A_s}{\tan \mathcal{B}}\right), \quad (9.1)$$

$$\mathrm{SPI} = A_S \tan \mathcal{B}, \quad (9.2)$$

where A_s is the specific catchment area ($m^2\ m^{-1}$), and $\mathcal{B}$ (radian) is the slope gradient (in degrees) (Regmi et al. 2010). TWI is the measure of moisture content at the site, and SPI is the measure of the erosive power of flowing water. SPI is based on the assumption that discharge is proportional to a specific catchment (Pourghasemi et al. 2013).

The study area is covered with five categories of soil types as shown in Fig. 9.3k. In general, the fine grain size of soil particles increases water absorption and subsequently landslide occurrence probability. The soil data of the study area were collected from the digitized drainage layer geomorphological map of Jalan Kota. Three different types of geology classes were identified in the study area as shown in Fig. 9.3i. This map and field observations showed that the study area was composed mainly of Belait, Pleistocene, and recent formations. However, the majority of the area is covered with Belait formations, whereas the southern, eastern, and northern borders of the area are covered with recent formations. The relationship between landslide occurrence

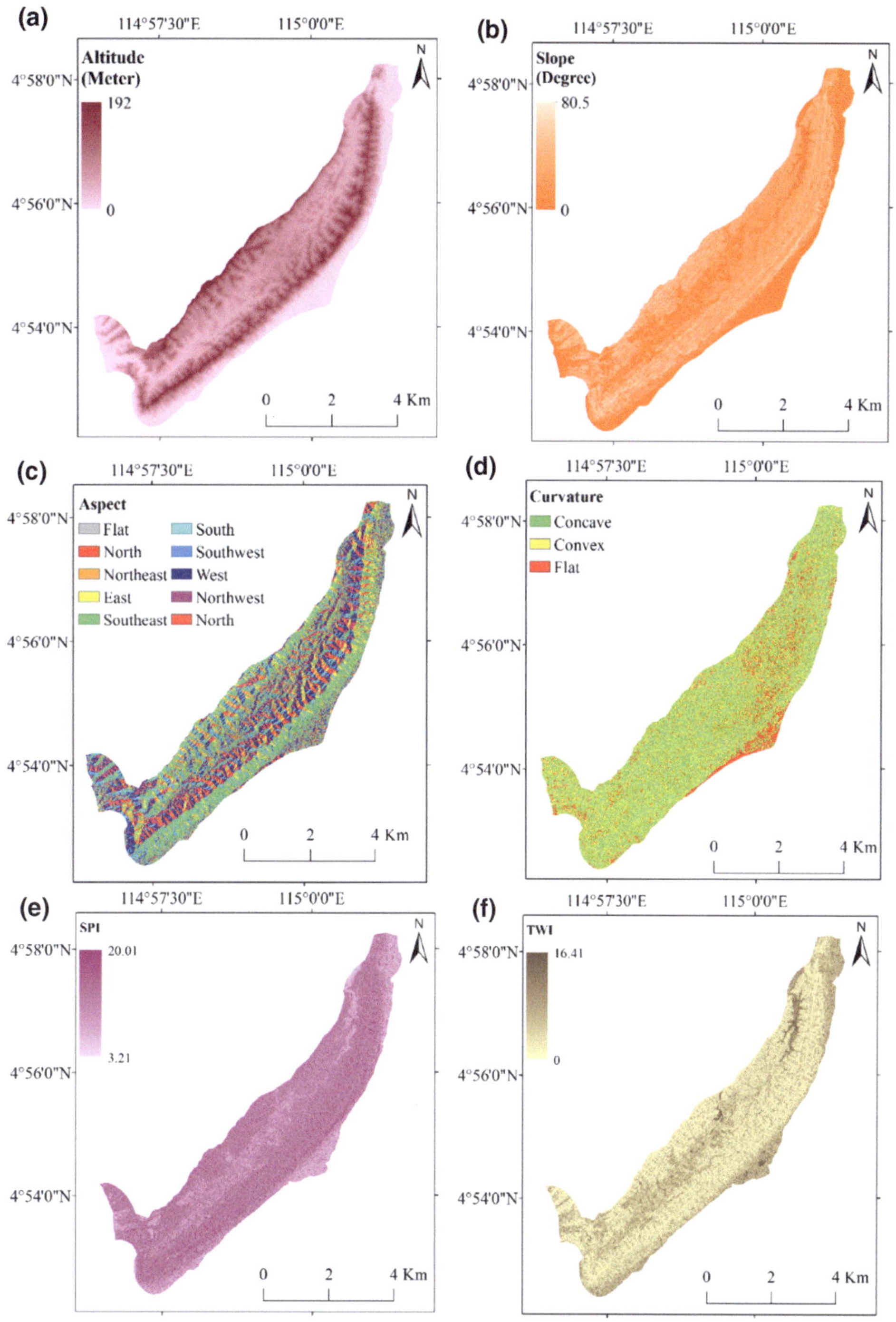

Fig. 9.3 Input thematic layers: **a** Altitude; **b** Slope; **c** Aspect; **d** Curvature; **e** SPI; **f** TWI; **g** TRI; **h** Distance from river; **i** Geology; **j** Distance from fault; **k** Soil; **l** LULC; **m** Rainfall; and **n** Distance from road

Fig. 9.3 (continued)

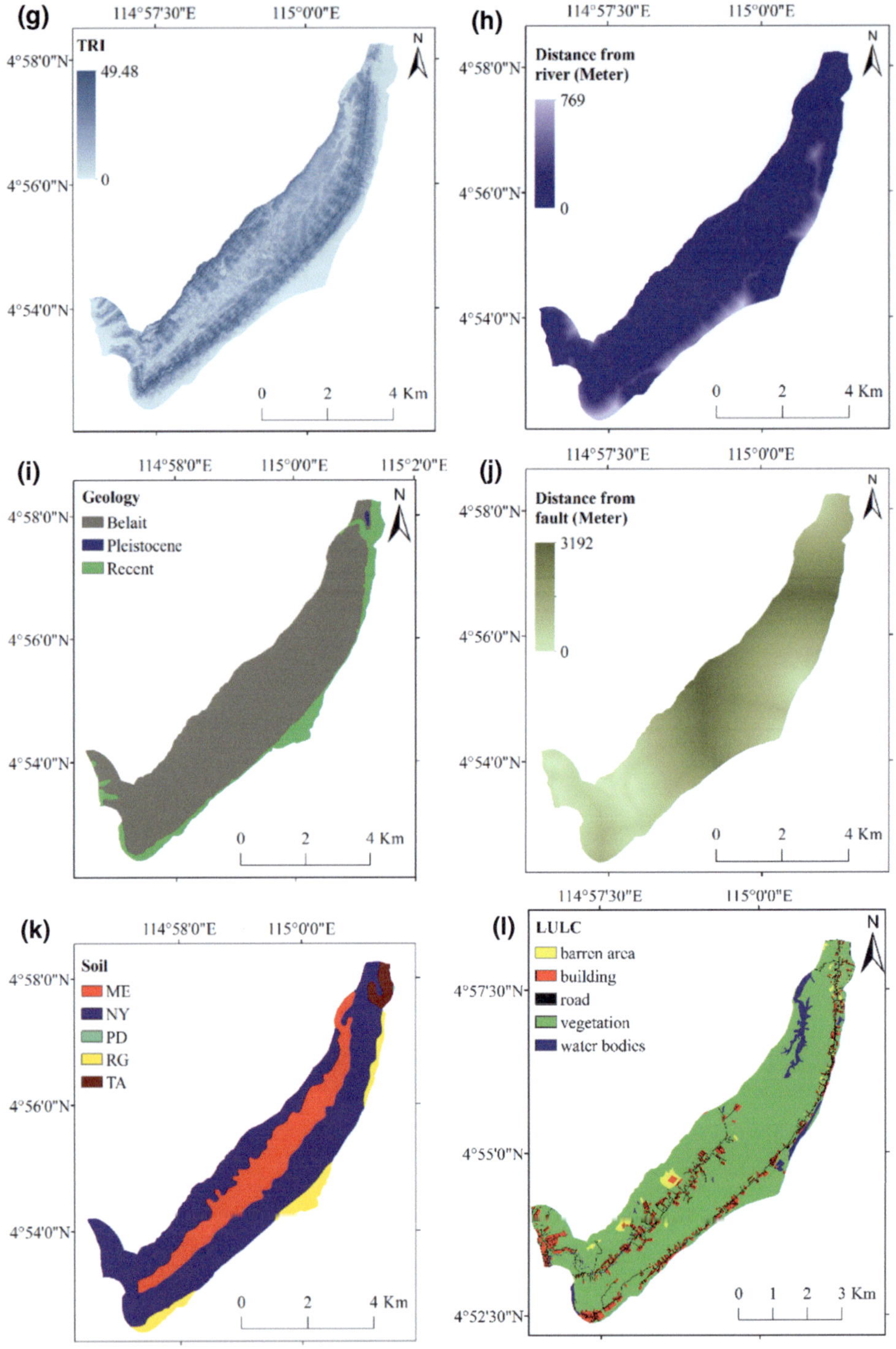

and land use type can be investigated by studying LULC in the study area. The land use map (Fig. 9.3 and Table 9.1) of Jalan Kota was produced using a topographic map. This area is mainly covered with vegetation, water bodies, built-up regions, roads, and few barren lands.

9.2.3 SVM Modeling

SVM is a popular learning machine based on the structural risk minimization principle (Yao et al. 2008; Wan and Lei 2009). Although this model is a supervised binary classifier,

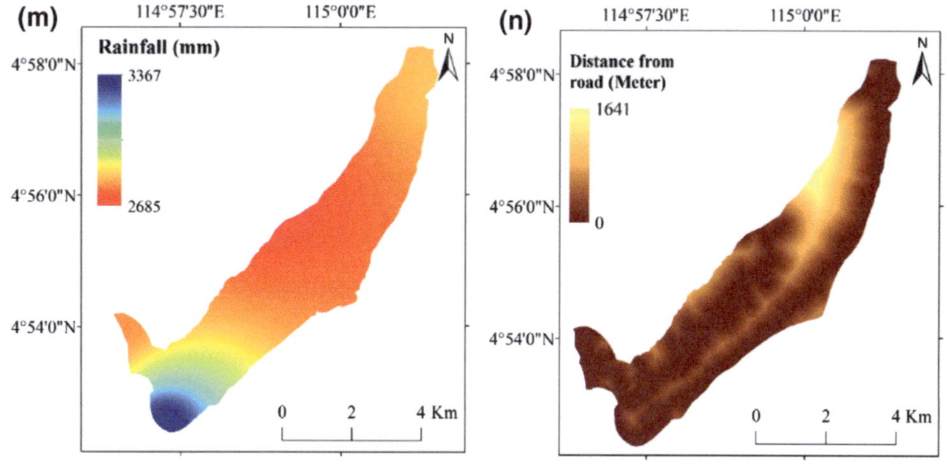

Fig. 9.3 (continued)

it can be easily transformed into several class problems (Belousov et al. 2002). The basis of this model is separating the hyperplane formation from the training data set, which is generated in the original space of n coordinates (xi parameters in vector x) between the points of two distinct classes (Marjanović et al. 2011). SVM extracts the maximum margin of separation between classes and builds a classification hyperplane in the center of the maximum margin (Pradhan 2013; Tehrany et al. 2014). If the point is located over the hyperplane, then it will be classified as +1; otherwise, it will be classified as −1. The training points closest to the optimal hyperplane are called support vectors. Once the decision surface is acquired, the classification of new data can be performed (Tien Bui et al. 2012).

For clarity, consider a training data set as pairs of (xi, yi) in which $xi \in Rn$, $yi \in [1, -1]$, and $i = 1, 2, \ldots, m$. The objective of SVM is to identify the n-dimensional hyperplane that discriminates the two classes based on landslide occurrence and non-occurrence $\{1, -1\}$ from the training data set. In case of linear separable data, a separating hyperplane can be defined as:

$$yi(w \cdot xi + b) \geq 1 - \xi i, \tag{9.3}$$

where w is a coefficient vector that defines the orientation of the hyperplane in feature space, b is the offset of the hyperplane from the origin, and ξi is the positive slack variables (Cortes and Vapnik 1995). Then, the solution for the coefficient vector (w) of SVM can be calculated using the Lagrange multiplier method (Samui 2008).

$$\text{Minimize} \sum_{i=1}^{n} \alpha_i\, y_i - \frac{1}{2}\sum_{i=1}^{n}\sum_{j=1}^{n} \alpha_i \alpha_j y_i y_j (x_i x_j), \tag{9.4}$$

$$\text{Subjected to} \begin{cases} \sum_{i=1}^{n} \alpha_i y_{j=0} \\ 0 \leq \alpha_i \leq C \end{cases}, \tag{9.5}$$

where α_i denotes the Lagrange multipliers, C is the penalty, and the slack variables ξi allow for penalized constraint violation. Therefore, the classification function can be defined as:

$$g(x) = \text{sign}\left(\sum_{i=1}^{n} y_i \alpha_i x_i + b\right). \tag{9.6}$$

In cases where the hyperplane cannot be separated using LN, a nonlinear function is applied as follows:

$$g(x) = \text{sign}\left(\sum_{i=1}^{n} y_i \alpha_i K(x_i, x_j) + b\right) \tag{9.7}$$

where $K(x_{i,}x_j)$ is the kernel function.

A successful SVM modeling with accurate results depends on the selection of kernel types and its parameters (Damaševičius 2010). Several types of kernel are most commonly used in SVM modeling, namely, LN, PL, RBF, and SIG (Pradhan 2013). PL and RBF (typically known as Gaussian kernels) are the most commonly used kernels in the literature (Marjanović et al. 2011). RBF, as an example of mapping function, exhibits good capability in generalization properties and in producing efficient interpolation. Compared with RBF, PL demonstrates better extrapolation capabilities at low-order degrees but requires high-order degrees to achieve good interpolation (Zhu et al. 2011). In general, LN is assumed as a specific case of RBF, whereas the performance SIG is similar to that of RBF in the case of parameters (Song et al. 2011). LN is applicable to linear separable circumstances; however, real-world problems are not linearly separable (Ali and Smith 2003). From these discussions, each kernel type has its own advantages and disadvantages with respect to applied analysis. Thus, the current study attempts to apply all four kernel types to assess the performance of each type in LSM. Different parameters should be considered for each kernel, as shown in Table 9.2.

Table 9.2 Equation and parameters of applied kernel for LSM

Kernel type	Equation	Kernel parameters
LN	$K(x_i, x_j) = x_i^T x_j$	–
PL	$K(x_i, x_j) = (-\gamma x_i^T x + 1)^d$	γ, d
RBF	$K(x_i, x_j) = \exp(-\gamma x_i - x_j^2)$	γ
SIG	$K(x_i, x_j) = \text{Tanh}\ (-\gamma x_i^T x + 1)^d$	γ

In SVM modeling, the trade-off between errors and margin, which helps avoid overfitting of the model, is typically achieved using regularization parameters (C) (Marjanović et al. 2011). Low C values increase training errors. By contrast, the accuracy of the results increases substantially with high C values, thereby generating a small margin. The appropriate value should be defined for all kernel types. Kernel width or gamma (γ) value, which controls the degree of nonlinearity, should also be considered in the model (Tien Bui et al. 2012). In addition, parameter "d" is the polynomial degree in the kernel function for PL, and "r" is the bias term in the kernel function for PL and SIG.

To select the optimal kernel parameters, cross-validation was performed using SPSS Modeler (Zhuang and Dai 2006; Pradhan 2013). From this process, n validation accuracies were achieved, and an average value was applied for the final landslide susceptibility model for each kernel type (Yao et al. 2008). As the first step, the standard ranges for all the parameters were defined. Each set of parameters was divided into n folds: one for the validation process and the remaining $n - 1$ for training purposes. In the next level, the data set was divided into five sample groups with the same number of samples. For the first iteration, the first fold was used in the validation process, whereas the rest of the four folds were applied in the training process. A similar procedure was used for other four models.

Finally, the area under curve (AUC) method was applied to evaluate the performance of each combination and eventually to select the best parameterization of SVM (Mu and Nandi 2007). AUC is a unitless summary metric that synthesizes the relationships between the reference Boolean feature and several diagnoses by the index. This technique is one of the most common validation techniques for evaluating the efficiency of model outputs with a success rate and prediction rate curve (Pradhan 2010). The rate curve illustrates the fitness of the probability maps with the actual landslide inventory map. In this manner, the process determines how well the method and the parameters produce the landslide susceptibility maps. This validation technique can range from 0 to 1, in which higher AUC values indicate a stronger positive correlation. To calculate AUC, the ordered pixel values based on the probability map were classified into 100 classes. These values were set on the vertical axis (y) with accumulated 1% intervals on the horizontal axis (x). The existence of landslide location training and testing in each interval was evaluated, and the resultant success and prediction rates were calculated. If the total area is one, then perfect matching and accuracy are achieved.

9.3 Results and Discussion

In this section, the effects of conditioning factors on the landslide probability map and the performance of each SVM kernel type are assessed in terms of predictive capability and mapping of future landslide occurrence. Cross-validation was performed on four kernel types based on the parameters given in Table 9.2. These parameters were used to run the final modeling for these kernels. The outputs of each model were converted into GIS format, and then, the probability index for each kernel type was computed. Figure 9.4 presents the maps derived for each kernel type. The efficiency and accuracy of these maps were evaluated using AUC.

The output-derived maps can be classified using various classification techniques, such as equal interval, quantile, standard deviation, and natural break, depending on research specifications. All the aforementioned techniques were applied, and the output maps were examined. Most of the results assigned a large part of the area with a high probability of landslide occurrence except for quantile classification, which presented an appropriate appearance of hazard locations. Finally, the probability maps were classified into five classes in terms of susceptibility to landslide occurrence: very low, low, moderate, high, and very high (Pradhan 2010).

As shown in Fig. 9.4, the probability map derived from the SIG kernel appears different from the other maps. This result indicates that this kernel type is less efficient in accurately identifying landslide occurrence sites. In addition, the SIG kernel is highly sensitive to input data. The difference among the other three maps (LN, PL, and RBF) is evident in the severity of a hazard; however, the extent and direction of a hazard location are nearly similar. The PL-derived map assigns a very high probability of landslide occurrence to a larger area than the other maps. By contrast, SIG designates the smallest area as a hazard location compared with the other kernel types. However, the southern part of the study area is a common susceptible zone in all the four derived maps.

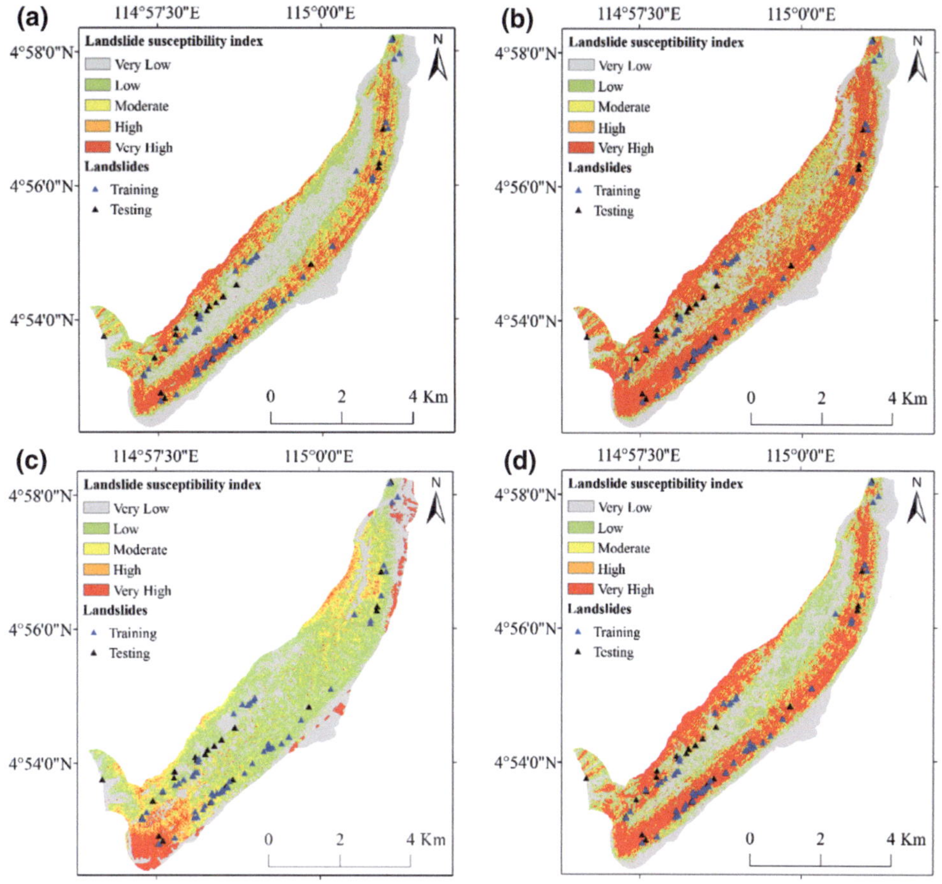

Fig. 9.4 Landslide susceptibility maps derived from SVM model using **a** RBF; **b** Polynomial; **c** Sigmoid; and **d** Linear

The comparison of Figs. 9.3 (conditioning factors) and 9.4 (landslide probability maps) indicates that the relationship between conditioning factors and landslide occurrence can be observed. In case of altitude, landslides mainly occur at an elevation higher than 150–190 m. The probability of landslide occurrence is evidently very low in low-elevation areas. Areas with a higher slope degree of nearly 80° are more susceptible to landslides than flat areas. Similarly, the flat areas in aspect and curvature maps (Figs. 9.3c, d) have minimum susceptibility to landslide occurrence. Convex areas are regarded as susceptible zones because of the concentration of loose debris on the inclined surface, which may slide during heavy rainfall. High SPI and TRI increase the probability of landslide occurrence, whereas high TWI decreases it. For geological type, landslides mainly occur in areas covered with Belait formations. Similarly, for soil type, most landslides occur in second class (NY), whereas landslide probability occurrence is minimal for other soil types. Rainfall did not exert a significant effect on probability maps; however, the southern part of the study area with the highest amount of rainfall has the maximum probability of landslide occurrence. The LULC map indicates that most landslides occur in vegetation areas, but a direct correlation cannot be observed in the case of road networks. Similarly, distance to fault appears to be correlated with landslide occurrence, although an insignificant effect can be observed in proximity to fault zones.

Despite these discussions, visual interpretation is not an appropriate means to evaluate the performance of kernel types. Hence, the probability maps were assessed using the AUC validation technique by measuring both success rate and prediction rate curves (Table 9.3). The rate curve illustrates the fitness of the probability maps produced from different kernel types with the actual landslide inventory map. Figure 9.5 and Table 9.3 show the success rate and prediction rate of all the probability maps derived from the

Table 9.3 Results of AUC for each kernel type

Kernel type	γ gamma	C	d	Success rate (%)	Prediction rate (%)
RBF	0.1	10	0.1	88.21	82.90
PL	1	10	0.1	82.66	78.33
SIG	1	10	0.1	47.93	49.10
LN		10	0.1	82.63	84.93

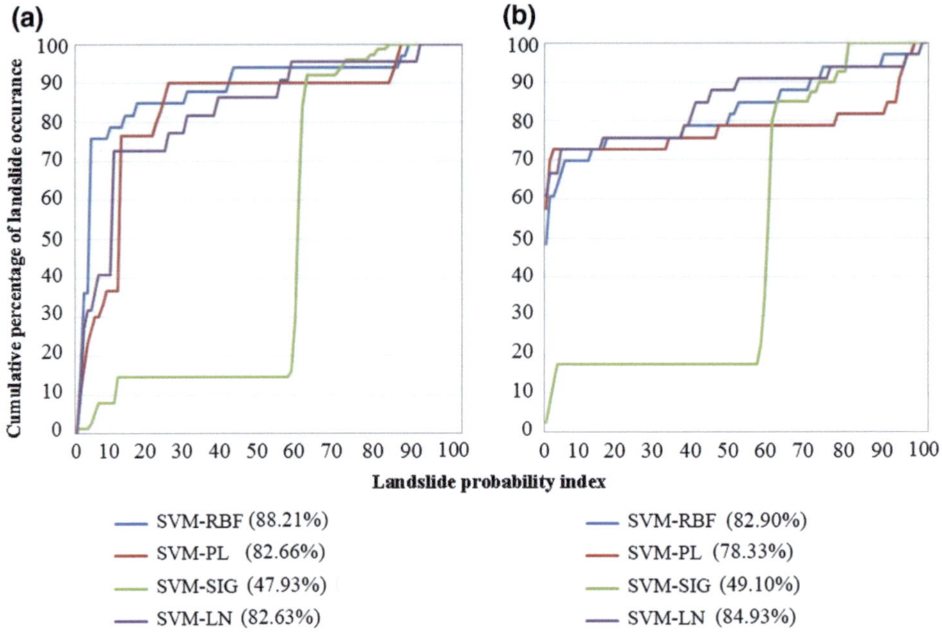

Fig. 9.5 Graphic representation of the cumulative frequency diagram presenting the cumulative landslide occurrence (%; *y*-axis) in landslide probability index rank (%; *x*-axis): **a** success rate; and **b** prediction rate

four kernel types. AUC ranges from 0.5 to 1.0. The value of 0.5 indicates random fit of the model. As shown in Table 9.3 and Fig. 9.5, RBF achieves the highest success rate value (88.21%) and the second highest prediction rate value (82.90%). As expected, SIG yields the lowest success rate and prediction rate values (47.93 and 49.10%, respectively). The LN and PL kernels also obtain reasonable results in both rating assessments with respect to the RBF model. These evaluations show that the selection of kernel type for SVM modeling can affect the output probability maps of landslide occurrence with similar input data. Therefore, RBF and LN exhibit better results in LSM. The higher accuracy of RBF may be attributed to the nature of performance of this function. In landslide susceptibility evaluation, the approximation capacity of RBF is higher than the other kernel functions. Landslide susceptibility assessment is a nonlinear process; therefore, nonlinear approaches perform better than linear approaches.

9.4 Conclusion

As a disaster management strategy, LSM is one of the main interests of planning authorities, particularly for landslide-prone areas in Brunei. The geographical condition in Jalan Kota is developed by mountains where landslide can easily occur after a heavy rainfall. Given the aforementioned reasons, numerous landslides have occurred in recent years and caused considerable human casualties, property losses, and environmental degradation. Therefore, the local government is investigating appropriate and rapid solutions for identifying areas prone to landslide occurrence. In recent years, RS and GIS have been proven to be useful tools for addressing geospatial issues, particularly natural hazard problems. However, sophisticated statistical and machine learning-based approaches are required to address serious natural hazards, such as landslides. Although several approaches are available in this field, the successful application and performance of SVM have made this method more common than other approaches. Moreover, various kernel types for SVM modeling provide a challenging area in evaluating the most effective kernel types for LSM. The main objective of this study is to evaluate the four main kernel types (RBF, PL, SIG, and LN) for SVM to select the most optimal kernel for LSM. Through a comprehensive field observation and literature review, the most effective landslide conditioning factors, such as altitude, slope, aspect, curvature, geological properties, soil type, and rainfall, were selected. The analysis showed the effects and relationship of each conditioning factor on landslide probability occurrence. The probability maps derived from each kernel type were compared with the actual landslide map to assess the efficiency and accuracy of their performance. The AUC technique was used to run the success rate and prediction rate processes. The comparison showed that the RBF kernel type exhibited better performance in terms of success and prediction rates. By contrast, the SIG kernel exhibited the weakest results because of its origin from neural networks. This study provides valuable information for identifying effective conditioning factors and their effects on landslide occurrence and the performance of different kernel types with respect to these factors. In addition, the output maps can assist government officials and planners in mitigating and even preventing losses in life and properties caused by landslide disasters by relocating new developments in this area.

References

Akgun, A., Sezer, E. A., Nefeslioglu, H. A., Gokceoglu, C., & Pradhan, B. (2012). An easy-to-use MATLAB program (MamLand) for the assessment of landslide susceptibility using a Mamdani fuzzy algorithm. *Computers & Geosciences, 38*(1), 23–34.

Ali, S., & Smith, K. A. (2003). Automatic parameter selection for polynomial kernel. In: *Information Reuse and Integration, 2003. IRI 2003. IEEE International Conference on* (pp. 243-249). IEEE.

Ayalew, L., & Yamagishi, H. (2005). The application of GIS-based logistic regression for landslide susceptibility mapping in the Kakuda-Yahiko mountains, Central Japan. *Geomorphology, 65*(1), 15–31.

Baeza, C., & Corominas, J. (2001). Assessment of shallow landslide susceptibility by means of multivariate statistical techniques. *Earth Surface Processes and Landforms, 26*(12), 1251–1263.

Ballabio, C., & Sterlacchini, S. (2012). Support vector machines for landslide susceptibility mapping: The Staffora river basin case study, Italy. *Mathematical Geosciences, 44*(1), 47–70.

Barredo, J., Benavides, A., Hervás, J., & van Westen, C. J. (2000). Comparing heuristic landslide hazard assessment techniques using GIS in the Tirajana basin, Gran Canaria Island, Spain. *International Journal of Applied Earth Observation and Geoinformation, 2*(1), 9–23.

Belousov, A., Verzakov, S., & Von Frese, J. (2002). Applicational aspects of support vector machines. *Journal of Chemometrics, 16* (8–10), 482–489.

Bonham-Carter, G. (1994). Geographic information systems for geoscientists: Modelling with GIS (Vol. 13). UK: Elsevier.

Brenning, A. (2005). Spatial prediction models for landslide hazards: Review, comparison and evaluation. *Natural Hazards and Earth System Science, 5*(6), 853–862.

Bui, D. T., Pradhan, B., Lofman, O., Revhaug, I., & Dick, Ø. B. (2013). Regional prediction of landslide hazard using probability analysis of intense rainfall in the Hoa Binh province, Vietnam. *Natural Hazards, 66*(2), 707–730.

Chacón, J., Irigaray, C., Fernandez, T., & El Hamdouni, R. (2006). Engineering geology maps: Landslides and geographical information systems. *Bulletin of Engineering Geology and the Environment, 65*(4), 341–411.

Chau, K., Wu, C., & Li, Y. (2005). Comparison of several flood forecasting models in Yangtze River. *Journal of Hydrologic Engineering, 10*(6), 485–491.

Chen, F.-W., & Liu, C.-W. (2012). Estimation of the spatial rainfall distribution using inverse distance weighting (IDW) in the middle of Taiwan. *Paddy and Water Environment, 10*(3), 209–222.

Chi, K. -H., Park, N. -W., & Chung, C. -J. (2002). Fuzzy logic integration for landslide hazard mapping using spatial data from Boeun, Korea. *International archives of photogrammetry remote sensing and spatial information sciences, 34*(4), 54–59.

Cortes, C., & Vapnik, V. (1995). Support-vector networks. *Machine Learning, 20*(3), 273–297.

Cui, X., Zhao, X., Ji, M., Wang, S., & Zhang, P. (2010). Research on landslide prediction based on support vector model. In: *Computer Design and Applications (ICCDA), 2010 International Conference on* (Vol. 3, pp. V3-540–V543-544): IEEE.

Dai, F., & Lee, C. (2002). Landslide characteristics and slope instability modeling using GIS, Lantau Island, Hong Kong. *Geomorphology, 42*(3), 213–228.

Damaševičius, R. (2010). Optimization of SVM parameters for recognition of regulatory DNA sequences. *Top, 18*(2), 339–353.

Ghalkhani, H., Golian, S., Saghafian, B., Farokhnia, A., & Shamseldin, A. (2013). Application of surrogate artificial intelligent models for real-time flood routing. *Water and Environment Journal, 27*(4), 535–548.

Gokceoglu, C., & Sezer, E. (2009). A statistical assessment on international landslide literature (1945–2008). *Landslides, 6*(4), 345–351.

Jebur, M. N., Pradhan, B., & Tehrany, M. S. (2014). Detection of vertical slope movement in highly vegetated tropical area of Gunung pass landslide, Malaysia, using L-band InSAR technique. *Geosciences Journal, 18*(1), 61–68.

Konadu, D., & Fosu, C. (2009). Digital elevation models and GIS for watershed modelling and flood prediction–a case study of Accra Ghana. In: *Appropriate Technologies for Environmental Protection in the Developing World* (pp. 325–332). The Netherlands: Springer.

Lee, S., & Pradhan, B. (2007). Landslide hazard mapping at Selangor, Malaysia using frequency ratio and logistic regression models. *Landslides, 4*(1), 33–41.

Lifeng, Y., & Youshui, Z. (2006). Debris flow hazard assessment based on support vector machine. *Wuhan University Journal of Natural Sciences, 11*(4), 897–900.

Lineback Gritzner, M., Marcus, W. A., Aspinall, R., & Custer, S. G. (2001). Assessing landslide potential using GIS, soil wetness modeling and topographic attributes, Payette River, Idaho. *Geomorphology, 37*(1), 149–165.

Liu, Y., & De Smedt, F. (2005). Flood modeling for complex terrain using GIS and remote sensed information. *Water Resources Management, 19*(5), 605–624.

Magliulo, P., Di Lisio, A., Russo, F., & Zelano, A. (2008). Geomorphology and landslide susceptibility assessment using GIS and bivariate statistics: A case study in southern Italy. *Natural Hazards, 47*(3), 411–435.

Marjanović, M., Kovačević, M., Bajat, B., & Voženílek, V. (2011). Landslide susceptibility assessment using SVM machine learning algorithm. *Engineering Geology, 123*(3), 225–234.

Micheletti, N., Foresti, L., Kanevski, M., Pedrazzini, A., & Jaboyedoff, M. (2011). *Landslide susceptibility mapping using adaptive support vector machines and feature selection* (Master Thesis submitted to University of Lausanne Faculty of Geosciences and Environment for the Degree of Master of Science in Environmental Geosciences, 99p).

Mu, T., & Nandi, A. K. (2007). Breast cancer detection from FNA using SVM with different parameter tuning systems and SOM–RBF classifier. *Journal of the Franklin Institute, 344*(3), 285–311.

Oh, H.-J., & Pradhan, B. (2011). Application of a neuro-fuzzy model to landslide-susceptibility mapping for shallow landslides in a tropical hilly area. *Computers & Geosciences, 37*(9), 1264–1276.

Park, S., Choi, C., Kim, B., & Kim, J. (2013). Landslide susceptibility mapping using frequency ratio, analytic hierarchy process, logistic regression, and artificial neural network methods at the Inje area, Korea. *Environmental Earth Sciences, 68*(5), 1443–1464.

Pourghasemi, H. R., Jirandeh, A. G., Pradhan, B., Xu, C., & Gokceoglu, C. (2013). Landslide susceptibility mapping using support vector machine and GIS at the Golestan Province, Iran. *Journal of Earth System Science, 122*(2), 349–369.

Pourghasemi, H. R., Mohammady, M., & Pradhan, B. (2012). Landslide susceptibility mapping using index of entropy and conditional probability models in GIS: Safarood Basin, Iran. *Catena, 97*, 71–84.

Pozdnoukhov, A. (2009). *Machine learning for spatial environmental data: Theory, applications, and software.* Switzerland: EPFL press.

Pradhan, B. (2010). Landslide susceptibility mapping of a catchment area using frequency ratio, fuzzy logic and multivariate logistic regression approaches. *Journal of the Indian Society of Remote Sensing, 38*(2), 301–320.

Pradhan, B. (2011). Use of GIS-based fuzzy logic relations and its cross application to produce landslide susceptibility maps in three test areas in Malaysia. *Environmental Earth Sciences, 63*(2), 329–349.

Pradhan, B. (2013). A comparative study on the predictive ability of the decision tree, support vector machine and neuro-fuzzy models in landslide susceptibility mapping using GIS. *Computers & Geosciences, 51,* 350–365.

Pradhan, B., & Buchroithner, M. F. (2010). Comparison and validation of landslide susceptibility maps using an artificial neural network model for three test areas in Malaysia. *Environmental and Engineering Geoscience, 16*(2), 107–126.

Pradhan, B., & Lee, S. (2010). Regional landslide susceptibility analysis using back-propagation neural network model at Cameron Highland, Malaysia. *Landslides, 7*(1), 13–30.

Ramani, S. E., Pitchaimani, K., & Gnanamanickam, V. R. (2011). GIS based landslide susceptibility mapping of Tevankarai Ar sub-watershed, Kodaikkanal, India using binary logistic regression analysis. *Journal of Mountain Science, 8*(4), 505–517.

Regmi, A. D., Yoshida, K., Pourghasemi, H. R., DhitaL, M. R., & Pradhan, B. (2014). Landslide susceptibility mapping along Bhalubang—Shiwapur area of mid-Western Nepal using frequency ratio and conditional probability models. *Journal of Mountain Science, 11*(5), 1266–1285.

Regmi, N. R., Giardino, J. R., & Vitek, J. D. (2010). Modeling susceptibility to landslides using the weight of evidence approach: Western Colorado, USA. *Geomorphology, 115*(1), 172–187.

Samui, P. (2008). Slope stability analysis: A support vector machine approach. *Environmental Geology, 56*(2), 255–267.

Sarkar, S., Kanungo, D. P., Patra, A., & Kumar, P. (2008). GIS based spatial data analysis for landslide susceptibility mapping. *Journal of Mountain Science, 5*(1), 52–62.

Song, S., Zhan, Z., Long, Z., Zhang, J., & Yao, L. (2011). Comparative study of SVM methods combined with voxel selection for object category classification on fMRI data. *PLoS ONE, 6*(2), e17191.

Subramanian, N., & Ramanathan, R. (2012). A review of applications of analytic hierarchy process in operations management. *International Journal of Production Economics, 138*(2), 215–241.

Tehrany, M. S., Pradhan, B., & Jebur, M. N. (2014). Flood susceptibility mapping using a novel ensemble weights-of-evidence and support vector machine models in GIS. *Journal of Hydrology, 512,* 332–343.

Tien Bui, D., Pradhan, B., Lofman, O., & Revhaug, I. (2012). Landslide susceptibility assessment in vietnam using support vector machines, decision tree, and Naive Bayes Models. *Mathematical Problems in Engineering, 2012*(974638), 26.

Tilmant, A., Vanclooster, M., Duckstein, L., & Persoons, E. (2002). Comparison of fuzzy and nonfuzzy optimal reservoir operating policies. *Journal of Water Resources Planning and Management, 128*(6), 390–398.

Umar, Z., Pradhan, B., Ahmad, A., Jebur, M. N., & Tehrany, M. S. (2014). Earthquake induced landslide susceptibility mapping using an integrated ensemble frequency ratio and logistic regression models in West Sumatera Province, Indonesia. *Catena, 118,* 124–135.

Van Westen, C. J. (2000). The modelling of landslide hazards using GIS. *Surveys In Geophysics, 21*(2–3), 241–255.

Wan, S., & Lei, T. C. (2009). A knowledge-based decision support system to analyze the debris-flow problems at Chen-Yu-Lan River, Taiwan. *Knowledge-Based Systems, 22*(8), 580–588.

Wang, Y.-T., Seijmonsbergen, A. C., Bouten, W., & Chen, Q.-T. (2015). Using statistical learning algorithms in regional landslide susceptibility zonation with limited landslide field data. *Journal of Mountain Science, 12*(2), 268–288.

Westen, C. V., & Terlien, M. (1996). An approach towards deterministic landslide hazard analysis in GIS. A case study from Manizales (Colombia). *Earth Surface Processes and Landforms, 21*(9), 853–868.

Xu, C., Dai, F., Xu, X., & Lee, Y. H. (2012). GIS-based support vector machine modeling of earthquake-triggered landslide susceptibility in the Jianjiang River watershed, China. *Geomorphology, 145,* 70–80.

Yao, X., Tham, L., & Dai, F. (2008). Landslide susceptibility mapping based on support vector machine: A case study on natural slopes of Hong Kong, China. *Geomorphology, 101*(4), 572–582.

Yilmaz, I. (2009). A case study from Koyulhisar (Sivas-Turkey) for landslide susceptibility mapping by artificial neural networks. *Bulletin of Engineering Geology and the Environment, 68*(3), 297–306.

Yin, Y., Wang, F., & Sun, P. (2009). Landslide hazards triggered by the 2008 Wenchuan earthquake, Sichuan, China. *Landslides, 6*(2), 139–152.

Youssef, A. M., Pradhan, B., Pourghasemi, H. R., & Abdullahi, S. (2015). Landslide susceptibility assessment at Wadi Jawrah Basin, Jizan region, Saudi Arabia using two bivariate models in GIS. *Geosciences Journal*, 1–21.

Zhu, X., Zhang, S., Jin, Z., Zhang, Z., & Xu, Z. (2011). Missing value estimation for mixed-attribute data sets. *Knowledge and Data Engineering, IEEE Transactions on, 23*(1), 110–121.

Zhuang, L., & Dai, H. (2006). Parameter optimization of kernel-based one-class classifier on imbalance learning. *Journal of Computers, 1* (7), 32–40.

Ensemble Disagreement Active Learning for Spatial Prediction of Shallow Landslide

10

Biswajeet Pradhan, Maher Ibrahim Sameen and Bahareh Kalantar

10.1 Introduction

In Malaysia, landslides are considered as the most frequent and devastating natural disaster that cause human life and property losses. The spatial prediction of landslides is the basic step required for hazard and risk assessments. Spatial prediction methods of landslides are established and documented in the literature. However, several research directions on this topic need to be developed and explored in depth. The current improvement in computer technology and laser scanning systems provide improved data processing capabilities and topographic datasets, as well as new trends in landslide modeling and methods that can deal with such advanced technologies and datasets.

Several techniques have been developed for landslide susceptibility modeling (or spatial prediction). Support vector machine (SVM) (Bui et al. 2012; Tien et al. 2012; Xu et al. 2012a, b; Peng et al. 2014; Pourghasemi et al. 2013; Yang et al. 2014), logistic regression (LR) (Akgun 2012; Ayalew and Yamagishi 2005), and frequency ratio (FR) (Pradhan 2010; Latif et al. 2012; Ozdemir and Altural 2013; Lee and Sambath 2006; Tazik et al. 2014) are among the common techniques used as benchmarks in several studies. Marjanović et al. (2011) compared SVM, LR, and decision tree models for landslide prediction. Results revealed that the SVM model outperformed the other models, as well as the knowledge-driven [i.e., analytical hierarchy process (AHP)] methods. Xu et al. (2012a, b) evaluated the efficiency of SVM model for landslide susceptibility mapping in their study, which achieved best results with a success rate of 79.20% and predictive accuracy of 79.13% using the radial basis function (RBF). Ballabio and Sterlacchin (2012) investigated landslide susceptibility mapping using the SVM model. The SVM procedure outperformed other techniques (i.e., LR, linear discriminant analysis, and naive Bayes) in terms of accuracy and generalization capacity. Moreover, Hong et al. (2016) used SVM model to assess landslide susceptibility in China. The results indicated that the prediction rates for the four SVM models are 81.0% (radial basis function), 71.0% (polynomial), 40.0% (sigmoid), and 63.0% (linear). Thus, the RBF-SVM model has the highest overall performance. Colkesen et al. (2016) assessed landslide susceptibility using the SVM and LR models. Results showed that the SVM models outperformed the traditional LR model by 18%.

However, landslide susceptibility modeling requires an adequate number of landslide inventories, and their qualities are verified before placing them into the models to ensure the quality and reliability of the outputs. Landslide inventories are collected from various sources such as field investigations and examinations of remote sensing images and orthophotographs, generating uncertainties in landslide historical data (Guzzetti et al. 2012). Owing to the uncertainty in landslide inventories, verifying the training data is needed before conducting any regression analysis to achieve accurate spatial prediction of landslides (Guzzetti et al. 2012). To verify landslide inventories, extensive manual works on remote sensing images or orthophotographs are necessary. However, this work is costly and time-consuming. One of the solutions to this is active learning, which is a subfield of machine learning that aims to select the most informative training samples among the available ones (Demir and Bruzzone 2015). In this manner, the active learner aims to achieve high accuracy using a fewer landslide inventories, thereby minimizing the cost of obtaining high-quality training data. The main concept of active learning is measuring the classification uncertainty of unlabeled samples. Active learning can be used with most machine learning algorithms such as SVM and multivariate statistical methods such as LR to improve the quality of training dataset before the learning process. Liao et al. (2005) presented an active learning method to improve the generalization ability of the LR model. Pasollie et al. (2014)

B. Pradhan (✉) · M.I. Sameen · B. Kalantar
Department of Civil Engineering, University Putra Malaysia, Serdang, Malaysia
e-mail: biswajeet24@gmail.com

B. Pradhan (ed.), *Laser Scanning Applications in Landslide Assessment*,
DOI 10.1007/978-3-319-55342-9_10

proposed a new active learning approach for SVM classification in urban areas. The experimental results showed the effectiveness of regularization in the spatial domain for active learning purpose. Di and Crawford (2012) investigated the principles and capabilities of several approaches for the view generation on hyperspectral data classification based on the multiview adaptive maximum disagreement active learning method. Results indicated that the proposed method outperforms the random sampling and the state-of-the-art SVM margin sampling techniques. Active learning showed efficient remote sensing applications (i.e., image classification, and feature extraction). However, the literature search shows that no previous works are used for active learning to train landslide susceptibility models.

Therefore, this study proposes a framework for predicting shallow landslides in a tropical region using the ensemble disagreement active learning strategy. This strategy by selecting informative training samples and employing multiple classifiers improves the accuracy of landslide prediction results.

The rest of this chapter is organized as follows. First, ensemble disagreement active learning is briefly explained. Second, the study area and the main datasets are described. Third, the proposed framework is presented and discussed in detail. Fourth, the results are presented, and the main findings are discussed. Finally, a brief conclusion is presented, including the main findings, limitations of the current study, and future directions.

10.2 Ensemble Disagreement Active Learning: A Preview

Active learning is a new strategy in machine learning algorithms, which aims to improve the prediction accuracy of regression models (Demir and Bruzzone 2015; Tuia et al. 2009). The main concept of active learning is a feedback loop between an expert (sample annotator) and machine (computer program) that eventually tunes the machine model. The model starts with a set of labeled data to judge received data. Experts label a sample of the machine's output, and their work is turned over into the model. Experts continue to label the data until the model achieves unexceptional accuracy.

Ensemble disagreement is a common strategy in active learning (Körner and Wrobel 2006), and it originates in the query-by-committee method by Seung et al. (1992). The use of multiple classifiers in improving the accuracy of the results is fundamental to this strategy. Hence, it allows it to learn from examples where the results of various methods disagree. For example, a landslide-type classifier may label a small landslide as shallow landslides; however, data from a second method might show that the small size landslide is actually a translational landslide; this helps the first classifier correctly judge whether the landslide is a shallow landslide. Ensemble disagreement is an iterative procedure that adds the queried training samples along with this label to the training dataset. The iteration continues until desired prediction accuracy is reached.

10.3 Study Area and Data

In this research, a part of Cameron Highlands, a tropical rainforest area in the Peninsular Malaysia at the northwestern tip of Pahang at approximately 200 km from Kuala Lumpur, was selected as the case study considering the frequent occurrences of landslides here. Reports from government agencies and previous studies indicate that several landslides occurred in Cameron Highlands and caused major damages to properties (Lateh et al. 2010). Quaternary, Devonian granite, and schist are the main lithology types in this region (Pradhan and Lee 2010). A 25-km^2 area is selected within the Cameron Highlands to perform the analysis and test the active learning model in the current work (Fig. 10.1). The lowest and highest altitudes are 1200 and 1800 m, respectively.

However, landslide inventories are collected and prepared using remote sensing data and geographic information system (GIS). Remote sensing data included archived 1:10,000–1: 50,000 aerial photographs, high-resolution satellite images such as SPOT, and high-resolution LiDAR-based orthophotographs were used for the visual identification of landslide occurrences in the study area. In addition, several landslide locations were digitized from historical landslide records and previous studies (Pradhan et al. 2010). Moreover, field observations were used to collect fresh landslide scars and to confirm the landslides collected from remote sensing data. Several landslides are shallow rotational and a few translational in type, which are not shallow in type, were removed from the analysis. In assembling a database to assess the surface area and a number of landslides in the study area, a total of 192 landslides were mapped in an area of 25 km^2 (Fig. 10.1).

10.3.1 Landslide Conditioning Factors

The LiDAR point clouds were collected on January 26, 2015, over a part of Cameron Highlands. The fight height of the airborne platform was 1510 m, the point density was 8 per m^2, and the frequency rate of laser sensor was 25,000 Hz. In addition, the absolute accuracy (measured by root mean square error [RMSE]) of the acquired data are 0.15 and 0.3 m in vertical and horizontal dimensions, respectively.

For the spatial prediction of shallow landslides, nine landslide conditioning factors were derived from LiDAR

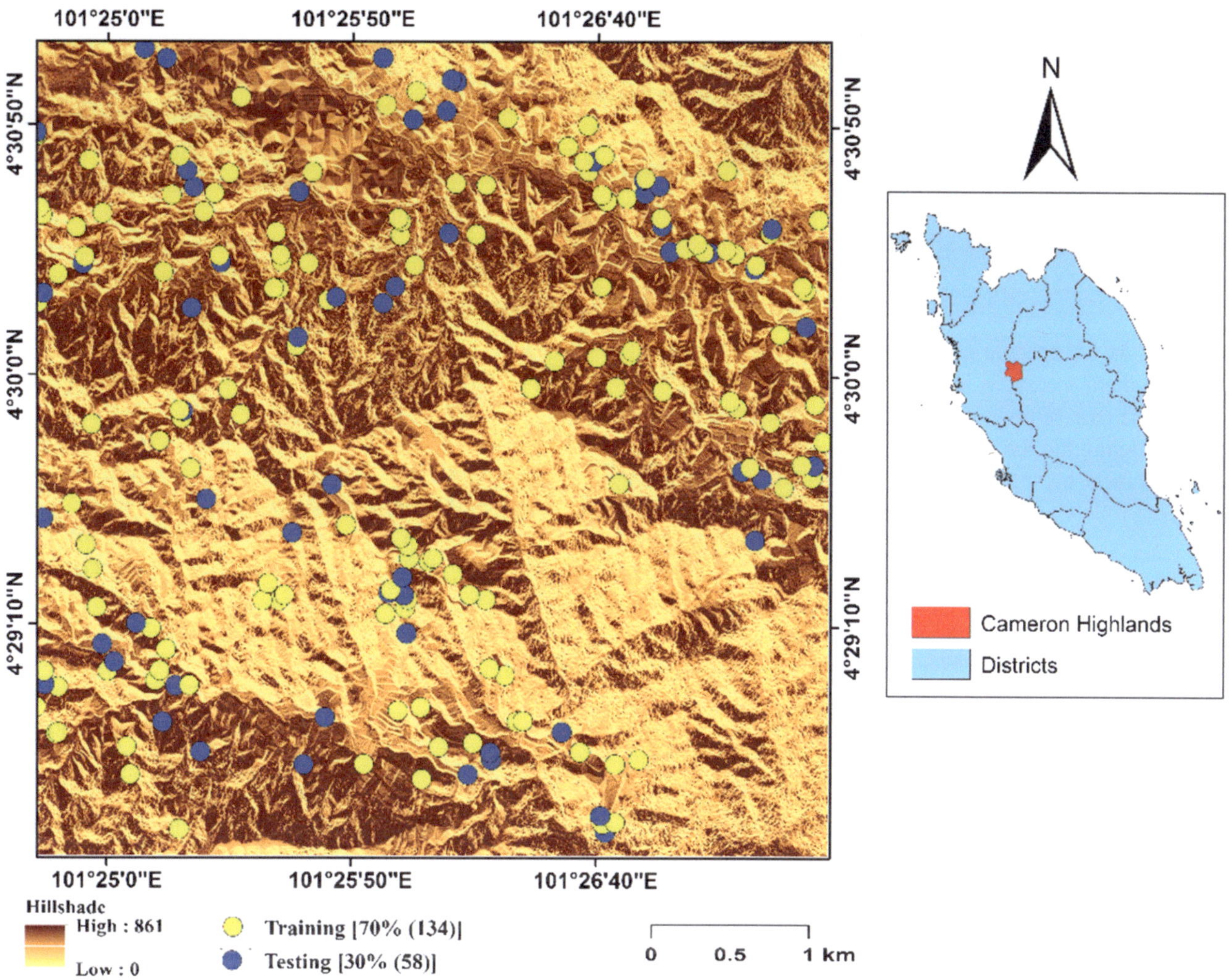

Fig. 10.1 Geographic location of the study area

point clouds. These factors are altitude, slope, aspect, plan curvature, profile curvature, topographic wetness index (TWI), topographic roughness index (TRI), sediment transport index (STI), and stream power index (SPI). These factors were reclassified by the quantile classification method in the ArcGIS software.

Altitude is controlled by several geologic and geomorphological processes. Generally, landslides occur at intermediate elevations as slopes tend to be covered by a layer of thin colluvium that is prone to landslides. The lowest and highest altitudes were observed at 1200 and 1800 m, respectively. The slope is another landslide conditioning factor, which measures the elevation change rate in the direction of steepest descent. The slope map of the study area was divided into five slope angle classes. In this study, the highest slope was observed at 82.4°. In addition, aspect is the slope direction measured in degrees from north in a clockwise direction, ranging from 0° to 360°. Parameters such as exposure to sunlight, rainfall, and drying winds control the soil moisture concentration, which in turn is determined landslide occurrence. Furthermore, plan curvature is described as the curvature of a contour line formed by the intersection of a horizontal plane with the surface. This influences the flow convergence and divergence across a surface. The profile curvature affects the acceleration and deceleration of downslope flows because of the erosion and deposition influences. Plan and profile curvature maps were reclassified into three classes: convex, flat, and concave lands from negative, zero, and positive values, respectively.

Moreover, four hydrological factors generated from the LiDAR data were used in the current study for landslide susceptibility mapping. The TWI describes the effect of topography on the location and size of saturated source areas of runoff generation (Beven and Kirkby 1979; Moore et al. 1991). Topographic wetness index (TWI) is calculated as follows:

$$\text{TWI} = \ln \frac{A_s}{\tan(\beta)} \quad (10.1)$$

where A_s is the specific catchment area of each cell and β is the slope gradient (in degrees) of the topographic heights. SPI, a measure of the erosion power of the stream, is also considered as a factor contributing toward the stability within the study area. The SPI is expressed as follows:

$$\text{SPI} = A_s \times \tan(\beta) \quad (10.2)$$

where A_s is the specific catchment area and β is the local slope gradient measured in degrees. However, STI, which reflects the erosive power of the overland flow, was derived by considering the transport capacity limiting sediment flux and catchment evolution erosion theories. STI is calculated using the following expression (Mohammady et al. 2012):

$$\text{STI} = \left(\frac{A_s}{22.13}\right)^{0.6} \left(\frac{\sin\beta}{0.0896}\right)^{1.2} \quad (10.3)$$

where A_s is the specific catchment area and β is the local slope gradient measured in degrees.

TRI is another important hydrological factor affecting landslide susceptibility. These hydrological factors were reclassified into five classes based on the quantile method and used for landslide susceptibility modeling. TRI is calculated using Eq. 10.4 (Jebur et al. 2014) (Fig. 10.2):

$$\text{TRI} = \sqrt{\text{Abs}(\max^2 - \min^2)} \quad (10.4)$$

10.4 Methods

10.4.1 Active Learning Landslide Susceptibility Mapping

The proposed framework of active learning for prediction shallow landslides is presented in Fig. 10.3. First, data processing starts by importing LiDAR point clouds and landslide inventory map into a GIS software with appropriate data storage format. Second, a very high-resolution digital elevation model is derived from the LiDAR point clouds at 0.5 m spatial resolution. In this step, the non-ground points were filtered by the multi-scale curvature algorithm (Evans and Hudak 2007). From the generated DEM, nine landslide conditioning factors were derived at the same resolution. These factors were then prepared in a GIS geodatabase along with the landslide inventory data. Third, the landslide inventory data was divided into two groups. The first group is composed of 34 landslides and saved as "Training T1," whereas the remaining landslides were saved as "Training T2" in the same geodatabase.

In the first group, landslides were confirmed by visual interpretation to remove the uncertain samples. However, the landslides in the second group were used in an iteration process to add only the correct samples to the first group. Furthermore, two models were built using the SVM and LR algorithms from the Training T1 data and the nine landslide conditioning factors. These models were then used to classify the subsets of landslides in the second group (Training T2). The 10 subsets are currently used as the total number of landslides is not large (192). However, a large number of landslide inventory data necessitates increased subsets. Afterward, the accuracy of the landslide prediction was measured by the success and prediction rates. A threshold was created to test the agreement or disagreement in the results of the two models. This threshold (T) is the absolute difference between the success rates of both models. Currently, T was selected as 0.1. This will allow samples with relatively small disagreement between the models to be relabeled, thereby improving the quality of the final training dataset. Furthermore, the subsets, where both models agree ($T \leq 0.1$) were added to the Training T1 data, whereas the subsets where the models failed to agree were relabeled by human interpretation and eventually added to the Training T1 data.

Therefore, once the high-quality training data (Tu) was generated, the models were rebuilt using the same landslide conditioning factors and their accuracies were measured and compared with the original models. The landslide susceptibility maps were generated from the models that were trained by the high-quality training dataset created by active learning.

10.4.2 Landslide Susceptibility Modeling

10.4.2.1 Support Vector Machine

Vapnik (1995) developed SVM, a nonlinear classification model, derived from machine learning techniques. SVM aims to determine an optimal separating hyper plane (maximizing the margin width) between two classes in a feature space. The training points near the hyperplane are called support vectors, and they are utilized for classification once the decision surface is obtained. The separating hyper plane is found as follows:

$$y_i(w \times x_i + b) \geq 1 - \xi_i \quad (10.5)$$

where w is the coefficient vector that defines the hyper plane orientation in the feature space; b is the offset of the hyper plane from the origin; and ξ_i is the positive slack variables (Cortes and Vapnik 1995). The optimal hyper plane is found by solving the following optimization problem (Jebur et al. 2014):

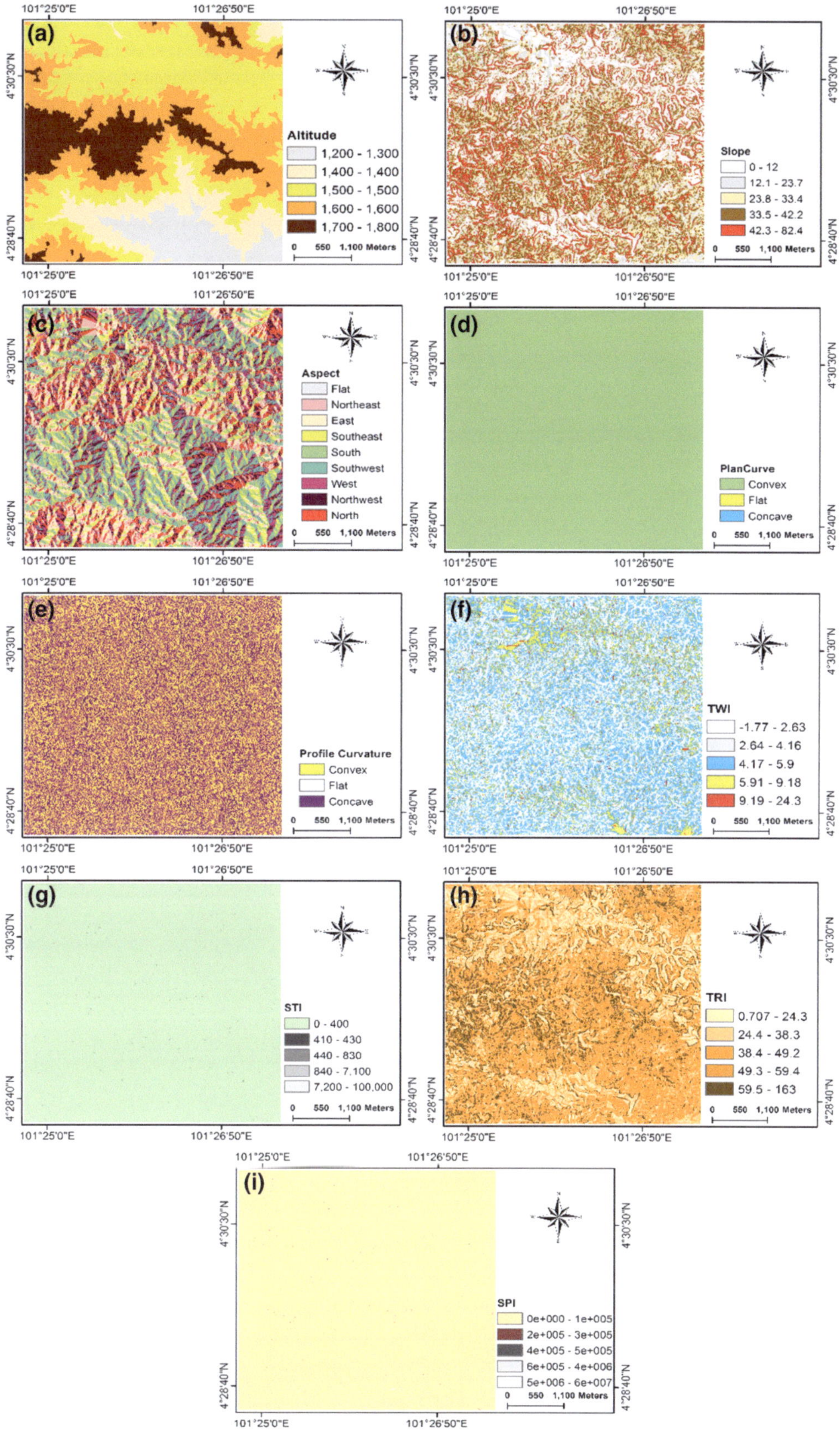

Fig. 10.2 Landslide conditioning factors derived from high-resolution airborne LiDAR, **a** altitude, **b** slope, **c** aspect, **d** plan curvature, **e** profile curvature, **f** topographic wetness index (TWI), **g** sediment transport index (STI), and **h** topographicroughness index (TRI), **i** stream power index (SPI).

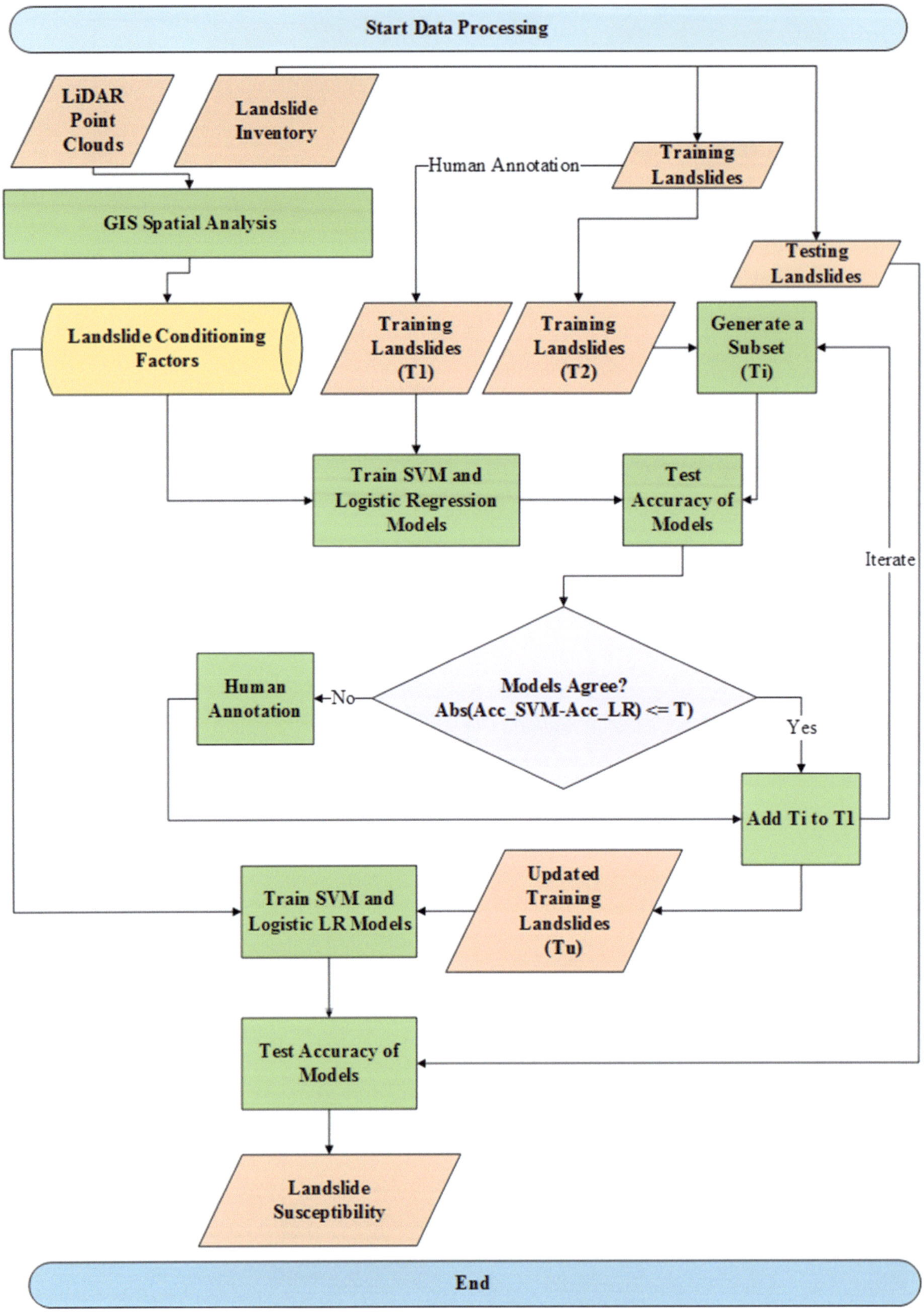

Fig. 10.3 Flowchart of the proposed spatial prediction method of shallow landslides by active learning

$$\begin{aligned} &\text{Minimize } \sum_{i=1}^{n} \alpha_i - \frac{1}{2}\sum_{i=1}^{n}\sum_{j=1}^{n} \alpha_i\alpha_j y_i y_j (x_i x_j) \\ &\text{subject to } \sum_{i=1}^{n} \alpha_i y_j = 0, \quad 0 \le \alpha_i \le C \end{aligned} \tag{10.6}$$

where α_i is the Lagrange multiplier and C is the penalty. For data classification, the following decision function is applied as follows:

$$g(x) = \text{sign}\left(\sum_{i=1}^{n} y_i \alpha_i x_i + b\right) \tag{10.7}$$

In case of nonlinearly separating samples, the decision function (Eq. 10.8) is rewritten as follows:

$$g(x) = \text{sign}\left(\sum_{i=1}^{n} y_i \alpha_i K(x_i, x_j) + b\right) \tag{10.8}$$

In this process, the original data are transformed into a higher dimensional space using a nonlinear kernel function (K). The common K functions used with SVM in landslide studies are linear, sigmoid, radial basis function, and polynomial (Bui et al. 2012). However, several studies showed that radial basis function is more suitable for landslide susceptibility application than other kernel functions (Pourghasemi et al. 2013; Xu et al. 2012a, b).

10.4.2.2 Logistic Regression

The logistic regression model is a widely used multivariate statistical technique to establish the relationship between explanatory factors and various types of targets such as landslide prediction (Bai et al. 2010; Das et al. 2010; Hyun-Joo et al. 2010). The predicted values which ranged from 0 to 1 are defined by the following expressions (Ozdemir 2016):

$$y = b_0 + b_1 x_1 + b_2 x_2 + \cdots + b_n x_n \tag{10.9}$$

$$y = \log_e\left(\frac{p}{1-p}\right) \tag{10.10}$$

$$p = \frac{e^y}{1+e^y}, \tag{10.11}$$

where y is the linear logistic model, b_0 is the model intercept, n is the number of landslide conditioning factors (5), b is the weight of each factor, x is the landslide conditioning factor, and P is the probability of landslide occurrence (landslide susceptibility index). The function y is represented as logit (P), i.e., the log (to base e) of the odds or likelihood ratio that the dependent variable is 1.

10.4.3 Accuracy Assessment

Results of the landslide susceptibility models are validated by estimating the area under the ROC curve (AUC), success, and prediction rates. These accuracy measures explain the percentage of detected landslides that fall into each defined susceptibility level and display as the cumulative frequency graph (Chung and Fabbri 2003; Intarawichian and Dasananda 2011). In conducting the ROC accuracy assessment, the results of landslide susceptibility models are compared with the training landslides created from various sources (i.e., aerial photograph, high-resolution satellite images, and previous records) (Pradhan and Kim 2014). In rate curve, the y-axis is normally considered as the cumulative percentage of observed landslide occurrences in various susceptibility classes, whereas the x-axis corresponds to the cumulative percentage of the area of the susceptibility classes. The total AUC is used to qualitatively determine the prediction accuracy of the susceptibility map in which a larger area refers to a higher accuracy achieved (Pourghasemi et al. 2013; Intarawichian and Dasananda 2011; Mathew et al. 2009).

10.5 Results and Discussion

This section presents the results of the analysis. The SVM penalty parameter is fine-tuned, and the landslide prediction models by the traditional and active learning approaches, the landslide susceptibility models, and the effects of active learning on multicollinearity statistics are presented and discussed.

10.5.1 Results of Fine-Tuning the Penalty Parameter of the SVM Model

The SVM model includes several user-defined parameters that require fine-tuning to achieve the highest possible accuracy (Yao et al. 2008). The kernel-type function and the penalty parameter or C-value are the most significant parameters that affect the accuracy of the model. Among landslide studies, the best kernel function was found to be the radial basis function, and this result is obtained in several studies (Pourghasemi et al. 2013; Zare et al. 2013; Xu et al. 2012a, b; Bui et al. 2012). According to Zhu et al. (2011), RBF is primarily advantageous because of its good interpolation abilities.

However, there is no specific conclusion on the best C-value for landslide studies. Therefore, C-value was fine-tuned by examining its sensitivity on the accuracy of

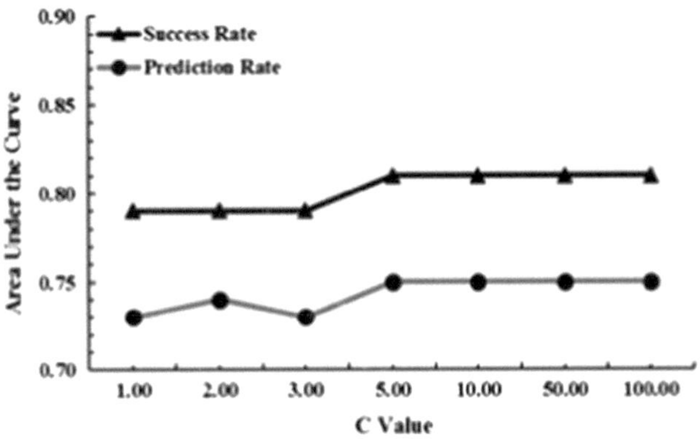

Fig. 10.4 Effects of *C* parameter on the accuracy of landslide prediction

landslide classification (Fig. 10.4). The graph suggests that the best *C*-value is 5 or greater and that it should be examined until 100. As a result, a *C*-value of 100 was selected in the current study, and it was used to build the subsequent SVM models.

10.5.2 Results of Spatial Prediction of Shallow Landslides by Active Learning

Table 10.1 shows the results of the landslide prediction accuracy of various iterations (*i*) in active learning. Overall, nine *i* were processed in this study, and this number is increased with the increasing landslide inventories in the training dataset. Success and prediction rates achieved by the SVM and LR models are presented. The lowest and highest success rates are 0.3 and 1, respectively. However, the lowest and highest prediction rates are 0.41 and 1, respectively. On the basis of these results, the difference between the success rates of the two methods was compared and judged to select the samples for the query. On the basis of selected threshold ($T \leq 0.1$), the samples that meet the condition was directly added to the Training T1 data, whereas the samples that fail to meet the condition were manually checked, relabeled, and added to the Training T1 data. This allowed the generation of high-quality Tu to be used in training the SVM and LR models to create landslide susceptibility maps. Results show that the *i* (*i* = 1, 3, 6, 9) were directly added to the Training T1 data. However, the remaining *i* (*i* = 2, 4, 5, 7, 8, 10) were relabeled through visual checking on high-resolution orthophotographs and then added to the Training T1 data.

In addition, Table 10.2 shows the results of the landslide susceptibility models trained by the Training T1 and updated T1 data. Results indicate that the active learning can improve the accuracy of the shallow landslide prediction from 0.75 to 0.93 in the case of LR model and from 0.84 to 0.89 in the case of SVM model. This result suggests that active learning is sufficient for landslide susceptibility modeling when the SVM and LR models are used.

10.5.3 Landslide Susceptibility Maps

Four landslide susceptibility maps were generated (Fig. 10.5). The first map was generated with the SVM model trained using all training datasets (T1). The second map was generated with the LR model trained by all training datasets. However, the third and fourth maps were generated by the SVM and LR models trained by the updated training dataset (the case of active learning), respectively. The susceptibility maps were reclassified into five classes: very low, low, moderate, high, and very high. The classification was performed using the quantile method (Jebur et al. 2014).

Landslide susceptibility maps show that most of the north and southeast areas are highly and very highly susceptible to shallow landslides, respectively. In contrast, the middle and

Table 10.1 Accuracy of landslide prediction of different iterations in active learning

Training landslides (Ti)	Accuracy of landslide prediction				T	Status
	SVM		Logistic regression			
	Succ. rate	Pred. rate	Succ. rate	Pred. rate		
Iteration 1	0.8	0.74	0.65	0.57	0.15	Added to T1
Iteration 2	0.85	0.81	0.87	0.82	0.02	Relabeled
Iteration 3	0.7	0.64	0.57	0.6	0.13	Added to T1
Iteration 4	0.9	0.83	1	1	0.1	Relabeled
Iteration 5	0.4	0.46	0.48	0.49	0.08	Relabeled
Iteration 6	0.6	0.55	0.74	0.66	0.14	Added to T1
Iteration 7	0.9	0.83	0.88	0.84	0.02	Relabeled
Iteration 8	0.85	0.78	0.91	0.85	0.06	Relabeled
Iteration 9	0.4	0.46	0.52	0.53	0.12	Added to T1
Iteration 10	0.3	0.44	0.3	0.41	0	Relabeled

Table 10.2 Accuracy of landslide prediction of the traditional learning and active learning methods

Training data	Accuracy of landslide prediction				*T*
	SVM		Logistic regression		
	Succ. rate	Pred. rate	Succ. rate	Pred. rate	
T1	0.81	0.75	0.84	0.84	0.03
Updated T1	0.88	0.93	0.87	0.89	0.01

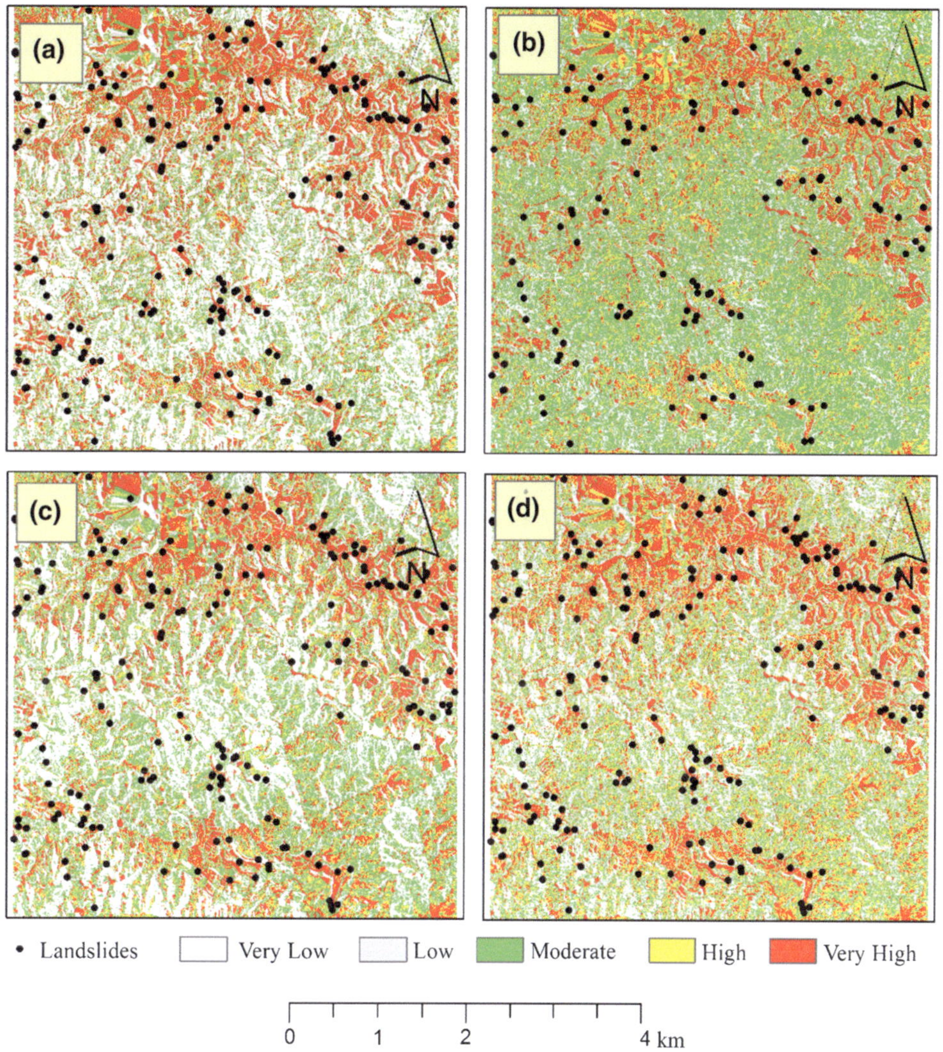

Fig. 10.5 Landslide susceptibility maps, **a** SVM with all the training dataset, **b** LR with all the training dataset, **c** SVM with active learning, and **d** LR with active learning

the west parts of the study area are classified as low and very low classes. The estimated weights for the landslide conditioning factors by the SVM and LR models indicated that the high and very high susceptibility are contributed to the corresponding steep slope and high TRI. Figure 10.6 shows the calculated number of landslides in the susceptibility classes for both models. A slight difference between the SVM and LR models was observed in terms of the number of landslides in the very low, low, and moderate classes. Therefore, a significant difference between the two models existed in terms of the number of landslides in the high and very high classes. The calculated numbers of landslides in the very high classes of the SVM and LR models are 109 and 67, respectively.

10.5.4 Effects on Multicollinearity Statistics

The collinearity among landslide conditioning factors can influence model performance (Ballabio and Sterlacchini 2012). The landslide factor was checked for collinearity using variance inflation factors (VIF) (Fox and Monette

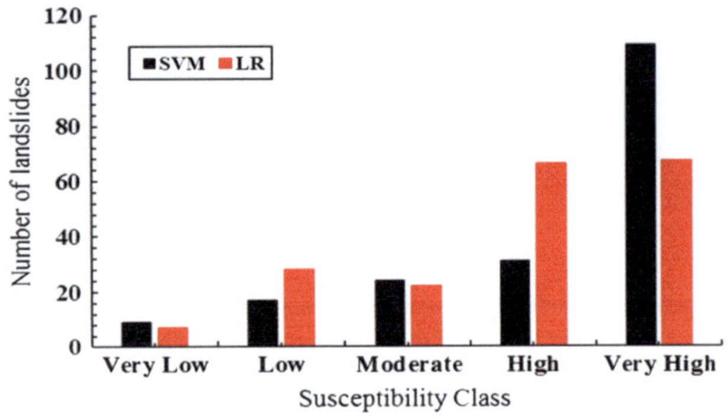

Fig. 10.6 A number of landslides in each susceptibility class for the SVM and LR models

1992). Generally, a VIF value of 10 or more is derived to prove high collinearity. Table 10.3 shows the results of the multicollinearity analysis for traditional and active learning. Results show the high collinearity (>10) of the two parameters SPI and STI. Given the result, these parameters were discarded from further analysis and excluded in the final models.

In addition, the Mann–Whitney test was used to evaluate the effects of active learning on the multicollinearity statistics (Table 10.4). This test the absence of a significant difference between the estimated VIF values for the model generated by the Training and updated T1 data. The following hypotheses were tested as follows:

$$\begin{cases} H_0\colon \Delta S = 0 \\ H_a\colon \Delta S \neq 0 \end{cases} \quad (10.12)$$

where ΔS is the difference of location between the samples. As the computed p-value is greater than the significance level alpha = 0.05, one cannot reject the null hypothesis H_0. The risk to reject the null hypothesis H_0 while it is true is 85.98%.

10.6 Discussion

The spatial prediction of landslides is a challenging task for several reasons including the spatial variations of landslide factors, the high-quality landslide inventories for model training, and the complex mechanisms of various types of landslides. Several landslide susceptibility modeling approaches have been proposed to handle a number of relevant aspects. The current study proposed an active learning-based landslide susceptibility mapping technique to semi-automatically evaluate the quality of landslide inventory prior to its use in developing regression models. The SVM and LR models were used for the evaluation of landslide samples-based disagreeing results. The landslide inventory dataset was evaluated by an iterative procedure, and a new landslide training subset was subsequently produced in which uncertain samples (i.e., landslides that SVM and LR failed to agree) were removed. The main advantages

Table 10.3 A number of landslides in each susceptibility class for the SVM and LR models

Landslide factor	VIF-T1	VIF-updated T1	Difference
Profile curvature	1.149	1.115	0.034
Slope	6.902	6.917	−0.015
SPI	14.010	14.230	−0.220
STI	16.890	17.190	−0.300
TRI	7.155	7.060	0.095
TWI	1.343	1.215	0.128
Aspect	1.008	1.007	0.001
DEM	1.129	1.081	0.048
Plan curvature	1.208	1.162	0.046

Table 10.4 Mann–Whitney test/two-tailed test

Parameter	Calculated value
U	43
Expected value	40.5
Variance (U)	128.25
p-value (Two-tailed)	0.859
Alpha	0.05

*An approximation has been used to compute the p-value

of this method are its effectiveness when the landslide inventory data are collected from various sources and when the data quality is unknown. In addition, the method enhances the inherent statistical properties of the training dataset. For example, the results (Sect. 3.4) showed reduced multicollinearity effects after removing the uncertain landslides from the training dataset. The other advantages of the proposed method include the improved success and prediction rates of SVM and LR models, as well as the updated landslide inventory datasets for further works.

Generally, there is a nonlinear relationship between landslide problems and their triggering factors (Yesilnacar and Topal 2005). LR and SVM provide the potential to overcome the limitations of statistical methods, which are distribution-based and inability to handle multisource data that are commonly collected from the field. SVM and LR satisfy more rigorous landslide susceptibility mapping requirements (Yesilnacar and Topal 2005). However, these models require high-quality training data to achieve accurate results. High-quality landslide inventory should include well-distributed landslides with an adequate number of landslides in the studied region. Therefore, selecting only high-quality landslide training data can improve the accuracy of landslide prediction by the SVM and LR methods. The accuracy assessment showed that the prediction rates increased by 0.18 and 0.5 in the SVM and LR models, respectively. Thus, the SVM method has more generalization ability than the LR model given that the former is based on risk minimization. Therefore, SVM attempts to maximize the margin between the closest support vectors, whereas LR maximizes the posterior class probability. This means that LR is more dependent on class probability than that of SVM, which makes it sensitive to the size, distribution, and the quality of landslide training data. In addition, LR converges to any decision boundary that can divide the training landslides into positive and negative targets, whereas the objective of SVM causes the decision boundary to lie geometrically between the support vectors. Furthermore, the literature shows that the accuracy of SVM and LR are comparable in landslide predictions, and the model selection for landslide studies remains data dependent, such that, various models should be tried. The selection of model is based on a certain accuracy measure (i.e., AUC, overall accuracy).

Compared with the LR model, SVM requires several user-defined parameters such as penalty parameter (c), kernel-type function, and gamma parameter of the kernel functions. The accuracy of produced landslide susceptibility maps using SVM significantly depends on these parameters. To improve the accuracy of landslide susceptibility outputs, a frequent practice is the fine-tuning of SVM parameters. The current study showed that the best C-value is 100, whereas RBF function was selected as the kernel-type function. This selection was based on suggestions in recent literature. Several studies showed that RBF is sufficient for landslide studies with no specific reason. The selection of SVM kernel function is automated using the cross-validation approach. However, this is tricky as it is easy to over-fit the SVM model and to end up with a worse model compared to random selection and to other types of kernel functions.

The analysis of factor importance by estimated SVM coefficients and the original landslide inventory showed that slope (90.63), TRI (2.56), and altitude (−0.65) are the most three important factors for landslide prediction in the study area. However, LR showed that slope (1.26), altitude (1.29), and TWI (1.260) are the most influential factors on landslide occurrence in the study area. Using the updated landslide training dataset, the estimated coefficients by SVM and LR were changed and the aspect (−1.5 by SVM and 1.3 by LR) was found to be more significant than altitude (0.47 by SVM and 1.04 by LR). The variations in estimating factor coefficient by the SVM and LR models produced landslide susceptibility maps in which susceptibility classes spatially vary within the study area. In addition, SVM was less affected through the reduced training samples than LR (Fig. 10.6). Therefore, the landslide susceptibility map generated by the SVM model and the active learning-based selected landslide training data were suggested for land-use planning and other risk mitigation analyses (Fig. 10.7).

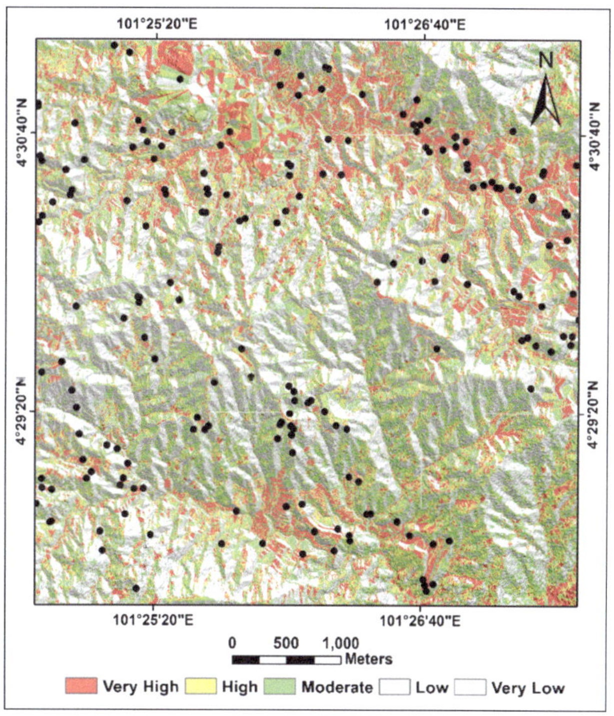

Fig. 10.7 Suggested landslide susceptibility map for land-use planning in Cameron Highlands

10.7 Conclusion

This study proposed and discussed a framework based on ensemble disagreement active learning, which aims to select the most informative landslide inventories for use in model training. This allowed building models with enhanced performance and generalization capacity. Experimental results showed that the active learning can improve the success and prediction accuracy of the SVM and LR models. The success rates of the SVM model with active learning and without active learning are 0.81 and 0.88, respectively. However, the success rates achieved by the LR model are 0.84 and 0.87 for the active learning and without active learning method, respectively. The prediction accuracy of the SVM with active learning was observed at 0.93, which was significantly improved compared with the SVM model trained through the traditional method. In addition, the prediction rates achieved by the LR model are 0.48 and 0.89 for the original training data and the updated training data, respectively. Results of the current study suggest that the combination of active learning and common landslide susceptibility modeling methods can increase accuracies. In addition, active learning allowed the removal of non-informative samples from the training data to achieve models with better performances. The use of active learning also reduced the effects of collinearity among the landslide conditioning factors. This finding was tested by the Mann–Whitney test (p-value two-tailed = 0.859).

However, the current study included certain limitations, which require further investigations in the future works. First, other active learning strategies need to be examined and their effectiveness for landslide susceptibility assessment should be studied. Second, the combination of ensemble disagreement active learning with other landslide susceptibility modeling approaches also needs to be explored. Finally, the same analysis should be repeated in various regions with different landslide conditioning factors, including LiDAR-based and other factors.

References

Akgun, A. (2012). A comparison of landslide susceptibility maps produced by logistic regression, multi-criteria decision, and likelihood ratio methods: A case study at İzmir, Turkey. *Landslides, 9*(1), 93–106.

Ayalew, L., & Yamagishi, H. (2005). The application of GIS-based logistic regression for landslide susceptibility mapping in the Kakuda-Yahiko Mountains, Central Japan. *Geomorphology, 65*(1), 15–31.

Bai, S. B., Wang, J., Lü, G. N., Zhou, P. G., Hou, S. S., & Xu, S. N. (2010). GIS-based logistic regression for landslide susceptibility mapping of the Zhongxian segment in the three Gorges area, China. *Geomorphology, 115*(1), 23–31.

Ballabio, C., & Sterlacchini, S. (2012). Support vector machines for landslide susceptibility mapping: the Staffora River Basin case study, Italy. *Mathematical Geosciences, 44*(1), 47–70.

Beven, K. J., & Kirkby, M. J. (1979). A physically based, variable contributing area model of basin hydrology/Un modèle à base physique de zone d'appel variable de l'hydrologie du bassin versant. *Hydrological Sciences Journal, 24*(1), 43–69.

Bui, D. T., Pradhan, B., Lofman, O., Revhaug, I., & Dick, O. B. (2012). *Application of support vector machines in landslide susceptibility assessment for the Hoa Binh province (Vietnam) with kernel functions analysis* (Doctoral dissertation, International Environmental Modelling and Software Society (iEMSs)).

Chung, C. J. F., & Fabbri, A. G. (2003). Validation of spatial prediction models for landslide hazard mapping. *Natural Hazards, 30*(3), 451–472.

Colkesen, I., Sahin, E. K., & Kavzoglu, T. (2016). Susceptibility mapping of shallow landslides using kernel-based Gaussian process, support vector machines and logistic regression. *Journal of African Earth Sciences, 118,* 53–64.

Cortes, C., & Vapnik, V. (1995). Support-vector networks. *Machine learning, 20*(3), 273–297.

Das, I., Sahoo, S., van Westen, C., Stein, A., & Hack, R. (2010). Landslide susceptibility assessment using logistic regression and its comparison with a rock mass classification system, along a road section in the northern Himalayas (India). *Geomorphology, 114*(4), 627–637.

Demir, B., & Bruzzone, L. (2015). A novel active learning method in relevance feedback for content-based remote sensing image retrieval. *IEEE Transactions on Geoscience and Remote Sensing, 53*(5), 2323–2334.

Di, W., & Crawford, M. M. (2012). View generation for multiview maximum disagreement based active learning for hyperspectral image classification. *IEEE Transactions on Geoscience and Remote Sensing, 50*(5), 1942–1954.

Evans, J. S., & Hudak, A. T. (2007). A multiscale curvature algorithm for classifying discrete return LiDAR in forested environments. *IEEE Transactions on Geoscience and Remote Sensing, 45*(4), 1029–1038.

Fox, J., & Monette, G. (1992). Generalized collinearity diagnostics. *Journal of the American Statistical Association, 87*(417), 178–183.

Guzzetti, F., Mondini, A. C., Cardinali, M., Fiorucci, F., Santangelo, M., & Chang, K. T. (2012). Landslide inventory maps: New tools for an old problem. *Earth-Science Reviews, 112*(1), 42–66.

Hong, H., Pradhan, B., Jebur, M. N., Bui, D. T., Xu, C., & Akgun, A. (2016). Spatial prediction of landslide hazard at the luxi area (china) using support vector machines. *Environmental Earth Sciences, 75* (1), 1–14.

Hyun-Joo, O., Saro, L., & Soedradjat, G. M. (2010). Quantitative landslide susceptibility mapping at Pemalang area. Indonesia. *Environmental Earth Sciences, 60*(6), 1317–1328.

Intarawichian, N., & Dasananda, S. (2011). Frequency ratio model based landslide susceptibility mapping in lower Mae Chaem watershed, Northern Thailand. *Environmental Earth Sciences, 64* (8), 2271–2285.

Jebur, M. N., Pradhan, B., & Tehrany, M. S. (2014). Optimization of landslide conditioning factors using very high-resolution airborne laser scanning (LiDAR) data at catchment scale. *Remote Sensing of Environment, 152,* 150–165.

Körner, C., & Wrobel, S. (2006). Multi-class ensemble-based active learning. In *European conference on machine learning* (pp. 687–694). Berlin, Heidelberg: Springer.

Lateh, H., Jefriz, Muhiyuddin, W. M., Taib, B., & Khan, Y. A. (2010). Monitoring of hill-slope movement due to rainfall at Gunung Pass of Cameron Highland district of Peninsular Malaysia. *International Journal of Earth Sciences and Engineering, 3,* 6–12.

Latif, Z. A., Aman, S. N. A., & Pradhan, B. (2012, March). Landslide susceptibility mapping using LiDAR derived factors and frequency ratio model: Ulu Klang area, Malaysia. In *2012 IEEE 8th International Colloquium on Signal Processing and its Applications (CSPA)* (pp. 378–382).

Lee, S., & Sambath, T. (2006). Landslide susceptibility mapping in the Damrei Romel area, Cambodia using frequency ratio and logistic regression models. *Environmental Geology, 50*(6), 847–855.

Liao, X., Xue, Y., & Carin, L. (2005, August). Logistic regression with an auxiliary data source. In *Proceedings of the 22nd international conference on Machine learning* (pp. 505-512). USA: ACM.

Marjanović, M., Kovačević, M., Bajat, B., & Voženílek, V. (2011). Landslide susceptibility assessment using SVM machine learning algorithm. *Engineering Geology, 123*(3), 225–234.

Mathew, J., Jha, V. K., & Rawat, G. S. (2009). Landslide susceptibility zonation mapping and its validation in part of Garhwal Lesser Himalaya, India, using binary logistic regression analysis and receiver operating characteristic curve method. *Landslides, 6*(1), 17–26.

Mohammady, M., Pourghasemi, H. R., & Pradhan, B. (2012). Landslide susceptibility mapping at Golestan Province, Iran: A comparison between frequency ratio, Dempster-Shafer, and weights-of-evidence models. *Journal of Asian Earth Sciences, 61,* 221–236.

Moore, I. D., Grayson, R. B., & Ladson, A. R. (1991). Digital terrain modelling: A review of hydrological, geomorphological, and biological applications. *Hydrological Processes, 5*(1), 3–30.

Ozdemir, A. (2016). Sinkhole susceptibility mapping using logistic regression in Karapınar (Konya, Turkey). *Bulletin of Engineering Geology and the Environment, 75*(2), 681–707.

Ozdemir, A., & Altural, T. (2013). A comparative study of frequency ratio, weights of evidence and logistic regression methods for landslide susceptibility mapping: Sultan Mountains, SW Turkey. *Journal of Asian Earth Sciences, 64,* 180–197.

Pasolli, E., Melgani, F., Tuia, D., Pacifici, F., & Emery, W. J. (2014). SVM active learning approach for image classification using spatial information. *IEEE Transactions on Geoscience and Remote Sensing, 52*(4), 2217–2233.

Peng, L., Niu, R., Huang, B., Wu, X., Zhao, Y., & Ye, R. (2014). Landslide susceptibility mapping based on rough set theory and support vector machines: A case of the Three Gorges area, China. *Geomorphology, 204,* 287–301.

Pourghasemi, H. R., Jirandeh, A. G., Pradhan, B., Xu, C., & Gokceoglu, C. (2013). Landslide susceptibility mapping using support vector machine and GIS at the Golestan Province, Iran. *Journal of Earth System Science, 122*(2), 349–369.

Pradhan, A. M. S., & Kim, Y. T. (2014). Relative effect method of landslide susceptibility zonation in weathered granite soil: A case study in Deokjeokri Creek, South Korea. *Natural hazards, 72*(2), 1189–1217.

Pradhan, B. (2010). Landslide susceptibility mapping of a catchment area using frequency ratio, fuzzy logic and multivariate logistic regression approaches. *Journal of the Indian Society of Remote Sensing, 38*(2), 301–320.

Pradhan, B., & Lee, S. (2010). Regional landslide susceptibility analysis using back-propagation neural network model at Cameron Highland, Malaysia. *Landslides, 7*(1), 13–30.

Pradhan, B., Sezer, E. A., Gokceoglu, C., & Buchroithner, M. F. (2010). Landslide susceptibility mapping by neuro-fuzzy approach in a landslide-prone area (Cameron Highlands, Malaysia). *IEEE Transactions on Geoscience and Remote Sensing, 48*(12), 4164–4177.

Seung, H. S., Opper, M., & Sompolinsky, H. (1992). Query by committee. In *Proceedings of the fifth annual workshop on Computational learning theory* (pp. 287–294). USA: ACM.

Tazik, E., Jahantab, Z., Bakhtiari, M., Rezaei, A., & Alavipanah, S. K. (2014). Landslide susceptibility mapping by combining the three methods fuzzy logic, frequency ratio and analytical hierarchy process in Dozain basin. *The International Archives of Photogrammetry, Remote Sensing and Spatial Information Sciences, 40*(2), 267.

Tien Bui, D., Pradhan, B., Lofman, O., & Revhaug, I. (2012). Landslide susceptibility assessment in Vietnam using support vector machines, decision tree, and Naive Bayes models. *Mathematical Problems in Engineering, 2012.*

Tuia, D., Ratle, F., Pacifici, F., Kanevski, M. F., & Emery, W. J. (2009). Active learning methods for remote sensing image classification. *IEEE Transactions on Geoscience and Remote Sensing, 47*(7), 2218–2232.

Vapnik, V. (1995). *The nature of statistical learning*. New York: Springer.

Xu, C., Dai, F., Xu, X., & Lee, Y. H. (2012a). GIS-based support vector machine modeling of earthquake-triggered landslide susceptibility in the Jianjiang River watershed, China. *Geomorphology, 145,* 70–80.

Xu, C., Xu, X., Dai, F., & Saraf, A. K. (2012b). Comparison of different models for susceptibility mapping of earthquake triggered landslides related with the 2008 Wenchuan earthquake in China. *Computers & Geosciences, 46,* 317–329.

Yang, B., Xu, W., & Yao, W. (2014). Extracting buildings from airborne laser scanning point clouds using a marked point process. *GIScience & Remote Sensing, 51*(5), 555–574.

Yao, X., Tham, L. G., & Dai, F. C. (2008). Landslide susceptibility mapping based on support vector machine: A case study on natural slopes of Hong Kong. *China. Geomorphology, 101*(4), 572–582.

Yesilnacar, E., & Topal, T. (2005). Landslide susceptibility mapping: a comparison of logistic regression and neural networks methods in a medium scale study, Hendek region (Turkey). *Engineering Geology, 79*(3), 251–266.

Zare, M., Pourghasemi, H. R., Vafakhah, M., & Pradhan, B. (2013). Landslide susceptibility mapping at Vaz Watershed (Iran) using an artificial neural network model: A comparison between multilayer perceptron (MLP) and radial basic function (RBF) algorithms. *Arabian Journal of Geosciences, 6*(8), 2873–2888.

Zhu, X., Zhang, S., Jin, Z., Zhang, Z., & Xu, Z. (2011). Missing value estimation for mixed-attribute data sets. *IEEE Transactions on Knowledge and Data Engineering, 23*(1), 110–121.

11 Performance Evaluation and Sensitivity Analysis of Expert-Based, Statistical, Machine Learning, and Hybrid Models for Producing Landslide Susceptibility Maps

Biswajeet Pradhan, Maher Ibrahim Seeni and Bahareh Kalantar

11.1 Introduction

Landslides are active natural hazards in many areas of the world. Landslides damage and destroy man-made structures and landforms, causing many deaths and injuries every year. In general, landslides cause higher losses annually than other types of natural disasters, such as floods, earthquakes, and windstorms (García-Rodríguez et al. 2008). Climate change, geologic conditions, and high tectonic events are the most important causes of landslides, consequently causing millions of financial and life losses worldwide. Rainfall-induced landslides are common in Malaysia, especially during the rainy seasons. Landslides are often triggered because of a single heavy thunderstorm or rains that last for days (Pradhan and Lee 2010). Malaysia has experienced many landslides in recent years. A major landslide occurred for over four days in Cameron Highlands, from December 4 to 7, 1994 (Shaluf and Ahmadun 2006). This landslide occurred due to farming activities, which involved indiscriminate clearing of land coupled with continuous downpours. Another landslide occurred on June 30, 1995 in Genting Highlands Resorts. Part of the hill came crashing down as flood waters washed tons of earth and fallen trees down the slope. Many vehicles, including cars, buses, and vans, on their way to Genting Highlands, were swept down (Shaluf and Ahmadun 2006).

The assessment of landslide-susceptible areas is essential to effective landuse management; these assessments can identify the best management strategies, such as reducing the effects of landuse activities in vulnerable slope areas (Gorsevski et al. 2006a, b). A landslide susceptibility assessment can reveal the areas that may experience future landslides based on historical records and several conditional factors. The methods for landslide susceptibility assessment can be divided into four groups: (1) statistical, (2) expert, (3) machine learning, and (4) hybrid methods (Wu et al. 2013). The statistical methods are further classified into bivariate and multivariate methods; the bivariate techniques analyze each landslide factor separately, whereas the multivariate methods analyze them simultaneously (Ghosh et al. 2011).

In this chapter, several landslide susceptibility models are developed and evaluated using different approaches: bivariate-, multivariate-, expert-, data mining-, and hybrid-based approaches. The main objective of this chapter is to compare the landslide susceptibility maps generated by the aforementioned methods. To achieve this purpose, frequency ratio (FR), statistical index (SI), weights-of-evidence (WoE), logistic regression (LR), partial least squares (PLS), discriminant analysis (DA), analytic hierarchy process (AHP), fuzzy AHP, support vector machine (SVM), random forest (RF), decision tree (DT), and hybrid models, including FR–SVM, LR–RF, and SI–LR, were developed. Overall, 37 landslide inventories and 10 landslides conditioning factors were prepared in the geographic information system (GIS) using very high-resolution light detection and ranging (LiDAR) point clouds and other relevant topographical databases.

11.2 Previous Work

Many studies have been conducted on landslide susceptibility assessment. These studies have used various methods, such as expert, statistical (bivariate and multivariate), machine learning, and hybrid techniques. The following sections review recent findings on these methods.

11.2.1 Expert-Based Methods

These methods attempt to describe the phenomenon by considering experts' knowledge about the landslide hazard (Sezer et al. 2017). The main advantages of these methods

B. Pradhan (✉) · M.I. Seeni · B. Kalantar
Department of Civil Engineering, University Putra Malaysia, Serdang, Malaysia
e-mail: biswajeet24@gmail.com

B. Pradhan (ed.), *Laser Scanning Applications in Landslide Assessment*,
DOI 10.1007/978-3-319-55342-9_11

are that the landslide inventory data are not necessary and that its generalization ability can be higher than that of data-driven methods (Sezer et al. 2017). However, these methods also have limitations, such as the determination of a number of factors to be included in the model and their sensitivity to expert opinions. Among the common expert-based methods used for landslide susceptibility mapping are the AHP (Ercanoglu et al. 2008; Althuwaynee et al. 2014a, b; Hasekioğulları and Ercanoglu 2012) and fuzzy AHP (Zhu et al. 2014). Hasekiogullari and Ercanoglu (2012) applied the AHP model and analyzed the effect of the conditioning factors on landslide susceptibility assessment in Turkey. Seven landslide susceptibility maps were produced based on the effect of different numbers of conditioning factors. The results showed an area under the curve (AUC) value of 0.797 using the AHP method with nine conditioning factors, whereas the other maps were less accurate. Zhu et al. (2014) showed that the knowledge-based model under the fuzzy concept can predict future landslides with a high accuracy and extrapolate to other study areas. In addition, Pourghasemi et al. (2012a, b, c) compared fuzzy logic and AHP models for landslide susceptibility mapping. They found that the fuzzy logic model (89.7%) performed better than the AHP method (81.1%). More recently, Sezer et al. (2017) developed a landslide susceptibility model based on the modified AHP (M–AHP) and compared it with a fuzzy-based method. They found the M–AHP to be more accurate (AUC = 0.82) than the fuzzy-based method (0.66). In addition, the results showed that the M–AHP was significantly faster than the fuzzy-based method. Other studies by Magliulo et al. (2009) and Bourenane et al. (2015) showed that the expert-based method provides a landslide susceptibility map that does not completely fit the surveyed spatial distribution of the landslides. Other expert-based methods include the spatial multicriteria evaluation model (SMCE) (Pourghasemi et al. 2012a, b, c) and weighted linear combination (Ahmed 2015a, b).

11.2.2 Statistical Methods

Bivariate and multivariate analyses are the most significant techniques among several statistical methods for landslide susceptibility mapping. The two succeeding sections review important studies on these methods.

11.2.3 Bivariate Methods

In bivariate statistical methods, each landslide factor map is combined with the landslide inventory map, and the weighting values based on landslide densities are calculated for each parameter class (Süzen and Doyuran 2004a, b). The bivariate analysis is easy to conduct, and the contribution of each factor can be directly computed (Kavzoglu et al. 2015a, b). On the other hand, the limitations of these methods are the restriction of input data to categorical/reclassified data and its high sensitivity to the accuracy of the thematic maps. Bivariate statistical methods include FR (Demir et al. 2015; Youssef et al. 2014, 2015; Pradhan et al. 2010), SI, and WoE. Raman and Punia (2012) used bivariate statistical modeling for landslide susceptibility mapping. The results were consistent with the inventory data; the method could predict 72% of the active landslides in the very high and high susceptibility zones. Pradhan et al. (2010) used the FR model for landslide susceptibility assessment in Cameron Highlands, Selangor, and Penang areas in Malaysia. The analysis showed reasonable results with an overall accuracy of 83 and 70% for Cameron and Selangor, respectively. Intarawichian and Dasananda (2011) applied the FR model for landslide susceptibility assessment in Thailand. They used 10 conditioning factors and a landslide inventory to produce a landslide susceptibility map. The results were evaluated using the area under the ROC curve method where 80.06 and 84.82% were achieved for the success and prediction rate, respectively. In addition, bivariate and GIS-based methods based on the area density concept for landslide hazard mapping were proposed by Kelarestaghi and Ahmadi (2009). Bijukchhen et al. (2013) compared the heuristic and bivariate statistical modeling methods for landslide susceptibility mapping. Their results showed that both heuristic and bivariate statistical methods could achieve success rates of over 80%. More recently, Youssef et al. (2016a, b) compared different probabilistic and bivariate models for landslide susceptibility assessment in Saudi Arabia. Their study showed that the AUC for success rates are 0.813, 0.815, 0.800, and 0.777, while the prediction rates are 0.95, 0.952, 0.946, and 0.934 for FR, WoE, index of entropy (IoE), and Dempster–Shafer (DS) models, respectively. Comparing the bivariate methods with expert-based methods, Bourenane et al. (2015) showed that the bivariate statistics-based method provided better results and appeared to be more accurate than the expert-based methods. In another comparative study, Shahabi et al. (2013) showed that bivariate methods (information value and density area) outperformed the multivariate methods (linear regression and DA).

11.2.4 Multivariate Methods

In multivariate analysis, all factor maps and landslide-occurred lands are simultaneously analyzed to determine the landslide susceptibility values. Unlike the bivariate model, the multivariate model evaluates the relative contribution of each factor by emphasizing the factors known to contribute to landslide occurrence (Nandi and Shakoor 2010). The most popularly used multivariate methods are LR (Pham et al. 2016a, b; Erener et al. 2016; Demir et al. 2015; Pradhan and Lee 2010; Umar et al. 2014; Youssef et al. 2015; Budimir et al. 2015; Pradhan 2010a, b) and DA (Baeza et al. 2010a, b; Santacana et al. 2003). Mousavi et al. (2011) applied the LR model for landslide susceptibility assessment in Iran. They used 95 landslide locations and five conditioning factors (elevation, slope, curvature, rainfall, and distance to the fault). Their results indicated that the LR method was 85.3% accurate. In a recent paper, Budimir et al. (2015) reviewed several studies that used the LR analysis for landslide susceptibility assessment. They concluded that although most studies used LR for landslide susceptibility assessment, more precision and consistency were needed in selecting covariates for the LR analysis. Kavzoglu et al. (2015a, b) compared LR with DT and SVM methods as well as several bivariate models (FR, WoE, and SI). Their results indicated that the multivariate method (i.e., SVM, LR, and DT) outperformed the bivariate methods (i.e., FR, SI, and WOE) by approximately 13%. Within the multivariate methods, the SVM model performed with the highest accuracy, while the FR method was the most effective and accurate bivariate method. Huang et al. (2015) showed that the LR model (84.05%) predicts more accurately than the FR model (76.64%) for the study area. In addition, Pradhan (2010a, b) compared the FR, FL, and LR models for landslide susceptibility mapping. The results showed that the FR model (89%) is better in prediction than the FL (accuracy is 84%) and LR (accuracy is 85%) models. Results showed that among the fuzzy operators, the gamma operator ($\lambda = 0.9$) showed the best accuracy (84%). Murillo-García et al. (2015) compared the DA, LR, and neural network (NN) models. The best result (AUC = 0.821) was achieved by combining all the models together. On the other hand, other multivariate methods, such as the partial least squares (PLS) method, were applied for various applications. For instance, Conforti et al. (2014a, b) applied PLS to predict soil organic matter (SOM) from the reflectance spectra.

11.2.5 Machine Learning Methods

Pradhan (2013) compared the SVM, DT, and neuro-fuzzy models for landslide susceptibility assessment. The results showed that the DT model has higher prediction performance (83.07%) on testing data, whereas the success rate showed that neuro-fuzzy model has better prediction (94.21%) capability on training data among all models. In addition, Hong et al. (2016) compared four kernel functions used in the SVM model for landslide susceptibility assessment. Their results revealed that the radial basis function achieved the best results with an accuracy of 0.738 (AUC of validation dataset). Similar results were also achieved by Chen et al. (2016) and Pourghasemi et al. (2013a, b, c). Xu et al. (2016)'s SVM model performed better than an NN model for earthquake-triggered landslide susceptibility mapping. Moreover, the SVM was also found to perform better than the NN model in terms of accuracy and generalization capacity.

Park et al. (2014) applied the DT model, which is a type of data mining method for landslide susceptibility assessment in the GIS environment. 3994 landslide inventories on 18 landslides conditioning factors were used to assess the susceptible area. Three algorithms were used to create DT, namely Chi-squared (χ^2), automatic interaction detector (CHAID), and the quick, unbiased, and efficient statistical tree (QUEST) algorithms. FR and DT were used to identify and measure the correlations between the conditioning factors and detected landslide locations. The AUC analysis was used to evaluate the landslide susceptibility assessment. The results show accuracies of 81.56 and 80.91% for the CHAID and QUEST algorithms, respectively. In general, the DT models using the CHAID and QUEST algorithms can be suitable for landslide susceptibility assessment. Sue et al. (2015) applied the SVM method for landslide susceptibility assessment in China. Seven conditioning factors and 354 landslide inventories were used for landslide susceptibility assessment. AUC analysis was used to evaluate the SVM model. The results showed that the SVM model performs well and is stable with an accuracy of 96%.

11.2.6 Hybrid Models

Bui et al. (2016) developed a hybrid integration method of least square SVM and differential evolution optimization for landslide susceptibility mapping. Their proposed model performed well (82% of AUC) on both training and validating datasets. In addition, the comparative study showed that the model outperformed the SVM, NN, and DT models. In addition, Pham et al. (2016a, b) proposed a novel ensemble model of rotation forest and naïve Bayes for landslide susceptibility assessment. The proposed model had a high prediction capability (AUC = 0.846) and a relatively high accuracy (ACC = 78.77%). The study also showed that the model outperformed the AdaBoost, Bagging, MultiBoost, and random forest. In another paper, Althuwaynee et al. (2016) proposed a novel CHAID and AHP hybrid model for landslide susceptibility modeling. Their results showed that the novel model (AUC = 0.80) outperformed the AHP model (0.76) and was more reliable. On the other hand, Sangchini et al. (2016) applied a hybrid method using bivariate and AHP models and compared their proposed method with the LR model. The results showed that the hybrid method performed better in comparison with the LR model, with an AUC of 0.914 and 0.865, respectively. Furthermore, Lee et al. (2015) applied a hybrid method using FR and an adaptive neuro-fuzzy inference system (ANFIS) for landslide susceptibility assessment in the GIS environment. Seventeen conditioning factors and landslide occurrence areas can assess landslide susceptibility. The results showed that the hybrid method (ANFIS) can effectively assess landslide susceptibility. Moosavi and Niazi (2016) developed the hybrid wavelet packet–statistical models (WP–SM) for landslide susceptibility mapping. Results showed that the transformed wavelet packet can be effectively used to produce precise landslide susceptibility maps; the best accuracy was achieved with the SVM model.

11.3 Pilot Test Site

In this chapter, the study area was selected in Cameron Highlands located in the near northern central part of Peninsular Malaysia. The area is located between 101° 24′ 00″E and 101° 25′ 10″E latitudes and 4° 30′ 00″N and 4° 30′ 55″N longitudes (Fig. 11.1). This area is prone to landslides and debris flows for several reasons. First, the area is undergoing developments and land clearing for housing and other infrastructures. Second, the annual rainfall of the tropical hilly regions, including the Cameron Highlands, is high, averaging between 2500 and 3000 mm per annum. Intense rainfalls often overflow the rivers and streams in Cameron Highlands, flooding the area, subsequently producing landslides with debris flowing along the river valleys. In addition, intense rainfalls affect the stability of the slope, causing gully erosions.

The area's topography is hilly mountainous with a slope angle ranging from 0° to 78°; a small part of the area is flat. On the other hand, the geology of Cameron Highlands comprises granite, which is classified as megacrystic biotite granite (Pradhan 2010a, b). In addition, the metasediments comprise schist, phyllite, slate, and limestone (Pradhan 2010a, b). Recently, many landslide events that have damaged housing and other properties have been reported. Therefore, this area was selected for studying the performance of different landslide susceptibility models. The goal was to determine the model that could most accurately predict the spatial distribution of landslide events in the area.

11.4 Datasets

11.4.1 Landslide Conditioning Factors

Several remotely sensed datasets were acquired to prepare the landslide conditioning factors for landslide susceptibility mapping. The LiDAR data of the study area were acquired to construct a high digital elevation model (DEM) of the 25 km^2 area. The average flight height of the LiDAR system during data acquisition was 1510 mm. The data were captured on January 26, 2015. The LiDAR data collection showed nearly eight-point clouds per square meter with a 25,000-Hz pulse rate frequency. In addition, the absolute accuracy of the LiDAR data met the root mean square errors of 0.15 m in the vertical axis and 0.3 m in the horizontal axis.

The DEM was constructed from the last return LiDAR point clouds at a 0.5-m spatial resolution. The last return was used because the last pulses reach the vegetated areas. The DEM was derived using the ESRIs ArcGIS 10.3 algorithm, which is based on the multiscale curvature method proposed by Evans and Hudak (2007). This method is based on a curvature filtering technique (Haugerud and Harding 2001) and iteratively classifies the LiDAR point clouds into ground and non-ground points at multiple scales. An inverse distance weighting interpolation technique was utilized to convert the point clouds classified as ground points into a raster format.

Once the DEM was derived, several topographical, geomorphological, and hydrological factors were obtained (Fig. 11.2). The factors derived from the LiDAR DEM were altitude, slope, aspect, curvature, topographic roughness index (TRI), and distance to stream. Altitude, slope, aspect, and curvature were derived using the spatial analysis tools, whereas TRI was derived by applying the following equation in the algebra map:

$$\mathrm{TRI} = \mathrm{Abs}\left(\sqrt{\overset{2}{\max} - \overset{2}{\min}}\right) \tag{11.1}$$

where max and min are the biggest and smallest values of the cells in the nine rectangular neighborhoods in the altitude raster.

On the other hand, the distance to the stream was derived in two steps. First, the streams were delineated using flow accumulation and converted into a vector format. In the second step, a Euclidean distance analysis was conducted to produce the distance to the stream factor. In addition, the distance to the road and the distance to the lineament factors were derived using the same distance analysis and existing topographic maps of the study area.

In addition, the normalized difference vegetation index (NDVI) was derived from a high-resolution SPOT 5 image (10 m) by applying the band ratio:

$$\mathrm{NDVI} = \frac{\mathrm{NIR} - \mathrm{RED}}{\mathrm{NIR} + \mathrm{RED}} \tag{11.2}$$

where the NIR and RED are the reflectance values of the near infrared and red bands of the SPOT image. The output

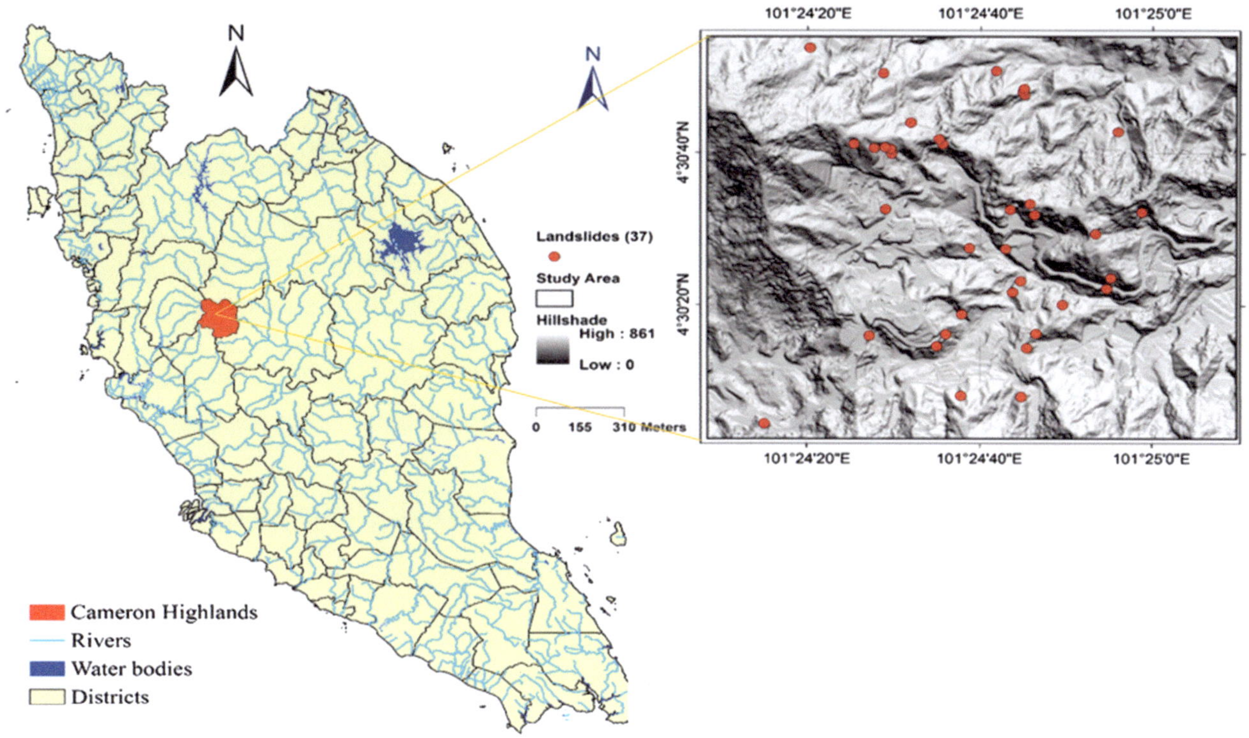

Fig. 11.1 Geographic location of the study area (Cameron Highlands)

NDVI raster often ranges from −1 to 1, where the negative values indicate non-vegetated areas and the high positive values indicate healthy green vegetation. Furthermore, to produce the vegetation density factor, the NDVI raster was reclassified into four classes [−1 to −0.5 (non-vegetation), −0.51 to 0 (low density), 0.1 to 0.5 (moderate density), and 0.51 to 1 (high density)] using the manual classification method. Moreover, the land cover map of the study area was produced by using the SPOT 5 image (10 m spatial resolution) acquired on July 8, 2004 was classified to map the different land cover classes (Pradhan and Lee 2010). A supervised classification method including the GIS-based refinements and field verifications was used to classify the SPOT 5 image into 11 classes.

The detail classes of landslide conditioning factors are shown in Fig. 11.2. The reclassification process was done by the quantile classification method in the ArcGIS software. The altitude of the area ranged from 1500 to 1800 m, where the northwest part of the area had the highest altitudes (Fig. 11.2a). The slope map was classified into five classes: 0–12, 13–24, 25–35, 36–47, and 48–78. The slope angles of the terrain ranged from flat lands to 78°. The aspect map was classified into nine classes, i.e., flat, northeast, east, southeast, south, southwest, west, northwest, and north. On the other hand, the curvature of the area was classified into two classes where the negative values were assigned as convex classes and the positive values were assigned as concave classes. As the study area is a rainforest area, high NDVI values (>0.18) were dominant. The highest NDVI values were found in the northern part of the area where mixed agriculture and forest were found. In addition, the vegetation density produced based on the NDVI values classified the area into four classes: non-vegetation, low density, moderate density, and high-density vegetation. The land cover map shows that the area is almost covered by agricultural lands and forests. In addition, the area included water bodies, industrial, infrastructure, facilities, as well as mixed, commercial, transportation, agriculture, residential, and vacant lands. Furthermore, the TRI map was classified into five classes: 0–24, 25–38, 39–50, 51–59, and 60–150. The TRI map shows that the agricultural areas have small TRI values; on the other hand, high TRI values were found in the forest areas. Distance to the lineament, distance to the road, and distance to the stream were classified into five classes; their details are shown in Fig. 11.2.

11.4.2 Landslide Inventory Dataset

The landslide inventory map of Cameron Highlands was produced by mapping the landslide area by visual interpretation of aerial photographs and SPOT 5 images; additionally, available previous reports were analyzed (Pradhan and Lee 2010; Pradhan et al. 2010; Pradhan 2011). Typically

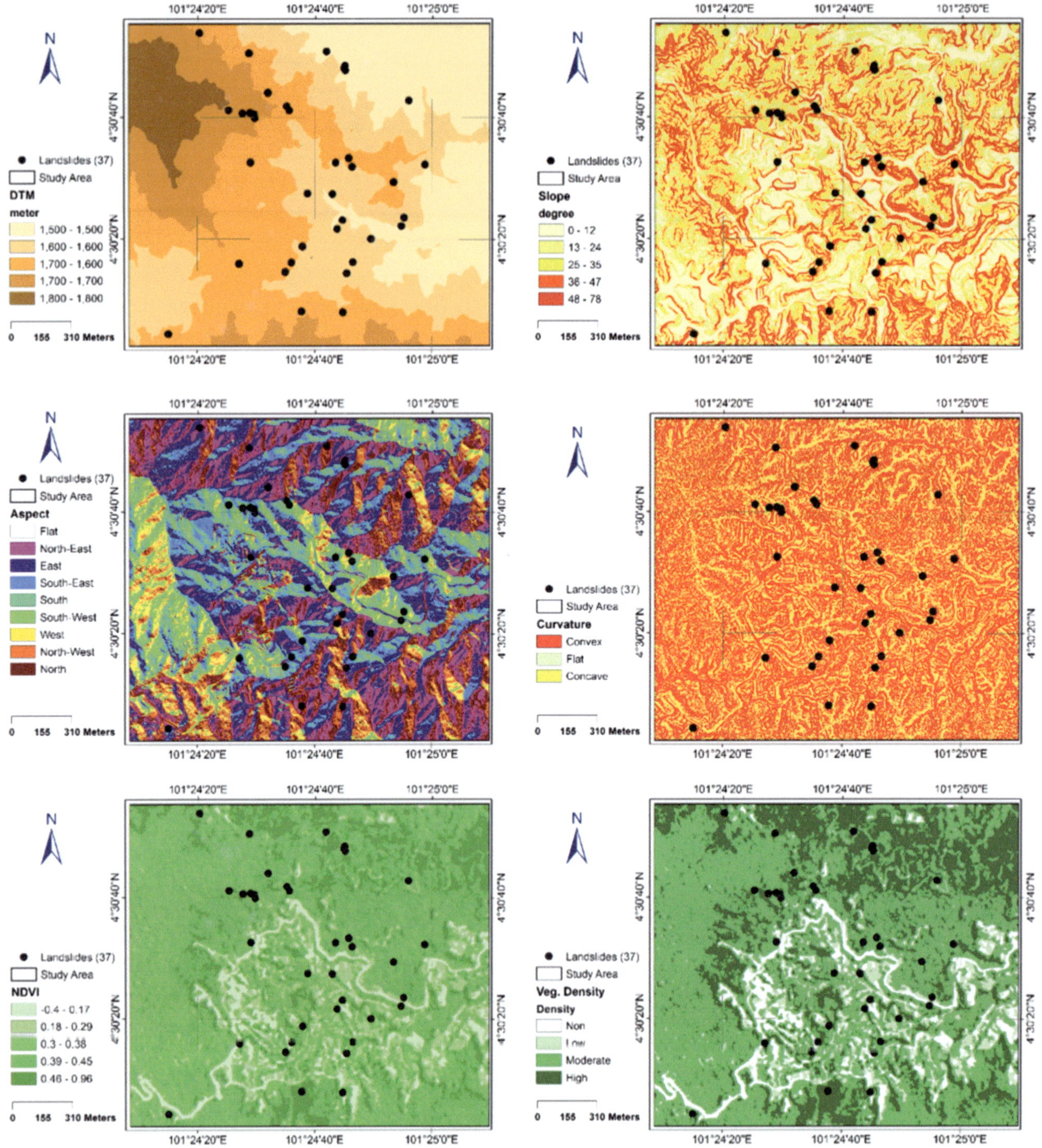

Fig. 11.2 Landslide conditioning factors

speaking, a comprehensive landslide inventory database should contain information about the date when the landslide occurred, the type of landslides, and the damages and injuries incurred because of the landslide event. However, such detail and complete landslide inventory maps are unavailable in Malaysia, and extracting landslide information from previous satellite images or orthophotographs alone is difficult. Therefore, this work used the best available landslide inventory for the study area. The landslide locations were extracted from archived aerial photographs, high-resolution satellite images, and previous records. A total of 37 landslides were identified in the study area. In addition, landslides were verified based on visual interpretation and field investigations.

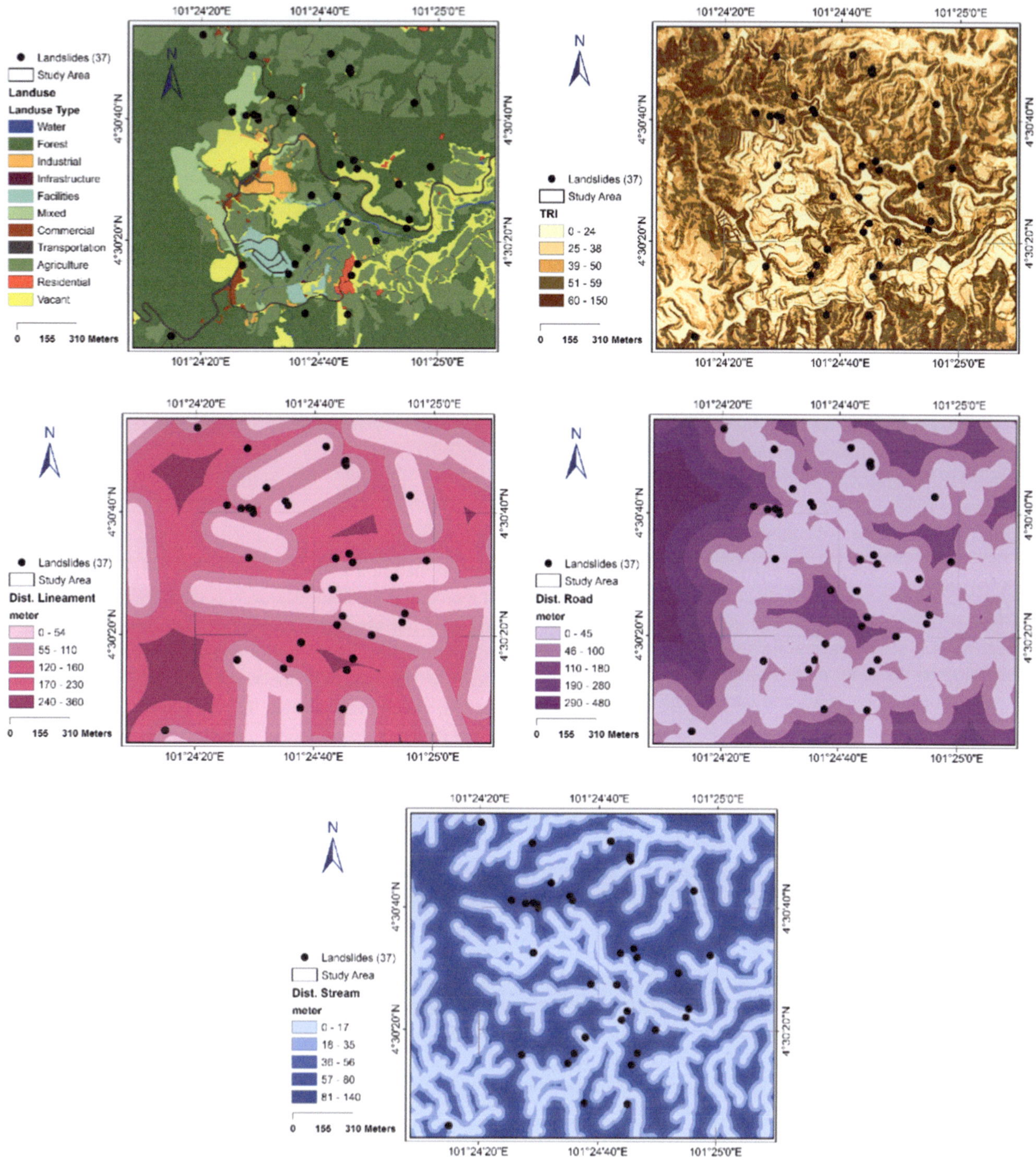

Fig. 11.2 (continued)

11.5 Theory of Models

11.5.1 Expert-Based Methods

11.5.1.1 Analytic Hierarchy Process

The AHP model was proposed by Saaty in 1980. It is an expert-based modeling method that mainly depends on expert opinions. The AHP is a multicriteria decision-making support approach that provides the user a scale of preferences drawn from a set of alternatives. In the last decades, AHP has gained much attention in many applications including site selection, suitability analysis, and landslide susceptibility mapping (Pourghasemi et al. 2012a, b, c; Kayastha et al. 2013). The first stage of AHP is breaking a complex problem into its component factors. Then these factors are arranged in a hierarchal order. After that, numerical values are assigned to subjective judgments on the relative importance of each factor; the judgments are synthesized to determine the priorities to be assigned to these factors (Saaty and Vargas 2001).

Next, the priorities are imported into the AHP matrixes. A pairwise comparison matrix is created by making dual comparisons made in this context. The weights are calculated from the pairwise comparison matrix undertaking eigenvalue and eigenvector calculations. The eigenvector corresponding to the largest eigenvalue of the matrix provides the relative priorities of the factors, i.e., if one factor has a preference, its eigenvector component is larger than that of the others. Thus, a vector of weights is obtained. This vector reflects the relative importance of the various factors from the matrix of paired comparisons.

In AHP, an index of consistency, known as the consistency ratio (CR), is used to indicate the probability that the matrix judgments were randomly generated (Saaty 1977).

$$\mathrm{CR} = \frac{\mathrm{CI}}{\mathrm{RI}} \tag{11.3}$$

where *RI*RI is the average of the resulting consistency index depending on the order of the matrix given by Saaty (1977); CI is the consistency index and can be expressed as

$$\mathrm{CI} = \frac{(\lambda_{\max} - n)}{n - 1} \tag{11.4}$$

where $\lambda_{\max}$ is the largest or principal eigenvalue of the matrix that can be easily calculated from the matrix, and n is the order of the matrix. If *CR*CR values are >0.1, the models are automatically discarded.

11.5.1.2 Fuzzy Analytic Hierarchy Process

Fuzzy AHP embeds the fuzzy theory to the basic AHP, which was developed by Saaty (1977). AHP takes the pairwise comparisons of different alternatives with respect to various criteria and provides a decision support tool for multicriteria decision problems. In a general AHP model, the objective is in the first level, and the criteria and subcriteria are in the second and third levels, respectively. Finally, the alternatives are found in the fourth level (Kilincci and Onal 2011). The AHP model cannot reflect human thinking. On the other hand, in the fuzzy AHP model, qualitative judgment can be qualified to compare more intuitionistic and reduce assessment bias in the pairwise comparison process (Roodposhti et al. 2014). Another advantage of including the fuzzy concepts in the AHP model is the capability of representing vague data. Further details of fuzzy AHP and its applications for landslide susceptibility modeling can be found in Roodposhti et al. (2014).

11.5.2 Bivariate Methods

11.5.2.1 Weights-of-Evidence

Weights-of-Evidence (WoE) is a nonlinear statistical technique based on the log-linear form of the Bayesian probability model (Bonham-Carter 1994). An early application of WoE in landslide susceptibility modeling was reported by Lee et al. (2002). The detailed mathematical formulations of this method are described in Bonham-Carter (1994). WoE calculates the weight of each landslide conditioning factor in the model based on the absence and presence of landslides in a given area. The calculations are done using the following equations (Bonham-Carter 1994):

$$W_i^+ = \ln \frac{P\{B|L\}}{P\{B|L\}} \tag{11.5}$$

$$W_i^- = \ln \frac{P\{\bar{B}|L\}}{P\{\bar{B}|\bar{L}\}} \tag{11.6}$$

where P indicates the probability, ln is the normal log, B is the presence of a potential landslide factor, $\bar{B}$ is the absence of a potential landslide factor, L is the presence of landslide, and $\bar{L}$ is the absence of landslide. In addition, W_i^+ is the positive weight, which indicates the presence of a landslide factor at the landslide locations; and W_i^- is the negative weight, which indicates that the absence of a landslide factor is the landslide location. As a result, $W_f = W_i^+ + W_i^-$ is the weight contrast reflecting the overall spatial association between landslide factors and landslide locations.

11.5.2.2 Statistical Index

The SI method is based on the statistical correlations that are often revealed by crossing the landslide inventory map with attributes of a different map factors (Bui et al. 2011). In the first stage, the density of landslides per parameter class is calculated from the map-crossing results (Yalcin 2008). Then, these densities are standardized by relating them to the overall density in the entire area (Oztekin and Topal 2005). The following equation was suggested by van Westen (1997) to calculate the landslide susceptibility index (LSI):

$$\mathrm{SI} = \mathrm{Ln}\left(\frac{\mathrm{Npix(Si)}}{\mathrm{Npix(Ni)}} / \frac{\mathrm{SNpix(Si)}}{\mathrm{SNpix(Ni)}}\right) \tag{11.7}$$

where Npix(Si) is the number of pixels contained in a landslide in a certain parameter class, Npix(Ni) is the total number of pixels in a certain parameter class, SNpix(Si) is the number of pixels in all landslides, and SNpix(Ni) is the total number of all pixels. Then, the SI value of each landslide conditioning factor is calculated. Finally, all layers are overlaid, and a resulting susceptibility map is obtained. Further details can be found in Van Westen (1997).

11.5.2.3 Frequency Ratio

The FR model depends on the observed relationship between the landslide inventory map and each landslide conditioning factor (Bonham-Carter 1994). The goal of these models is to establish relationships between the landslide locations and the conditioning factors in the study area (Lee and Pradhan 2007). The advantages of the FR model are that it is easy to calculate and run, and the landslide susceptibility maps that are produced by the FR models often spatially match the landslide inventories. In addition, many researchers have proposed this method (Yilmaz 2009; Intarawichian and Dasananda 2011; Schleier et al. 2014).

Frequency is calculated from the analysis of the relationship between landslides and the attributed factors. Therefore, the frequency ratios of each factor's type or range were calculated from their relationship with landslide events. The frequency ratio was calculated for the subcriteria of the parameter, and the frequency ratios were summed to calculate the LSI (Lee and Talib 2005).

$$\mathrm{LSI} = \mathrm{Fr}_1 + \mathrm{Fr}_2 + \cdots + \mathrm{Fr}_n \quad (11.8)$$

where Fr is the rating of each factor's type or range. Further details on FR can be found in Bonham-Carter (1994) and Yalcin et al. (2011).

11.5.3 Multivariate Methods

11.5.3.1 Logistic Regression

The LR is a multivariate statistical technique that works under the probability concepts and is widely used for landslide susceptibility assessment throughout the world (Ayalew et al. 2005; Schicker and Moon 2012; Devkota et al. 2013). In LR, the landslide conditioning factors were first transformed into a logit variable. Then, the maximum likelihoods were computed from the logit variables (Bai et al. 2010). The main advantage of the LR model is that through the addition of a suitable link function to the usual linear regression model, the variables may be either continuous, discrete, or of any composition of both types; the variables do not necessarily have normal distributions (Pradhan 2011).

The LR model is widely accepted in landslide susceptibility assessments and is computed using the following equations:

$$Y = \mathrm{Logit}(p) = \mathrm{Ln}\left(\frac{p}{p-1}\right) \quad (11.9)$$

$$Y = C_0 + C_1X_1 + C_2X_2 + \cdots + C_nX_n \quad (11.10)$$

where p is the probability that has a dependent variable; $Y = 1$; $p/(1-p)$ is the so-called odd or frequency ratio; C_0 is the intercept; and $C_1, C_2, \ldots, C_n$ are the coefficients, which measure the contribution of the independent factors $(X_1, X_2, \ldots, X_n)$ to the variations in Y (Lee 2005). Further information about LR can be found in Lee and Pradhan (2007), Yalcin et al. (2011), Pradhan and Lee (2010), Pradhan (2010a, b), and Althuwaynee et al. (2014).

11.5.3.2 Discriminant Analysis

Discriminant Analysis (DA) predicts membership in a group or category based on observed values of several continuous variables. Specifically, DA predicts a classification (*X*) variable (nominal or ordinal) based on known continuous responses (*Y*). The data for DA comprise a sample of observations with known group membership together with their values on the continuous variables.

These are two of the univariate statistical techniques that are used to build a predictive model based on groups that consider the observed explanatory variables (McLachlan 2004). They are used to classify each mapping as either landslide or non-landslide. In DA, the estimated values (i.e., *L*) are determined by a linear combination of a set of explanatory variables, such as $L = \beta X + c$ $(c = constant\mathrm{constant})$, which best differentiates the group of a case by finding the β coefficients. In QDA, the task is to determine a quadratic surface that best separates the group of a case. Unlike LDA, QDA finds a group membership comprising a square $n \times n$ matrix (n = number of explanatory variables) and a linear combination of these variables such that $= x^{\mathrm{T}}Ax + b^{\mathrm{T}}x + c$, where A is the $n \times n$ coefficient matrix, b is the linear combination coefficient, and c is a constant.

11.5.3.3 Partial Least Squares

Partial Least Squares (PLS) regression is a technique that generalizes and combines features of a principal component analysis and multiple regressions (Abdi 2003). It breaks down predictors into a smaller set of uncorrelated components and then performs least-squares regression on these components rather than on the original data. PLS regression is especially useful when predictors are highly collinear or when more predictors exist than observations, in which ordinary least-squares regression either produces coefficients with high standard errors or fails completely. Detailed mathematical formulations and discussions on PLS can be found in Geladi and Kowalski (1986).

11.5.4 Machine Learning Methods

11.5.4.1 Support Vector Machine

Vapnik (1995) developed SVM, a nonlinear classification model derived from machine learning techniques. SVM aims to determine an optimal separating hyper plane (maximizing the margin width) between two classes in a feature space (Pham et al. 2016a, b). The training points near the hyper plane are called support vectors and are utilized for classification once the decision surface is obtained (Pourghasemi et al. 2013a, b, c). The separating hyper plane is found as follows:

$$y_i(w \times x_i + b) \geq 1 - \xi_i \quad (11.11)$$

where w is the coefficient vector that defines the hyper plane orientation in the feature space; b is the offset of the hyper plane from the origin; and ξ_i is the positive slack variables (Cortes and Vapnik 1995). The optimal hyper plane is found by solving the following optimization problem (Jebur et al. 2014):

$$\begin{aligned} &\text{Minimize} \sum_{i=1}^{n} \alpha_i - \frac{1}{2}\sum_{i=1}^{n}\sum_{j=1}^{n} \alpha_i\alpha_j y_i y_j (x_i x_j) \\ &\text{subject to} \sum_{i=1}^{n} \alpha_i y_j = 0, \quad 0 \leq \alpha_i \leq C \end{aligned} \quad (11.12)$$

where α_i is the Lagrange multiplier, and C is the penalty for data classification. The following decision function is applied as follows:

$$g(x) = \text{sign}\left(\sum_{i=1}^{n} y_i \alpha_i x_i + b\right) \quad (11.13)$$

In the case of nonlinearly separating samples, the decision function (Eq. 11.13) is rewritten as follows:

$$g(x) = \text{sign}\left(\sum_{i=1}^{n} y_i \alpha_i K(x_i, x_j) + b\right) \quad (11.14)$$

In this process, the original data are transformed into a higher dimensional space using a nonlinear kernel function (K). The common K functions used with SVM in landslide studies are linear, sigmoid, radial basis, and polynomial (Bui et al. 2012). However, several studies showed that the radial basis function is more suitable for landslide susceptibility application than the other kernel functions (Pourghasemi et al. 2013a, b, c; Xu et al. 2012). A thorough discussion on the SVM method for landslide susceptibility assessment is presented in Marjanović et al. (2011).

11.5.4.2 Random Forest

Random Forest (RF) is a model building strategy that provides estimators of the Bayes classifier. Unlike other models, the RF model has several measures of variable importance. One of the most dependable measures is based on the decrease in classification accuracy when values of a variable in a node of a tree are permuted randomly (Breiman 2001). For landslide susceptibility mapping, RF models take advantage of the high variance among individual trees, allowing each tree to vote for the class membership and assigning respective classes according to the majority of the votes. Such ensembles demonstrate robust and accurate performance on complex datasets with little need for fine-tuning. Complex datasets include the presence of many noisy variables, such as factors that could possibly, but not plausibly, affect the dependent variable Stumpf and Kerle (2011). Moreover, in this study, an OOB (out-of-bag) sample is used considering that the observations are not used for building the current tree and the two types of errors in which mean decreases in accuracy and mean decreases in Gini are calculated (Breiman 2001; Calle and Urrea 2010). Such types of errors can be used to rank and select variables (Liaw and Wiener 2002; Biau et al. 2008).

11.5.4.3 Decision Tree

Decision Tree (DTs) are a class of predictive data mining tools that predict either a categorical or continuous response variable. A DT comprises nodes and splits in the data. The tree starts with all the training data residing in the first node. An initial split is made using a predictor variable, which segments the data into two or more child nodes. Splits can then be made from the child nodes, and no more splits are made from a terminal node. Predictions are made based on the composition of terminal nodes. DTs offer many advantages. One important advantage is the ease of interpretation of a DT and its ability to model complex relationships between variables (Bui et al. 2012). In addition, making model predictions with DTs does not involve mathematical calculations as in general linear models. However, these methods have disadvantages, such as their susceptibility to noisy data and that multiple output attributes are not allowed. An in-depth discussion and the mathematical equations used in DT modeling can be found in Bui et al. (2012).

11.6 Methods

11.6.1 Factor Analysis

Given several issues such as multicollinearity, outliers, and spatial variations of landslide inventories and the conditioning factors, factor analysis is an important step required in susceptibility assessments (Pradhan and Lee 2010; Jebur et al. 2014). The aim of this analysis is to detect and remove the problematic factors, which can mislead the coefficient estimations and eventually reduce the performance of regression models (Mancini et al. 2010). In addition, inclusion of problematic factors in the analysis can lead to a poor model design, which subsequently limits the use of the models in producing landslide susceptibility maps for the target area. Multicollinearity refers to landslide conditioning factors that are correlated with other factors in the model. Severe multicollinearity can increase the variance of the coefficients estimates. It also makes the model sensitive to small changes, thereby making it difficult to interpret. Multicollinearity is often detected through cross-correlation matrix construction among landslide conditioning factors. The use of advanced methods to detect multicollinearity among predictors enables the estimation of the variance-inflated factor (VIF), which can be calculated using the following expression:

$$\mathrm{VIF} = \frac{1}{1 - R'^2} \quad (11.15)$$

where R' is the multiple correlation coefficient between a factor and the remaining factors in the model.

Furthermore, researchers have recently suggested several solutions to address multicollinearity (Van Den Eeckhaut et al. 2010; Verachtert et al. 2011; Jebur et al. 2014). Common practices include removing the highly correlated factors, linearly combining the highly correlated factors, and running more advanced models that account for multicollinearity, such as a ride regression. In the present study, highly correlated factors (VIF => 4.00) are removed.

11.6.2 Factor Optimization

With the increase in the number of landslide conditioning factors, the required training sample size also grows; optimizing landslide conditioning factors in landslide susceptibility mapping has become important. In addition, increasing the number of landslide conditioning factors increases the computational costs, and having a number of input landslide factors may not necessarily improve the prediction accuracy of landslide occurrences. Furthermore, increasing the number of landslide factors can mislead the estimated regression coefficients and then reduce the generalizability of the susceptibility models into different subsets within the study area or within different sites. In this study, two methods of factor optimization, the Chi-square statistic and Gini importance, are used to remove nonsignificant factors at a 95% confidence level.

11.6.2.1 Factor Optimization Using the Chi-Square Statistic

The significant landslide factors can be identified using the Chi-square statistic (Sarkar and Kanungo 2004). This approach aims to test the significance of the relationship between categorical variables. It is based on the computation of the expected and observed frequencies. The Chi-square value and its significance level depend on the overall number of landslide events and the conditioning factors. The hypothesis test is defined as follows:

H_0 The landslide factor (e.g., slope) and landslide occurrence are independent

H_1 The landslide factor (e.g., slope) and landslide occurrence are independent

$$\chi^2 = \sum_{i=1}^{n} \frac{(O_i - E_i)^2}{E_i}, \quad (11.16)$$

where O_i is the observed frequency count at level i for landslide factor, and E_i is the expected frequency count at level i for landslide factor.

Once the Chi-square value and the p-value for each landslide conditioning factor are computed, the p-value is evaluated against the significance level of (0.05), which allows us to determine the significance of the relationship between the landslide factor and the landslide occurrence. A higher Chi-square value implies that the factor more ably detects landslide probabilities.

11.6.2.2 Factor Optimization Using the Information Value (IV) Method

The information value (IV) is a statistical indicator of the overall predictive power of the characteristics. It is calculated using the following expression:

$$IV = \left[\sum_{i=1}^{k} (g_i - b_i) \,.\, \ln\left(\frac{g_i}{b_i}\right)\right] .\, 100 \quad (11.17)$$

where k is the number of classes in a landslide conditioning factor, g_i is the column-wise percentage distribution of the total "good" cases in the ith bin, and b_i is column-wise percentage distribution of the total "bad" cases in the ith bin. The optimum landslide conditioning factors are selected using the IV calculated by the above formula (Eq. 11.2). In addition, the Gini coefficient and the Gramer's V statistic are computed for each factor. The Gini coefficient is a summary statistic of the Lorenz curve and a measure of inequality in a population. The Gini coefficient ranges from a minimum value of zero, given that all individuals are equal, to a theoretical maximum of one in an infinite population in which every individual except for one has a size of zero. On the other hand, the Cramer's V measures the correlation of landslide factors. It ranges from 0 (no correlation exists) to 1 (ideal correlation) and can be calculated using the following formula:

$$V = \sqrt{\frac{\chi^2}{n.\min(w - 1, k - 1)}} \quad (11.18)$$

where χ^2 is the Chi-square statistic, n is the number of cases available in the dataset, w is the number of categories in the dependent variable (=2 "Landslide," "Non-landslide"), and k

is the number of categories in the independent variables (landslide conditioning factors).

11.6.3 Landslide Susceptibility Modeling

Multiple regression analysis can be applied to spatially predict landslide occurrences in an area. In this method, several independent variables (i.e., landslide conditioning factors) and a dependent variable (i.e., the presence/absence of landslides) are used to establish a linear equation via the estimation of coefficients for the independent variables. The literature contains several techniques for various performances. The performance of a regression model depends on a number of issues, including the quality and quantity of training data, the spatial distribution of a dependent variable, the topographic and geomorphology of the study area, and the model concept and its assumptions. In the present study, 14 models are developed and evaluated for landslide susceptibility mapping in the study area. The models are grouped into five categories: bivariate, multivariate, expert-based, data manning, and hybrid (Fig. 11.3). The details of these models are explained in Sect. 11.2 and are listed in Table 11.1.

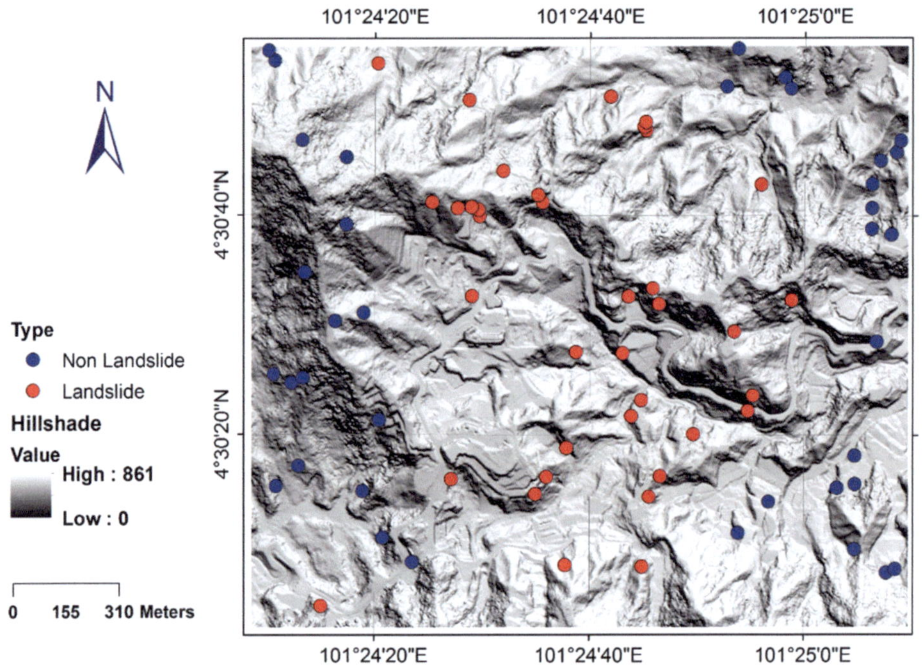

Fig. 11.3 Landslide and non-landslide samples used for susceptibility modeling

Table 11.1 Landslide susceptibility models and their abbreviations used in the current study

Category	Model	Code
Bivariate	Weight of evidence	WoE
	Statistical index	SI
	Frequency ratio	FR
Multivariate	Logistic regression	LR
	Discriminant analysis	DA
	Partial least squares	PLS
Expert-based	Analytic hierarchy process	AHP
	Fuzzy analytic hierarchy process	FAHP
Data mining	Support vector machine	SVM
	Random forest	RF
	Decision tree	DT
Hybrid	Frequency ratio–logistic regression	FR–LR
	Frequency ratio–support vector machine	FR–SVM
	Weight of evidence–random forest	WoE–RF

11.6.4 Accuracy Assessment

11.6.4.1 Success and Prediction Rates

The success and prediction rate curves explain the percentage of known landslides that fall into susceptibility level ranks and show the cumulative frequency graph (Chung and Fabbri 2003; Intarawichian and Dasananda 2011). The developed landslide susceptibility maps are compared with the spatial distribution of the landslide inventory. The success rate curve is created from the landslides used for training, whereas the prediction rate curve is created from the landslides used for validation (Pradhan and Kim 2014). Constructing the success and prediction rate curves, the cumulative percentage of observed landslide occurrences in different landslide susceptibility classes and the cumulative percentage of the area of the susceptibility classes are plotted in the y- and x-axis, respectively. In addition, the total AUC can be used to qualitatively determine the prediction accuracy of the susceptibility map, in which a larger area implies a higher accuracy (Lee 2005; Mathew et al. 2009; Intarawichian and Dasananda 2011; Pourghasemi et al. 2013a, b, c).

11.6.4.2 Spatial Agreement

A Spearman correlation analysis is used to evaluate the spatial agreements between landslide-susceptible maps (Hühnerbach and Masson 2004). The Spearman correlation between two landslide susceptibility maps assesses the monotonic relationships (linear and nonlinear). A perfect Spearman correlation of +1 or −1 occurs when each of the landslide conditioning factors is a perfect monotone function of the other (Setchi and Anuar 2016). The Spearman correlation is appropriate for continuous and discrete variables. In the present study, it is applied to discrete landslide susceptibility maps to assess the spatial agreements of each landslide susceptibility class in two different models. The Spearman correlation measures the strength and the direction of the relationship between two variables. In this study, the Spearman correlation measures the strength and the direction of the relationship between the two landslide susceptibility maps. The Spearman correlation is calculated using the following formula (MacFarland and Yates 2016):

$$\rho = 1 - \frac{6\sum d_i^2}{n(n^2 - 1)} \tag{11.19}$$

where ρ is the Spearman rank correlation, d_i is the difference between the ranks of corresponding values x_i and y_i, and n is the number of values in each dataset.

The Spearman rank correlation matrix was generated using all pixels in the study area. For each landslide susceptibility map, the pixel values were extracted and stored in a table, and the pair landslide susceptibility maps were assessed using the Spearman rank correlation estimated using the Eq. (11.19).

11.6.4.3 Landslide Density Graphs

A landslide density graph is often produced to investigate the predicted landslide densities in each landslide susceptibility class (Galli et al. 2008). This graph is constructed by plotting the ratio of occurred landslides pixels over the ratio of non-occurred landslide pixels for each classified susceptible zone in a diagram. This graph is an important aspect of an accuracy assessment of landslide susceptibility models, as it provides information on the distribution of landslide events in the study area and in each landslide susceptibility class.

11.7 Results

11.7.1 Results of Factor Analysis

Tables 11.2 and 11.4 show the results of factor analysis. The estimated correlation coefficients of landslide conditioning factors indicate that the NDVI and vegetation density as well as the slope and TRI have the two highest correlation coefficients of 0.92 and 0.84, respectively. Examining only the correlation coefficients is insufficient to detect multicollinearity because the pairwise correlations may be small while linear correlations among three or more landslide conditioning factors may exist. Therefore, calculating the VIF is essential to detecting the multicollinearity in the dataset. Table 11.3 shows the estimated VIF values of all landslide conditioning factors. The VIF values indicate that the NDVI (VIF = 4.45) and vegetation density (VIF = 4.16) are the two factors that suffer multicollinearity in the dataset. On the other hand, the calculated VIF of TRI factor is 2.30, indicating low linear dependence with other factors. The VIF explains how much the estimated regression coefficients are inflated by the correlations among the landslide conditioning factors in the model. Therefore, keeping the factors that suffer from multicollinearity will inflate the regression coefficients, producing an unreliable model. A common practice in removing the multicollinear factors in a model is investigating factors with a VIF greater than 4.00. In the present study, the NDVI factor is removed because its correlation coefficient than that of the vegetation density. As a result, only 10 factors are used in the landslide susceptibility models developed in the present study.

11.7.2 Results of Factor Optimization

Factor optimization is an important step in regression analysis, as it reduces the redundant information contained in multiple factors. Table 11.5 shows the results of the factor analysis in the present study using Chi-square and Gini importance methods. The first method indicated that the three most important factors for predicting landslides in the

Table 11.2 Correlation matrix between landslide conditioning factors

Factor	Slope	Curvature	Aspect	Distance to lineament	Distance to road	Distance to stream	NDVI	Vegetation density	Landuse	Altitude	TRI
Slope	1.00	−0.08	0.00	−0.04	0.10	0.01	0.29	0.25	0.12	0.23	**0.84**
Curvature	−0.08	1.00	−0.12	−0.01	−0.03	0.14	−0.15	−0.22	0.10	0.31	0.00
Aspect	0.00	−0.12	1.00	−0.07	0.09	−0.10	0.29	0.24	−0.20	−0.07	0.02
Distance to lineament	−0.04	−0.01	−0.07	1.00	0.27	0.13	−0.22	−0.22	−0.24	0.24	−0.15
Distance to road	0.10	−0.03	0.09	0.27	1.00	−0.01	0.11	0.03	−0.48	0.47	0.07
Distance to stream	0.01	0.14	−0.10	0.13	−0.01	1.00	−0.04	−0.05	0.31	0.43	0.04
NDVI	0.29	−0.15	0.29	−0.22	0.11	−0.04	1.00	**0.92**	0.07	0.04	0.35
Vegetation density	0.25	−0.22	0.24	−0.22	0.03	−0.05	**0.92**	1.00	0.07	−0.04	0.27
Landuse	0.12	0.10	−0.20	−0.24	−0.48	0.31	0.07	0.07	1.00	0.02	0.20
Altitude	0.23	0.31	−0.07	0.24	0.47	0.43	0.04	−0.04	0.02	1.00	0.20
TRI	**0.84**	0.00	0.02	−0.15	0.07	0.04	0.35	0.27	0.20	0.20	1.00

Table 11.3 Estimated VIF for each landslide conditioning factor

Variable	Summary statistics and multicollinearity			
	Means	Std. Devs	Multiple	VIF
Slope	32.16	14.96	**0.74**	2.20
Curvature	−1.05	19.61	0.23	1.05
Aspect	180.18	107.93	0.16	1.03
Distance to lineament	108.43	75.37	0.21	1.05
Distance to road	74.19	80.28	0.51	1.36
Distance to stream	37.92	27.73	0.32	1.11
NDVI	0.37	0.10	**0.88**	**4.45**
Vegetation density	3.20	0.83	**0.87**	**4.16**
Landuse	4.92	2.32	0.43	1.23
Altitude	1581.65	76.86	0.54	1.42
TRI	49.53	14.64	**0.75**	2.30

study area are landuse (Chi-square value = 34.172), distance to road (Chi-square value = 22.581), and altitude (Chi-square value = 20.554). On the other hand, the Gini method showed that the landuse (information value = 1.766), altitude (information value = 1.445), and slope (information value = 0.963) are the three most important factors for predicting the areas which are most susceptible to landslides. Although the two methods disagree on the three most important factors in predicting landslides in the Cameron Highlands, they agree on the two most irrelevant factors, NDVI and vegetation density. The estimated Chi-square values of NDVI and the vegetation density are 3.906 and 0.949, respectively.

We found in the multicollinearity analysis that NDVI and the vegetation density are highly correlated ($R^2 = 0.92$); we thus removed the NDVI from further analysis. The aim of factor optimization is to analyze the importance of each landslide conditioning factor regardless of its correlation with the remaining factors. Therefore, none of the factors were removed based on their importance value estimated either by the Chi-square method or by the Gini method.

Human activities and changes in environmental conditions affect the landuse and can thus affect landslide activity. Deforestation, forest logging, road construction, and cultivation on steep slopes are the main activities that have significant effects on landslide activity in the focused area. However, in most of the landslide studies as well as the present study, landuse is often viewed as a static factor extracted from a single date satellite image. In addition, landslides may occur on the road and on the side of the slopes that are affected by the road. A road that is constructed beside steep slopes causes a decrease in the load

Table 11.4 Factor loadings shows the first three factors calculated by the Varimax normalized method

Variable	Factor loadings (varimax normalized)		
	Factor 1	Factor 2	Factor 3
Slope	**0.86**	0.00	−0.08
Curvature	0.02	0.01	0.20
Aspect	−0.01	0.05	−0.30
Distance to lineament	−0.11	0.32	0.18
Distance to road	0.10	**0.96**	−0.15
Distance to stream	0.05	0.00	0.07
NDVI	0.27	−0.06	**−0.93**
Vegetation density	0.19	−0.13	**−0.93**
Landuse	0.19	−0.51	0.03
Altitude	0.23	0.46	0.01
TRI	**0.98**	−0.04	−0.09
Expl. Var	1.91	1.53	1.93
Prp. Totl	0.17	0.14	0.18

Table 11.5 Estimated factor importance and ranking using the Chi-square and Gini methods

Factor	Chi-square method		Gini method		
	Chi-square	*p*-value	Gini index	Information value	Cramer's V
Landuse	**34.172**	0.000	0.317	1.766	0.606
Distance to road	**22.581**	0.001	0.383	0.180	0.483
Altitude	**20.554**	0.004	0.353	1.445	0.543
Aspect	18.596	0.010	0.433	0.568	0.365
Distance to lineament	16.431	0.021	0.270	0.665	0.678
Slope	15.800	0.045	0.414	0.963	0.414
TRI	14.810	0.096	0.414	0.778	0.414
Distance to stream	13.541	0.060	0.380	0.518	0.490
Curvature	6.148	0.292	0.375	0.918	0.500
NDVI	3.906	0.918	0.430	0.128	0.374
Vegetation density	0.949	0.814	0.494	0.052	0.113

both on the topography and on the heel of the slope. This phenomenon increases the stress on the back of the slope because of the changes in topography. In previous studies, the distance to the road was found to be an unimportant factor (Yalcin 2008). This is mainly because the design of roads plays a significant role in determining whether the road is a contributing factor. A drop-down road section may behave like a wall, a net source, a net sink, or a corridor for water flow (Yalcin 2008). Furthermore, altitude is found to positively affect landslide proneness (Pachauri and Pant 1992). Resistant lithological units and rainfall often underlie higher altitudes.

On the other hand, vegetation density effects on landslide occurrence may be classified as hydrological or mechanical in nature. The hydrological factors include the loss of precipitation by interception, the removal of soil moisture by evapotranspiration, and the effects of hydraulic conductivity (Ercanoglu et al. 2004). The mechanical factors comprise soil reinforcement by roots, surcharge, wind-loading, and surface protection (Ercanoglu et al. 2004). However, in the present study, the vegetation density was found to be non-significant, as the training landslides are mostly located in other landuse classes. Furthermore, the study area is mostly covered by dense vegetation in which the non-landslide samples are generated randomly.

11.7.3 Results of Bivariate Statistical Models

Three bivariate statistical landslide susceptibility models were developed, FR, SI, and WoE. In bivariate statistical models, a landslide conditioning factor was compared to the landslide inventory map to determine the influence of the factors. Table 11.6 contains the estimated coefficients for each landslide conditioning factor and its sub-level classes. A high coefficient value does not indicate the importance of the factor given that units are different. For example, the degree or radian measures the slope, whereas meters measure the altitude. However, the coefficients can be used to explain the contribution of the factors to the landslides occurrence in the study area. The positive and the negative signs (variables in the equation) in Table 11.6 indicate the positive and negative relationships of a specific factor and the landslide occurrence. In addition, the bivariate analysis showed that most of the landslides occurred at an altitude ranging from 1527 to 1639 m. In the record, 22 (60%) landslides occurred in areas with slopes greater than 36 degrees. In addition, the southeast section contained almost 27% of the slope failure records. Meanwhile, the curvature has three faces: flat, convex, and concave. The analysis showed that 18 landslides occurred in convex curvature and that four landslides occurred in concave areas; no landslide occurred in flat areas. For the other factors, the large number of landslides occurred in areas with a TRI greater than 51. Moreover, the mixed landuse, high vegetation density, distance to the road from 40 to 100 m, distance to lineament from 55 to 110 m, and distance to stream from 36 to 56 m are some of the other factors. The slopes above 36° tend to be more unstable than the less steep ones, although the analysis showed that landslides (7) occurred in areas with a slope of less than 13 degrees.

Landslide susceptibility maps were generated using the developed bivariate statistical models, as shown in Fig. 11.4. The landslide susceptibility maps were reclassified into five classes using the ArcGIS quantile classification method. The classes are very low, low, moderate, high, and very high. The ROC accuracy assessment showed that the highest accuracy could be achieved with FR considering both the success and the prediction rates (Fig. 11.5). The FR model could achieve success and prediction rates of 0.70 and 0.69, respectively. The success rate of the SI and WoE models was similar (0.68), whereas the prediction rate of SI (0.68) was higher than the WoE model by 2% of ROC.

11.7.4 Results of Multivariate Statistical Models

Three multivariate statistical models, PLS, LR, and DA, were used to produce landslide susceptibility maps of the study area using the 37 landslide inventories. PLS and DA were implemented in Minitab 17 using the leave one out cross-validation approach and the 10 landslide conditioning factors. Furthermore, a binary logistic regression (BLR) using a logit function as a link function was used to establish the nonlinear relationship between the 10 landslide conditioning factors and landslide occurrence. Table 11.7 shows the estimated coefficients. The coefficients were interpreted as the influence weight of the specific factor on landslide occurrence. The weights of the specific factors were used in the process of the sum of weights in GIS to produce the landslide susceptibility maps. Afterward, the produced maps were reclassified into five classes using the quantile classifier method (Fig. 11.6).

The estimated coefficients from the PLS method showed that the slope aspect, distance to lineament, and distance to the road had a negative relationship with landslide occurrence. According to the standardized coefficients, landuse was found to have the highest positive influence on landslide occurrence, whereas the distance to the road was found to have the highest negative influence on slope failure occurrence. In the DA model, factors with the largest coefficients were the ones that contributed most to the prediction of landslide occurrence (Table 11.8). On the other hand, in the LR model, the influence of a specific landslide conditioning factor could be quantified by observing the odds ratios. If p is the proportion of observations with an outcome of 1 (landslide presence), then $1 - p$ is the probability of an outcome of 0 (landslide absent). Then, the ratio $p/(1 - p)$ is called the odds of it ranging from 0 to positive infinity. In addition, the confidence interval indicates the level of uncertainty around the measure of effect, which in this case is expressed as an odds ratio. The odds ratios in Table 11.9 show that the distance to stream (odds ratio = 1.0229), landuse (odds ratio = 1.0608), altitude (odds ratio = 1.0145), and TRI (odds ratio = 1.1921) have the highest effects on increasing the probability of landslide occurrence in the study area.

The AOCs calculated from the ROC curves indicate that BLR has higher success (0.81) and prediction (0.77) rates than the PLS and DA models (Fig. 11.7). In addition, the accuracies also show that the PLS model can more accurately predict the landslides in the test dataset, with a success and prediction rate of 0.71 and 0.65, respectively. The DA model has a low success (0.42) and prediction (0.33) rate. Pourghasemi et al. (2013a, b, c), Lombardo et al. (2015), and Patriche et al. (2016) reported high success and prediction rates of the BLR model.

11.7.5 Results of Machine Learning Models

Data mining models, such as SVM, RF, and DT, have become popular in many geoscience applications for their high prediction capability. In the present study, three models are built using the SVM, RF, and DT algorithms with 37 landslide inventories and 10 landslide conditioning factors. Table 11.10 shows the estimated coefficients. The SVM model shows that factors such as slope aspect, distance to lineament, and distance to the road had a negative relationship with landslide occurrence, whereas the remaining

Table 11.6 Results of bivariate statistical modeling

Data layers	Classes	Pixels in classes	Landslide pixels	FR	SI	W^+	W^-	Wf
Altitude	1451–1526	2,804,061	5	0.591	−0.526	−0.526	0.114	−0.640
	1527–1582	3,441,544	12	1.149	0.145	0.145	−0.063	0.208
	1583–1639	3,640,730	15	1.273	0.311	0.312	−0.167	0.479
	1640–1718	1,628,053	5	0.906	0.018	0.018	−0.003	0.021
	1719–1839	747,338	0	0.000	0.000	0.000	0.000	0.000
Slope	0–12	2,729,980	1	0.121	2.109	−2.108	0.225	−2.333
	13–24	2,682,317	6	0.577	0.299	−0.299	0.070	−0.369
	25–35	3,125,291	8	0.619	0.165	−0.164	0.051	−0.215
	36–47	2,676,565	17	1.368	0.744	0.745	−0.370	1.114
	48–78	1,051,347	5	0.918	0.455	0.455	−0.056	0.511
Aspect	Flat	359	0	0.000	0.000	0.000	−0.084	0.084
	North–East	2,100,307	3	0.331	−0.748	−0.748	0.011	−0.759
	East	2,142,655	6	0.565	−0.075	−0.074	−0.123	0.049
	South–East	1,846,774	10	0.939	0.585	0.585	0.049	0.536
	South	1,354,505	4	0.493	−0.022	−0.021	0.093	0.072
	South–West	1,129,248	7	1.833	0.720	0.720	0.070	0.651
	West	987,710	1	0.268	−1.092	−1.092	0.029	−1.120
	North–West	1,098,257	2	0.737	−0.505	−0.505	−0.020	−0.484
	North	1,605,685	4	1.046	−0.192	−0.191	0.000	−0.191
Curvature	Convex	6,117,114	18	1.290	−0.025	−0.025	0.024	−0.049
	Flat	0	0	0.000	0.000	0.000	0.000	0.000
	Concave	6,148,386	19	1.406	0.024	0.024	−0.025	0.049
TRI	0–24	1,657,409	0	0.000	0.000	0.000	0.000	0.000
	25–38	2,297,400	2	0.251	−1.243	−1.243	0.090	−1.333
	39–50	3,160,497	8	0.993	−0.176	−0.175	−0.037	−0.139
	51–59	3,566,593	13	1.139	0.189	0.189	−0.135	0.324
	60–150	1,593,877	14	1.981	1.068	1.069	−0.132	1.201
Landuse	Water	39,631	0	0.000	0.000	0.000	0.000	0.000
	Forest	6,161,626	4	0.184	−1.536	−1.536	−0.111	−1.425
	Industrial	202,633	1	0.724	0.492	0.492	0.670	−0.178
	Infrastructure	347,197	6	2.442	1.745	1.746	−0.160	1.906
	Facilities	3,716,563	0	0.000	0.000	0.000	0.000	0.000
	Mixed	60,137	25	19.397	4.926	4.926	−1.097	6.023
	Commercial	1,101,967	0	0.000	0.000	0.000	0.000	0.000
	Transportation	2438	0	0.000	0.000	0.000	0.000	0.000
	Agriculture	177,062	0	0.000	0.000	0.000	0.000	0.000
	Residential	384,818	0	0.000	0.000	0.000	0.000	0.000
	Vacant	74,678	1	4.467	1.490	1.490	0.334	1.157
Vegetation Density	Non	731,850	1	0.649	−0.792	−0.792	0.034	−0.826
	Low	1,339,275	4	1.642	−0.010	−0.010	0.001	−0.011
	Moderate	6,450,950	15	1.321	−0.261	−0.260	0.226	−0.487
	High	3,746,675	17	1.895	0.408	0.408	−0.251	0.659

(continued)

Table 11.6 (continued)

Data layers	Classes	Pixels in classes	Landslide pixels	FR	SI	W^+	W^-	Wf
Distance to Road	0–45	5,299,275	23	1.439	0.363	0.364	−0.406	0.770
	46–100	3,504,632	11	0.862	0.039	0.040	−0.016	0.056
	101–180	2,021,571	3	0.405	−0.710	−0.709	0.095	−0.805
	181–280	882,963	0	0.000	0.000	0.000	0.000	0.000
	281–480	557,059	0	0.000	0.000	0.000	0.000	0.000
Distance to lineament	0–54	3,195,812	12	1.245	0.219	0.219	−0.090	0.309
	55–110	3,440,293	16	1.611	0.433	0.433	−0.237	0.670
	111–160	2,857,921	6	0.717	−0.363	−0.362	0.088	−0.450
	161–230	2,011,560	3	0.500	−0.705	−0.704	0.094	−0.799
	231–360	759,914	0	0.000	0.000	0.000	0.000	0.000
Distance to Stream	0–17	3,748,358	8	0.708	−0.346	−0.346	0.121	−0.467
	18–35	3,265,472	8	0.564	−0.208	−0.208	0.066	−0.274
	36–56	2,619,826	9	0.488	0.130	0.130	−0.039	0.169
	57–80	1,818,382	8	0.313	0.377	0.377	−0.083	0.461
	81–140	813,462	4	0.108	0.488	0.489	−0.046	0.535

factors had positive effects on landslide occurrence in the study area. In addition, the highest positive coefficient of 1.796 was found for the landuse factor, and then, the TRI and altitude factors were found to have high coefficients compared with the remaining factors. The estimated coefficients of the SVM model for TRI and altitude were 1.606 and 0.822, respectively. Among the factors that show negative effects, the distance to the road had the lowest coefficient value (−1.881), as estimated by the SVM method.

Figure 11.8 presents the OOB predictions of the RF model, and Table 11.10 indicates the estimated coefficients. In general, the OOB predictions indicate that when the resulting model is applied to new observations (i.e., test data), the answer will obtain an error of 20% of the time. On the other hand, the error on the training dataset is ~10% of the time. The RF model computes the importance of the factors and is represented in values ranging from 0 to 100. The results show that the most important factor is landuse (importance value = 100) and that the least important factor is the vegetation density (importance value = 0.184).

A DT is a knowledge-based modeling approach that can be used to establish relationships between a set of independent variables (i.e., landslide conditioning factors) and a dependent variable (i.e., landslide occurrence) in terms of crisp rules. In the present study, the DT model is developed using the same data in the SVM and RF models explained earlier. Analyzing the estimated coefficients by the DT method shows that the landuse is the most important factor; its coefficient is equal to 1. The DT model is consistent with the RF model on the less important factor; the estimated coefficient of the vegetation density from the DT model is 0.033 (Fig. 11.9).

The calculated ROC-based accuracies, which are represented in success and prediction rates, indicated that the SVM model has the highest prediction capability on both the training and testing datasets (Fig. 11.10). The success and the prediction rates of the SVM model are 0.83 and 0.77, respectively. The success and the prediction rates of the DT and RF models show that the DT model outperforms the RF model. The estimated success rates of the DT and RF models are 0.76 and 0.72, respectively. The prediction rates of the two models are 0.73 and 0.71, respectively. The dataset of the present study shows that the RF has more generalization capability than the DT and SVM model.

11.7.6 Results of Expert-Based Models

Expert-based models such as AHP and fuzzy AHP were also applied to produce the landslide susceptibility maps for the study area. The same 10 landslide conditioning factors were used in the modeling. However, the landslide inventory data were not used, considering that these techniques were not required because they depended on expert opinions. For each factor, a specific weight was given by a number of experts. Then, the geometric mean was applied to calculate the final weights of the factors. In the present study, the weights were gathered from the previous studies applied in areas that have similar characteristics. Table 11.11 shows the pairwise comparison matrix for the landslide conditioning factors. The information is obtained from various researches on AHP modeling for landslide susceptibility. The main references are (Yalcin et al. 2011; Hasekioğulları and Ercanoglu 2012; Pourghasemi et al. 2013a, b, c; Kayastha

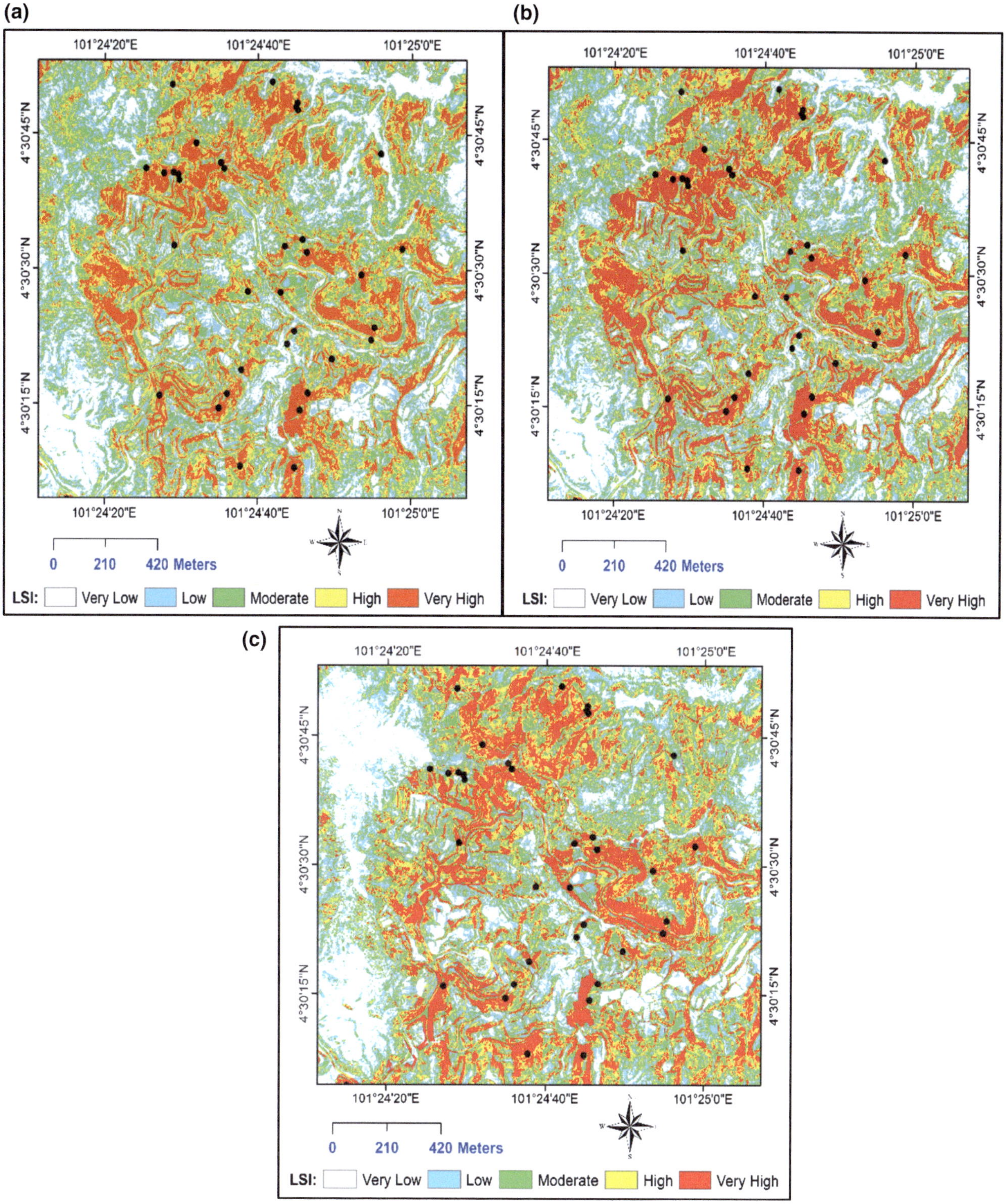

Fig. 11.4 Landslide susceptibility maps produced by the bivariate methods: **a** WoE, **b** SI, and **c** FR

et al. 2013; Park et al. 2013). The pairwise comparison matrix is then used to calculate the weights of the factors and to ensure the reliability of the model in which the inconsistency ratio is calculated. Table 11.12 shows the estimated weights for the landslide conditioning factors, where the inconsistency ratio is 9.7%. In addition, the fuzzy AHP method is applied to evaluate the necessity of fuzzy integration on expert-based landslide susceptibility models.

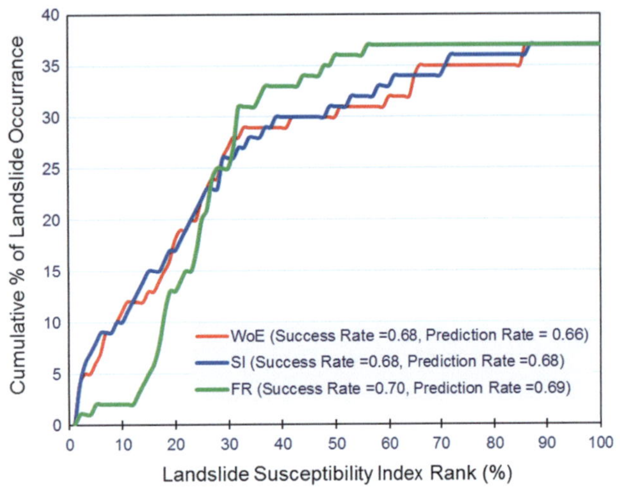

Fig. 11.5 Success and prediction rates of bivariate methods

Table 11.7 Estimated coefficients of landslide conditioning factors by the PLS method

Factor	Coefficients	Standardized coefficients
Slope	0.0053	0.1587
Curvature	0.0020	0.0768
Aspect	−0.0006	−0.1225
Distance to lineament	−0.0011	−0.1611
Distance to road	−0.0015	−0.2446
Distance to stream	0.0023	0.1269
Vegetation density	0.0260	0.0428
Landuse	0.0568	0.2620
Altitude	0.0001	0.0217
TRI	0.0076	0.2211

Table 11.12 shows the estimated weights of fuzzy AHP method, and the ranks are calculated based on the weights.

The AHP method showed that the most important factor is the TRI (weight = 0.25, rank = 1), which the fuzzy AHP also determined to be the most influential factor. In addition, both models agreed that the slope and landuse were the second and the third important factors contributing to the landslide occurrence in the Cameron Highlands. The linear regression analysis between the estimated weights of the AHP and fuzzy AHP showed that the two models have a correlation of 0.96 (Fig. 11.11). The difference between the AHP and fuzzy AHP was found in the calculation of ranks of altitude, slope aspect, and distance to stream. Furthermore, both models agreed that the least important factor is the distance to the road, whose estimated weights were 0.024 and 0.027, respectively (Fig. 11.12)

Figure 11.13 shows the calculated success and the prediction rates from the ROC curves for the AHP and fuzzy AHP models. The results showed that the incorporation of fuzzy concept in the expert-based modeling of landslide susceptibility does not significantly increase the success rate. In addition, the prediction rate of the traditional AHP was found to be higher than the fuzzy AHP method. The estimated prediction rates of the AHP and fuzzy AHP were 0.71 and 0.67, respectively.

11.7.7 Result of Hybrid Models

To improve the prediction rate of the landslide susceptibility models, recent studies have suggested combining two or more models in an integrated or hybrid model (Gorsevski and Jankowski 2010; Althuwaynee et al. 2016; Aghdam et al. 2016). Three hybrid models are developed and evaluated in the present study. The models are FR–SVM, FR–LR, and WoE–RF. Table 11.13 shows the estimated weights of the landslide conditioning factors obtained by the hybrid models. The three models disagreed on the influence of the factors related to landslide occurrence in Cameron Highlands. For example, the factors with negative relationships with landslide occurrence in the FR–SVM model are slope, distance to road, and distance to lineament. On the other hand, these factors were found to have positive effects on landslide occurrence by the FR–LR and WoE–RF models. In addition, the FR–SVM showed that landuse has the highest positive effect on landslide occurrence, whereas the distance to road and altitude were found to have the highest coefficient values in the FR–LR and WoE–RF models, respectively (Fig. 11.14).

The ROC accuracies of the hybrid models are shown in Fig. 11.15. The highest success and prediction rates were achieved by FR–LR model. Similarly, the FR–SVM model outperformed the WoE–RF model both in success and in prediction rates. The success rates of the FR–LR, FR–SVM, and WoE–RF were 0.84, 0.76, and 0.68, respectively. On the other hand, the prediction rates of the three models were 0.83, 0.75, and 0.65, respectively. Differences in the success and prediction rates of the FR–LR and FR–SVM models indicated that these models could more ably generalize findings than the WoE–RF model; however, this finding requires further study. In addition, the OOB predictions of the WoE–RF model showed that when the resulting model is applied to new observations, an error occurs 23% of the time (Fig. 11.16).

11.7.8 Landslide Density Graphs

Landslide density graphs, which are useful for assessing the validity of landslide susceptibility maps with the existing slope instability conditions, show the landslide percentage that occurred in landslide susceptibility classes. For the

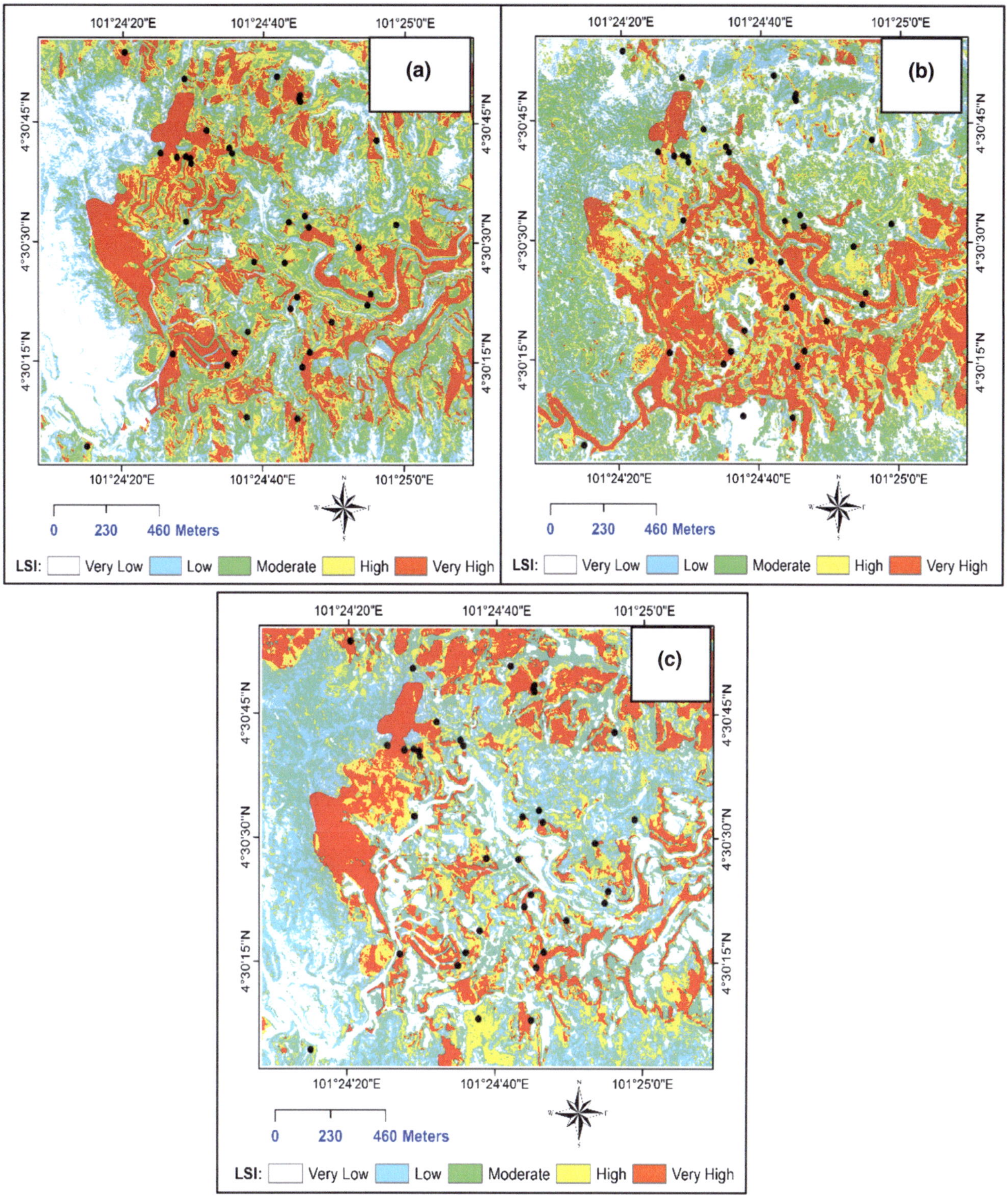

Fig. 11.6 Landslide susceptibility maps produced by the multivariate methods: **a** PLS, **b** DA, and **c** LR

bivariate models, the WoE model showed that ~75% of the landslides occur at a very high susceptibility class, whereas the FR and SI models showed lower percentages of 63 and 70%, respectively (Fig. 11.17). The percentage of landslide occurrence in the very high susceptibility zone is remarkably higher than other zones in all bivariate models, indicating the validity of the produced susceptibility maps with the existing landslide inventory map.

Table 11.8 Estimated discriminant function for landslide and non-landslide samples by the DA method

Factor	Discriminant function for groups	
	Landslides	Non-landslides
Slope	−0.91	−0.94
Curvature	−0.55	−0.54
Aspect	0.01	0
Distance to lineament	−0.02	−0.03
Distance to road	−0.28	−0.31
Distance to stream	−0.47	−0.45
Vegetation density	4.46	4.55
Landuse	−0.93	−0.79
Altitude	0.55	0.55
TRI	0.73	0.87

Table 11.9 Estimated regression coefficients and odds ratios for landslide conditioning factors by the LR method

Factor	Coefficients	Standardized coefficients	Odds ratio	95% CI
Slope	−0.0524	0.0573	0.9489	(0.8482,1.0616)
Curvature	−0.0252	0.0295	0.9751	(0.9204,1.0332)
Aspect	−0.0069	0.0050	0.9931	(0.9833,1.0029)
Distance to lineament	−0.0145	0.0080	0.9855	(0.9701,1.0012)
Distance to road	−0.0546	0.0178	0.9469	(0.9144,0.9805)
Distance to stream	0.0226	0.0174	1.0229	(0.9886,1.0583)
Vegetation density	−0.002	0.626	0.9979	(0.2924,3.4060)
Landuse	0.059	0.23	1.0608	(0.6761,1.6645)
Altitude	0.0144	0.0085	1.0145	(0.9977,1.0316)
TRI	0.1757	0.0672	1.1921	(1.0451,1.3598)

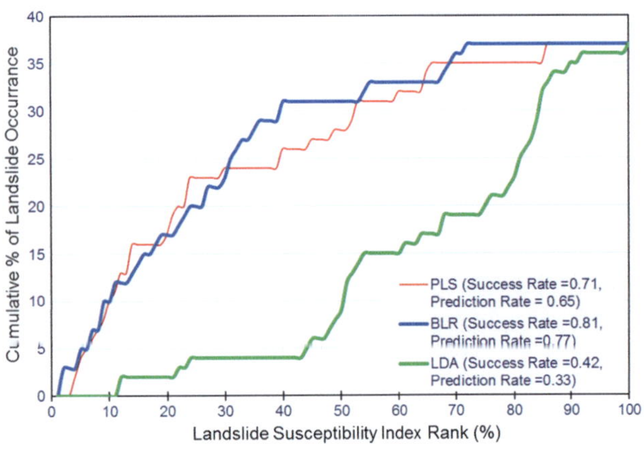

Fig. 11.7 Success and prediction rates of the multivariate methods

In addition, Fig. 11.18 shows that the landslide density values of very high susceptibility zone in maps produced by multivariate statistical approaches vary from one model to another. In the LR model, the percentage value of the very high susceptibility class was ~62%, whereas the percentage value in the PLS and DA models were 77 and 28%, respectively. These values indicate that LR and PLS models are much more valid than the DA model with the existing landslide inventory map. This also suggests the use of LR and PLS models for producing landslide susceptibility maps for landuse planning.

In landslide susceptibility maps produced by data mining models, the landslide percentages in the very high susceptibility zone were 77, 50, and 71% for the SVM, RF, and DT models, respectively (Fig. 11.19). In addition, the SVM, RF, and DT models had significantly higher landslide percentages in the high susceptibility zone than the other zones, corresponding to 14, 27, and 20%, respectively. As a result, the SVM model can produce a better landslide susceptibility map for the landuse planning because of the validity of the map relative to existing field conditions.

The analysis of landslide susceptibility maps produced by expert-based models demonstrated that landslide percentage in the very high susceptibility zone in the AHP (68%) was higher than that of the fuzzy AHP (63%) (Fig. 11.20). Similarly, the landslide percentage of the AHP model in the high susceptibility zone was 23%, which was slightly higher than that of fuzzy AHP model.

Figure 11.21 shows the landslide percentages in the landslide susceptibility zones in maps produced by hybrid models. The graph shows that the FR–SVM and FR–LR models are more valid than the WoE–RF model with the

Table 11.10 Estimated coefficients of landslide conditioning factors by the SVM, RF, and DT method

Factor	Coefficients		
	SVM	RF	DT
Slope	0.624	0.571	0.467
Curvature	0.548	0.407	0.152
Aspect	−0.672	0.632	0.363
Distance to lineament	−0.867	0.415	0.389
Distance to road	−1.881	0.760	0.636
Distance to stream	0.811	0.612	0.377
Vegetation density	0.263	0.184	0.033
Landuse	1.796	1.000	1.000
Altitude	0.822	0.668	0.423
TRI	1.606	0.702	0.467

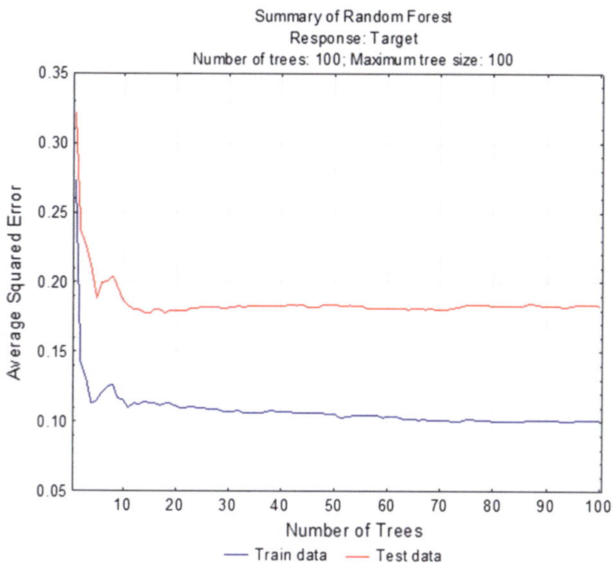

Fig. 11.8 OOB predictions of the RF model

existing landslide inventory map. This result occurred because the landslide percentages in the very high susceptibility zone were 79, 76, and 47% for the FR–SVM, FR–LR, and WoE–RF models, respectively. In addition, the landslide percentage in the high susceptibility zone using the WoE–RF model was 35%, which was almost twice that of the other two models.

11.7.9 Sensitivity of the Models

High-quality data are not always available, and their acquisition is expensive for small-scale projects. Therefore, evaluating model sensitivities with respect to the number of landslide conditioning factors is important. Four sets of data were evaluated; the first set included three factors: altitude, slope, and curvature. The second dataset included 50% of the available data (i.e., altitude, slope, curvature, TRI, and aspect). The third dataset included all factors except vegetation density, whereas the fourth dataset comprised of all the factors. The datasets were evaluated based on the success and prediction rates estimated from the ROC curves.

Table 11.14 shows the results of the model sensitivity evaluation using the success rate metric. When only three factors (i.e., altitude, slope, curvature) were included in the models, the best success rate was achieved by the FR model (0.88), which belongs to the bivariate category. In addition, success rates of 0.85 and 0.72 were achieved by the WoE and SI models, respectively, which were higher than the values achieved by other models in other categories. In the multivariate category, all models performed similarly; the highest success rate of 0.62 was achieved by the PLS method. The AHP model outperformed the fuzzy AHP model by 2%. In the data mining category, the best model was the RF model, which had a success rate of 0.63. Furthermore, the evaluation of hybrid models showed that the combination of the FR and LR models can result in a higher success rate than other combinations. The success rate of the FR–LR model was 0.64, whereas the success rates of the FR–SVM and WoE–RF models were 0.57 and 0.58, respectively.

When 50% of the landslides were used as training data for the models, the accuracy evaluation showed a different result compared to when only three factors were used. The best model according to the success rate (0.81) was the WoE model. In addition, the FR and SI models, with success rates of 0.79 and 0.74, respectively, achieved higher success rates than other models in other categories. Among the multivariate models, the best model was the PLS model, with a success rate of 0.68. In addition, the LR model performed slightly better than the DA model. Among the expert-based methods, the fuzzy AHP achieved a success rate of 0.67, which was higher than AHP by 1%. On the other hand, the best model among data mining approaches was found to be the RF model, with a success rate of 0.70. The success rates of the SVM and DT models were 0.61 and 0.56,

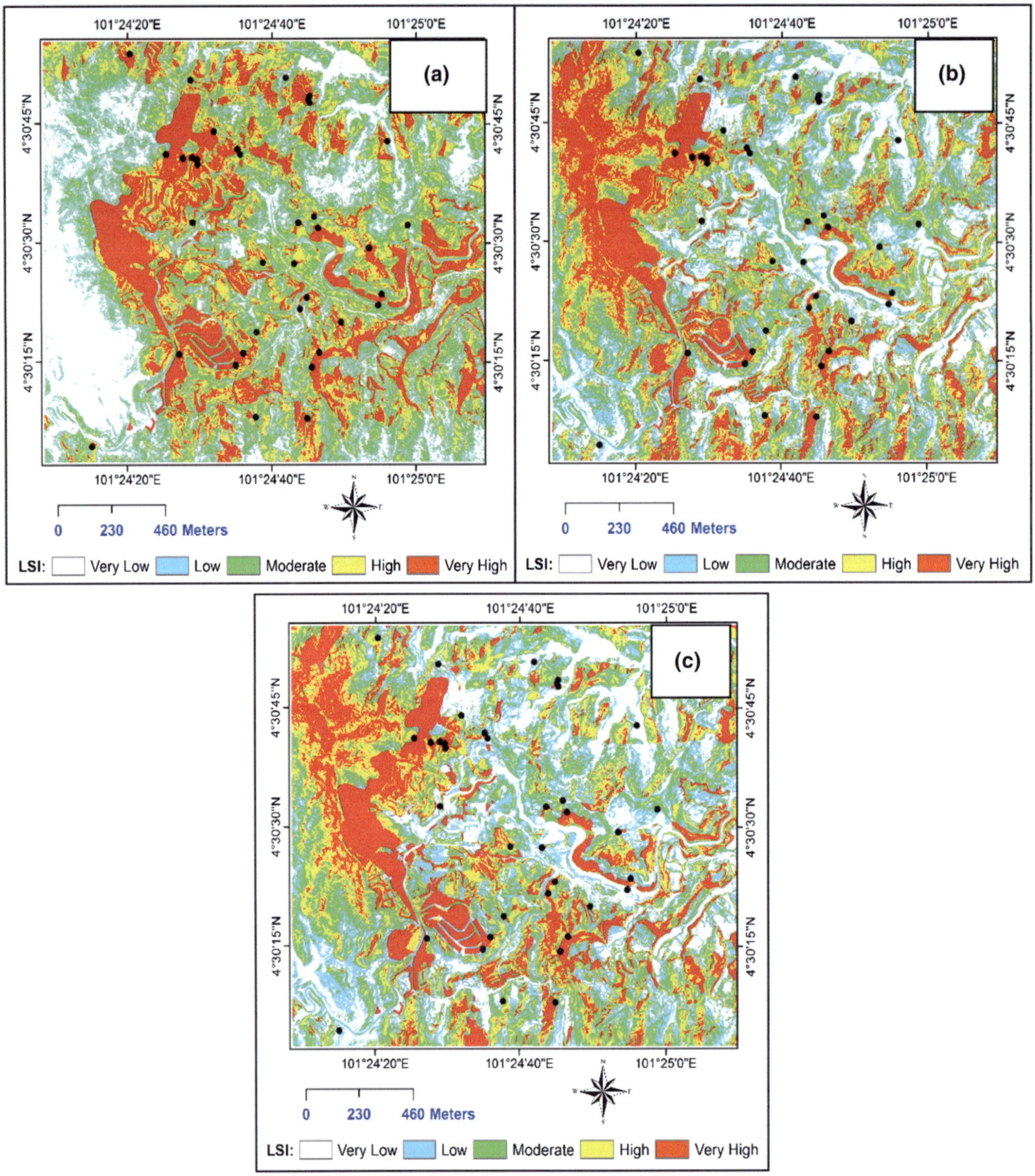

Fig. 11.9 Landslide susceptibility maps produced by the data mining methods: **a** SVM, **b** RF, and **c** DT

respectively. The assessment of the hybrid models showed that the best model was the WoE–RF model, which had a success rate of 0.69. The success rates of the FR–SVM and FR–LR models were 0.58 and 0.67, respectively.

The same set of models was also evaluated using eight landslide conditioning factors where the vegetation density factor was excluded. The sensitivity evaluation showed that the best model based on success rate was the WoE model (0.89). The FR and SI models also achieved success rates higher than ~0.80. The multivariate models, including LR, DA, and PLS, achieved success rates of 0.75, 0.64, and 0.71, respectively, indicating that the LR model could outperform

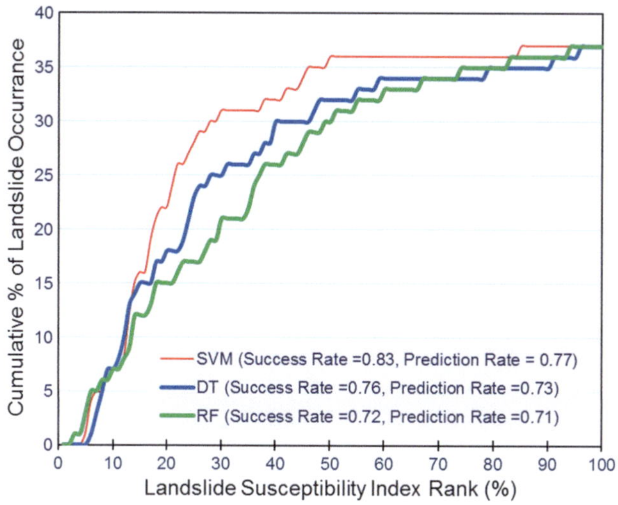

Fig. 11.10 Success and prediction rates of the data mining methods

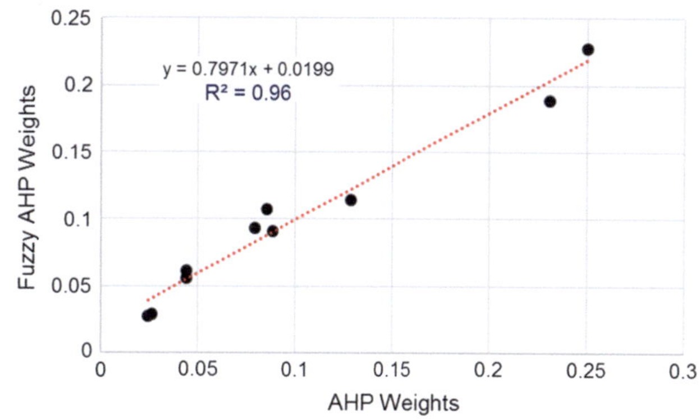

Fig. 11.11 Linear regression between the estimated coefficients by the AHP and fuzzy AHP methods

Table 11.11 Pairwise comparison matrix using in the AHP method

	Altitude	Slope	Aspect	Curvature	Vegetation density	Distance to lineament	Distance to road	Distance to stream	TRI	Landuse
Altitude	1									
Slope	4.00	1								
Aspect	0.25	0.25	1							
Curvature	1.00	0.25	0.25	1						
Vegetation density	0.50	0.11	0.25	2.00	1					
Distance to lineament	0.25	0.25	0.33	0.25	0.25	1				
Distance to road	0.25	0.17	0.20	0.33	0.33	1.00	1			
Distance to stream	2.00	0.33	1.00	4.00	3.00	3.00	2.00	1		
TRI	4.00	1.00	4.00	9.00	9.00	4.00	6.00	3.00	1	
Landuse	2.00	0.33	4.00	3.00	3.00	5.00	4.00	2.00	0.33	1

Inconsistency ratio = 9.7%

Table 11.12 Estimated coefficients for the landslide conditioning factors by the AHP and fuzzy AHP methods

Factor	AHP priority	AHP rank	FAHP priority	FAHP rank
Altitude	0.085	5	0.107	4
Slope	0.231	2	0.189	2
Aspect	0.079	6	0.093	5
Curvature	0.044	7	0.061	7
Vegetation density	0.044	8	0.056	8
Distance to lineament	0.026	9	0.029	9
Distance to road	0.024	10	0.027	10
Distance to stream	0.088	4	0.091	6
TRI	0.250	1	0.228	1
Landuse	0.128	3	0.114	3

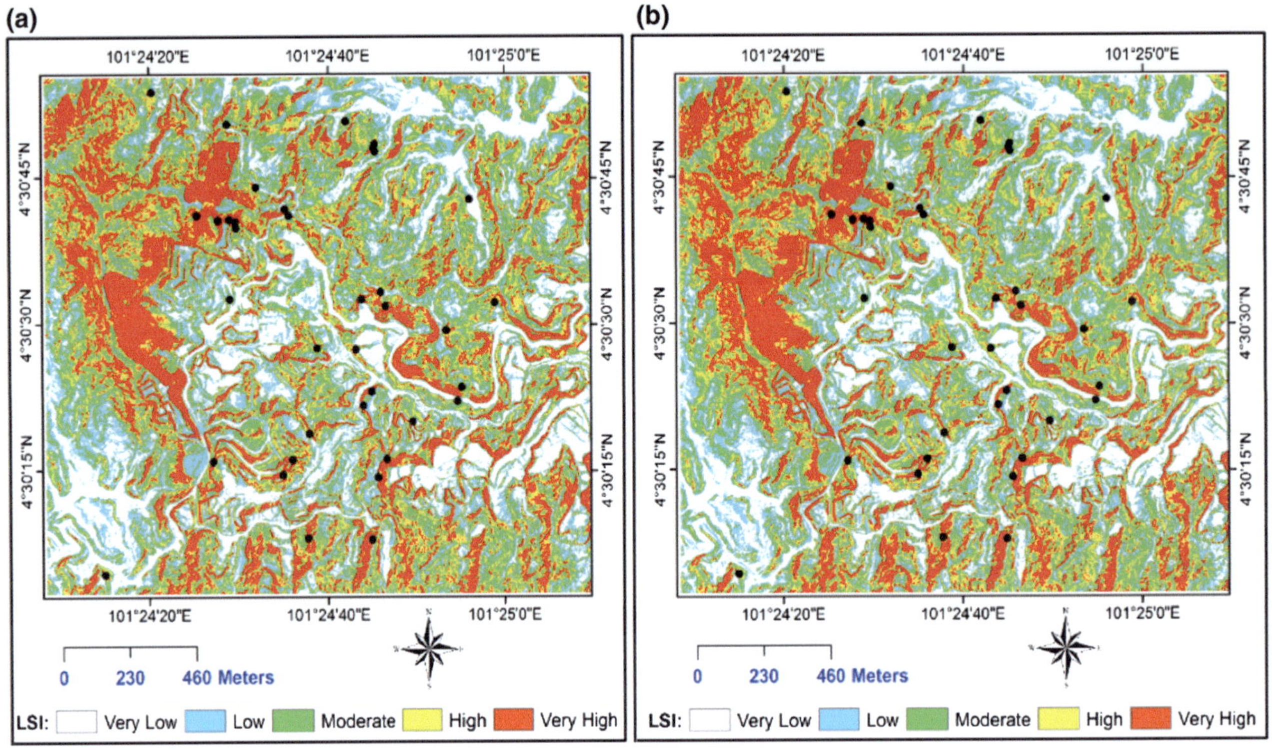

Fig. 11.12 Landslide susceptibility maps produced by the AHP **a** and fuzzy AHP **b** methods

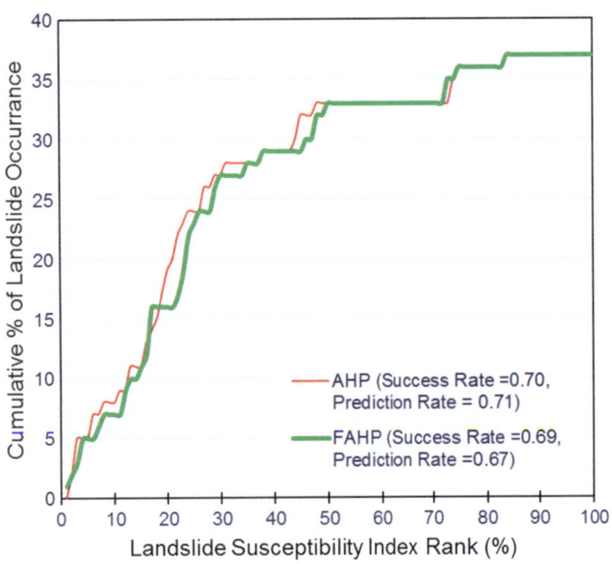

Fig. 11.13 Success and prediction rates of the expert-based methods

the DA and PLS models. Among the expert-based models, the AHP and fuzzy AHP models performed equally, with a success rate of 0.72. Furthermore, the success rates of data mining methods indicated that the RF outperformed the SVM and DT methods. The SVM and DT models performed almost similarly, with a success rate of ~0.76. Among the hybrid models, the best model was the WoE–RF model, which achieved a success rate of 0.77. The FR–LR model considerably outperformed the FR–SVM model by a 0.06 difference in success rate.

When vegetation density was included with other factors in the modeling, the performance of the models changed remarkably. The best model was the FR–LR hybrid model, with a success rate of 0.83. Bivariate models performed almost equally, with success rates of 0.66, 0.69, and 0.68 corresponding to the WoE, FR, and SI models, respectively. Among the multivariate models, the LR model significantly outperformed the DA and PLS models. The success rate of the LR model was 0.77, whereas the DA and PLS models had success rates of 0.33 and 0.65, respectively. The traditional AHP achieved a better success rate than the fuzzy AHP; the difference in their success rates was 0.04. In addition, the success rates of the SVM, RF, and DT models were 0.77, 0.70, and 0.73, respectively, indicating that the SVM model achieved a better success rate than the two other data mining methods. A comparison of the hybrid models showed that the FR–SVM model significantly (by 10% of success rate) outperformed the WoE–RF model, while FR–LR achieved the best success rate, as previously mentioned.

The sensitivity analysis based on success rates showed that the bivariate models were strongly influenced by vegetation density. The average success rate of the three models decreased by 11% when comparing the models developed using only three factors with the models developed by

Table 11.13 Estimated factor coefficients by the hybrid models

Factor	Coefficients		
	FR-SVM	FR-LR	WoE-RF
Slope	0.625	0.052	0.576
Curvature	0.548	0.025	0.167
Aspect	−0.673	0.007	0.426
Distance to lineament	−0.868	0.014	0.407
Distance to road	−1.881	0.054	0.714
Distance to Stream	0.812	−0.022	0.513
Vegetation density	0.264	0.002	0.279
Landuse	1.796	−0.059	0.795
Altitude	0.822	−0.014	1
TRI	1.606	−0.175	0.703

including all of the available factors. With respect to multivariate models, the success rate of LR model increased as the number of included landslide conditioning factors increased. The success rate of the LR model with three factors was 0.54, and a success rate of 0.81 was achieved when all factors were included. The DA and PLS models showed an increase in performance by increasing the number of factors from three to eight. In contrast, their performance decreased when vegetation density was added as an additional independent variable. Among expert-based models, a slight increment in performance was observed with respect to different numbers of landslide conditioning factors used. In general, the performance of both the AHP and fuzzy AHP models were improved by adding additional factors when the vegetation density was incorporated in the modeling. The success rates of the SVM and DT models were increased by increasing the number of landslide conditioning factors. Consequently, their success rates improved by ~20% when the number of factors increased from three to 10. Meanwhile, the RF model was affected by adding the vegetation density to the model. In general, the success rate of hybrid models improved as the number of factors increased. The success rates of the FR–SVM and FR–LR models significantly increased when 10 factors were used instead of three. Although the success rate of the RF model improved by 20% when the landslide factors increased from three to eight, the success rate of the model decreased by 13% when vegetation density was included in the model.

Table 11.15 shows the sensitivity of the models to landslide factors based on the prediction rate metric. Using three landslide factors (altitude, slope, and curvature), the best model was the WoE model (0.78). Furthermore, among the bivariate models, the FR model outperformed the SI model by 13%. Among the multivariate models, the PLS method, which achieved a prediction rate of 0.63, outperformed the LR and DA methods. Moreover, the LR model performed better than the DA model, as observed through their prediction rates of 0.57 and 0.44, respectively. Expert-based models, including the AHP and fuzzy AHP models, performed almost equally, with an average prediction rate of 0.58. In the data mining category, the RF model achieved the best prediction rate (0.67), while the SVM and DT models achieved a prediction rate of 0.46 and 0.52, respectively. Furthermore, among the hybrid models, the best model was found to be FR–LR model with a prediction rate of 0.63. The FR–SVM and WoE–RF models achieved prediction rates of 0.52 and 0.57, respectively.

Furthermore, using 50% of the landslide factors showed a different performance compared to using only three. Among the bivariate models, the FR and SI methods performed almost equally (~0.67), whereas the WoE model achieved the best prediction rate (0.77). On the other hand, the PLS model achieved a better performance than the other two multivariate methods. The prediction rates of the PLS, LR, and DA methods were 0.70, 0.62, and 0.55, respectively. The fuzzy AHP model, as an expert-based method, outperformed the traditional AHP technique by a 0.05 prediction rate. Furthermore, the RF (0.69) and SVM (0.56) data mining methods achieved higher prediction rates than the DT approach (0.53). Moreover, hybrid models performed differently, and the best model was found to be the WoE–RF mode, which a prediction rate of 0.70. The prediction rates of the FR–SVM and FR–LR methods were 0.53 and 0.67, respectively.

When 80% of the factors were used to develop the landslide susceptibility models, the WoE method achieved the best prediction rate, 0.82. Within the bivariate category, the SI model performed better than the FR model. On the other hand, the LR achieved a better prediction rate than the PLS and DA multivariate methods. The prediction rates of the LR, PLS, and DA methods were 0.73, 0.67, and 0.57, respectively. Moreover, the fuzzy AHP outperformed the traditional AHP by 3%, observing a similar result with that when 50% of the factors were used. In the data mining category, the best model was the RF model, which achieved a prediction rate of 0.80; the prediction rates of the SVM and DT models were 0.71 and 0.73, respectively.

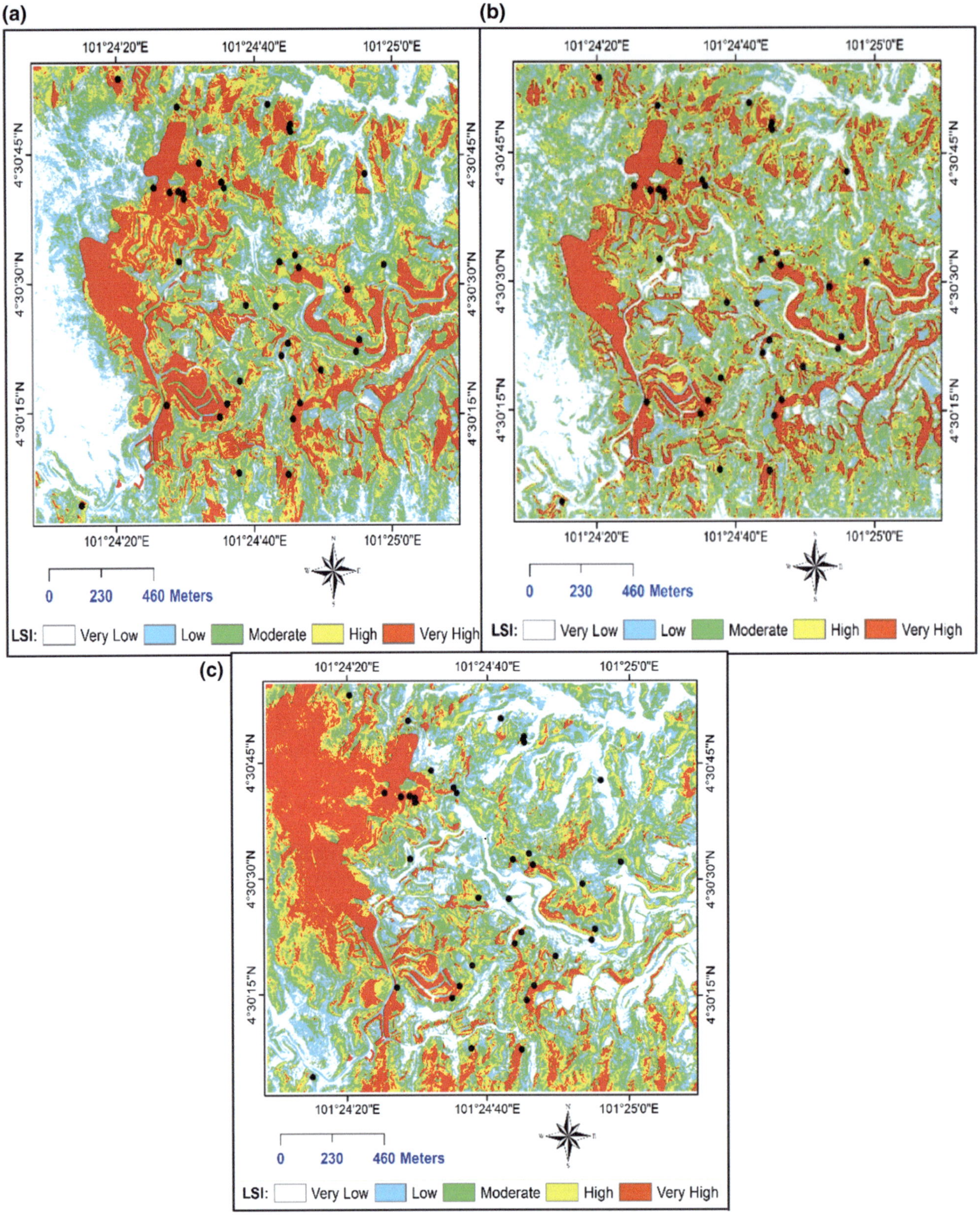

Fig. 11.14 Landslide susceptibility maps produced by the hybrid methods

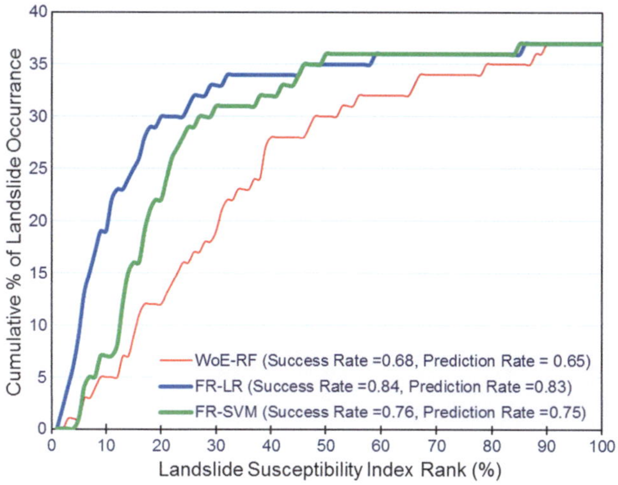

Fig. 11.15 Success and prediction rates of the hybrid models

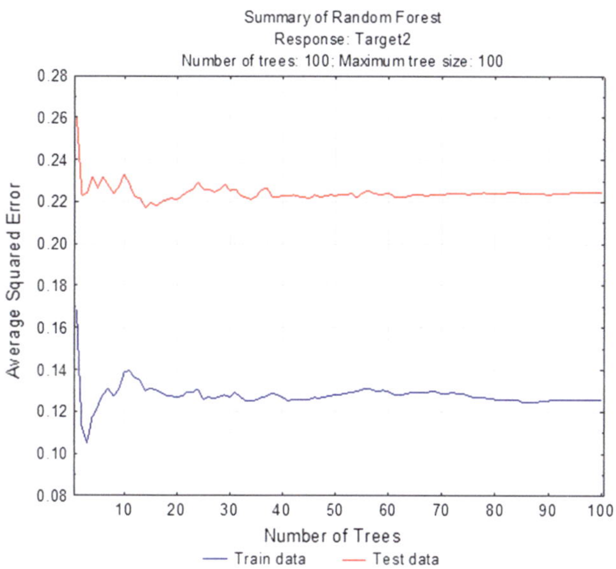

Fig. 11.16 OOB predictions of the RF WoE-RF model

Including all available landslide conditioning factors in the models, the accuracy assessment showed that the best model was the FR–LR hybrid model, with a prediction rate of 0.83. The prediction capability of bivariate models was reduced by ~9% after incorporating vegetation density in the models. Among the multivariate methods, the highest prediction rate (0.77) was achieved by the LR model. The DA method achieved the lowest prediction rate (0.33) among all other models. Furthermore, including vegetation density into the expert-based models yielded an improved performance of the traditional AHP method but reduced the prediction rate of the fuzzy AHP model by 6%. The SVM model, which achieved a prediction rate of 0.77, performed better than the RF and DT models. The RF and DT methods performed almost equally, and their average prediction rate was 0.72. As with the hybrid models, the FR–SVM and FR–LR models achieved higher prediction rates than the WoE–RF method.

The sensitivity analysis based on the number of landslide factors and prediction rate showed that bivariate models were significantly affected when all factors were included in the models. The WoE model was strongly affected by the incorporation of vegetation density, whereas the FR method was less affected, such that its prediction rate was reduced by only 1%. Increasing the number of factors included in the model improved the prediction rate of the LR model, whereas the best prediction rate of PLS method was found to be 0.70 when 50% of the factors were included in the model. The highest prediction rate of the DA (0.57) was achieved when 80% of the landslide conditioning factors were included in the model. Expert-based models were less affected by increasing the number of factors. The performance of the AHP model increased due to the increase in the number of factors and the incorporation of vegetation density, whereas the prediction capability of the fuzzy AHP decreased by 6%. The prediction rates of the data mining methods were improved by increasing the number of landslide factors; however, the RF model was affected by including the vegetation density factor into the model. Similarly, the hybrid models showed an increase in their prediction capability when the number of landslide factors increased. The WoE–RF model was affected by a more than 10% prediction rate when the vegetation density was considered.

11.7.10 Spatial Agreements of the Models

Despite using the same landslide susceptibility model with an identical prediction rate, the spatial agreement of the produced classification maps was often inconsistent, and their spatial patterns were considerably different. Therefore, analyzing the spatial agreements between the developed models in a specific category and between categories is important. Spatial agreements were analyzed by using the Spearman correlations where nonlinear relationships were examined. Table 11.16 shows the Spearman correlations calculated for the developed models that included all factors. Perfect spatial agreements (1.00) are highlighted in white color, high agreements (>0.80) in blue color, and red and yellow colors indicate the moderate (0.50–0.79) and low (<0.50) spatial agreements, respectively, between the models.

The Spearman correlations show near perfect spatial agreement between the SVM and FR–SVM models. Spatial agreements between the maps produced by the bivariate models showed high agreements, and the lowest Spearman correlation of 0.74 was found between the FR and SI

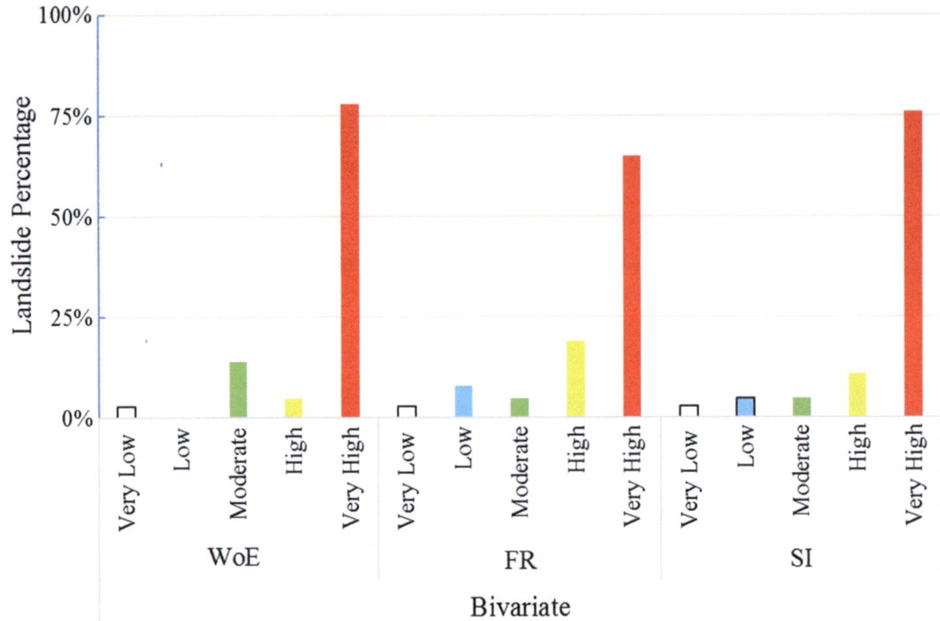

Fig. 11.17 Landslide density graph of the bivariate methods

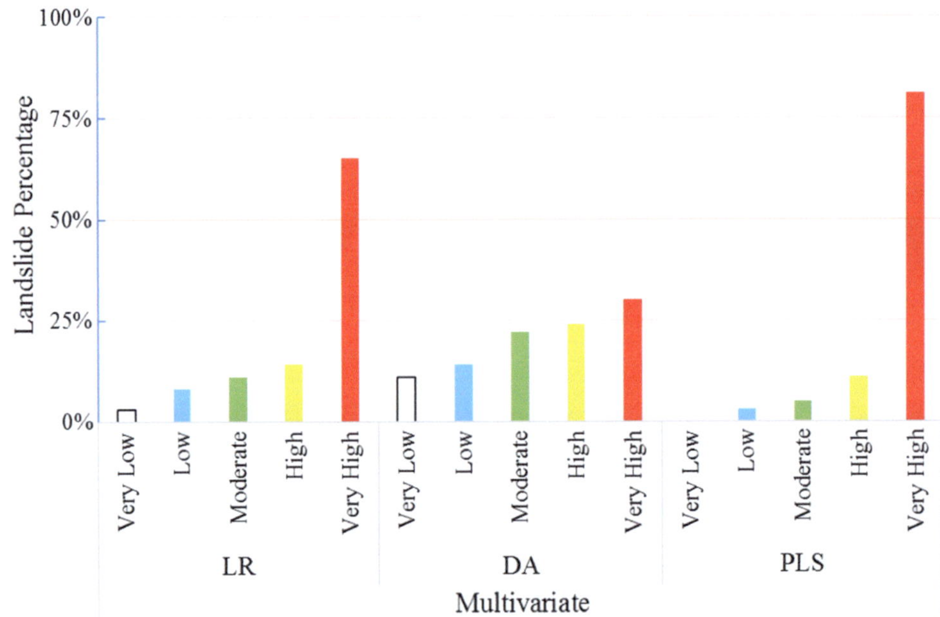

Fig. 11.18 Landslide density graph of the multivariate methods

models. On the other hand, among multivariate models, a negative correlation (i.e., disagreement) of −0.27 was found between the LR and PLS models. In addition, a low spatial agreement of 0.24 was observed between the LR and DA models. Among maps produced by expert-based methods, a very high spatial agreement (0.97) was found between the AHP and fuzzy AHP models. The spatial patterns in landslide susceptibility maps produced by the data mining approaches varied from one model to another. In general, a high agreement was found between the DT and RF models (0.93); however, the SVM model showed a low agreement between the DT and the RF models (0.37 and 0.27, respectively). Furthermore, hybrid models showed high spatial agreements between the FR–LR and FR–SVM models (0.77), whereas the agreement between the WoE–RF and other models was all less than 0.50.

The spatial agreements between the bivariate models and other categories showed moderate agreement (∼0.60). The agreements of the WoE and SI models were higher than the agreements between the FR model and other models. The highest correlations were found between the bivariate and hybrid models. In addition, correlations between bivariate

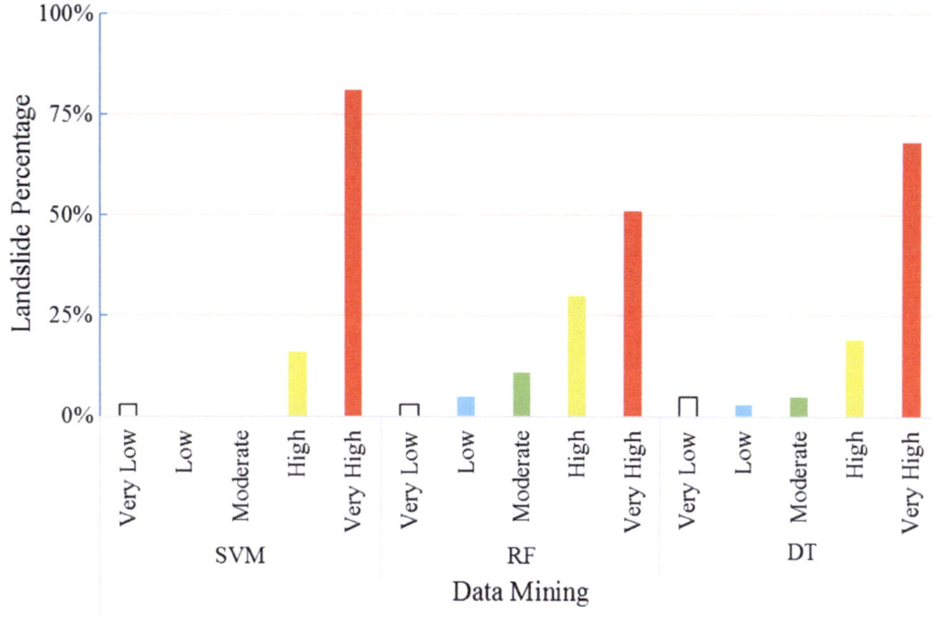

Fig. 11.19 Landslide density graph of the data mining methods

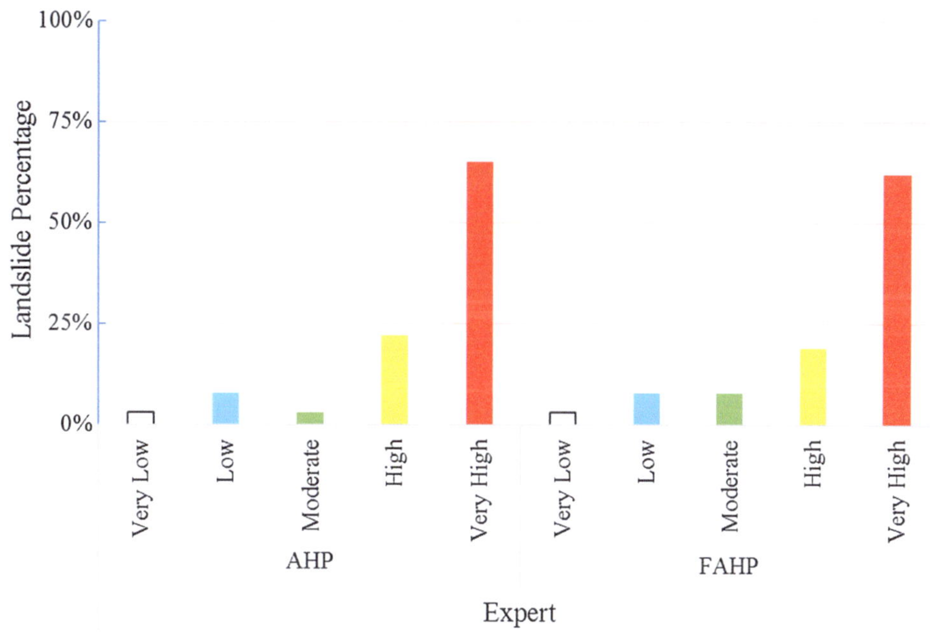

Fig. 11.20 Landslide density graph of the AHP and fuzzy AHP methods

and the SVM models were moderate (0.42–0.64); however, lower correlations (0.11–0.42) were observed when they were compared with RF and DT models.

On the other hand, Spearman correlations showed that the spatial agreements between the multivariate methods and the bivariate techniques ranged from low to moderate. The lowest agreement (−0.06) was found between the DA and FR models, whereas the highest agreement was observed between the PLS and SI methods (0.63). Furthermore, low correlations were found between multivariate and expert-based models, which indicated low spatial agreements and patterns in landslide susceptibility maps produced by these methods. In general, low spatial agreements were found between multivariate and data mining models, except for those of the PLS and SVM models, whose correlation value was 0.96. Furthermore, low to very high spatial agreements were observed between the multivariate and hybrid techniques. The lowest agreement (−0.01) was found between the DA and WoE–RF models, whereas the highest agreement (0.96) belonged to the PLS and FR–SVM models.

Furthermore, the expert-based models had high to very high spatial agreements with the RF and DT data mining approaches. However, they showed lower agreement when compared with the SVM model. In addition, the AHP and fuzzy AHP models showed good spatial agreements with the hybrid models; the highest correlation was found to be

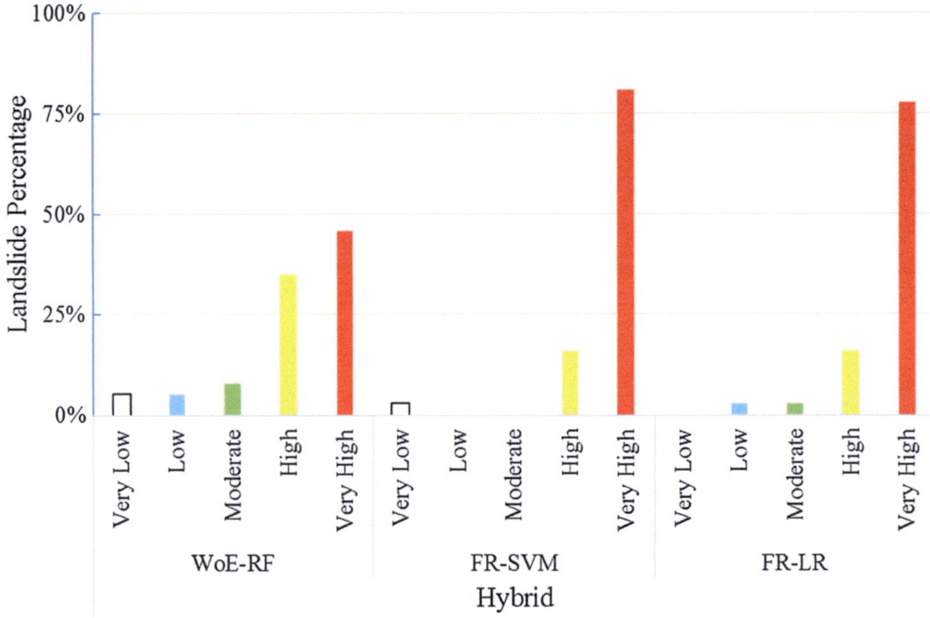

Fig. 11.21 Landslide density graph of the hybrid methods

Table 11.14 Success rates of various models using different subsets of landslide conditioning factors

Category	Method	Percent of factors			
		30%	50%	80%	100%
		Altitude slope curvature	Altitude slope curvature TRI aspect	Altitude slope curvature aspect distance to road distance to stream distance to lineament landuse	Altitude slope curvature aspect distance to road distance to stream distance to lineament landuse vegetation density
Bivariate	WoE	0.85	0.81	0.89	0.68
	FR	0.88	0.79	0.80	0.70
	SI	0.72	0.74	0.83	0.68
Multivariate	LR	0.54	0.63	0.75	0.81
	DA	0.57	0.61	0.64	0.42
	PLS	0.62	0.68	0.71	0.71
Expert-based	AHP	0.63	0.66	0.72	0.70
	FAHP	0.61	0.67	0.72	0.69
Data mining	SVM	0.49	0.61	0.76	0.83
	RF	0.63	0.70	0.8	0.72
	DT	0.54	0.56	0.77	0.76
Hybrid	FR–SVM	0.57	0.58	0.65	0.76
	FR–LR	0.64	0.67	0.71	0.84
	WoE–RF	0.58	0.69	0.77	0.68

Success rates

0.86 for the fuzzy AHP and WoE–RF models. Moreover, the Spearman correlations showed low to very high spatial agreements between the data mining methods, which included the SVM, DT and RF, and hybrid models. The DT and RF models had a correlation of 0.89 and 0.94, respectively, with the WoE–RF model. The SVM and WoE–RF models had a low correlation (0.26), indicating fewer agreements on spatial patterns of landslide susceptibility zones. In contrast, the spatial agreements between the SVM and FR–LR and FR-SVM models were 0.77 and 1.00, respectively. Other details are listed in Table 11.16.

Table 11.15 Prediction rates of various models developed by different landslide factor subsets

Category	Method	Percent of factors			
		30%	50%	80%	100%
		Altitude slope curvature	Altitude slope curvature TRI aspect	Altitude slope curvature aspect distance to road distance to stream distance to lineament landuse	Altitude slope curvature aspect distance to road distance to stream distance to lineament landuse vegetation density
Bivariate	WoE	0.78	0.77	0.82	0.66
	FR	0.71	0.68	0.70	0.69
	SI	0.58	0.67	0.77	0.68
Multivariate	LR	0.57	0.62	0.73	0.77
	DA	0.44	0.55	0.57	0.33
	PLS	0.63	0.70	0.67	0.65
Expert-based	AHP	0.59	0.61	0.70	0.71
	FAHP	0.58	0.66	0.73	0.67
Data mining	SVM	0.46	0.56	0.71	0.77
	RF	0.67	0.69	0.80	0.71
	DT	0.52	0.53	0.73	0.73
Hybrid	FR–SVM	0.52	0.53	0.59	0.75
	FR–LR	0.63	0.67	0.71	0.83
	WoE–RF	0.57	0.70	0.78	0.65

11.7.11 Discussion and Model Comparison

The complex nature of landslide mechanisms and the difficulty of forecasting triggering factors make landslide prediction and risk assessment challenging. Several factors affecting the spatial distribution of landslides include human activities, material properties, and geological and geomorphological conditions. Several attempts were made to improve our understanding of how machine learning-based approaches can predict spatial distribution of landslides in an area; however, no agreements have been made on which algorithm should be used. Therefore, the main objective of the current work was to evaluate the performance and the sensitivity of several commonly used landslide susceptibility models. The models were grouped into five categories, including bivariate and multivariate statistical, expert-based, data mining, and hybrid methods. The models were run on 37 landslide inventories and 10 landslide conditioning factors. The models were evaluated based on a dataset collected over the Cameron Highlands, which is located in the northern part of Peninsular Malaysia and is highly prone to landslides.

Factor analysis showed that the NDVI and vegetation density are highly correlated with the remaining factors. After analyzing the VIF values of these factors, the NDVI factor was removed from further analysis because it had a VIF that exceeded the selected threshold (4.00). On the other hand, results of factor optimization indicated that landuse, altitude, slope, and distance to the road were the most important factors, whereas NDVI and vegetation density were found to be statistically nonsignificant. The logical explanations of these findings are as follows. Human activities and changes in environmental conditions affect the landuse and, as a result, can affect landslide activity. Deforestation, forest logging, road construction, and cultivation on steep slopes are the main activities that affect landslide activity in the focused area. However, in most landslide studies as well as the current study, landuse is used as a static factor extracted from a single date satellite image. In addition, landslides may occur on the road and on the side of slopes affected by the road. Roads constructed beside steep slopes decrease the load on both the topography and the heel of the slope. The changes in topography increase the stress on the back of the slope. In several works, the distance to the road is considered as an unimportant factor (Yalcin 2008). This consideration results from the design of roads determining whether the road will be a contributing factor. A drop-down road section may behave like a wall, a net source, a net sink, or a corridor for water flow (Yalcin 2008). Furthermore, the altitude was found to have a positive effect on landslide proneness (Pachauri and Pant 1992). Higher altitudes are often underlain by resistant lithological units and rainfall.

On the other hand, the effects of vegetation density on landslide occurrence may be classified as hydrological or mechanical. The hydrological factors include the loss of

Table 11.16 Calculated Spearman correlations among developed models

	WoE–RF	SVM	DA	DT	FAHP	FR–LR	FR–SVM	FR	PLS	LR	SI	AHP	RF
WoE	0.39	0.63	−0.01	0.39	0.58	0.62	0.63	0.77	0.63	0.52	0.96	0.62	0.39
WoE–RF	1.00	0.26	−0.01	0.89	0.86	0.43	0.26	0.15	0.20	0.45	0.42	0.82	0.94
SVM	0.26	1.00	−0.29	0.37	0.36	0.77	1.00	0.42	0.96	0.57	0.64	0.41	0.27
DA	−0.01	−0.29	1.00	−0.21	0.03	−0.11	−0.29	0.11	−0.27	0.24	−0.06	0.02	−0.10
DT	0.89	0.37	−0.21	1.00	0.77	0.48	0.37	0.11	0.32	0.52	0.43	0.75	0.93
FAHP	0.86	0.36	0.03	0.77	1.00	0.57	0.36	0.40	0.34	0.49	0.58	0.97	0.85
FR–LR	0.43	0.77	−0.11	0.48	0.57	1.00	0.77	0.47	0.75	0.51	0.62	0.61	0.42
FR–SVM	0.26	1.00	−0.29	0.37	0.36	0.77	1.00	0.42	0.96	0.57	0.64	0.41	0.27
FR	0.15	0.42	0.11	0.11	0.40	0.47	0.42	1.00	0.44	0.34	0.74	0.44	0.15
PLS	0.20	0.96	−0.27	0.32	0.34	0.75	0.96	0.44	1.00	0.57	0.63	0.40	0.22
LR	0.45	0.57	0.24	0.52	0.49	0.51	0.57	0.34	0.57	1.00	0.53	0.50	0.47
SI	0.42	0.64	−0.06	0.43	0.58	0.62	0.64	0.74	0.63	0.53	1.00	0.62	0.42
AHP	0.82	0.41	0.02	0.75	0.97	0.61	0.41	0.44	0.40	0.50	0.62	1.00	0.81
RF	0.94	0.27	−0.10	0.93	0.85	0.42	0.27	0.15	0.22	0.47	0.42	0.81	1.00

precipitation by interception, the removal of soil moisture by evapotranspiration, and the effects of hydraulic conductivity (Ercanoglu et al. 2004). The mechanical factors comprise the reinforcement of soil by roots, surcharge, wind-loading, and surface protection (Ercanoglu et al. 2004). However, the vegetation density was found to be nonsignificant in the current study. This result occurred because the training landslides were mostly located in other landuse classes. Moreover, the study area was mostly covered by dense vegetation where non-landslide samples were generated randomly. Therefore, selecting non-landslide samples should be done meticulously to improve the factor optimization in landslide susceptibility assessments. We suggest that non-landslide samples be selected by distributing spatially balanced points using landuse layers in the random process.

The accuracy of the bivariate models was almost similar both in terms of success and prediction rates. This result occurred because they work under the same concept of statistically comparing the landslide conditioning factors with the landslide inventory map. These methods are sensitive to the reclassification of landslide conditioning factors and the spatial distribution of landslide inventories in the study area. However, little difference was observed between their success and prediction rates, indicating a good generalization ability of the models. Models with good generalization ability are essential to producing susceptibility maps that are appropriate for the landslide inventory map and the field conditions. In addition, bivariate models agreed strongly on the spatial distribution of landslide susceptibility classes. On the other hand, their sensitivity analysis showed that the accuracy of the FR model is less affected by selecting different subsets of landslide factors. The prediction capability of these models was found to be affected by reducing the number of landslide conditioning factors, and they performed best when 80% of the factors were included in the model (Table 11.17). Including less important factors such as vegetation density into the bivariate models reduced their success and prediction rates. Thus, factor optimization is an important step to improving the accuracy of bivariate models.

In contrast, multivariate models were found to have different success and prediction rates, and the best model was the LR model. Increasing the number of landslide factors improved the accuracy of these models; however, including the vegetation density as an additional factor in the model reduced their performance. They had low spatial agreements with many other models, but a very high agreement was found between the PLS and WoE–RF models. On the other hand, expert-based models showed a good performance even with fewer landslide conditioning factors. In addition, a

Table 11.17 Average success and prediction rates of the bivariate, multivariate, expert-based, data mining, and hybrid models with different landslide factor subsets

Accuracy	Category	Percent of factors			
		30%	50%	80%	100%
Success rate	Bivariate	0.82	0.78	0.84	0.69
	Multivariate	0.58	0.64	0.70	0.65
	Expert-based	0.62	0.67	0.72	0.70
	Data mining	0.55	0.62	0.78	0.77
	Hybrid	0.60	0.65	0.71	0.76
Prediction rate	Bivariate	0.69	0.71	0.76	0.68
	Multivariate	0.55	0.62	0.66	0.58
	Expert-based	0.59	0.64	0.72	0.69
	Data mining	0.55	0.59	0.75	0.74
	Hybrid	0.57	0.63	0.69	0.74

small difference was observed between their success and prediction rates in increasing the usability of these models for producing a landslide susceptibility map that is effective for landuse planning in environments with scarce landuse inventories. Moreover, they had a high spatial agreement among each other as well as with the models from the data mining and hybrid approaches.

Among data mining approaches, the SVM model outperformed both the DT and RF models in terms of success and prediction rates. The accuracy of the RF model showed that the SVM and DT methods had a lower generalization ability than the RF model. Data mining models also had a high spatial agreement with other models in different categories. On the other hand, the sensitivity analysis of these models revealed that using a higher number of landslide conditioning factors can improve accuracy even when the vegetation density factor is included. Furthermore, hybrid models, especially the FR–LR model, showed high success and prediction rates. Their sensitivity analysis indicated that increasing the number of landslide factors could improve their success and prediction rates. In general, they showed a high spatial agreement with the other models evaluated in the current study.

While each regression model has its own advantages and disadvantages, our findings demonstrated the different prediction capabilities of the developed models as well as different sensitivities to input data. However, having a guideline in selecting models for rapid landslide susceptibility assessments is important for emergency cases where models cannot be selective based on thorough evaluations. The most important factor for model selection is the validity of the model with the existing inventory map and field conditions. This procedure makes the use of landslide susceptibility maps effective for decision making and landuse planners. Other factors that need to be considered are the prediction rate and sensitivity of the model. Accounting for these factors, the best model among the ones developed in the current study is the FR–LR hybrid model. This model could produce a landslide susceptibility map in more than 75% of the landslide inventories located in the very high susceptible class (Fig. 11.22). In addition, it achieved the best prediction rate of 0.83 when all factors were included, and it achieved a prediction rate of 0.63 when only altitude, slope, and curvature were considered. Moreover, the FR–LR model had a systematic sensitivity to the number of landslide factors, where its prediction rate increased by increasing the number of factors. Several studies have likewise found that the hybrid models can achieve better accuracy than single models. Sangchini et al. (2016) reported that the hybrid bivariate AHP models achieved better accuracy than the LR model. Wang et al. (2016) showed that the LR model could achieve more accurate results than bivariate models and artificial neural networks. Meng et al. (2015) indicated that hybrid models produce subjectivity and more accurate results than single models. Furthermore, Youssef et al. (2015) reported a similar prediction rate of the FR–LR model (0.83), and their comparison indicated the superior effectiveness of the hybrid FR–LR model over single bivariate models. Thus, the results of the present study and most of the published studies suggest paying more attention to hybrid models to compensate for the shortcomings of single models.

11.8 Conclusion

Landslide susceptibility assessment methods have progressed significantly in the last decade; however, the performance and sensitivity of these models are not well understood, thereby making the model selection a challenge.

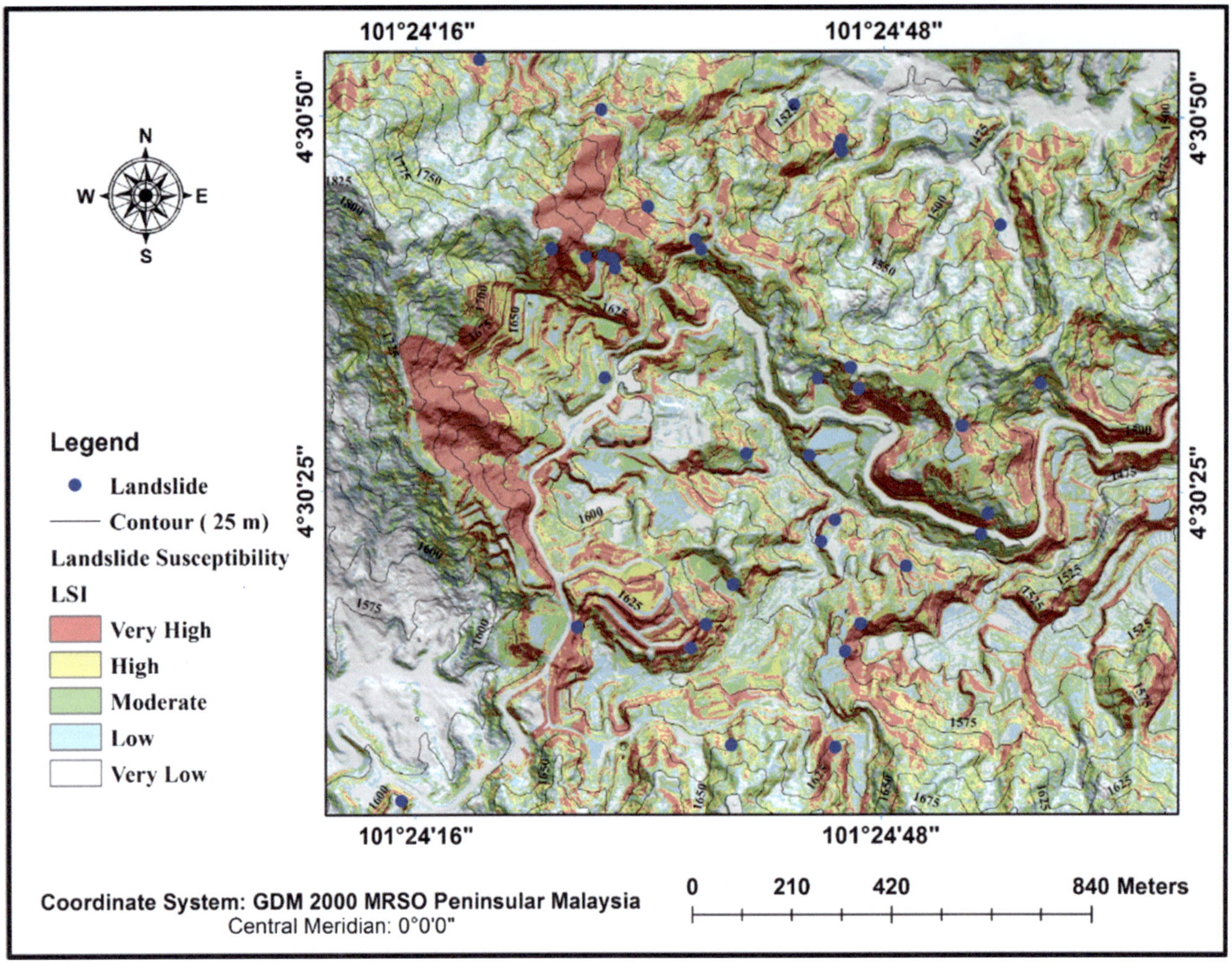

Fig. 11.22 Proposed landslide susceptibility map for landuse planning in Cameron Highlands

This study contributed comprehensive assessments of the accuracy and sensitivity of 14 landslide susceptibility models with the aim of selecting a model that (1) can accurately predict the spatial distribution of landslide inventories, (2) is valid with the existing inventory map, and (3) is not excessively sensitive to input data. The analysis showed that the best model corresponds to the hybrid FR–LR model, which had a prediction accuracy of 0.83 when using 10 factors and 0.63 when using only three factors. Landslide density graphs showed the robustness of this model in predicting landslide inventories in the very high susceptible class. It could predict ~77% of the landslides in the very high susceptible class of the landslide susceptibility map. Other advantages of this model are its ease of use, high spatial agreements with several other models, and its logical representation of the spatial distribution of landslides. This study suggests that future works should pay more attention to hybrid models, study their sensitivity and reusability in more detail, and aim to optimize such models for better spatial predictions of landslides.

References

Abdi, H. (2003). Partial least square regression (PLS regression). *Encyclopedia for Research Methods for the Social Sciences*, 792–795.

Aghdam, I. N., Varzandeh, M. H. M., & Pradhan, B. (2016). Landslide susceptibility mapping using an ensemble statistical index (Wi) and adaptive neuro-fuzzy inference system (ANFIS) model at Alborz Mountains (Iran). *Environmental Earth Sciences, 75*(7), 1–20.

Ahmed, B. (2015a). Landslide susceptibility mapping using multi-criteria evaluation techniques in Chittagong Metropolitan Area, Bangladesh. *Landslides*, *12*(6), 1077–1095.

Ahmed, B. (2015b). Landslide susceptibility modelling applying user-defined weighting and data-driven statistical techniques in Cox's Bazar Municipality, Bangladesh. *Natural Hazards*, *79*(3), 1707–1737.

Althuwaynee, O. F., Pradhan, B., & Lee, S. (2016). A novel integrated model for assessing landslide susceptibility mapping using CHAID and AHP pair-wise comparison. *International Journal of Remote Sensing, 37*(5), 1190–1209.

Althuwaynee, O. F., Pradhan, B., Park, H. J., & Lee, J. H. (2014a). A novel ensemble bivariate statistical evidential belief function with knowledge-based analytical hierarchy process and multivariate statistical logistic regression for landslide susceptibility mapping. *Catena, 114*, 21–36.

Althuwaynee, O. F., Pradhan, B., Park, H. J., & Lee, J. H. (2014b). A novel ensemble decision tree-based CHi-squared Automatic Interaction Detection (CHAID) and multivariate logistic regression models in landslide susceptibility mapping. *Landslides, 11*(6), 1063–1078.

Ayalew, L., Yamagishi, H., Marui, H., & Kanno, T. (2005). Landslides in Sado Island of Japan: Part II. GIS-based susceptibility mapping with comparisons of results from two methods and verifications. *Engineering Geology, 81,* 432–445.

Baeza, C., Lantada, N., & Moya, J. (2010a). Validation and evaluation of two multivariate statistical models for predictive shallow landslide susceptibility mapping of the Eastern Pyrenees (Spain). *Environmental Earth Sciences, 61*(3), 507–523.

Baeza, C., Lantada, N., & Moya, J. (2010b). Influence of sample and terrain unit on landslide susceptibility assessment at La Pobla de Lillet, Eastern Pyrenees, Spain. *Environmental Earth Sciences, 60* (1), 155–167.

Bai, S. B., Lü, G., Wang, J., Zhou, P., & Ding, L. (2010). GIS-based rare events logistic regression for landslide-susceptibility mapping of Lianyungang, China. *Environmental Earth Sciences, 62*(1), 139–149.

Biau, G., Devroye, L., & Lugosi, G. (2008). Consistency of random forests and other averaging classifiers. *Journal of Machine Learning Research, 9*(Sep), 2015–2033.

Bijukchhen, S. M., Kayastha, P., & Dhital, M. R. (2013). A comparative evaluation of heuristic and bivariate statistical modelling for landslide susceptibility mappings in Ghurmi–Dhad Khola, east Nepal. *Arabian Journal of Geosciences, 6*(8), 2727–2743.

Bonham-Carter, G. F. (1994). Geographic information systems for geoscientists-modeling with GIS. *Computer Methods in the Geoscientists, 13,* 398.

Bourenane, H., Bouhadad, Y., Guettouche, M. S., & Braham, M. (2015). GIS-based landslide susceptibility zonation using bivariate statistical and expert approaches in the city of Constantine (Northeast Algeria). *Bulletin of Engineering Geology and the Environment, 74*(2), 337–355.

Breiman, L. (2001). Random forests. *Machine learning, 45*(1), 5–32.

Budimir, M. E. A., Atkinson, P. M., & Lewis, H. G. (2015). A systematic review of landslide probability mapping using logistic regression. *Landslides, 12*(3), 419–436.

Bui, D. T., Lofman, O., Revhaug, I., & Dick, O. (2011). Landslide susceptibility analysis in the Hoa Binh province of Vietnam using statistical index and logistic regression. *Natural hazards, 59*(3), 1413.

Bui, D. T., Pradhan, B., Lofman, O., Revhaug, I., & Dick, O. B. (2012). *Application of support vector machines in landslide susceptibility assessment for the Hoa Binh province (Vietnam) with kernel functions analysis* [Doctoral dissertation, International Environmental Modelling and Software Society (iEMSs)].

Bui, D. T., Tuan, T. A., Klempe, H., Pradhan, B., & Revhaug, I. (2016). Spatial prediction models for shallow landslide hazards. a comparative assessment of the efficacy of support vector machines, artificial neural networks, kernel logistic regression, and logistic model tree. *Landslides, 13*(2), 361–378.

Calle, M. L., & Urrea, V. (2011). Letter to the editor: stability of random forest importance measures. *Briefings in bioinformatics, 12* (1), 86–89.

Calle, M. L., Urrea, V., Malats, N., & Van Steen, K. (2010). mbmdr: an R package for exploring gene–gene interactions associated with binary or quantitative traits. *Bioinformatics, 26*(17), 2198–2199.

Chen, W., Li, W., Chai, H., Hou, E., Li, X., & Ding, X. (2016). GIS-based landslide susceptibility mapping using analytical hierarchy process (AHP) and certainty factor (CF) models for the Baozhong region of Baoji City, China. *Environmental Earth Sciences, 75*(1), 63.

Chung, C. J. F., & Fabbri, A. G. (2003). Validation of spatial prediction models for landslide hazard mapping.*Natural Hazards, 30*(3), 451–472.

Conforti, M., Muto, F., Rago, V., & Critelli, S. (2014a). Landslide inventory map of north-eastern Calabria (South Italy). *Journal of maps, 10*(1), 90–102.

Conforti, M., Pascale, S., Robustelli, G., & Sdao, F. (2014b). Evaluation of prediction capability of the artificial neural networks for mapping landslide susceptibility in the Turbolo River catchment (northern Calabria, Italy). *Catena,113*, 236–250.

Cortes, C., & Vapnik, V. (1995). Support-vector networks. *Machine Learning, 20*(3), 273–297.

Demir, G., Aytekin, M., & Akgun, A. (2015). Landslide susceptibility mapping by frequency ratio and logistic regression methods: an example from Niksar–Resadiye (Tokat, Turkey). *Arabian Journal of Geosciences, 8*(3), 1801–1812.

Devkota, K. C., Regmi, A. D., Pourghasemi, H. R., Yoshida, K., Pradhan, B., Ryu, I. C., et al. (2013). Landslide susceptibility mapping using certainty factor, index of entropy and logistic regression models in GIS and their comparison at Mugling-Narayanghat road section in Nepal Himalaya. *Natural Hazards, 65*(1), 135–165.

Ercanoglu, M., Gokceoglu, C., & Van Asch, T. W. (2004). Landslide susceptibility zoning north of Yenice (NW Turkey) by multivariate statistical techniques. *Natural Hazards, 32*(1), 1–23.

Ercanoglu, M., Kasmer, O., & Temiz, N. (2008). Adaptation and comparison of expert opinion to analytical hierarchy process for landslide susceptibility mapping. *Bulletin of Engineering Geology and the Environment,67*(4), 565–578.

Erener, A., Mutlu, A., & Düzgün, H. S. (2016). A comparative study for landslide susceptibility mapping using GIS-based multi-criteria decision analysis (MCDA), logistic regression (LR) and association rule mining (ARM). *Engineering Geology, 203*, 45–55.

Evans, J. S., & Hudak, A. T. (2007). A multiscale curvature algorithm for classifying discrete return LiDAR in forested environments. *IEEE Transactions on Geoscience and Remote Sensing, 45*(4), 1029–1038.

Galli, M., Ardizzone, F., Cardinali, M., Guzzetti, F., & Reichenbach, P. (2008). Comparing landslide inventory maps. *Geomorphology, 94*(3), 268–289.

García-Rodríguez, M. J., Malpica, J. A., Benito, B., & Díaz, M. (2008). Susceptibility assessment of earthquake-triggered landslides in El Salvador using logistic regression. *Geomorphology, 95*(3), 172–191.

Geladi, P., & Kowalski, B. R. (1986). Partial least-squares regression: a tutorial. *Analytica chimica acta, 185*, 1–17.

Ghosh, S., Carranza, E. J. M., van Westen, C. J., Jetten, V. G., & Bhattacharya, D. N. (2011). Selecting and weighting spatial predictors for empirical modeling of landslide susceptibility in the Darjeeling Himalayas (India).*Geomorphology, 131*(1), 35–56.

Gorsevski, P. V., Gessler, P. E., Boll, J., Elliot, W. J., & Foltz, R. B. (2006a). Spatially and temporally distributed modeling of landslide susceptibility.*Geomorphology, 80*(3), 178–198.

Gorsevski, P. V., Gessler, P. E., Foltz, R. B., & Elliot, W. J. (2006b). Spatial prediction of landslide hazard using logistic regression and ROC analysis.*Transactions in GIS, 10*(3), 395–415.

Gorsevski, P. V., & Jankowski, P. (2010). An optimized solution of multi-criteria evaluation analysis of landslide susceptibility using fuzzy sets and Kalman filter. *Computers & Geosciences, 36*(8), 1005–1020.

Gorsevski, P. V., Jankowski, P., & Gessler, P. E. (2006). Heuristic approach for mapping landslide hazard integrating fuzzy logic with analytic hierarchy process. *Control and Cybernetics, 35*(1), 121.

Hasekioğulları, G. D., & Ercanoglu, M. (2012). A new approach to use AHP in landslide susceptibility mapping: A case study at Yenice (Karabuk, NW Turkey). *Natural Hazards, 63*(2), 1157–1179.

Haugerud, R. A., & Harding, D. J. (2001). Some algorithms for virtual deforestation (VDF) of LIDAR topographic survey data. *International Archives of Photogrammetry Remote Sensing and Spatial Information Sciences, 34*(3/W4), 211–218.

Hong, H., Pradhan, B., Jebur, M. N., Bui, D. T., Xu, C., & Akgun, A. (2016). Spatial prediction of landslide hazard at the Luxi area (China) using support vector machines. *Environmental Earth Sciences, 75*(1), 40.

Hühnerbach, V., & Masson, D. G. (2004). Landslides in the North Atlantic and its adjacent seas: an analysis of their morphology, setting and behaviour. *Marine Geology*, 213(1), 343–362.

Intarawichian, N., & Dasananda, S. (2011). Frequency ratio model based landslide susceptibility mapping in lower Mae Chaem watershed, Northern Thailand. *Environmental Earth Sciences, 64* (8), 2271–2285.

Jebur, M. N., Pradhan, B., & Tehrany, M. S. (2014). Optimization of landslide conditioning factors using very high-resolution airborne laser scanning (LiDAR) data at catchment scale. *Remote Sensing of Environment, 152*, 150–165.

Kavzoglu, T., Sahin, E. K., & Colkesen, I. (2015a). An assessment of multivariate and bivariate approaches in landslide susceptibility mapping: a case study of Duzkoy district. *Natural Hazards, 76*(1), 471–496.

Kavzoglu, T., Sahin, E. K., & Colkesen, I. (2015b). Selecting optimal conditioning factors in shallow translational landslide susceptibility mapping using genetic algorithm. *Engineering Geology, 192*, 101–112.

Kayastha, P., Dhital, M. R., & De Smedt, F. (2013). Application of the analytical hierarchy process (AHP) for landslide susceptibility mapping: A case study from the Tinau watershed, west Nepal. *Computers & Geosciences, 52,* 398–408.

Kelarestaghi, A., & Ahmadi, H. (2009). Landslide susceptibility analysis with a bivariate approach and GIS in Northern Iran. *Arabian Journal of Geosciences, 2*(1), 95–101.

Kilincci, O., & Onal, S. A. (2011). Fuzzy AHP approach for supplier selection in a washing machine company. *Expert systems with Applications, 38*(8), 9656–9664.

Lee, S. (2005). Application of logistic regression model and its validation for landslide susceptibility mapping using GIS and remote sensing data. *International Journal of Remote Sensing, 26* (7), 1477–1491.

Lee, S., & Pradhan, B. (2007). Landslide hazard mapping at Selangor, Malaysia using frequency ratio and logistic regression models. *Landslides, 4*(1), 33–41.

Lee, S., & Talib, J. A. (2005). Probabilistic landslide susceptibility and factor effect analysis. *Environmental Geology, 47*(7), 982–990.

Lee, S., Choi, J., & Min, K. (2002). Landslide susceptibility analysis and verification using the Bayesian probability model. *Environmental Geology, 43*(1–2), 120–131.

Liaw, A., & Wiener, M. (2002). Classification and regression by randomForest. *R news, 2*(3), 18–22.

Lombardo, L., Cama, M., Conoscenti, C., Märker, M., & Rotigliano, E. (2015). Binary logistic regression versus stochastic gradient boosted decision trees in assessing landslide susceptibility for multiple-occurring landslide events: Application to the 2009 storm event in Messina (Sicily, southern Italy). *Natural Hazards, 79*(3), 1621–1648.

MacFarland, T. W., & Yates, J. M. (2016). Spearman's rank-difference coefficient of correlation. *Introduction to nonparametric statistics for the biological sciences using R* (pp. 249–297). Berlin: Springer International Publishing.

Magliulo, P., Di Lisio, A., & Russo, F. (2009). Comparison of GIS-based methodologies for the landslide susceptibility assessment. *Geoinformatica, 13*(3), 253–265.

Mancini, F., Ceppi, C., & Ritrovato, G. (2010). GIS and statistical analysis for landslide susceptibility mapping in the Daunia area, Italy. *Natural Hazards and Earth System Sciences, 10*(9), 1851.

Marjanović, M., Kovačević, M., Bajat, B., & Voženílek, V. (2011). Landslide susceptibility assessment using SVM machine learning algorithm.*Engineering Geology, 123*(3), 225–234.

Mathew, R., Karp, C. M., Beaudoin, B., Vuong, N., Chen, G., Chen, H. Y., & DiPaola, R. S. (2009). Autophagy suppresses tumorigenesis through elimination of p62. *Cell, 137*(6), 1062–1075.

McLachlan, G. (2004). *Discriminant analysis and statistical pattern recognition* (Vol. 544). Hoboken: John Wiley & Sons.

Meng, Q., Miao, F., Zhen, J., Wang, X., Wang, A., Peng, Y., et al. (2015). GIS-based landslide susceptibility mapping with logistic regression, analytical hierarchy process, and combined fuzzy and support vector machine methods: A case study from Wolong Giant Panda Natural Reserve, China. *Bulletin of Engineering Geology and the Environment, 75,* 1–22.

Mohamed Shaluf, I., & Ahmadun, F. L. R. (2006). Disaster types in Malaysia: An overview. *Disaster Prevention and Management: An International Journal, 15*(2), 286–298.

Moosavi, V., & Niazi, Y. (2016). Development of hybrid wavelet packet-statistical models (WP-SM) for landslide susceptibility mapping. *Landslides, 13*(1), 97–114.

Mousavi, S. Z., Kavian, A., Soleimani, K., Mousavi, S. R., & Shirzadi, A. (2011). GIS-based spatial prediction of landslide susceptibility using logistic regression model. *Geomatics, Natural Hazards and Risk, 2*(1), 33–50.

Murillo-García, F. G., Alcántara-Ayala, I., Ardizzone, F., Cardinali, M., Fiourucci, F., & Guzzetti, F. (2015). Satellite stereoscopic pair images of very high resolution: a step forward for the development of landslide inventories.*Landslides, 12*(2), 277–291.

Nandi, A., & Shakoor, A. (2010). A GIS-based landslide susceptibility evaluation using bivariate and multivariate statistical analyses. *Engineering Geology, 110*(1), 11–20.

Neuhäuser, B., & Terhorst, B. (2007). Landslide susceptibility assessment using "weights-of-evidence" applied to a study area at the Jurassic escarpment (SW-Germany). *Geomorphology, 86*(1), 12–24.

Oztekin, B., & Topal, T. (2005). GIS-based detachment susceptibility analyses of a cut slope in limestone, Ankara—Turkey. *Environmental Geology, 49,* 124–132.

Pachauri, A. K., & Pant, M. (1992). Landslide hazard mapping based on geological attributes. *Engineering geology, 32*(1-2), 81–100.

Park, I., & Lee, S. (2014). Spatial prediction of landslide susceptibility using a decision tree approach: a case study of the Pyeongchang area, Korea. *International Journal of Remote Sensing, 35*(16), 6089–6112.

Park, S., Choi, C., Kim, B., & Kim, J. (2013). Landslide susceptibility mapping using frequency ratio, analytic hierarchy process, logistic regression, and artificial neural network methods at the Inje area, Korea. *Environmental Earth Sciences, 68*(5), 1443–1464.

Patriche, C. V., Pirnau, R., Grozavu, A., & Rosca, B. (2016). A comparative analysis of binary logistic regression and analytical hierarchy process for landslide susceptibility assessment in the Dobrov River Basin, Romania. *Pedosphere, 26*(3), 335–350.

Pham, B. T., Bui, D. T., Pham, H. V., Le, H. Q., Prakash, I., & Dholakia, M. B. (2016a). Landslide hazard assessment using random subspace fuzzy rules based classifier ensemble and probability analysis of rainfall data: a case study at Mu Cang Chai District, Yen Bai Province (Viet Nam). *Journal of the Indian Society of Remote Sensing*, 1–11.

Pham, B. T., Pradhan, B., Bui, D. T., Prakash, I., & Dholakia, M. B. (2016b). A comparative study of different machine learning methods for landslide susceptibility assessment: a case study of Uttarakhand area (India). *Environmental Modelling & Software, 84*, 240–250.

Pourghasemi, H. R., Jirandeh, A. G., Pradhan, B., Xu, C., & Gokceoglu, C. (2013a). Landslide susceptibility mapping using support vector machine and GIS at the Golestan Province, Iran. *Journal of Earth System Science, 122*(2), 349–369.

Pourghasemi, H. R., Moradi, H. R., & Aghda, S. F. (2013b). Landslide susceptibility mapping by binary logistic regression, analytical hierarchy process, and statistical index models and assessment of their performances. *Natural Hazards, 69*(1), 749–779.

Pourghasemi, H. R., Pradhan, B., Gokceoglu, C., Mohammadi, M., & Moradi, H. R. (2013c). Application of weights-of-evidence and certainty factor models and their comparison in landslide susceptibility mapping at Haraz watershed, Iran. *Arabian Journal of Geosciences, 6*(7), 2351–2365.

Pourghasemi, H. R., Mohammady, M., & Pradhan, B. (2012a). Landslide susceptibility mapping using index of entropy and conditional probability models in GIS: Safarood Basin, Iran. *Catena, 97*, 71–84.

Pourghasemi, H. R., Pradhan, B., & Gokceoglu, C. (2012b). Application of fuzzy logic and analytical hierarchy process (AHP) to landslide susceptibility mapping at Haraz watershed, Iran. *Natural hazards, 63*(2), 965–996.

Pourghasemi, H. R., Pradhan, B., Gokceoglu, C., & Moezzi, K. D. (2012c). Landslide susceptibility mapping using a spatial multi criteria evaluation model at Haraz Watershed, Iran. In Terrigenous mass movements (pp. 23–49). Springer Berlin Heidelberg.

Pradhan, B. (2010a). Landslide susceptibility mapping of a catchment area using frequency ratio, fuzzy logic and multivariate logistic regression approaches. *Journal of the Indian Society of Remote Sensing, 38*(2), 301–320.

Pradhan, B. (2010b). Remote sensing and GIS-based landslide hazard analysis and cross-validation using multivariate logistic regression model on three test areas in Malaysia. *Advances in space research, 45*(10), 1244–1256.

Pradhan, B. (2011). Manifestation of an advanced fuzzy logic model coupled with Geoinformation techniques to landslide susceptibility mapping and their comparison with logistic regression modelling. *Environmental Ecology Statistics, 18*(3), 471–493.

Pradhan, B. (2013). A comparative study on the predictive ability of the decision tree, support vector machine and neuro-fuzzy models in landslide susceptibility mapping using GIS. *Computers & Geosciences, 51*, 350–365.

Pradhan, A. M. S., & Kim, Y. T. (2014). Relative effect method of landslide susceptibility zonation in weathered granite soil: a case study in Deokjeok-ri Creek, South Korea. *Natural hazards, 72*(2), 1189–1217.

Pradhan, B., & Lee, S. (2010). Delineation of landslide hazard areas on Penang Island, Malaysia, by using frequency ratio, logistic regression, and artificial neural network models. *Environmental Earth Sciences, 60*(5), 1037–1054.

Pradhan, B., Sezer, E. A., Gokceoglu, C., & Buchroithner, M. F. (2010). Landslide susceptibility mapping by neuro-fuzzy approach in a landslide-prone area (Cameron Highlands, Malaysia). *IEEE Transactions on Geoscience and Remote Sensing, 48*(12), 4164–4177.

Raman, R., & Punia, M. (2012). The application of GIS-based bivariate statistical methods for landslide hazards assessment in the upper Tons river valley, Western Himalaya, India. Georisk: Assessment and Management of Risk for Engineered Systems and Geohazards, 6(3), 145–161.

Roodposhti, M. S., Rahimi, S., & Beglou, M. J. (2014). PROMETHEE II and fuzzy AHP: an enhanced GIS-based landslide susceptibility mapping. *Natural hazards, 73*(1), 77–95.

Saaty, T. L. (1977). A scaling method for priorities in hierarchical structures. *Journal of Mathematical Psychology, 15*, 234–281.

Saaty, T. L., & Vargas, G. L. (2001). *Models, methods, concepts, and applications of the analytic hierarchy process*. Boston: Kluwer Academic Publisher.

Sarkar, S., & Kanungo, D. P. (2004). An integrated approach for landslide susceptibility mapping using remote sensing and GIS. Photogrammetric Engineering & Remote Sensing, 70(5), 617-625.

Sangchini, E. K., Emami, S. N., Tahmasebipour, N., Pourghasemi, H. R., Naghibi, S. A., Arami, S. A., et al. (2016). Assessment and comparison of combined bivariate and AHP models with logistic regression for landslide susceptibility mapping in the Chaharmahal-e-Bakhtiari Province, Iran. *Arabian Journal of Geosciences, 9*(3), 1.

Santacana, N., Baeza, B., Corominas, J., De Paz, A., & Marturiá, J. (2003). A GIS-based multivariate statistical analysis for shallow landslide susceptibility mapping in La Pobla de Lillet area (Eastern Pyrenees, Spain). *Natural hazards, 30*(3), 281–295.

Schicker, R., & Moon, V. (2012). Comparison of bivariate and multivariate statistical approaches in landslide susceptibility mapping at a regional scale. *Geomorphology, 161*, 40–57.

Schleier, M., Bi, R., Rohn, J., Ehret, D., & Xiang, W. (2014). Robust landslide susceptibility analysis by combination of frequency ratio, heuristic GIS-methods and ground truth evaluation for a mountainous study area with poor data availability in the Three Gorges Reservoir area, PR China. *Environmental earth sciences, 71*(7), 3007–3023.

Setchi, R., & Anuar, F. M. (2016). Multi-faceted assessment of trademark similarity. *Expert Systems with Applications, 65*, 16–27.

Sezer, E. A., Nefeslioglu, H. A., & Osna, T. (2017). An expert-based landslide susceptibility mapping (LSM) module developed for Netcad Architect Software. *Computers & Geosciences, 98*, 26–37.

Shahabi, H., Ahmad, B. B., & Khezri, S. (2013). Evaluation and comparison of bivariate and multivariate statistical methods for landslide susceptibility mapping (case study: Zab basin). *Arabian Journal of Geosciences, 6*(10), 3885–3907.

Stumpf, A., & Kerle, N. (2011). Object-oriented mapping of landslides using Random Forests. *Remote Sensing of Environment, 115*(10), 2564–2577.

Süzen, M. L., & Doyuran, V. (2004a). A comparison of the GIS based landslide susceptibility assessment methods: multivariate versus bivariate.*Environmental geology, 45*(5), 665–679.

Süzen, M. L., & Doyuran, V. (2004b). Data driven bivariate landslide susceptibility assessment using geographical information systems: a method and application to Asarsuyu catchment, Turkey. *Engineering Geology, 71*(3), 303–321.

Umar, Z., Pradhan, B., Ahmad, A., Jebur, M. N., & Tehrany, M. S. (2014). Earthquake induced landslide susceptibility mapping using an integrated ensemble frequency ratio and logistic regression models in West Sumatera Province, Indonesia. *Catena, 118*, 124–135.

Van Den Eeckhaut, M., Marre, A., & Poesen, J. (2010). Comparison of two landslide susceptibility assessments in the Champagne–Ardenne region (France). *Geomorphology, 115*(1), 141–155.

van Westen, C. J. (1997). *Statistical landslide hazard analysis. ILWIS 2.1 for windows application guide* (pp. 73–84). ITC Publication: Enschede.

Vapnik, V. (1995). *The nature of statistical learning*. New York: Springer.

Verachtert, E., Van Den Eeckhaut, M., Poesen, J., Govers, G., & Deckers, J. (2011). Prediction of spatial patterns of collapsed pipes in loess-derived soils in a temperate humid climate using logistic regression.*Geomorphology,130*(3), 185–196.

Wang, L. J., Guo, M., Sawada, K., Lin, J., & Zhang, J. (2016). A comparative study of landslide susceptibility maps using logistic regression, frequency ratio, decision tree, weights of evidence and artificial neural network. *Geosciences Journal, 20*(1), 117–136.

Wu, X., Niu, R., Ren, F., & Peng, L. (2013). Landslide susceptibility mapping using rough sets and back-propagation neural networks in the Three Gorges, China. *Environmental earth sciences*, *70*(3), 1307–1318.

Wu, X., Ren, F., & Niu, R. (2014). Landslide susceptibility assessment using object mapping units, decision tree, and support vector machine models in the Three Gorges of China. *Environmental earth sciences*, *71*(11), 4725–4738.

Xu, C., Dai, F., Xu, X., & Lee, Y. H. (2012). GIS-based support vector machine modeling of earthquake-triggered landslide susceptibility in the Jianjiang River watershed, China. *Geomorphology, 145*, 70–80.

Yalcin, A. (2008). GIS-based landslide susceptibility mapping using analytical hierarchy process and bivariate statistics in Ardesen (Turkey): Comparisons of results and confirmations. *Catena, 72*(1), 1–12.

Yalcin, A., Reis, S., Aydinoglu, A. C., & Yomralioglu, T. (2011). A GIS-based comparative study of frequency ratio, analytical hierarchy process, bivariate statistics and logistics regression methods for landslide susceptibility mapping in Trabzon, NE Turkey. *Catena, 85*(3), 274–287.

Yilmaz, I. (2009). Landslide susceptibility mapping using frequency ratio, logistic regression, artificial neural networks and their comparison: a case study from Kat landslides (Tokat—Turkey). *Computers & Geosciences*, *35*(6), 1125–1138.

Youssef, A.M., Pradhan, B., Maerz, N.H., (2014) Debris flow impact assessment caused by 14 April 2012 rainfall along the Al-Hada Highway, Kingdom of Saudi Arabia using high resolution satellite imagery. Arabian Journal of Geosciences, *7*(7), 2591–2601. http://dx.doi.org/10.1007/s12517-013-0935-0

Youssef, A. M., Al-Kathery, M., & Pradhan, B. (2015). Landslide susceptibility mapping at Al-Hasher area, Jizan (Saudi Arabia) using GIS-based frequency ratio and index of entropy models. *Geosciences Journal*, *19*(1), 113–134.

Youssef, A. M., Pourghasemi, H. R., Pourtaghi, Z. S., & Al-Katheeri, M. M. (2016). Landslide susceptibility mapping using random forest, boosted regression tree, classification and regression tree, and general linear models and comparison of their performance at Wadi Tayyah Basin, Asir Region, Saudi Arabia. *Landslides*, *13*(5), 839–856.

Youssef, A. M., Pourghasemi, H. R., El-Haddad, B. A., & Dhahry, B. K. (2016). Landslide susceptibility maps using different probabilistic and bivariate statistical models and comparison of their performance at Wadi Itwad Basin, Asir Region, Saudi Arabia. *Bulletin of Engineering Geology and the Environment*, *75*(1), 63–87.

Youssef, A. M., Pradhan, B., Jebur, M. N., & El-Harbi, H. M. (2015). Landslide susceptibility mapping using ensemble bivariate and multivariate statistical models in Fayfa area, Saudi Arabia. *Environmental Earth Sciences, 73*(7), 3745–3761.

Zhou, J. W., Cui, P., & Hao, M. H. (2016). Comprehensive analyses of the initiation and entrainment processes of the 2000 Yigong catastrophic landslide in Tibet, China. *Landslides*, *13*(1), 39–54.

Zhu, A. X., Wang, R., Qiao, J., Qin, C. Z., Chen, Y., Liu, J., & Zhu, T. (2014). An expert knowledge-based approach to landslide susceptibility mapping using GIS and fuzzy logic. *Geomorphology*, 214, 128–138.

Part V

Landslide Vulnerability and Risk Modelling

Slope Vulnerability and Risk Assessment Using High-Resolution Airborne Laser Scanning Data

12

Biswajeet Pradhan and Norbazlan Mohd Yusof

12.1 Introduction

Natural hazards, such as landslides, earthquakes, and floods, result in considerable losses of lives and properties (Calil et al. 2015; Tierney et al. 2001). Natural disasters are in fact the main cause of irrecoverable damages worldwide (Varoonchotikul 2003). In different regions worldwide, landslides pose a threat to infrastructure especially in the transportation sector. Many studies have agreed that the transportation sector is probably the most affected sector during and after landslides because it loses billions of dollars annually as a result of such natural disaster (Jaiswal et al. 2010, 2011).

Consequently, numerous tools have been developed over the years to address the effects of natural hazards. Geographic information systems (GIS) are known as a powerful set of tools that facilitate the gathering, storing, retrieval, analysis, and exhibition of spatial information (Opolot 2013). Remote sensing (RS) is commonly described as a technique for obtaining information about the earth's surface without physical interaction (Joshi et al. 2004). The efficiency of RS and GIS has revolutionized natural hazard management by fulfilling all the requirements at each stage (Dou et al. 2015). Currently, light detection and ranging (LiDAR) data are used by researchers in most landslide hazard and risk mapping studies (Kobal et al. 2015). The accessibility of very high-resolution digital elevation models (DEMs), obtained by high-resolution LiDAR sensors, makes it possible for researchers to recognize and map slope failures. Moreover, various conditioning factors can be extracted from high-resolution DEMs for susceptibility and hazard mapping. The use of accurate and optimized conditioning factors has a direct effect on the final maps. A significant advantage of LiDAR lies in its capability to penetrate vegetation areas and acquire valuable information on topographic conditions.

Regarding the disadvantages of LiDAR, the cost of LiDAR data acquisition and data processing is extremely high. Another disadvantage of LiDAR data is its poor morphological quality; sharp linear features such as building boundaries cannot be captured. The consequence is that it is hard to get high-accuracy building models only from LiDAR data if its point density is not high. Moreover, algorithms for segmentation of LiDAR data are application-dependent. Meaning that there is no fixed algorithm that can be pre-set to apply to all kinds of LiDAR data under any circumstances. Large non-ground objects are common in LiDAR sets so that a large window size is needed in order to correctly eliminate non-ground objects. Simultaneously, a larger window will create a smoother result. Usually, this will eliminate fine objects and change the topography dramatically. This will make it very complex and difficult, even impossible, to recover the topography (Ma 2004).

Landslide management typically comprises several of stages: prediction, prevention, and damage assessment (Dai et al. 2002). The spatial prediction of landslides is important for urban planning because its results assist the public and emergency departments in making early preparations for the occurrence of landslides. Landslide-prone areas can be identified through a susceptibility analysis; in this way, early warning and emergency response can be performed to facilitate early preparations and decrease disaster effects (Kia et al. 2012). Landslide susceptibility mapping, using digital spatial information, has been studied extensively since the early 1980s (Van Westen et al. 2008). Most susceptibility mapping techniques in landslide studies are based on the statement, "The past is key to the future," which indicates the potential of using previous landslide records in predicting future landslides with consideration of certain conditioning factors.

A hazard map can be used as a main guide for infrastructure development and urban management at a specific time and location (Crozier and Glade 2006). Spatial and

B. Pradhan (✉) · N.M. Yusof
Department of Civil Engineering, University Putra Malaysia, Serdang, Malaysia
e-mail: biswajeet24@gmail.com

B. Pradhan (ed.), *Laser Scanning Applications in Landslide Assessment*,
DOI 10.1007/978-3-319-55342-9_12

temporal probabilities are the two main components of landslide hazard assessment. A landslide hazard is commonly defined as "the probability of occurrence within a specified period of time and within a given area of a potentially damaging phenomenon" (Althuwaynee et al. 2014a, b). This definition usually answers the question on location, "where a landslide will occur," and that on time, "when and how frequent a landslide will occur." In most cases, researchers consider susceptibility maps as the only component of hazard maps because of data scarcity (Chau et al. 2004). Some researchers directly multiply the three components of hazard probability (i.e., spatial, temporal, and magnitude) because they presume an independent relationship among these variables.

Vulnerability is one of the most significant concepts that have been broadly studied in hazard risk management. Landslide vulnerability has been defined in several ways by several studies (Fuchs et al. 2007; Muthukumar 2013; Uzielli et al. 2008), but it is commonly used to describe physical, social, economic, and environmental circumstances that can make particular inhabitants highly susceptible to the effects of landslide hazards. Risk is defined as the uncertain product of a hazard and the probable extent of damage. Moreover, risk refers to the expected damage to lives, belongings, and economy caused by a specific hazard in a particular region and reference period (Opolot 2013). Risk includes the concepts of the danger to lives, challenge in evacuating residents and their properties during a landslide, possible damages to buildings, social interruption, loss of production, and damage to public property (Dang et al. 2011).

12.2 Landslide Hazard, Vulnerability, and Risk Assessment: A Preview

Landslide susceptibility mapping methods are generally categorized as qualitative or quantitative in nature (Guzzetti et al. 1999). Statistical methods are preferred over qualitative methods because the latter involves expert knowledge, which may cause uncertainty in the outcomes (Ayalew and Yamagishi 2005; Feizizadeh and Blaschke 2013). Thus, quantitative methods have become very popular in recent years (Yilmaz 2009). Various statistical methods are available, and those used in landslide susceptibility mapping mainly include logistic regression (LR) (Demir et al. 2015; Yalcin et al. 2011), frequency ratio (Pradhan and Lee 2010), artificial neural network (ANN) (Park et al. 2013; Conforti 2014), decision tree (Saito et al. 2009), and support vector machine (Moonjun 2007).

Current research utilizes the popular LR method for landslide susceptibility mapping because its efficiency and proficiency in landslide mapping have been proven by numerous studies (Althuwaynee et al. 2014a, b; Devkota et al. 2013; Pradhan 2010, 2011). A landslide probability index represents the predicted probabilities of a landslide for each pixel in the presence of a given set of conditioning factors. A susceptibility map can be prepared through the popular method adopted in the literature by dividing the probability map into a specific number of classes (Ayalew and Yamagishi 2005). Probabilistic analysis considers the statistical relationships between historical landslide locations and conditioning factors. The probability index ranges from 0 to 0.99, and producing a susceptibility map requires the division of a probability map into different categories (Ohlmacher and Davis 2003). Different categorization methods have been examined in the GIS environment; examples of such methods include standard deviation, natural break, equal interval, and quantile.

Several methods for landslide hazard mapping are available; they include heuristic algorithms (judgmental method) (Bulut et al. 2000), empirical probability (Crovelli 2000), magnitude–frequency relations (Stoffel 2010), rational methods (geomechanical approach) (Alonso and Pinyol 2010), and indirect approaches (Corominas et al. 2014). The heuristic approach is used when a judgment is based on the opinion obtained from a group of experts or specialists (Akgun 2012). In rational methods, the probability varies according to the results of the stability analyses and mathematical modeling (Romeo et al. 2013). Among all the listed approaches, the GIS platform proves to be useful tools in obtaining an output that shows the potential zonation of landslides (Haneberg 2004). Empirical probabilistic models are employed on the basis of past records of an event using the annual probability of occurrence and hydrological analyses. The magnitude of other contributing factors, such as earthquakes and rainfall, is integrated in analyses using indirect approaches. In this process, the critical values of rainfall or earthquake events are determined to define the return period for landslides (Umar et al. 2014). In this correlation analysis, the areas with high probability of landslide occurrence are recognized without any requirement for information on the size or type of failure. The quantitative prediction of landslide hazards is aimed at detecting the relationship between magnitude and frequency (Guzzetti et al. 2008).

The most common approach for hazard analysis starts with the creation of a landslide susceptibility map using conditioning factors such as distance from drainages, roads, and land cover, which are not expected to change within the specific prediction periods. Other factors, such as distance from drainages and roads and land cover, are impervious to significant changes after a few years. The next stage involves the preparation of the temporal triggering factors. The susceptibility map is then multiplied with the prepared temporal rainfall or earthquake. Althuwaynee et al. (2014a, b) performed a hazard analysis using the susceptibility map developed for Kuala Lumpur, Malaysia, on the basis of the evidential belief function model (Althuwaynee et al. 2012). They multiplied the rainfall factor with the derived susceptibility map and acquired a hazard map. Subsequently, they classified the hazard index into four geometrical interval hazardous groups: high, medium, low, and very low hazard regions. The drawback of their research is related to the lack of information about classification of the slope material failure and updated land use/land cover maps. Moreover, in Malaysia, landslides are recorded only when they are noticeable in terms of infrastructure damage or human fatality. Hence, many small landslides that covered an approximate area of 100 m^2 were not included. Guzzetti et al. (1999) summarized several landslide hazard evaluation studies.

Vulnerability is defined as the total damage caused by a particular natural hazard to a specific object or element at risk at a specific scale. Vulnerability is measured in a continuous scale ranging from 0 to 1, where 0 represents no loss and 1 represents total destruction (Galli and Guzzetti 2007). A full understanding of the interaction between elements at risk and slope failure is an essential step in assessing vulnerability. Four parameters are considered in vulnerability assessment: (1) sliding volume and its velocity, (2) distance of run out, (3) risk on people (their distance from the landslide; the construction elements where they are, such as roads or buildings; and their locations in roads or buildings), and (4) risk on structures and buildings (this involves the design and the distance from the failure) (Dai et al. 2002).

Vulnerability changes into a different form when used to deal with lives and properties. For example, a certain house can have a high vulnerability to both a slow-moving landslide and a rapid landslide. As for the people actually residing in this house, they may exhibit extremely high vulnerability to a rapid landslide and moderate vulnerability to a slow-moving landslide (Sattenpalli and Parkash 2013). Historic records serve as basis for the assessment of vulnerability. The vulnerability of a house varies according to the location of a given landslide. For instance, the vulnerability level may be high for a house at the base of a steep slope. By contrast, a house at a landslide deposition area exhibits low vulnerability. Expert judgment can be used to assess vulnerability levels and the probable depth of debris for specific facility types (Winter et al. 2014). Another method involves the use of statistical data comprising historical landslide records as basis for the analysis of the vulnerability of certain properties and people (Dai et al. 2002). For example, some cities (e.g., Hong Kong) maintain a record of landslides, including their size, type, and effects. Vulnerability can also be assessed by using the matrix algorithm proposed by Leone et al. (1996). This algorithm is applicable to many scenarios because of its flexibility. Moreover, this algorithm features minimal subjectivity relative to other methods. Two factors are involved in the use of a matrix algorithm: (1) the types and characteristics of landslides involved in the analysis and (2) the design of the buildings and other elements at risk, such as nature, age, and type (Dai et al. 2002). Muthukumar (2013) analyzed the vulnerability of landslide zonation mapping for Nilgiri Mountains, Western Ghats, South India, using the landslide per unit area method. As far as the geosystem parameters were concerned, five vital parameters (lithology, lineament, geomorphology, slope, and land use/land cover) were used for such vulnerability mapping.

The term element at risk covers all parameters, such as the environment, properties, and population that are affected by the occurrence of any phenomena such as a landslide (Corominas et al. 2014). The amount of risk varies according to the elements in the area affected by a hazard such as a landslide (Kappes 2012). Data on elements at risk should be collected for certain basic spatial units, which may be grid cells, administrative units, or homogeneous units with similar characteristics in terms of type and density of elements at risk. Other features, such as transportation lines and particular locations (e.g., dam site), can also be considered in relevant analyses (Corominas et al. 2014).

Landslide risk is defined as the product of landslide hazards and landslide vulnerability (Van Westen et al. 2008). A high landslide hazard and vulnerability value equate to a high landslide risk. Landslide risk assessment can be carried out on different scales using different methods for susceptibility and hazard assessment. It can also be qualitative or quantitative in nature. Various risks, such as distributed landslide risk, site-specific landslide risk, and global landslide risk, should be addressed in landslide risk assessment (Corominas et al. 2014).

Landslide risk assessment aims to create a thematic map that shows the level of risk in terms of lives and economic loss in a specific area either in a quantitative or in a qualitative manner (Cui et al. 2009). The spatial information for risk level can be obtained by multiplying the landslide probability index, LULC, properties, and population within the affected zone. These calculations may be easily obtained in the GIS environment. Risk assessment involves generating the risk level of a potential hazard (i.e., landslide) for the element at risk (people and property) in a particular area (Kanungo et al. 2008). This method can be used to determine whether the risk degree is dangerous on the basis of cost–benefit analyses. Quantitative risk assessment features a high degree of refinement for specific sites. Global risk assessment is aimed at calculating the total risk for lives and properties per year. Its results are helpful in managing the resources in landslide-prone areas in terms of allocation and policymaking.

Selecting the appropriate assessment method completely depends on the availability of spatial data. A specific data layer is needed to continue each analysis, and the absence of one parameter renders an analysis method useless. Risk analysis is a valid technique if and only if it fulfills a series of appropriate criteria. Bell and Glade (2004), Calvo and Savi (2009), Huang and Lyamin (2013), and other researchers have implemented several analysis techniques in landslide risk assessment. For instance, Pradhan and Lee (2009) performed a study on landslide hazard and risk assessment using an ANN algorithm in a GIS environment in Penang Island. Van Westen et al. (2008) stated that the risk and hazard assessment for any phenomena is limited by the availability of spatial information. Akgun (2012) studied Izmir City (west Turkey) and performed a landslide risk assessment. First, the LR method was applied to derive a susceptibility map. Earthquake and rainfall factors were used as triggering factors to derive a hazard map. Afterward, the settlement areas were derived from remote sensing data and considered as the elements at risk in order to derive a vulnerability map. Finally, a risk thematic map for Izmir City was derived by combining the hazard and vulnerability layers.

A comprehensive understanding of the landslide hazard phenomenon and its probable effects on society are vital for defining landslide control polices, risk mitigation projects, and other landslide management strategies. Numerous landslides have occurred in Malaysia in recent years. Most of these landslides threatened the lives and properties of the country's residents. Moreover, landslides often occurred near highways or in cut slopes in mountainous areas. The current study aims to perform landslide susceptibility, hazard, vulnerability, and risk modeling in the Gua Tempurung area, Malaysia, particularly along a selected stretch of the North–South Expressway. Given the significant reputation of the LR method in hazard studies, it is used to perform landslide susceptibility mapping in this study. The produced landslide susceptibility map will be used as basis for hazard, vulnerability, and risk assessment in the future.

12.3 Study Area

The Gua Tempurung Corridor of the North–South Expressway was chosen for the landslide hazard, vulnerability, and risk analysis because of the importance of this highway and the frequent occurrences of landslides in this region (Fig. 12.1). Many major cities are linked by this expressway in Western Peninsular Malaysia. Therefore, this highway represents the backbone of the west coast of Peninsular Malaysia. With a total length of 772 km, this expressway passes through seven states, namely Kedah, Perak, Negeri Sembilan, Johor, Penang, Selangor, and Malacca. This area is approximately located at 4°23′34.387″ N to 4°26′39.153″ N latitude and 101°11′18.607″ E to 100°13′33.626″ E longitude. The study area experiences frequent mass movements that cause erosion and landslides. The average annual rainfall ranges from 1170 to 1950 mm per year. Wet seasons in this area start from February to May and from September to December. Maximum rainfall occurs from March to May and from November to December. Undulating plateaus and hilly terrains are the pronounced geomorphologic characteristics of the Gua Tempurung area. Gunung Tempurung is the largest surface limestone mass (Kinta Limestone) of the Kinta Valley. Gua Tempurung is located in the middle part of Gunung Tempurung (Muhammad 2010). Sungai Tempurung which is a tributary of the Sungai Kampar flows westward from the granitic hill of the Main Range and carves its way through Gua Tempurung to the east. Limestone hills are characteristically steep-sided, with subvertical to overhanging cliffs. The base of limestone hills also often exhibits deep horizontal notches or undercuts due to dissolution by streams, groundwater, or swamp water. Though they might appear massive when viewed from the side, most limestone hills are actually "riddled" with numerous caverns and cave systems. The geology of the area mostly consists of Devonian granite and Quaternary. The limestone bedrock in this area rises above the alluvial plains to form limestone hills with steep to vertical slopes (mogote or tower karst). In recent years, many landslides have occurred along PLUS highways, roads, and streams, the sides of which have thus suffered from scouring.

12.4 Data

12.4.1 Landslide Inventory

Landslide inventory maps are the basis and first requirement of most landslide susceptibility mapping methods (Pradhan et al. 2014). Inventory maps are crucial for assessing the correlation between landslide occurrences and conditioning factors. Several field measurements were conducted in the study area to create an inventory map. The data were captured from the PLUS maintenance division, which holds field maintenance data. All the landslides were similar and they were shallow type. Moreover, they have occurred for the first time. A total of 17 landslides were identified in the Gua Tempurung area, and their related data were subsequently divided into two data sets for training and testing. Following the literature, 60% of the data on landslide occurrences were employed for modeling; the rest of the data were utilized for validation (Fig. 12.1). Training slope failure locations were used to create a dependent layer. The produced layer consisted of two values, namely 0 and 1,

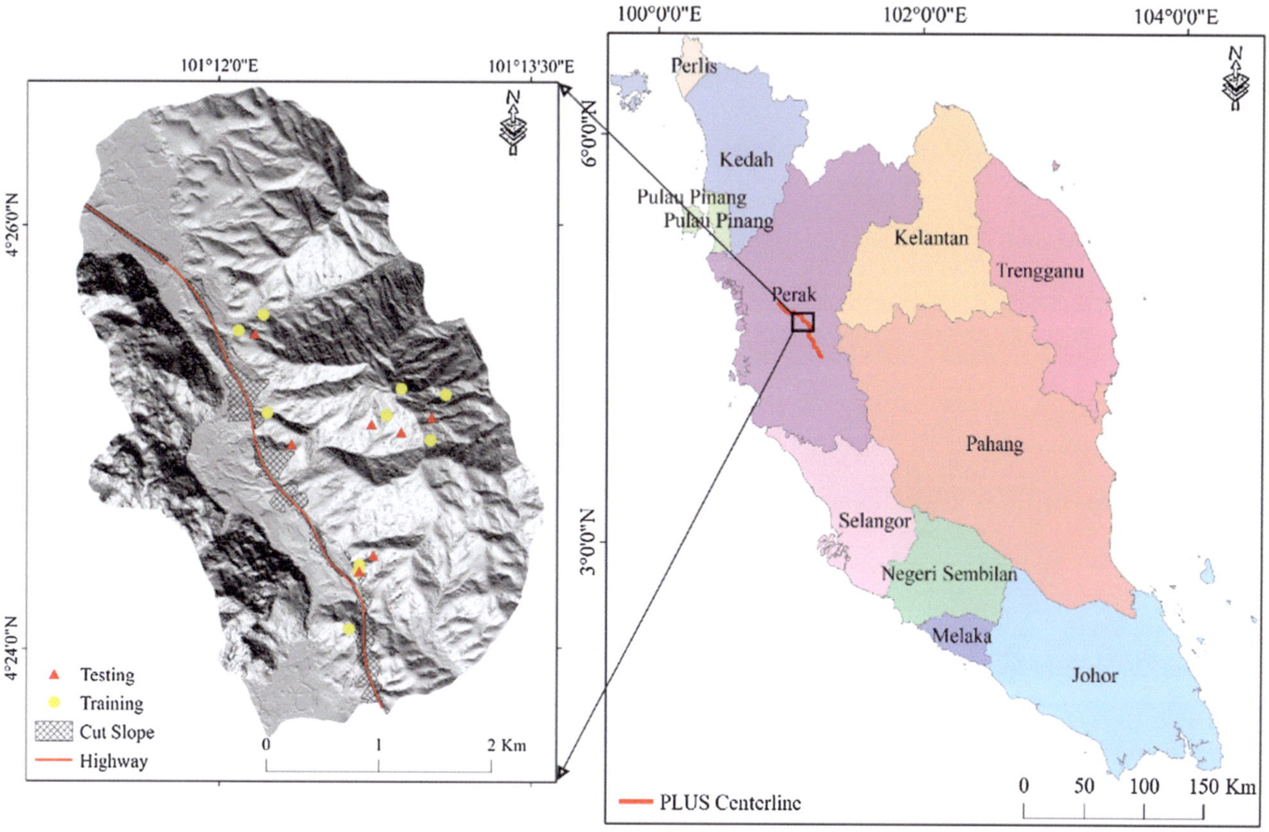

Fig. 12.1 Landslide locations map with a hill-shaded map of Gua Tempurung, Malaysia

where 1 denotes the presence of a landslide and 0 indicates the absence of landslide. The remaining slope failure locations were utilized to test the outcomes. Both layers were created in ArcGIS and then rasterized.

12.4.2 Landslide Conditioning Factors

The LiDAR data used in this study were collected on March 12, 2013, by Riegl LMS Q5600 and Camera Hassleblad 39Mp. The device has a spatial resolution of 13 cm, laser scanning angle of 60°, and camera angle −45°. In addition, the posting density of the LiDAR data was 3–4 pts/m^2.

Susceptibility maps are defined by qualitatively and quantitatively studying the conditioning factors in affected areas (Jebur et al. 2014). Choosing the appropriate data set that comprises conditioning factors is a challenging task; no defined precept can be adopted when choosing the number of conditioning factors sufficient for a specific susceptibility analysis (Wang 2013). These factors are essentially picked up on the basis of literature and expert knowledge. In the current study, conditioning factors were chosen with knowledge derived from the literature. The sole use high-resolution LiDAR data were proven to be sufficient (Jebur et al. 2014). Therefore, only airborne laser scanning data were used as the main data source. The LiDAR data were received as point cloud (LAS file) which was subsequently converted to LAS dataset in ArcGIS software. LAS dataset to raster tool was used to create DEM, and triangulation method was used for interpolation (Kovač and Žalik 2010). The conditioning factor data set included altitude, slope, aspect, curvature, SPI, TWI, TRI, and river factors (Fig. 12.2). All the factors were resampled to a 1 m grid (based on the point density of LiDAR data), and the grid of the Gua Tempurung region was built with 4181 columns and 5665 rows (15,263,455 pixels; 15.26 km^2).

The altitude, slope, aspect, curvature, SPI, TWI, and TRI maps were derived from a DEM, as shown in Fig. 12.2a–g. At an altitude of 0–1339.24 m, the elevation layer was established. Slope is an influential conditioning factor in landslide occurrence. This factor directly affects slope failure occurrence and is typically considered in landslide susceptibility analysis (Alimohammadlou et al. 2013). As a slope becomes steep, the vertical component of gravity rises (Tournadour et al. 2015). The slope in the study area ranges from 0° to 89.38° (Fig. 12.2b). This layer was also used in the analysis as a continuous layer, where each cell represents the actual calculated slope. Slope aspect is an important conditioning factor (Budimir et al. 2015). Aspect can effect on the extent of rainfall and sunlight that influence the

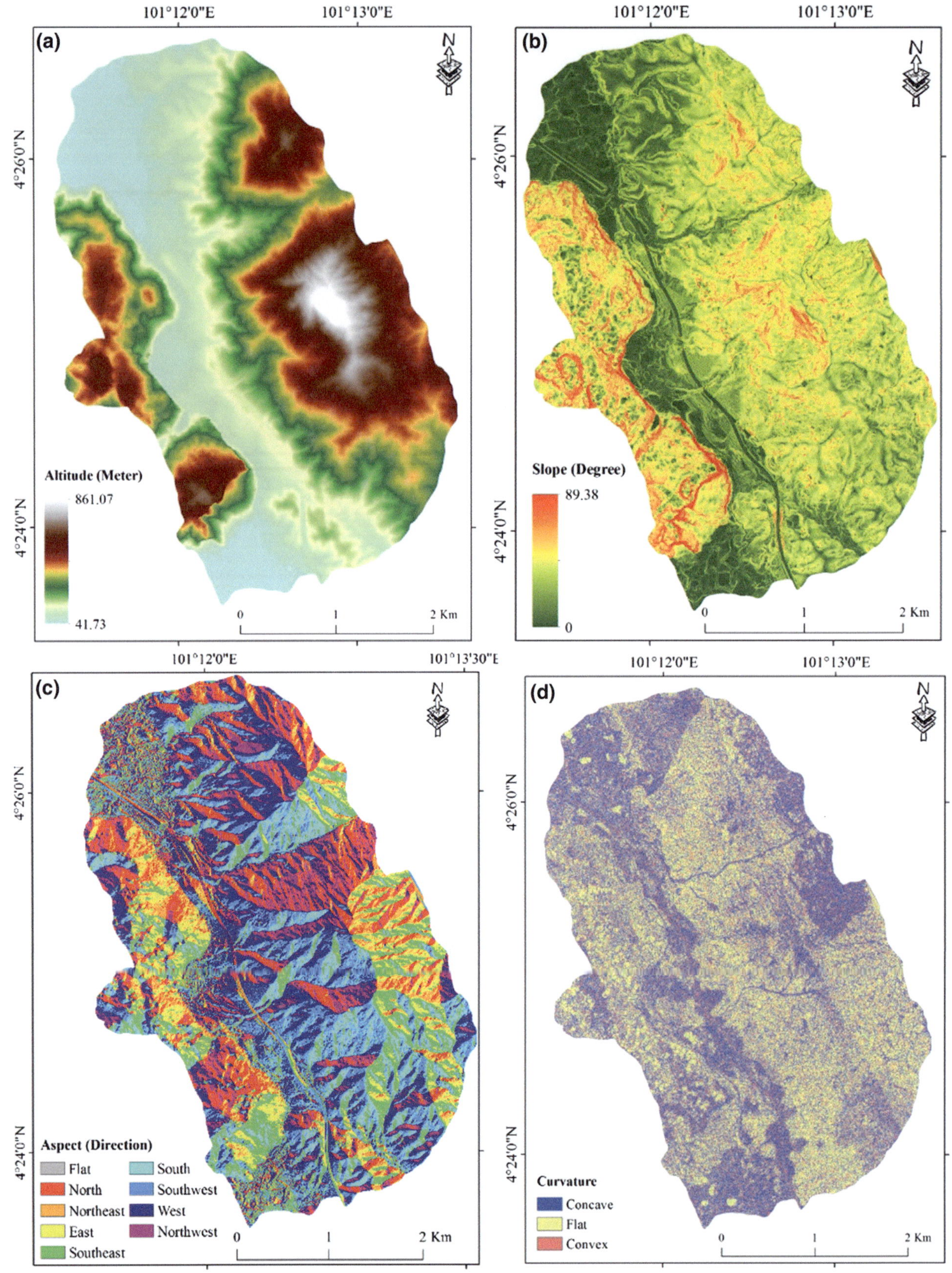

Fig. 12.2 Input conditioning factors: **a** altitude, **b** slope, **c** aspect, **d** curvature, **e** SPI, **f** TWI, **g** TRI, and **h** distance from river

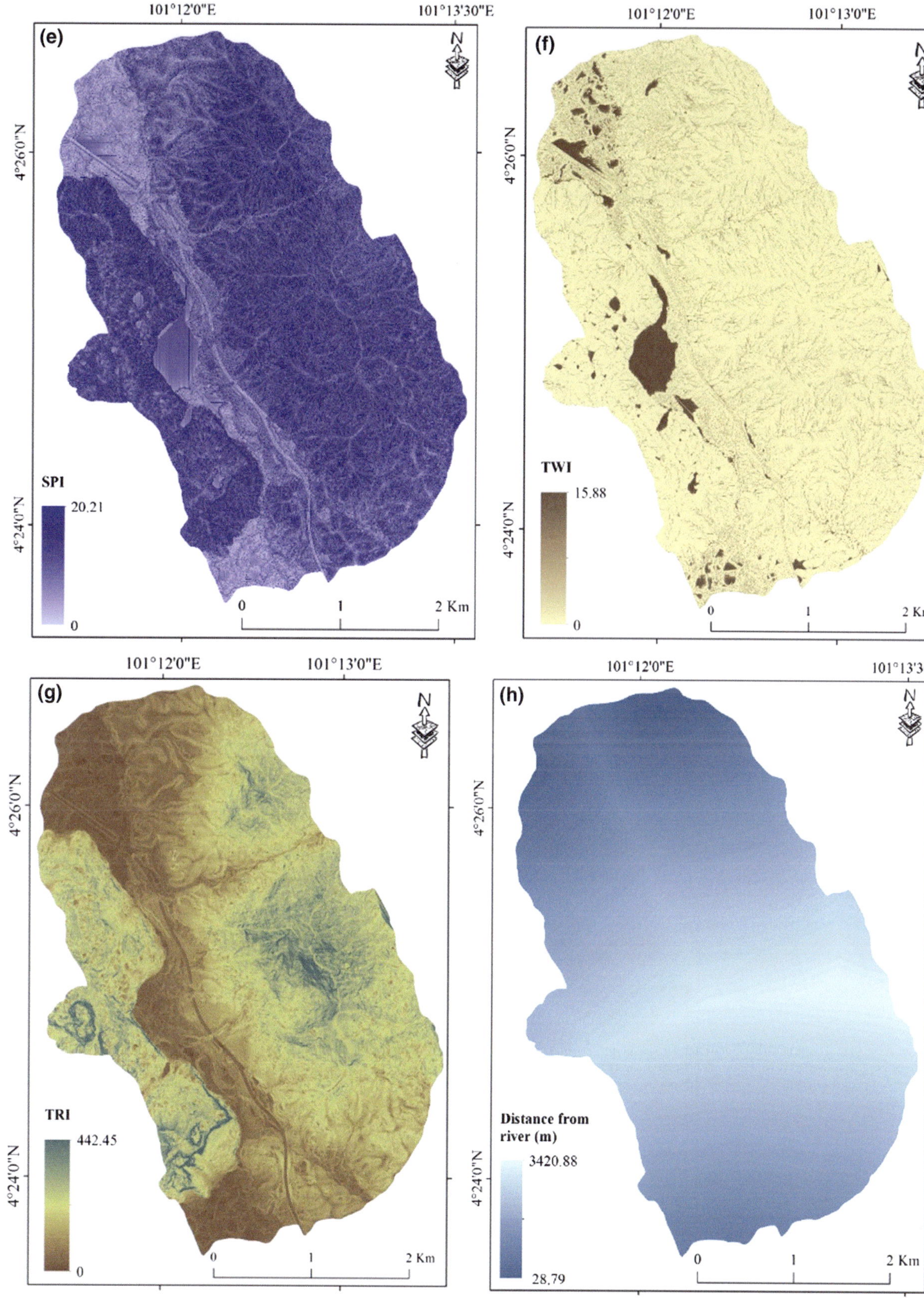

Fig. 12.2 (continued)

occurrence of slope failure. Moreover, it influences weathering and thus indirectly affects the sheer power of object mass. Although the relationship between landslide occurrence and slope has been proven, no exact rule exists with regard to the effect of this factor on slope failure (Pedrazzini et al. 2015). The aspect map utilized to recognize the association between aspect and slope failure occurrence is displayed in Fig. 12.2c. Nine classes were developed for the aspect map (flat, north, northeast, east, southeast, south, southwest, west, northwest, and north). The effect of curvature on slope failure reflects the convergence or divergence of water during downhill movement (Dou et al. 2014). Thus, this factor is another conditioning factor involved in landslide occurrence. In this study, curvature was derived from a DEM using spatial tool in ArcGIS and subsequently categorized into three classes: concave, convex, and flat.

Hydrological factors, such as SPI and TWI, were calculated using Eqs. (12.1) and (12.2). Some researchers consider these two factors as secondary topographic characteristics in landslide susceptibility mapping (Dragićević et al. 2015). Additional information on SPI and TWI can be found in the work of Yusof et al. (2015). SPI represents the power of water flow in terms of erosion (Yesilnacar and Topal 2005). A topographic wetness index (TWI) has been used to know the effect of topography on the location and size of saturated source areas of run-off generation (Yilmaz 2009). The water-related factors SPI and TWI were calculated using the following equations:

$$\mathrm{TWI} = \ln(A_s/\tan\beta) \qquad (12.1)$$

$$\mathrm{SPI} = A_s \tan\beta \qquad (12.2)$$

where A_s is the specific catchment area ($m^2\ m^{-1}$) and β (radian) is the slope gradient (in degrees). Another influential factor is TRI which is generally used in mass movement studies. In the current work, this factor is calculated using Eq. (12.3):

$$\mathrm{TRI} = \sqrt{\max^2 - \min^2} \qquad (12.3)$$

where max and min are the highest and minimum values of the cells in the nine rectangular neighborhoods of altitude, respectively. For the distance from river factor, only the undercutting of the side slopes of rivers might cause slope failure initiation.

12.5 Methodology

The methodology applied in this study is shown in Fig. 12.3. Since this research is done through a collaborative project funded by PLUS BERHAD, they have given the full permission as the owner and concessionary of the North–South Expressways to collect the required data at the Gua Tempurung Corridor of the North–South Expressway. As can be seen in Fig. 12.3, landslide inventory is divided into two classes of training and testing. Training dataset was used for performing the LR method, and testing dataset was kept to be used in validation stage. Conditioning factors and training dataset were used as an input for LR analysis. The correlation among the landslide occurrence and each condition factor was assessed. Landslide probability index was produced, and susceptibility map was created by classifying the probability index using

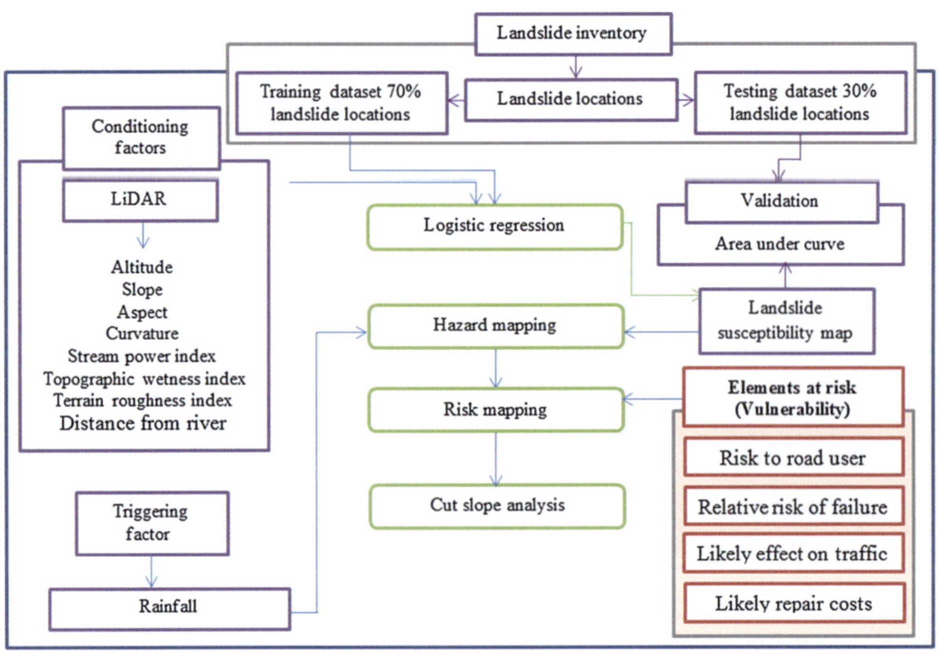

Fig. 12.3 Method applied in the current research

quantile method. In the next stage, hazard analysis was done and rainfall factor was used as triggering factor. Elements at risk factors were used to perform vulnerability analysis. Finally, risk map was produced using hazard and vulnerability maps. Finally, the cut slopes along the highway were classified based on their risk to the future landslide occurrence. Each stage will be described in more detail in relevant subsection.

12.5.1 Susceptibility Analysis

LR was applied to calculate the probability of landslide occurrence in the Gua Tempurung area using Eqs. (12.4) and (12.5) and the conditioning factors. The first requirement of this algorithm is a dependent layer (landslide inventory) that consists of two values, namely 1 and 0, which indicate the presence and absence of a landslide, respectively. Each conditioning factor (Fig. 12.2) was converted from raster into ASCII format. The SPSS software was utilized to implement a multivariate statistical analysis. The regression coefficients were subsequently calculated, as shown in Table 12.1. When the LR coefficient is high, its probable influence on landslide occurrence is large. Using the measured LR coefficients, the landslide probability index was calculated with Eq. (12.4):

$$P = 1/(1 + e^{-z}) \tag{12.4}$$

where p denotes the landslide probability ranging from 0 to 1 in an S-shaped curve and z denotes the lean combination, which can be calculated using Eq. (12.5):

$$Z = b^0 + b_1x_1 + b_2x_2 + b_3x_3 + b_nx_n \tag{12.5}$$

where b^0 is the intercept of the model which derives by LR as a constant value. b_i (i = 0, 1, 2, …, n) represents the coefficients of the LR model which also measured by LR calculations, and x_i (i = 0, 1, 2, …, n) denotes the conditioning factors such as slope, altitude, and rainfall (Demir et al. 2015).

Table 12.1 The coefficient values obtained from LR model

Landslide conditioning parameter	Logistic coefficient of Gua Tempurung
Altitude	0.255213
Slope	1.339465
Aspect	0.035935
Curvature	0.007980
SPI	1.904285
TWI	1.676523
TRI	5.143918
Distance from river	0.012420
Constant	56.172860

12.5.2 Validation

An efficiency validation method should be employed to verify the used algorithm or validate its prediction capabilities (Xu et al. 2012). In this study, the resulting landslide probability map was assessed by comparing it with the landslide inventory data using the area under the curve (AUC) method (Jebur et al. 2014). The reliability of the results can be assessed using AUC because this method defines prediction and success rates (Kritikos and Davies 2014). The success rate in this study was calculated using the same landslide inventory data used to train the model; these data made up 60% of the total available data. However, these training data cannot be used to evaluate the prediction efficiency of the model. Hence, the remaining 40% of the data was used to measure the prediction rate. Additional information about AUC and its equations can be found in (Tehrany et al. 2014). AUC shows the percentage of testing points that fall within the highest probability range. The measured probability index was sorted in descending order to compute the relative ranks for each prediction pattern. Consequently, the cell values were partitioned into 100 classes and were set on the vertical axis (y), along with accumulated 1% intervals on the horizontal axis (x). The existence of landslide locations (training and testing) in each interval was assessed, and the resultant success and prediction rates were measured.

12.5.3 Hazard Analysis

In hazard analysis, two factors should be considered: landslide susceptibility map and landslide triggering factor (Althuwaynee et al. 2014a, b). Triggering factors such as rainfall (Glade et al. 2000), earthquakes (Xu et al. 2014; Yuan ct al. 2015), and snowfall (Moreiras 2015) can trigger a landslide. The landslide hazard in Malaysia is mainly triggered by heavy rainfall (Althuwaynee et al. 2014a, b; Lee et al. 2014). Figure 12.4 shows the rainfall factor considered as a triggering factor in the current research. Many extreme events, such as flooding and overflowing, are caused by the heavy precipitation in the Gua Tempurung area. Thus, the annual long-term average precipitation values for the year 2014 were analyzed in this work. Various rain gauge stations (15 in total) were considered in the rainfall to derive the density rainfall map.

Equation (12.6) Xu et al. 2014) was calculated to create the final hazard map:

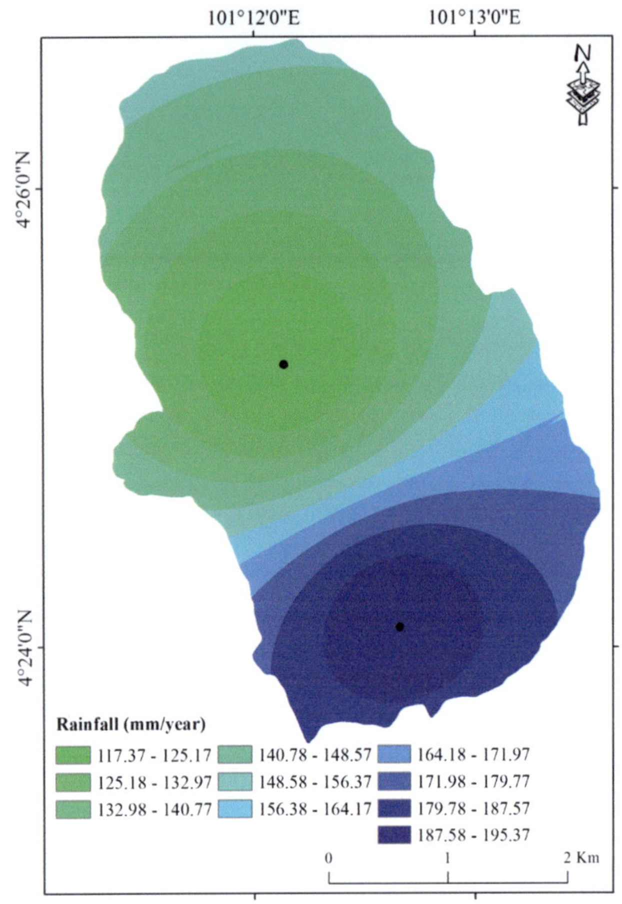

Fig. 12.4 Annual average precipitation map for the Gua Tempurung area

$$H = P_S \times P_T \tag{12.6}$$

where H represents the hazard probability, P_S represents the probability acquired from the LR analysis, and P_T represents the rainfall density layer. Before overlaying the layers, they must be reformatted to a unique dimension scale because of their different ranges. The same scales for the layers were achieved with Eq. (12.7).

$$X_{ij} = X_j - X_{ij}/X_{\max-j} - X_{\min-j} \tag{12.7}$$

where X_{ij} is the standardized score for the ith alternative and jth attribute, X_{ij} is the raw score, and $X_{\max-j}$ and $X_{\min-j}$ are the maximum and minimum scores for the jth attribute, respectively. In the new format, 0 and 1 represent the minimum and maximum values for each layer. Consequently, the resulting layer from the overlaying process must also be scaled from 0 to 1, where 0 represents very low hazard and 1 represents very high hazard.

The hazard map was then reclassified into five classes, namely very low, low, moderate, high, and very high. Each cut slope was analyzed separately. The maximum area analysis was carried out and slopes categorized according to the maximum area class (Sarkar et al. 2008).

12.5.4 Vulnerability Assessment

Vulnerability to a certain phenomenon should be assessed in performing risk analysis (Papathoma Köhle et al. 2015). Vulnerability is equivalent to the complete loss of assets or the total destruction of elements at risk (Murillo-García et al. 2015). Given the lack of data, vulnerability was simplified in this study to facilitate the computation. Landslide vulnerability (VL) is defined mathematically as follows (He et al. 2014):

$$V_L = P[D_L \geq 0|L], \quad (0 \leq D_L \leq 1) \tag{12.8}$$

where D_L is the element at risk for a given phenomenon (landslide, L) that is expected to or will definitely be damaged and V_L is the degree of loss given to a landslide for a particular element or the percentage of damage to such element. The probability of vulnerability can be defined as a scale ranging from 0 to 5, where 5 represents complete damage and 0 represents no damage (Pan et al. 2014).

Economic and heuristic approaches are used to represent the factors that trigger a landslide (Shaharom et al. 2014). In a heuristic (qualitative and descriptive) scale analysis, the expected damage of a particular element at risk is assessed. In this work, a vulnerability analysis for the Gua Tempurung area was carried out using the following criteria for each cut slope:

(a) **Risk to Road User**

Possibility of inflicting injury or damage to property and road users

(b) **Relative Risk of Failure**

Possibility of an existing failure that can grow and affect other parts of the slope or its stability

(c) **Likely Effect on Traffic**

Possibility of failure encroaching onto the expressway; existence of an alternative route to bypass that particular location

(d) **Likely Repair Costs**

Relates to the possible volume of earthwork, sufficient space for access/construction, and complexity of earthwork. Each criterion carries its own weightage against the level of

<Criteria in Consideration>

5 =
4 =
3 =
2 =
1 =

↑ In increasing level of severity, with level 5 being most severe

Fig. 12.5 Level of each criterion in landslide vulnerability analysis

severity and other vulnerability factors, as shown in Fig. 12.5.

12.5.5 Risk Analysis

In this research, risk analysis was conducted to calculate the expected amount of loss caused by landslides in the Gua Tempurung area and the effect of each cut slope. Risk is determined with the formula for hazard and vulnerability, as shown in Eq. (12.9), where R is the expected risk, H is the estimated hazard, and V is the assessed vulnerability (Althuwaynee et al. 2014a, b).

$$R = H \times V \tag{12.9}$$

Quantitative (probabilistic) and qualitative (heuristic) approaches are two popular methods for estimating landslide risk (Althuwaynee et al. 2014a, b). The quantitative approach is used to estimate the probability of expected damage or loss of life (Jakob et al. 2012). In this analysis, an index for the type of landslide and the related consequences are required (Kavzoglu et al. 2014). In the present study, a qualitative approach was used by combining the hazard and vulnerability results for each cut slope using Eq. (12.9). To ensure effective visual interpretation, the output risk map was categorized into five classes (very low, low, moderate, high, and very high) using an equal interval algorithm that is designed for such kind of application in an equal range of values given to each class.

12.6 Results and Discussion

The SPSS software was used to perform LR in order to evaluate the correlation between the conditioning factors with landslide occurrence. All the parameters used showed an acceptable significant probability (sig) value of below 0.05. The use of the sig factor indicates the significance of a parameter in landslide occurrence (Das et al. 2010). When the sig value is below 0.05, the conditioning factor has a significant mathematical effect on slope failure. Equation (12.10) was calculated to obtain the landslide probability index as follows:

$$\begin{aligned} Z = {} & (0.035935 \times \text{“aspect”}) - (1.676523 \times \text{“TWI”}) \\ & - (1.339465 \times \text{“slope”}) + (1.904285 \times \text{“SPI”}) \\ & + (5.143918 \times \text{“TRI”}) - (0.007980 \times \text{“curvature”}) \\ & - (0.255213 \times \text{“DEM”}) + (0.012420 \times \text{“river”}) \\ & - 56.172860 \end{aligned} \tag{12.10}$$

where −56.172860 is the intercept of the model. This value was derived from LR analysis representing as a constant value. Numbers multiplied by conditioning factors represent the coefficients of the LR model. The positive LR coefficient shows the positive correlation between the factor and flood occurrence. However, negative LR coefficient weights delineate that the conditioning factor has negative relationship with landslide occurrence. For instance, the acquired LR coefficient for TWI is −1.676523, which shows that as TWI increases, the probability of landslide occurrence will be decreased.

The probability map was then resampled using Eq. (12.4). The correlation between the conditioning factors and the landslide can be observed in the acquired results. The landslide probability index was acquired and categorized using a quantile classification method. Each pixel in the probability map showed the predicted value of existing landslides that is correlated with the used conditioning factor.

Probability index has the range from 0 to 0.99, and to produce a susceptibility map, the probability map needs to be divided into different categories (Regmi et al. 2014). Different classification algorithms can be found in the literature, and each algorithm is appropriate for a particular application. The most popular ones include natural breaks (Kritikos and Davies 2014), equal interval (Nandi and Shakoor 2010), and quantile (Jebur et al. 2015; Tehrany et al. 2013). Usually, the method for classification is selected by the objective of the research and the nature of the data (Zare et al. 2013). For instance, natural break is suitable for the case that there is a sudden change in the data values. Standard deviation is mostly used for categorizing the population density that produces the fixed number of classes not based on the users' desire. In most of the studies, the best results are achieved through the quantile method (Papathoma-Köhle et al. 2015). The quantile algorithm was chosen for this study because it is widely used in probability classification. The mathematical process of this algorithm includes dividing the chosen classes, where each class has an equal number of pixels (Martha et al. 2013; Xu et al. 2012; Yusof et al. 2015). The outcome of this method is more suitable than that of other methods when comparing the produced susceptibility map and the landslide distribution. The susceptibility map was created and divided into five classes, as shown in Fig. 12.6.

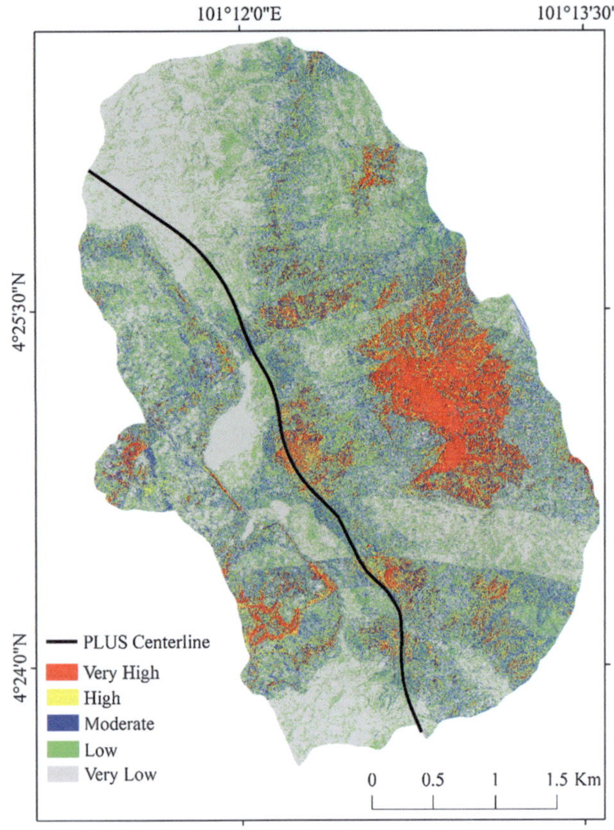

Fig. 12.6 Landslide susceptibility map derived from LR

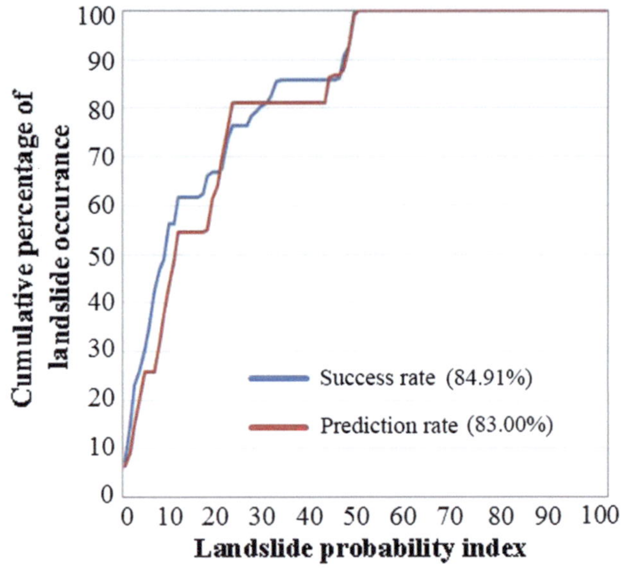

Fig. 12.7 AUC **a** success rate and **b** prediction rate

As can be seen in Fig. 12.6, the eastern part of the study area is highly susceptible to landslides because the characteristics of the area are prone to such hazard. Some parts near the highway are also highly susceptible, especially in the southern and center parts. Planners can use this information and avoid more landslide occurrences in those locations. AUC was utilized to calculate the success and prediction rates of the algorithm used in this study (Fig. 12.7). The generated success and prediction rates for LR were 84.91 and 83.00%, respectively. LR achieved an acceptable accuracy according to the acquired results.

After producing the landslide susceptibility map using topographic and geomorphologic factors, the same year rainfall (2014) was used to generate the hazard map. In Malaysia, the only recorded slope failures are the visible ones in terms of infrastructure damage or human casualty. Therefore, Eq. (12.6) was applied to generate the hazard map as shown in Fig. 12.8. Although the eastern part of the study area is clearly susceptible to landslides, the hazard was moderate. This finding is attributed to the fact that the rainfall in that year was less than the average. By contrast, the southern part featured high susceptibility and hazard values. The hazard was very low in the middle of the highway because of the low rainfall average in that area in 2014 relative to the other parts of the study area. The significant difference between hazard and susceptibility can be attributed to the amount of rainfall in 2014.

A vulnerability map was derived according to the following criteria: risk to road user, relative risk of failure, likely effect on traffic, and likely repair costs. These criteria were applied only on the cut slopes of the highway because this study aims to calculate the risk of these slopes. As shown in Fig. 12.5, each slope was assigned with a number (1–5) associated with the condition of the slope under particular criteria. Subsequently, the outcome was resampled using Eq. (12.6) to normalize the vulnerability with a range between 0 and 1. The vulnerability map was generated as shown in Fig. 12.9. Most of the cut slopes have a vulnerability of more than 0.5, with the exception of the seven slopes with low degree of vulnerability. The highest vulnerability concentrated in the center of the study area. The northern part showed moderately high vulnerability slopes.

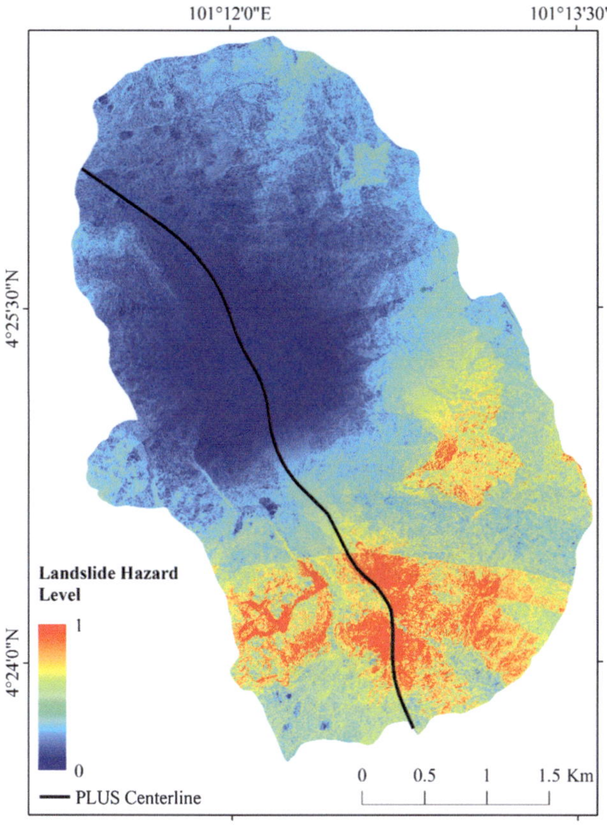

Fig. 12.8 Hazard map for the Gua Tempurung area

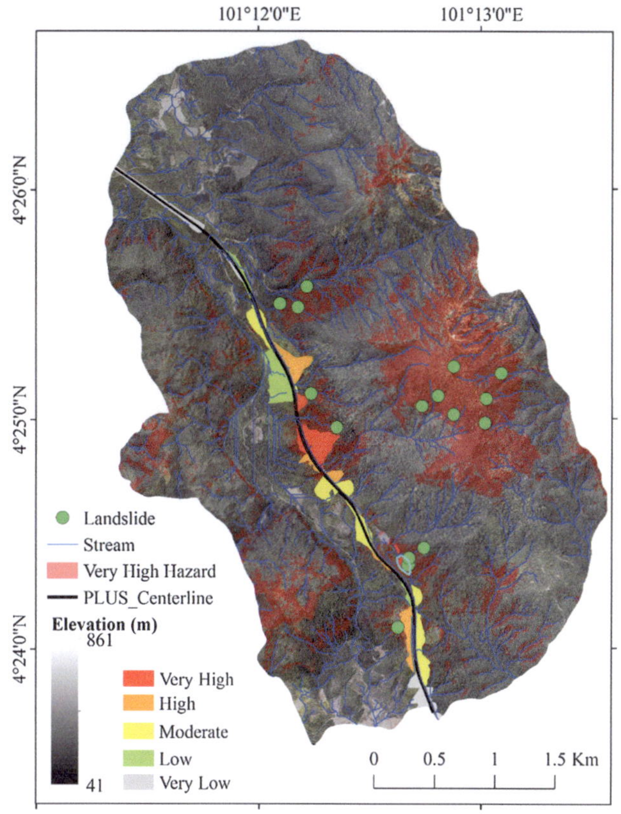

Fig. 12.10 Generated risk map for Gua Tempurung for each cut slope

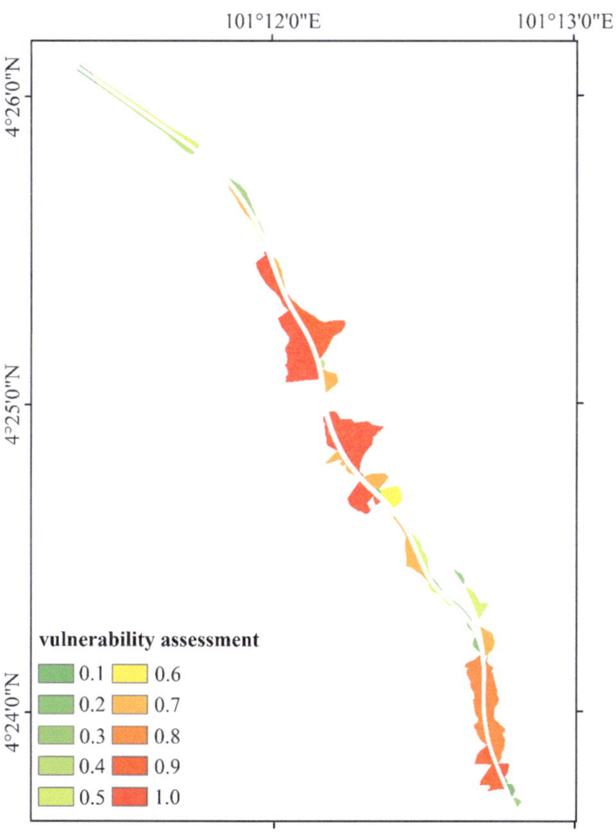

Fig. 12.9 Vulnerability map associated with each cut slope

Using Eq. (12.10), the hazard map for each slope was generated and categorized into five groups using an equal interval algorithm, as shown in Fig. 12.10. Some of the slopes that showed high vulnerability were categorized as moderate risk because of the conditions of the resulting hazard, as shown in the southern part of the study area. Meanwhile, some cut slopes were estimated as high risk although they reflected moderate vulnerability, as shown in the middle part of the area. Finally, some cut slopes were categorized as very high risk because the hazard and vulnerability of these slopes were very high.

12.7 Conclusion

This study sought to provide a comprehensive analysis of landslide vulnerability and risk mapping. The specific objective was to detect the most risky slope cuts along the PLUS highway, specifically in the Gua Tempurung area. Once cut slopes with a high degree of risk are recognized, safety and maintenance actions can be implemented to increase the security of the expressway. The selected study area is highly susceptible to landslides, as evidenced by the large number of landslides reported for this region each year. In this work, the LR method was utilized to map landslide-susceptible areas. The conditioning factor data set consisted of altitude, slope, aspect, curvature, SPI, TWI, TRI, and distance from river. A DEM was generated using airborne LiDAR images measuring 1 m × 1 m pixels and was then used as the basis in deriving an altitude map. All the conditioning factors featured a pixel size of 1 m × 1 m. LR was applied, and the landslide-prone regions were recognized. The probability index was classified into the following five classes with the quantile method: very high, high, moderate, low, and very low susceptibility. Hazardous areas were then detected by multiplying the probability map with the rainfall factor. A risk map was produced by combining the landslide hazard and vulnerability maps. On the basis of the defined criteria (elements at risk), five vulnerability levels were assigned to each cut slope. To implement risk analysis, each cut slope must belong to one class of vulnerability and one class of hazard. The maximum area was considered in the case where one cut slope contained various hazard weights. That is, the cut slope belonged to the hazard class that covered most of its area. Finally, AUC was used to assess the susceptibility map. The success and prediction rates were 84.91 and 83.00%, respectively. On the basis of the acquired accuracies, we can conclude that the LR method is an efficient statistical method that can be used in hazard studies because of its simple structure and robust performance.

The advantages of this research, focusing on a specific area and particular cut slopes, can lead to more efficient outcomes than that produced when the focus is on a vast area. Highway authorities can use the proposed method to avoid future damages to highways and save the lives of road users and residents. Current highway studies focus more on vulnerability than on an accurate risk assessment. This study can improve accuracy when data of the temporal rainfall for a specific landslide is available. In this analysis, the vulnerability of the infrastructure around the highway was not considered. Hence, detailed information can be used in future studies to achieve a highly comprehensive assessment. Temporal risk assessment can also be conducted when temporal hazard maps are produced. The produced maps may be helpful for planners, decision makers, and local governments in landslide management and planning in the study area. Moreover, the produced risk map can be used for the future maintenance of the PLUS highway.

References

Akgun, A. (2012). A comparison of landslide susceptibility maps produced by logistic regression, multi-criteria decision, and likelihood ratio methods: A case study at İzmir, Turkey. *Landslides, 9*(1), 93–106.

Alimohammadlou, Y., Najafi, A., & Yalcin, A. (2013). Landslide process and impacts: A proposed classification method. *Catena, 104,* 219–232.

Alonso, E. E., & Pinyol, N. M. (2010). Criteria for rapid sliding I. A review of Vaiont case. *Engineering Geology, 114*(3), 198–210.

Althuwaynee, O. F., Pradhan, B., & Ahmad, N. (2014a). Estimation of rainfall threshold and its use in landslide hazard mapping of Kuala Lumpur metropolitan and surrounding areas. *Landslides, 12,* 1–15.

Althuwaynee, O. F., Pradhan, B., & Lee, S. (2012). Application of an evidential belief function model in landslide susceptibility mapping. *Computers & Geosciences, 44,* 120–135.

Althuwaynee, O. F., Pradhan, B., Park, H. J., & Lee, J. H. (2014b). A novel ensemble bivariate statistical evidential belief function with knowledge-based analytical hierarchy process and multivariate statistical logistic regression for landslide susceptibility mapping. *Catena, 114,* 21–36.

Ayalew, L., & Yamagishi, H. (2005). The application of GIS-based logistic regression for landslide susceptibility mapping in the Kakuda-Yahiko Mountains, Central Japan. *Geomorphology, 65*(1), 15–31.

Bell, R., & Glade, T. (2004). Quantitative risk analysis for landslides? Examples from Bíldudalur, NW-Iceland. *Natural Hazard Earth System, 4*(1), 117–131.

Budimir, M., Atkinson, P., & Lewis, H. (2015). A systematic review of landslide probability mapping using logistic regression. *Landslides, 12*(3), 1–18.

Bulut, F., Boynukalin, S., Tarhan, F., & Ataoglu, E. (2000). Reliability of landslide isopleth maps. *Bulletin of Engineering Geology and the Environment, 58*(2), 95–98.

Calil, J., Beck, M. W., Gleason, M., Merrifield, M., Klausmeyer, K., & Newkirk, S. (2015). Aligning natural resource conservation and flood hazard mitigation in California. *PLoS ONE, 10*(7), e0132651. doi:10.1371/journal.pone.0132651

Calvo, B., & Savi, F. (2009). A real-world application of Monte Carlo procedure for debris flow risk assessment. *Computers & Geosciences, 35*(5), 967–977.

Chau, K., Sze, Y., Fung, M., Wong, W., Fong, E., & Chan, L. (2004). Landslide hazard analysis for Hong Kong using landslide inventory and GIS. *Computers & Geosciences, 30*(4), 429–443.

Conforti, M., Pascale, S., Robustelli, G., & Sdao, F. (2014). Evaluation of prediction capability of the artificial neural networks for mapping landslide susceptibility in the Turbolo River catchment (northern Calabria, Italy). *Catena, 113,* 236–250.

Corominas, J., Van Westen, C., Frattini, P., Cascini, L., Malet, J. P., Fotopoulou, S., et al. (2014). Recommendations for the quantitative analysis of landslide risk. *Bulletin of Engineering Geology and the Environment, 73*(2), 209–263.

Crovelli, R. A. (2000). *Probability models for estimation of number and costs of landslides.* Reston, VA: US Geological Survey.

Crozier, M. J., & Glade, T. (2006). Landslide hazard and risk: Issues, concepts and approach. *Landslide Hazard and Risk (Wiley, West Sussex), 14,* 1–40.

Cui, P., Zhu, Y. Y., Han, Y. S., Chen, X. Q., & Zhuang, J. Q. (2009). The 12 May Wenchuan earthquake-induced landslide lakes: Distribution and preliminary risk evaluation. *Landslides, 6*(3), 209–223.

Dai, F., Lee, C., & Ngai, Y. Y. (2002). Landslide risk assessment and management: An overview. *Engineering Geology, 64*(1), 65–87.

Dang, N. M., Babel, M. S., & Luong, H. T. (2011). Evaluation of food risk parameters in the Day River flood diversion area, Red River delta, Vietnam. *Natural Hazards, 56*(1), 169–194.

Das, I., Sahoo, S., van Westen, C., Stein, A., & Hack, R. (2010). Landslide susceptibility assessment using logistic regression and its comparison with a rock mass classification system, along a road section in the northern Himalayas (India). *Geomorphology, 114*(4), 627–637.

Demir, G., Aytekin, M., & Akgun, A. (2015). Landslide susceptibility mapping by frequency ratio and logistic regression methods: An example from Niksar-Resadiye (Tokat, Turkey). *Arabian Journal of Geosciences, 8*(3), 1801–1812.

Devkota, K. C., Regmi, A. D., Pourghasemi, H. R., Yoshida, K., Pradhan, B., Ryu, I. C., et al. (2013). Landslide susceptibility mapping using certainty factor, index of entropy and logistic regression models in GIS and their comparison at Mugling-Narayanghat road section in Nepal Himalaya. *Natural Hazards, 65*(1), 135–165.

Dou, J., Bui, D. T., Yunus, A. P., Jia, K., Song, X., Revhaug, I., et al. (2015). Optimization of causative factors for landslide susceptibility evaluation using remote sensing and GIS data in parts of Niigata, Japan. *PloS One, 10*(7), e0133262. doi:10.1371/journal.pone.0133262

Dou, J., Oguchi, T., Hayakawa, Y. S., Uchiyama, S., Saito, H., & Paudel, U. (2014). *GIS-based landslide susceptibility mapping using a certainty factor model and its validation in the Chuetsu Area, Central Japan. Landslide Science for a Safer Geoenvironment* (pp. 419–424). Berlin: Springer.

Dragićević, S., Lai, T., & Balram, S. (2015). GIS-based multicriteria evaluation with multiscale analysis to characterize urban landslide susceptibility in data-scarce environments. *Habitat International, 45,* 114–125.

Feizizadeh, B., & Blaschke, T. (2013). GIS-multicriteria decision analysis for landslide susceptibility mapping: Comparing three methods for the Urmia lake basin, Iran. *Natural Hazards, 65*(3), 2105–2128.

Fuchs, S., Heiss, K., & Hübl, J. (2007). Towards an empirical vulnerability function for use in debris flow risk assessment. *Natural Hazard Earth Systems, 7*(5), 495–506.

Galli, M., & Guzzetti, F. (2007). Landslide vulnerability criteria: A case study from Umbria, Central Italy. *Environmental Management, 40* (4), 649–665.

Glade, T., Crozier, M., & Smith, P. (2000). Applying probability determination to refine landslide-triggering rainfall thresholds using an empirical "Antecedent Daily Rainfall Model". *Pure and Applied Geophysics, 157*(6–8), 1059–1079.

Guzzetti, F., Carrara, A., Cardinali, M., & Reichenbach, P. (1999). Landslide hazard evaluation: A review of current techniques and their application in a multi-scale study, Central Italy. *Geomorphology, 31*(1), 181–216.

Guzzetti, F., Peruccacci, S., Rossi, M., & Stark, C. P. (2008). The rainfall intensity–duration control of shallow landslides and debris flows: An update. *Landslides, 5*(1), 3–17.

Haneberg, W. C. (2004). A rational probabilistic method for spatially distributed landslide hazard assessment. *Environmental and Engineering Geoscience, 10*(1), 27–43.

He, X., Hong, Y., Yu, X., Cerato, A. B., Zhang, X., & Komac, M. (2014). Landslides susceptibility mapping in Oklahoma state using GIS-based weighted linear combination method. In *Landslide science for a safer geoenvironment* (pp. 371–377).

Huang, J., Lyamin, A., Griffiths, D., Krabbenhoft, K., & Sloan, S. (2013). Quantitative risk assessment of landslide by limit analysis and random fields. *Computers and Geotechnics, 53,* 60–67.

Jaiswal, P., van Westen, C. J., & Jetten, V. (2010). Quantitative landslide hazard assessment along a transportation corridor in southern India. *Engineering Geology, 116*(3), 236–250.

Jaiswal, P., van Westen, C. J., & Jetten, V. (2011). Quantitative assessment of landslide hazard along transportation lines using historical records. *Landslides, 8*(3), 279–291.

Jakob, M., Stein, D., & Ulmi, M. (2012). Vulnerability of buildings to debris flow impact. *Natural Hazards, 60*(2), 241–261.

Jebur, M. N., Pradhan, B., & Tehrany, M. S. (2014). Optimization of landslide conditioning factors using very high-resolution airborne laser scanning (LiDAR) data at catchment scale. *Remote Sensing of Environment, 152,* 150–165.

Jebur, M. N., Pradhan, B., & Tehrany, M. S. (2015). Manifestation of LiDAR-derived parameters in the spatial prediction of landslides using novel ensemble evidential belief functions and support vector machine models in GIS. *IEEE J-STARS, 8*(2), 674–690.

Joshi, C., de Leeuw, J., & van Duren, I. C. (Eds.) (2004). Remote sensing and GIS applications for mapping and spatial modelling of invasive species. In *Proceedings of ISPRS*.

Kanungo, D., Arora, M., Gupta, R., & Sarkar, S. (2008). Landslide risk assessment using concepts of danger pixels and fuzzy set theory in Darjeeling Himalayas. *Landslides, 5*(4), 407–416.

Kappes, M. S., Keiler, M., von Elverfeldt, K., & Glade, T. (2012). Challenges of analyzing multi-hazard risk: A review. *Natural Hazards, 64*(2), 1925–1958.

Kavzoglu, T., Sahin, E. K., & Colkesen, I. (2014). Landslide susceptibility mapping using GIS-based multi-criteria decision analysis, support vector machines, and logistic regression. *Landslides, 11*(3), 425–439.

Kia, M. B., Pirasteh, S., Pradhan, B., Mahmud, A. R., Sulaiman, W. N. A., & Moradi, A. (2012). An artificial neural network model for flood simulation using GIS: Johor River Basin, Malaysia. *Environmental Earth Science, 67*(1), 251–264.

Kobal, M., Bertoncelj, I., Pirotti, F., Dakskobler, I., & Kutnar, L. (2015). Using LiDAR data to analyse Sinkhole characteristics relevant for understory vegetation under forest cover-case study of a high Karst Area in the Dinaric Mountains. *PLoS ONE, 10*(3), e0122070. doi:10.1371/journal.pone.0122070

Kovač, B., & Žalik, B. (2010). Visualization of LiDAR datasets using point-based rendering technique. *Computers & Geosciences, 36* (11), 1443–1450.

Kritikos, T., & Davies, T. (2014). Assessment of rainfall-generated shallow landslide/debris-flow susceptibility and runout using a GIS-based approach: application to western Southern Alps of New Zealand. *Landslides, 15,* 1–25.

Lee, M. L., Ng, K. Y., Huang, Y. F., & Li, W. C. (2014). Rainfall-induced landslides in Hulu Kelang area, Malaysia. *Natural Hazards, 70*(1), 353–375.

Leone, F., Asté, J., & Leroi, E. (1996). Vulnerability assessment of elements exposed to mass-movement: Working toward a better risk perception. *Landslides-Glissements de Terrain Balkema, Rotterdam, 7,* 263–270.

Ma, R. (2004). *Building model reconstruction from LiDAR data and aerial photographs*. The Ohio State University.

Martha, T. R., van Westen, C. J., Kerle, N., Jetten, V., & Kumar, K. V. (2013). Landslide hazard and risk assessment using semi-automatically created landslide inventories. *Geomorphology, 184,* 139–150.

Moonjun, R. (2007). Application of artificial neural network and decision tree in a GIS–based predictive soil mapping for landslide vulnerability study: A case study of Hoi Num Rin sub–Watershed, Thailand, ITC. *Enschede, 104,* 16–32.

Moreiras, S. M. (2015). Landslides as climate indicators in Argentinean Central Andes (32° S). In *Engineering geology for society and territory* (Vol. 1, pp. 443–447).

Muhammad, R. F. (2010). The geomorphology and origin of Gua Tempurung, Perak, West Malaysia. *Bulletin of the Geological Society of Malaysia, 56,* 127–132.

Murillo-García, F., Rossi, M., Fiorucci, F., & Alcántara-Ayala, I. (2015). Population Landslide vulnerability evaluation: The case of the indigenous population of Pahuatlán-Puebla, Mexico. In *Engineering geology for society and territory* (Vol. 2, pp. 1793–1797).

Muthukumar, M. (2013). GIS based geosystem response modelling for landslide vulnerability mapping parts of Nilgiris, South India. *Disaster Advances, 6*(7), 58–66.

Nandi, A., & Shakoor, A. (2010). A GIS-based landslide susceptibility evaluation using bivariate and multivariate statistical analyses. *Engineering Geology, 110*(1), 11–20.

Ohlmacher, G. C., & Davis, J. C. (2003). Using multiple logistic regression and GIS technology to predict landslide hazard in northeast Kansas, USA. *Engineering Geology, 69*(3), 331–343.

Opolot, E. (2013). Application of remote sensing and geographical information systems in flood management: A review. *Research Journal of Applied Sciences Engineering and Technology, 6,* 1884–1894.

Pan, W. S., Lu, Y. D., & Guo, J. Y. (2014). Risk assessment and management of geological disaster based on risk period analysis and GIS in Loess Areas. In *Applied mechanics and materials* (Vol. 675, pp. 1184–1191).

Papathoma-Köhle, M., Zischg, A., Fuchs, S., Glade, T., & Keiler, M. (2015). Loss estimation for landslides in mountain areas–An integrated toolbox for vulnerability assessment and damage documentation. *Environment Modelling and Software, 63,* 156–169.

Park, S., Choi, C., Kim, B., & Kim, J. (2013). Landslide susceptibility mapping using frequency ratio, analytic hierarchy process, logistic regression, and artificial neural network methods at the Inje area, Korea. *Environmental Earth Sciences, 68*(5), 1443–1464.

Pedrazzini, A., Humair, F., Jaboyedoff, M., & Tonini, M. (2015). Characterisation and spatial distribution of gravitational slope deformation in the Upper Rhone catchment (Western Swiss Alps). *Landslides, 8,* 1–19.

Pradhan, B. (2010). Landslide susceptibility mapping of a catchment area using frequency ratio, fuzzy logic and multivariate logistic regression approaches. *Journal of the Indian Society of Remote Sensing, 38*(2), 301–320.

Pradhan, B. (2011). Manifestation of an advanced fuzzy logic model coupled with Geo-information techniques to landslide susceptibility mapping and their comparison with logistic regression modelling. *Environmental and Ecological Statistics, 18*(3), 471–493.

Pradhan, B., Hagemann, U., Tehrany, M. S., & Prechtel, N. (2014). An easy to use ArcMap based texture analysis program for extraction of flooded areas from TerraSAR-X satellite image. *Computers & Geosciences, 63,* 34–43.

Pradhan, B., & Lee, S. (2009). Landslide risk analysis using artificial neural network model focusing on different training sites. *International Journal of Physical Sciences, 3*(11), 1–15.

Pradhan, B., & Lee, S. (2010). Delineation of landslide hazard areas on Penang Island, Malaysia, by using frequency ratio, logistic regression, and artificial neural network models. *Environmental Earth Sciences, 60*(5), 1037–1054.

Regmi, N. R., Giardino, J. R., McDonald, E. V., & Vitek, J. D. (2014). A comparison of logistic regression-based models of susceptibility to landslides in western Colorado, USA. *Landslides, 11*(2), 247–262.

Romeo, R. W., Mari, M., & Pappafico, G. (2013). A performance-based approach to landslide risk analysis and management. In C. Margottini, P. Canuti, & K. Sassa (Eds.), *Landslide science and practice* (pp. 91–95). Berlin: Springer.

Saito, H., Nakayama, D., & Matsuyama, H. (2009). Comparison of landslide susceptibility based on a decision-tree model and actual landslide occurrence: The Akaishi Mountains, Japan. *Geomorphology, 109*(3), 108–121.

Sarkar, S., Kanungo, D. P., Patra, A., & Kumar, P. (2008). GIS based spatial data analysis for landslide susceptibility mapping. *Journal of Mountain Science-England, 5*(1), 52–62.

Sattenpalli, S., & Parkash, S. (2013). Landslides, land use systems and food security. In C. Margottini, P. Canuti, & K. Sassa (Eds.), *Landslide science and practice* (pp. 151–161). Berlin: Springer.

Shaharom, S., Huat, L. T., & Othman, M. A. (2014). Area based landslide hazard and risk assessment for Penang Island Malaysia. In K. Sassa, P. Canuti, & Y. Yin (Eds.), *Landslide Science for a Safer Geoenvironment* (pp. 513–519).

Stoffel, M. (2010). Magnitude–frequency relationships of debris flows —A case study based on field surveys and tree-ring records. *Geomorphology, 116*(1), 67–76.

Tehrany, M. S., Pradhan, B., & Jebur, M. N. (2013). Spatial prediction of flood susceptible areas using rule based decision tree (DT) and a novel ensemble bivariate and multivariate statistical models in GIS. *Journal of Hydrology, 504,* 69–79.

Tehrany, M. S., Pradhan, B., & Jebur, M. N. (2014). Flood susceptibility mapping using a novel ensemble weights-of-evidence and support vector machine models in GIS. *Journal of Hydrology, 512,* 332–343.

Tierney, K. J., Lindell, M. K., & Perry, R. W. (2001). *Facing the unexpected: Disaster preparedness and response in the United States*. USA: Joseph Henry Press.

Tournadour, E., Mulder, T., Borgomano, J., Hanquiez, V., Ducassou, E., & Gillet, H. (2015). Origin and architecture of a mass transport complex on the northwest slope of Little Bahama Bank (Bahamas): Relations between off-bank transport, bottom current sedimentation and submarine landslides. *Sediment Geology, 317,* 9–26.

Umar, Z., Pradhan, B., Ahmad, A., Jebur, M. N., & Tehrany, M. S. (2014). Earthquake induced landslide susceptibility mapping using an integrated ensemble frequency ratio and logistic regression models in West Sumatera Province, Indonesia. *Catena, 118,* 124–135.

Uzielli, M., Nadim, F., Lacasse, S., & Kaynia, A. M. (2008). A conceptual framework for quantitative estimation of physical vulnerability to landslides. *Engineering Geology, 102*(3), 251–256.

Van Westen, C. J., Castellanos, E., & Kuriakose, S. L. (2008). Spatial data for landslide susceptibility, hazard, and vulnerability assessment: An overview. *Engineering Geology, 102*(3), 112–131.

Varoonchotikul, P. (2003). *Flood forecasting using artificial neural networks*. The Netherlands: CRC Press.

Wang, L. J., Sawada, K., & Moriguchi, S. (2013). Landslide susceptibility analysis with logistic regression model based on FCM sampling strategy. *Computers & Geosciences, 57,* 81–92.

Winter, M., Smith, J., Fotopoulou, S., Pitilakis, K., Mavrouli, O., Corominas, J., et al. (2014). An expert judgement approach to determining the physical vulnerability of roads to debris flow. *Bulletin of Engineering Geology and the Environment, 73*(2), 291–305.

Xu, C., Dai, F., Xu, X., & Lee, Y. H. (2012). GIS-based support vector machine modeling of earthquake-triggered landslide susceptibility in the Jianjiang River watershed, China. *Geomorphology, 145,* 70–80.

Xu, C., Xu, X., Yao, X., & Dai, F. (2014). Three (nearly) complete inventories of landslides triggered by the May 12, 2008 Wenchuan Mw 7.9 earthquake of China and their spatial distribution statistical analysis. *Landslides, 11*(3), 441–461.

Yalcin, A., Reis, S., Aydinoglu, A., & Yomralioglu, T. (2011). A GIS-based comparative study of frequency ratio, analytical hierarchy process, bivariate statistics and logistics regression methods for landslide susceptibility mapping in Trabzon, NE Turkey. *Catena, 85*(3), 274–287.

Yesilnacar, E., & Topal, T. (2005). Landslide susceptibility mapping: A comparison of logistic regression and neural networks methods in a medium scale study, Hendek region (Turkey). *Engineering Geology, 79*(3), 251–266.

Yilmaz, I. (2009). Landslide susceptibility mapping using frequency ratio, logistic regression, artificial neural networks and their comparison: A case study from Kat landslides (Tokat-Turkey). *Computers & Geosciences, 35*(6), 1125–1138.

Yuan, R. M., Tang, C. L., & Deng, Q. H. (2015). Effect of the acceleration component normal to the sliding surface on earthquake-induced landslide triggering. *Landslides, 12*(2), 335–344.

Yusof, N. M., Pradhan, B., Shafri, H. Z. M., Jebur, M. N., & Yusoff, Z. (2015). Spatial landslide hazard assessment along the Jelapang Corridor of the North–South expressway in Malaysia using high resolution airborne LiDAR data. *Arabian Journal of Geoscience, 8,* 1–12.

Zare, M., Pourghasemi, H. R., Vafakhah, M., & Pradhan, B. (2013). Landslide susceptibility mapping at Vaz Watershed (Iran) using an artificial neural network model: A comparison between multilayer perceptron (MLP) and radial basic function (RBF) algorithms. *Arabian Journal of Geoscience, 6*(8), 2873–2888.

13 Landslide Risk Assessment Using Multi-hazard Scenario Produced by Logistic Regression and LiDAR-Based DEM

Biswajeet Pradhan and Waleed M. Abdulwahid

13.1 Introduction

The rapid urban development and population growths worldwide push threats to the people because of landslides and other mass movements. Landslide is one of the natural disasters causing significant damages to lives and properties. Approximately 820,000 km^2 inhabited with a population of ∼ 66 million are relatively classified as high-risk areas (Dilley 2005). Landslide management programs start by predicting the future potential landslides, designing strategies for preventing or at least reducing the risk of potential landslides and estimating the risks at elements as well as the damages caused by the materials moved from high slopes to lower slopes. Each component of landslide management programs has been extensively studied in literature with various techniques. Recently, Light Detection and Ranging (LiDAR) remote sensing has been proven to be an effective solution for landslide prediction and risk assessments. This is due to the advantages of these systems that include the capability of penetrating through the vegetation canopies and producing high to very high-resolution digital elevation models (DEM). Subsequently, detailed topographic and geomorphologic information can be extracted, and potential correlations with landslide occurrence could be established. On the other hand, various statistical and machine learning methods have been developed and proposed for landslide modeling and risk assessments. Overall, high-quality data and efficient modeling and data analysis methods are the basic requirements to improve the landslide management programs.

The first step in landslide risk assessment is the preparation of landslide inventory data. These data can be collected from various sources such as aerial orthophotos, satellite images, previous records and newspapers, and LiDAR data. The database of a landslide inventory often contains the spatial information (where) the landslides occurred, the type of mass movements, and the date (when) of the events. Once the landslide inventory map is produced, a landslide susceptibility assessment is often the next step. A landslide susceptibility mapping is a process of predicting future landslides from the use of previous landslide records with consideration of certain conditioning factors. Landslide susceptibility mapping is essential for land use planning and risk assessments. On the other hand, risk assessment requires the hazard and vulnerability analysis. Landslide hazard is defined as the probability of occurrences of a potential landslide in a specified period of time considering the landslide susceptibility (spatial probability) and triggering factors (temporal probability) (Althuwaynee et al. 2014a, b). In addition, vulnerability is another concept used in landslide risk assessments. It is one of the important concepts that have been extensively explored in landslide risk management. It describes the physical, social, economic, and environmental factors that can make particular inhabitants highly susceptible to the effects of landslide hazards. As a result of the mentioned concepts, landslide risk is defined as the product of landslide hazards and landslide vulnerability (van Westen et al. 2008). The output of risk analysis is thematic maps representing the uncertain product of a hazard and the probable extent of different types of damages. In other words, landslide risk refers to the expected damages to lives, economy, and assets caused by a specific hazard in a given area (Opolot 2013). Risk assessment involves generating the risk level of a potential hazard (i.e., landslide) for the element at risk (people and property) in a particular area (Kanungo et al. 2008; Althuwaynee and Pradhan 2016).

13.2 Previous Work

Extensive works have been done on landslide susceptibility and risk assessments. For the purpose of finding the research problem, only important and recent studies were

B. Pradhan (✉) · W.M. Abdulwahid
Department of Civil Engineering, University Putra Malaysia, Serdang, Malaysia
e-mail: biswajeet24@gmail.com

B. Pradhan (ed.), *Laser Scanning Applications in Landslide Assessment*,
DOI 10.1007/978-3-319-55342-9_13

reviewed and discussed in this section. Early works in landslide susceptibility mapping have come a long way as the GIS software and hardware become more available in 1970s. In addition, in the beginning, the methods were simple that included heuristic and simple statistical-based models. However, later many statistical and machine learning methods have been developed and improved the landslide susceptibility assessment techniques. The advancement of remote sensing systems was also another reason for improving the landslide susceptibility assessment techniques. Data become more available and accessible; recently, LiDAR has been widely used to prepare high-quality landslide inventory maps. Among the popular techniques and widely accepted methods for landslide susceptibility mapping are support vector machine (SVM) and logistic regression (LR). SVM belongs to the machine learning methods, and it has been used by many authors for landslide susceptibility modeling showed promising success and prediction rates (Yao and Dai 2006; Yao et al. 2008).

On the other hand, LR has been demonstrated to be successful in various case studies from different geographic regions (Bai et al. 2010; Falaschi et al. 2009; Pradhan and Youssef 2010). LR is a multivariate statistical model that employs the maximum likelihood estimation method. It is used to establish a relationship between a number of landslide predictors and the landslide occurrence (absence/absent). Pourghasemi et al. (2013) applied SVM model in landslide susceptibility mapping at the Golestan Province, Iran. Results of prediction rates indicated that radial basis function (RBF) (85%) models performed better than other types of kernel. Kavzoglu et al. (2014) compared GIS-based multi-criteria decision making, SVM and LR models for landslide susceptibility mapping. The results showed that the GIS-based multi-criteria and SVM outperformed the LR model in terms of success and prediction rates. Pradhan (2013) compared the predictive ability of the decision tree (DT), SVM, and neuro-fuzzy models. SVM was found to have the best performance (area under the curve = 0.95) compared to DT and neuro-fuzzy models. Wu et al. (2014) compared SVM and DT models for landslide susceptibility assessment in Three Gorges of China. The results showed that the object-based SVM model had the highest correct rate of 89.36 compared with other models. In Italy, Ballabio and Sterlacchini (2012) evaluated the SVM model for landslide susceptibility mapping. They found that SVM model can outperform other techniques in terms of accuracy and generalization capacity. In a recent paper, Hong et al. (2016) evaluated the SVM model with four kernel types for landslide susceptibility mapping in China. Their results showed that the RBF function is the most suitable kernel type for landslide susceptibility mapping in the study area.

On the other hand, Devkota et al. (2013) compared LR with other techniques such as certainty factor, index of entropy for landslide susceptibility assessment in Nepal Himalaya. Their results suggested that LR can produce accuracy more than 80%; however, other methods also performed well and comparable to LR in terms of accuracy. On the other hand, Akgun (2012) produced a landslide susceptibility map at İzmir, Turkey, with LR and compared the result with models such as multi-criteria decision making and likelihood ratio methods. According to the results of the study, LR was found to be the best model with an accuracy of 0.810 (AUC). On the contrary, Ozdemir and Altural (2013) found that frequency ratio and weights-of-evidence outperformed the LR model in Sultan Mountains, SW Turkey. However, the accuracy of LR (0.937) was acceptable for producing landslide susceptibility maps for land use planning. In addition, an analysis by Xu et al. (2013) over an area in China suggested that LR can also achieve accurate results with incomplete landslide inventory. Costanzo et al. (2014) developed a forward LR model for earth-flow landslide susceptibility mapping in Platani river basin (southern Sicily, Italy). Their results indicated the efficiency of LR for landslide susceptibility assessment with an accuracy over 0.80 (AUC). Furthermore, Solaimani et al. (2013) compared frequency ratio and LR models for landslide susceptibility assessment. They found that frequency ratio (79.48%) is more accurate than LR (77.4%) model. In a case study area in the USA, Regmi et al. (2014) showed that LR can produce high-quality landslide susceptibility maps with various scenarios. The accuracy of LR was above 80% using different sampling techniques and different mass movements (landslides, debris flows). In more recent works such as Zhang et al. (2016), Bui et al. (2016), Erener et al. (2016), Bornaetxea et al. (2016), LR was found to be suitable for landslide susceptibility assessment and it has stable accuracy.

In addition, according to several studies, spatial and temporal probabilities are the two main components of landslide hazard assessment (Althuwaynee et al. 2014b; Guzzetti et al. 2006). On the other hand, some other researchers used three components instead including the magnitude of the landslide events (Erener and Düzgün 2013). Landslide hazard is commonly defined as "the probability of occurrence within a specified period of time and within a given area of a potentially damaging phenomenon" (Varnes 1984). For landslide hazard mapping, several methods have been used, which include heuristic algorithms (judgmental method), empirical probability, magnitude–frequency relations, rational methods (geomechanical approach), and indirect approaches (Corominas et al. 2014). The heuristic methods of landslide hazard assessment are based on opinions obtained from a group of experts (Akgun 2012). In empirical methods, the probability is analyzed based on the results of stability analysis and mathematical modeling. Empirical probabilistic models are employed on the basis of historical records of a landslide event using the annual probability of occurrence and hydrological analysis. The magnitude of other contributing factors, such as earthquakes and rainfall, is integrated into the analyses using indirect approaches. In this process, the critical values of rainfall or earthquake events are determined to define the return period for landslides (Umar et al. 2014). In this correlation analysis, areas with high probability of landslide occurrence are recognized without information on the size or type of failure. The quantitative prediction of landslide hazards is aimed at detecting the relationship between magnitude and frequency (Fausto Guzzetti et al. 2008). Overall, GIS-based methods show the potential zonation of landslide hazards (Haneberg 2004).

Furthermore, in terms of landslide vulnerability and risk assessment, several studies have developed various approaches. The expert decision could be used to assess vulnerability levels and the probable depth of landslides for specific facility types. Muthukumar (2013) analyzed the vulnerability of landslide susceptibility mapping for Nilgiri Mountains in the Western Ghats, South India, using the landslide per unit area method. Lithology, lineament, geomorphology, slope, and land use/land cover were used for vulnerability mapping. On the other hand, Althuwaynee and Pradhan (2016) developed a semi-quantitative landslide risk analysis using GIS-based exposure analysis in Kuala Lumpur city, Malaysia. With their approach, the number of elements affected by landslides and the population density under landslide risk could be quantified. Akgun et al. (2012) used GIS and remote sensing data for landslide risk assessment in Izmir city (west Turkey). Several landslide conditioning factors were modeled by LR method, and hazard map was then produced by considering earthquakes and precipitation as main triggering factors in the study area. Then, an accurate land use map produced at 94% of accuracy was used to assess the landslide risks in the area. Their results suggested that the methods applied could be useful for land use planner and local government authorities. In another paper, Vranken et al. (2015) showed a case study of a landslide risk assessment in a densely populated hilly area. Their method could quantify the affected areas by landslides and showed that the annual risk values vary depending on the spatial probability of landslides produced by different methods. Several other authors developed methods of landslide risk assessments using spatial data such as Calvo and Savi (2009), Huang et al. (2013), and Pradhan and Lee (2009).

The main objective of this chapter is to develop landslide risk maps for the Ringlet area located in Cameron Highlands using logistic regression and multi-hazard scenarios constructed by analyzing 15-year rainfall data. With regard to the advantages of this study, focusing on a specific area can lead to more efficient outcomes than when the focus is on a vast area. Planning authorities can use the proposed method to prevent future damage to land parcels and to protect the lives of residents. This study can improve its accuracy when data on the return periods and landslide intensity for a specific landslide occurrence are available. Such maps may be used as a reference by planners, decision makers, and local governments in landslide management and planning in the study area.

13.3 Methodology

13.3.1 Study Area

The study area is the Ringlet area, a township in Cameron Highlands, Malaysia. The area is home to many farmers cultivating various types of crops such as strawberry, vegetables, and flowers. Due to roadside cut and modified slopes, the area has experienced many landslide events where significant risks threatened to the farmers and other communities in the area. The area is bounded by the following coordinates: 101° 21′ 54″ to 101° 24′ 51″ longitude and 4° 22′ 52″ to 4° 25′ 48″ latitude (Fig. 13.1). The average altitude of the area is 1,200 m above mean sea level, and the total land area is 24.38 km^2.

The Ringlet district is located on the eastern flank of the main range and is mainly composed of granite. However, scattered outliers (roof pendants) of meta-sediments are also present. The granite in the Ringlet area is classified as megacrystic biotite granite (Pradhan and Lee 2010). The meta-sediments consist of schist, phyllite, slate, and limestone (Pradhan and Lee 2010. Post-Triassic–Mesozoic granite comprises most of the granite rocks. However, there are few patches of metamorphic rocks, mostly comprised of Silurian–Ordovician Schist, phyllite, limestone, and sandstone.

The average daytime temperature is 24 °C and average nighttime temperature is 14 °C. The average annual rainfall is 2660 mm. The intensity of the rain is one factor that affects the fill slopes, causing severe sheet, rill, and gully erosion. The climate is conducive to a wide range of subtropical crops. However, the land is steeply sloping with 66% of the slopes having gradients of more than 20°. Approximately 80% of the area is forested.

13.4 Spatial Database

13.4.1 Landslide Inventory Map

The most important step in landslide susceptibility mapping is the preparation of landslide inventory maps (Pradhan et al.

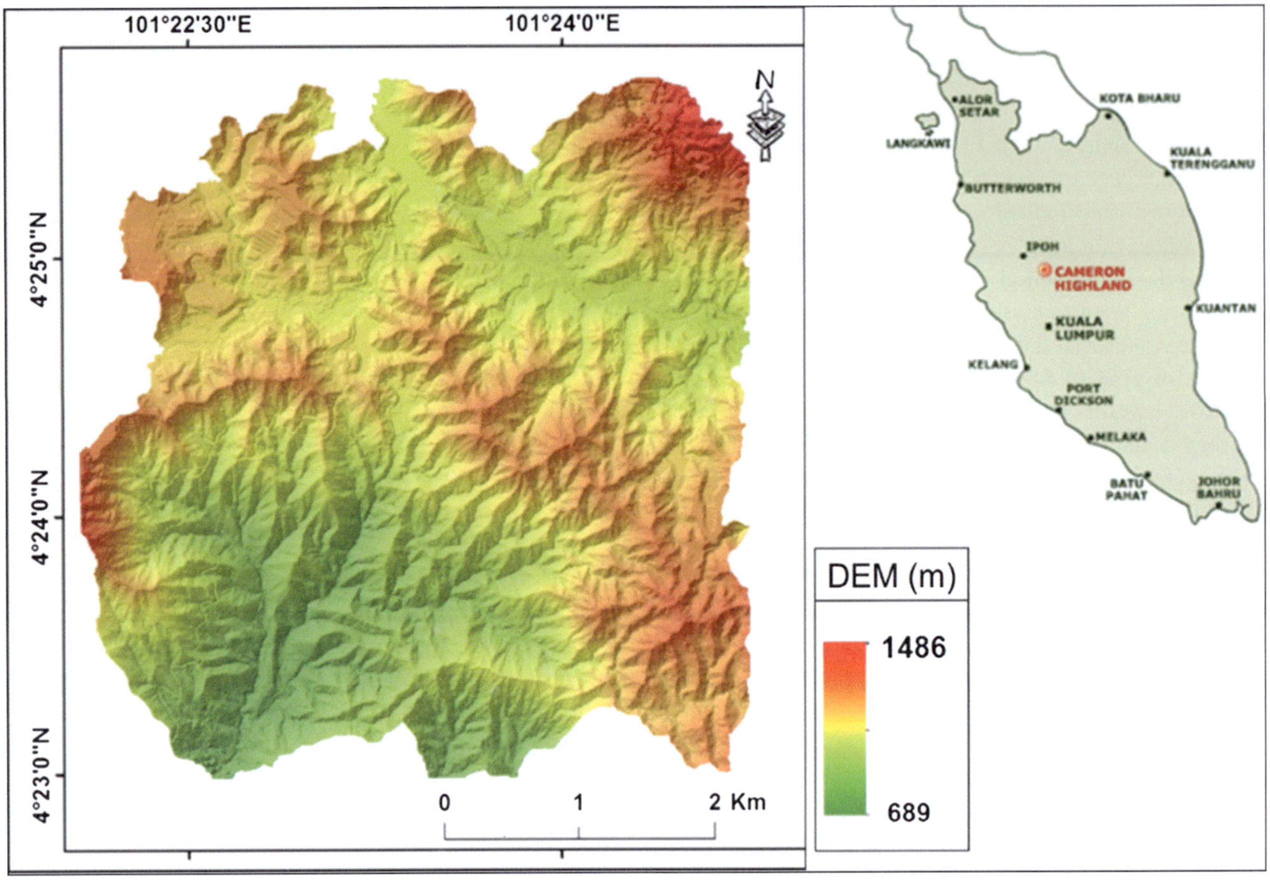

Fig. 13.1 The DEM map of the study area (Ringlet area, Cameron Highlands)

2014). Inventory maps are vital for evaluating the spatial relationship between landslide events and conditioning factors. Several measurements were conducted in the study area to create an inventory map. Remote sensing methods were used to obtain historical records of the landslides over the past 11 years. Archived 1:10,000–1: 50,000 aerial photographs and SPOT 5 panchromatic satellite images were used for the visual detection of landslide occurrences in the study area. The source material varies in quality with respect to describing the precise location of the landslide event. Based on aerial photo interpretation, the locations of the individual landslides were drawn on 1: 25,000 maps and the location plotted as closely as possible. Field observations were done using GPS to detect the exact location of the landslide and to confirm the fresh landslide scars. In the aerial photographs and high-resolution satellite images, historical landslides could be observed as breaks in the forest canopy, bare soil, or geomorphologic features like head and side scarps, flow tracks, and soil and debris deposits below a scar. A total of 164 landslides were identified in the study area (Fig. 13.2); these landslides were characterized as shallow rotational landslides with an area of 39,243 m^2 for the largest landslide and 14.6 m^2 for the smallest landslide. Then, the landslides were divided into two sets. According to the literature, 70% of the data on landslide occurrences were used for modeling, and the remaining landslide data were used for validation. Training landslide locations were used to create a dependent layer. The produced layer consisted of two values, namely 0 and 1, where 1 denotes the presence of a landslide and 0 indicates the absence of a landslide. The remaining landslide locations were used to test the outcomes. Both layers were created in ArcGIS and then rasterized.

13.4.2 Landslide Conditioning Factors

Susceptibility maps are defined by qualitatively and quantitatively studying the conditioning factors in affected areas (Jebur et al. 2014). Choosing the appropriate data set that comprises conditioning factors is a challenging task; no standard rules can be adopted when selecting the number of conditioning factors that are sufficient for a specific susceptibility analysis (Wang et al. 2013). These factors are essentially chosen on the basis of the literature and expert knowledge. In the current study, conditioning factors were

Fig. 13.2 Landslide inventory map of the study area

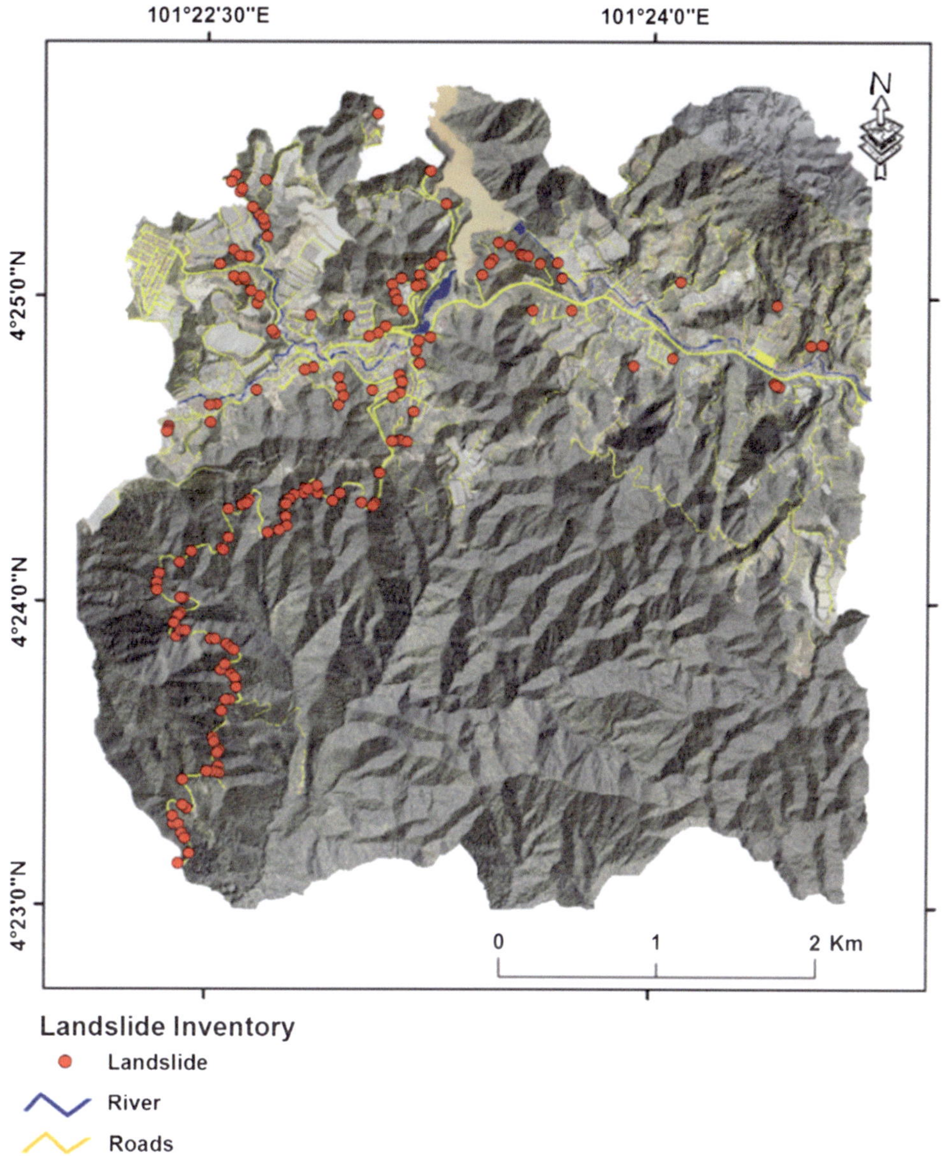

chosen with knowledge derived from the literature. The sole use of high-resolution LIDAR data was proven to be sufficient (Jebur et al. 2014). The conditioning factor data set included altitude, slope, aspect, curvature, stream power index (SPI), topographic wetness index (TWI), terrain roughness index (TRI), distance from a river, distance from roads, distance from lineament, sediment transport index, and geology (Fig. 13.3). All the factors were resampled to a 2 m grid, and the grid of the Ringlet area was built with 2484 columns and 2499 rows (6,207,516 pixels; 24.38 km^2).

The LiDAR data were used in constructing the altitude. The LiDAR vector point data were recorded over 25 km^2 of the Ringlet area and nearby area with flight height around 1510 mm on January 15, 2015, and resulted in nearly eight points per square meter with a 25,000 Hz pulse rate frequency. The absolute accuracy of the LiDAR data should meet the root-mean-square errors of 0.15 m in the vertical axis and 0.3 m in the horizontal axis.

The altitude, slope, aspect, curvature, SPI, TWI, and TRI maps were derived from a DEM, as shown in Fig. 13.3. Scale data can be defined as a scale to evaluate the influence of a layer on slope failure occurrence (Althuwaynee et al. 2014a, b). The slope is an influential conditioning factor in landslide occurrence. This factor directly affects landslide occurrence and is typically considered in landslide susceptibility analysis (Alimohammadlou et al. 2013). As the slope becomes steep, the vertical component of gravity increases (Tournadour et al. 2015). The slope in the study area ranges from 0° to 87.18° (Fig. 13.3b). This layer was also used in the analysis as a continuous layer, where each cell represents the calculated slope. Slope aspect is an important conditioning factor (Budimir et al. 2015). The morphological state

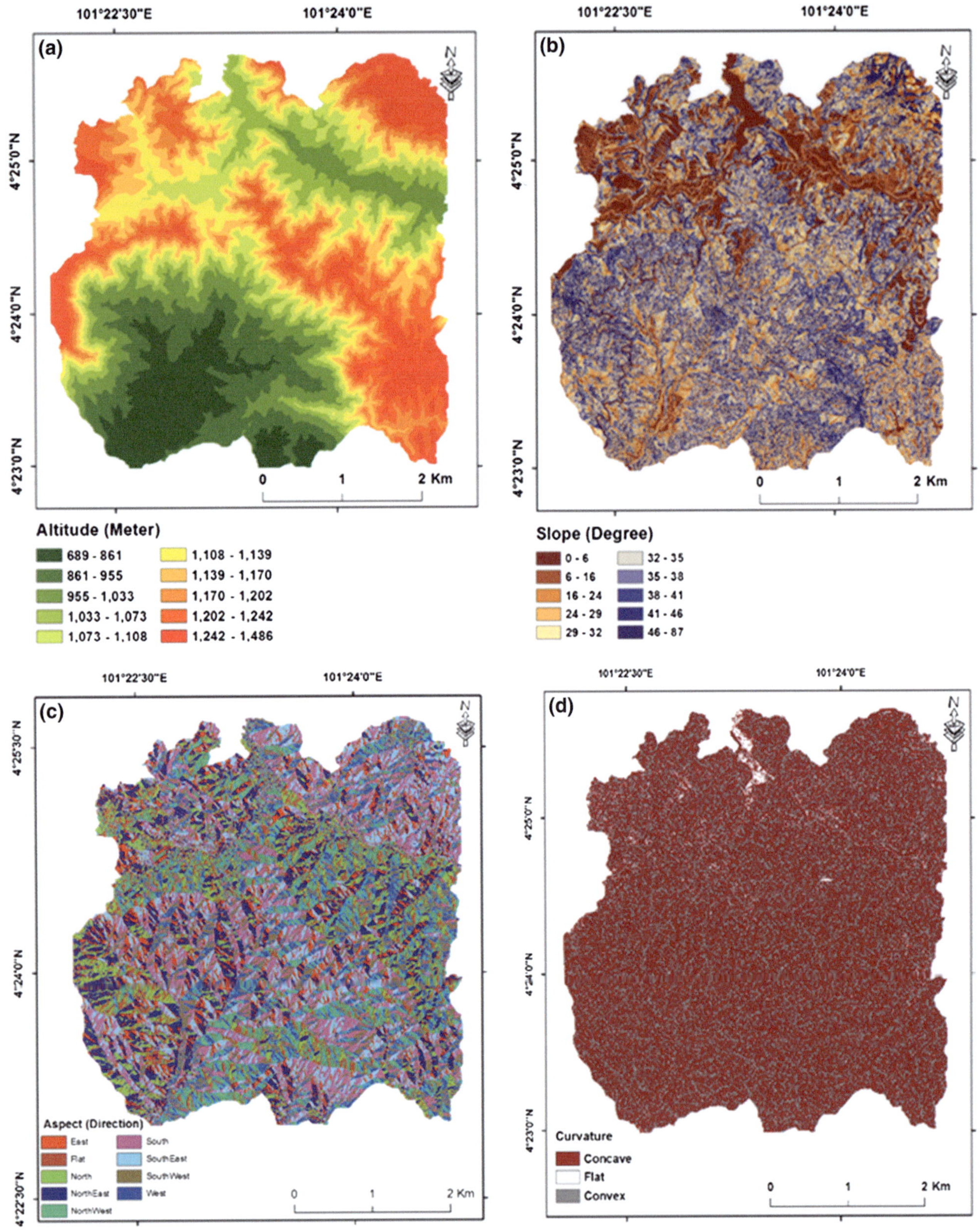

Fig. 13.3 Input conditioning factors: **a** altitude, **b** slope, **c** aspect, **d** curvature, **e** SPI, **f** TWI, **g** TRI, **h** STI, **i** distance from rivers, **j** distance from roads, **k** distance from lineament, and **l** geology

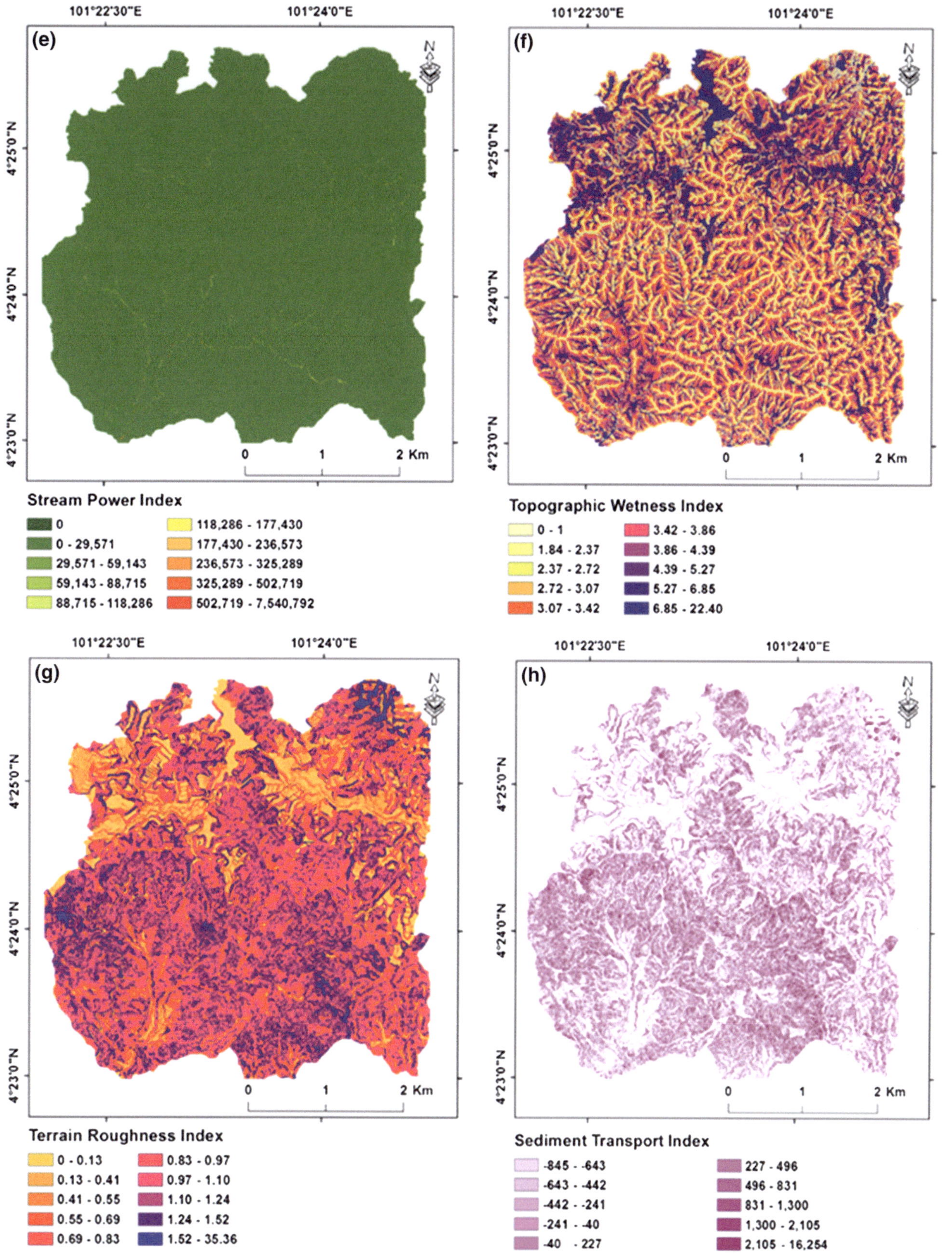

Fig. 13.3 (continued)

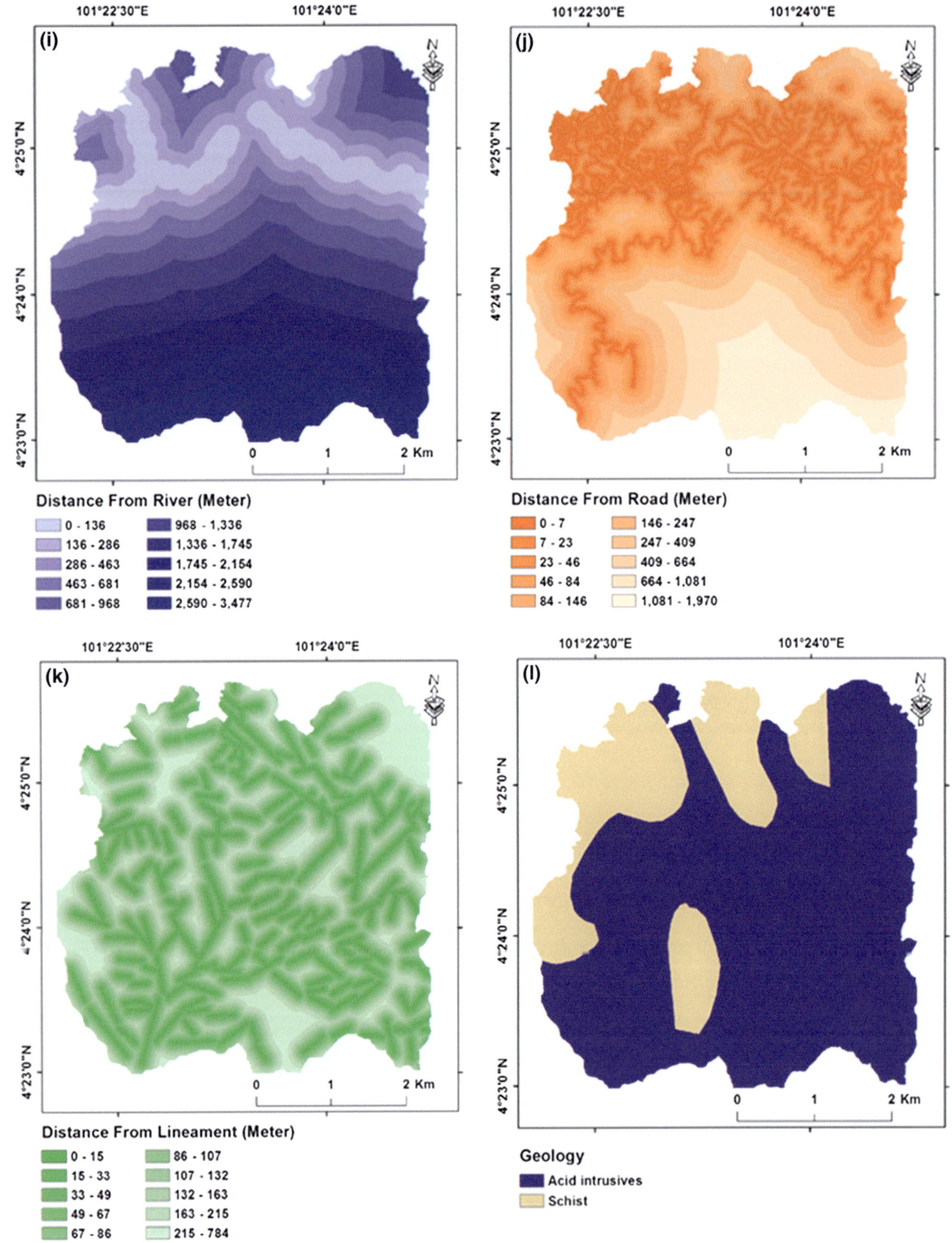

Fig. 13.3 (continued)

of the study range and the degree of precipitation and daylight are the meteorological circumstances that influence the event of landslide occurrence. Aspect influences weathering and, by implication, the sheer force of the object. Although the relationship between landslide occurrence and slope has been demonstrated, no accurate rule exists with respect to the effect of this element on slope failure (Pedrazzini et al. 2015). The aspect map was employed to draw the relationship and is shown in Fig. 13.3c. Ten classes were produced for the aspect map (flat, north, northeast, east, southeast, south, southwest, west, northwest, and north). The effect of curvature on slope failure reflects the convergence or divergence of water during downhill movement (Dou et al. 2014). Thus, this factor is another conditioning factor involved in landslide occurrence. In this study, curvature was derived from a DEM and subsequently categorized into three classes: concave, convex, and flat (Fig. 13.3d).

Hydrological factors, such as SPI and TWI, were calculated using Eqs. 13.1 and 13.2. Many scientists consider these two components as auxiliary geographic attributes in landslide susceptibility mapping (Dragićević et al. 2015). Additional information on SPI and TWI can be found in the work of Yusof et al. (2015). The water-related factors SPI and TWI were calculated using the following equations:

$$\mathrm{TWI} = \ln(A_s/\tan\beta), \tag{13.1}$$

$$\mathrm{SPI} = A_s \tan\beta, \tag{13.2}$$

where A_s is the specific catchment area ($\mathrm{m}^2\ \mathrm{m}^{-1}$) and β (radian) is the slope gradient (in degree) (Regmi et al. 2010). Another influential factor is TRI, which is generally used in mass movement studies. In the current study, this factor is calculated with Eq. 13.3:

$$\mathrm{TRI} = \sqrt{\max^2 - \min^2}, \tag{13.3}$$

where max and min are the highest and minimum values of the cells in the nine rectangular neighborhoods of altitude, respectively. TRI was also categorized into 10 classes using the quantile technique. STI defines the procedure of the slope failure and deposition (Fig. 13.3h) and is computed by using Eq. 13.4 as follows:

$$\mathrm{STI} = \left(\frac{A_S}{22.13}\right)^{0.6}\left(\frac{\sin\beta}{0.0896}\right)^{1.3}, \tag{13.4}$$

where β is the slope at each pixel and A_S is the upstream area. For the distance from the river factor, only the undercutting of the side slopes of rivers might cause slope failure initiation (Yang et al. 2014). The distance from road (Fig. 13.3j) is considered as an important factor because constructing roads in hilly areas weakens the stability of the slope structure and therefore increases the area's susceptibility to landslides. The distance from roads was calculated using the Euclidean distance in the spatial analyst tool.

Topographic structures (lineaments) are tectonic breaks that usually decrease rock strength. These structures comprise faults, overlays, and shear areas. They are responsible for triggering a large number of landslides in the study area. Lineaments were acquired from the topographic map and DEM of the territory. In this way, the distance from the lineaments was computed (Fig. 13.3k).

Geology influences the shear strength of rock mass, penetrance, and accordingly, the probability of an increase in neutral pressure in the subsoil. Geological data were generated by digitizing geological boundaries, fieldwork, and interpretation of aerial photos rasterized and resampled to a 2 m grid. In several studies, geology is one of the most significant conditioning factors in the distribution of landslides. In the present study, two geological types are given: acid intrusive and schist (Fig. 13.3l).

For the factors: altitude, slope, aspect, distance from a river, distance from roads, distance from lineament, sediment transport index, SPI, TWI, and TRI, all categorized into 10 classes using the quantile technique (Youssef et al. 2015).

13.5 Landslide Susceptibility and Hazard Assessments

13.5.1 LR Model

LR model is used to produce the landslide susceptibility map. The LR was used to determine the landslide probability in the Ringlet area. The landslide susceptibility index was calculated using the following Eqs. (13.5) and (13.6) of LR multivariate statistical modeling method:

$$P = \frac{1}{1 + \mathrm{e}^{-z}} \tag{13.5}$$

where P denotes the landslide, probability ranging from 0 to 1 in an S-shaped curve, and z denotes the lean combination, which can be calculated using Eq. (13.6) as follows:

$$Z = b_0 + b_1x_1 + b_2x_2 + b_3x_3 + b_nx_n, \tag{13.6}$$

where b_0 is the intercept, b_i (i = 0,1,2, …, n) represents the LR coefficients, and x_i (i = 0, 1, 2, …, n) represents the conditioning layers (Demir et al. 2015).

The LR was implemented in SPSS statistical software. The steps included: converting the landslide conditioning factors into ASCII files that had a similar coordinate system and geographic extents. Then, the dependent variable was prepared by giving a value of 1 to the pixels where the

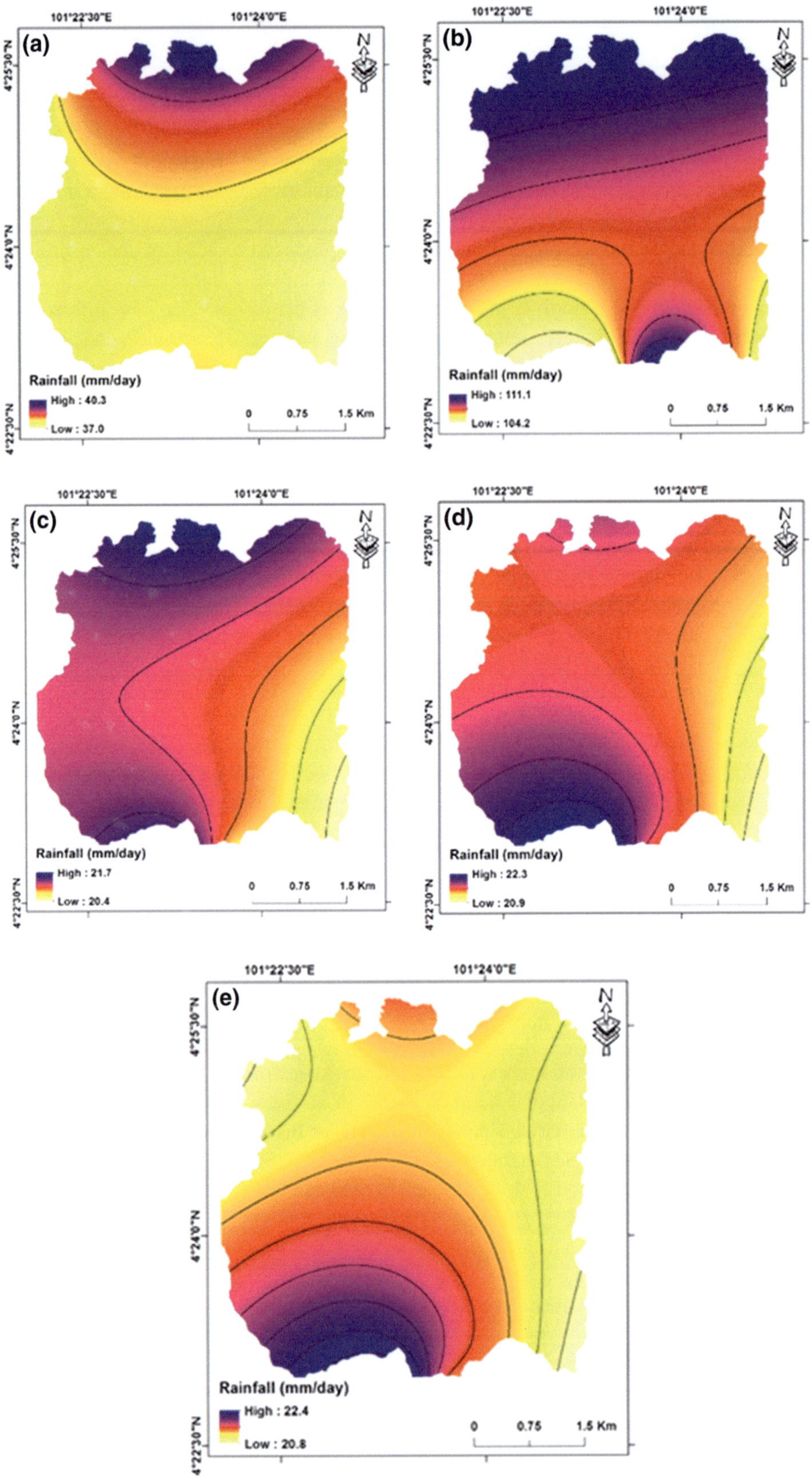

Fig. 13.4 Rainfall intensity for 15 years' period: **a** average intensity (any day in year), **b** abnormal value intensity recorded, **c** average rainfall intensity for 5-year return period, **d** average rainfall intensity for 10-year return period, and **e** average rainfall intensity for 15-year return period

landslide inventory data are, and a value of 0 to the remaining pixels. After that, the LR model was run and the coefficients of the landslide conditioning factors were calculated. These coefficients were used in GIS software to produce the landslide susceptibility map for the study area.

13.5.2 Validation of Landslide Susceptibility Map

Landslide susceptibility models often validated by ROC curves and by calculating the area under the curve (AUC) using the training and testing landslide inventory data. To create the ROC curves, the produced landslide susceptibility map was compared with the landslide inventory data (Chung and Fabbri 2003). The AUC validation method defines the prediction and success rates (Kritikos and Davies 2014). The success rate was calculated using the landslide inventory data used to train the LR model, the data accounted for 70% of the whole landslide inventory data. On the other hand, the prediction rate was calculated using the remaining 30% of landslide inventory data, the data that the LR model did not see before, and thus, the prediction rate can show the prediction efficiency of the model. Additional information on AUC and its equations can be found in the work of Tehrany et al. (2014). AUC shows the percentage of testing points that fall within the highest probability range. Practically, the first step of creating ROC curves was that the landslide susceptibility index was sorted in the downward arrangement. Then, the cell values were divided into 100 intervals (classes) to the horizontal axis (x), and the y-axis represented the cumulative landslides.

13.5.3 Landslide Hazard Assessment

Landslide hazard refers to the temporal probability of occurrence of a landslide event with a given intensity. In addition, hazards also have a spatial component related to the initiation of the hazard and the spreading of the landslides (Van Westen et al. 2008). In the hazard analysis, two factors were considered: landslide susceptibility map and landslide rainfall triggering factor (Althuwaynee et al. 2014a, b). Factors such as rainfall (Glade et al. 2000), earthquakes (Xu et al. 2014; Yang et al. 2015), and snowfall (Moreiras and Sepúlveda 2015) trigger a landslide. Landslides in Malaysia are mainly triggered by heavy rainfall (Althuwaynee et al. 2014a, b; Lee et al. 2014). Figure 13.4 shows the rainfall factors considered as triggering factors in the current study. Many extreme events, such as flooding and overflowing, are caused by heavy precipitation in the study area. Thus, the average (any day in the year) and abnormal intensity recorded in any day for 15 years, as well as three return periods (15, 10, and 5 years for average intensity per day), were analyzed in this study. A total of eight rain gauge stations were considered to derive the rainfall density maps by implementing the inverse distance weighting (IDW) for interpolation, and this method is fit for the type of a scattered points (rain gauge stations).

The final hazard maps were produced in GIS after calculating the hazard using the following expression (Eq. 13.7) (Xu et al. 2014):

$$H = P_S \times P_T, \tag{13.7}$$

where H represents the hazard probability, P_S represents the probability acquired from the LR modeling, and P_T represents the rainfall density layer.

13.5.4 Landslide Vulnerability Assessment

The vulnerability is an important step in landslide risk assessment. According to Fell et al. (2008), the vulnerability is defined as the level of loss to a given component in the area affected by the landslide. The authors demonstrated the vulnerability levels from 0 (no misfortune) to 1 (absolute misfortune). For property, the damage is estimated as the damage with respect to the estimation of the property; for people, the probability of loss is estimated for a given person affected by the landslide. In addition, vulnerability is also defined in terms of susceptibility, exposure, and coping capacity which is mathematically expressed in Eq. 13.8:

$$\text{Vulnerability} = \frac{\text{Susceptibility} \times \text{Exposere}}{\text{Coping Capacity}} \tag{13.8}$$

The exposure-based landslide vulnerability analysis defines the proportion of elements at risk expected to be influenced by the landslide event (Lee and Jones 2004). Exposure has sometimes used a component of vulnerability (Li et al. 2010) and sometimes used a separate component in the risk equation (Fell et al. 2005). The exposure-based vulnerability analysis was categorized under the quantitative-based approaches (Lee and Jones 2004). Ghosh et al. (2012) used 12 landslides hazard scenarios to calculate the number of the element at risk or those that are likely to be affected per mapping unit. In addition, Burns et al. (2013) performed the exposure analysis through a series of spatial and tabular queries between hazard zones, to find and quantify the community buildings that exposed to landslide hazard in prone areas. In general, the exposure-based approach is capable of identifying the number of buildings, population density, and road networks at risk for the preparation of risk maps despite a data-scarce environment.

In the current study, the landslide vulnerability is defined as the following expression:

$$V_L = P[D_L \geq 0|L], (0 \leq D_L \leq 1), \qquad (13.9)$$

where D_L is the element of risk for a given phenomenon (landslide, L) that is expected to or will definitely be damaged, and V_L is the probability of loss given to a landslide for a particular element or the percentage of damage to such element. The probability of vulnerability can be defined as a scale ranging from 0 to 5, where 5 represents complete damage and 0 represents no damage.

The following criteria for each type of LULC (Table 13.1) are:

- The cost of each type of LULC by square meter.
- The cost value was obtained from the literature (Table 13.2).
- The time required for reconstruction.
- The time required as indicated by relevant agencies in the literature for each type of LULC (Table 13.2).
- The relative risk of the landslide (surrounding area).
- The possibility of existing damage to any property affects neighboring properties.
- The risk to the population.
- The possibility of inflicting damage to properties and people in certain locations of a landslide.
- General effect of certain damage (overall area)

Table 13.1 The vulnerability value assessment for each type of LULC

LULC type	Cost (%)	Time to repair (%)	Relative risk (%)	Risk to population (%)	General effect of risk (%)	Vulnerability (%)
Building of worship areas	1.00	1.00	0.20	1.00	0.00	0.60
Dam areas	1.00	1.00	1.00	1.00	1.00	1.00
Industrial building areas	1.00	0.80	1.00	1.00	0.50	0.90
Forest areas	0.00	0.00	0.00	0.00	0.00	0.00
Grass areas	0.00	0.20	0.00	0.50	0.00	0.20
Mixed perennial crop areas	0.00	0.80	0.00	0.50	0.50	0.50
Open land areas	0.00	0.00	0.00	0.00	0.00	0.00
Pond areas	0.00	0.20	0.00	0.50	0.00	0.20
Power line areas	1.00	1.00	1.00	0.50	1.00	0.90
Residential building areas	1.00	1.00	1.00	1.00	0.50	0.90
Roads	1.00	1.00	1.00	1.00	1.00	1.00
River	0.00	0.20	0.00	0.00	0.50	0.20

Table 13.2 The cost value and time to repair

LULC type	Cost (RM per m^2)	Time to repair (per day)
Building of worship areas	396.00	60.00
Dam areas	237.00	90.00
Industrial building areas	300.00	30.00
Forest areas	11.00	0.00
Grass areas	0.10	2.00
Mixed perennial crop areas	0.20	10.00
Open land areas	0.00	0.00
Pond areas	12.00	2.00
Power line areas	300.00	10.00
Residential building areas	300.00	30.00
Roads	250.00	10.00
River	12.00	2.00

13.5.5 Landslide Risk Assessment

Landslide risk is the amount of the negative impact to well-being, property, or the environment. Risk assessments include three stages, determination of risks in a certain environment, risk evaluation, and risk management which involve the final decision and the product of the combination of the two previous stages in producing the best possible solution (Bell and Glade 2004). (Xu et al. 2012) defined the risk as the probability of harm caused by a particular hazard to a specific element. In the present study, risk analysis was conducted to calculate the expected amount of loss caused by landslides in the study area. The risk is determined on the basis of the formula for hazard and vulnerability, as shown in Eq. 13.10, where R is the expected risk, H is the estimated hazard, and V is the assessed vulnerability (Varnes 1984).

$$R = H \times V. \tag{13.10}$$

In the present study, a semi-quantitative approach was used by combining the hazard and vulnerability results using Eq. 13.8. To ensure effective visual interpretation, and also, by exposure spatial overlay of hazard map (for given return periods) and elements at risk to estimating loss and annualized risk. In addition, the loss analysis has to be conducted for each combination of hazard maps and an element at risk map. Then, the annualized risk was calculated using the following equation:

$$Risk = \frac{1}{P1} \times S1 + \left(\frac{1}{P2} - \frac{1}{P1}\right) \times \frac{S1 + S2}{2} + \left(\frac{1}{P3} - \frac{1}{P2}\right) \times \frac{S2 + S3}{2} \tag{13.11}$$

where $P1$ and $P2$ are the return period used, and $S1$ and $S2$ are the losses.

13.6 Results

This section presents the results of landslide susceptibility, hazard, vulnerability, and risk assessments. It also briefly highlights the findings, whereas the discussion is given in a separate section.

13.6.1 Results of Landslide Susceptibility

The landslide susceptibility map for the study area was produced by the LR model. The LR model established a relationship between the landslide conditioning factors and the landslide occurrence (absence/presence) as shown in Eq. (13.6). In this study, the landslide susceptibility index (LSI) was reclassified into five susceptibility classes according to the quantile classification method of ArcGIS 10.2, very low, low, moderate, high, and very high classes (Fig. 13.5). The quantile classification method was utilized because of its popularity (Jebur et al. 2015; Tehrany et al. 2013); however, other methods (i.e., natural breaks, equal intervals) could also be used depending on the application types and the case study being investigated (Kritikos and Davies 2014; Nandi and Shakoor 2010).

$$\begin{aligned} LSI = {} & (0.010 \times \text{``Altitude''}) + (0.053 \times \text{``Slope''}) \\ & + (0.286 \times \text{``Aspect''}) \\ & + (0.030 \times \text{``Curvature''}) - (0.51 \times \text{``SPI''}) \\ & - (0.022 \times \text{``TWI''}) + (0.021 \times \text{``TRI''}) \\ & + (0.076 \times \text{``STI''}) - (0.017 \times \text{``Lineament''}) \\ & + (0.184 \times \text{``Geology''}) + (0.149 \\ & \times \text{``river''}) + (0.004 \times \text{``road''}) - 0.816 \end{aligned} \tag{13.12}$$

The generated map (Fig. 13.5) reflects the potential of landsliding in Ringlet area, Malaysia. The AUC-based validation of this map showed that the success and prediction rates of the LR model were 86.22% and 84.87%, respectively. LR achieved an acceptable accuracy according to the ROC curves as shown in Fig. 13.6.

13.6.2 Results of Landslide Hazard Mapping

After the landslide susceptibility map was produced based on the topographic and geomorphologic factors, the values of rainfall for 15 years were used to generate the hazard maps. The hazard maps were categorized using a quantile classification method. The hazard maps were divided into five classes, namely very low, low, medium, high, and very high as shown in Fig. 13.7. The available rainfall information (2000–2015) has made the production of hazard maps.

Several scenarios of hazards were evaluated based on the results of landslide susceptibility and rainfall triggering factor. For different hazard types, determining if the hazard types are dependent or not is important, which means finding out whether or not the hazard types are related to the same triggering event. In this study, all of the hazards are related to the same trigger that is rainfall. This fact is important in estimating risk, where we take the maximum loss per LULC type of the hazards. Therefore, we would not add the losses for different hazards but would take the maximum losses for each hazard.

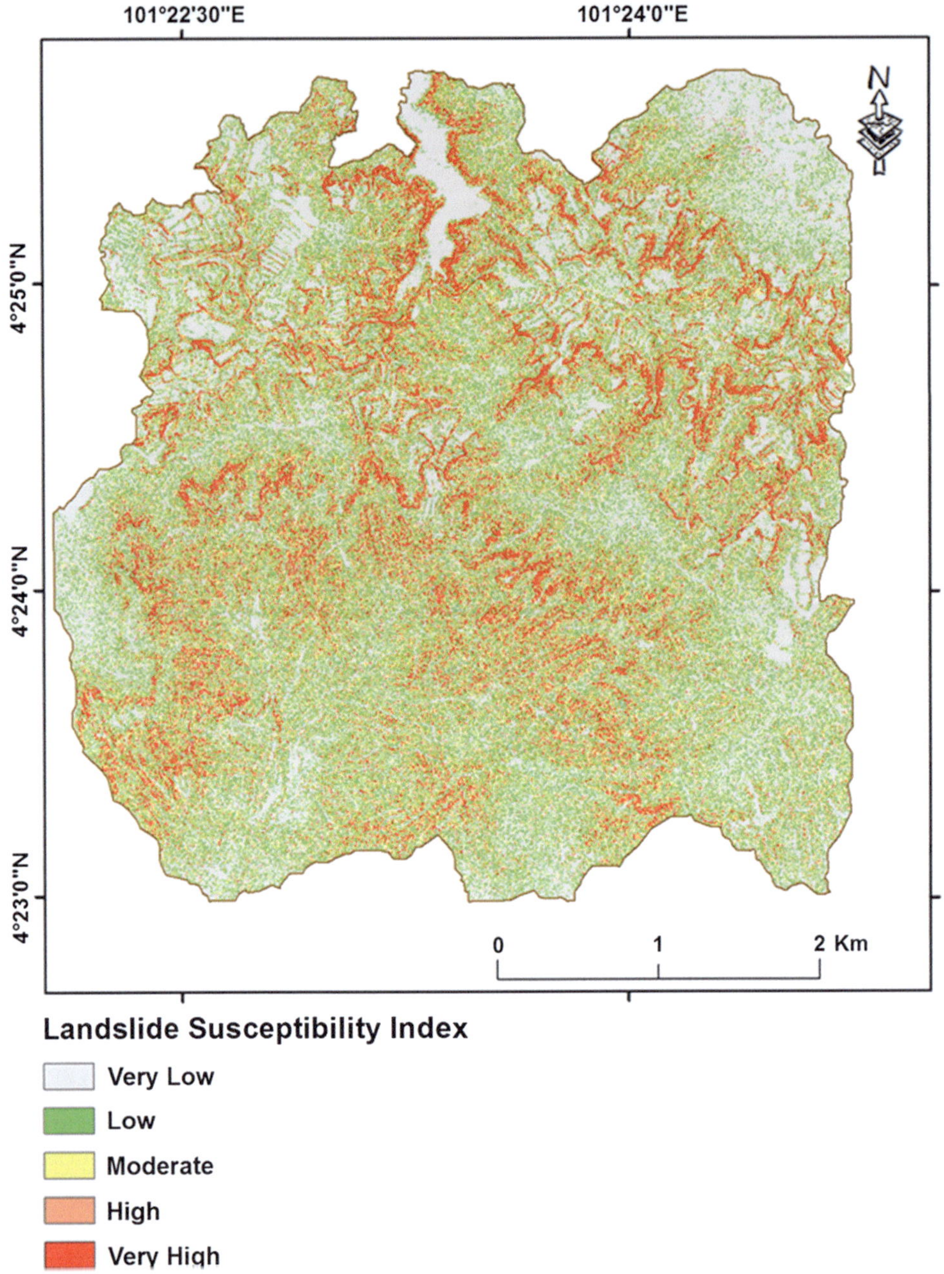

Fig. 13.5 Landslide susceptibility map of the study area produced by LR model

Although most parts of the study area were generally susceptible to landslides, the hazard was high and very high. By contrast, the significant difference between hazards of return periods, average, abnormal can be attributed to the amount of rainfall intensity.

13.6.3 Results of Vulnerability Assessment

A vulnerability map was derived according to the LULC criteria because of the lack of information on landslide intensity: cost, the time required for reconstruction, the relative risk of landslide, the risk to the population, and general effect on certain damage. The LULC map of the study area was produced by a supervised classification of high-resolution SPOT image (10 m spatial resolution) supplemented by very high-resolution aerial orthophotos and manual editing. Overall, 12 classes were identified and mapped as shown in Fig. 13.8. Forest area is considered as the main LULC type in the study area. Each LULC type was assigned a number (1–5) associated with the condition of the LULC type under particular

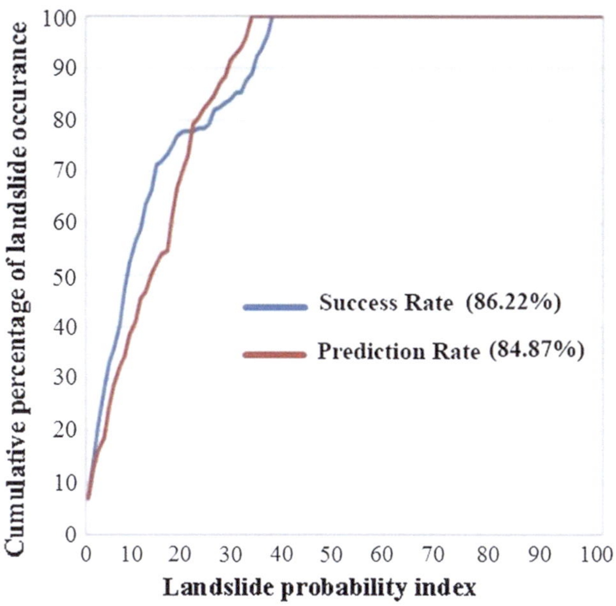

Fig. 13.6 AUC **a** success rate curve and **b** prediction rate curve

criteria. Subsequently, the outcome was resampled using Eq. 13.13.

$$X_{ij} = X_j - X_{ij}/X_{\max-j} - X_{\min-j,} \tag{13.13}$$

where X_{ij} is the standardized score for the ith alternative and jth attribute, X_{ij} is the raw score, and $X_{\max-j}$ and $X_{\min-j}$ are the maximum and minimum scores for the jth attribute, respectively. To normalize vulnerability with a range between 0 and 1, the vulnerability map was generated and divided into five classes, namely very low, low, medium, high, and very high. Most of the LULC types have a vulnerability of more than 0.5. The highest vulnerability was concentrated in certain parts of the study area (Fig. 13.9).

The results show that some of the LULC types that showed high vulnerability were categorized as low risk because of the conditions of the resulting hazard, as shown in the northern part of the study area. Meanwhile, some roads and power lines were categorized as high risk because the hazard and vulnerability of these facilities were very high. Lastly, the dam area was categorized as low risk when it shows high vulnerability because of a hazard condition.

13.6.4 Results of Risk Assessment

Thereafter, risk analysis was conducted. The LULC map is cross-matched with the results of return-period hazard maps with respect to vulnerability index (Fig. 13.10) to show the risk maps, while the losses were calculated for the three return periods by implementing just the area of each type of LULC crossed with high and very high hazard, and then, the losses are computed depending on the cost of each type of LULC in meter and the vulnerability index. The losses are aggregated for the LULC in the type of monetary loss.

Figure 13.11 shows the generated risk maps produced with two different hazard scenarios. The first scenario included any day in the year (average rainfall values), whereas the second scenario analyzed the rain values with abnormal intensity.

Table 13.3 shows the total risk computed over the Ringlet area for three different return periods in the absence of mitigation measures. It is worth noting that, due to the nonlinear increase in hazard with time, the increase in expected losses with time is not linear.

$$\begin{aligned}\text{Annual risk} &= 0.06 \times 35{,}922{,}250 + (0.1 - 0.06) \\ &\quad \times (35{,}922{,}250 + 29{,}147{,}100)/2 \\ &\quad + (0.2 - 0.1) \times (29{,}147{,}100 + 21{,}346{,}050)/2 \\ &= 5{,}981{,}379.00 \text{ MYR.}\end{aligned}$$

The area under risk is calculated and defined as annual risk (Fig. 13.12). The annual risk is used in the cost–benefit analysis, where the difference in annual risk before and after the implementation of risk reduction measures (benefit) is compared with the cost of implementation.

13.7 Discussion and Conclusion

The results of this study showed that the average annual economic risk of landslides is 5,981,379.00 MYR in the study area. The uncertainty in the calculated losses related to a number of factors including the quality and the availability of data, the accuracy of landslide susceptibility and hazard maps, and the definition of multi-hazard scenarios. The quantity and the spatial distributions of

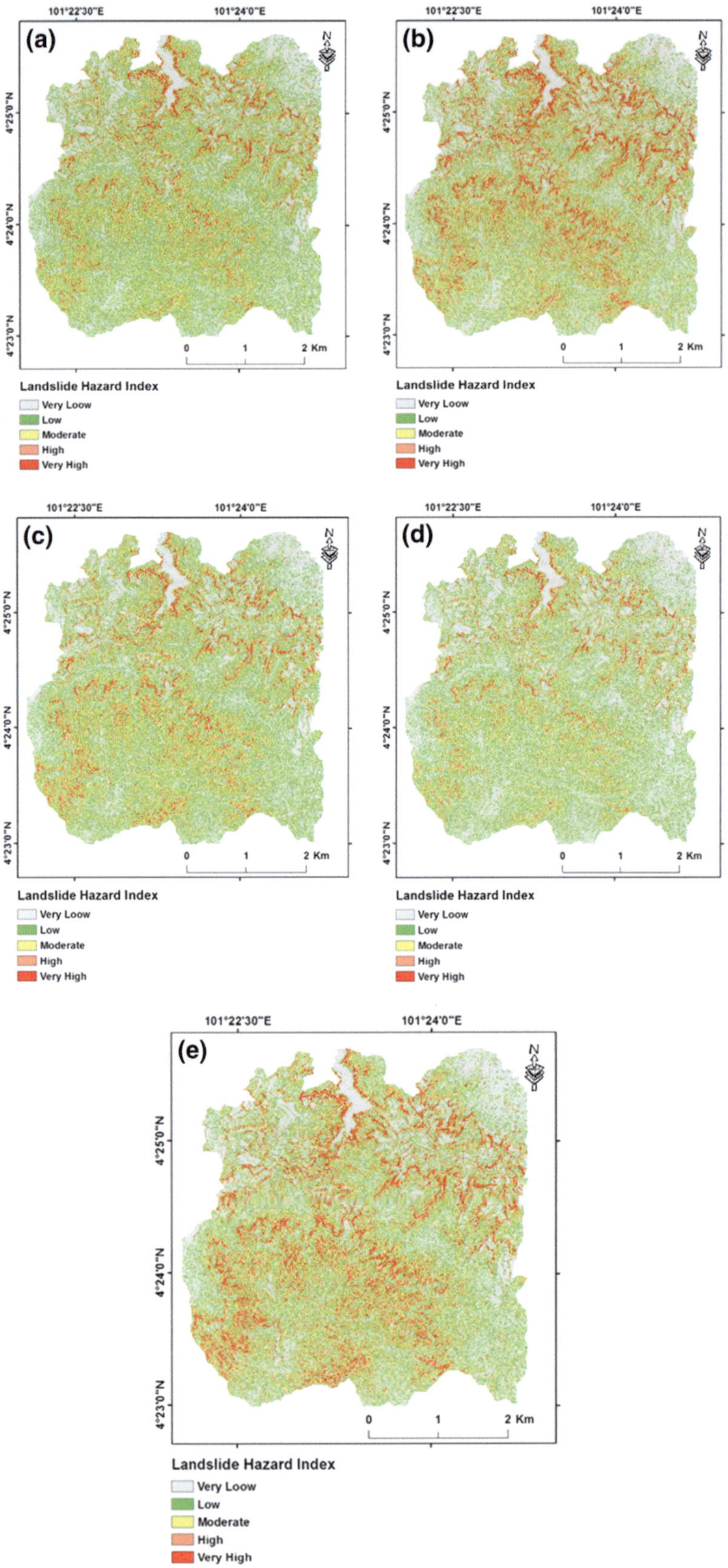

Fig. 13.7 Hazard maps for the study area: **a** hazard map when average intensity of rain (in any day), **b** hazard map when abnormal intensity of rain recorded (in day), **c** hazard map for 5-year return period with average intensity of rainfall per day, **d** hazard map for 10-year return period, and **e** hazard map for 15-year return period

Fig. 13.8 LULC map of the study area

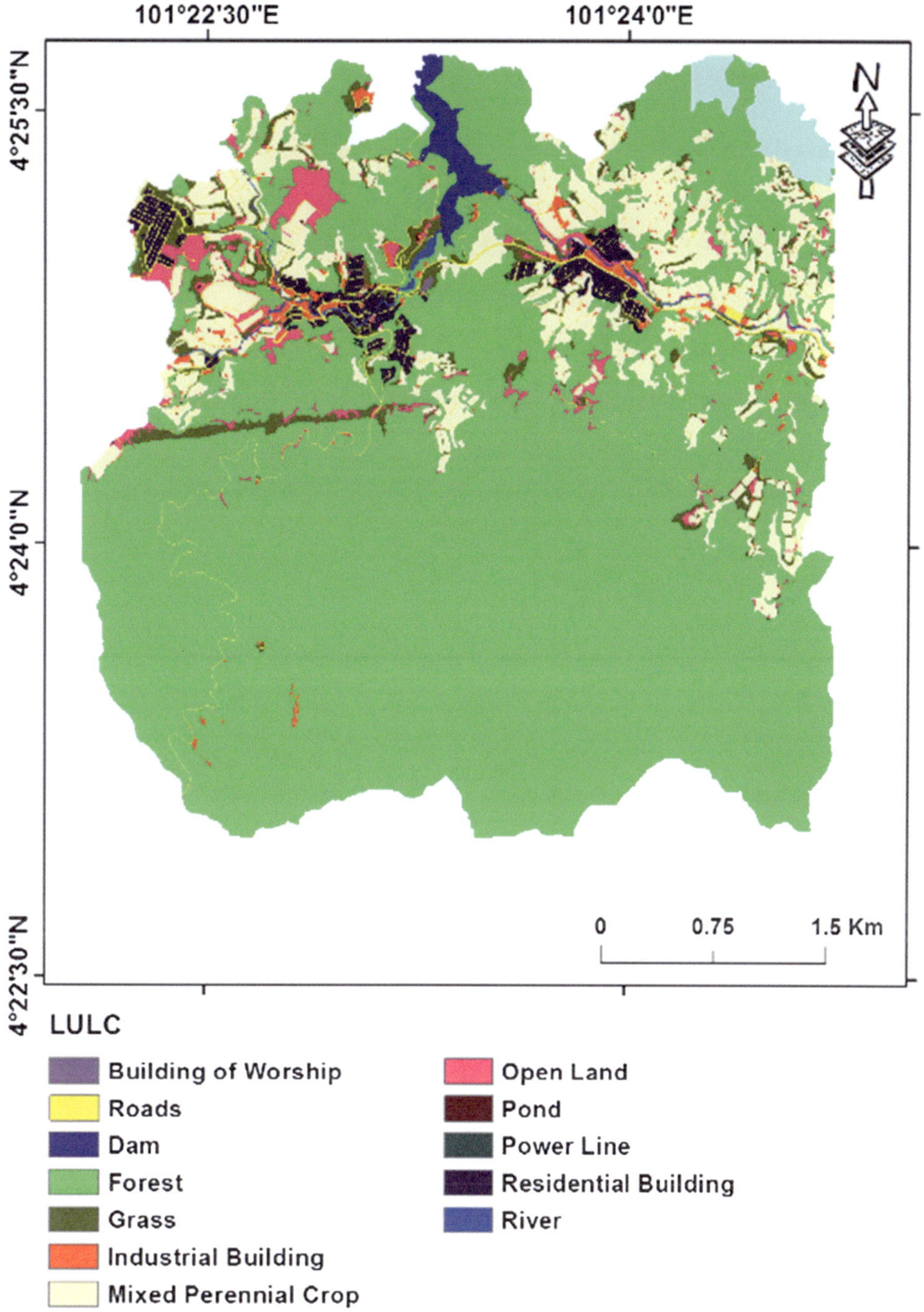

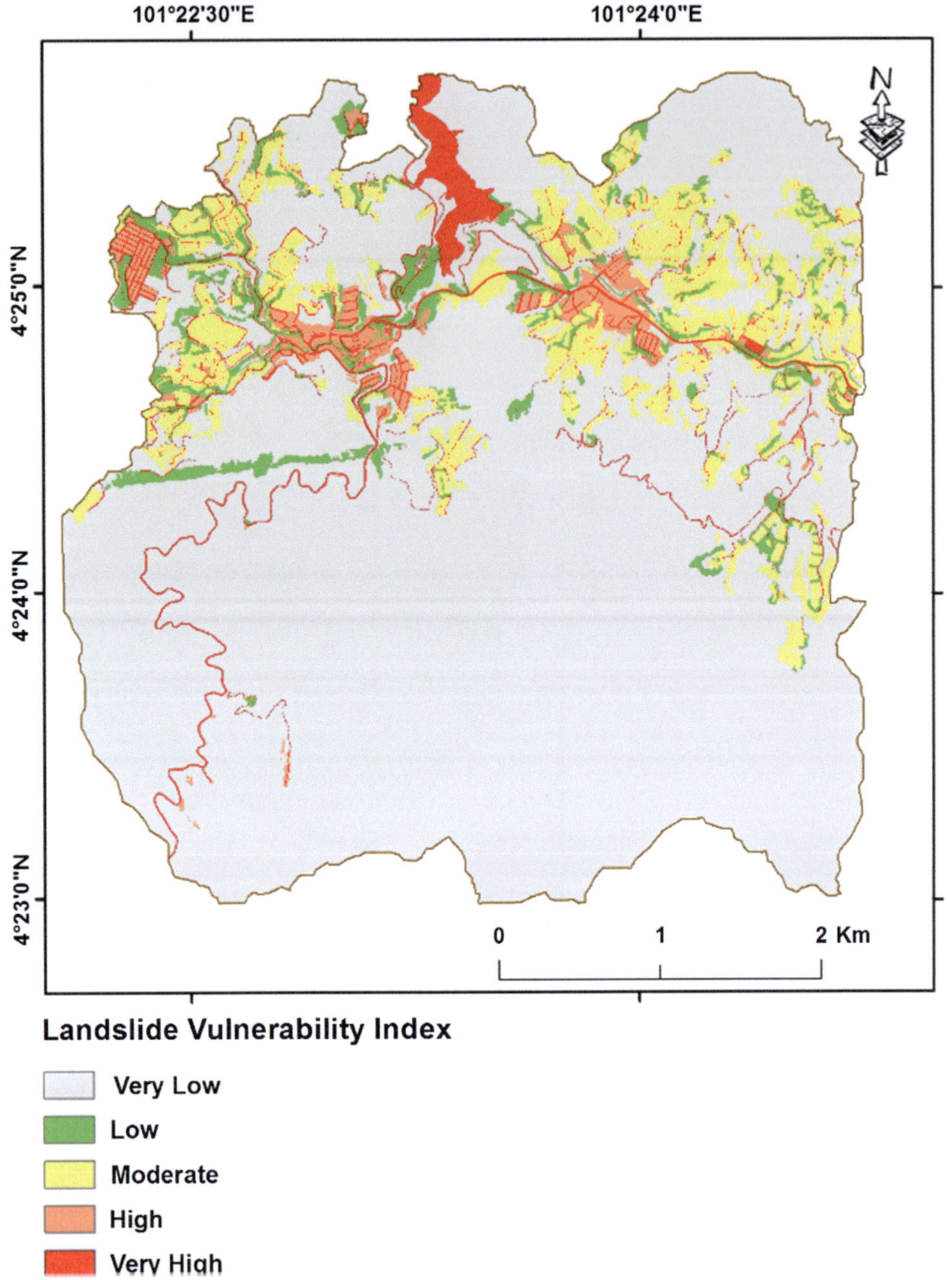

Fig. 13.9 Landslide vulnerability map

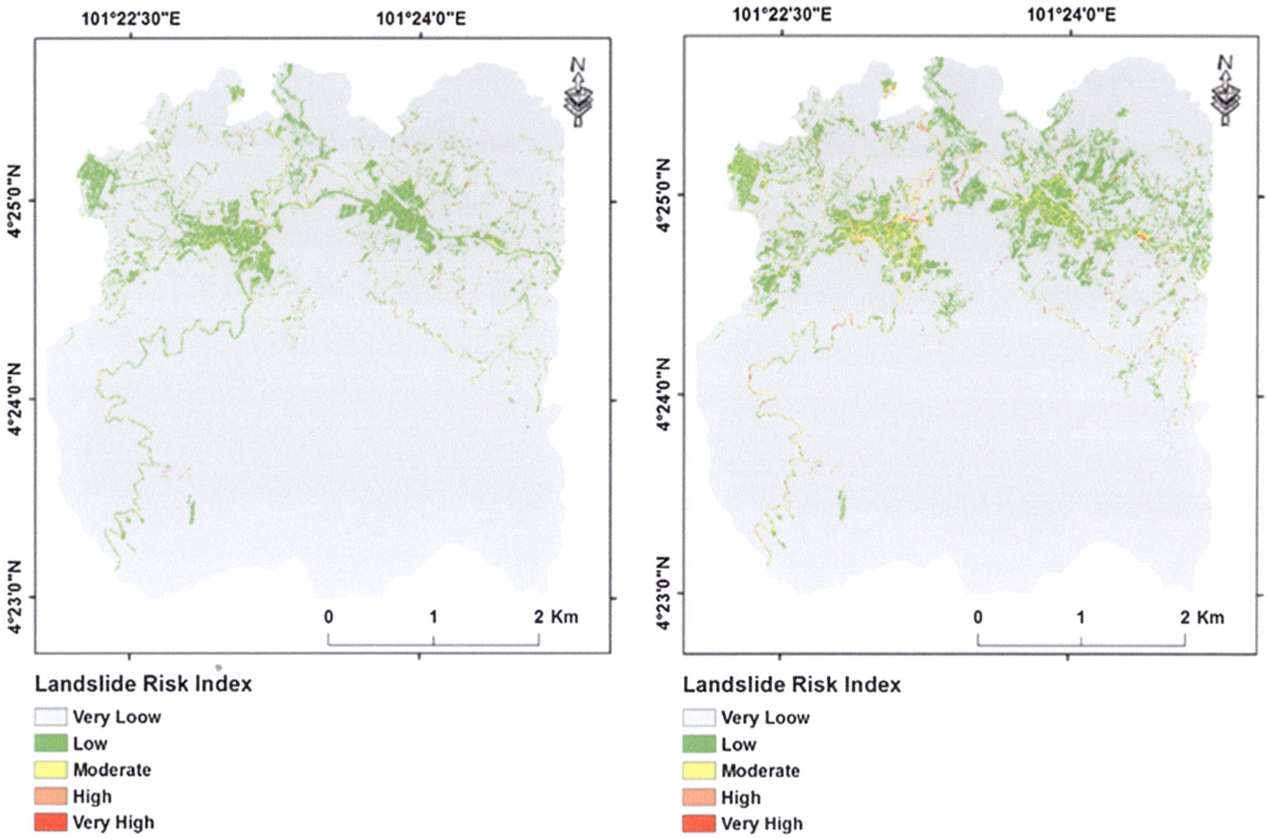

Fig. 13.10 Generated risk maps for the study area for different scenarios: (*left*) risk map for any day in the year, (*right*) risk map when the rain value record abnormal intensity

historical landslides, as well as the missing information such as date of landslide events and their magnitudes, affected on the accuracy of the calculated losses. In addition, the prediction accuracy of LR model (84.87%) can be considered as acceptable for landslide susceptibility assessment, but still there were uncertainties in the estimated losses due to the error rate of the LR model. Furthermore, during extreme rainfall events, multiple hazards (landslides, flash floods) may affect the same locations. To reduce the uncertainties due to this problem, only maximum losses were considered. However, this cannot solve the problem completely, and further investigations and modeling with high-quality data are required. Another uncertainty is due to the quality of vulnerability curves which mostly based on expert opinion and previous works. To improve the landslide risk modeling, spatial libraries that contain building types, the number of people in each building, and hazard intensities will be necessary.

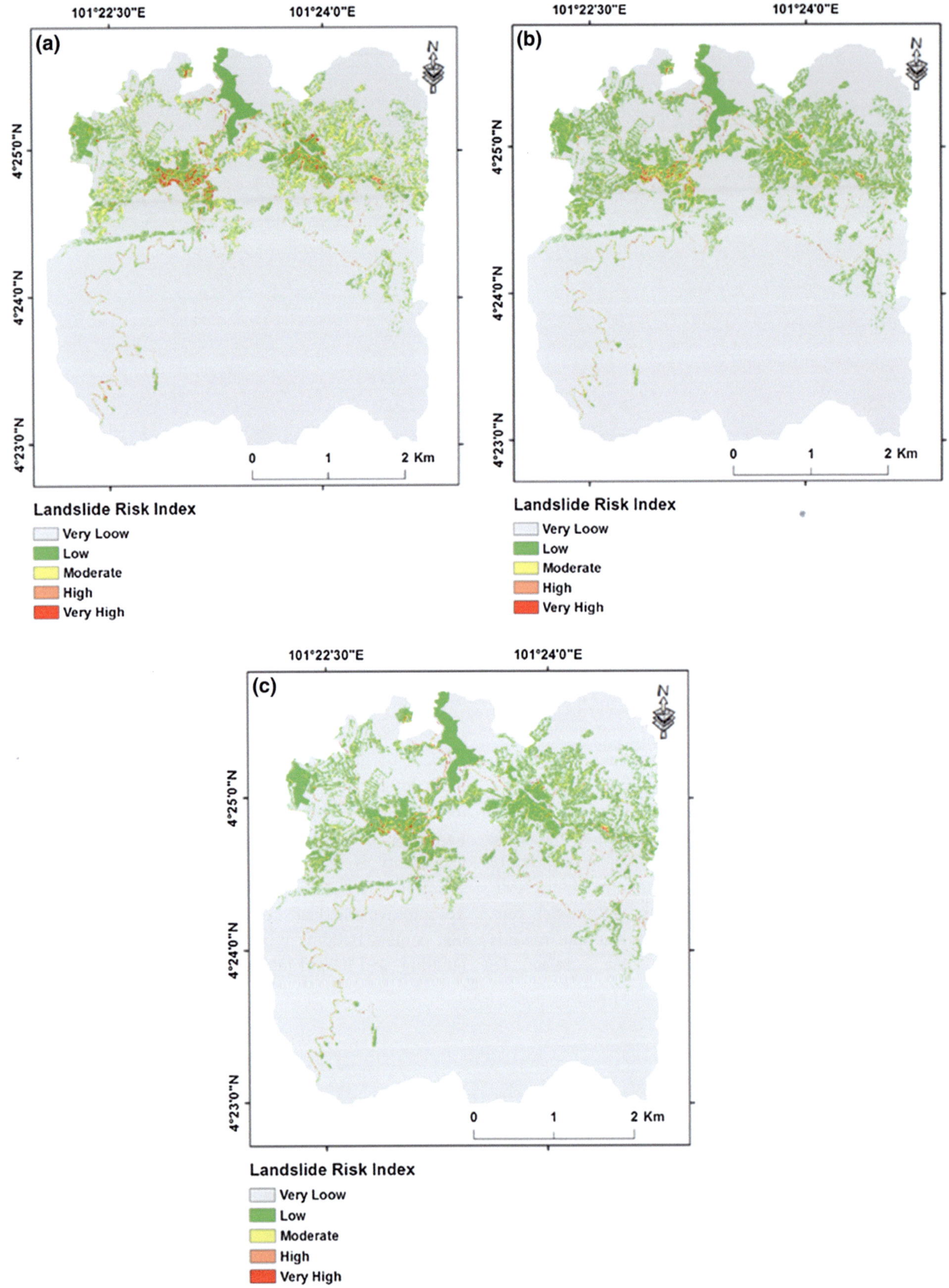

Fig. 13.11 Generated risk maps for the study area for different return periods: **a** risk map for 15 years, **b** risk map for 10 years, and **c** risk map for 5 years

Table 13.3 Loss estimation for each return period

Loss	Return period	Annual probability
21,346,050	5	0.2
29,147,100	10	0.1
35,922,250	15	0.06

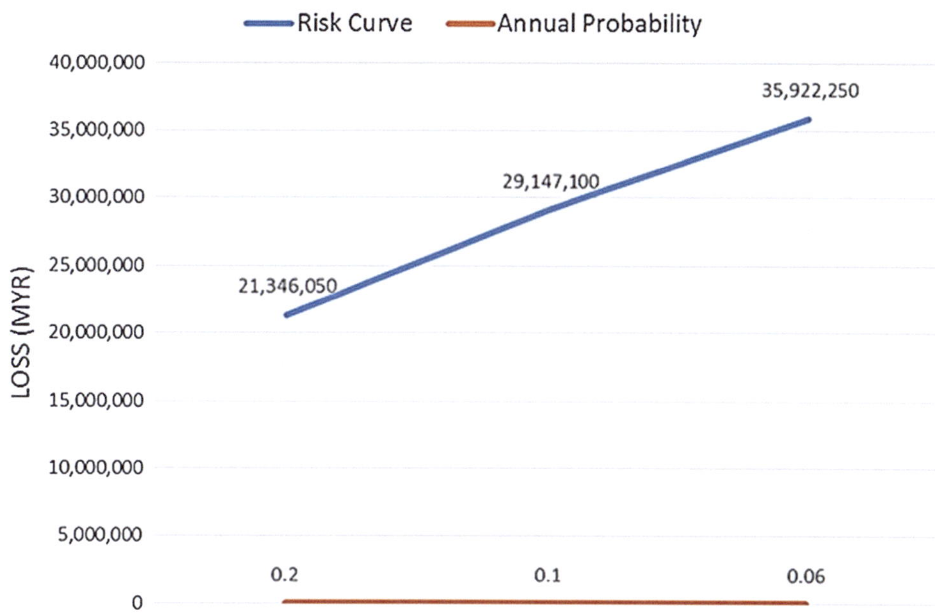

Fig. 13.12 Risk curve for the study area

There are many challenges to accurately model multi-hazard landslide risks which are important to be further investigated in future works. Building comprehensive landslide inventory databases is the most important research problem as it creates uncertainties in all the subsequent analysis, establishing relationships between hazards intensities and historical building damages and losses in other properties. Finally, validation methods of landslide risk assessment are at early stages. The traditional method is to compare the calculated losses with those of similar events that have occurred in the past. However, to improve the landslide risk modeling approaches, validation methods should be improved so that better comparison can be done with the available models.

References

Akgun, A. (2012). A comparison of landslide susceptibility maps produced by logistic regression, multi-criteria decision, and likelihood ratio methods: a case study at İzmir, Turkey. *Landslides, 9*(1), 93–106.

Akgun, A., Kıncal, C., & Pradhan, B. (2012). Application of remote sensing data and GIS for landslide risk assessment as an environmental threat to Izmir city (west Turkey). *Environmental Monitoring and Assessment, 184*(9), 5453–5470.

Alimohammadlou, Y., Najafi, A., & Yalcin, A. (2013). Landslide process and impacts: A proposed classification method. *Catena, 104,* 219–232.

Althuwaynee, O. F., & Pradhan, B. (2016). Semi-quantitative landslide risk assessment using GIS-based exposure analysis in Kuala Lumpur City.

Althuwaynee, O. F., Pradhan, B., & Ahmad, N. (2014a). Estimation of rainfall threshold and its use in landslide hazard mapping of kuala lumpur metropolitan and surrounding areas. *Landslides,* 1–15.

Althuwaynee, O. F., Pradhan, B., Park, H.-J., & Lee, J. H. (2014b). A novel ensemble bivariate statistical evidential belief function with knowledge-based analytical hierarchy process and multivariate statistical logistic regression for landslide susceptibility mapping. *Catena, 114,* 21–36.

Bai, S., Wang, J., Lü, G., Zhou, P., Hou, S., & Xu, S. (2010). GIS-based logistic regression for landslide susceptibility mapping of the Zhongxian segment in the three gorges area, China. *Geomorphology, 115*(1), 23–31.

Ballabio, C., & Sterlacchini, S. (2012). Support vector machines for landslide susceptibility mapping: the Staffora River Basin case study, Italy. *Mathematical Geosciences, 44*(1), 47–70.

Bell, R., & Glade, T. (2004). Quantitative risk analysis for landslides? Examples from Bíldudalur, NW-Iceland. *Natural Hazards and Earth System Science, 4*(1), 117–131.

Bornaetxea, T., Antigüedad, I., & Ormaetxea, O. (2016, April). Shallow landslide susceptibility model for the Oria river basin, Gipuzkoa province (North of Spain). Application of the logistic regression and comparison with previous studies. In *EGU General Assembly Conference Abstracts* (Vol. 18, p. 7715).

Budimir, M., Atkinson, P., & Lewis, H. (2015). A systematic review of landslide probability mapping using logistic regression. *Landslides*, 1–18.

Bui, D. T., Tuan, T. A., Klempe, H., Pradhan, B., & Revhaug, I. (2016). Spatial prediction models for shallow landslide hazards: A comparative assessment of the efficacy of support vector machines, artificial neural networks, kernel logistic regression, and logistic model tree. *Landslides, 13*(2), 361–378.

Calvo, B., & Savi, F. (2009). A real-world application of monte carlo procedure for debris flow risk assessment. *Computers & Geosciences, 35*(5), 967–977.

Chung, C. F., & Fabbri, A. G. (2003). Validation of spatial prediction models for landslide hazard mapping. *Natural Hazards, 30*(3), 451–472.

Corominas, J., Van Westen, C., Frattini, P., Cascini, L., Malet, J.-P., Fotopoulou, S., et al. (2014). Recommendations for the quantitative analysis of landslide risk. *Bulletin of Engineering Geology and the Environment, 73*(2), 209–263.

Costanzo, D., Chacón, J., Conoscenti, C., Irigaray, C., & Rotigliano, E. (2014). Forward logistic regression for earth-flow landslide susceptibility assessment in the Platani river basin (southern Sicily, Italy). *Landslides, 11*(4), 639–653.

Demir, G., Aytekin, M., & Akgun, A. (2015). Landslide susceptibility mapping by frequency ratio and logistic regression methods: An example from Niksar-Resadiye (Tokat, Turkey). *Arabian Journal of Geosciences, 8*(3), 1801–1812.

Devkota, K. C., Regmi, A. D., Pourghasemi, H. R., Yoshida, K., Pradhan, B., Ryu, I. C., et al. (2013). Landslide susceptibility mapping using certainty factor, index of entropy and logistic regression models in GIS and their comparison at Mugling–Narayanghat road section in Nepal Himalaya. *Natural Hazards, 65* (1), 135–165.

Dilley, M. (2005). Natural disaster hotspots: A global risk analysis. World Bank Publications.

Dou, J., Oguchi, T., Hayakawa, Y. S., Uchiyama, S., Saito, H., & Paudel, U. (2014). GIS-based landslide susceptibility mapping using a certainty factor model and its validation in the Chuetsu area, Central Japan. In *Landslide science for a safer geoenvironment* (pp. 419–424). Berlin: Springer.

Dragićević, S., Lai, T., & Balram, S. (2015). GIS-based multicriteria evaluation with multiscale analysis to characterize urban landslide susceptibility in data-scarce environments. *Habitat International, 45,* 114–125.

Erener, A., & Düzgün, H. S. (2013). A regional scale quantitative risk assessment for landslides: Case of Kumluca watershed in Bartin, Turkey. *Landslides, 10*(1), 55–73.

Erener, A., Mutlu, A., & Düzgün, H. S. (2016). A comparative study for landslide susceptibility mapping using GIS-based multi criteria decision analysis (MCDA), logistic regression (LR) and association rule mining (ARM). *Engineering Geology, 203*, 45–55.

Falaschi, F., Giacomelli, F., Federici, P., Puccinelli, A., Avanzi, G., Pochini, A., et al. (2009). Logistic regression versus artificial neural networks: Landslide susceptibility evaluation in a sample area of the Serchio river valley, Italy. *Natural Hazards, 50*(3), 551–569.

Fell, R., Corominas, J., Bonnard, C., Cascini, L., Leroi, E., & Savage, W. Z. (2008). Guidelines for landslide susceptibility, hazard and risk zoning for land-use planning. *Engineering Geology, 102*(3), 99–111.

Fell, R., Ho, K. K., Lacasse, S., & Leroi, E. (2005). A framework for landslide risk assessment and management. *Landslide Risk Management*, 3–25.

Ghosh, S., van Westen, C. J., Carranza, E. J. M., Jetten, V. G., Cardinali, M., Rossi, M., et al. (2012). Generating event-based landslide maps in a data-scarce Himalayan environment for estimating temporal and magnitude probabilities. *Engineering Geology, 128,* 49–62.

Glade, T., Crozier, M., & Smith, P. (2000). Applying probability determination to refine landslide-triggering rainfall thresholds using an empirical "Antecedent Daily Rainfall Model". *Pure and Applied Geophysics, 157*(6–8), 1059–1079.

Guzzetti, F., Ardizzone, F., Cardinali, M., Galli, M., Reichenbach, P., & Rossi, M. (2008). Distribution of landslides in the Upper Tiber River basin, Central Italy. *Geomorphology, 96*(1), 105–122.

Guzzetti, F., Galli, M., Reichenbach, P., Ardizzone, F., & Cardinali, M. (2006). Landslide hazard assessment in the Collazzone area, Umbria, Central Italy. *Natural Hazards and Earth System Science, 6*(1), 115–131.

Haneberg, W. C. (2004). A rational probabilistic method for spatially distributed landslide hazard assessment. *Environmental and Engineering Geoscience, 10*(1), 27–43.

Hong, H., Pradhan, B., Bui, D. T., Xu, C., Youssef, A. M., & Chen, W. (2016). Comparison of four kernel functions used in support vector machines for landslide susceptibility mapping: A case study at Suichuan area (China). *Geomatics, Natural Hazards and Risk*, 1–26.

Huang, J., Lyamin, A., Griffiths, D., Krabbenhoft, K., & Sloan, S. (2013). Quantitative risk assessment of landslide by limit analysis and random fields. *Computers and Geotechnics, 53*, 60–67.

Jebur, M. N., Pradhan, B., & Tehrany, M. S. (2014). Optimization of landslide conditioning factors using very high-resolution airborne laser scanning (LiDAR) data at catchment scale. *Remote Sensing of Environment, 152,* 150–165.

Jebur, M. N., Pradhan, B., & Tehrany, M. S. (2015). Using ALOS PALSAR derived high-resolution DInSAR to detect slow-moving landslides in tropical forest: Cameron Highlands, Malaysia. *Geomatics, Natural Hazards and Risk, 6*(8), 741–759.

Kanungo, D., Arora, M., Gupta, R., & Sarkar, S. (2008). Landslide risk assessment using concepts of danger pixels and fuzzy set theory in Darjeeling Himalayas. *Landslides, 5*(4), 407–416.

Kavzoglu, T., Sahin, E. K., & Colkesen, I. (2014). Landslide susceptibility mapping using GIS-based multi-criteria decision analysis, support vector machines, and logistic regression. *Landslides, 11*(3), 425–439.

Kritikos, T., & Davies, T. (2014). Assessment of rainfall-generated shallow landslide/debris-flow susceptibility and runout using a GIS-based approach: Application to western southern Alps of New Zealand. *Landslides*, 1-25.

Lee, E., & Jones, D. (2004). Landslide risk assessment.

Lee, M. L., Ng, K. Y., Huang, Y. F., & Li, W. C. (2014). Rainfall-induced landslides in Hulu Kelang area, Malaysia. *Natural hazards, 70*(1), 353–375.

Li, Z., Nadim, F., Huang, H., Uzielli, M., & Lacasse, S. (2010). Quantitative vulnerability estimation for scenario-based landslide hazards. *Landslides, 7*(2), 125–134.

Moreiras, S. M., & Sepúlveda, S. A. (2015). Megalandslides in the Andes of central Chile and Argentina (32°–34° S) and potential hazards. *Geological Society, London, Special Publications, 399*(1), 329–344.

Muthukumar, M. (2013). GIS based Geosystem response modeling for landslide vulnerability mapping parts of Nilgiris, South India. *Disaster Advances, 6*(7), 58–66.

Nandi, A., & Shakoor, A. (2010). A GIS-based landslide susceptibility evaluation using bivariate and multivariate statistical analyses. *Engineering Geology, 110*(1), 11–20.

Opolot, E. (2013). Application of remote sensing and geographical information systems in flood management: A review. *Research Journal of Applied Sciences, Engineering and Technology, 5*(10), 1884–1894.

Ozdemir, A., & Altural, T. (2013). A comparative study of frequency ratio, weights of evidence and logistic regression methods for landslide susceptibility mapping: Sultan Mountains, SW Turkey. *Journal of Asian Earth Sciences, 64*, 180–197.

Pedrazzini, A., Humair, F., Jaboyedoff, M., & Tonini, M. (2015). Characterisation and spatial distribution of gravitational slope deformation in the upper rhone catchment (western Swiss Alps). *Landslides*, 1–19.

Pourghasemi, H. R., Jirandeh, A. G., Pradhan, B., Xu, C., & Gokceoglu, C. (2013). Landslide susceptibility mapping using support vector machine and GIS at the Golestan Province, Iran. *Journal of Earth System Science, 122*(2), 349–369.

Pradhan, B. (2013). A comparative study on the predictive ability of the decision tree, support vector machine and neuro-fuzzy models in landslide susceptibility mapping using GIS. *Computers & Geosciences, 51*, 350–365.

Pradhan, B., Abokharima, M. H., Jebur, M. N., & Tehrany, M. S. (2014). Land subsidence susceptibility mapping at Kinta Valley (Malaysia) using the evidential belief function model in GIS. *Natural Hazards, 73*(2), 1019–1042.

Pradhan, B., & Lee, S. (2009). Landslide risk analysis using artificial neural network model focusing on different training sites. *International Journal of Physical Sciences, 3*(11), 1–15.

Pradhan, B., & Lee, S. (2010). Delineation of landslide hazard areas on Penang island, Malaysia, by using frequency ratio, logistic regression, and artificial neural network models. *Environmental Earth Sciences, 60*(5), 1037–1054.

Pradhan, B., & Youssef, A. M. (2010). Manifestation of remote sensing data and GIS on landslide hazard analysis using spatial-based statistical models. *Arabian Journal of Geosciences, 3*(3), 319–326.

Regmi, N. R., Giardino, J. R., McDonald, E. V., & Vitek, J. D. (2014). A comparison of logistic regression-based models of susceptibility to landslides in western Colorado, USA. *Landslides, 11*(2), 247–262.

Regmi, N. R., Giardino, J. R., & Vitek, J. D. (2010). Modeling susceptibility to landslides using the weight of evidence approach: Western Colorado, USA. *Geomorphology, 115*(1), 172–187.

Solaimani, K., Mousavi, S. Z., & Kavian, A. (2013). Landslide susceptibility mapping based on frequency ratio and logistic regression models. *Arabian Journal of Geosciences, 6*(7), 2557–2569.

Tehrany, M. S., Pradhan, B., & Jebur, M. N. (2013). Spatial prediction of flood susceptible areas using rule based decision tree (DT) and a novel ensemble bivariate and multivariate statistical models in GIS. *Journal of Hydrology, 504*, 69–79.

Tehrany, M. S., Pradhan, B., & Jebuv, M. N. (2014). A comparative assessment between object and pixel-based classification approaches for land use/land cover mapping using SPOT 5 imagery. *Geocarto International, 29*(4), 351–369.

Tournadour, E., Mulder, T., Borgomano, J., Hanquiez, V., Ducassou, E., & Gillet, H. (2015). Origin and architecture of a mass transport complex on the northwest slope of little Bahama Bank (Bahamas): Relations between off-bank transport, bottom current sedimentation and submarine landslides. *Sedimentary Geology, 317*, 9–26.

Umar, Z., Pradhan, B., Ahmad, A., Jebur, M. N., & Tehrany, M. S. (2014). Earthquake induced landslide susceptibility mapping using an integrated ensemble frequency ratio and logistic regression models in West Sumatera Province, Indonesia. *Catena, 118*, 124–135.

van Westen, C. J., Castellanos, E., & Kuriakose, S. L. (2008). Spatial data for landslide susceptibility, hazard, and vulnerability assessment: An overview. *Engineering Geology, 102*(3), 112–131.

Varnes, D. J. (1984). Landslide hazard zonation: A review of principles and practice (No. 3).

Vranken, L., Vantilt, G., Van Den Eeckhaut, M., Vandekerckhove, L., & Poesen, J. (2015). Landslide risk assessment in a densely populated hilly area. *Landslides, 12*(4), 787–798.

Wang, L., Sawada, K., & Moriguchi, S. (2013). Landslide susceptibility analysis with logistic regression model based on FCM sampling strategy. *Computers & Geosciences, 57*, 81–92.

Wu, X., Ren, F., & Niu, R. (2014). Landslide susceptibility assessment using object mapping units, decision tree, and support vector machine models in the Three Gorges of China. *Environmental Earth Sciences, 71*(11), 4725–4738.

Xu, C., Dai, F., Xu, X., & Lee, Y. H. (2012). GIS-based support vector machine modeling of earthquake-triggered landslide susceptibility in the Jianjiang River watershed, China. *Geomorphology, 145*, 70–80.

Xu, J., Nyerges, T. L., & Nie, G. (2014). Modeling and representation for earthquake emergency response knowledge: perspective for working with geo-ontology. *International Journal of Geographical Information Science, 28*(1), 185–205.

Xu, C., Xu, X., Dai, F., Wu, Z., He, H., Shi, F., et al. (2013). Application of an incomplete landslide inventory, logistic regression model and its validation for landslide susceptibility mapping related to the May 12, 2008 Wenchuan earthquake of China. *Natural Hazards, 68*(2), 883–900.

Yang, W., Peng, Z., Wang, B., Li, Z., & Yuan, S. (2015). Velocity contrast along the rupture zone of the 2010 Mw6. 9 Yushu, China, earthquake from fault zone head waves. *Earth and Planetary Science Letters, 416*, 91–97.

Yang, M., Wu, M., & Liu, J. (2014). Estimating landslide-induced riverbed roughness variation by using lidar data. *Journal of Marine Science and Technology, 22*(4), 424–429.

Yao, X., & Dai, F. (2006). Support vector machine modeling of landslide susceptibility using a GIS: A case study. *IAEG, 2006*, 793.

Yao, X., Tham, L., & Dai, F. (2008). Landslide susceptibility mapping based on support vector machine: A case study on natural slopes of Hong Kong, China. *Geomorphology, 101*(4), 572–582.

Youssef, A. M., Al-Kathery, M., & Pradhan, B. (2015). Landslide susceptibility mapping at Al-Hasher Area, Jizan (Saudi Arabia) using GIS-based frequency ratio and index of entropy models. *Geosciences Journal, 19*(1), 113–134.

Yusof, N. M., Pradhan, B., Shafri, H. Z. M., Jebur, M. N., & Yusoff, Z. (2015). Spatial landslide hazard assessment along the jelapang corridor of the north-south expressway in Malaysia using high-resolution airborne LiDAR data. *Arabian Journal of Geosciences*, 1–12.

Zhang, M., Cao, X., Peng, L., & Niu, R. (2016). Landslide susceptibility mapping based on global and local logistic regression models in Three Gorges Reservoir area, China. *Environmental Earth Sciences, 75*(11), 1–11.

Part VI
Debris Flow Modelling

Debris Flow Susceptibility Assessment Using Airborne Laser Scanning Data

14

Biswajeet Pradhan, Bahareh Kalantar, Waleed M. Abdulwahid and Bui Tien Dieu

14.1 Introduction

Debris flows and related landslide failure phenomena occur in many mountainous areas worldwide and pose significant hazards to settlements, human lives, and transportation corridors (Jaedicke et al. 2008, 2009). Debris flows occur on different terrains where sufficient debris materials are available and the angle of slope is steep enough. Flow behavior is of different types, namely confined, unconfined, and transition (Fannin and Wise 2001). Flow behavior can be identified according to the specifications of the source and evidence area (Lorente et al. 2007). Given the heavy precipitation in upper catchments, debris flows become frequent in highly susceptible areas. However, a susceptibility map does not directly represent the temporal probability occurrence because temporal data are not used in debris flow modeling. Debris flow susceptibility mapping is generally implemented to delineate potential danger areas with minimum data requirements (Glade 2005; Jakob 2005; van Westen et al. 2006). Susceptibility analyses can identify the most vulnerable and potentially unstable areas, especially in down-slope regions, that could be affected by debris flow. Susceptibility analysis is the main assessment tool to identify potential impact zones because it determines where further studies need to be conducted in detail (van Westen et al. 2006).

Debris flow processes are controlled by many factors, such as geological setting, relief, substratum type, and availability of debris, and strongly influenced by triggering factors, such as rainfall and heavy snowmelt events (Fischer et al. 2012). In recent years, the use of high-resolution airborne laser scanning data from light detection and ranging (LiDAR) has become extremely popular in debris flow modeling. LiDAR-derived data can be used as main inputs in debris flow source identification and susceptibility assessment to produce highly accurate and useful results. LiDAR-derived parameters can aid in the accurate modeling of depths and velocities and representation of field conditions. Accurate and reliable debris flow susceptibility maps can be produced with physical modeling methods. Physical models incorporate hydrological models and infinite slope stability in their analysis (Frattini et al. 2010). These models describe the geometry evolution of a finite mass of granular materials and the velocity distributions as the mass slides down an inclined plane. These models also simulate maximum flow height, momentum, and pressure, which are useful in assessing potential risks in an area (Christen et al. 2012).

This paper presents debris flow susceptibility mapping with a physical modeling approach implemented in Flow-R and rapid mass movement simulation (RAMMS) environments. Several topographical parameters (e.g., slope, aspect, and curvature) and thematic layers (e.g., land use and lithology) were examined. These parameters and thematic layers were then used to identify debris flow sources in the study area. Afterward, a susceptibility map of debris flow was produced with Flow-R and RAMMS models. The affected areas were evaluated by overlaying the produced susceptibility map and land use.

14.2 Previous Studies

In mountain environments, debris flows are the most dangerous natural hazards that threaten human lives. Designing debris flow barriers is necessary to mitigate the potential hazard and impact force of debris flows. Currently, developing theoretical and mechanistic models are insufficient (Proske et al. 2008; Hübl et al. 2009). Therefore, debris flow impact models for engineering purposes are based on empirical and rational models. Examples of such case studies were performed by Egli (2005) and Wendeler (2008).

B. Pradhan (✉) · B. Kalantar · W.M. Abdulwahid
Department of Civil Engineering, University Putra Malaysia, Serdang, Malaysia
e-mail: biswajeet24@gmail.com

B.T. Dieu
Department of Business Administration and Computer Science, Faculty of Art and Sciences, Telemark University College, Hallvard Eikas Plass 1, 3800 Bø Telemark, Norway

B. Pradhan (ed.), *Laser Scanning Applications in Landslide Assessment*,
DOI 10.1007/978-3-319-55342-9_14

Estimation of impact force can be performed under real-world conditions or in laboratories through experimental studies. Measurements of impact forces under real-world conditions were performed by Zhang (1993) and Hu et al. (2011) in China, König (2006) in Austria, and Wendeler et al. (2007) in Switzerland. An advantage of observing real debris flows is that scaling consideration is unnecessary (Scheidl et al. 2013). However, measuring additional indicators, such as flow height, speed, and density, is complex. Moreover, predicting the time of debris flow is difficult. Therefore, laboratory experiments, which involve small-scale measurements, were developed. However, physical modeling presents possible scale effects (Armanini and Scotton 1992; Ishikawa et al. 2008; Hübl and Holzinger 2003; Monney et al. 2007; Wendeler 2008). Given that a scale-dependent interaction exists between solid and fluid phases in debris flow behavior (Iverson and Denlinger 2001; Iverson et al. 2011), the hydrodynamic method was developed to extrapolate from the experimental scale to the field scale (Hübl and Holzinger 2003). This approach considers both geometric and simple kinematic similarities. A dimensionless number called the Froude number is used to characterize kinematic similarity and reveal the performance of a model in relation to a real system (Scheidl et al. 2013). Therefore, the values of the Froude number and natural debris flow events should be similar.

Hübl et al. (2009) investigated the relationship between the impact force of debris flow and the Froude number for both field data and data from miniaturized laboratory tests. They reported that the Froude number of the miniaturized tests ranged from 1.2 to 2.0; however, the range of 0–2 was used in the field measurement. They concluded that the model presents a systematic error when the ranges of input and field data do not agree. Therefore, hydrodynamic models do not perform well with low velocities and low Froude numbers. According to this literature review, the approaches for the prediction of run-out distance of landslide debris can be categorized into analytical, numerical, and empirical models (Dai and Lee 2002). Analytical models define the dynamic behavior of debris movement and depend on lumped mass methods in which the debris mass is assumed to be a single point (e.g., the sled model) (Sassa 1989). Numerical models define the dynamic motion of debris, and rheological models describe the material behavior of debris (Dai and Lee 2002). From a mathematical point of view, numerical models, such as RAMMS (Christen et al. 2010), are utilized to investigate the variation in debris flow. These models are pivotal tools for risk assessment and designing measures related to mitigation processes (Deubelbeiss and Graf 2013). However, models that use a numerical analysis consist of friction parameters that need to be adjusted through different means, such as matching past events' run-out distances, flow paths, flow heights, and velocities. Significant challenges, including a large variety of debris, different flow types, and different material compositions, are encountered when the calibration process is utilized; these challenges considerably influence the choice of friction factors (Naef et al. 2006; Rickenmann et al. 2006). To deal with this issue, the software platform RAMMS is used. RAMMS has a user-friendly graphical interface and flexible configurations, which make the software simple to understand (Christen et al. 2012). RAMMS can deal with debris flow but overlooks current erosion and entrainment (Berger et al. 2011; Schürch et al. 2011). However, "bulking" techniques are currently under development. Many applications that show the good performance of RAMMS in debris flow are found in Scheuner et al. (2011) and Berger et al. (2012).

Empirical methods for landslide run-out computation have been established based on several parameters, such as run-out distance, damage corridor width, velocity, depth of the moving mass, and depth of deposits (Dai and Lee 2002). Empirical approaches are simple and user friendly; therefore, they are normally used for assessment in landslide run-out modeling. Debris flow empirical models, such as Flow-R, have been designed to process geographic information system (GIS)-based regions in susceptibility assessments (for in-depth information on Flow-R, readers can refer to Carrara et al. 1995; Chung and Fabbri 1999; Hofmeister and Miller 2003; Melelli and Taramelli 2004; and Guinau et al. 2007). This technique demonstrates superior performance in the identification of potential source areas and equivalent propagation extent. Furthermore, Flow-R is free of charge for different operating systems, such as Windows and Linux, available at www.flow-r.org. It is used in many geographical areas to produce regional debris flow susceptibility maps with high efficiency levels. Initially, it was used in Switzerland for the Canton de Vaud (Horton et al. 2008) and the Val de Bagnes (Jaboyedoff et al. 2012). Subsequently, it was utilized by academic centers, universities, and local geological services in France (Kappes et al. 2011), Italy (Blahut et al. 2010; Lari et al. 2010, 2011), Norway (Fischer et al. 2012), Argentina (Baumann et al. 2011), and Pakistan (Horton et al. 2011). Previous studies have shown that insufficient work has been performed on debris flow modeling in tropical areas because of the thick vegetation and complicated environments in these areas.

14.3 Overall Methodological Flow

The main goal of this study is to assess the areas affected by potential debris flows in Cameron Highlands by integrating Flow-R and RAMMS models. The detailed methodological flowchart is presented in Fig. 14.1. First, data were

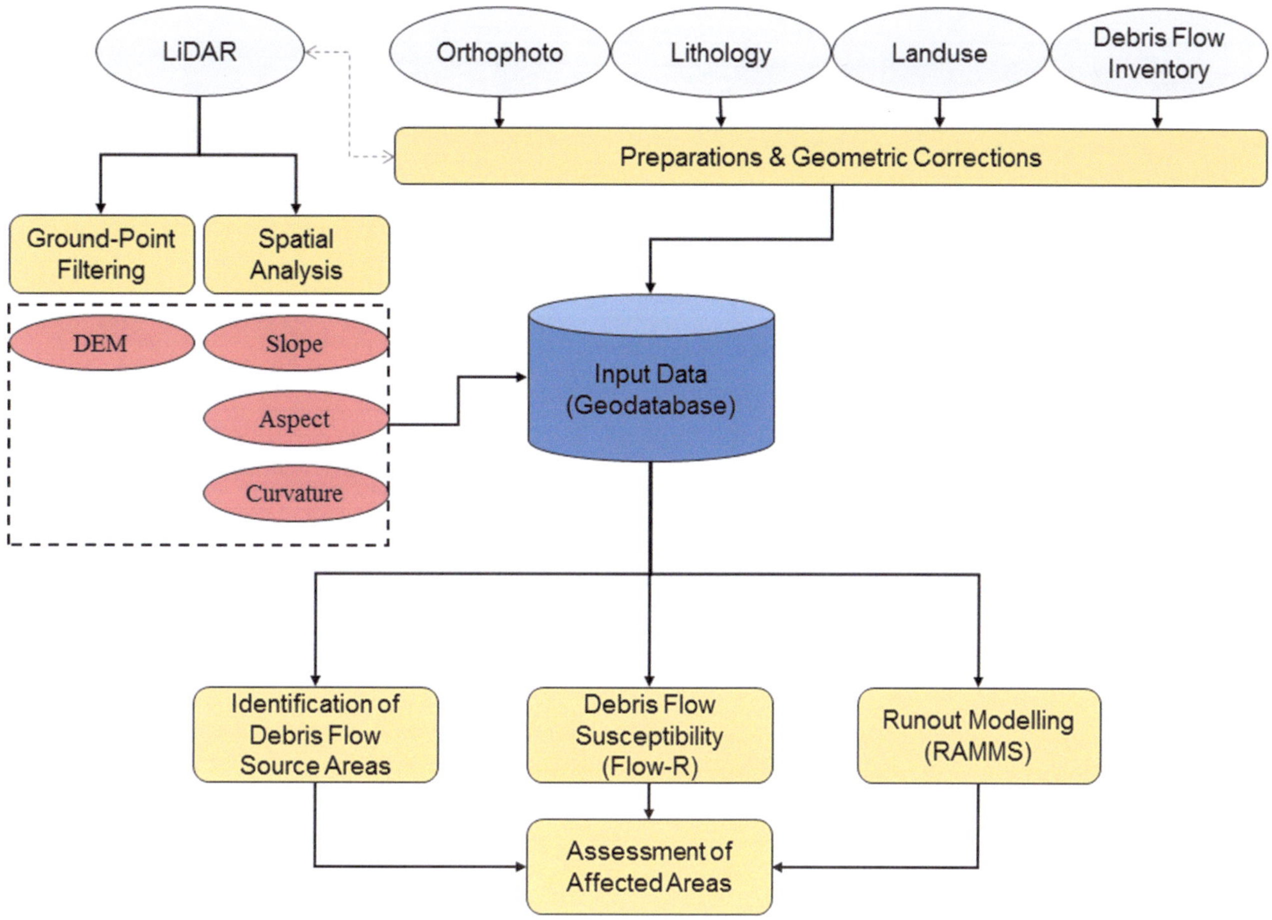

Fig. 14.1 Flowchart of the methodology

preprocessed, prepared, and assembled in a geodatabase. Second, the sources of debris flows were identified by using LiDAR data and other thematic layers, such as land use and lithology. The results of this step were validated with existing debris flow inventories. Third, the Flow-R model was used to produce a debris flow susceptibility map of the study area. Fourth, run-out distances were modeled with the physical RAMMS model. Finally, the areas affected by potential debris flows were assessed by using the generated debris flow susceptibility and land use data.

14.4 Study Area

In this research, a part of Cameron Highlands was selected as the case study because of its frequent occurrences of landslides and debris flow. This location is a tropical rain-forest area in Peninsular Malaysia at the northwestern tip of Pahang approximately 200 km from Kuala Lumpur. Reports from government agencies and previous studies indicate that several landslides have occurred in Cameron Highlands and caused major damage to properties. Quaternary, Devonian granite, and schist are the main lithology types in this region (Pradhan and Lee 2010). A 25-km^2 area was selected within Cameron Highlands for the analysis and testing of the Flow-R and RAMMS models in the current work.

14.5 Data Preprocessing

Airborne LiDAR-derived data and other thematic layers, such as land use, lithology, and debris flow inventory data, were acquired. LiDAR point clouds were collected on January 26, 2015, over the study area. The flight height of the airborne platform was 1510 m, the point density was 8/m^2, and the frequency rate of the laser sensor was 25,000 Hz. The absolute accuracy (measured by the root mean-square error) of the acquired data was 0.15 and 0.3 m in vertical and

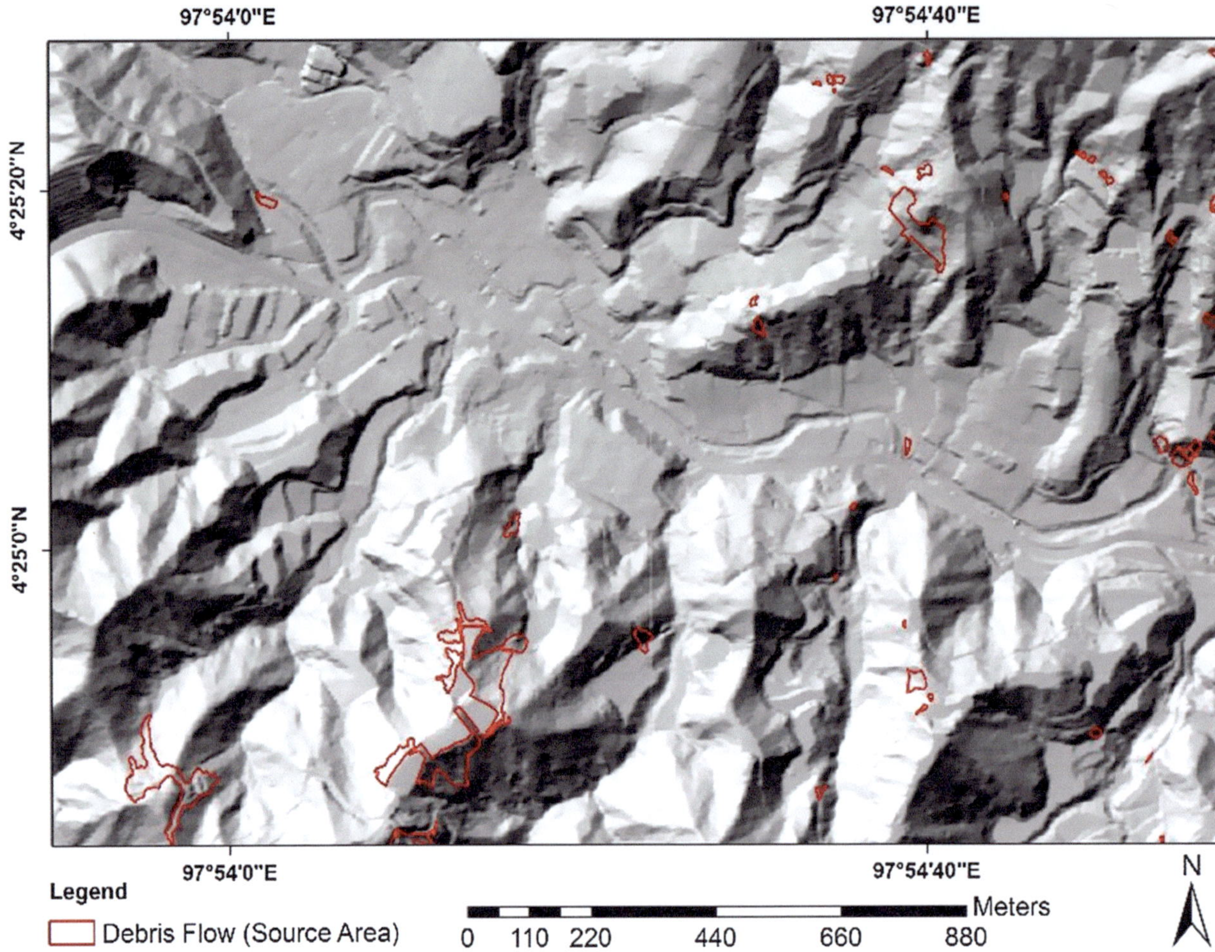

Fig. 14.2 Debris flow inventory map

horizontal dimensions, respectively. Along with LiDAR point clouds, orthophotos were captured with a Nikon D800 camera. The lithology map of the study area was obtained from the Department of Geology and Mineral Sciences, Malaysia. The lithology of the area consists of two main types: schist and phyllite (Fig. 14.3h). The land use map was derived from a SPOT 5 satellite image with a supervised classification technique and then manually refined and labeled in GIS (Fig. 14.3g). This procedure resulted in an accurate and detailed land use map that helps in the study of areas affected by potential debris flows. An inventory of debris flow was obtained through various means, such as interpretation of multi-temporal aerial photographs and satellite imageries, previous studies, and multiple field visits to the site (Fig. 14.2).

To prepare the data for modeling and further analysis, first, LiDAR point clouds were filtered to produce an accurate digital elevation model using the ArcGIS 10.3 LiDAR filtering algorithm (Fig. 14.3a). A digital elevation model (DEM) was produced at 2 m spatial resolution using the last return points. Second, four terrain factors, namely slope (Fig. 14.3d), aspect (Fig. 14.3b), plan curvature (Fig. 14.3e), and profile curvature (Fig. 14.3f), as well as a hydrological factor [i.e., flow accumulation (Fig. 14.3c)] were derived from the DEM model at the same resolution by spatial analysis tools available in the ArcGIS 10.3 software. Third, the LiDAR factors, debris flow inventory, and thematic layers (i.e., lithology and land use) were geometrically corrected and assembled in a geodatabase. Lastly, the prepared data were used for the identification of debris flow source areas, debris flow susceptibility assessment, run-out modeling, and estimation of areas affected by potential debris flow.

14.6 Identification of Debris Flow Source Areas

A good-quality debris flow inventory map is required for detailed and accurate debris flow modeling and run-out distance estimation. Given that the debris flow inventories for the study area were incomplete and not up to date, the

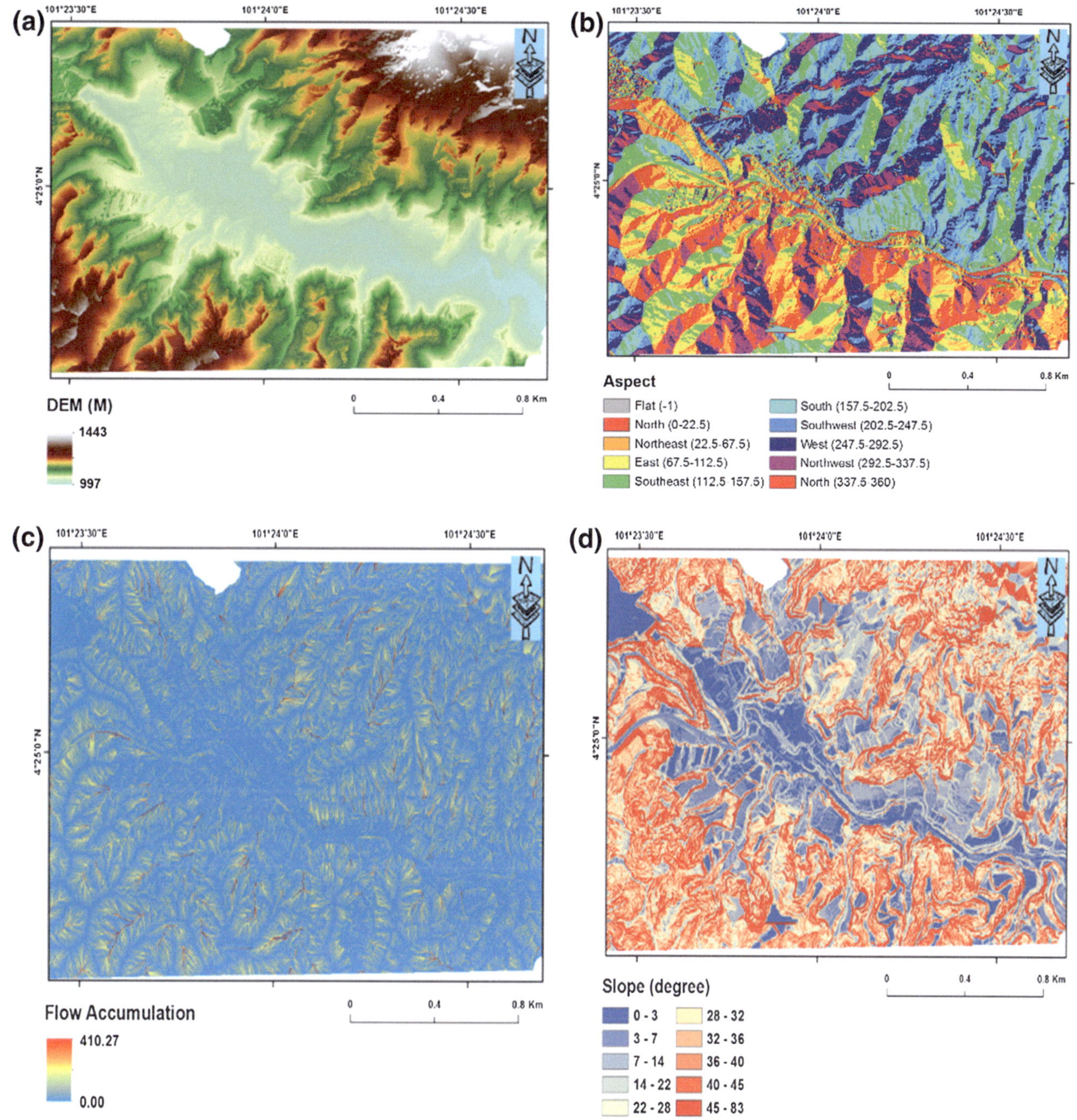

Fig. 14.3 Factors used in defining the source areas: **a** altitude, **b** aspect, **c** flow accumulation, **d** slope, **e** plan curvature, **f** profile curvature, **g** land use/cover, and **h** lithology

source areas were delineated using an indexed approach (Fischer et al. 2012). In this approach, LiDAR-derived factors, such as altitude, slope, aspect, and curvature, as well as thematic maps, such as land use and lithology, were applied (Fischer et al. 2012). The approach is a knowledge-based method, in which the input factors are reclassified into two classes based on a specified threshold for each factor (Fischer et al. 2012). In this step, the resulting raster data have two classes: potential areas for debris flow and non-debris flow areas. Threshold selection is challenging because the factors spatially vary from one geographical area to another. However, literature (Guinau et al. 2007; Kappes et al. 2011)

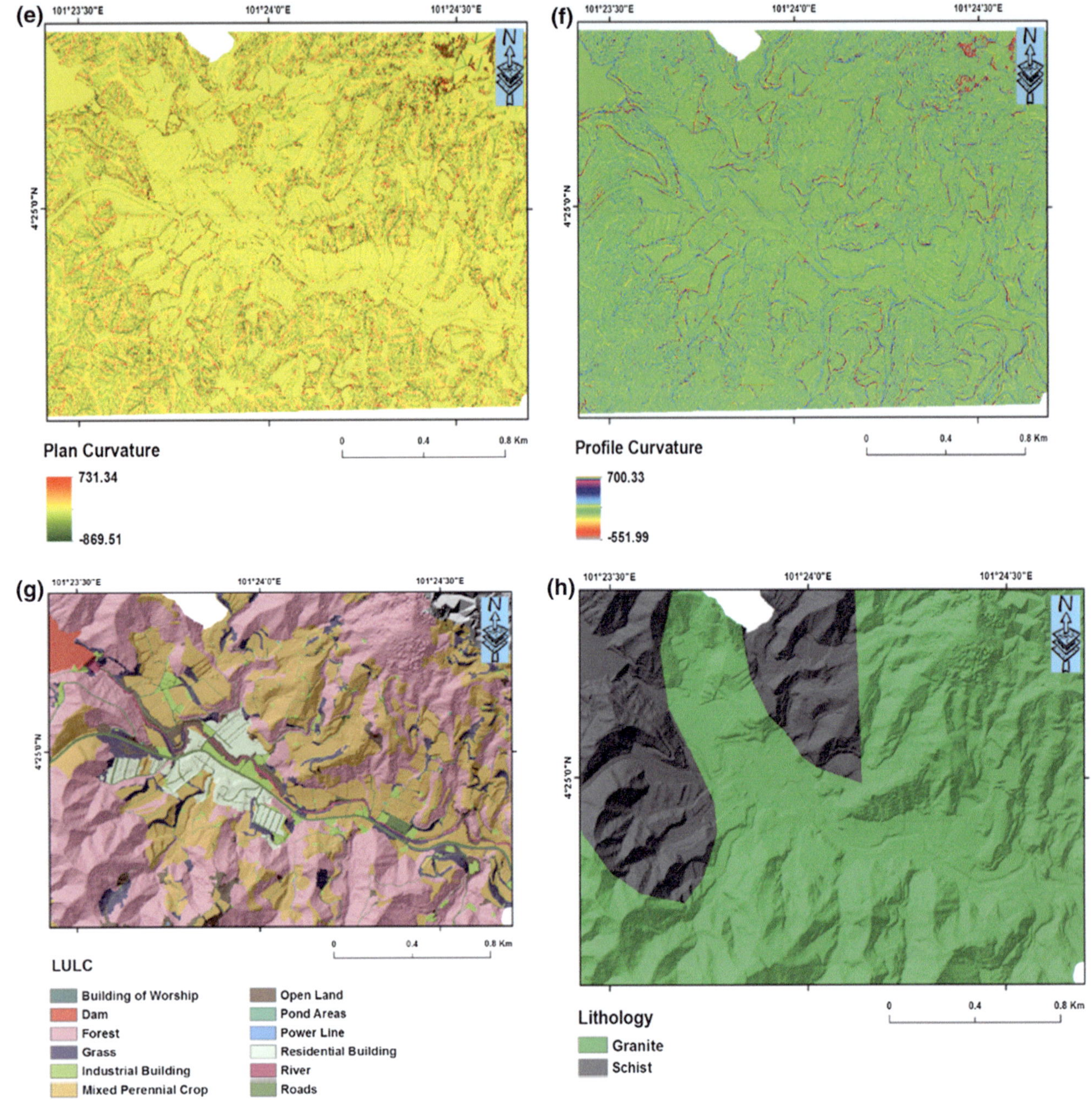

Fig. 14.3 (continued)

indicates that several thresholds can be used with slight modifications for optimization for a local study area. In the current study, modification of factor thresholds was performed by interpreting the orthophotos and associated classification raster (Hürlimann et al. 2006). Thresholds were specified for both confined and unconfined debris flows, given that both are common in Malaysia. The threshold value used in debris flow studies was $-2/100\ \text{m}^{-1}$ for curvature (Horton et al. 2008; Blahut et al. 2010). The slope angles for grassland were 25° to 41° (mean at 37°) and 25° to 47° (mean at 39°) for forest areas. In addition, given that the study area is mainly granitic bedrock where major debris flows occur, granitic bedrock areas were used as a potential area for debris flow. Geological setting is the main factor that affects the variability of debris flows in the study area. Bedrock type mainly defines weathering type and intensity, thus influencing loose sediment availability, type, and grain sizes, which in turn control the disposition of debris flows. After reclassifying all the available factors with their corresponding selected thresholds, the debris flow source areas in the study area were identified a. Good agreement was observed between the results of the modeled source areas and the debris flow tracks mapped in the inventory map.

14.7 Modeling Debris Flow Susceptibility (Flow-R Model)

Flow-R is an empirical model developed in MATLAB for debris flow susceptibility assessment. This model is used by different countries to produce debris flow susceptibility maps with acceptable accuracy. Initially, Flow-R was used by Switzerland for the Canton de Vaud (Horton et al. 2008) and the Val de Bagnes (Jaboyed-off et al. 2012) regions. The model regards DEM as input data and processes it in two steps. In the first step, the source areas of debris flow are identified by utilizing morphological and other user-defined criteria. The next step includes propagating the debris flow from the initial sources identified in the first step on the basis of frictional laws and flow direction algorithms. The Flow-R tool is supported by a user graphical interface to simplify the identification of source areas of debris flow and to evaluate the model parameters. Selection of model parameters in Flow-R is an empirical process, which means that the user should determine suitable parameters based on hazard types (Horton et al. 2013).

14.8 Run-Out Modeling with RAMMS

RAMMS is a modified model based on the Navier–Stokes equation and Voellmy–Salm (VS) frictional relationship (Christen et al. 2012). Debris flow is described as a hydraulic-based depth-averaged continuum model (Christen et al. 2010). The environment is described in three dimensions using the RAMMS model. The direction of the flow down a surface is shown by x and y, and the elevation perpendicular to the profile is provided by $z(x, y)$ (Fig. 14.4). A gravitational acceleration vector and a time factor are defined for each direction. The main parameters of non-uniform motion are flow height (H) and mean velocity (U).

The RAMMS model requires several input data, starting conditions, and model friction parameters (Christen et al. 2012). DEM is the main input data because it characterizes

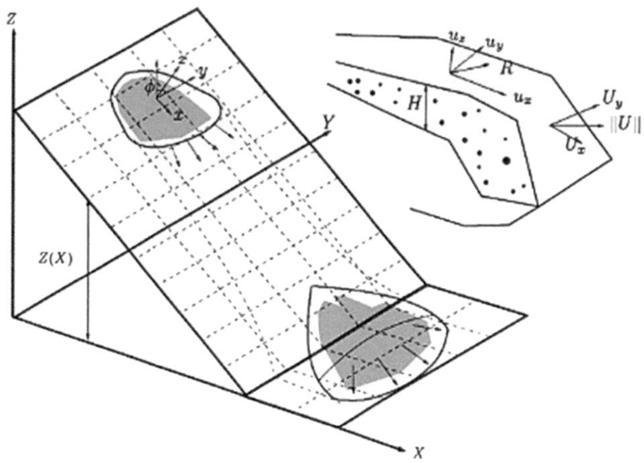

Fig. 14.4 Three dimensions of environment using RAMMS model (modified after Christen et al. 2010)

natural terrains. The starting conditions are often defined as a release area extracted from the identified debris flow sources. The results of the model are sensitive to the definition of the initial conditions (Bartelt et al. 2012). The maximum flow height, velocities, pressure, and momentum can be predicted with the VS model. The geophysical mass movement of debris flow is calculated with the RAMMS model.

A minimum travel angle, which defines the slope between the start and end points of debris flow, is used to characterize the maximum run-out distance. The minimum travel angle is often set to 10° based on Huggel et al. (2003). In this study, to determine the susceptibility of debris flow, three run-out scenarios were analyzed (high-, medium-, and low-frequency scenarios). Travel angle has a negative relationship with debris volume. In other words, a high angle value indicates a low volume of debris and a short run-out distance. Finally, the susceptibility map was produced with three classes by considering the three scenarios with travel angles of 10° for the high-frequency scenario, 15° for the medium-frequency scenario, and 30° for the low-frequency scenario.

14.9 Mapping of LULC Areas Affected by Potential Debrls Flows

Another aspect of the use of run-out models is the possibility to estimate the LULC areas affected by debris flows. To prepare maps, the land use layer is usually overlaid with debris flow deposition depths. With additional field data, such as number of causalities and cost per square meter, transfer functions can be developed. These transfer functions generally translate the affected areas into losses. In this study, affected LULC areas were mapped to show the spatial or geographical distribution of expected losses from debris flow in the study area (Fig. 14.11). However, because of the limitations in field data, quantitative assessment of these losses was not achieved in this study. With such field data, performing a complete risk assessment by using the loss index to define the vulnerability of each element at risk (cost per m^2) is possible.

14.10 Results

14.10.1 Debris Flow Source Area Map

The debris flow sources were mapped with the Flow-R model, which combines a set of thematic layers with user-defined thresholds. Five thematic layers were used (altitude, slope, plan curvature, land use, and geology). However, because the authors wanted to use the inventory data available for the study area, these inventory data were added as an additional layer to improve the accuracy of the source areas. In addition, the spatial variations in the controlling factor (i.e., slope) make the delineation of the source areas with the thresholds recommended by previous studies a challenge and may not be valid with the field conditions in other areas. For example, Fischer et al. (2012) used a slope angle of 25° to 45° as a slope threshold in Norway; however, in the current study, inventory statistics showed that 8 out of 34 debris flows occurred in areas with slopes beyond this range. Therefore, the use of debris flow inventory data in Flow-R modeling is an important step. Figure 14.5a shows the debris flow source map produced with the Flow-R model. The overall accuracy of source identification is 82%. Figure 14.5c shows an example of a correctly identified source, and Fig. 14.5b shows an example of misidentified sources. Accuracy was estimated by overlaying the produced source area map from the Flow-R model and the available inventory map at the pixel level. Overall, six sources in the inventory map were not identified for the following reason. After resampling DEM to 2 m spatial resolution as suggested by other researchers, the geomorphological parameters of small debris flows (<100 m^2) were affected by the neighborhood pixels. In small debris flows, the details of geomorphological features can only be accurately detected at fine DEM resolutions, and resampling leads to averaging the pixel values. This resampling most probably changed the actual values of several samples in the study. Meanwhile, using a DEM spatial resolution of 2 m reduces the effects of surface roughness on propagation extent and channelization. In addition, although several researchers have recommended the use of 10 m DEM, the use of large spatial resolutions in the current study led to artifact propagation and difficulty in the interpretation of run-out distance susceptibility because the susceptibility map of small debris flows were pixelated.

The Flow-R model identified approximately 0.05 km^2 out of 3 km^2 of potentially unstable slopes. The percentage of pixels identified as debris flow sources were located in land use and lithology classes, as shown in Table 14.1. Most of the debris flows were identified in vacant areas (98.9%) and in areas with granite-type lithology (99.1%). The identified debris flow sources were distributed in other land use types as follows: forest (0.6%), industrial areas (0.02%), transportation (0.21%), and agriculture (0.24%). Meanwhile, only 0.9% of the pixels were found in areas that are of schist-, phyllite-, and slate-type lithology.

Land use is considered an important conditioning factor for various natural hazards, including debris flows. Land uses, such as built-up and outcropping bedrocks, control the availability of materials. Vegetation and dense forests reduce the initiation potentials because they reduce surface runoff. Another advantage of using land use in the identification of debris flow initiations is the removal of some uncertain individual or small group pixels, in which the conditions of the thresholds of geomorphological parameters set by the user are met. Meanwhile, the lithology layer contributes to sediment availability.

14.10.2 Results of Run-Out Modeling

The identified debris flow initiations were used to model the run-out distances using the method proposed by Horton et al. (2013) that is mainly based on simple frictional laws. The model consecutively processed the active cells in the dataset. The path and spreading of the debris flows were modeled with a spreading algorithm, whereas run-out distances were determined with the friction law algorithm. The model produced two main outputs: maximum probability of susceptibility and maximum kinetic energy. Maximum probability of susceptibility ranged from 0 to 1, as shown in Fig. 14.6a. In the figure, a surface with a red color has a higher probability to be reached by a debris flow than a blue one. This high probability of run-out propagation is related to drainage channels. This relation is more obvious in the 3D view (Fig. 14.6b), which shows the probability of debris flow paths increases (yellow to red) in the drainage channels.

The probability of the run-out distance of debris flow paths is used to reproduce debris flow events. The quality of such simulations depends largely on model parameterization

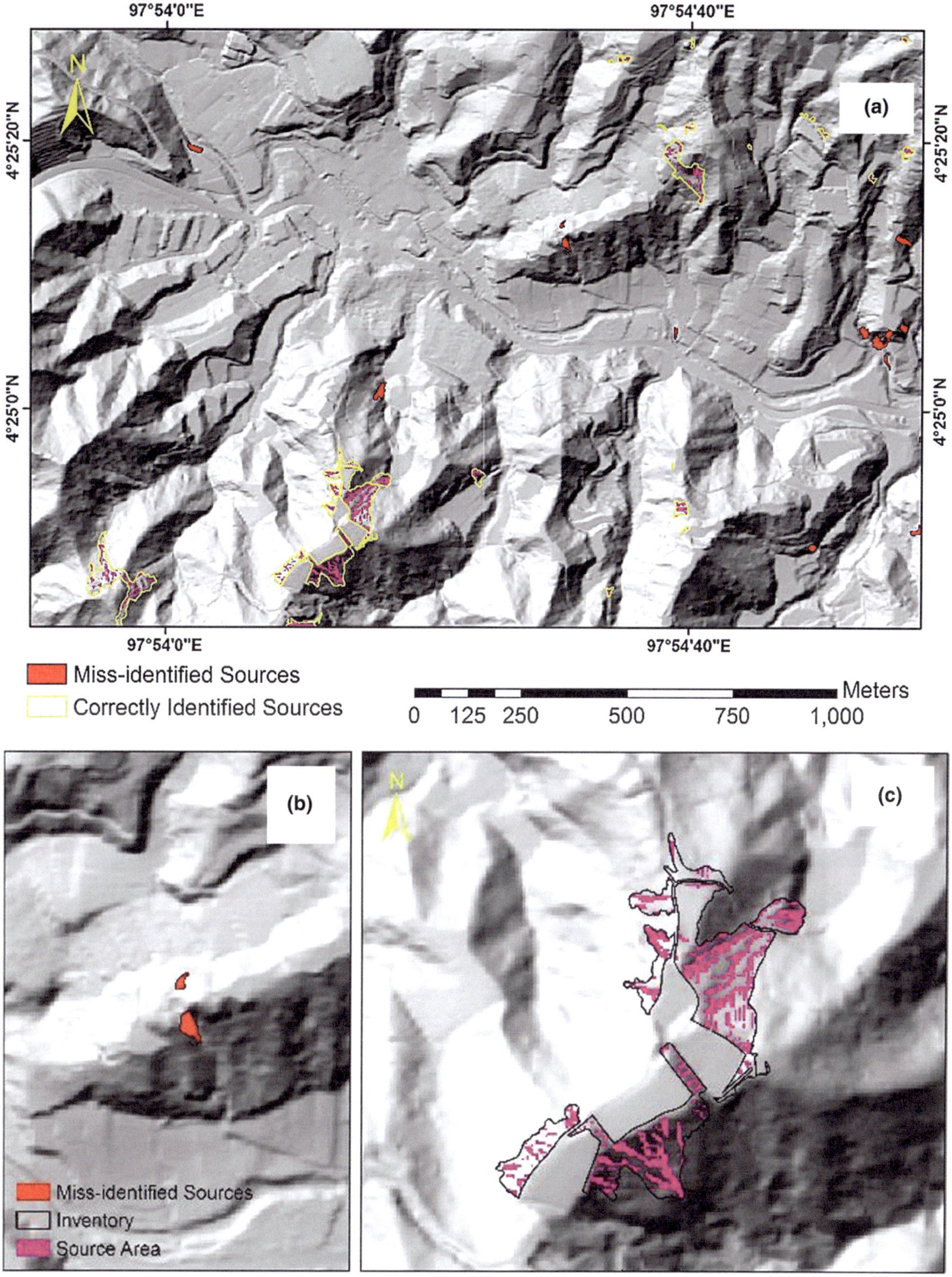

Fig. 14.5 Identified debris flow source areas. **a** Source area of the study area shows the correctly identified and miss-identified sources based on the available inventory map, **b** an example on the miss-identified sources, and **c** an example on the correctly identified source area

Table 14.1 Percent of the pixels identified as debris flow sources in each land use and lithology types

	Class	Percent of pixels identified as debris flow sources
Land use	Forest	0.60
	Industrial areas	0.02
	Transportation	0.21
	Agriculture	0.24
	Vacant land	98.9
Lithology	Schist, phyllite, slate	0.90
	Granite	99.10

and identification of initiations (Table 14.2). Extensive field measurements are necessary to calibrate models for accurate simulations of run-out distances (Graf and McArdell 2009). The model parameters directly affect the simulation results, and the location and size of debris flow sources affect the length and extent of the simulated run-out distances (Iovine et al. 2003). Therefore, to improve the quality of the simulation of the run-out distance of debris flow paths, accurate field measurements and robust debris flow source area identification algorithms are needed.

Although the accuracy assessment and inventory debris flow data revealed the effectiveness of the source area identification approach (i.e., overall accuracy of 82%), the uncertainty in the run-out distance simulations is caused by model parameterization. The quality of DEM also plays a significant role in identifying the source areas and simulating the run-out distances of debris flows. A spatial resolution of 2 m was used in this study because the simulations were carried out in a relatively small area ($\sim 3\ \text{km}^2$) that contained urban and dense forest locations. This resolution produced reasonable results; however, fine resolutions were not tried because they result in surface roughness and channelization of propagation extent.

Figure 14.7a shows the debris flow run-out distance susceptibility map produced with Flow-R and several examples of the calculated distances in GIS for the simulated debris flow paths. The distances were measured from the source area to the farthest point of the propagation extent. The calculated run-out distances for the 24 debris flows simulated in Flow-R showed minimum and maximum values of 92 and 6692 m, respectively. Five of the debris flows, including the one with the longest simulated run-out distance, could exert significant potential damage to urban and the surrounding areas (Fig. 14.7b). The three most highly susceptible debris flows are highlighted in red circles; the two other debris flows mostly affect agricultural lands (Fig. 14.7b). The run-out simulations did not show the quantitative values for debris flow effects on urban or other vulnerable features. The details of the debris flow effects on different land use types will be discussed later based on the disposition of the debris.

The maximum velocity threshold of 15 ms^{-1} was selected for the run-out distance simulations according to previous studies (Fischer et al. 2012), given that no velocity measurements exist for debris flow events in the study area. Figure 14.8 shows the angle of reach and run-out distance profile of the debris flow event with the longest run-out distance (692 m). The elevation of the source area ranged from 1100 to 1160 m, and the elevation of the farthest deposit reached by the debris flow is 1030 m. The calculated angle of reach is 11°. Several studies implemented the travel angle of 11° for spreading; these studies include Haeberli (1983), Rickenmann and Zimmermann (1993), and Bathurst et al. (1997). Selection of the angle of reach threshold is important because it supports the important role of event volume in run-out distance. Meanwhile, the run-out distance profile shows that the peak probability (0.9) is located just at the end of the source area. Therefore, the probability generally decreased with several fluctuations because of the difference in the morphological characteristics of the terrain and surface roughness.

Figure 14.9 shows the reclassified run-out distance susceptibility modeled in Flow-R. The map was reclassified into four classes of low, moderate, high, and very high based on the travel angles calculated in GIS. The remaining areas were considered non susceptible. The percentages of the classes are as follows: 23.5% for low, 21.80% for moderate, 36.5% for high, and 18.20% for very high. The simulated debris flow susceptibility is important for land use planning and for estimating the potential damages during any debris flow event in the study area. However, in this map, temporal probability is assumed to be constant over the entire hazard initiation probability class. Therefore, the produced susceptibility map can serve as a preliminary tool in the mitigation of debris flow risks in the study area.

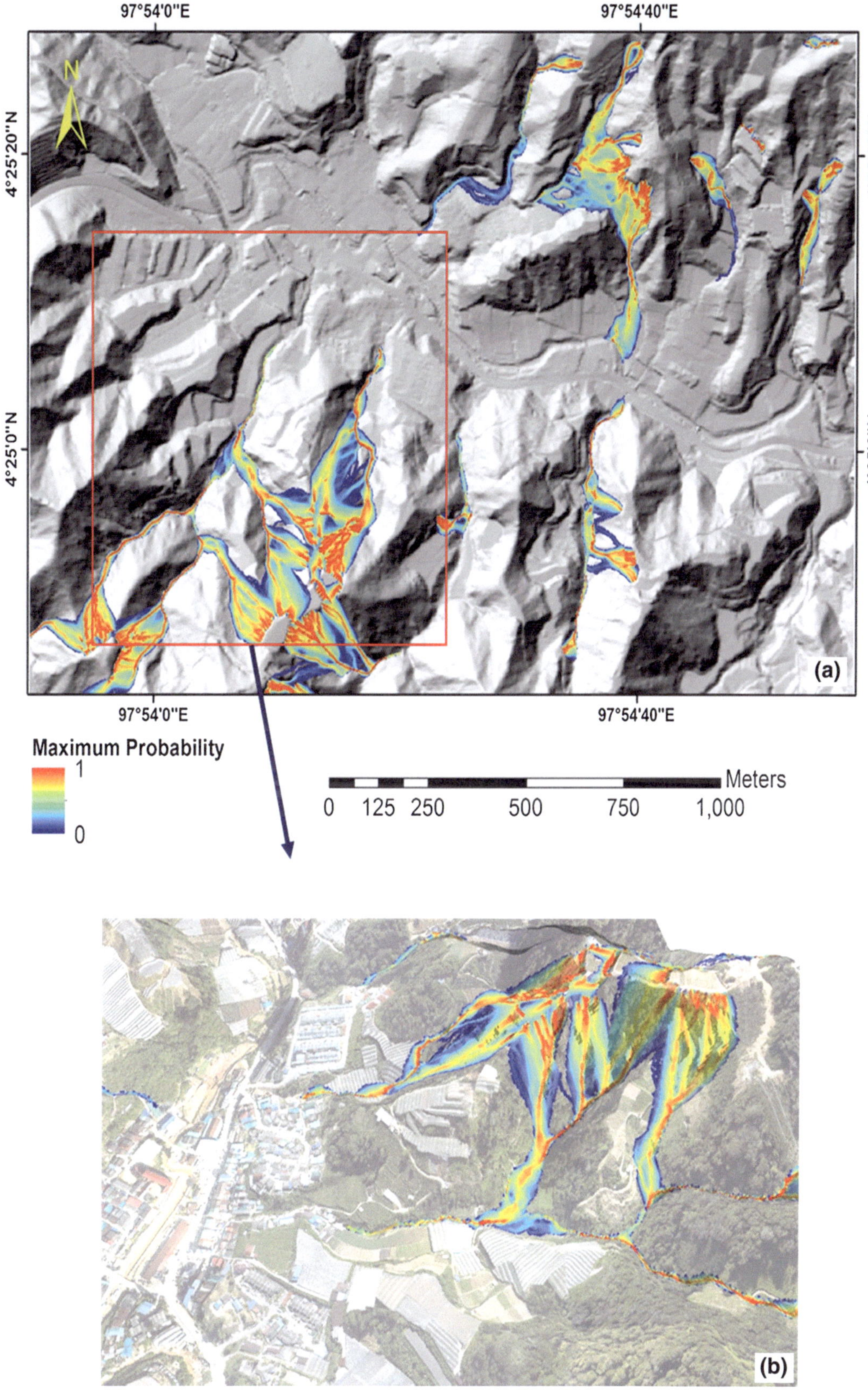

Fig. 14.6 Maximum probability of debris flow occurrence **a** shows the probability distribution in the study area and **b** shows a 3D view of the maximum probability for specific debris flows highlighting their details

Table 14.2 Model parameterization for the debris flow source identification and susceptibility analysis

Source		Run out	
Parameter	Values chosen	Parameter	Values chosen
Planar curvature threshold	<−2/100 m^{-1}	Flow direction algorithm	Modified Holmgren (1994)
Slope angle	>18°	Inertial algorithm	Weights
Altitude	>1100 m	Holmgren exponent	1
Land use/cover	Forest, bare land, agriculture included.	Angle of reach (=constant friction loss angle)	7°
Lithology	Schist excluded		
Inventory data	Converted into probability (>0.8)		

14.10.3 Results of Geophysical Parameter Simulation with RAMMS

Five geophysical parameters for each debris flow source area were simulated with the RAMMS model. However, only the two most important ones are discussed. The velocity of the debris flow varied across the area; the minimum velocity was close to 0 ms^{-1}, and the maximum velocity was 15 ms^{-1} (Fig. 14.10a). Velocity increased in the steep terrains and contributed to the downward movement of debris flow. The velocity of debris flow also increased because of the volume of materials entrained as the debris moved and the basal friction component of the flow at various points. Furthermore, velocity increased at the confluence of flows in the channel because of the narrowing of the channel. Meanwhile, velocity was low in the source areas at the top of catchments mainly because debris materials have not gained enough energy to flow because of the inertial force that tends to restrict the flow of movement. However, flow depth remained fairly constant but increased at the mouth of the debris flow fan where all the debris flow materials tend to accumulate, thus forming a debris dam. The simulated dispositions ranged from 0 m to 0.75 m.

Figure 14.11 shows the areas of LULC affected by debris flow deposition depth. Exact prediction of building vulnerability to debris flow impact is challenging because debris flow velocities and flow depths at the point of impact can be estimated with uncertainties for different debris flow scenarios (Jakob et al. 2012). In addition, land use information is too general, which limits the simulation of the exact elastic, plastic, or rigid behavior of buildings. Therefore, in this study, only qualitative assessment was performed for the land use effects of debris flow simulated with the Flow-R and RAMMS models. The results showed that highways, residential areas, and agricultural lands are at risk. The average deposition depth predicted for highways is 0.5 m. The predicted velocities at these points reach 8 m/s, which can pose significant risks for passing vehicles and pedestrians. For agricultural lands, although the depth is less than 0.5 m, the potential debris flow could exert significant effects on crops, especially during harvest time. The middle area of the south part shows that several residential buildings are at risk of debris flow with different magnitudes depending on the type of materials and structural behavior of the buildings. Velocity and flow depths also play a role in the extent of the effect of debris flow on buildings. However, to estimate the accurate effect of debris flow on residential areas and transportation infrastructures, accurate input data are necessary.

This preliminary investigation of the potential effects of debris flow on different infrastructures can help local government agencies and the national government create land use plans and engineered structures.

14.11 Conclusion

This study conducted preliminary debris flow modeling and run-out susceptibility assessment by using Flow-R and RAMMS models. A 2 m DEM was prepared from LiDAR point clouds, and other parameters were used for the modeling and susceptibility assessment. The Flow-R model was used to produce debris flow initiations and a susceptibility map, and the RAMMS model was utilized to simulate the geophysical parameters of the identified debris flow sources.

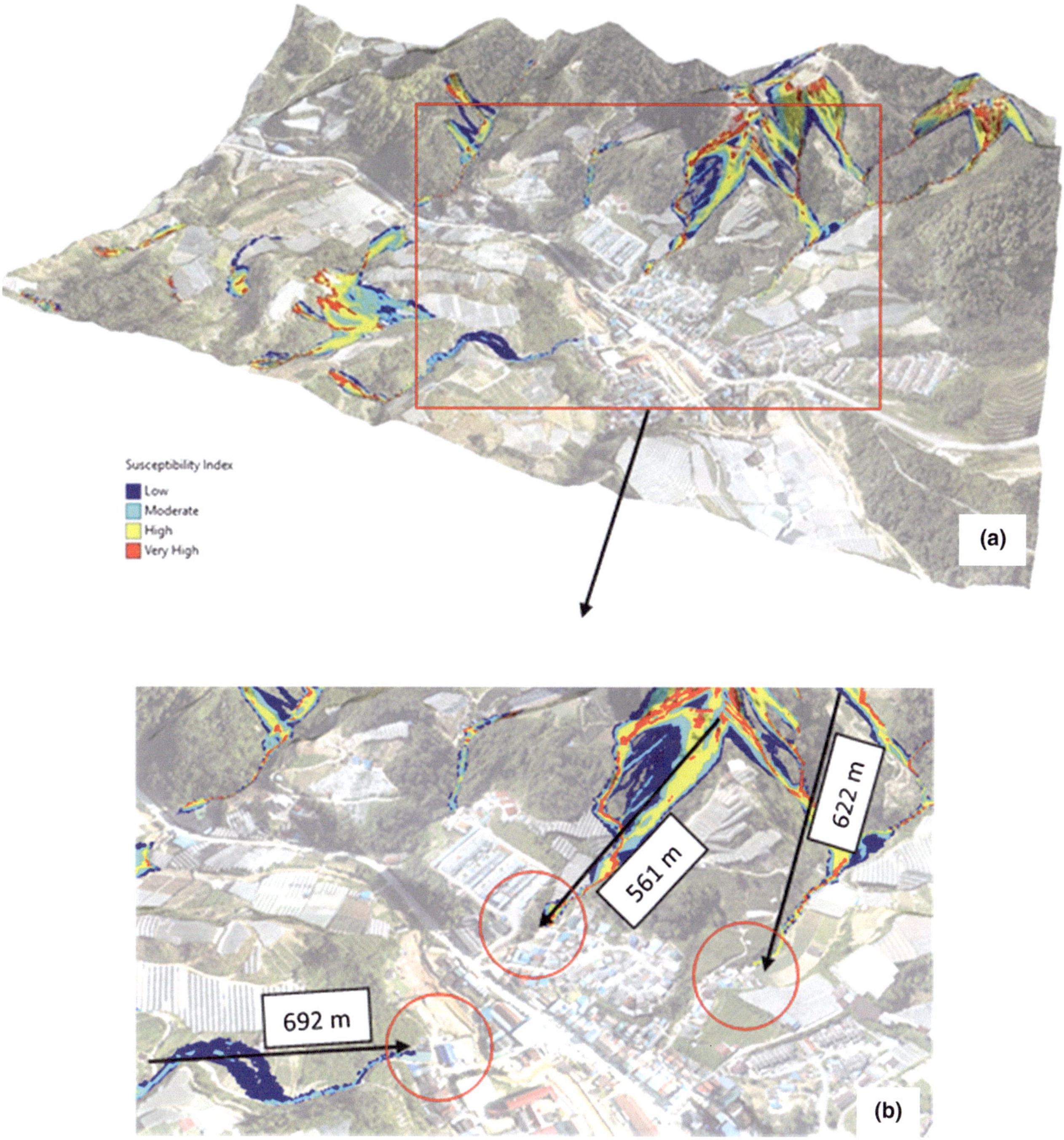

Fig. 14.7 **a** Run-out distance susceptibility map and **b** the calculated run-out distances based on the debris flow susceptibility map

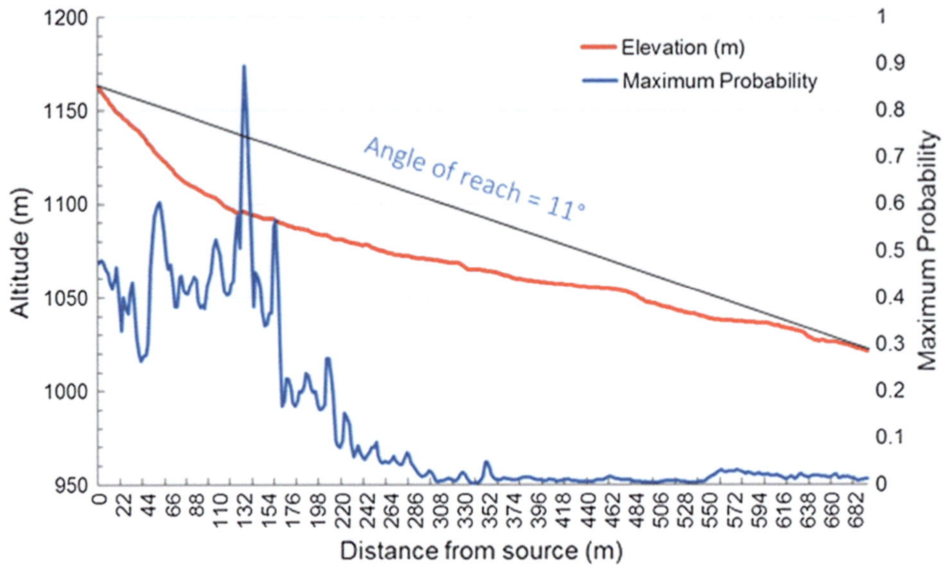

Fig. 14.8 Profile of the debris flow with the longest run-out distance (692 m) identified in the study area. A *line* for the identification of the angle of reach was positioned between the source area and the furthest point of the run-out

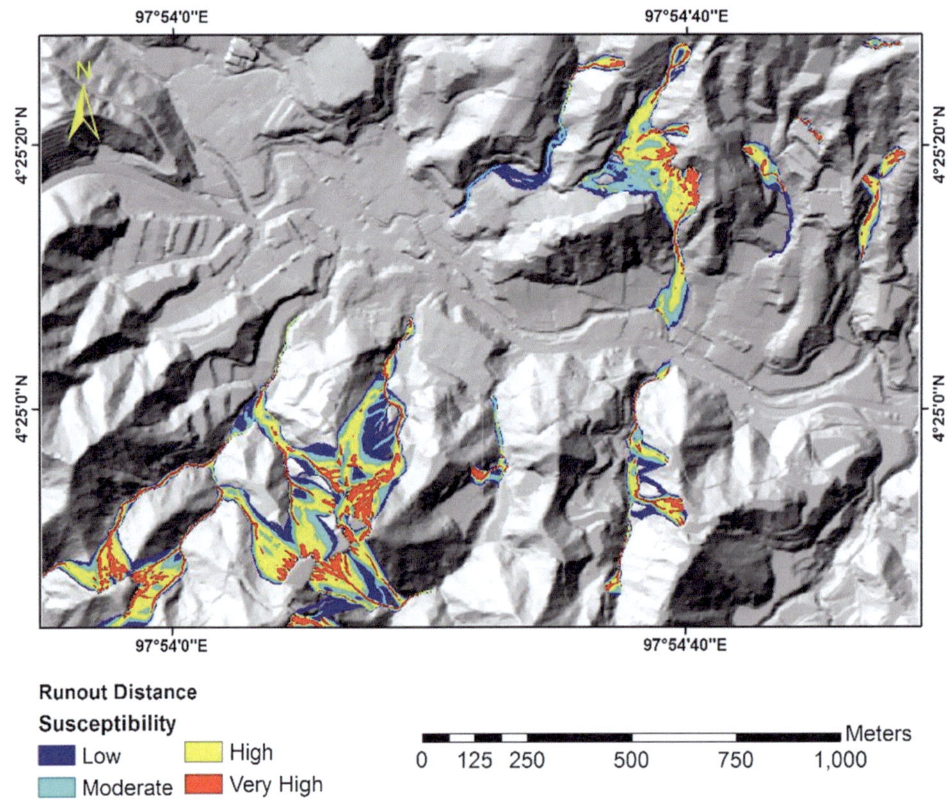

Fig. 14.9 Reclassified debris flow susceptibility map

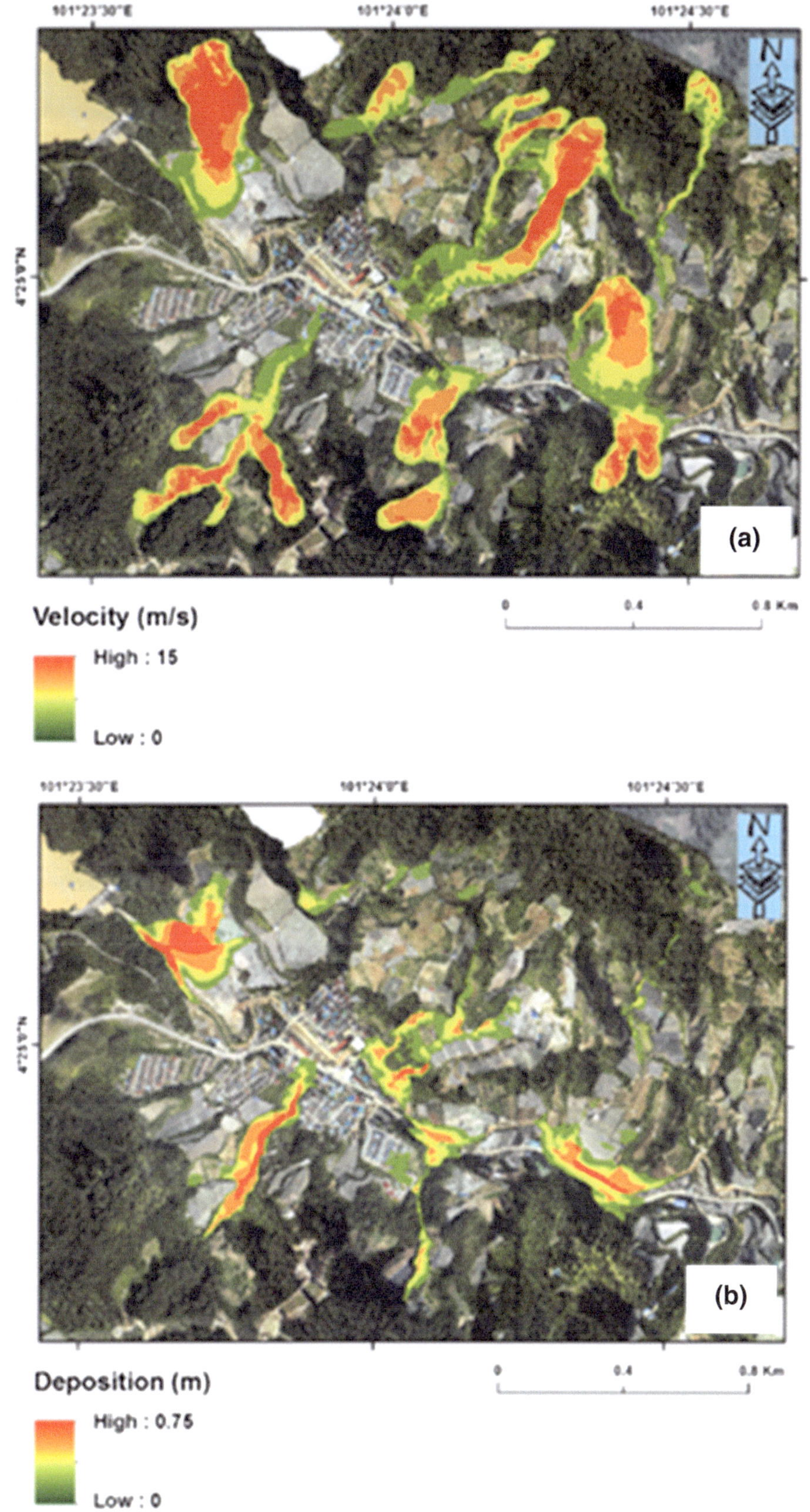

Fig. 14.10 RAMMS derived geophysical parameters **a** velocity, **b** deposition

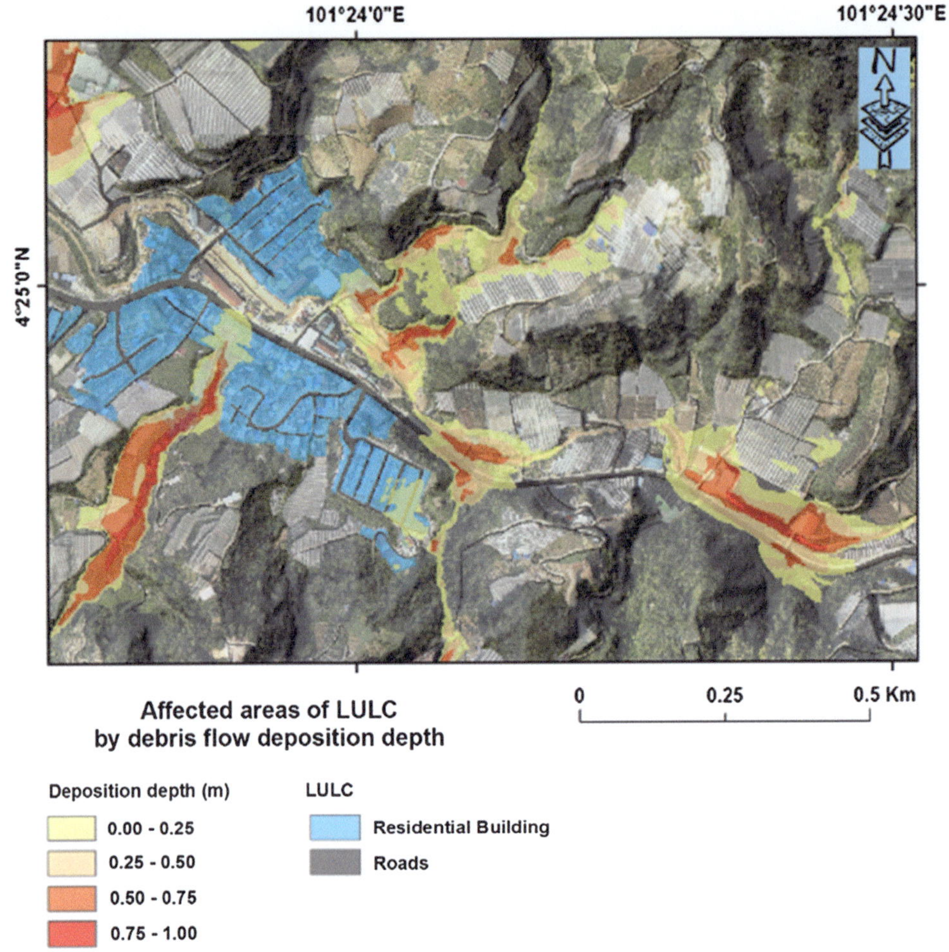

Fig. 14.11 Affected areas of LULC by debris flow deposition depths

The overall accuracy of source identification was 82%. The 2 m DEM was suitable for modeling debris flow events in the study area. The Flow-R model identified approximately 0.05 km^2 out of 3 km^2 of potentially unstable slopes. The percentage of pixels identified as debris flow sources were mostly located in vacant areas (98.9%). The run-out simulations showed that the 24 debris flows had minimum and maximum values of 92 and 6692 m, respectively. Five of the debris flows, including the one with the longest simulated run-out distance, showed a significant potential damage to urban and the surrounding areas. Furthermore, the geophysical parameter simulations indicated that the maximum velocity is 15 ms^{-1}, and the deposition depth could reach 0.75 m. Future work should focus on detailed sensitivity analyses of physical models to improve the accuracy of the estimated parameters for potential debris flows. These analyses will allow for an accurate quantitative assessment of debris flow effects on land use features, and improved risk management strategies can be established.

References

Armanini, A., & Scotton, P. (1992). Experimental analysis on the dynamic impact of a debris flow on structures. *Internationales symposion interpraevent, 6,* 107–116.

Bartelt, P., Bühler, Y., Buser, O., Christen, M., & Meier, L. (2012). Modeling mass-dependent flow regime transitions to predict the stopping and depositional behavior of snow avalanches. *Journal of Geophysical Research: Earth Surface, 117*(F1).

Bathurst, J. C., Burton, A., & Ward, T. J. (1997). Debris flow run-out and landslide sediment delivery model tests.*Journal of Hydraulic Engineering, 123*(5), 410–419.

Baumann, V., Wick, E., Horton, P., & Jaboyedoff, M. (2011, October). Debris flow susceptibility mapping at a regional scale along the National Road N7, Argentina. In *Proceedings of the 14th Pan-American Conference on Soil Mechanics and Geotechnical Engineering* (pp. 2–6).

Berger, C., McArdell, B. W., & Schlunegger, F. (2011). Direct measurement of channel erosion by debris flows, Illgraben, Switzerland. *Journal of Geophysical Research: Earth Surface, 116*(F1).

Berger, C., McArdell, B., & Lauber, G. (2012). Murgangmodellierung im Illgraben, Schweiz. mit dem numerischen 2D-Modell RAMMS.

Murgangmodellierung in der Praxis. In *12th Congress Interpraevent*.

Blahut, J., Horton, P., Sterlacchini, S., & Jaboyedoff, M. (2010). Debris flow hazard modelling on medium scale: Valtellina di Tirano, Italy. *Natural Hazards and Earth System Sciences, 10*(11), 2379–2390.

Carrara, A., Cardinali, M., Guzzetti, F., & Reichenbach, P. (1995). GIS technology in mapping landslide hazard. In *Geographical information systems in assessing natural hazards* (pp. 135–175). Netherlands: Springer.

Christen, M., Bühler, Y., Bartelt, P., Leine, R., Glover, J., Schweizer, A., et al. (2012). Integral hazard management using a unified software environment. In *12th Congress Interpraevent* (pp. 77-86).

Christen, M., Kowalski, J., & Bartelt, P. (2010). RAMMS: Numerical simulation of dense snow avalanches in three-dimensional terrain. *Cold Regions Science and Technology, 63*(1), 1–14.

Chung, C. J. F., & Fabbri, A. G. (1999). Probabilistic prediction models for landslide hazard mapping. *Photogrammetric Engineering and Remote Sensing, 65*(12), 1389–1399.

Dai, F. C., & Lee, C. F. (2002). Landslide characteristics and slope instability modeling using GIS, Lantau Island, Hong Kong. *Geomorphology, 42*(3), 213–228.

Deubelbeiss, Y., & Graf, C. (2013). Two different starting conditions in numerical debris flow models–Case study at Dorfbach, Randa (Valais, Switzerland). GRAF, C.(Red.) Mattertal–ein Tal in Bewegung. *Publikation zur Jahrestagung der Schweizerischen Geomorphologischen Gesellschaft, 29*, 125–138.

Egli, T. (2005). Wegleitung Objektschutz gegen gravitative Naturgefahren. VKF.

Fannin, R. J., & Wise, M. P. (2001). An empirical-statistical model for debris flow travel distance. *Canadian Geotechnical Journal, 38*(5), 982–994.

Fischer, L., Rubensdotter, L., Sletten, K., Stalsberg, K., Melchiorre, C., Horton, P., et al. (2012, June). Debris flow modeling for susceptibility mapping at regional to national scale in Norway. In *Proceedings of the 11th International and 2nd North American Symposium on Landslides* (pp. 3–8).

Frattini, P., Crosta, G., & Carrara, A. (2010). Techniques for evaluating the performance of landslide susceptibility models. *Engineering Geology, 111*(1), 62–72.

Glade, T. (2005). Linking debris-flow hazard assessments with geomorphology. *Geomorphology, 66*(1), 189–213.

Graf, C., & McArdell, B. W. (2009, September). Debris-flow monitoring and debris-flow runout modelling before and after construction of mitigation measures: an example from an instable zone in the Southern Swiss Alps. In La géomorphologie alpine: entre patrimoine et contrainte. Actes du colloque de la Société Suisse de Géomorphologie (pp. 3–5).

Guinau, M., Vilajosana, I., & Vilaplana, J. M. (2007). GIS-based debris flow source and runout susceptibility assessment from DEM data? A case study in NW Nicaragua. *Natural Hazards and Earth System Science, 7*(6), 703–716.

Haeberli, W. (1983). Frequency and characteristics of glacier floods in the Swiss Alps. *Annals of Geophysics, 4*: 85–90.

Hofmeister, R. J., & Miller, D. J. (2003). GIS-based modeling of debris-flow initiation, transport and deposition zones for regional hazard assessments in western Oregon, USA. In *Debris-flow hazards mitigation: Mechanics, prediction, and assessment* (pp. 1141–1149). Rotterdam: Millpress.

Holmgren, P. (1994). Multiple flow direction algorithms for runoff modelling in grid based elevation models: an empirical evaluation. *Hydrological processes, 8*(4), 327–334.

Horton, P., Jaboyedoff, M., Bardou, E., Locat, J., Perret, D., Turmel, D., et al. (2008). Debris flow susceptibility mapping at a regional scale.

Horton, P., Jaboyedoff, M., Rudaz, B. E. A., & Zimmermann, M. (2013). Flow-R, a model for susceptibility mapping of debris flows and other gravitational hazards at a regional scale. *Natural Hazards and Earth System Sciences, 13*(4), 869.

Horton, P., Jaboyedoff, M., Zimmermann, M., Mazotti, B., & Longchamp, C. (2011). Flow-R, a model for debris flow susceptibility mapping at a regional scale–some case studies. *Italian Journal of Engineering Geology, 2*, 875–884.

Hu, K., Wei, F., & Li, Y. (2011). Real-time measurement and preliminary analysis of debris-flow impact force at Jiangjia Ravine, China. *Earth Surface Processes and Landforms, 36*(9), 1268–1278.

Huggel, C., Kääb, A., Haeberli, W., & Krummenacher, B. (2003). Regional-scale GIS-models for assessment of hazards from glacier lake outbursts: evaluation and application in the Swiss Alps. *Natural Hazards and Earth System Science, 3*(6), 647–662.

Hübl, J., & Holzinger, G. (2003). Entwicklung von Grundlagen zur Dimensionierung kronenoffener Bauwerke für die Geschiebebewirtschaftung in Wildbächen: Kleinmaßstäbliche Modellversuche zur Wirkung von Murbrechern. WLS Report, 50.

Hübl, J., Suda, J., Proske, D., Kaitna, R., & Scheidl, C. (2009, September). Debris flow impact estimation. In Eleventh International Symposium on Water Management and Hydraulic Engineering (Vol. 1, pp. 137–148).

Hürlimann, M., Copons, R., & Altimir, J. (2006). Detailed debris flow hazard assessment in Andorra: A multidisciplinary approach. *Geomorphology, 78*(3), 359–372.

Iovine, G., Di Gregorio, S., & Lupiano, V. (2003). Debris-flow susceptibility assessment through cellular automata modeling: an example from 15? 16 December 1999 disaster at Cervinara and San Martino Valle Caudina (Campania, southern Italy). *Natural Hazards and Earth System Science, 3*(5), 457–468.

Ishikawa, N., Inoue, R., Hayashi, K., Hasegawa, Y., & Mizuyama, T. (2008). Experimental approach on measurement of impulsive fluid force using debris flow model.

Iverson, R. M., & Denlinger, R. P. (2001). Mechanics of debris flows and debris-laden flash floods. In *Seventh Federal Interagency Sedimentation Conference* (pp. IV-1–IV-8).

Iverson, R. M., Reid, M. E., Logan, M., LaHusen, R. G., Godt, J. W., & Griswold, J. P. (2011). Positive feedback and momentum growth during debris-flow entrainment of wet bed sediment. *Nature Geoscience, 4*(2), 116–121.

Jaboyedoff, M., Oppikofer, T., Abellán, A., Derron, M. H., Loye, A., Metzger, R., et al. (2012). Use of LIDAR in landslide investigations: A review. *Natural Hazards, 61*(1), 5–28.

Jaedicke, C., Lied, K., & Kronholm, K. (2009). Integrated database for rapid mass movements in Norway. *Natural Hazards and Earth System Sciences, 9*(2), 469–479.

Jaedicke, C., Solheim, A., Blikra, L. H., Stalsberg, K., Sorteberg, A., Aaheim, A., et al. (2008). Spatial and temporal variations of Norwegian geohazards in a changing climate, the GeoExtreme Project. *Natural Hazards and Earth System Sciences, 8*(4), 893–904.

Jakob, M. (2005). Debris-flow hazard analysis. In *Debris-flow hazards and related phenomena* (pp. 411–443). Berlin: Springer.

Jakob, M., Stein, D., & Ulmi, M. (2012). Vulnerability of buildings to debris flow impact. *Natural Hazards, 60*(2), 241–261.

Kappes, M. S., Malet, J. P., Remaître, A., Horton, P., Jaboyedoff, M., & Bell, R. (2011). Assessment of debris-flow susceptibility at medium-scale in the Barcelonnette Basin, France. *Natural Hazards and Earth System Sciences, 11*(2), 627–641.

König, U. (2006). *Real scale debris flow tests in the Schesatobel-valley* (Doctoral dissertation. Master's thesis, University of Natural Resources and Life Sciences, Vienna, Austria).

Lari, S., Crosta, G. B., Frattini, P., Horton, P., & Jaboyedoff, M. (2011, June). Regional-scale debris-flow risk assessment for an alpine valley. In Proceedings *of the 5th International Conference on*

Debris-Flow Hazards Mitigation: Mechanics, Prediction, and Assessment, Padua, Italy (pp. 14–17).

Lari, S., Frattini, P., Crosta, G. B., Jaboyedoff, M., & Horton, P. (2010). Rockfall and debris flow societal and economic risk assessment at the regional scale. In *Acts 10th World Water Day, Accademia Nazionale dei Lincei, Rome, 22* (pp. 179–187).

Lorente, A., Beguería, S., Bathurst, J. C., & García-Ruiz, J. M. (2007). Debris flow characteristics and relationships in the Central Spanish Pyrenees.

Melelli, L., & Taramelli, A. (2004). An example of debris-flows hazard modeling using GIS. *Natural Hazards and Earth System Science, 4* (3), 347–358.

Monney, J., Herzog, B., Wenger, M., Wendeler, C., & Roth, A. (2007). Einsatz von multiplen Stahlnetzbarrieren als Murgangruckhalt. *Wasser Energie Luft, 3,* 255–259.

Naef, D., Rickenmann, D., Rutschmann, P., & McArdell, B. W. (2006). Comparison of flow resistance relations for debris flows using a one-dimensional finite element simulation model. *Natural Hazards and Earth System Science, 6*(1), 155–165.

Pradhan, B., & Lee, S. (2010). Landslide susceptibility assessment and factor effect analysis: Backpropagation artificial neural networks and their comparison with frequency ratio and bivariate logistic regression modelling. *Environmental Modelling and Software, 25* (6), 747–759.

Proske, D., Kaitna, R., Suda, J., & Hübl, J. (2008). Abschätzung einer Anprallkraft für murenexponierte Massivbauwerke. *Bautechnik, 85* (12), 803–811.

Rickenmann, D., Laigle, D. M. B. W., McArdell, B. W., & Hübl, J. (2006). Comparison of 2D debris-flow simulation models with field events. *Computational Geosciences, 10*(2), 241–264.

Rickenmann, D., & Zimmermann, M. (1993). The 1987 debris flows in Switzerland: documentation and analysis.*Geomorphology, 8*(2-3), 175–189.

Sassa, K. (1989). Special lecture: Geotechnical model for the motion of landslides. *International Journal of Rock Mechanics and Mining Sciences and Geomechanics Abstracts, 26*(2), 88.

Scheidl, C., Rickenmann, D., & McArdell, B. W. (2013). Runout prediction of debris flows and similar mass movements. In *Landslide science and practice* (pp. 221–229). Berlin: Springer.

Scheuner T., Schwab S., & McArdell B. (2011). Application of a two-dimensional numerical model in risk and hazard assessment in Switzerland. In 5th DFHM, Padua, Italy.

Schürch, P., Densmore, A. L., Rosser, N. J., & McArdell, B. W. (2011). Dynamic controls on erosion and deposition on debris-flow fans. *Geology, 39*(9), 827–830.

Van Westen, C. J., Van Asch, T. W., & Soeters, R. (2006). Landslide hazard and risk zonation—Why is it still so difficult? *Bulletin of Engineering Geology and the Environment, 65*(2), 167–184.

Wendeler, C. S. I. (2008). Murgangrückhalt in Wildbächen: Grundlagen zu Planung und Berechnung von flexiblen Barrieren.

Wendeler, C., Volkwein, A., Roth, A., Denk, M., & Wartmann, S. (2007). Field measurements and numerical modelling of flexible debris flow barriers. In *Debris-flow hazards mitigation: Mechanics, prediction, and assessment* (pp. 681–687). Rotterdam: Millpress.

Zhang, S. (1993). A comprehensive approach to the observation and prevention of debris flows in China. *Natural Hazards, 7*(1), 1–23.

Part VII

Rockfall Hazard Assessment

Rockfall Hazard Assessment: An Overview 15

Biswajeet Pradhan and Ali Mutar Fanos

15.1 Introduction

Rockfalls are landslides that exhibit mass movements and highly varied volume and that involve rock masses ranging from several cubic centimeters to thousands of cubic meters. Rockfalls happen when rock masses are detached from a cliff face and freely fall under the effect of gravity (Blahut et al. 2013; Youssef et al. 2015; Varnes 1984). Even small-magnitude events can be extremely destructive because of their high velocities reaching up to tens of meters per second; they can potentially damage roads and cause fatalities. After a landslide occurs, rockfalls are among the natural hazards that mostly affect roads with steep roadside cuttings through brittle rock masses (Kharel and Dhakal 2013). Therefore, a rockfall is a serious natural disaster in mountainous regions and poses a major threat to infrastructure, transportation networks, and people.

Rockfalls are composed of detached rocks from a cliff face, with subsequent free-falling (flying), bouncing, sliding, and rolling motions along a slope surface with high velocity (Arbanas et al. 2012; Ferrari et al. 2013; Leine et al. 2014). Rock detachment is basically attributed to discontinuities, as well as to relevant weathering and deterioration, along surfaces. Major triggering factors of rockfalls include saturation, erosion, freezing temperatures, weakening caused by water runoff, earthquakes, wildfires, the presence of vegetation roots, frequent freeze–thaw cycles, thermal expansion–contraction, and high rainfalls (Asteriou et al. 2012; Wyllie 2014; Sabatakakis et al. 2015).

Traditional surveying techniques restrict the gathering of spatial datasets to generate digital elevation models (DEMs) required for rockfall modeling (Salvini et al. 2013). In the last decade, new remote sensing technologies and powerful geographical information systems (GIS) have increased topographic information, thereby providing a basis for developing new methodologies to analyze Earth surfaces; moreover, new techniques, such as light detection and ranging (LiDAR), have risen rapidly in the fields of geohazard assessment and modeling. At present, both airborne and ground-based LiDAR surveys are essential for analyzing detailed topographies (Fanos et al. 2016; Youssef et al. 2015; Barbarella et al. 2013; Pradhan et al. 2005; Tonini and Abellan 2014; Stephenne et al. 2014; Fanos and Pradhan 2016).

15.2 Slope Failure Problem

Rockfalls pose considerable threats to public transportation networks and properties located in hilly regions and rock cuttings. However, rockfalls are not as economically dangerous as large-scale failures that can block vital roads for days. Rockfall fatalities tend to be of the same order as those in other types of rock slope instabilities. Martin (1988) reported that rockfalls, small rockslides, and ravelings are the most frequent problems on road transportation networks in the mountainous regions of North America. Hungr and Evans (1988) reported 13 rockfall fatalities in the last 87 years in the mountain motorways of British Columbia, Canada. Over the last decades, increasing incidents of slope failure has been observed in Malaysia. Most of these incidents have occurred on cut slopes or embankments alongside roads and highways in mountainous areas (Pradhan et al. 2010). Shu and Lai (1980) recorded a major rockfall event in Gunung Cheroh, Ipoh, Malaysia. This event involved the collapse of the entire face of a cliff as a single plate weighing approximately 23,000 tons and measuring 33 m in length. It resulted in 40 fatalities, and numerous cattle were also killed. Among the most recent disastrous slope failures occurred on August 7, 2011, in Kampung Sungai Ruil, the Cameron Highlands. Another incident occurred on May 21, 2011, in Hulu Langat. Moreover, a rockfall buried the back portion of

B. Pradhan (✉) · A.M. Fanos
Department of Civil Engineering, University Putra Malaysia, Serdang, Malaysia
e-mail: biswajeet24@gmail.com

B. Pradhan (ed.), *Laser Scanning Applications in Landslide Assessment*,
DOI 10.1007/978-3-319-55342-9_15

an illegal factory located at the foothills of a limestone hill in Bercham, Ipoh, Perak, western Malaysia in December 2004. This incident caused two deaths. Some rockfall incidents have not resulted in fatalities but have caused major inconveniences. Examples include the Athenaeum Condominium in Ulu Kelang in May 1999 and the rocky slope failure in Bukit Lanjan on the New Klang Valley Expressway in 2003. Both events resulted in traffic disruptions that lasted for six months.

15.3 Rockfall

A "rockfall" is a slope process that involves rock fragment detachment and their subsequent falling, bouncing, rolling, sliding, and deposition (Varnes 1978). In certain cases, rockfalls are quantitatively measured by describing the insignificant phenomenon of falling blocks of rocks of a few cubic meters up to 10,000 m^3. Meanwhile, "rockslides" are characterized by falling blocks of over 100,000 m^3, whereas "rock avalanches" may extend to a few million cubic meters (Dussauge-Peisser et al. 2002). Rockfalls occur regularly when one or multiple blocks fall, bounce, slide, or roll down a slope. In a scree slope, a falling block may move beyond the slope edge and stop at a certain distance from the base of the slope. Falling blocks pose the largest hazard to the surrounding areas of a slope, and their uncertain behavior is a major challenge in assessing rockfall hazards (Evans and Hungr 1993).

15.4 Rockfall Definitions

"Rockfalls" or "rockfalls" refer to rock quantities that fall freely from a cliff face. Rockfalls are rock fragments (blocks) that detached by sliding, falling, or toppling, and then fall off a steep cliff face (vertically or a sub-vertically), moving downslope by flying, bouncing over ballistic trajectories, or rolling over talus or debris slopes (Varnes 1978). Chen et al. (1994) defined a "rockfall" as a sudden independent block movement or a complex of continuous rock detachments from a steep slope. Lee and Elliot (1998) defined "rockfall" as "the downslope boulders movement (from natural slopes) or blocks (from cut faces) which, when not correctly restrained, have the potential to damage or destroy structures along their trajectory or creating an impediment to the public transportation networks."

Richards (1988) provided a summary of commonly accepted properties of rockfalls as follows:

- A rockfall event comprises one block or a set of blocks that detached from a cliff face.
- Each falling rock behaves independently of other rocks.
- A temporary loss of earth contact and high downhill acceleration occur.
- Blocks gain significant kinetic energy during their descent.

Rockfall failures vary from slipping failures that form on the slipping surface of rocky slopes. Rockfalls, which include small individual rock blocks, should be differentiated from rock avalanches, which is characterized by a huge amount of mass motion and a portion of the entire slope (which can include the bed rock and the slope face) collapsing suddenly.

15.5 Methods of Data Collection for Rockfall Hazard Analysis

The essential variables of rockfall hazard analysis are shown in Fig. 15.1. Susceptibility, magnitude, rockfall run-out, and exposure are frequently assessed using well-set mapping and measurement methods or directly defined by in situ specialists using a heuristic approach. Current developments in digital data collection platforms and widely available computing resources have allowed digital and indirect assessments of rock mass stability. Such remote sensing techniques involve LiDAR and photogrammetry.

15.5.1 Heuristic or Experience-Based Approaches

The heuristic or experience-based approach is frequently used for rock mass assessment when rock assessment experts are available, and failure modes and geological settings are well understood by the experts. Mining environments depend heavily on heuristic assessment because in situ experts are regularly exposed to geological, structural, and failure models. Nevertheless, in spatially different circumstances, such as transportation corridors where complicated and varying geologies and failure modes can be estimated, heuristic approaches are only utilized at the primary level to identify rock masses that require further assessment and possible mitigation techniques (Ruff and Czurda 2008).

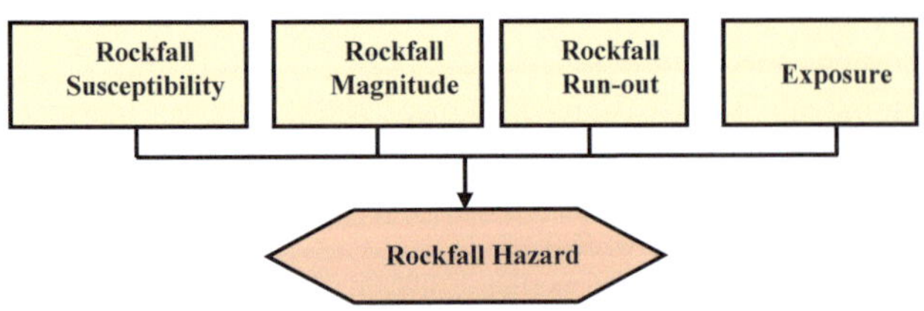

Fig. 15.1 Rockfall hazard evaluation framework

15.5.2 Mapping and Measurement Techniques

Mapping and measurement methods include the immediate physical exposure of the assessing engineer to a possibly unstable rock mass. The engineer typically measures visibly accessible structural characteristics, such as discontinuities involving joints, beds, and faults. The engineer will also assess positional parameters, such as physical setting (height, slope length, and face angle), and how a rock mass interacts with a highway in man-made/natural obstacles (e.g., barriers) and the presence of shoulders and ditches. These measurements and geologic mapping results form the bases for a rockfall hazard evaluation system (Crosta and Agliardi 2004).

15.5.3 Photogrammetric Analysis

Photogrammetry methods for the rock mass assessment of potentially unstable slopes include the alignment and 3D projection of two stereo photographs. The output 3D stereo photographs enable the assessment of rock mass geometry and structure. Discontinuity orientation is measured, faults are detected, and kinematic instability is computed. Photogrammetry methods for rock mass assessment are widely published and adopted in the geological community. Comprehensive examples of processing and data collection techniques are provided in Kemeny and Post (2003) and Haneberg (2007).

15.5.4 Light Detection and Ranging (LiDAR)

LiDAR is a range-based imagery technique that can create an accurate 3D model of the Earth surface within a short period. LiDAR data are basically gathered via mobile aerial surveying (e.g., using helicopters and airplanes) or static terrestrial (using a tripod) methods. The resulting datasets include millions to billions of points in a space coordinate (XYZ) that can be converted into geographical coordinates, such as the Universal Transverse Mercator (UTM). Every point information group typically includes a “color” value associated with the measured intensity of a returning beam as detected by a scanner or associated with the true colors derived from a combination of photographic techniques (Höfle and Rutzinger 2011).

Remote geomechanical assessment of structural discontinuity has conventionally involved the use of photogrammetry techniques. In recent years, technological advancements have led to the assessment and use of LiDAR-based techniques in remote geomechanical analyses. Discontinuity mapping accuracy using LiDAR data should be assessed and compared with conventional compass-based methods prior to implementing LiDAR in engineering workflows (James et al. 2007).

15.6 Rockfall Research Background

Initial research on rockfall behavior was conducted by Ritchie (1963). He stated the necessity for a prediction method for the material stability of rock cut surfaces. Moreover, he developed standards for designing ditches and cut slopes (Fig. 15.2) by performing hundreds of full-scale rockfall tests. These tests are still extensively used in rockfall protection design at present. Ritchie (1963) investigated the movements and trajectory of rocks and attempted to formulate an analytical solution for rockfall based on movement laws.

Subsequent to Ritchie’s work, substantial progress has been achieved in rockfall behavior analysis. Most of these studies are relevant to the highway projects. Rockfall research has been conducted by empirical investigations, physical simulation, and computer simulation (Dorren 2003). Initial research was typically accomplished via empirical approaches, whereas computer modeling has been widely used in the past two decades.

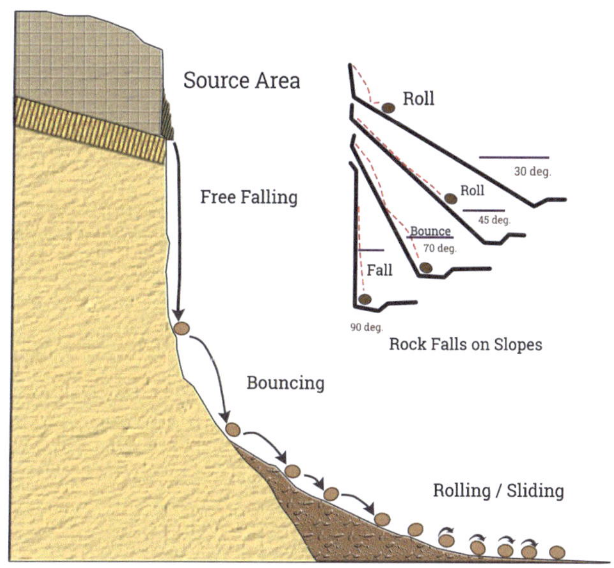

Fig. 15.2 A typical rockfall process and the rockfall design standards based on Ritchie’s (1963) work

15.7 Rockfall Mechanics

Ritchie (1963) provided design criteria for describing relationships among the variables of cliffs, namely slope angle, ditch depth, and fallout area width. Rockfall mechanics for cliffs and slopes have been considered to propose different solutions, including fences or barriers and ditches to accommodate rockfalls. With regard to problematic slope gradients, Ritchie (1963) noted that a large rock had a long run-out distance from the origin. He observed that a falling rock would spend more time in the air and would stop when the slope became sufficiently flat and slope irregularity became sufficiently high to reduce rock velocity. Ritchie also proposed a mechanical approach to describe rockfall trajectory. He explained rock path as a sequence of parabolic trajectories affected by the impact angle that determined the velocity along and perpendicular to the effect plane. Ritchie also suggested that the shape and size of a rock have minimal effect on its falling or rolling characteristics. He inferred that a falling rock must follow particular laws of energy, mass, restitution, velocity, and impact, although it would be affected by friction, time, and gravity. The mechanical considerations of Ritchie are associated with a single rock and its movement is unaffected by neighboring rock fragments involved in a rockfall. Potential energy due to gravity is transformed into kinetic energy in rockfalls. The line that connects the rockfall origin and the final deposition of the rockfall is known as the "energy line," and the gradient is known as the "energy line angle" (Salvini et al. 2013).

15.8 Rockfall-Triggering Factors

Rockfalls begin with the detachment of rocks from a cliff face in a rockfall source region (Youssef et al. 2015). Rocky slopes are subjected to varying weathering degrees that can cause joint opening and cracking, thereby promoting rockfall. Rockfall promotion degree relies on the elements of the environment that cause weathering, i.e., chemical and physical, and on bedrock type (Day 1997). The triggering mechanism determines the occurrence of a rockfall regardless of the weathering rate. The triggering factors of rockfall conditions and mechanisms have been widely characterized in the literature. Rockfall-triggering mechanisms can be classified into rockfall motivators and movement causes. Nevertheless, differentiating between movement causes and rockfall motivators is complicated because a particular process, such as frost shattering, typically motivates weathering that leads to rockfall. Moreover, slope morphology and the direct neighborhood of probable falling rocks are significant elements for determining whether rocks will fall.

Gardner (1983) observed rockfalls in a mountainous region and concluded that such phenomena occurred particularly on glaciers over steep rocky slopes that were alternately subjected to thawing and freezing. Such rockfalls occur frequently, have small magnitudes, and are common in steep regions (Jomelli and Francou 2000). Similarly, Douglas (1980) examined frequent and small-magnitude rockfall events and proved that such events were caused by frost. Nevertheless, he declared that the geotechnical characteristics of the bedrock play a significant role. These findings corroborated the opinion of Luckman (1976), who demonstrated that the geological and morphological natures of cliffs and the variations in temperature of rock surfaces controlled rockfalls. Vidrih et al. (2001) characterized different rockfall causes and explored the correlation between earthquake activities and rockfalls. They inferred that earthquakes would trigger rockfalls. Wieczorek et al. (2000) reported that rockfalls could be triggered by various causes, such as seismic activities, water freezing–thawing cycles in joints, rapid snow thawing, rainstorms, root wedging and permeation, and stress relief deglaciation. In most studies on slope movements, factors that triggered the movements were either unnoticed or unreported. Reported rockfall events have indicated that the rapid melting of snow, earthquakes, and extensive winter rainstorms have caused more movements than human activities and freezing–thawing conditions. Human activities that reduce slope stability in hard rocks remain the main element compared with geological elements, but may vary significantly, such as in the undercutting of slopes through excavations or quarrying for infrastructure. Moreover, animals can also cause rockfalls, such as goats climbing on steep rock faces.

The overall view demonstrates that diverse elements have been recorded as rockfall-triggering parameters. In most cases, however, geological, topographical, and climatic factors combined with time determine whether a rockfall will occur. A dynamic analysis by Salvini et al. (2013) concluded that water saturation and the implemented acceleration of earthquakes could affect the stability of nearly all blocks. A gradual decrease in the stability of steep rocky slopes is one of the potential effects of warming in high mountain areas. Lately, the possible direct role of warm temperatures in triggering rockfalls has been studied (Allen and Huggel 2013). Rockfalls in Malaysia are mainly triggered by tropical rainfall and flash floods that cause failure of the rock surface along fractures, joints, and cleavage planes (Pradhan 2010; Pradhan and Lee 2010).

15.9 Motion Modes of Falling Rocks

After a rock detaches and proceeds downslope, it descends the slope in various motion types. The type of movement highly depends on the mean of the slope incline (Fig. 15.3). The most significant motion modes are free-falling or flying through the air, bouncing over the surface of a slope, and sliding or rolling on the slope surface.

15.9.1 Free-Falling of Rocks

The free-falling of rocks occurs when slopes are extremely steep. Ritchie (1963) stated that the free-falling of rocks would occur if the gradient of the slope was greater than 76°. However, this value varies in different field conditions. Figure 15.1 illustrates that rock movements of approximately 70° transform gradually from bounce to fall.

Azzoni et al. (1995) mentioned that two types of movements could occur during the free-falling of rocks. The first movement is the translation of the rock center, whereas the second is the rotation of the block around the center. Rotation and translation are significant because falling blocks are never round in shape. After a rock rotates in the air, it can bounce in various directions following impact compared with its previous direction. The velocity of a free-falling rock is affected by air friction. Nevertheless, Bozzolo and Pamini (1986) noted that air friction would not affect rock movement. Azzoni et al. (1995) reported that a rock colliding with other falling rocks also influenced free-falling rocks and their trajectories. However, this effect is difficult to analyze during rockfall events or field surveys.

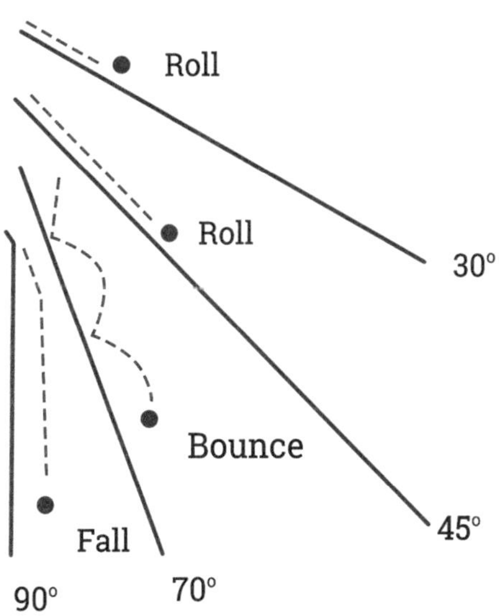

Fig. 15.3 Motion modes of rock during their fall on slopes based on the mean gradients of slope (Ritchie 1963)

15.9.2 Bouncing, Rolling, and Sliding of Rocks

Rock movement occurs on or close to the surface of a slope when the mean incline of the slope is reduced in the downslope section. After free-falling, a rock collides with the surface of the slope; this movement is defined as rock bouncing. Rocks, particularly weak ones, tend to break down into fragments at first bounce (Bozzolo and Pamini 1986). Evans and Hungr (1993) stated that 75–85% of the energy from the first fall was lost during the first collision regardless of whether a rock broke or not. When the mean gradient of the slope is lower than approximately 45°, rock movement transforms gradually from bouncing to rolling because of the rotational momentum collected by the rock. Moreover, a rolling rock is nearly permanently in touch with the surface of a slope (Hungr and Evans 1988). During the transition from bouncing to rolling, a rock revolves rapidly and only the edges with a high radius come in contact with the slope surface. Thus, the center of gravity moves along a nearly direct path, which is an effective movement mode with regard to energy loss. Erismann (1986) stated that the combination of bouncing and rolling was among the major mechanisms of displacement. Sliding is another type of movement on a slope surface. However, sliding typically occurs only during the first and final phases of a rockfall event. As a sliding rock begins to fall, it bounces or rolls as the mean incline of a slope increases. A rock normally stops because of energy loss due to friction if the mean gradient of a slope does not change during sliding (Bozzolo and Pamini 1986). Basson (2012) reported that a falling rock could exhibit four types of movement along its track: free-falling, rolling, bouncing, and sliding. Typically, a rockfall incident experiences more than one of these movements. During free-falling, no interaction occurs between the slope and the falling body; however, an interaction occurs for the other types of motion, during which the rock can be fractured into smaller portions.

15.10 Lateness of Moving Rocks

A moving rock stops after experiencing various modes of motion. The velocity and stopping of a falling rock rely primarily on the mean incline of a slope because a falling rock normally decelerates on a flat slope and accelerates on a steep slope. In addition to the mean incline of the slope, velocity depends on the material that covers the slope, such as soil, vegetation, and scree. Small rocks are easier to stop than large ones because their kinetic energy that aggregates during a rockfall is less than that of the large rocks. Small rocks can be easily stopped by huge obstacles, such as trees. Moreover, they can be easily impeded in the depressions among larger boulders on a slope surface. These reasons are the major sorting effects on falling rocks over a slope (Statham and Francis 1986). In general, the sorting effect should only be considered for the upper portion of scree slopes because rocks with varying sizes in avalanches are mainly deposited at the base (Jomelli and Francou 2000). The stopping of falling rocks is a gradual process rather than a sudden one. Rocks stop because of energy loss from collision forces and friction along the surface of the slope. The frictional force of moving rocks does not only depend on their shape, but also on the characteristics of the slope surface (Statham and Francis 1986). These characteristics can vary considerably within short distances. Thus, the frictional force between the slope surface and a rock can be described using the dynamic angle of friction. The dynamic angle of friction is associated with surface roughness, which has been defined by Pfeiffer and Bowen (1989) as the height variation perpendicular to the slope within a particular distance of the slope. The dynamic angle of friction of falling rocks is described by Kirkby and Statham (1975) as shown in Eq. (15.1):

$$\tan \phi_{\mathrm{ud}} = \tan \phi_0 + c * d(2 * r) \tag{15.1}$$

where ϕ_{ud} is the friction dynamic angle (°); ϕ_0 is the internal friction angle (°), which ranges from 20.3° to 33.7°; c is a constant ranging from 0.16 to 0.25; d is the mean scree diameter on the surface of the slope (m); and r is the rock radius (m).

Forest cover also affects the transportation of scree or large rocks. Zinggeler et al. (1991) studied the significance of trees in stopping falling boulders and inferred that the topography of a slope surface was equally significant; moreover, the collision of falling rocks with tree trunks led to energy loss, thereby ultimately causing rocks to stop in flat regions of a slope surface. Hétu and Gray (2000) noted the influence of a forest on scree movement on a slope surface. They observed that the concentration of rocks over forest edges on scree slopes increased with increasing forest density. Moreover, they mentioned the permanent struggle between forest settlement and active scree slope development. The front area of an active scree slope moves downslope when a forest is disrupted by fire or a large-scale mass movement. Their research elucidated the incapability of forests to stop large-scale destruction from rockfall incidents; however, forests provide efficient protection for small-scale and high-frequency rockfall events.

15.11 Rockfall Modeling and Analysis

Several models can calculate run-out areas of rockfall incidents and their characteristics in terms of trajectory, frequency, velocity, bouncing height, and kinetic energy (Volkwein et al. 2011). All existing rockfall models can be divided into three major types: (1) process-based, (2) empirical, and (3) GIS-based rockfall models.

15.11.1 Process-Based Rockfall Models

Process-based models simulate or explain rockfall movement modes over slope surfaces. Gigli et al. (2014) used 2D and 3D rockfall simulation models to calculate bounce height, rock velocity, and kinetic energy based on rock position along the profiles or on the slope. A 3D rockfall model was utilized to simulate the effect of slope morphology on rockfall trajectories at the regional scale, whereas a 2D rockfall model enabled implementation of a larger number of simulations along the slope profiles specified by the 3D modeling. The lumped mass approach was applied in the two models. Each rock was symbolized by a simple point with its mass settled at the center, and rockfall trajectories were simulated by considering the physical laws that controlled the sequence of various rockfall motion modes (free-falling or flying, bouncing, rolling, and sliding). As an effective and rational technique for protection measures and performance-based design, a 3D rockfall simulation technique assists in depicting rockfall motion on a slope and in probabilistically considering vegetation impact.

Masuya et al. (2009) elaborated a typical evaluation technique and analyzed the manner in which a rockfall combined with vegetation interference and other elements. As an application, a real slope where a rockfall occurred because of an earthquake activity was examined. The advantages and validity of the proposed technique were used as bases for the measurement planning and hazard mapping of a rockfall. Ma et al. (2011) simulated actual rockfall via discontinuous deformation analysis. In the simulation, rockfall energy losses were classified into three types: friction loss, collision loss, and loss by vegetation. The result of the in situ experiments illustrated that energy loss resulting from collision was among the most significant elements. Rockfall impact force is defined by its movement velocity and behavior, which are conditioned by slope incline, rock shape, height, and surface roughness of the rockfall trajectory.

Undulating and rough slopes tend to cause changes in rockfall trajectories. An irregular slope easily changes the behavior of a rockfall movement from sliding or rolling to bouncing. Moreover, a large slope incline increases bouncing movement behavior, whereas a small slope incline easily initiates sliding and rolling. Furthermore, slope surface undulation immediately influences rock collision angle, and rockfall behavior easily changes from sliding or rolling modes to bouncing mode (Wang and Lee 2012).

15.11.2 Empirical Rockfall Models

Empirical rockfall models are typically based on the correlation between the topographical factors and trajectory length of rockfall incidents. Such models are occasionally defined as statistical models (Keylock and Domaas 1999). Leine et al. (2013) developed a complete 3D simulation method for rockfall dynamics. The simulation of a rockfall was performed using hard contact laws based on the non-smooth contact dynamic technique. The rock was modeled similar to that of an arbitrarily convex polyhedron, and the terrain was modeled using a high-resolution DEM. Leine et al. (2013) proposed a specialized law of friction for rockfall that provided scarring behavior description (i.e., a falling rock tended to slide before bouncing on a slope surface). The geometry of rock effect on rockfall dynamic has been examined using two numerical simulations. Topal et al. (2007) devised a 2D rockfall assessment that was performed over several slope profiles. Rockfall characteristics in terms of run-out distance, bouncing height, kinetic energy, and rock velocity over each profile were evaluated using the 2D rockfall model. The outcomes of the simulation were utilized to outline the regions at risk. Mikoš et al. (2006) used a 2D rockfall simulation program to analyze rockfall in two longitudinal profiles. First, the program was calibrated in a previous rockfall event in two longitudinal profiles using different numbers of blocks. The initial values of the associated model parameters were obtained from the literature, and various combinations were tested.

Rockfall run-out has been largely determined based on terrain roughness and surface characteristics. The number of released blocks affects run-out distance. In particular, when the roughness of a slope surface is high, a relatively large number of released blocks should be used. Large blocks have a larger bounce height and higher total kinetic energy but lower run-out distance than small blocks. A forest may virtually stop blocks that are less than 0.2 m, but has no effect on 6 m blocks. The calibrated model has been applied to another rockfall event in two longitudinal profiles without and with a gallery for rockfalls. The results of the simulation (bounce height, total kinetic energy) confirmed the appropriateness of the gallery location. They concluded that silent witnesses, such as released blocks and tree damages, should be used in the case of an active rockfall; otherwise, more than one profile should be simulated. They also mentioned that the upper scar on the rock face should be considered in the calculation. In the case of active rockfalls, rock face color indicates the release points. Silent witnesses may help to a certain extent. Therefore, a rockfall model should be calibrated with another rockfall event under same the field conditions before it can be used.

Ahmad et al. (2013) studied various numerical simulations using rockfall characteristics in terms of maximum rebound height, translational velocity, and total kinetic energy. They also performed a comparative assessment by increasing the rock mass and slope height. Their analysis result showed that varying angles of slope geometry produced more problems than the rock mass in the rockfall scenario. Moreover, these researchers stated that nearly all of the rockfalls occurred because of the orientation and nature of discontinuities in the blocks. In the case of varying slope geometry, bounce height is more variable than the other parameters. However, as rock mass increases, bounce height increases with the same trajectory. Bounce heights exhibit complicated behavior as height increases. Consequently, the geometry of a slope is a more crucial parameter for rockfall compared with the mass of rockfall blocks.

15.11.3 Rockfall Analysis Using GIS-Based Models

Conventional information management related to rockfalls has typically been presented in report form, and photographs are generally organized into file folders and kept in filing cabinets. Data are arranged using indexing techniques to facilitate information search. At present, non-digitized methods cannot match our applications, and thus, introducing new techniques for information management is necessary; these techniques should consider information technology that comprises storage, acquisition, analysis, and distribution of information through a variety of electronic software and equipment products (Antoniou 2013). Information technology involves more than replacing file folders with electronic media; it completely changes the manner in which information is viewed and used. In geotechnical engineering, GIS application has focused on areas where data are defined spatially (Antoniou et al. 2008). GIS technologies and databases have been adapted for information storage associated with major geotechnical issues and their management (Fish and Lane 2002). At present, modern solutions for information technologies used in geotechnical engineering are not limited to stand-alone applications that have been developed in the past decades. However, integrating other sophisticated technologies, such as Web-based applications using the GIS environment and electronic data gathering, has produced modern techniques for the method, in which information is viewed and used in programming interfaces and applications to create maps and reports.

In the past decades, GIS has become a common technique utilized in managing and calculating natural hazards, including rockfalls (Pradhan 2010). GIS analysis has been widely proposed for generating rockfall hazard maps (Antoniou 2013). Rockfall models based on GIS are either raster-based modeling, for which input information are supplied via GIS analysis, or run within a GIS environment. Such rockfall modeling consists of three steps: identifying the rockfall source region in the zone of interest, determining the rockfall trajectory, and computing the length of run-out distance (Hegg and Kienholz 1995).

Lan et al. (2007, 2010) utilized a 3D extension for GIS to determine rockfall characteristics in terms of run-out distance, energy, and velocity. Inventory data were utilized to calibrate the mechanical parameters of the rockfall process. They proposed comprehensive methods for rockfall hazard assessment that considered the characteristics of rockfall source regions, the rockfall physical process, and the spatial attribution of rockfall energy and frequency. To evaluate the potential effect of rockfalls on railway operations, rockfall hazard distribution was investigated using rockfall frequency and energy-simulated distribution. They concluded that 3D rockfall modeling provides a fast framework for rockfall hazard assessment and for understanding the rockfall geomorphic process because it deals with 3D rockfall physical processes and the interaction of rockfall with slope topography. Moreover, it elucidates rockfall processes in terms of trajectory and dissipation as well as predicts their energy and frequency spatial distribution. To assess potential rockfall trajectories, Salvini et al. (2013) used the ArcHydro module of ArcMap and assumed that a rockfall would follow the direction of the steepest gradient. The morphological profile of rockfall trajectories was derived by interpolating 3D points obtained using a method developed in ArcInfo Workstation combined with the ArcMap Easy Profiler tool.

Jaboyedoff and Labiouse (2011) demonstrated that rockfall distribution regions could be specified by using a geometric rule known as the energy line or shadow angle technique based on a simple model of Coulomb friction performed in CONEFALL software. Run-out zones are evaluated from a DEM and a grid or cell file that represents probable rockfall source regions. Moreover, CONEFALL enables evaluation of maximum and mean rock energies and velocities in the rockfall distribution region. The identification of probable rockfall source areas is among the major difficulties in rockfall hazard assessment at a regional scale. Loye et al. (2009) studied probable rockfall source regions based on the distribution of the slope angle derived from a high-resolution DEM combined with other data obtained from topographic maps and geological GIS formats. The results showed that the predicted probable rockfall source areas match in situ observations conducted on test areas and derived from orthophotograph analysis.

Jaboyedoff et al. (2012a) used CONEFALL, which could simply implement a GIS environment, to assess run-out zones from potential source areas. Blahut et al. (2013) used both CONEFALL and RockFall Analyst (RA) codes in quantitative rockfall hazard and risk analysis to identify rockfall hazard regions. They concluded that RA could map rockfall hazard more realistically than CONEFALL in a variety of natural conditions, particularly within the studied region and provide realistic input data for risk assessment. The difference is attributed to the complex input information used in RA, which represents the local slope and energy loss coefficients of falling boulders. Moreover, CONEFALL calculations simplify the modeled rockfall by considering the sliding of rock blocks rather than their falling and bouncing.

15.12 Rockfall Trajectory Modeling Approaches

Rockfall trajectory codes can be categorized into 2D, 2.5D, and 3D rockfall trajectories models and adopt one of the simulation approaches. The analysis using the selected model can be performed probabilistically or deterministically.

15.12.1 2D Rockfall Trajectories Models

A 2D trajectory model simulates rockfall trajectory in a spatial domain determined by two axes. Such models can compute along the slope profile with user input based on distance axes (x, y) and an elevation axis (z) (Azzoni et al. 1995). Such profiles frequently follow the steepest descent line (Basson 2012). Another type of 2D model is rockfall trajectory, which is computed in a spatial framework determined through two distance axes (x, y), such as the contour lines of a map or the elevation values of a raster. This model typically computes the rockfall path using the run-out distance and velocity with sliding block and topographic–hydrologic methods. Gigli et al. (2014) used a 2D model to implement numerous rockfall simulations over the most crucial slope profiles specified through 3D modeling. Youssef et al. (2015) used 2D rockfall simulation software to elucidate the simulation of rockfall and define the main effect of falling rocks on neighboring regions. Antoniou and Lekkas (2010) used a 2D model under seismic and static loading circumstances in run-out distance analysis. Moreover, the selection of the 2D slope profile is critical to derive practical analysis outcomes using 2D rockfall models. Such models only produce a spatial representation of rockfall trajectory distribution because they require selecting critical 2D cross sections. In addition, such models are unable to determine 3D rockfall process characteristics (Lan et al. 2007, 2010).

15.12.1.1 2.5D Rockfall Trajectory Models

A 2.5D model, also known as a quasi-3D model, is the second type of trajectory model and is basically a GIS-supported 2D model used to obtain pre-specified falling trajectories. This model separates the rockfall path direction in the (*x*, *y*) domain from the kinematics of falling rocks and rock trajectory along the vertical plane. To calculate the horizontal direction of falling in the (*x*, *y*) domain, this model divides the kinematics of rockfall calculation, bouncing heights, and positions. This condition indicates that this model performs two different 2D computations: defining the slope profile location in the (*x*, *y*) domain and simulating rockfall over the slope profile in 2D. An example of this model is one that calculates rockfall kinematics over a slope profile that follows the steepest gradient specified using digital surface information, such as the Rocky3 rockfall model (Dorren and Seijmonsbergen 2003).

15.12.2 3D Rockfall Trajectory Models

A 3D rockfall model is defined as a trajectory model that calculates rockfall path along a 3D plane (*x*, *y*, *z*) in each part of the calculation. Moreover, an interrelationship exists among rockfall trajectory direction in the (*x*, *y*) domain, the kinematics of a falling rock, its rebound height and position, and, if included, the influence of trees. The main advantages of 3D models include the converging and diverging effects of terrain and extraordinary or unexpected trajectories. However, 3D models require spatially evident parameter maps of the site, which are more time-consuming to prepare than the definition of parameter values for simulating rockfall trajectories based on slope profile (Volkwein et al. 2011). Examples of 3D models include rapid mass movements used by Leine et al. (2013) for rockfall dynamics, RA, a 3D rockfall process model integrated into GIS that enables effective handling of numerous geospatial information related to rockfall behavior used by Lan et al. (2007, 2010), Macciotta et al. (2011), Blahut et al. (2013), and Samodra et al. (2013, 2014) to assess rockfall characteristics in terms of run-out distance, energy, and velocity. Gigli et al. (2014) used RA to assess the effects of slope morphology on rock trajectory at a regional scale. Lopez-Saez et al. (2016) used a 3D rockfall trajectory model for four various dates to assess the effects of land use and land cover changes on rockfall propagation. The simulation permitted determination of return periods and rockfall kinetic energy, and consequently, the definition of associated hazards at the urban front for each time step.

15.13 Simulation Approaches

Rockfall simulation methods can be classified into three approaches: (i) lumped mass, (ii) rigid body (Hungr and Evans 1988), and (iii) the hybrid approach (Frattini et al. 2008). The following subsections discuss each approach.

15.13.1 Lumped Mass Approach

The lumped mass approach, which is the most widely used rockfall simulation method, considers falling rocks as point masses. Lumped mass models disregard the size and shape of a falling rock; moreover, the mass of the falling rock does not influence its trajectory but is used only to calculate energy. Lumped mass models simulate rockfall with different motion modes (flying, bouncing, sliding, rolling, and final deposition). They require two input parameters: the coefficients of normal and tangential restitutions (Rn and Rt) to compensate for the lack of physics applied in simplified models. The two coefficients of restitution parameters depend on several factors, such as the friction characteristics of falling rocks, incident angle, slope friction, and collision point in a falling rock with a non-spherical shape (Basson 2012).

15.13.2 Rigid Body Approach

Rigid body or rigorous models consider the volume and shape of a falling rock. However, considering the size and shape of individual rocks results in exaggerated computational demands that complicate the evaluation of a rockfall hazard at a regional scale (Guzzetti et al. 2002). The two input parameters in rockfall simulation based on a rigid body approach are dynamic friction (μ) and the normal restitution coefficient (Rn). Dynamic friction is the tangent line of the frictional angle that can be derived from empirical data. Chai et al. (2013) demonstrated that the empirical parameter Rt could be derived via rigid body impact mechanics using only the material parameters Rn and μ. They also introduced the influences of rock size and shape, as well as their interactions with the slope, to compute rockfall trajectory and derive Rt.

15.13.3 Hybrid Approach

The hybrid approach combines with the other two approaches such as using rigid body approach in order to simulate rolling, impact and bounce and lumped mass approach in order to simulate free fall (Frattini et al. 2008).

15.14 Parameters for Rockfall Analysis

The parameters required to simulate rockfalls can be divided into two groups: geometric parameters (seed point identification, topography, outcropping material limit, location of points of interest, or elements at risk) and mechanical parameters (coefficient of restitution (COR), roughness, and friction angle) (Gigli et al. 2014). The most significant parameter is COR. Geometric properties are derived from field elevation observations, whereas mechanical–physical characteristics can be obtained from in situ and laboratory tests or from the implementation of back analysis (Firpo et al. 2011). The trajectories of rockfall, bouncing height, and impact energy rely on slope surface roughness, slope geometry, and rockfall block characteristics (Arbanas et al. 2012). The mass of a rock considerably influences impact energy, such as the kinetic energy of a block, which consists of a smaller rotational component and a translational component. Therefore, rigorously and accurately characterizing a potentially unstable rocks mass or size in a particular field is important (Spadari et al. 2013). Moreover, rock slope geometry significantly affects the post-impact behavior of falling rocks. Slope geometry is vital in any rockfall analysis. Its effect adds a dimension to the final impact distance variation (Vijayakumar et al. 2011). The identification of probable rockfall source regions is a difficult task in rockfall prediction. Source zones are frequently obtained from apparent evidence, such as the deposition of a talus slope below a cliff face, field measurements, and historical register information. Rocky outcrops, and consequently, unstable rockfall source areas are mostly found on steep slopes (Jaboyedoff et al. 2012a).

15.14.1 COR

COR describes the kinematic behavior of a falling rock as it hits the slope surface. Every time a rock hits a slope surface, its movement characteristics are changed. Hoek (2007) described COR as the mathematical expression of the retarding capacity of a surface material when dealing with falling rocks. Each slope has unique properties that vary among regions along the slope. Each falling rock also has unique properties. Therefore, characterizing COR is difficult because each case has a unique set of properties. To simplify this process, COR is generalized to fit the behavior of similar falling rocks downslope with known parameters. Rn is a classic parameter. It denotes material characteristics, which are determined by contacting slope rigidity. Meanwhile, Rt is an experimental parameter that is measured using slope material and vegetation. The range of the proposed Rt value is relatively larger than the range of the proposed Rn value. For example, the proposed Rn values of firm soil and talus slope range from 0.1 to 0.2, whereas their proposed Rt values range from 0.5 to 1.0 (Chai et al. 2013). In another paper, Vijayakumar et al. (2012) used a simple mechanical model to demonstrate that the computed Rn had a value greater than 1.0; this trend, which is evident in certain rockfall field data, is caused by the eccentricity of rock shape and its rotational energy. Their group also demonstrated that the computed coefficient can become negative in some cases. Although such cases seem to break the law of energy conservation, the appropriate description is found in the definition of COR itself. In most situations, the rock body is a point mass; thus, Rn must be based on the incoming and outgoing velocities of the center of mass. Otherwise, the rotating energy is unaccounted for in the point mass model.

The values of COR may vary considerably, depending on the site conditions. Therefore, the two components of COR (normal and tangential) have to be determined separately for each field. The in situ test or back analysis of falling rocks can be used to derive COR. The most critical input parameters for simulating rockfall phenomena are CORs, which control the bouncing of rocks (Asteriou et al. 2012). Chau et al. (2002) presented the results of an experimental study on COR for spherical blocks that affected a rock slope. A plaster modeling material was used to cast both the slopes and the blocks. A positive correlation was noted between slope angle and Rn. However, no apparent relation was detected between slope angle and Rt. When the resultant velocity ratio and kinetic energy ratio before and after impact were utilized to define COR, COR evidently increased with slope angle.

The physical or mechanical parameters (COR and friction angle) are particularly significant inputs for rockfall simulation; these factors control block bouncing, velocity magnitude, and rockfall trajectory analysis (Asteriou et al. 2012; Lato et al. 2012). In addition, the loss of rock boulder energy upon impact is controlled by COR (Keskin 2013; Samodra et al. 2014; Sabatakakis et al. 2015). In principle, hard materials exhibit higher CORs than soft materials. Moreover, Rt increases with Rn. Slight changes in COR values cause entirely different trajectories.

COR is one of the most significant and most difficult parameters for assessment in rockfall analysis (Papathanassiou et al. 2013). However, these values may vary dramatically, depending on the circumstances of each site (Topal et al. 2007). These values also differ for diverse materials and various types of vegetation covering the slope; in addition, the values are dissimilar within the same environment (Macciotta et al. 2014). To avoid obtaining irrelevant values, coefficient distribution can be truncated between the important minimum and maximum values (Frattini et al. 2013). Attempts to model rock-ground effects using a single COR do not sufficiently capture rockfall variability (Glover et al. 2015). Moreover, the slope friction angle cannot be derived via field testing (Ku 2012).

The most commonly used definitions of COR components (normal and tangential) are

$$Rt = V_{tr}/V_{ti}; \quad Rn = V_{nr}/V_{ni} \tag{15.2}$$

where V_{nr} and V_{ni} are the quantities of the rebounding and incoming velocities of the normal component, respectively; and V_{tr} and V_{ti} are the quantities of the rebounding and incoming velocities of the tangential component, respectively (Fig. 15.4).

Chiessi et al. (2010) performed rockfall hazard assessment using two individual approaches. The analyzed rockfall trajectories is strongly affected by the input parameters, particularly the COR values. Asteriou et al. (2012) performed in situ and laboratory tests to determine the parameters that affected rockfall trajectories. CORs are the most critical parameters in rockfall modeling. Wyllie (2014) documented rockfalls in five locations, including the effects on rock, talus, colluvium, asphalt, and concrete. The values of Rn and Rt were calculated for these locations. The field results showed that Rn was related to the impact angle. The Rn values are essentially independent of the slope material. The Rt values ranged from 0.3 to 0.8. This coefficient is related to the friction coefficient at the impact point. Its value is independent of the velocity and normal force. The calculated field values for CORs are consistent with the principles of impact mechanics.

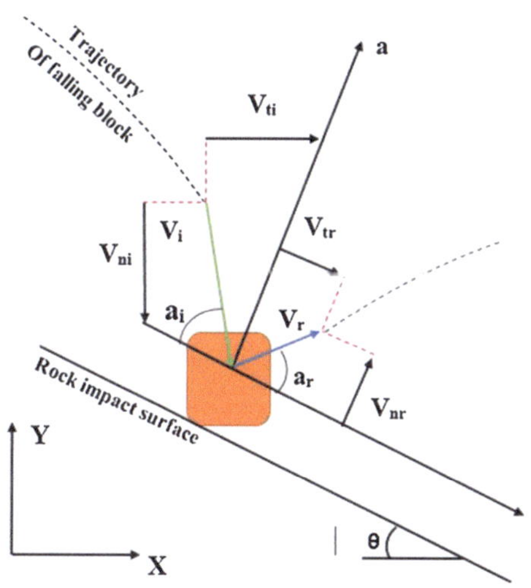

Fig. 15.4 Components of translational velocities before and after impact (Asteriou et al. 2012)

15.15 Possible Measures for Mitigation of Rockfall Hazard

15.15.1 Potential Rockfall Problem Identification

The identification of all probable rockfall hazards using common techniques for rockfall hazard assessment is neither practical nor possible. For example, when studying the blocks on the highest slope portion, rockfall hazard is apparent. Nevertheless, the most hazardous types of rockfall occur when a rock is suddenly detached from a cliff face through comparatively small deformations in neighboring rocks. This event may happen when a force affects discontinuity across planes, thereby separating a rock from its surrounding. A change in discontinuity is attributed to water pressure or the reduced shear strength of planes because of long-term damage after weathering. This phenomenon can sometimes trigger rockfalls of considerable sizes or, in excessive cases, large-scale slope failures. Rock faces should be accurately examined for probable rockfall problems. However, not all rockfall hazards will be revealed through this examination.

15.15.2 Decrease in Energy Level Related to Excavation

Conventional excavation techniques for rocky slopes include blasting. Even with controlled and planned explosions, high-intensity forces affect rock masses for a short period. Wedges and blocks may be triggered by such strong forces. Therefore, to reduce rockfall hazard caused after excavation by explosion, another method should be used; for example, ripping requires concentrated vibrations or short-period forces on rock masses. Manual and mechanical excavation techniques can also be utilized. When an enormous amount of rocks need to be destroyed, chemical expansion may be used.

15.15.3 Physical Restraint of Rockfalls

Rockfalls vary spatially and temporarily; thus, detecting all rockfall hazards is impossible. Techniques for reducing the effects of these rockfalls hazard must be considered. These techniques are elucidated in Fig. 15.5.

A berm is a relatively efficient method for catching rockfalls; this structure is commonly utilized on a permanent slope. Nevertheless, berms may be excavated from the top down. During construction, the use of berms is limited in

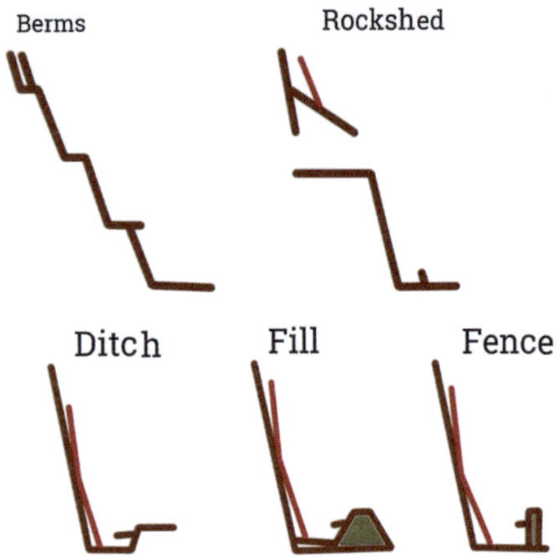

Fig. 15.5 Possible techniques to reduce the damage due to rockfall, after Spang and Rautenstrauch (1988)

minimizing rockfall hazards. Avalanche shelters or rocksheds are frequently utilized over a steep slope above a narrow roadway or railway. A steep slope roof coverage with a comparatively narrow extent is required for efficient shelter. For a wide multi-lane expressway, designing a rockshed structure with an adequate capability to resist large rockfalls may be impossible. In general, a fill of soil or gravel is recommended at the top of a rockshed to function as a rockfall deflector and retarder. Rock traps can effectively catch rockfalls by providing an adequate room at the slope toe to accommodate such trap. In the case of a relatively narrow highway at a steep slope toe, adequate room for rock trap accommodation is difficult to find. Frequently utilized barriers or catch fences are estimated to have an energy absorption capacity of 100 kN/m^2. This value is equal to a 250 kg rock mass falling with a velocity of approximately 20 m/s. However, a robust barrier fence can have an energy absorption capacity of up to 2500 kN/m^2, which can stop a 6200 kg rock moving at a velocity of approximately 20 m/s. The use of a mesh draped over a rock face is another restraint technique that deserves further consideration. Meshes extend along a rock face and are attached at several positions over the slope surface. A mesh is not used to stop rockfalls but is intended to trap the falling blocks between the rock face and the mesh, thereby reducing rock velocity that causes rocks to bounce out onto the highway. The construction of a catch ditch at a slope toe is probably the most efficient permanent rockfall protection technique for most expressways. To increase the efficiency of ditches, a ditch base is normally covered with a gravel layer for the energy absorption of falling boulders; a robust barrier fence is placed between the highway and the ditch. The location of a barrier fence can be assessed based on rockfall analysis, such as rockfall trajectories and their characteristics.

15.16 DEM

A DEM is a 3D representation of a topographical terrain. Current geomorphometry focuses on the parameters obtained for a terrain surface (slope, slope aspect, and curvature) and the spatial features or land surface objects (cirque, watershed boundary, and drainage network) from DEM. This characterization depends on the general and specific geomorphometric analysis modes. Specific modes describe discrete surface objects, such as landforms, whereas general modes describe a continuous terrain surface. The most typical data format is the DEM square grid, where the gridding sets of points in Cartesian spaces are assigned with elevation values that characterize the terrain surface (Wilson 2012).

A DEM provides basic information about topographic relief. The resolution of this model significantly affects modeling outcomes, thereby indicating that its selection is a critical step in the numerical modeling of rockfalls (Salvini et al. 2013; Bühler et al. 2014). This finding is attributed to the reliable prediction of such events, which is highly relevant to the 3D characteristic of real slope geometry (Ku 2012). In particular, LiDAR techniques may be applied in rockfall hazard assessment because of their capacity to produce precise and accurate ground surface DEMs (Rayburg et al. 2009; Barbarella et al. 2013). Bühler et al. (2014) used high-resolution LiDAR (50 cm) to sample the land morphology of an extremely active rockfall region. Their group resampled the obtained DEM into various resolutions. Rockfall simulation was conducted while the terrain effect parameters of the model were kept fixed. In addition, the release orientation was varied to mimic the naturally stochastic initial circumstances of boulder fall detachment, whereas potential energy was kept fixed. The various results of rockfall simulation were compared to assess the effect of DEM resolution on completely 3D rockfall simulation. DEM resolution significantly affects the results of rockfall simulation, thereby demonstrating that DEM selection is a crucial part of numerical rockfall simulation.

15.16.1 DEM Data Acquisition

DEM production integrates three correlated functions: (i) terrain surface sampling, such as the collection of altitudes; (ii) surface model generation from the sampled altitudes; and (iii) error correction in the generated DEM (Hengl et al. 2010). Data sources and processing techniques for creating DEMs have developed rapidly from a topographic map and land survey transformation to passive remote sensing techniques and more recently to active remote sensing techniques using radar and LiDAR (Wilson 2012). DEM can be derived from different sources with various spatial resolutions ranging from a few centimeters to 90 m.

DEM spatial resolution that symbolizes surface topography can considerably influence the results of rockfall simulation. In particular, the terrain roughness of a boulder field or a scree slope is included when the spatial resolution of DEM (centimeters to meters) is fine (Bühler et al. 2014). Nelson et al. (2009) classified DEM data into three general classes based on collection methods: (i) land survey methods involving theodolite, electronic distance measurement (EDM), total stations, and global navigation satellite system (GNSS) instruments; (ii) existing topographic maps, which are in hardcopy form, including the elements of contour lines, lakes, rivers, and spot heights; and (iii) remote sensing techniques, including airborne and satellite photogrammetric techniques, airborne and terrestrial laser scanning, and airborne and satellite radar.

15.16.2 Pre-Processing of Data and DEM Construction

The preparation of elevation data for geomorphometric analyses is a complex process because elevation itself is generally not the attribute of concern. The actual geomorphological accuracy may be evaluated using terrain parameters and feature measurement, including landforms or in-site drainage lines and the comparison of their locations, distributions, and shapes with data derived from geomorphometric analyses (Wilson et al. 2008). Reuter et al. (2009) proposed that the actual application of DEMs in geomorphometric analyses could be evaluated by answering the following questions. (i) What is the terrain roughness representation accuracy? (ii) What is the accuracy of representation of the ground surface shape (i.e., convex and concave shapes, water divergence or convergence, deposition, and erosion)? (iii) What is the accuracy of detection of streamlines and world ridgelines? (iv) How regularly is elevation measured over the entire concern area? The responses to these questions and other comparable queries are interconnected. Errors will be mostly present in the preferred or accessible DEM despite the responses to these significant queries. Error magnitude and frequency depend on the methods and techniques utilized for data gathering, the implemented algorithms in pre-processing, and ground surface characteristics.

Elevation data resolution (horizontal and vertical) was utilized to describe ground topography. These data definitely have a major effect on the information level and the accurate description of terrain objects, as well as on the values of the terrain surface parameters, which are calculated from a DEM (Bühler et al. 2014). Grid spacing also influences the accuracy and values of landform objects and the parameters of land surface (Raaflaub and Collins 2006). The rapid development of mass-produced sources and remote sensing DEMs over the last two decades requires new techniques for DEM pre-processing. Webster and Dias (2006) and Reuter et al. (2009) described varied approaches and possibilities for orthorectifying DEMs, reducing local noise and outliers, filtrating water surface, filtrating clear noise, filtrating forests in DEM, filling sinks and voids, mosaic neighboring DEM, and filtrating LiDAR DEM.

Before producing the triangulated mesh required to generate a DEM of the cliff surface from the obtained point cloud, a pre-processing step has to be performed, including two major functions: (1) eliminating vegetation cover and (2) differentiating rock outcrops from the construct surroundings and the detritus at the slope toe within the point clouds. The decimation and segmentation stages are implemented using manual and automated methods; point cloud filtering based on the various intensity of the pulse return is reflected from the scanned features (Fanti et al. 2013). This tedious pre-processing of point clouds is justified by the eventual objective, i.e., the generation of a dependable DEM that is appropriate for rock mass discontinuity characterization. In general, non-geological and vegetation points, as well as registration error, can contribute to noise in point clouds and influence the procedure of automatic meshing (Buckley et al. 2008). After point extraction on a rock mass out-crop, the resultant points can be resampled to obtain a uniform and regular spatial distribution. To create continuous surfaces from discrete information, the sampled point clouds can be eventually triangulated by considering the fill holes to produce homogeneous surfaces.

15.16.3 Computation of Terrain Parameters

In the regular workflow of digital modeling, the focus will be on obtaining spatial features (land surface objects) and the values of terrain parameters after an appropriate DEM is produced. The parameters of a terrain surface are directly obtained without further input from the DEM. To characterize these parameters, various terms are utilized in the literature. Olaya (2009) defined these parameters as "basic" parameters of terrain surfaces; these parameters could be computed from a DEM without additional comprehension of the explained region. Florinsky (1998) differentiated local elementary attributes that were computed as a function of their neighboring and regional elementary attributes, which required a wide regional ground surface area analysis from a computing approach. Wilson and Burrough (1999) demonstrated the differences between the regional and local surface characteristics of the presence of local interactions between "action-at-a-distance" forces and surrounding points. The parameters of the local land surface included aspect, slope, and curvatures; the parameters of regional ground surface included the extent of flow trajectory, downslope contribution area, dispersion area, and the proximity to the closest ridgeline. Most local parameters are calculated through a (3 × 3) window approach moving over a grid and computing the parameters of the land surface for the concerned cell, i.e., the centric cell of the (3 × 3) window (Fig. 15.6). Specific principles are available on how to deal with this approach; the edges create a new raster for each parameter, which has the same size as the DEM. Various formulas have been presented for calculating aspect and slope as the first extraction and curvatures as the second extraction (Florinsky 1998; Shary et al. 2002).

The computation and interpretation of the aspect and slope grids are reasonably uncomplicated. The vertical curvature or profile and the horizontal curvature or tangential are frequently utilized to differentiate locally concave and convex shapes. In geosciences, the curvature sign is written by convention as negative and positive for concave and convex shapes, respectively (Olaya 2009). That is, concave and convex tangential curvatures represent the convergence and divergence of flow paths, respectively. The concave profile curvature represents slope flattening, and thus, a decrease in potential energies. By contrast, the convex profile curvature represents flow acceleration, and consequently, a local increment in potential energies. Finally, the scale issues with the framework of the land surface parameters should be addressed. Local land shapes frequently exhibit a constant difference in altitude values from point to point over land surface; these shapes significantly influence regional and local ground surface characteristics. However, this role is affected by data and processing factors. Florinsky (1998) reported that local attributes, such as aspect, slope incline, and curvature, were mathematical variables instead of actual values. This assertion can be expanded to all local ground surface characteristics because of two reasons. First, the shape of the local ground surface can depends on various mathematical descriptions; thus, computed local attributes depend on the selection of the algorithm. Second, the terrain shape represented by DEM is a scale function, which is combined with terrain complexity, spatial scale, and resolution or scale of data from which the land surface is observed.

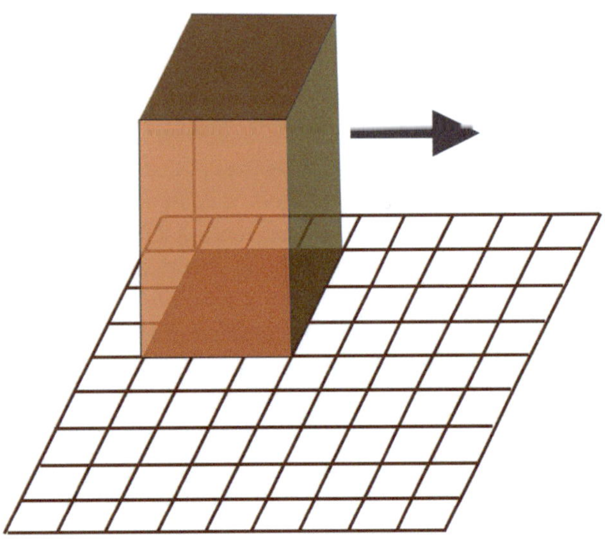

Fig. 15.6 Terrain parameters are typically computed by moving a 3 × 3 grid across a DEM (Olaya 2009)

15.16.4 Error Calculation in DEM

Errors in DEM are variable, depending on the sensor selection or particular application (distribution technique); thus, a DEM will be deduced from another to present elevation difference for evaluating deposition, erosion, and change (Burns et al. 2010). The other group of difficulties is related to altitude error propagation in terrain parameters, and considerable effort is frequently required for error identification. Nevertheless, errors associated with the data source are mostly difficult to remove; any individual concerned with utilizing terrain surface parameters derived from DEM should be aware of these errors and how these errors may influence workflow and interpretation, and consequently, the results. Several methods have been presented to evaluate the accuracy of DEM altitude values (Temme et al. 2009).

Several researchers have compared a series of altitudes obtained from a DEM and the actual altitude values derived from the most precise sources of topographic information; the root-mean-square error (RMSE) of altitude, which represents the variation between true and derived values, has also been calculated (Wise 2000). The only issue with this method is that it disregards the spatial distribution pattern and the existence of systematic bias of errors, which are crucial to these terrain surface parameters (Hutchinson and Gallant 2000). The aforementioned parameters are significantly affected by the shape of the terrain surface. Carara et al. (1997) proposed multi-criteria with broader significance to assess the quality of a DEM created from contour lines. The values of the DEM and contour lines should be similar and close to the contours. The values of the DEM should be within the range specified throughout an interval of contours. The values of the DEM should nearly linearly differ with the values of the contour line interval. Artifacts have to be restricted to within a small portion of the data set, and DEM patterns should reflect the actual shape in flat areas. Hutchinson and Gallant (2000) measured the quality of a DEM, which was created from surface contour lines; point elevation and streamlines data suggested a wide and more varied list of simple metrics that included some of the same schemes. Abellán et al. (2009, 2010) applied a nearest neighbor (NN) averaging technique to minimize the error in

RAW data obtained using the LiDAR technique. The NN method is composed of three steps: (1) interpolation of data to a square grid or cell, (2) search for the n neighboring points, and (3) the NN average value calculation for each point, except for the edges. The selection of the DEM interpolation technique can strongly influence DEM surface properties (Wise 2011; Gallay et al. 2013). Bater and Coops (2009) examined seven interpolation procedures using airborne laser scanning (ALS) data. These data were categorized from random subsets into a verification data set and a prediction dataset. A series of DEMs was generated through the natural neighbor, linear, regularized spline, quintic, spline with tension, inverse distance weighted, and finite difference approach interpolation algorithms. These researchers concluded that the natural neighbor interpolation algorithm provided the best outcomes at minimal effort among all the algorithms.

15.16.5 Remote Sensing Techniques for Capturing DEM

The topography of complex landscapes is challenging to obtain in remote regions; moreover, ground-based survey techniques can be difficult, time-consuming, and less/insignificant landscape features. DEM can be generated from various data sources, but this diversity can result in different precision and accuracy degrees. Remote sensing and GIS have revolutionized hazard studies because of their efficient data collection, analysis, and validation processes (Pradhan et al. 2011). During the last two decades, the rapid development of data sources for produced DEMs, such as photogrammetry, shuttle radar topographic mission (SRTM), radar interferometry (InSAR), and LiDAR surveying, has considerably improved DEM resolution because of the highly accurate and precise data that they can provide.

15.16.5.1 Photogrammetric-Based DEM Generation

Photogrammetry provides 3D point coordinates with expected accuracy from stereo- or multi-photographs, such as from photographs taken for the same scene from various viewpoints. The accuracy of the coordinates relies on the number of elements that should be considered in designing the steps for any photogrammetric surveying: the calibration of the camera, the orientation of photographs, and the restitution of objects. Object-based photograph construct and matching from the movement, extraction, and matching of tie points can be automatically performed; thus, the orientation of a photograph can be derived without any manual measures. The parameters of orientation can also be specified immediately through the fixation of a GNSS device, incorporated with an inertial measurement unit (IMU) and a camera. This combination allows processing without requiring ground control points (GCPs). The restitution of an object can be performed automatically or manually by a technician. Photogrammetry has a long history among remote sensing technologies for DEM production. Photogrammetry techniques have confirmed their efficiency for an extensive range of mapping applications, including the production of DEMs, cartographic maps, and orthophotographs. Photogrammetry is commonly utilized as a multipurpose spatial data-capturing technique given the rapid development in the utilization and maturation of GIS. Figure 15.7 illustrates a 1 arc-second (30 m) photogrammetric DEM. DEM generation based on photogrammetry principles has two operating phases: the initial measuring phase and the second phase for DEM derivation. The primary data sources are from aerial photographs (film-based or digital). Digital image-processing techniques are applied by interacting (user-based) measured techniques or an automatic technique. The DEM points identified by the interpolation process from the aerial photographs (stereo pairs) are based on object matching (Chang et al. 2004).

Photogrammetry based on repeated aerial photography is considered an adequate remote sensing technique for long-term monitoring of small deformation rates along large regions (Strozzi et al. 2010). Terrestrial photogrammetry is another type of photogrammetry; this method is beneficial for performing elaborate surveys of geometric–structural environments, even in unreachable areas. Firpo et al. (2011) used various terrestrial photogrammetric techniques for rockfall simulation. Their result demonstrated that the quantity and accuracy of geometric and geological information acquired from a photogrammetric survey permitted the numerical assessment of the correlation between rock factors as a function of their load conditions and mechanical or physical properties. In addition, potential shadow can be removed by varying the focal length and shooting position. Nevertheless, this survey technique has drawbacks, such as in extremely high slopes, where a full photogrammetric survey of the highest regions of a rocky wall is impossible to perform.

15.16.5.2 SRTM DEM Generation

The 3 arc-second SRTM DEM, which has been developed based on satellite data gathered during a nine-day window in 2000, includes a considerable portion of the world (58°S to 60°N). This dataset is recognized as one of the most proportionate, perfect, and common environment datasets worldwide (Zandbergen 2008; Nelson et al. 2009). The spacing of 3 arc-second grids (~90 m) is preferable compared with the worldwide GTOPO30 DEM spacing of 1 km. An accuracy evaluation using kinematic GNSS data demonstrated elevation accuracy, where 90% of the errors were less than 5 m (Rodriguez et al. 2006). Recent studies

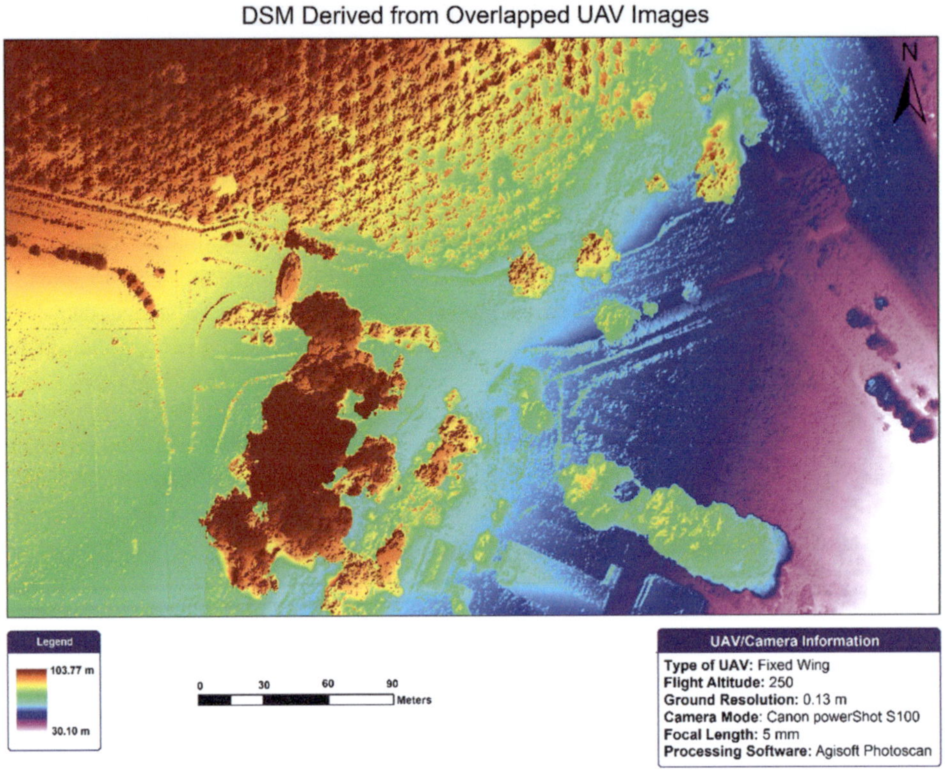

Fig. 15.7 A 1 arc-second photogrammetric DEM (Chang et al. 2004)

have demonstrated a positive correlation between elevation error and canopy height (Hofton et al. 2006; Berry et al. 2007). The relatively new technique of Advanced Spaceborne Thermal Emission and Reflectance Radiometer Global DEM (ASTER GDEM) was released in 2009. The ASTER GDEM technique provides greater spatial covering (83°N to 83°S vs. 60°N to 58°S) and higher resolution (1 arc-second vs. 3 arc-seconds) as well as comparison of the horizontal and vertical accuracies for SRTM (Nelson et al. 2009; Slater et al. 2009). Furthermore, the issues of missing data due to cloud are considerably easier to fill. Nevertheless, a 30-m resolution is still inadequate for supporting vegetation, soil mapping, and related phenomena in most terrain.

15.16.5.3 InSAR DEM Generation

Synthetic aperture radar (SAR) is defined as a side-looking active radar range technique. SAR utilizes the microwave part of the electromagnetic spectrum, including frequencies ranging from 0.3 to 300 GHz or from 1 mm to 1 m in the wavelength range. InSAR requires two SAR photographs for the same location. Those photographs can be obtained separately at the time of revisiting the same location with a single antenna, such as in a typical spaceborne radar system or while simultaneously utilizing two antennas hanging on a platform, such as several satellite and typical airborne systems. Both photographs are then registered accurately with each other to calculate the phase variance between the pixels in the two photographs. This interferogram or phase variance can be utilized to obtain the DEM of the captured region. Figure 15.8 illustrates an InSAR DEM obtained from the images of the European remote sensing satellite (ERS)-2 and ERS-1 captured during the "tandem" mission (Chang et al. 2004). ERS SAR works with an incidence angle $\theta = 23°$ from the vertical direction at the center of the swath, which is 100 km wide, and at the wavelength $\lambda = 5.65$ cm. SAR revisits a particular location every 35 days.

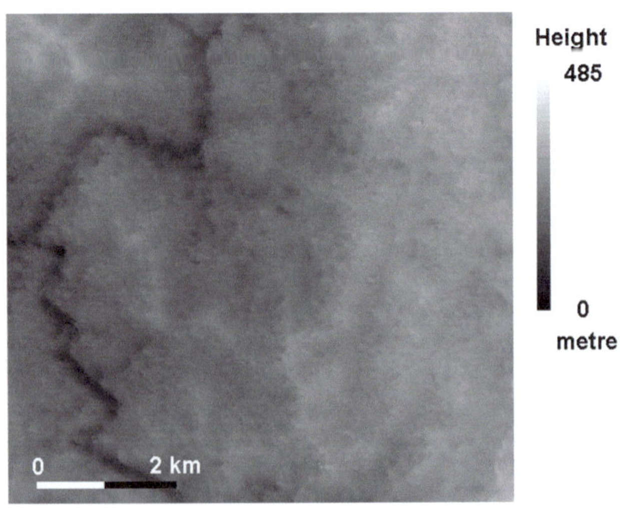

Fig. 15.8 InSAR tandem DEM

15.16.5.4 Unmanned Aerial System (UAS)

A UAS is known by various terms and names, such as aerial robot, drone, or unmanned aerial vehicle (UAV); the most popular acronym is UAV. A UAS is composed of three basic components, which are frequently defined as the unmanned aerial vehicle, the communication and data link, and the ground control station. Furthermore, other components of UAS are considered crucial, such as navigation sensors, autopilots, mechanical servos, imaging sensors, and a wireless system. DEMs and orthophotographs are two major UAS products. At present, a UAV can be utilized to accelerate the external direction phase and minimize operational cost. The main usefulness of UAV-based remote sensing applications for dangerous environments such as rockfalls, landslides, or mudslides is their capability to obtain data in hazardous regions of concern. Direct measurements in such areas are frequently impossible. The flying motion of a UAV is autonomously based on a programmed plan of flight that utilizes compound dynamic automated systems and GNSS/INS to guide external directions. The processing of a DEM creation basically relies on several elements, such as flight height, overlapping, and camera resolution. The differences in these factors influence the final accuracy of the obtained results. The standard algorithms for DEM creation suffer from normal errors caused by GNSS/INS instruments, particularly in location measurement related to each obtained photograph. The difference between real locations and these measurements is approximately a few meters. The complete flowchart of DEM generation algorithms is depicted in Fig. 15.9 (Ruiz et al. 2013).

The regular products of UAVs are dense DEMs of approximately 50 points per square meter, 2 cm seamless orthophotographs, and 3D vector maps with a 2 cm point precision in XY and 4 cm in altitude (Haarbrink and Eisenbeiss 2008). Hand-launched and simple UAVs that operate autonomously by utilizing its GPS-driven autopilot and, typically, an IMU sensor, are the most affordable systems. Nevertheless, most of these systems are significantly influenced by cold and wind, and thus, they are rarely utilized or difficult to utilize in mountainous topography (Niethammer et al. 2012). The most stable systems, generally those with a gasoline engine, have greater payloads and allow a more professional camera onboard or even conduct surveys using LiDAR instruments. Fritz et al. (2013) compared LiDAR point clouds and UAS-based point clouds by utilizing a frame camera for tree trunk reconstruction. The resulting points with reconstruction were less dense and less accurate than with LiDAR. In addition to the frame camera, the LiDAR technique plays a major role in point cloud creation. However, the use of LiDAR techniques in UAS platforms for DEM generation remains infrequent. Niethammer et al. (2010) used a UAV for high-resolution acquisitions of landslides. Digital surface models (DSMs) have been generated from the airborne photographs of a landslide using a modern features-based surface reconstruction technique that does not require any GCP information.

Giordan et al. (2014) used a micro-UAV in an emergency scenario related to rockfall phenomenon. The 3D photographs derived were utilized to create the first-order DSM, which provides quantitative information about the orientation and the dimension of the main discontinuity identified in the rock mass. The LiDAR technique was used to acquire a high-resolution DEM of the study area to improve and validate the results derived from micro-drone surveying.

15.17 Detection and Characterization of Rock Mass Movement

ALS and terrestrial laser scanning (TLS) technologies are responsible for the outstanding development in the characterization of rock slopes, primarily because rock instability are mainly dominated by a structure that is at least locally

Fig. 15.9 DEM generation algorithms (Ruiz et al. 2013)

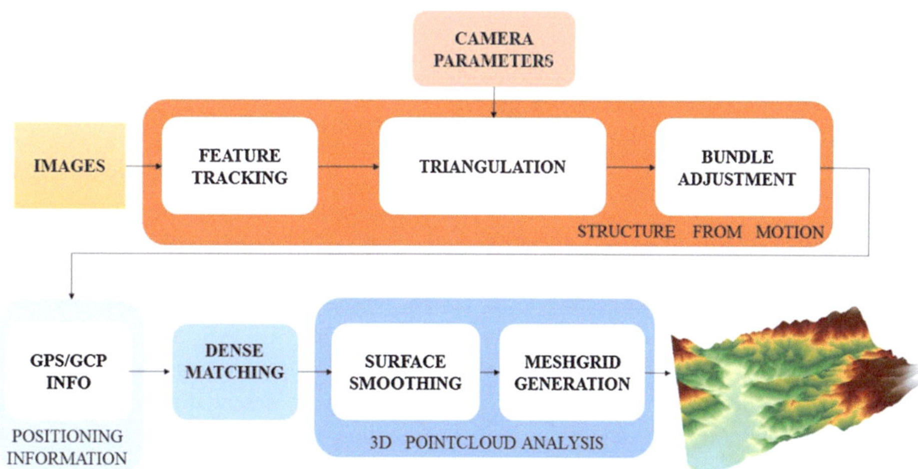

plane surfaces. The applications of TLS in rock mass characterization are countless (Oppikofer et al. 2009; Sturzenegger and Stead 2009; Lato et al. 2009a, b; Armesto et al. 2009). By contrast, the applications of ALS in generating DEMs are still uncommon.

15.17.1 Using ALS

Large-scale feature analyses are required to derive DEMs by utilizing ALS based on the COLTOP principle (Jaboyedoff et al. 2009b), which permits assigning a distinctive color to each topographical direction. This technique also enables rapid feature characterization of inaccessible areas (Brideau et al. 2009) and the reinterpretation of previous rockfalls (Froese et al. 2009). Oppikofer (2009) investigated a fiord valley at the regional scale to characterize the instability of former and present rock slopes, i.e., their mechanisms and volume.

15.17.2 Using TLS

The TLS characterization of a rock slope is one of the first applications that utilize TLS for slope mass movement, mostly to acquire accurate discontinuous orientations and slope profile (Bornaz et al. 2002; Slob et al. 2002). The various methods for characterizing discontinuity sets can be divided into three parts. The first part involves the use of the appropriate plane (Abellan et al. 2006; Sturzenegger and Stead 2009). The second part utilizes the triangular irregular network (TIN) surface as an indicator of plane orientation (Feng et al. 2001; Slob et al. 2002). Kemeny and Post (2003) provided a description of this methodology for rock mass characterization using Split-FX software. Lato et al. (2009a, b) demonstrated the optimum point number for point clouds to obtain a realistic result based on the precision level of the instrument. Furthermore, a high-density TIN is influenced by data noises. Finally, the other methods enable the automatic delineation of a set of neighborhood points, which are characterized by the same normal vector. Consequently, the computation of the orientation of plane discontinuities is acquired.

The COLTOP method allows visualization of each discontinuity orientation set by using a distinctive color that makes the method similar to in situ data capturing (Jaboyedoff et al. 2009b). The obtained data allow the analysis of rock instability mechanisms (Oppikofer et al. 2009; Janeras et al. 2004). Lato et al. (2009a, b) and Sturzenegger and Stead (2009) noted two types of bias in the definition of discontinuity orientation: (a) a scale bias is observed when spatial resolution (point spacing) is larger than the discontinuity sets and (b) an orientation bias is observed when the spatial resolution is influenced by the incidence angle of a given data set. Roughness determination is another aspect of the application that is integrated into rock slope characteristics. Haneberg (2007) and Tatone and Grassel-li (2009) attempted to improve the quantification of popular techniques for rock mass rating systems. However, certain limitations are associated with the achievement of this goal, such as the inherent instrumental accuracy and the resolution of various scanner locations on the eventual outcomes (Sturzenegger et al. 2007).

15.18 Rockfall Monitoring

The monitoring of surface displacements in a rocky slope is easier than in soil slopes because the displacement can be regarded as rigid body transformation (Monserrat and Crosetto 2008; Oppikofer et al. 2009; Abellan et al. 2009). Thus, the movements are considered a combination of translation and rotation in various slope portions. Detailed movement analyses that utilize the rigid body transformation method combined with comprehensive structural analyses enable the determination of potential rockfall mechanisms (Oppikofer et al. 2009). The possibility linkage of temporal and spatial rockfall predictions represents a considerable challenge in the monitoring of rockfalls. Recently, two various preliminary indicators were investigated: (a) the increase in rockfall activities prior to final collapses (Rosser et al. 2007) for a hard rock cliff and (b) preliminary displacement detection was investigated for a large rockfall (Oppikofer et al. 2008) and fragmental rockfalls as Abellan et al. (2009, 2010) illustrated that few centimeters displacement prior rockfall with few to hundred cubic meters. Although the magnitude of preliminary displacement can be same as that of instrument errors in some cases, several researchers have noted that these errors can be considerably reduced by considering the information from the adjacent points, such as in cases of filtering noise reduction or averaging (Monserrat and Crosetto 2008; Abellan et al. 2009). Abellan et al. (2009) demonstrated the possibility of millimetric terrain displacements detection in outdoor experiments, even with a single point, which had the highest standard deviation.

15.19 Rockfall Analysis

Traditional surveying techniques present significant restrictions in acquiring spatial datasets required for rockfall assessment. The use of modern technologies, such as LiDAR, has rapidly improved in the field of geohazard evaluation. ALS and TLS surveys are currently considered essential tools for accurate and dense information collection to assist detailed topographical analysis. From a deterministic perspective, rockfall trajectory depends on (1) the location point at which the rock detaches, (2) the rock that is detached, (3) the slope properties, and (4) how the rock behaves over slope surface (Salvini et al. 2013). The first elements correlate with the source position (i.e., plane coordinates and altitude) (Li and Lan 2015).

DEM resolution significantly influences detection of rockfall sources: a rough DEM tends to smooth the values of the slope angle (Michoud et al. 2012). Loye et al. (2009) reported that the higher the DEM resolution, the smaller the probable source regions. Moreover, the rougher the DEM, the lower the evident slope angle. For example, for a 10 m DEM, the apparent vertical cliff slope angle is 55°, whereas it is 83° for a 2 m DEM. The accessibility of a high-resolution DEM at a regional scale facilitates detailed terrain analysis. Thus, a DEM-based geomorphometric approach accurately detects probable rockfall sources; such regions are defined depending on the distribution of the slope angle derived from crossing very high-resolution DEM with other information obtained from topographic maps and land cover in GIS format. Nevertheless, a major challenge encountered at a regional scale rockfall hazard mapping is the identification of these regions (Loye et al. 2009). Novel remote sensing techniques, such as high-resolution photography and laser scanners, will quantitatively characterize rockfall source regions safely and efficiently (Stock et al. 2011).

15.19.1 Using ALS

Rockfall hazard assessment at a regional scale from source areas to rockfall distribution using ALS–DEM is not regularly implemented. The first challenge is determining rockfall source regions. This step is typically conducted by utilizing the slope angle threshold (Frattini et al. 2008). However, Loye et al. (2009) showed that further details could probably be obtained from slope angle distribution. The threshold depends on the type of bedrock, the DEM resolution, and the absence or presence of a land cover (Loye et al. 2009). This technique allows effectively distinction between a real cliff and one drawn on the topographic map. As proposed by Günther (2003), a structural analysis performed on a DEM acquired via ALS could be the foundation for kinetic tests. Janeras et al. (2004) showed that the result accuracy was significantly enhanced by utilizing a high-resolution DEM.

Rockfall hazard assessment requires frequently executing trajectory modeling for delineating rockfall distribution areas. Rockfall model has been significantly enhanced by using a DEM obtained via ALS (Lan et al. 2007) by offering further dispersal in propagation as indicated by Agliardi and Crosta (2003). Moreover, the profile of kinetic energy is significantly modified with increasing DEM resolution. These parameters are essential for rockfall hazard mapping and for eliminating measurements. Airborne LiDAR was utilized with rockfall spatial modeling by Lan et al. (2010) for a rockfall assessment strategy along a section of a railway. Their group concluded that utilizing LiDAR could explain the usefulness of rockfall hazard assessment along a portion of the railway. ALS allowed them to achieve accurate modeling of terrain geomorphology and to acquire the geometry of significant infrastructure. The simulation results from the high-resolution ALS–DEM present better correspondence with the historic rockfall than the results from coarse DEM. The simulation offers logical rockfall frequency distribution over the railway corridor and the accurate positions of the high possibility of rockfall compared with the historical observations. Topographical analyses utilizing the ALS dataset can also determine possible rockfall source areas based on the slope angle and topographic contrast.

15.19.2 Using TLS

High-resolution DEM is required to implement rockfall simulations and kinematic analyses. In particular, LiDAR techniques are interesting in rockfall hazard assessment because of their capability to produce highly precise and accurate DEMs of the Earth's surface (Barbarella et al. 2013). TLS can provide spatial resolution, high accuracy, and rapid information gathering, and thus, it is becoming increasingly utilized in rockfall research for small areas (Fanti et al. 2013; Gigli et al. 2012; Lato et al. 2012). Gigli et al. (2014) used the TLS technique to provide all the geometric parameters required for implementing rockfall simulations (DEM, main source areas, and the outcropping material limits). Tonini and Abellan (2014) presented a method for extracting features from terrestrial LiDAR point clouds that focused on the recognition of a single rockfall event. The spatial distribution of these events has been analyzed, thereby demonstrating that detected rockfalls are clustered within a well-defined distance ranging from 1 to 3.5 m.

Salvini et al. (2013) and Lato et al. (2013) integrated data from various sources (topographical observations, photogrammetry, and laser scanning) to obtain a DEM of a slope surface, define possible run-out trajectories, and characterize rock mass. They showed that digital terrestrial photogrammetry and TLS provided a powerful analytical tool and model for studying rockfall hazard and the stability of rocky slopes. Gigli et al. (2014) applied TLS in combination with geomechanical surveys at three different areas to cover a wide range of features and examine a proposed approach. TLS was used to structure 3D surface model of the entire slopes to be investigated for kinetic analyses and rockfall modeling and to rebuild the geomechanical characteristics of the block masses and ultimately identify major rockfall source regions. Each slope was observed from various acquisition points to mitigate the shadow areas as much as possible given slope roughness and scan location limits.

Abellán et al. (2006) used TLS for a comprehensive rockfall research in a test site. High-resolution DEM and the reconstruction of joint geometry were consequently obtained. The DEM was utilized for rockfall inventory and for the accurate simulation of rockfall trajectories and velocities. By contrast, joint geometry enabled modeling of the volume and geometry of the source region in a current rockfall. Their group reported that the TLS technique could be used as a reference tool in rockfall studies in the near future. Janke (2013) compared DEM derivatives (aspect, slope, elevation, curvature, and hillshade) obtained from LiDAR and the US Geological Survey (USGS) DEMs for evaluating rock glaciers. He concluded that the USGS DEMs might suitable for analysis to characterize landform topographical setting at a regional scale. At a fine scale, however, rock glacier topography was illustrated more clearly on the LiDAR DEM, thereby making it a perfect tool for feature acquisition.

15.20 Conclusion

The following conclusions can be drawn from this chapter:

1. Rockfall modeling is frequently implemented via 2D or 3D rockfall model.
2. The selection of a 2D slope profile is crucial to derive practical results of rockfall analysis using 2D models. Such models are restricted in their capability to provide rockfall trajectories spatial distribution, as well as flying or bouncing height and kinetic energy.
3. Several stand-alone 3D rockfall models are accessible for 3D rockfall simulation. These models normally utilize topographic information that are transformed from other information sources, such as geospatial information in regular GRID or TIN, which are typical information constructs in most GIS programs. This information transformation is usually tedious and time-consuming, particularly for a large study area.
4. The accuracy of a DEM is crucial for rockfall assessment. Remote sensing techniques, particularly LiDAR techniques, offer the most accurate DEM among the various techniques for DEM generation.
5. The LiDAR techniques of ALS and TLS are responsible for considerable developments in rock slope characterization. The applications of TLS to characterize rock mass are countless, whereas those of ALS–DEM are still infrequent.
6. Rockfall modeling has two approaches: lumped mass and rigid body. The most popular approach is lumped mass because of its capability to model rockfalls in various motion modes, including flying or free-falling, impacting and rebounding, and rolling or sliding. In addition, the use of rigid body can produce immoderate computational demands that make evaluating rockfall hazard difficult at a regional scale.
7. The identification of rockfall source areas is a challenging task in rockfall simulation.
8. Mechanical parameters significantly affect rockfall trajectories and their characteristics.
9. Numerous studies have been conducted in rockfall hazard assessment. However, most of these studies used specific mechanical parameters but disregarded the uncertainty of these parameters.

References

Abellán, A., Calvet, J., Vilaplana, J. M., & Blanchard, J. (2010). Detection and spatial prediction of rockfalls by means of terrestrial laser scanner monitoring. *Geomorphology, 119*(3), 162–171.

Abellán, A., Jaboyedoff, M., Oppikofer, T., & Vilaplana, J. (2009). Detection of millimetric deformation using a terrestrial laser scanner: Experiment and application to a rockfall event. *Natural Hazards and Earth System Science, 9*(2), 365–372.

Abellán, A., Vilaplana, J., & Martínez, J. (2006). Application of a long-range terrestrial laser scanner to a detailed rockfall study at Vall de Núria (Eastern Pyrenees, Spain). *Engineering Geology, 88*(3), 136–148.

Agliardi, F., & Crosta, G. (2003). High resolution three-dimensional numerical modelling of rockfalls. *International Journal of Rock Mechanics and Mining Sciences, 40*(4), 455–471.

Ahmad, M., Umrao, R., Ansari, M., Singh, R., & Singh, T. (2013). Assessment of rockfall hazard along the road cut slopes of state highway-72, Maharashtra, India. *Geomaterials, 3*(1), 15–23.

Allen, S., & Huggel, C. (2013). Extremely warm temperatures as a potential cause of recent high mountain rockfall. *Global and Planetary Change, 107,* 59–69.

Antoniou, A. A. (2013). GIS-based evaluation of rockfall risk along routes in Greece. *Environmental Earth Sciences, 70*(5), 2305–2318.

Antoniou, A. A., & Lekkas, E. (2010). Rockfall susceptibility map for Athinios port, Santorini island, Greece. *Geomorphology, 118*(1), 152–166.

Antoniou, A., Papadimitriou, A., & Tsiambaos, G. (2008). A geographical information system managing geotechnical data for Athens (Greece) and its use for automated seismic microzonation. *Natural Hazards, 47*(3), 369–395.

Arbanas, Ž., Grošić, M., Udovič, D., & Mihalić, S. (2012). Rockfall hazard analyses and rockfall protection along the Adriatic coast of Croatia. *Journal of Civil Engineering and Architecture, 6*(3), 344–355.

Armesto, J., Ordóñez, C., Alejano, L., & Arias, P. (2009). Terrestrial laser scanning used to determine the geometry of a granite boulder for stability analysis purposes. *Geomorphology, 106*(3), 271–277.

Asteriou, P., Saroglou, H., & Tsiambaos, G. (2012). Geotechnical and kinematic parameters affecting the coefficients of restitution for rock fall analysis. *International Journal of Rock Mechanics and Mining Sciences, 54,* 103–113.

Azzoni, A., La Barbera, G., & Zaninetti, A. (1995). Analysis and prediction of rockfalls using a mathematical model. *International Journal of Rock Mechanics and Mining Sciences & Geomechanics Abstracts*, 709.

Barbarella, M., Fiani, M., & Lugli, A. (2013). Application of LiDAR-derived DEM for detection of mass movements on a landslide. *ISPRS-International Archives of the Photogrammetry, Remote Sensing and Spatial Information Sciences, 1*(3), 89–98.

Basson, F. (2012). Rigid body dynamics for rock fall trajectory simulation. In *46th US Rock Mechanics/Geomechanics Symposium*. American Rock Mechanics Association.

Bater, C. W., & Coops, N. C. (2009). Evaluating error associated with LiDAR-derived DEM interpolation. *Computers & Geosciences, 35*(2), 289–300.

Berry, P., Garlick, J., & Smith, R. (2007). Near-global validation of the SRTM DEM using satellite radar altimetry. *Remote Sensing of Environment, 106*(1), 17–27.

Blahut, J., Klimeš, J., & Vařilová, Z. (2013). Quantitative rockfall hazard and risk analysis in selected municipalities of the České Švýcarsko National Park, Northwestern Czechia. *Geografie, 118*(3), 205–220.

Bornaz, L., Lingua, A., & Rinaudo, F. (2002). Engineering and environmental applications of laser scanner techniques. *International Archives of Photogrammetry Remote Sensing and Spatial Information Sciences, 34*(3/B), 40–43.

Bozzolo, D., & Pamini, R. (1986). Simulation of rock falls down a valley side. *Acta Mechanica, 63*(1–4), 113–130.

Brideau, M., Yan, M., & Stead, D. (2009). The role of tectonic damage and brittle rock fracture in the development of large rock slope failures. *Geomorphology, 103*(1), 30–49.

Buckley, S. J., Howell, J., Enge, H., & Kurz, T. (2008). Terrestrial laser scanning in geology: Data acquisition, processing and accuracy considerations. *Journal of the Geological Society, 165*(3), 625–638.

Bühler, Y., Glover, J., Christen, M., & Bartelt, P. (2014). Digital elevation models in numerical rockfall simulations. In *EGU General Assembly Conference Abstracts* (p. 2109).

Burns, W. J., Coe, J. A., Kaya, B. S., & Ma, L. (2010). Analysis of elevation changes detected from multi-temporal LiDAR surveys in forested landslide terrain in western Oregon. *Environmental and Engineering Geoscience, 16*(4), 315–341.

Carrara, A., Bitelli, G., & Carla, R. (1997). Comparison of techniques for generating digital terrain models from contour lines. *International Journal of Geographical Information Science, 11*(5), 451–473.

Chai, S., Yacoub, T., Charbonneau, K., & Curran, J. (2013). The effect of rigid body impact mechanics on tangential coefficient of restitution. *Geo Montreal*.

Chang, H., Ge, L., Rizos, C., & Milne, T. (2004). Validation of DEMs derived from radar interferometry, airborne laser scanning and photogrammetry by using GPS-RTK. In *Geoscience and Remote Sensing Symposium (IGARSS'04)* (p. 2815). IEEE International.

Chau, K., Wong, R., & Wu, J. (2002). Coefficient of restitution and rotational motions of rockfall impacts. *International Journal of Rock Mechanics and Mining Sciences, 39*(1), 69–77.

Chen, H., Chen, R., & Huang, T. (1994). An application of an analytical model to a slope subject to rockfalls. *Bulletin of the Association of Engineering Geologists, 31*(4), 447–458.

Chiessi, V., D'Orefice, M., Mugnozza, G. S., Vitale, V., & Cannese, C. (2010). Geological, geomechanical and geostatistical assessment of rockfall hazard in San Quirico Village (Abruzzo, Italy). *Geomorphology, 119*(3), 147–161.

Crosta, G., & Agliardi, F. (2004). Parametric evaluation of 3D dispersion of rockfall trajectories. *Natural Hazards and Earth System Science, 4*(4), 583–598.

Day, R. W. (1997). Case studies of rockfall in soft versus hard rock. *Environmental and Engineering Geoscience, 3*(1), 133–140.

Dorren, L. K. (2003). A review of rockfall mechanics and modelling approaches. *Progress in Physical Geography, 27*(1), 69–87.

Dorren, L. K., & Seijmonsbergen, A. C. (2003). Comparison of three GIS-based models for predicting rockfall runout zones at a regional scale. *Geomorphology, 56*(1), 49–64.

Douglas, G. (1980). Magnitude frequency study of rockfall in Co., Antrim, N. Ireland. *Earth Surface Processes, 5*(2), 123–129.

Dussauge-Peisser, C., Helmstetter, A., Grasso, J., Hantz, D., Desvarreux, P., Jeannin, M., et al. (2002). Probabilistic approach to rock fall hazard assessment: Potential of historical data analysis. *Natural Hazards and Earth System Science, 2*(1/2), 15–26.

Erismann, T. (1986). Flowing, rolling, bouncing, sliding: Synopsis of basic mechanisms. *Acta Mechanica, 64*(1–2), 101–110.

Evans, S., & Hungr, O. (1993). The assessment of rockfall hazard at the base of talus slopes. *Canadian Geotechnical Journal, 30*(4), 620–636.

Fanti, R., Gigli, G., Lombardi, L., Tapete, D., & Canuti, P. (2013). Terrestrial laser scanning for rockfall stability analysis in the cultural heritage site of Pitigliano (Italy). *Landslides, 10*(4), 409–420.

Fanos, A.M., & Pradhan, B. (2016). Multi-scenario Rockfall Hazard Assessment Using LiDAR Data and GIS. *Geotechnical and Geological Engineering, 34*(5), 1375–1393. http://dx.doi.org/10.1007/s10706-016-0049-z.

Fanos, A.M., Pradhan, B., Aziz, A.A., Jebur, M.N., & Park, H. J. (2016). Assessment of multi-scenario rockfall hazard based on mechanical parameters using high-resolution airborne laser scanning data and GIS in a tropical area. *Environmental Earth Sciences, 75*, 1129. http://dx.doi.org/1007/s12665-016-5936-3.

Feng, Q., Sjögren, P., Stephansson, O., & Jing, L. (2001). Measuring fracture orientation at exposed rock faces by using a non-reflector total station. *Engineering Geology, 59*(1), 133–146.

Ferrari, F., Giani, G. P., & Apuani, T. (2013). Why can rockfall normal restitution coefficient be higher than one? *Rendiconti online Società Geologica Italiana*, 122.

Firpo, G., Salvini, R., Francioni, M., & Ranjith, P. (2011). Use of digital terrestrial photogrammetry in rocky slope stability analysis by distinct elements numerical methods. *International Journal of Rock Mechanics and Mining Sciences, 48*(7), 1045–1054.

Fish, M., & Lane, R. (2002). Linking New Hampshire's rock cut management system with a geographic information system. *Transportation Research Record: Journal of the Transportation Research Board, no., 1786*, 51–59.

Florinsky, I. V. (1998). Accuracy of local topographic variables derived from digital elevation models. *International Journal of Geographical Information Science, 12*(1), 47–62.

Frattini, P., Crosta, G. B., Agliardi, F., & Imposimato, S. (2013). Challenging calibration in 3D rockfall modelling. In *Landslide science and practice* (pp. 169–175). Berlin: Springer.

Frattini, P., Crosta, G., Carrara, A., & Agliardi, F. (2008). Assessment of rockfall susceptibility by integrating statistical and physically-based approaches. *Geomorphology, 94*(3), 419–437.

Fritz, A., Kattenborn, T., & Koch, B. (2013). UAV-based photogrammetric point clouds-tree stem mapping in open stands in comparison to terrestrial laser scanner point clouds. *International Archives of the Photogrammetry, Remote Sensing and Spatial Information Sciences, XL, 1*, W2.

Froese, C. R., Moreno, F., Jaboyedoff, M., & Cruden, D. M. (2009). 25 years of movement monitoring on South Peak, Turtle Mountain: Understanding the hazard. *Canadian Geotechnical Journal, 46*(3), 256–269.

Gallay, M., Lloyd, C. D., McKinley, J., & Barry, L. (2013). Assessing modern ground survey methods and airborne laser scanning for digital terrain modelling: A case study from the Lake District, England. *Computers & Geosciences, 51,* 216–227.

Gardner, J. (1983). Rockfall frequency and distribution in the Highwood Pass area, Canadian Rocky Mountains. *Zeitschrift für Geomorphologie, 27*(3), 311–324.

Gigli, G., Frodella, W., Mugnai, F., Tapete, D., Cigna, F., Fanti, R., et al. (2012). Instability mechanisms affecting cultural heritage sites in the Maltese Archipelago. *Natural Hazards and Earth System Sciences, 12,* 1883–1903.

Gigli, G., Morelli, S., Fornera, S., & Casagli, N. (2014). Terrestrial laser scanner and geomechanical surveys for the rapid evaluation of rock fall susceptibility scenarios. *Landslides, 11*(1), 1–14.

Giordan, D., Manconi, A., Facello, A., Baldo, M., Allasia, P., & Dutto, F. (2014). "Brief communication" the use of UAV in rock fall emergency scenario. *Natural Hazards and Earth System Sciences Discussions, 2*(6), 4011–4029.

Glover, J., Bartelt, P., Christen, M., & Gerber, W. (2015). Rockfall-simulation with irregular rock blocks. In *Engineering geology for society and territory* (Vol. 2, pp. 1729–1733). Berlin: Springer.

Günther, A. (2003). SLOPEMAP: Programs for automated mapping of geometrical and kinematical properties of hard rock hill slopes. *Computers & Geosciences, 29*(7), 865–875.

Guzzetti, F., Crosta, G., Detti, R., & Agliardi, F. (2002). STONE: A computer program for the three-dimensional simulation of rock-falls. *Computers & Geosciences, 28*(9), 1079–1093.

Haarbrink, R., & Eisenbeiss, H. (2008). Accurate DSM production from unmanned helicopter systems. *The International Archives of the Photogr ammetry, Remote Sensing and Spatial Information Sciences., 37,* 1259–1264.

Haneberg, W. C. (2007). Directional roughness profiles from three-dimensional photogrammetric or laser scanner point clouds. In E. Eberhardt, D. Stead, & T. Morrison (Eds.), *Rock mechanics: Meeting society's challenges and demands* (p. 101).

Hegg, C., & Kienholz, H. (1995). Determining paths of gravity-driven slope processes: The 'vector tree model'. In *Geographical information systems in assessing natural hazards* (pp. 79–92). Berlin: Springer.

Hengl, T., Heuvelink, G., & Van Loon, E. (2010). On the uncertainty of stream networks derived from elevation data: The error propagation approach. *Hydrology and Earth System Sciences, 14*(7), 1153–1165.

Hétu, B., & Gray, J. T. (2000). Effects of environmental change on scree slope development throughout the postglacial period in the Chic-Choc Mountains in the northern Gaspé Peninsula, Québec. *Geomorphology, 32*(3), 335–355.

Höfle, B., & Rutzinger, M. (2011). Topographic airborne LiDAR in geomorphology: A technological perspective. *Zeitschrift für Geomorphologie, Supplementary Issues, 55*(2), 1–29.

Hofton, M., Dubayah, R., Blair, J. B., & Rabine, D. (2006). Validation of SRTM elevations over vegetated and non-vegetated terrain using medium footprint lidar. *Photogrammetric Engineering & Remote Sensing, 72*(3), 279–285.

Hungr, O., & Evans, S. (1988). Engineering evaluation of fragmental rockfall hazards. In *Proceedings of the Fifth International Symposium on Landslides, Lausanne, AA Balkema, Rotterdam, Netherlands* (p. 685).

Hutchinson, M. F., & Gallant, J. C. (2000). Digital elevation models and representation of terrain shape. In *Terrain analysis: Principles and applications* (pp. 29–50).

Jaboyedoff, M., Choffet, M., Derron, M., Horton, P., Loye, A. Longchamp, C., et al. (2012a). Preliminary slope mass movement susceptibility mapping using DEM and LiDAR DEM. In *Terrigenous mass movements* (pp. 109–170). Berlin: Springer.

Jaboyedoff, M., & Labiouse, V. (2011). Technical Note: Preliminary estimation of rockfall runout zones. *Natural Hazards and Earth System Science, 11*(3), 819–828.

Jaboyedoff, M., Oppikofer, T., Abellán, A., Derron, M., Loye, A., Metzger, R., et al. (2012b). Use of LIDAR in landslide investigations: A review. *Natural Hazards, 61*(1), 5–28.

Jaboyedoff, M., Oppikofer, T., Locat, A., Locat, J., Turmel, D., Robitaille, D., et al. (2009b). Use of ground-based LIDAR for the analysis of retrogressive landslides in sensitive clay and of rotational landslides in river banks. *Canadian Geotechnical Journal, 46,* 1379–1390.

James, L. A., Watson, D. G., & Hansen, W. F. (2007). Using LiDAR data to map gullies and headwater streams under forest canopy: South Carolina, USA. *Catena, 71*(1), 132–144.

Janeras, M., Navarro, M., Arnó, G., Ruiz, A., Kornus, W., Talaya, J., et al. (2004). LiDAR applications to rock fall hazard assesment in Vall de Núria. In *4th ICA Mountain Cartography Workshop, Vall de Núria, Catalonia, Spain* (p. 1).

Janke, J. R. (2013). Using airborne LiDAR and USGS DEM data for assessing rock glaciers and glaciers. *Geomorphology, 195,* 118–130.

Jomelli, V., & Francou, B. (2000). Comparing the characteristics of rockfall talus and snow avalanche landforms in an Alpine environment using a new methodological approach: Massif des Ecrins, French Alps. *Geomorphology, 35*(3), 181–192.

Kemeny, J., & Post, R. (2003). Estimating three-dimensional rock discontinuity orientation from digital images of fracture traces. *Computers & Geosciences, 29*(1), 65–77.

Keskin, İ. (2013). Evaluation of rock falls in an urban area: The case of Boğaziçi (Erzincan/Turkey). *Environmental Earth Sciences, 70*(4), 1619–1628.

Keylock, C., & Domaas, U. (1999). Evaluation of topographic models of rockfall travel distance for use in hazard applications. *Arctic, Antarctic, and Alpine Research, 31*(3), 312–320.

Kirkby, M., & Statham, I. (1975). Surface stone movement and scree formation. *The Journal of Geology, 83*(3), 349–362.

Ku, C. (2012). Assessing rockfall hazards using a three-dimensional numerical model based on high resolution DEM. In *The Twenty-second International Offshore and Polar Engineering ConferenceInternational Society of Offshore and Polar Engineers* (p. 790).

Lan, H. Derek, Martin, C., & Lim, C. (2007). RockFall analyst: A GIS extension for three-dimensional and spatially distributed rockfall hazard modeling. *Computers & Geosciences, 33*(2), 262–279.

Lan, H., Martin, C. D., Zhou, C., & Lim, C. H. (2010). Rockfall hazard analysis using LiDAR and spatial modeling. *Geomorphology, 118* (1), 213–223.

Lato, M., Diederichs, M. S., Hutchinson, D. J., & Harrap, R. (2009a). Optimization of LiDAR scanning and processing for automated structural evaluation of discontinuities in rockmasses. *International Journal of Rock Mechanics and Mining Sciences, 46*(1), 194–199.

Lato, M. J., Diederichs, M. S., Hutchinson, D. J., & Harrap, R. (2012). Evaluating roadside rockmasses for rockfall hazards using LiDAR data: Optimizing data collection and processing protocols. *Natural Hazards, 60*(3), 831–864.

Lato, M., Hutchinson, J., Diederichs, M., Ball, D., & Harrap, R. (2009b). Engineering monitoring of rockfall hazards along transportation corridors: Using mobile terrestrial LiDAR. *Natural Hazards and Earth System Sciences, 9*(3), 935–946.

Lato, M., Kemeny, J., Harrap, R., & Bevan, G. (2013). Rock bench: Establishing a common repository and standards for assessing rockmass characteristics using LiDAR and photogrammetry. *Computers & Geosciences, 50,* 106–114.

Lee, K., & Elliott, G. (1998). *Rockfall: Application of computer simulation to design of preventive measures, planning, design and implementation of debris flow and rockfall hazards mitigation measures* (pp. 47–65). Hong Kong: Association of Geo-Technical Specialists & Hong Kong Institution of Engineers.

Leine, R., Schweizer, A., Christen, M., Glover, J., Bartelt, P., & Gerber, W. (2013). Simulation of rockfall trajectories with consideration of rock shape. *Multibody System Dynamics, 32*(2), 1–31.

Li, L., & Lan, H. (2015). Probabilistic modeling of rockfall trajectories: A review. *Bulletin of Engineering Geology and the Environment, 74* (4), 1163–1176.

Lopez-Saez, J., Corona, C., Eckert, N., Stoffel, M., Bourrier, F., & Berger, F. (2016). Impacts of land-use and land-cover changes on rockfall propagation: Insights from the Grenoble conurbation. *Science of the Total Environment, 547,* 345–355.

Loye, A., Jaboyedoff, M., & Pedrazzini, A. (2009). Identification of potential rockfall source areas at a regional scale using a DEM-based geomorphometric analysis. *Natural Hazards and Earth System Science, 9*(5), 1643–1653.

Luckman, B. (1976). Rockfalls and rockfall inventory data: Some observations from Surprise Valley, Jasper National Park, Canada. *Earth Surface Processes, 1*(3), 287–298.

Ma, G., Matsuyama, H., Nishiyama, S., & Ohnishi, Y. (2011). Practical studies on rockfall simulation by DDA. *Journal of Rock Mechanics and Geotechnical Engineering, 3*(1), 57–63.

Macciotta, R., Cruden, D., Martin, C., & Morgenstern, N. (2011). Combining geology, morphology and 3D modelling to understand the rock fall distribution along the railways in the Fraser River Valley, between Hope and Boston Bar. In *BC International Symposium on Rock Slope Stability in Open Pit Mining and Civil Engineering, Vancouver, BC, Canada.*

Macciotta, R., Martin, C. D., & Cruden, D. M. (2014). Probabilistic estimation of rockfall height and kinetic energy based on a three-dimensional trajectory model and Monte Carlo simulation. *Landslides, 12*(4), 1–16.

Martin, D. C. (1988). Rockfall control: An update (technical note). *Bulletin Association Engineering Geologists, 13*(14), 329–335.

Masuya, H., Amanuma, K., Nishikawa, Y., & Tsuji, T. (2009). Basic rockfall simulation with consideration of vegetation and application to protection measure. *Natural Hazards and Earth System Science, 9*(6), 1835–1843.

Michoud, C., Derron, M., Horton, P., Jaboyedoff, M., Baillifard, F., Loye, A., et al. (2012). Rockfall hazard and risk assessments along roads at a regional scale: Example in Swiss Alps. *Natural Hazards and Earth System Sciences, 12*(3), 615–629.

Mikoš, M., Petje, U., & Ribičič, M. (2006). *Application of a rockfall simulation program in an alpine valley in Slovenia* (pp. 199–211). Slopes, Failures and Landslides: Disaster Mitigation of Debris Flows.

Monserrat, O., & Crosetto, M. (2008). Deformation measurement using terrestrial laser scanning data and least squares 3D surface matching. *ISPRS Journal of Photogrammetry and Remote Sensing, 63*(1), 142–154.

Nelson, A., Reuter, H. I., & Gessler, P. (2009). Chapter 3 DEM production methods and sources. *Developments in Soil Science*, *33*, 65–85.

Niethammer, U., James, M., Rothmund, S., Travelletti, J., & Joswig, M. (2012). UAV-based remote sensing of the Super-Sauze landslide: Evaluation and results. *Engineering Geology, 128,* 2–11.

Niethammer, U., Rothmund, S., James, M., Travelletti, J., & Joswig, M. (2010). UAV-based remote sensing of landslides. *International Archives of Photogrammetry, Remote Sensing and Spatial Information Sciences, 38*(Part 5), 496–501.

Olaya, V. (2009). Chapter 6 basic land-surface parameters. *Developments in Soil Science, 33,* 141–169.

Oppikofer, T., Jaboyedoff, M., Blikra, L., Derron, M., & Metzger, R. (2009). Characterization and monitoring of the Åknes rockslide using terrestrial laser scanning. *Natural Hazards and Earth System Science, 9*(3), 1003–1019.

Oppikofer, T., Jaboyedoff, M., & Keusen, H. (2008). Collapse at the eastern Eiger flank in the Swiss Alps. *Nature Geoscience, 1*(8), 531–535.

Papathanassiou, G., Marinos, V., Vogiatzis, D., & Valkaniotis, S. (2013). A rock fall analysis study in Parnassos area, Central Greece. In *Landslide science and practice* (pp. 67–72). Berlin: Springer.

Pfeiffer, T. J., & Bowen, T. (1989). Computer simulation of rockfalls. *Bulletin of the Association of Engineering Geologists, 26*(1), 135–146.

Pradhan, B., & Lee, S. (2010). Regional landslide susceptibility analysis using back-propagation neural network model at Cameron Highland, Malaysia. *Landslides, 7*(1), 13–30.

Pradhan, B., Mansor, S., Pirasteh, S., & Buchroithner, M. F. (2011). Landslide hazard and risk analyses at a landslide prone catchment area using statistical based geospatial model. *International Journal of Remote Sensing, 32*(14), 4075–4087.

Pradhan, B., Mansor, S., Ramli, A. R., Mohamed Sharif, A. R. B., & Sandeep, K. (2005). LiDAR data compression using wavelets. *Remote Sensing International Society for Optics and Photonics,* 598305.

Pradhan, B., OHb, H., & Buchroithner, M. (2010). Use of remote sensing data and GIS to produce a landslide susceptibility map of a landslide prone area using a weight of evidence model. *Assessment, 11,* 13.

Raaflaub, L. D., & Collins, M. J. (2006). The effect of error in gridded digital elevation models on the estimation of topographic parameters. *Environmental Modelling and Software, 21*(5), 710–732.

Rayburg, S., Thoms, M., & Neave, M. (2009). A comparison of digital elevation models generated from different data sources. *Geomorphology, 106*(3), 261–270.

Reuter, H. I., Hengl, T., Gessler, P., & Soille, P. (2009). Chapter 4 Preparation of DEMs for geomorphometric analysis, *Developments in Soil Science, 33,* 87–120.

Richards, L. (1988). Rockfall protection: A review of current analytical and design methods. *Secondo ciclo di conferenze di meccanica ed ingegneria delle rocce, MIR, Politecnico di Torino, 11,* 1–13.

Ritchie, A. M. (1963). Evaluation of rockfall and its control. *Highway research record, no., 17,* 13–28.

Rodriguez, E., Morris, C. S., & Belz, J. E. (2006). A global assessment of the SRTM performance. *Photogrammetric Engineering & Remote Sensing, 72*(3), 249–260.

Rosser, N., Lim, M., Petley, D., Dunning, S., & Allison, R. (2007). Patterns of precursory rockfall prior to slope failure. *Journal of Geophysical Research: Earth Surface (2003–2012), 112*(F4), 1–14.

Ruff, M., & Czurda, K. (2008). Landslide susceptibility analysis with a heuristic approach in the Eastern Alps (Vorarlberg, Austria). *Geomorphology, 94*(3), 314–324.

Ruiz, J., Diaz-Mas, L., Perez, F., & Viguria, A. (2013). Evaluating the accuracy of DEM generation algorithms from UAV imagery. *International Archives of the Photogrammetry, Remote Sensing and Spatial Information Sciences, 40,* 333–337.

Sabatakakis, N., Depountis, N., & Vagenas, N. (2015). Evaluation of rockfall restitution coefficients. In *Engineering geology for society and territory* (Vol. 2, pp. 2023–2026). Berlin: Springer.

Salvini, R., Francioni, M., Riccucci, S., Bonciani, F., & Callegari, I. (2013). Photogrammetry and laser scanning for analyzing slope stability and rock fall runout along the Domodossola-Iselle railway, the Italian Alps. *Geomorphology, 185,* 110–122.

Samodra, G., Chen, G., Sartohadi, J., Hadmoko, D., & Kasama, K. (2014). Automated landform classification in a rockfall-prone area, Gunung Kelir, Java. *Earth Surface Dynamics, 2*(1), 339–348.

Samodra, G., Sartohadi, J., Chen, G., & Kasama, K. (2013). *Application of supervised landform classification of 9-unit slope model for preliminary rockfall risk analysis in Gunung Kelir, Java* (pp. O-8-1–O-8-4), Geo morphometry.org.

Shary, P. A., Sharaya, L. S., & Mitusov, A. V. (2002). Fundamental quantitative methods of land surface analysis. *Geoderma, 107*(1), 1–32.

Slob, S., Hack, H., & Turner, A. K. (2002). An approach to automate discontinuity measurements of rock faces using laser scanning techniques. *Proceedings of ISRM EUROCK, 2002,* 87–94.

Spadari, M., Kardani, M. De, Carteret, R., Giacomini, A., Buzzi, O., Fityus, S., et al. (2013). Statistical evaluation of rockfall energy ranges for different geological settings of New South Wales, Australia. *Engineering Geology, 158,* 57–65.

Spang, R., & Rautenstrauch, R. (1988). Empirical and mathematical approaches to rockfall protection and their practical applications. In *5th International Symposium on Landslides* (p. 1237).

Statham, I., & Francis, S. (1986). *Influence of scree accumulation and weathering on the development of steep mountain slopes* (pp. 245–267). Hillslope Processes. (Abrahams AD).

Stephenne, N., Frippiat, C., Veschkens, M., Salmon, M., & Pacyna, D. (2014). Use of a LiDAR high resolution digital elevation model for risk stability analysis. *EARSeL eProceedings, 13*(S1), 24–29.

Stock, G. M., Bawden, G. W., Green, J. K., Hanson, E., Downing, G., Collins, B. D., et al. (2011). High-resolution three-dimensional imaging and analysis of rock falls in Yosemite Valley, California. *Geosphere, 7*(2), 573–581.

Strozzi, T., Delaloye, R., Kääb, A., Ambrosi, C., Perruchoud, E., & Wegmüller, U. (2010). Combined observations of rock mass movements using satellite SAR interferometry, differential GPS, airborne digital photogrammetry, and airborne photography interpretation. *Journal of Geophysical Research: Earth Surface (2003–2012), 115*(F1).

Sturzenegger, M., & Stead, D. (2009). Quantifying discontinuity orientation and persistence on high mountain rock slopes and large landslides using terrestrial remote sensing techniques. *Natural Hazards and Earth System Science, 9*(2), 267–287.

Sturzenegger, M., Yan, M., Stead, D., & Elmo, D. (2007). Application and limitations of ground-based laser scanning in rock slope characterization. In *Proceedings of the First Canadian US Rock Mechanics Symposium* (p. 29).

Tatone, B. S., & Grasselli, G. (2009). A method to evaluate the three-dimensional roughness of fracture surfaces in brittle geomaterials. *Review of Scientific Instruments, 80*(12), 125110.

Temme, A. J. A. M., Heuvelink, G. B. M., Schoorl, J. M., & Claessens, L. (2009). Chapter 5 geostatistical simulation and error propagation in geomorphometry. *Developments in Soil Science, 33,* 121–140.

Tonini, M., & Abellan, A. (2014). "Rockfall detection from terrestrial LiDAR point clouds: A clustering approach using R", Journal of Spatial Information Science, no. *To appear JOSIS issue, 8,* 95–101.

Topal, T., Akin, M., & Ozden, U. A. (2007). Assessment of rockfall hazard around Afyon Castle, Turkey. *Environmental Geology, 53* (1), 191–200.

Varnes, D. J. (1978). Slope movement types and processes. *Transportation Research Board Special Report, no., 176,* 11–33.

Varnes, D. J. (1984). *Landslide hazard zonation: A review of principles and practice* (p. 64).

Vidrih, R., Ribičič, M., & Suhadolc, P. (2001). Seismogeological effects on rocks during the 12 April 1998 upper Soča Territory earthquake (NW Slovenia). *Tectonophysics, 330*(3), 153–175.

Vijayakumar, S., Yacoub, T., & Curran, J. H. (2011). *On the effect of rock size and shape in rockfall analyses*. USA: Proceedings of the US Rock Mechanics Symposium (ARMA) San Francisco CA.

Vijayakumar, S., Yacoub, T., Ranjram, M., & Curran, J. (2012). Effect of rockfall shape on normal coefficient of restitution. In *46th US Rock Mechanics/Geomechanics Symposium*. American Rock Mechanics Association.

Volkwein, A., Schellenberg, K., Labiouse, V., Agliardi, F., Berger, F., Bourrier, F., et al. (2011). Rockfall characterisation and structural protection-a review. *Natural Hazards and Earth System Sciences, 11,* 2617–2651.

Wang, I., & Lee, C. (2012). Simulation and statistical analysis of motion behavior of a single rockfall. *International Journal of Civil and Environmental Engineering, 61,* 853–862.

Webster, T. L., & Dias, G. (2006). An automated GIS procedure for comparing GPS and proximal LiDAR elevations. *Computers & Geosciences, 32*(6), 713–726.

Wieczorek, G. F., Snyder, J. B., Waitt, R. B., Morrissey, M. M., Uhrhammer, R. A., Harp, E. L., et al. (2000). Unusual July 10, 1996, rock fall at Happy Isles, Yosemite National Park, California. *Geological Society of America Bulletin, 112*(1), 75–85.

Wilson, J. P. (2012). Digital terrain modeling. *Geomorphology, 137*(1), 107–121.

Wilson, J. P., Aggett, G., & Yongxin, D. (2008). Water in the landscape: A review of contemporary flow routing algorithms. In *Advances in digital terrain analysis* (pp. 213–236). Berlin: Springer.

Wilson, J. P., & Burrough, P. A. (1999). Dynamic modeling, geostatistics, and fuzzy classificaiton: New sneakers for a new geography? *Annals of the Association of American Geographers, 89* (4), 736–746.

Wise, S. (2000). Assessing the quality for hydrological applications of digital elevation models derived from contours. *Hydrological Processes, 14*(11–12), 1909–1929.

Wise, S. (2011). Cross-validation as a means of investigating DEM interpolation error. *Computers & Geosciences, 37*(8), 978–991.

Wyllie, D. C. (2014). Calibration of rock fall modeling parameters. *International Journal of Rock Mechanics and Mining Sciences, 67,* 170–180.

Yoshimatsu, H., & Abe, S. (2006). A review of landslide hazards in Japan and assessment of their susceptibility using an analytical hierarchic process (AHP) method. *Landslides, 3*(2), 149–158.

Youssef, A. M., Pradhan, B., Al-Kathery, M., Bathrellos, G. D., & Skilodimou, H. D. (2015). Assessment of rockfall hazard at Al-Noor Mountain, Makkah city (Saudi Arabia) using spatio-temporal remote sensing data and field investigation. *Journal of African Earth Sciences, 101,* 309–321.

Zandbergen, P. (2008). Applications of shuttle radar topography mission elevation data. *Geography Compass, 2*(5), 1404–1431.

Zinggeler, A., Krummenacher, B., & Kienholz, H. (1991). Steinschlagsimulation in Gebirgswäldern. *Berichte und Forschungen, 3,* 61–70.

Application of LiDAR in Rockfall Hazard Assessment in Tropical Region

16

Biswajeet Pradhan and Ali Mutar Fanos

16.1 Introduction

Rockfall is one of the catastrophes which threaten the human's life and properties in mountainous and hilly regions such as Malaysia with steep and high-elevation topography. Prediction and mitigation of such phenomenon can be carried out via the identification of rockfall source areas (seeder points) and modelling of rockfall trajectories and their characteristics. Therefore, a proper rockfall analysis method is required in order to map and thus to understand the characteristics of rockfall catastrophe. This research adopted various methods to investigate, analyse and assess rockfall in terms of identification of rockfall source areas, modelling of rockfall trajectories and their characteristics and consequently rockfall hazard map. Geographic information system (GIS) and light detection and ranging (LiDAR) techniques can support comprehensive rockfall management as they can provide rapid data gathering and analysis for hazard research. Therefore, current research is divided into two general aspects. The first aspect that basically employed RS technique is to gather data of the study area using light detection and ranging (LiDAR) airborne laser scanner. Traditional survey techniques show significant restriction for gathering spatial information needed for the modelling of rockfall. The utilizing of recently developed techniques such as LiDAR has quickly increased in the field of geohazard evaluation. The two LiDAR surveys techniques (airborne and ground-based) are presently considered as the essential tools for accurate and dense information capturing to simplify comprehensive topographical analysis. A portion of North-South Expressways at Jelapang, Malaysia, was used as a study area for rockfall hazard assessment.

16.2 Methodology

In this study, multi-criteria method was used to identify rockfall source areas (seeder points) based on DEM derivatives (slope, aspect and curvature) in addition to topographic contrast and terrain type or land use/cover (LULC). DEM was obtained from high-resolution airborne laser scanning data (LiDAR). Terrain type or LULC was extracted from the high-resolution aerial photograph. Kinematical modelling of rockfall process was carried out through discrete time steps which allow the modelling of free falling or flying, impacting and bouncing, and rolling or sliding motions in a 3D frame. Mechanical parameters (coefficients of restitution (Rn and Rt) and friction angle) were considered the crucial parameters for rockfall analysis. Multi-rockfall scenarios were conducted based on a range of mechanical parameters. Many raster surfaces are generated utilizing raster modelling to represent the spatial distribution of the rockfall characteristics containing spatial frequency, flying or bouncing height and kinematical energy. These are crucial for the final prediction of rockfall hazard map. Eventually, a raster surface predicting rockfall hazard was produced utilizing a spatial modelling that takes into consideration all raster data of the rockfall characteristics. Expert's opinion was included in this step, and geometric mean was applied and then AHP to give appropriate weight for each rockfall characteristics (frequency, bouncing height and energy). In order to mitigate rockfall hazard, a barrier location was suggested based on the rockfall characteristics (height and energy), and the assessment was performed again with a barrier to show the efficiency of barrier eliminating rockfall hazard. Figure 16.1 shows the overall methodology applied in this study.

16.3 Study Area

The Jelapang corridor of the NSE is also known as the PLUS Expressway in Malaysia (Fig. 16.2) and has been chosen for rockfall analysis because of the common happening of mass

B. Pradhan (✉) · A.M. Fanos
Department of Civil Engineering, University Putra Malaysia, Serdang, Malaysia
e-mail: biswajeet24@gmail.com

B. Pradhan (ed.), *Laser Scanning Applications in Landslide Assessment*,
DOI 10.1007/978-3-319-55342-9_16

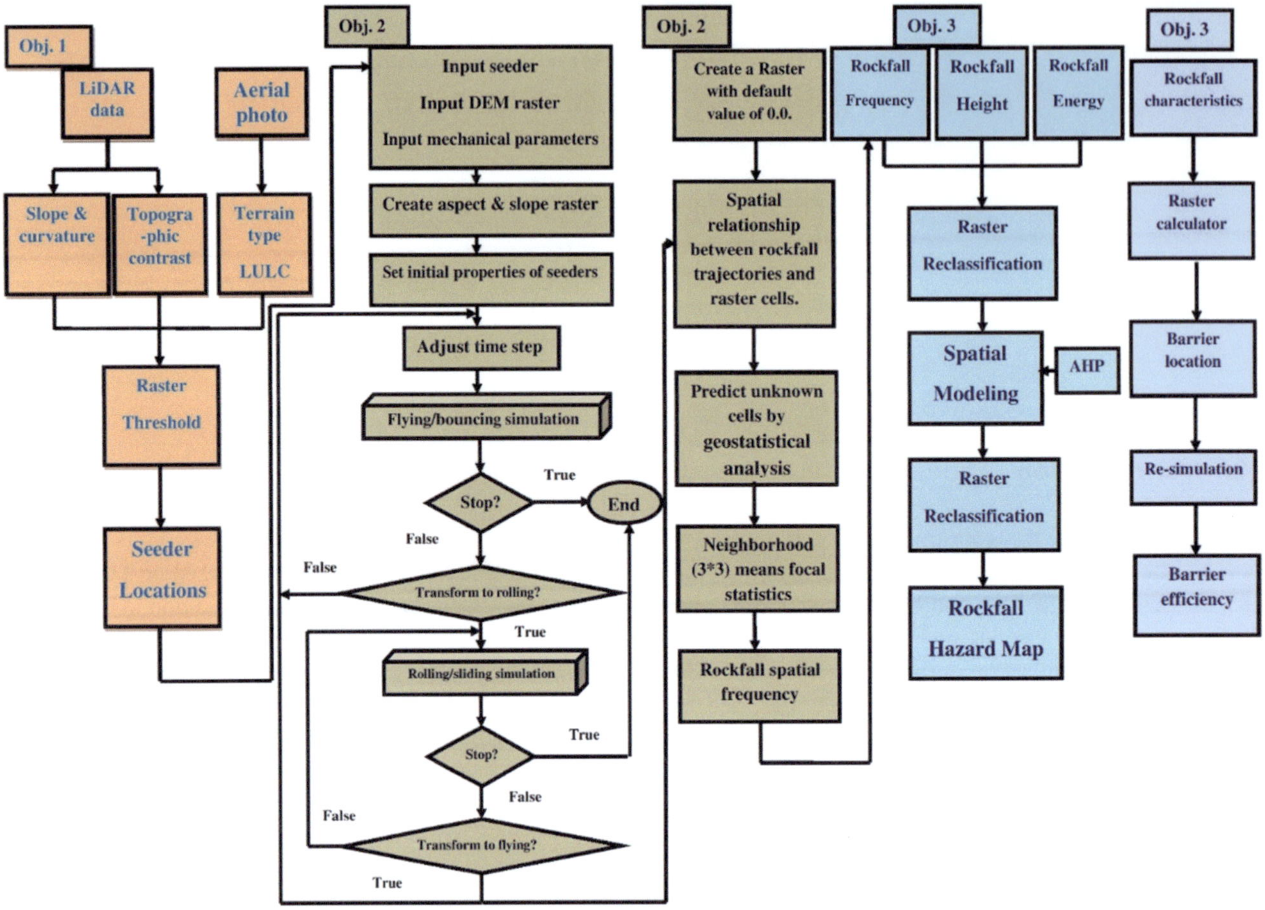

Fig. 16.1 Overall methodology

movements in this area. The highway connects numerous significant towns and cities in the west coast of Peninsular Malaysia; this path also serves as the backbone of this coast. Moreover, this highway is the longest in Malaysia at around 772 km and crosses seven major states in the Peninsula, namely Kedah, Penang, Perak, Selangor, Negeri Sembilan, Malacca and Johor. This area of study is roughly located in the zone of 4° 41′ 14″N to 4° 41′ 41″N latitude and 101° 0′ 14″E to 101° 1′ 8″E longitude near the NSE Tunnel (Jelapang). This area experiences frequent mass movements that cause rockfall; furthermore, the level of annual precipitation is high in this area, ranging from 2500 to 3000 mm. This area experiences two noticeable rainy seasons from February to May and from September to December; precipitation is maximized between March to May and November to December. The geomorphology of the region is composed of a hilly terrain and an undulating plateau. The geology of the area mainly consists of Devonian and Quaternary granite (Yusof et al. 2015). Nonetheless, many rockfall events have occurred recently along the PLUS highway. According to the inventory data, three rockfall events have occurred in different regions within the study area. In addition, the slope of the study area is cut slope and categorized as a high risk area.

16.4 Data Used

16.4.1 LiDAR

Airborne light detection and ranging (LiDAR) is a popular remote sensing method is utilized to attain digital presentation of the topographic surface for regions with small to very large coverage. This method uses a laser sensor which is positioned on an airplane to record the distance from the device and various points on the earth. In each square, 100 points can be recorded based on some conditions such as elevation, speed and type of the sensor. Moreover, the condition of the terrain is also an important factor. In this research, LiDAR vector point data were gathered over ~6 km^2 of Jelapang corridor and the region around it on 8 March 2013 using airborne laser scanning (Table 16.1).

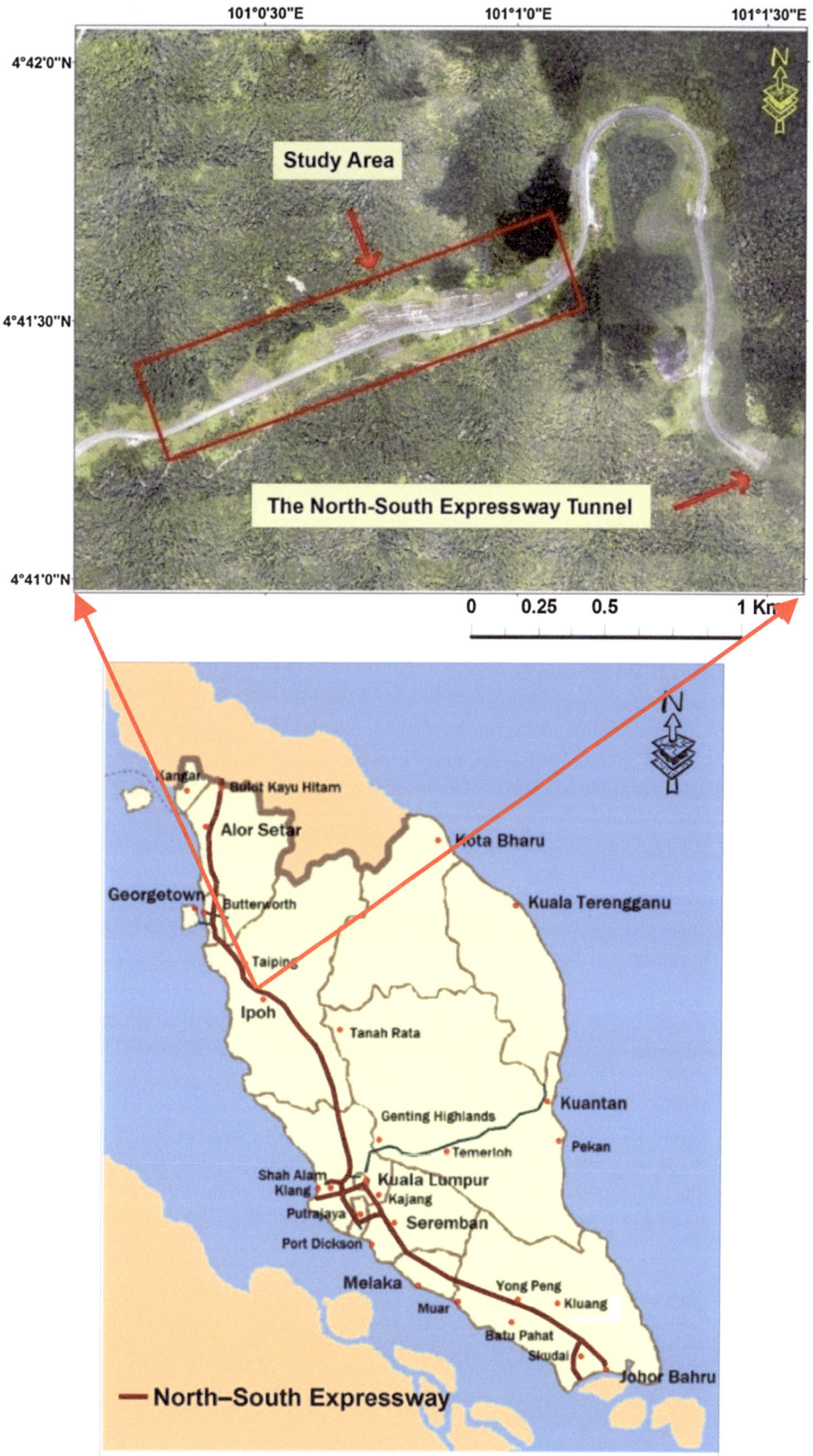

Fig. 16.2 Location of study area in Jelapang, Malaysia

Table 16.1 LiDAR Information

Scanner type	Riegl LM Q5600
Incident angle	60°
Height of flight	1000 m
Point density	3–4 pts./m^2
Wavelength	1550 nm
Pulse repetition rate	200 kHz
Scan frequency	111.1 Hz

Table 16.2 Aerial photograph information

Camera type	Hassleblad 39Mp
Camera angle	45°
Spatial resolution	13 cm

The data were recorded with approximately 10,343,246 data points. The LiDAR data have been provided by PLUS Highway Berhad in Malaysia.

16.4.2 Aerial Photograph

The use of high-resolution aerial photograph has become a famous choice for ground feature mapping. In this research, high-resolution aerial photograph (Table 16.2) was used for two purposes: first, to extract the terrain type or land use/cover (LULC) of the study area; and second, for the identification of rockfall source areas (seeder points).

16.4.3 Mechanical Parameters

In the present study, the mechanical parameters (coefficients of restitution and friction angle) are considered crucial for rockfall simulation. The rockfall simulation was conducted with the maximum and minimum range values of the restitution coefficient (Rn and Rt) and the friction angle for different terrain types (Table 16.3). These values were obtained from the most common values reported in the literature related to the study area properties (geological setting and terrain type) (Ma et al. 2011; Azzoni et al. 1995; Topal et al. 2007; Rammer et al. 2010; Asteriou et al. 2012; Ansari et al. 2014; Ku 2012; Ahmad et al. 2013; Akin et al. 2013; Chen et al. 2013; Keskin 2013; Singh et al. 2013; Wang et al. 2014; Samodra et al. 2014; Sabatakakis et al. 2015) and assigned to the terrain-type layers for each scenario. Consequently, five scenarios of rockfall have been prepared based on these values (minimum to maximum) with an interval of 0.05 for coefficients of restitution (bare earth and vegetated area), but for road the interval is 0.04 and 5° for friction angle. For instance, in first scenario, the values of the coefficient of restitution (Rn and Rt) are 0.35 and 0.70 for bare earth, 0.30 and 0.65 for vegetated area, and 0.36 and 0.86 for road, and the friction angle is 20°. Whereas in the fifth scenario, the values of coefficients of restitution (Rn and Rt) are 0.55 and 0.90 for bare earth, 0.50 and 0.85 for vegetated area, and 0.44 and 0.94 for road, and 40° for friction angle. The others scenarios are between these two scenarios.

16.5 DEM Extraction

ALS is used in this study to obtain a high-density point cloud; on 8 March 2013, LiDAR vector point data were gathered from over ~6 km^2 of the Jelapang corridor and the surrounding region using this technique. The data were recorded with approximately 10,343,246 data points. In almost all applications of LiDAR, the filtering process is essential to separate the LiDAR (pulse) returns derived from non-ground features (DSM) and those from the ground surface (DEM). Distinguishing non-ground features from ground features can be significantly challenging in areas with significant terrain variation; however, accurate DEM can be achieved if non-ground surface points are extracted before interpolation to a DEM raster. The ALS point cloud was filtered based on wave return to obtain bare earth from the ground (DEM) and to eliminate undesirable features. A number of various interpolation algorithms have been developed to interpolate ground surface LiDAR data. In the current study, the natural neighbour interpolation algorithm is implemented with a filtered ALS point cloud to fill cells that lack adequate data points (Kenner et al. 2014). Subsequently, the DEM was constructed. To enhance the DEM,

Table 16.3 Values of coefficient of restitution (Rn and Rt) and friction angle

Scenario	Bare earth		Vegetated area		Road		Friction angle (Degree)
	Rn	Rt	Rn	Rt	Rn	Rt	
1	0.35	0.70	0.30	0.65	0.36	0.86	20
2	0.40	0.75	0.35	0.70	0.38	0.88	25
3	0.45	0.80	0.40	0.75	0.40	0.90	30
4	0.50	0.85	0.45	0.80	0.42	0.92	35
5	0.55	0.90	0.50	0.85	0.44	0.94	40

fill process through GIS is applied to avoid unrealistic results that resulted from unconscionable cells, such as in the case of a sink. Then, a high-resolution DEM (0.5 m) was derived. Then the DEM derivatives (slope, aspect, curvature and topographic contrast) have been obtained using 3D Analyst in ArcMap.

16.6 Rockfall Sources Identification

In the present research, the rockfall source (seeder points) is identified by applying the popular approach involving slope threshold angles (Heckmann and Schwanghart 2013; Loye et al. 2009; Chai et al. 2015) and by setting a slope angle greater than 45° in addition to another criterion (Fig. 16.3), such as obtaining detailed topographic information of ALS and high-resolution (13 cm) aerial photograph. Information process was done via GIS, and raster was derived for the DEM and its derivatives (slope angle, curvature and topographic contrast) and terrain type. A sharp topographic contrast is associated with unexpected changing in the slope angle that assists in determining the existence of steep slopes with probable block detach. Topographic contrast can be accessed through the cooperation of the DEM (original DEM) with a smoothed DEM through a mean filter. Smoothed DEM is subtracted from the original one, and positive value denotes the higher segment of steep slopes (Lan et al. 2010; Macciotta et al. 2014).

16.7 Rockfall 3D Modelling in GIS

The performing of rockfall modelling processes consists of two main sections: first, the simulation of 3D rockfall trajectories; and second, the raster modelling of rockfall spatial distribution. The most popular approach for rockfall simulation models, a "lumped mass" or point mass technique is utilized in the rockfall model to assess the trajectories of rockfall. Nevertheless, due to the spatial auto relation of elements impacting rockfall such as, slope geometry, vegetation, geology, etc., control the spatial relation of rockfall incidents in terms of their run-out span, energy distribution and velocity, the raster modelling based on spatial geostatistics is utilized in dealing with the rockfall spatial distribution in terms of frequency, height and energy, and the uncertainty associated with them. Since the rockfall model is incorporated into the environment of GIS, all the geostatistical functions available in the GIS can be used. This allows an easy use for exploring the spatial information relevant to rockfall such as rockfall frequency, height and energy that can be discretized in cell format and assess their spatial autocorrelation and directional difference. All of these three elements must be taken into consideration together to achieve a realistic assessment of the rockfall hazard spatial distribution (Jaboyedoff et al. 2005). Realistic estimation of surfaces which depict the rockfall spatial distribution in terms of frequency, height and energy can be generated through different geostatistical techniques. Eventually, the assessment of rockfall hazard is implemented by considering all of this raster of rockfall characteristics in addition to AHP approach.

16.7.1 3D Rockfall Trajectories Simulation

Once the probable rockfall sources (seeder points) have been determined; rockfall physical process was implemented through the rockfall 3D model by taking into account terrain topographic and range of physical parameters (Table 3.3), to provide multi-scenario of rockfall simulation. In the analysis, 10 blocks have been thrown at each seeder point (Table 16.4). The mass of a falling rock was set to be 250 kg, which is based on field observations. In addition, a normal barrier has an energy absorption capacity of 100 kJ equivalent to a 250-kg rock moving at about 20 m/s. Default initial orientation of boulders is the cell raster dip orientation in which the source points are situated. Several boulders are thrown in every rock source point with different beginning orientations to simulate the uncertainty of rockfall sources. A number of initial orientations have been allocated for every rockfall source point that result in about 2400 simulated rockfall events. Since the DEM has been generated from the ALS information and the terrain topography spatial characteristics allocated, potential rockfall has been simulated from every potential source regions. It includes rock separation and fall or fly, bouncing, sliding, rolling and final stopping subsequently.

Rockfall kinematical modelling process is carried out using discrete time phases that automatically determined using both particle velocity and cell size. The used algorithms enable rockfall simulation in different motion modes including flying or free falling, bouncing, and sliding or rolling movements in a 3D frame (Fig. 16.4). The 3D vector operation is one of most significant elements in rockfall simulation. 3D vector indicates the rock physical quantities that are directional in 3D space. It can also be utilized for representing the boulder location, acceleration, force, displacement, momentum and velocity. It can also be utilized for predicting the boulder behaviour upon the impact.

Generally, vector can describe rockfall behaviour as in Eq. (16.1)

$$v = xr + ys + zt, \quad (16.1)$$

where r, s and t are the unit vectors in the x-, y- and z-directions, respectively.

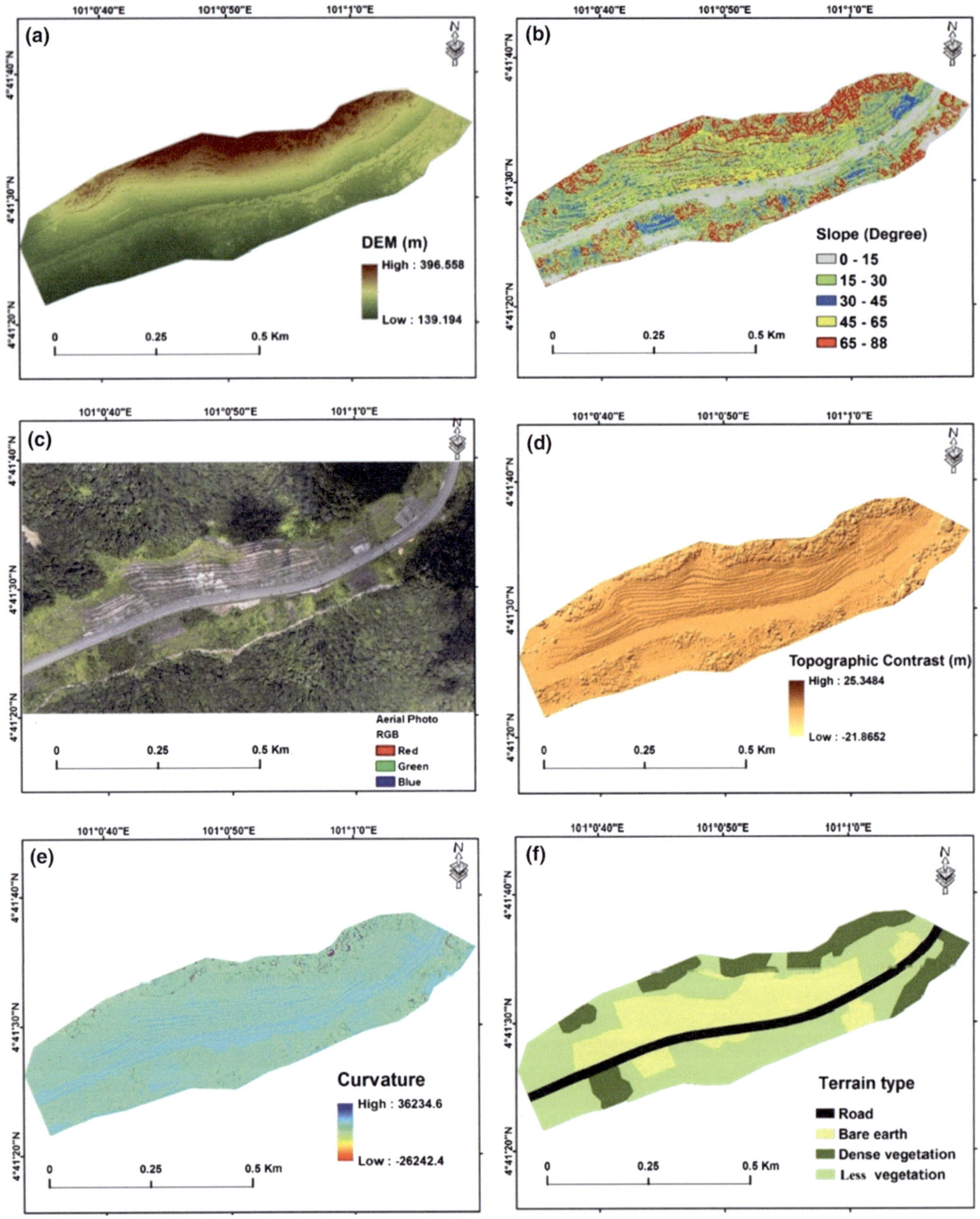

Fig. 16.3 **a** DEM, **b** slope angle, **c** aerial photograph, **d** topographic contrast, **e** curvature and **f** terrain type

Table 16.4 Seeder points properties

Mass of seeder	250 kg
Horizontal velocity (V_{xy})	3 m/s
Vertical velocity (V_z)	0 m/s
Offset height from surface	3 m
Seeder number at one location	10
Angle interval	3°

The falling rock motion in the lumped-mass approach is composed essentially of the projectile, bouncing and sliding algorithms, respectively (Ku 2012).

16.7.1.1 The Projectile Algorithm

The projectile algorithm computes the rockfall trajectory in space. The calculation result relies on whether the falling boulder has an initial velocity or not. When it has no initial velocity, a falling boulder falls on a straight path because of the gravity; whereas if it has initial velocity, a falling boulder falls on a parabolic curve. The rockfall trajectory in a 3D space can normally be represented as in Eq. (16.2)

$$\begin{bmatrix} x(t) \\ y(t) \\ z(t) \end{bmatrix} = \begin{bmatrix} V_{x0}t + x_0 \\ V_{y0}t + y_0 \\ -\frac{1}{2}gt^2 + V_{z0}t + z_0 \end{bmatrix} \quad (16.2)$$

Equation (16.2) can be rewritten as

$$\begin{bmatrix} x(t) \\ y(t) \\ z(t) \end{bmatrix} = \begin{bmatrix} 0 \\ 0 \\ -\frac{1}{2}gt^2 \end{bmatrix} + \begin{bmatrix} V_{x0} \\ V_{y0} \\ V_{z0} \end{bmatrix} t + \begin{bmatrix} x_0 \\ y_0 \\ z_0 \end{bmatrix} \quad (16.3)$$

in which (x_0, y_0, z_0) is a rock starting location, t is the time, g is the gravitational acceleration and (V_{x0}, V_{y0}, V_{z0}) is the starting velocity components for a mass particle in directions of x, y and z, respectively.

The velocity can be represented as

$$\begin{bmatrix} V_x(t) \\ V_y(t) \\ V_z(t) \end{bmatrix} = \begin{bmatrix} V_{x0} \\ V_{y0} \\ V_{z0} - gt \end{bmatrix} = \begin{bmatrix} 0 \\ 0 \\ -gt \end{bmatrix} + \begin{bmatrix} V_{x0} \\ V_{y0} \\ V_{z0} \end{bmatrix} \quad (16.4)$$

Determining of impact point at the end of a rock flying is the significant challenge in utilizing of the projectile algorithm. The point of impact is the intersection point of the grid cell surface and the flight path parabola. The plane surface can be determined for each cell as the Eq. (16.5)

$$Cx + Dy + Ez + F = 0 \quad (16.5)$$

In which C, D and E are the plane normal vector coefficients, and F is the distance from the origin.

The normal vector is vector orthogonal to the plane surface and utilized for constructing of the cell plane 3D geometry (Fig. 16.5). Normally, the plane normal is determined through the cross-products of two vectors characterizing the two plane edges. The calculation of aspect and slope grid from DEM is one of the ordinary functions in GIS, aspect angle φ and slope angle θ are used to structure each of cell plane normal vectors. The unit normal vector can be defined in the global Cartesian system as in Eq. (16.6)

$$u_n = (\sin\theta\sin\varphi, \sin\theta\cos\varphi, \cos\theta) \quad (16.6)$$

16.7.1.2 The Bouncing Algorithm

After the intersection or impact point is determined, the bouncing vector is determined through bouncing algorithm. Since the bouncing algorithm is computed on the slope surface, it requires a local coordinate system for computation. This, in turn, requires a coordinate transformation in this study. The transformation between a local coordinate (X', Y', Z') to a global coordinate (X, Y, Z) in a three-dimensional space can be achieved by rotating the coordinate system horizontally and vertically. After rotating an angle φ horizontally from the global coordinate to the coordinate (X^p, Y^p, Z^p), the transformation equation can be expressed as in Eq. (16.7)

$$\begin{bmatrix} X^p \\ Y^p \\ Z^p \end{bmatrix} = \begin{bmatrix} \cos\varphi & \sin\varphi & 0 \\ -\sin\varphi & \cos\varphi & 0 \\ 0 & 0 & 1 \end{bmatrix} \begin{bmatrix} X \\ Y \\ Z \end{bmatrix} \quad (16.7)$$

Similarly, after rotating an angle θ vertically from the global coordinate to the coordinate (X^v, Y^v, Z^v), the transformation equation can be expressed as in Eq. (16.8)

$$\begin{bmatrix} X^v \\ Y^v \\ Z^v \end{bmatrix} = \begin{bmatrix} \cos\theta & 0 & -\sin\theta \\ 0 & 1 & 0 \\ \sin\theta & 0 & \cos\theta \end{bmatrix} \begin{bmatrix} X \\ Y \\ Z \end{bmatrix} \quad (16.8)$$

Upon rotating both φ and θ from the global coordinate to the coordinate (X', Y', Z'), the transformation equation can be expressed as

$$\begin{bmatrix} X' \\ Y' \\ Z' \end{bmatrix} = \begin{bmatrix} \cos\theta & 0 & -\sin\theta \\ 0 & 1 & 0 \\ \sin\theta & 0 & \cos\theta \end{bmatrix} \begin{bmatrix} \cos\varphi & \sin\varphi & 0 \\ -\sin\varphi & \cos\varphi & 0 \\ 0 & 0 & 1 \end{bmatrix} \begin{bmatrix} X \\ Y \\ Z \end{bmatrix} \quad (16.9)$$

Equation (16.9) can be rewritten as

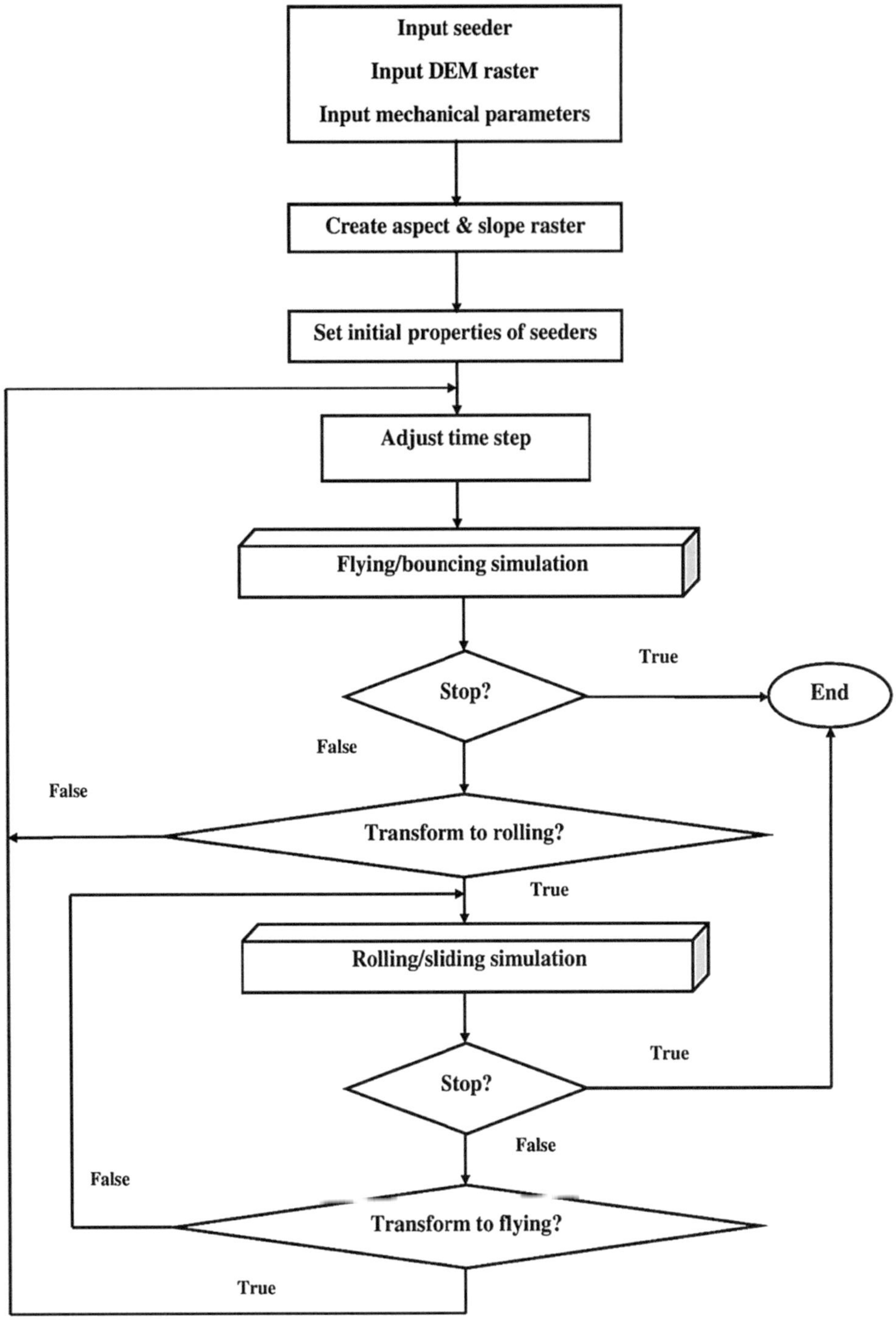

Fig. 16.4 Kinematics algorithm of rockfall trajectory creation adopted from Lan et al. (2007)

$$\begin{bmatrix} X' \\ Y' \\ Z' \end{bmatrix} = \begin{bmatrix} \cos\theta\cos\varphi & \cos\theta\sin\varphi & -\sin\theta \\ -\sin\varphi & \cos\varphi & 0 \\ \sin\theta\cos\varphi & \sin\theta\sin\varphi & \cos\theta \end{bmatrix} \begin{bmatrix} X \\ Y \\ Z \end{bmatrix} \tag{16.10}$$

Since the collision of a falling rock with the slope surface occurs in the bouncing algorithm, the corresponding velocity vector must be transformed into the local coordinate using Eq. (16.11)

$$\begin{bmatrix} V_x^{\mathrm{in}} \\ V_y^{\mathrm{in}} \\ V_z^{\mathrm{in}} \end{bmatrix} = \begin{bmatrix} \cos\theta\cos\varphi & \cos\theta\sin\varphi & -\sin\theta \\ -\sin\varphi & \cos\varphi & 0 \\ \sin\theta\cos\varphi & \sin\theta\sin\varphi & \cos\theta \end{bmatrix} \begin{bmatrix} V_x(t) \\ V_y(t) \\ V_z(t) \end{bmatrix} \tag{16.11}$$

Next, consider the energy dissipation during the collision. The coefficient of restitution is defined as the after-collision

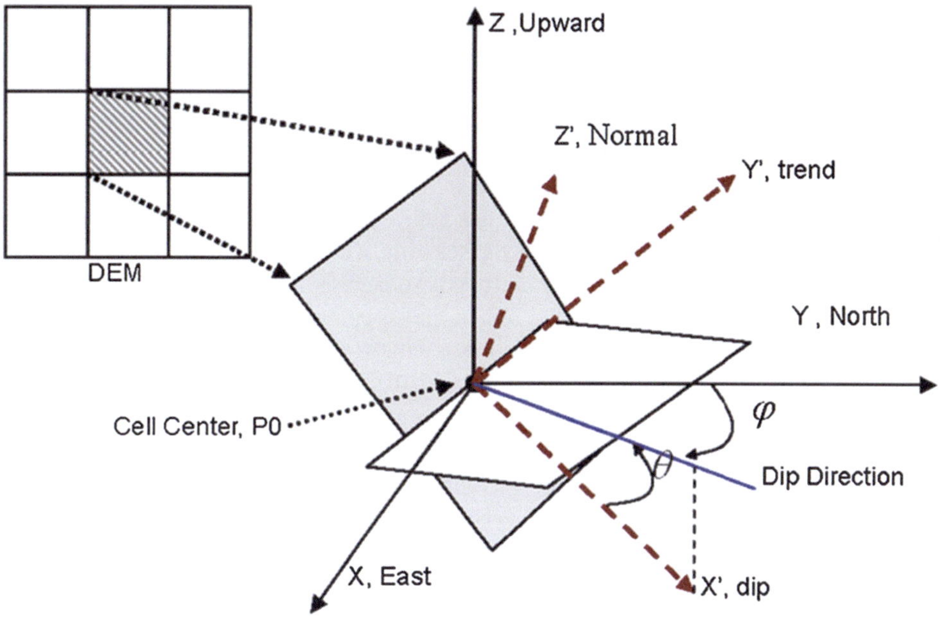

Fig. 16.5 Definition of cell plane and coordinate system cell plane is defined using cell centre (P0), slope angle θ and aspect angle φ

velocity components $(V_x^{\mathrm{out}}, V_y^{\mathrm{out}}, V_z^{\mathrm{out}})$ along slope normal and tangential directions divided by the before-collision velocity components $(V_x^{\mathrm{in}}, V_y^{\mathrm{in}}, V_z^{\mathrm{in}})$ along slope normal and tangential directions. The vector of the coefficient of restitution is expressed as:

$$\begin{bmatrix} V_x^{\mathrm{out}} \\ V_y^{\mathrm{out}} \\ V_z^{\mathrm{out}} \end{bmatrix} = \begin{bmatrix} \mathrm{Rt}V_x^{\mathrm{in}} \\ \mathrm{Rt}V_y^{\mathrm{in}} \\ \mathrm{Rn}V_z^{\mathrm{in}} \end{bmatrix} \tag{16.12}$$

where Rn and Rt are the coefficients of restitution in the normal and tangential directions, respectively. After re-bouncing from the slope, Eq. (16.13) transforms the velocity from local to global coordinates.

$$\begin{bmatrix} V_x(t) \\ V_y(t) \\ V_z(t) \end{bmatrix} = \begin{bmatrix} \cos\varphi\cos\theta & \sin\varphi & \cos\varphi\sin\theta \\ -\sin\varphi\cos\theta & \cos\varphi & -\sin\varphi\sin\theta \\ -\sin\theta & 0 & \cos\theta \end{bmatrix} \begin{bmatrix} V_x^{\mathrm{out}} \\ V_y^{\mathrm{out}} \\ V_z^{\mathrm{out}} \end{bmatrix} \tag{16.13}$$

The projectile or bouncing algorithm is continued and repeated along the rock movement in space until the velocity of moving rock reaches a particular velocity, such as 0.5 m/s (Lan et al. 2007) after the rock particle impacts the slope plane. The rock motion then transfers to the sliding algorithm.

16.7.1.3 Sliding/Rolling Algorithm

The segment of rolling/sliding in each cell plane is composed of a single 3D line. The length of this line is defined essentially through friction angle ϕ and slope angle θ. In each cell, the falling boulder starts rolling/sliding at one cell border and ends at another cell border over the sliding trajectory. The starting velocity of the moving rock in the next cell is the same magnitude as the end velocity of the sliding rock in the first cell. When the friction is the same as the angle the slope angle, the moving rock will slide on a 3D plane and its velocity stay fixed until it moves to another 3D plane. When the starting sliding velocity accesses zero, the boulder stops falling. When the friction angle is smaller than the slope angle, the boulder will continue sliding on a 3D plane with increasing of sliding velocity. The increasing velocity can be computed using Eq. (16.14).

$$\begin{bmatrix} V_x^{\mathrm{out}} \\ V_y^{\mathrm{out}} \\ V_z^{\mathrm{out}} \end{bmatrix} = \begin{bmatrix} \sqrt{V_{x0} + 2s_x gk} \\ V_{y0} + 2s_y gk \\ V_{z0}^{\mathrm{in}} \end{bmatrix} \tag{16.14}$$

where $k = \pm\sin(\theta) - \cos(\theta)\tan(\phi)$. The positive sign refers to the initial velocity of a falling rock moving on the upward slope, whereas the negative sign refers to the initial velocity of a falling rock moving on the downward slope. (V_{x0}, V_{y0}, V_{z0}) are the initial velocities for a rock mass point in the x-, y- and z-directions. g is the gravitational acceleration, θ is the slope gradient angle, ϕ is the friction angle with slope, s_x and s_y are the moving distance for a falling rock mass on a local coordinate system based on its projection on x-axis and y-axis, respectively.

Rock sliding or rolling calculation equations include both upslope and downslope motion. The rolling or sliding movement (upslope or downslope) is specified through the interaction between the normal vector of cell plane and the rock velocity vector. When the rock moves downslope, it follows the steepest slope path. The falling rock slides with increasing velocity when the acceleration is greater than zero, whereas the falling rock slides with the same velocity as the incoming velocity, when the acceleration is equal to zero. In both cases, the exiting velocity is calculated at the end of steepest trajectory segment. The length of the steepest segment will equal the sliding distance in this cell. Eventually, the falling rock slides with decreasing velocity, when the acceleration is smaller than zero. A zero exit velocity is used to compute the stopping distance. When the stopping distance is smaller than the steepest segment length, the simulation will stop and the exiting velocity will be zero. When the stopping distance is greater than the steepest path length, the exit velocity will be recomputed at the segment end. For flat cell or upslope sliding or rolling, first the stopping distance of rock is computed. When the stopping distance is more than the segment length, the sliding will continue to the next cell and the exit velocity is computed. When the stopping distance is smaller than the segment length, the simulation will stop and the exit velocity is zero.

When the angle between Normal 1 and Normal 2 vectors Fig. 16.6 is more than a crucial angle such as 45° and the falling velocity is more than a crucial value, for example 5 m/s (Lan et al. 2007), the moving mode will be converted from sliding/rolling into a flying. The results of all algorithms are 3D rockfalls trajectories and velocity associated with them that describe the whole processes of rockfall behaviour. These also will use as essential inputs for the next step of rockfall raster modelling.

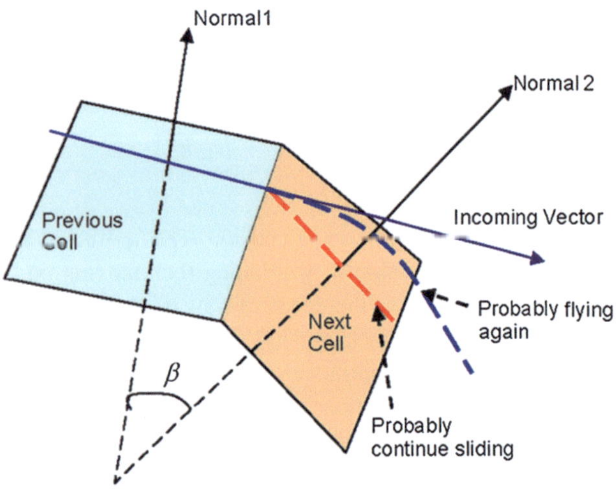

Fig. 16.6 Schematics for determining if movement of rock would transform from rolling to flying

16.7.2 Raster Modelling for Spatial Distribution of Rockfalls

Raster distribution modelling technique was performed in five scenarios to represent the spatial distribution of rockfall frequency, flying or bouncing height and kinetic energy for each scenario based on the result of 3D rockfall trajectory modelling. Figure 16.7 shows the workflow to create the raster of rockfall spatial frequency. It consists of four steps:

1. First, a raster with 0.0 cell value is generated. This raster is basically similar to the original input DEM raster in terms of cell size, extent and georeference system.
2 Second, topological analyses are conducted by determining the spatial relationship between each DEM cell and the 3D rockfall trajectories. A sample value is allocated to each cell as a rock falling down passing the slope in this step. For instance, this value remains 0.0 if no

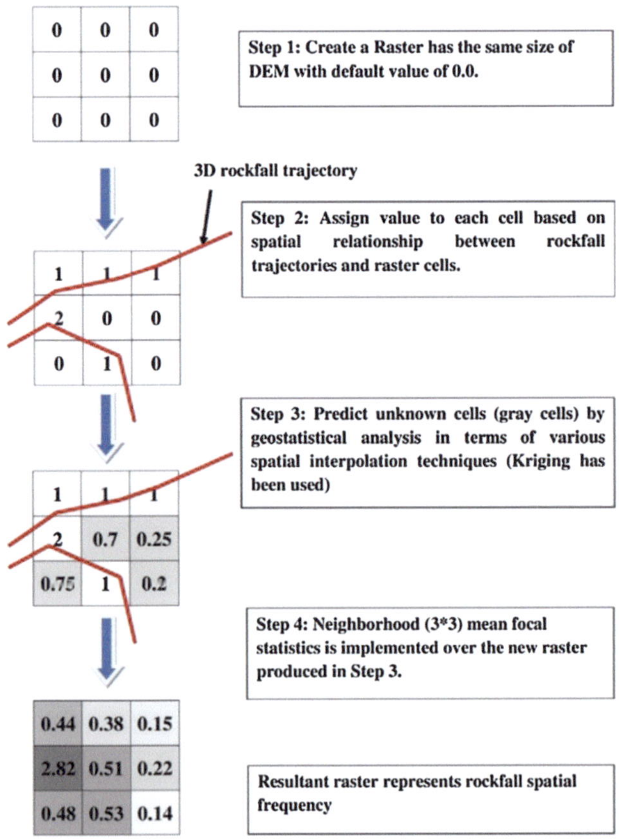

Fig. 16.7 Workflow involving four steps of a rockfall spatial frequency raster creation adopted from Lan et al. (2007)

rockfall trajectory surpasses its value. If one rockfall trajectory exceeds this value, then the cell is assigned a value of 1.0; when two rockfall trajectories surpass a certain value, a value of 2.0 is allocated to the cells.

3. After Step (2), a number of cells retain a value of 0.0. Nevertheless, this outcome does not indicate that these cells escape the rockfall effect; for instance, the cells near the closest cells with 2.0 cell value are certainly at risk. In this step, spatial statistical interpolation is performed to estimate cell value when the risk is unknown. Several spatial interpolation techniques are available for GIS including Kriging, global or local polynomial and inverse distance-weighted (IDW). The kriging technique relies on statistical and mathematical models associated with a probability. Kriging has an advantage in exploring the spatial relation, directional and structural variations, and performing error analysis of the rockfall-related spatial data. Kriging technique was used in the generation of rockfall characteristics raster. A new raster is generated in this step.
4. A focal analysis of the neighbourhood is conducted on the raster obtained in Step 3 to derive a continuous surface of estimated frequency of rockfall. The cell value identifies the spatial potential of facing a rockfall. The dark-coloured cells indicate high rockfall frequency, and light-coloured cells stay safe.

The same procedure was utilized to create a kinematic energy raster and a flying or bouncing height raster, with the only difference occurring in Step 2. The kinematic energy raster is created based on rockfall velocity and mass. The maximum energy of the rock is assigned to each cell. For a flying or bouncing height raster that indicates potential energy to the ground, the maximum height of trajectory is assigned to the cells in which the flight or bounce occurs. From this point, the procedure is identical to that of creating a spatial frequency raster. In order to calculate the probability of rockfall velocity, height and energy, a raster has been classified into different classes with an equal interval. Then, the raster elements have been calculated for each class and compared with a total number of raster elements.

16.7.3 Rockfall Hazard Map Generation

Eventually, rockfall hazard maps are generated through a spatial modelling that considers all aspects of the rockfall characteristics raster in five scenarios. Figure 16.8 shows the needed procedure to generate the rockfall hazard map. First, input the rockfall characteristics raster including kinematic energy, frequency and flying or bouncing height; and second, reclassify those characteristics raster by the same classification criterion. Different reclassification techniques are provided in GIS, such as natural breaks and equal interval. As a result, a new raster with changed grid values is generated which indicate the rockfall physical characteristics level. For example, when the kinematic energy raster is reclassified into five classes, a value of 5 represents the maximum kinematic energy and a value of 1 represents the minimum kinematic energy. Third, a spatial modelling is implemented through assigning various weights, which are obtained from the performing of AHP approach to the reclassified raster of rockfall characteristics. As a result, rockfall hazard maps are created. Finally, reclassify the rockfall hazard map to indicate various levels of hazard along the expressway. The area along the expressway was divided into different stations based on the kilometre of the expressway. The hazard percentage has been calculated along the expressway for each area between two stations.

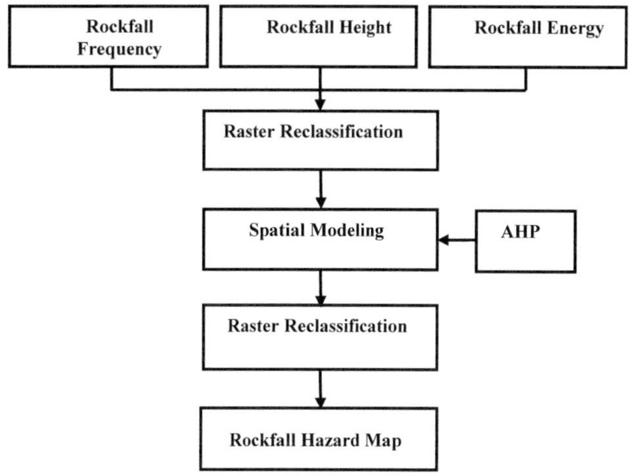

Fig. 16.8 Workflow for rockfall hazard map generation

16.7.3.1 The Weight of Rockfall Characteristics

Rockfall characteristics (Frequency, height and energy) were considered in the implementation of spatial modelling to derive rockfall hazard map. However, each element has a different effect in rockfall hazard. Therefore, expert's opinion was included in order to obtain an appropriate weight for each rockfall characteristic raster. Analytic hierarchy process (AHP) is a useful tool to supply a simple decision-making technique for complex decision-making issues with multiplied targets, multiple standards, and no structural characteristic by mathematizing the decision maker's decision-making by processes utilizing minimum quantitative data (Vaidya and Kumar 2006; Siqiao et al. 2010). AHP was applied in current research in order to obtain an appropriate weight for each rockfall characteristics. First, a questionnaire was prepared including the main goal (rockfall hazard assessment) and the criteria (energy, frequency and height). Since the hierarchy has been structured, the decision factors are differentiated by pairwise comparison in terms of their significance with respect to their dominate criteria. In particular, the decision makers are requested to answer pairwise comparisons in which two factors at a time are in contrast with regard to their contributions to rockfall hazard assessment. The comparative significance values are specified on a nine-point scale known as "Saaty's Fundamental Scale" (Saaty 1980). The questionnaire was sent to some of the experts in the field of geological engineering. The opinion of six experts was taken into account and geometric mean was applied to derive appropriate intensity of importance for rockfall characteristics. The geometric mean is a kind of average or mean that specifies the typical value or central tendency of a set of numbers through utilizing of their values product against the arithmetic mean that utilizes their sum. The geometric mean is known as the *n*th root of the product of *n* numbers. Using a geometric mean "normalizes" the range being averaged so that no range controls the weight, and change of a given percentage in any of the characteristics has a similar influence on the geometric mean. Geometric mean can be defined as in Eq. 16.15:

$$\overline{w}_{\text{geom}} = \sqrt[n]{\prod_{i=1}^{n} w_i} = \sqrt[n]{w_1 * w_2 * \ldots * w_n} \tag{16.15}$$

where w_i is the weightage for *i* number, and *n* is the number of opinion (Table 16.5).

The weights computing steps in AHP were as follows (Table 16.6):

Table 16.5 Sample of pairwise comparisons among objectives/alternatives

No.	Criteria	Energy	Frequency	Height
1	Energy	1		
2	Frequency		1	
3	Height			1

Table 16.6 Intensity of importance definition (Saaty, 1980)

Intensity of importance	Definition
1	Equal importance
2	Equal-to-moderate importance
3	Moderate importance
4	Moderate-to-strong importance
5	Strong importance
6	Strong-to-very strong importance
7	Very strong importance
8	Very strong-to-extremely strong importance
9	Extremely importance

First, by normalizing each line of vector in judgment matrix, we get

$$w'_{ij} = a_{ij} / \sum_{i=1}^{n} a_{ij} \tag{16.16}$$

Second, by summing the rows of w'_{ij}, it can be obtained that

$$w'_i = \sum_{i=1}^{n} w'_{ij} \tag{16.17}$$

Third, normalizing $w'_i, w_i = w'_i/n$ the obtained

$$w = (w_1, w_2, \ldots, w_n)^T \tag{16.18}$$

is the approximate eigenvector.

Fourth, by calculating λ as the approximate value of the maximum latent root, we get

$$\lambda = \frac{1}{n} \sum_{i=1}^{n} \frac{(Aw)_i}{w_i} \tag{16.19}$$

In this formula $\lambda_{\max}$ was the maximum latent root of the consistent matrix and *n* was the number of the paired comparison factors.

The consistency index is calculated as in Eq. (16.20)

$$\text{CI} = \frac{\lambda - n}{n - 1} \tag{16.20}$$

Finally, a consistency check was made on the obtained weight vector, as shown below:

$$\text{CR} = \text{CI}/\text{RI} \tag{16.21}$$

where RI is the average of the resultant consistency index depending on the order of the matrix provided (for 3 × 3 matrix is 0.58), and CI is the consistency index.

16.8 Eliminating of Rockfall Hazard

Some mitigation measures have been taken previously in the study area, such as berms in order to reduce rockfall velocity thus reduce rockfall energy and nets to catch small rocks (Fig. 16.9). However, these measures do not eliminate the rock fall hazard. In the design of protective measures against rockfalls, such as barriers, fences and sheds, information on impact energy must be collected to determine the strength of the structure in addition to data on trajectories to identify location and size (Wyllie 2014). Mitigation techniques have been used extensively to reduce rockfall hazards, and the design of these methods is based on the trajectory estimation, bouncing heights and kinetic energies of unstable blocks. These parameters can be obtained through kinematic simulations, which are among the most popular approaches to assess rockfall hazards at present (Ferrari et al. 2013). To mitigate the rockfall phenomenon and for designing mitigate measures, simulation programs are frequently utilized based on probabilistic lumped-mass assessment models (Asteriou et al. 2012). Therefore, the optimal barrier location (Fig. 16.10) is determined in this study based on limited energy and bouncing height to reduce rockfall hazards. The simulation procedures described above are repeated with a (5 m) height barrier to highlight its efficiency at the chosen location.

16.9 Results and Discussion

16.9.1 Digital Elevation Model (DEM)

There are many techniques to derive DEM. However, each technique has different accuracy and limitation. On the other hand, remote sensing techniques such as LiDAR provide very accurate and detailed DEM over traditional techniques (Stephenne et al. 2014). In addition, the selection of the DEM interpolation technique can have a remarkable influence on the DEM surface properties. In this research, high-resolution DEM (0.5 m) was derived (Fig. 16.11) from

Fig. 16.9 Mitigation measures that have been taken in the study area

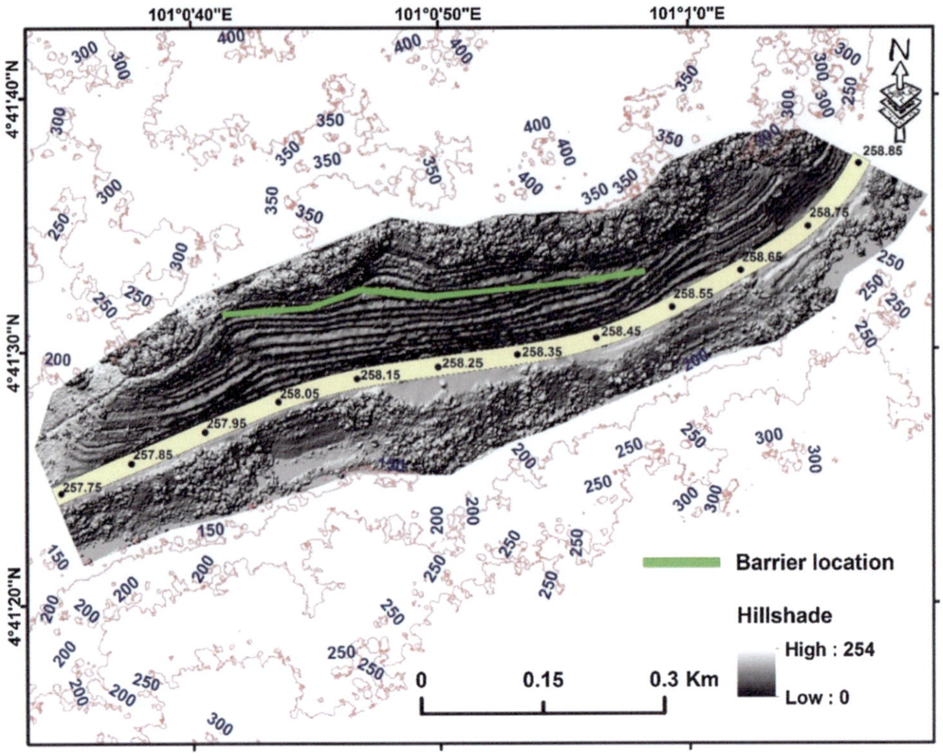

Fig. 16.10 Suggested barrier location

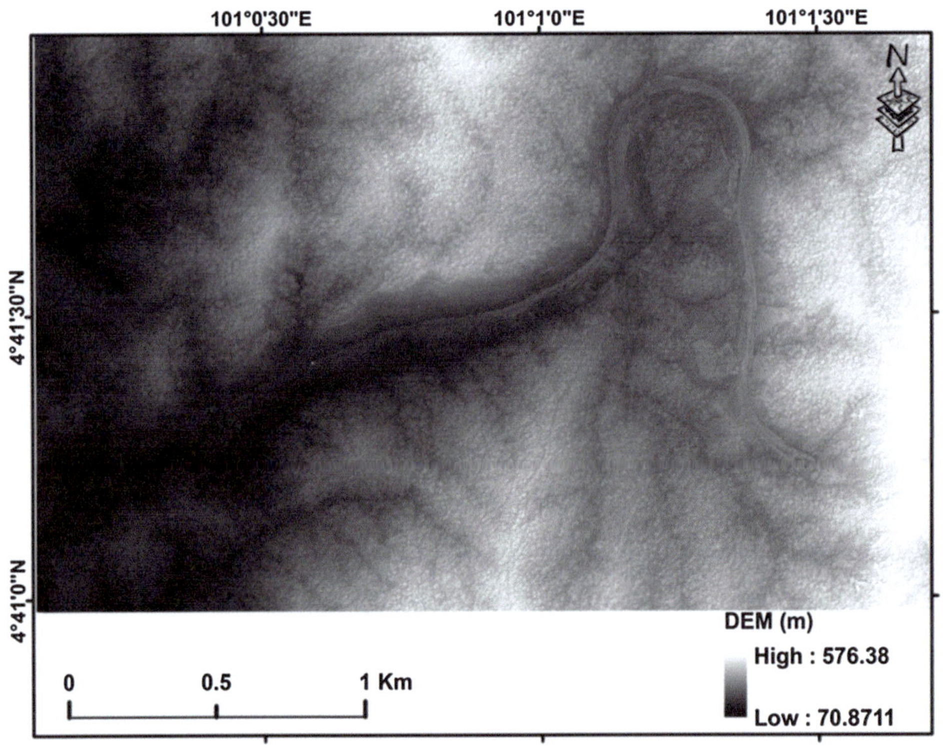

Fig. 16.11 High-resolution LiDAR DEM (0.5 m)

high-density LiDAR point cloud (up to 20 cm) using natural neighbour interpolation algorithm. One of the most common methods to measure DEM quality/accuracy is the root-mean-square error (RMSE). The RMSE of LiDAR data was (0.19 m). Since the DEM accuracy or resolution strongly affects the rockfall modelling results, the obtained

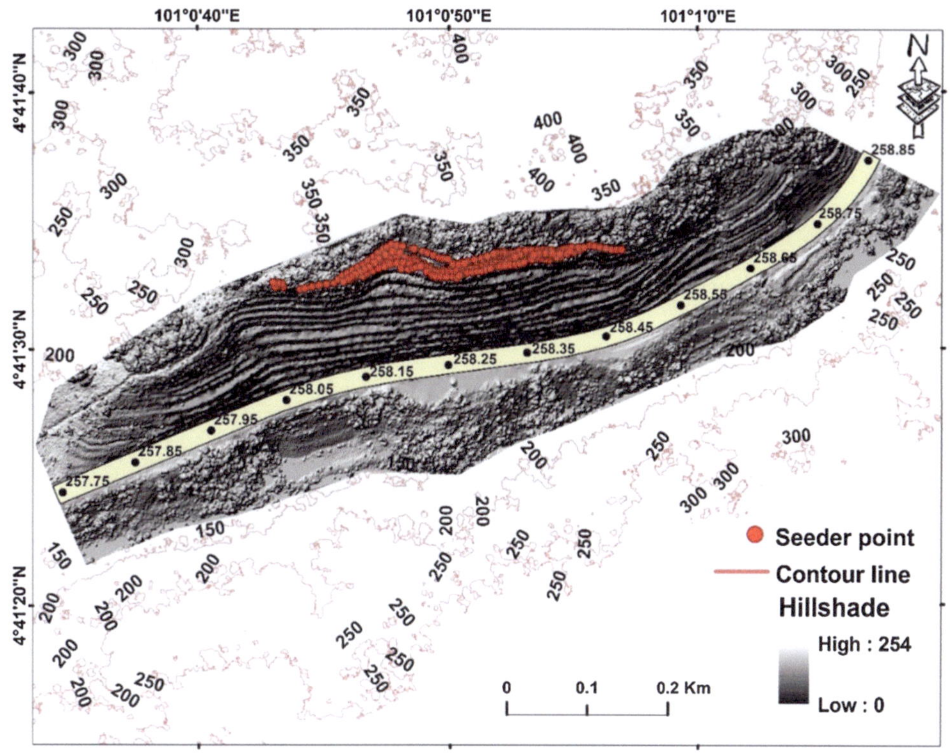

Fig. 16.12 Locations of seeder points

DEM can assist a lot in rockfall simulation providing very detailed terrain morphology. Consequently, very accurate DEM derivatives (aspect, slope and curvature) were derived, which are essential in rockfall modelling.

16.9.2 Rockfall Source Areas (Seeder Points)

The determination/identification of rockfall source areas is the main problem in the characterization of a region affected by rockfall. Various positions of the rockfall source areas along a cliff lead to various possibilities of propagation and a variety of interaction with passive countermeasures existing in the area (Assali et al. 2014). Rockfall source areas have been defined based on multi-criteria as shown in Fig. 16.3, and the seeder point locations have been assigned where steepest areas, highest elevation, positive topographic contrast and where terrain type is bare earth or less vegetation. Figure 16.12 illustrates the locations of rockfall source areas. It shows also the density of seeder points is different from area to another, and the seeder points are distributed along the cliff face, that corresponds to the field observation as shown in Fig. 16.13, which illustrates the weakness of rocks and the discontinuity along the cliff face.

16.9.3 Rockfall Trajectories

Seeder points yielded approximately 2400 trajectories for each scenario of the five scenarios which are differentiated based on the values of mechanical parameters in each scenario; some of these trajectories stopped on the slope surface, another on the road and the remainder crossed the road (Fig. 16.14a). The end of the trajectories varies for each scenario due to the differences in the mechanical parameters (coefficients of restitution [Rn and Rt] and friction angle). Rockfall simulation is extremely challenging because when the rock moves downslope, various potential motion modes are induced: free fall (flying), bouncing, rolling and sliding (Leine et al., 2013). The restitution coefficients and the friction angle are considered to be crucial parameters in the modelling of rockfall. As shown in Fig. 16.13a, the characteristics of rockfall trajectory are strongly affected by the mechanical parameters (Rn, Rt and friction angle), especially run-out distance and bouncing height.

The finding suggests that these two variables increase with the values of the mechanical parameters. Therefore, some rockfall trajectories were stopped on the slope surface, another on the road and a few crossed the road in the first scenario. This occurrence is attributed to the minimum values of the mechanical parameters. By contrast, most of the

Fig. 16.13 Field observation of rockfall source areas (**a**–**c**)

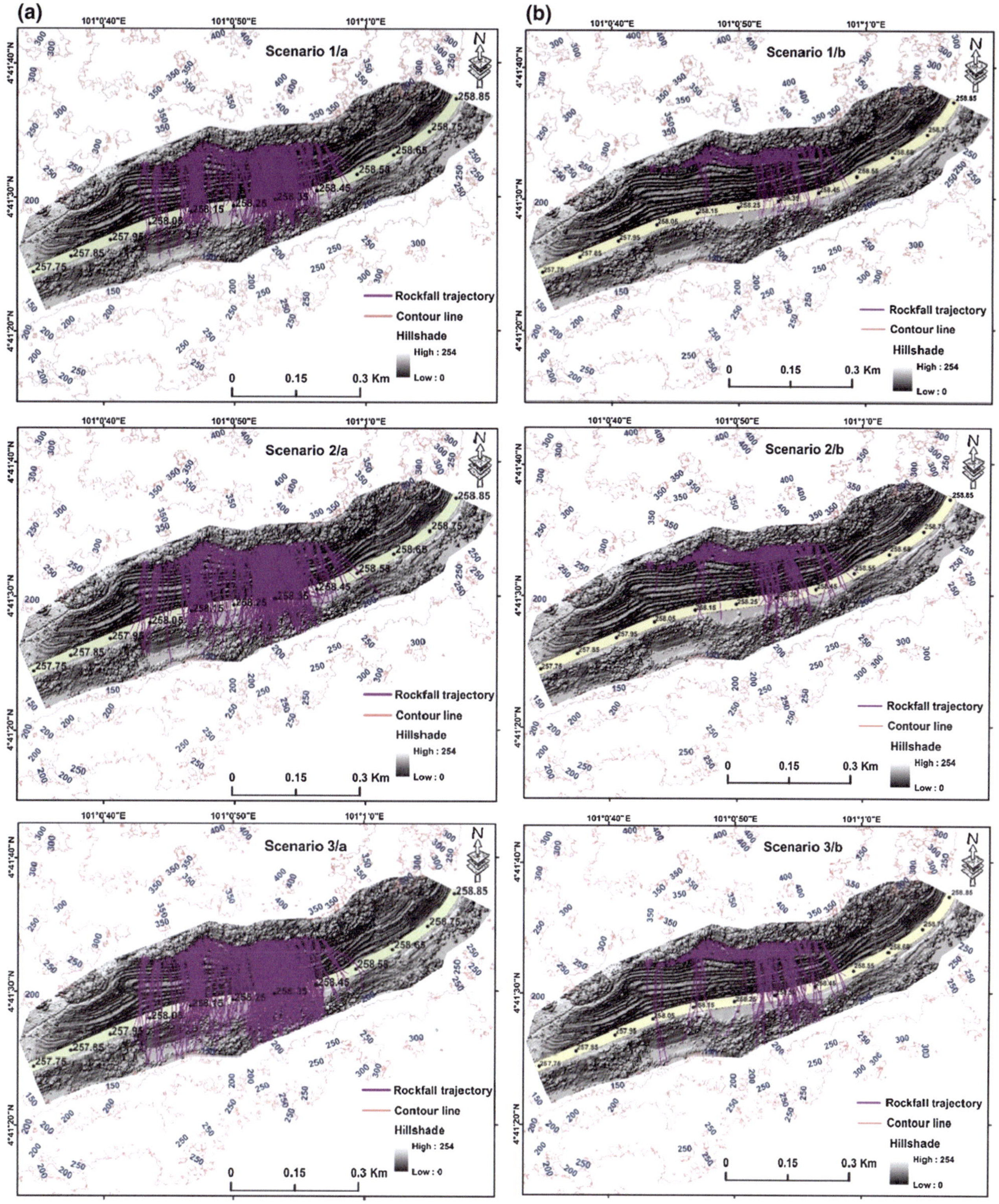

Fig. 16.14 Rockfall trajectories in five scenarios; **a** without barrier; and **b** with barrier

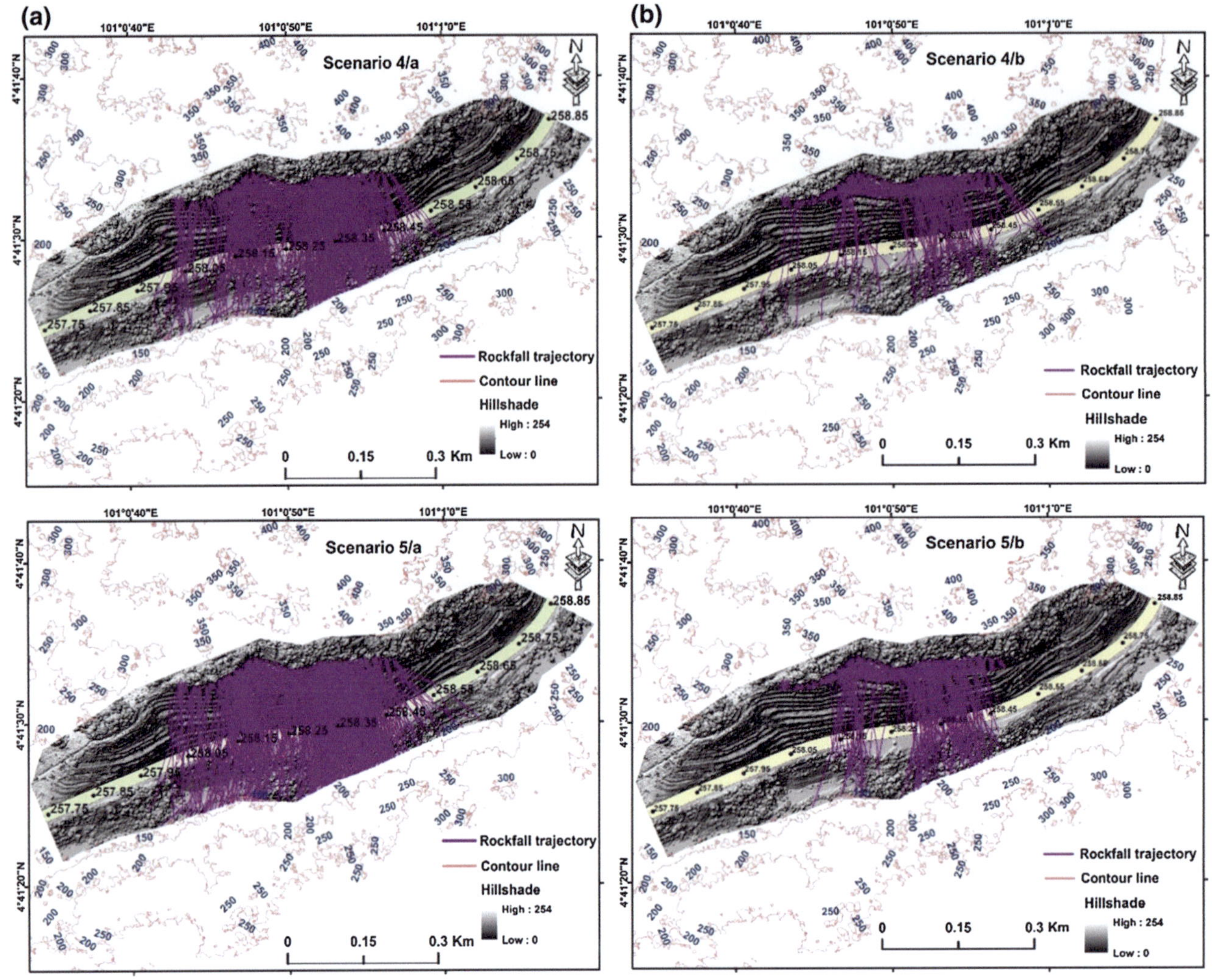

Fig. 16.14 (continued)

rockfall trajectories crossed the road in the fifth scenario due to the maximum values of these parameters. The highest density of rockfall trajectories in all scenarios was from station (kilometre of the expressway) 258.15 to 258.18 and from 258.30 to 258.40. However, this density varies from scenario to another due to the difference in mechanical parameters that affect the rockfall trajectory stopping distance. Therefore, the lowest density was in scenario (1), and the highest density was in scenario (5). The simulation results of rockfall trajectory with the suggested barrier (Fig. 16.10), showed that the suggested barrier location can eliminate the rockfall effect (Fig. 16.14b). Nonetheless, a few trajectories crossed the barrier, and barrier efficiency varies for each scenario due to the differences in the characteristics of the rockfall trajectories in each scenario; the outcome indicates that an increase in the values of the mechanical parameters corresponded to a decrease in barrier efficiency. This is because due to various characteristics of rockfall trajectories in terms of height, velocity and energy. With an increase in mechanical parameters, the behaviour of rockfall trajectories becomes more complicated especially the bouncing height and energy which affect barrier efficiency.

Table 16.7 illustrates the barrier efficiency in each scenario based on the percentage of the trajectories that either arrived or crossed the road to the number of trajectories that

Table 16.7 Efficiency percentage of barrier

Scenario	Efficiency percentage (%)
1	94.86
2	93.75
3	92.73
4	89.63
5	84.88

stopped by the barrier. It also shows that the barrier efficiency in scenarios 1, 2 and 3 was almost same but it decreased in scenarios 4 and 5. This indicates that the barrier efficiency is decreased with an increase in the values of mechanical parameters (Rn, Rt and friction angle). However, the barrier efficiency is still acceptable (since the lowest efficiency percentage is 84.86%), and the barrier can aid to eliminate rockfall hazard.

Figure 16.15 demonstrates the rockfall profiles of one trajectory and the same trajectory in each scenario without and with a barrier. The location of start or release point of the profile is 101° 0′ 53.238″E 4° 41′ 33.796″N. These

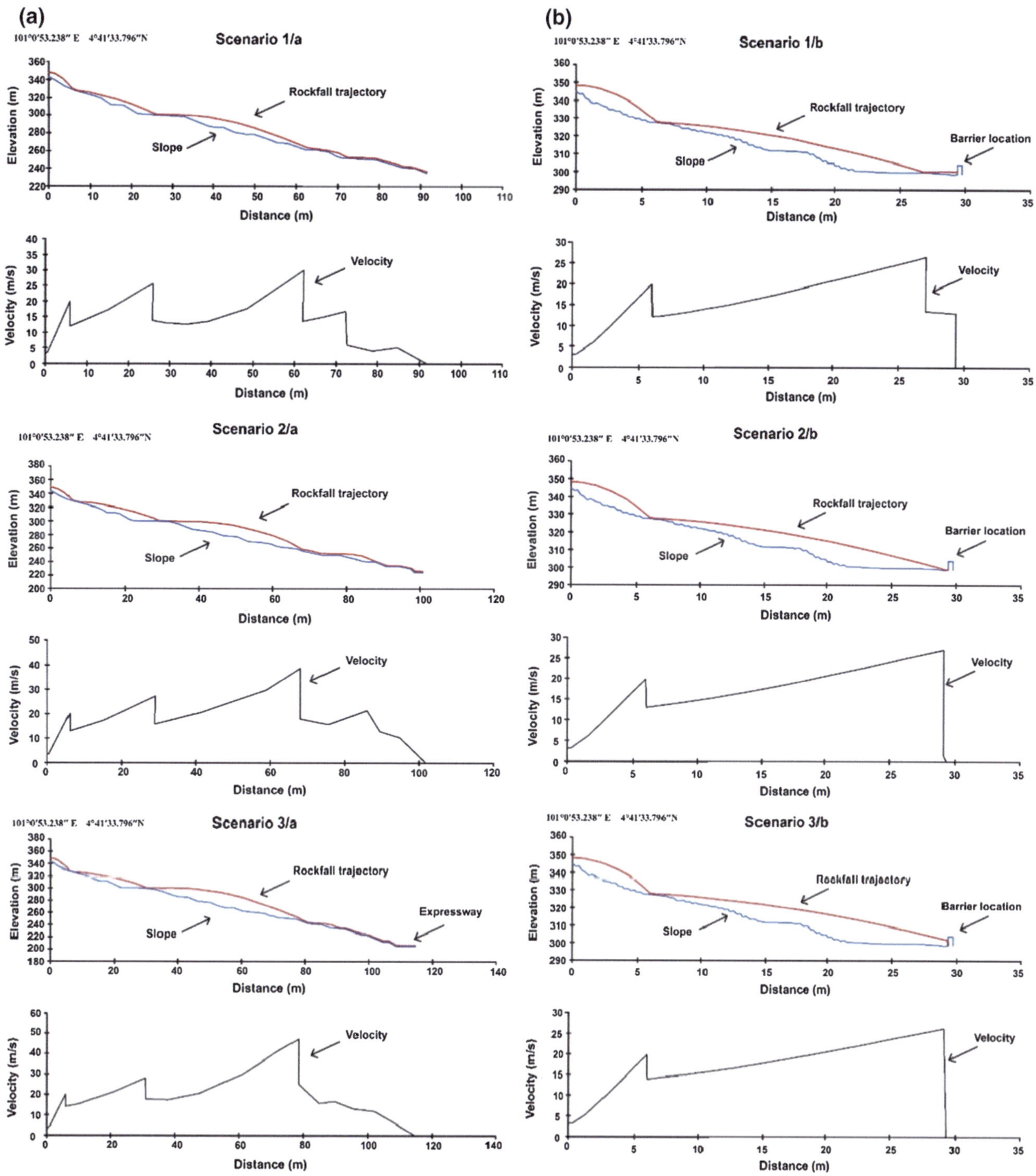

Fig. 16.15 Profiles of a rockfall trajectory and velocity in each scenario **a** without; and **b** with barrier

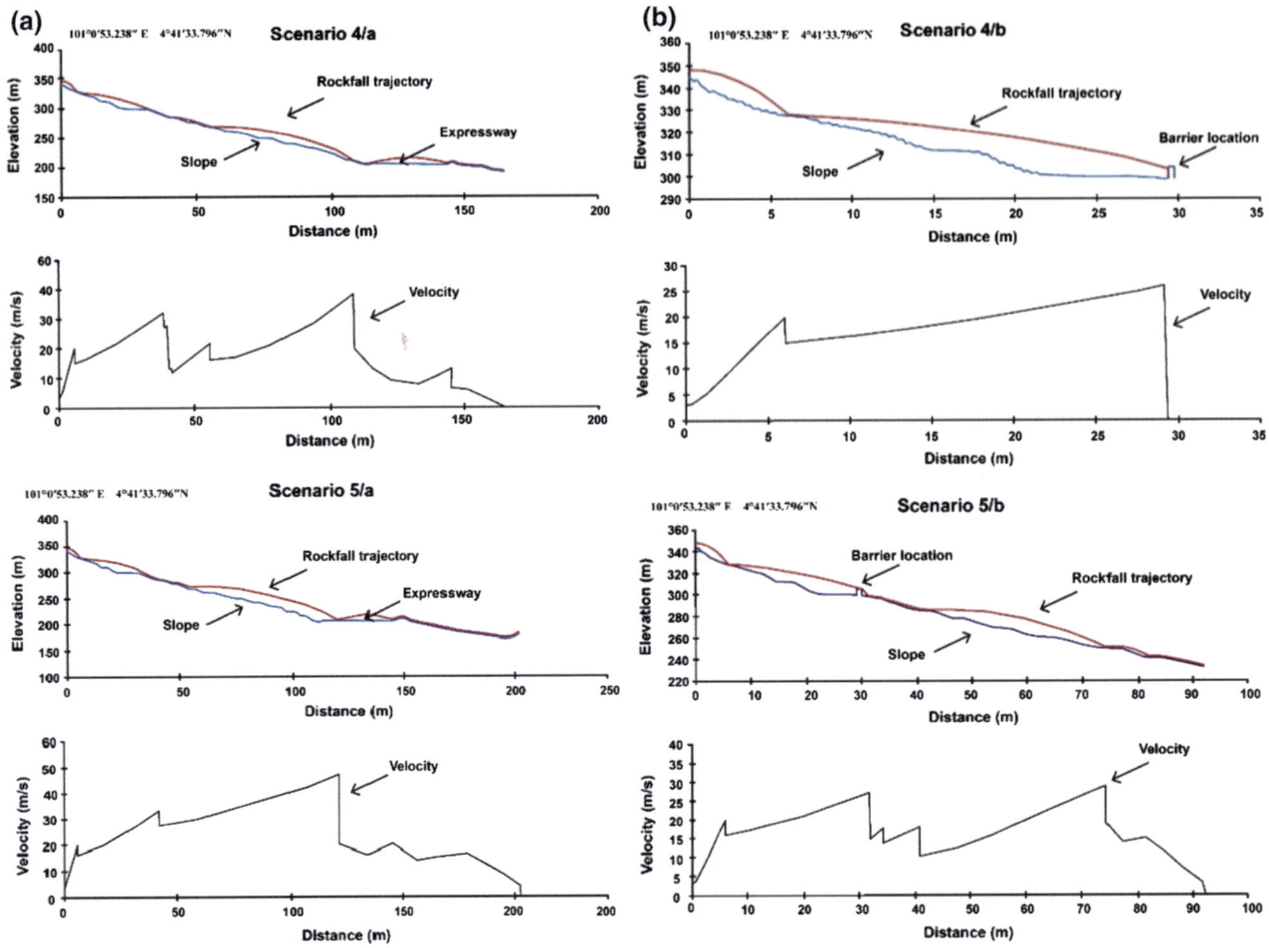

Fig. 16.15 (continued)

profiles include rockfall trajectory, slope surface and velocity. They indicate the difference in rockfall characteristics with respect to run-out distance and bouncing height. These two variables increase when the values of Rn, Rt and the friction angle are increased; thus, the stopping distances (final deposition) in the first and fifth scenarios are approximately 90 and 200 m, respectively. The change in rockfall behaviour from flying/bouncing to sliding/rolling and vice versa in the same trajectory is clarified as well. This occurrence is attributed to a change in acceleration and slope surface. The velocity also changes based on the motion mode. This factor was increased in the flying/bouncing mode and decreased in the sliding/rolling mode, especially at the impact point (intersection point between flight path and slope plane). It also shows that the rock was stopped on slope surface in the first and second scenarios but in the third it stopped on the expressway, whereas the rock was crossed the expressway in the fourth and fifth scenarios. The velocity reflects the efficiency with which the barrier stops the trajectory. The barrier stopped the rockfall trajectory in the first, second, third and fourth scenarios. The barrier has effectively reduced the velocity of the trajectory after the collision, and as a result the rock stopped within a distance of around 90 m instead of 200 m in the fifth scenario. In the fifth scenario, the barrier did not stop the rockfall trajectory because the bouncing height was more than the barrier height. It is clear that the bouncing or the flying height is increased in each scenario due to the increasing in the values of mechanical parameters.

16.9.4 Rockfall Spatial Distribution

16.9.4.1 Rockfall Frequency

Figure 16.16 illustrates the spatial frequency of the simulated rockfall for the entire study area. Raster of rockfall frequency is extracted for each scenario without and with a barrier. There are many classification schemes available for raster classification, such as natural breaks, equal intervals, quantiles and standard deviations. However, natural breaks are recommended for map classification (Evans 1977; Maceachren 1994; Ayalew and Yamagishi 2005; Pradhan et al. 2014).

Therefore, the frequency raster is classified into five classes (ranging from very low to very high) based on natural breaks classification algorithm. Frequency basically relies on the path of rockfall trajectories and varies for each scenario due to the difference in these trajectories. Therefore, the number of areas at risk is increased with the mechanical parameters.

The maximum frequency in each scenario was observed at stations 258.15 and 258.35 (ranging from high to very

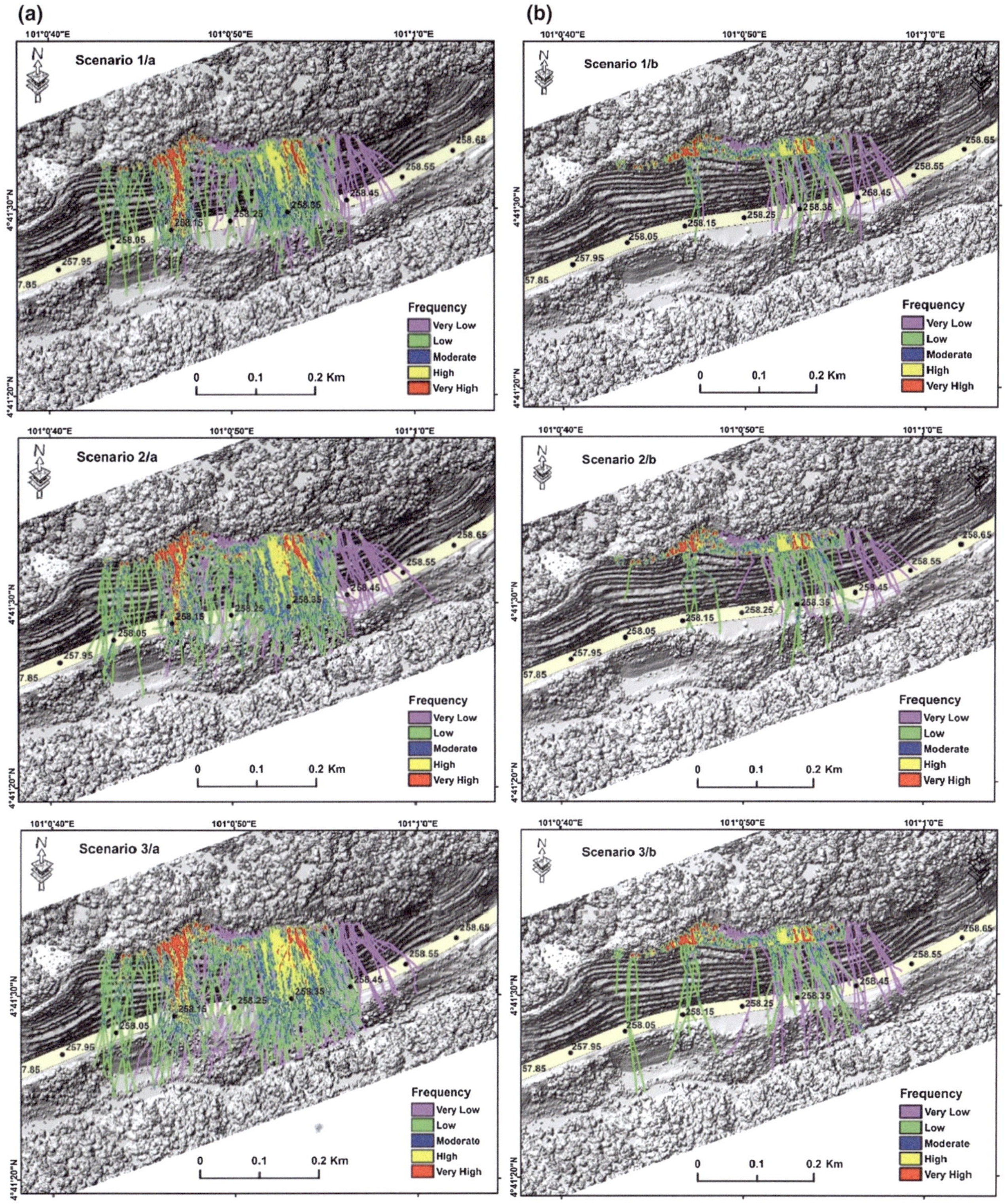

Fig. 16.16 Rockfall frequency in five scenarios; **a** without; and **b** with barrier

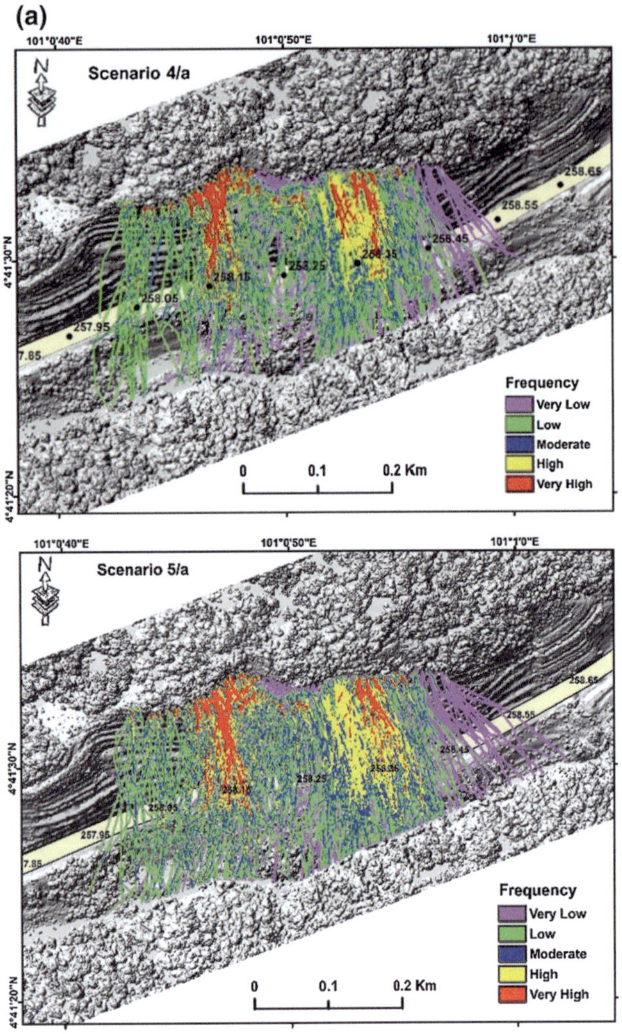

Fig. 16.16 (continued)

high) in the first scenario. In the second scenario, the maximum frequency was at station 258.15, 258.35 and 258.40 (ranging from high to very high). In the third and fourth scenarios, the maximum frequency was from station 258.15 to 258.18 and from 258.33 to 258.40. In the fifth scenario, the maximum frequency was from station 258.15 to 258.20 and from 258.30 to 258.42 (ranging from high to very high). Furthermore, the effect of the suggested barrier location eliminates the rockfall hazard and illustrates how the barrier modifies the frequency levels. For instance, the barrier reduced the frequency from very high to very low in the first scenario and from very high to low in the second scenario at station 258.15 and from high to low at station 258.35 in the first and second scenarios. In the third and fourth scenarios, the frequency deteriorated from very high and high to low. In the fifth scenario, the frequency was reduced but not the same amount as in the first, second, third and fourth scenarios. However, the barrier still shows an acceptable efficiency eliminating the rockfall frequency (since it is effectively decreased the frequency rate).

16.9.4.2 Rockfall Height

Figure 16.17 demonstrates the rockfall height in each scenario without and with the barrier. The height raster was classified into five classes ranging from very low to very high based on natural breaks classification algorithm. The rockfall height varies from scenario to another due to various mechanical parameters. The highest rockfall height was with the highest values of mechanical parameters (fifth scenario). In general, the highest bouncing height was observed over

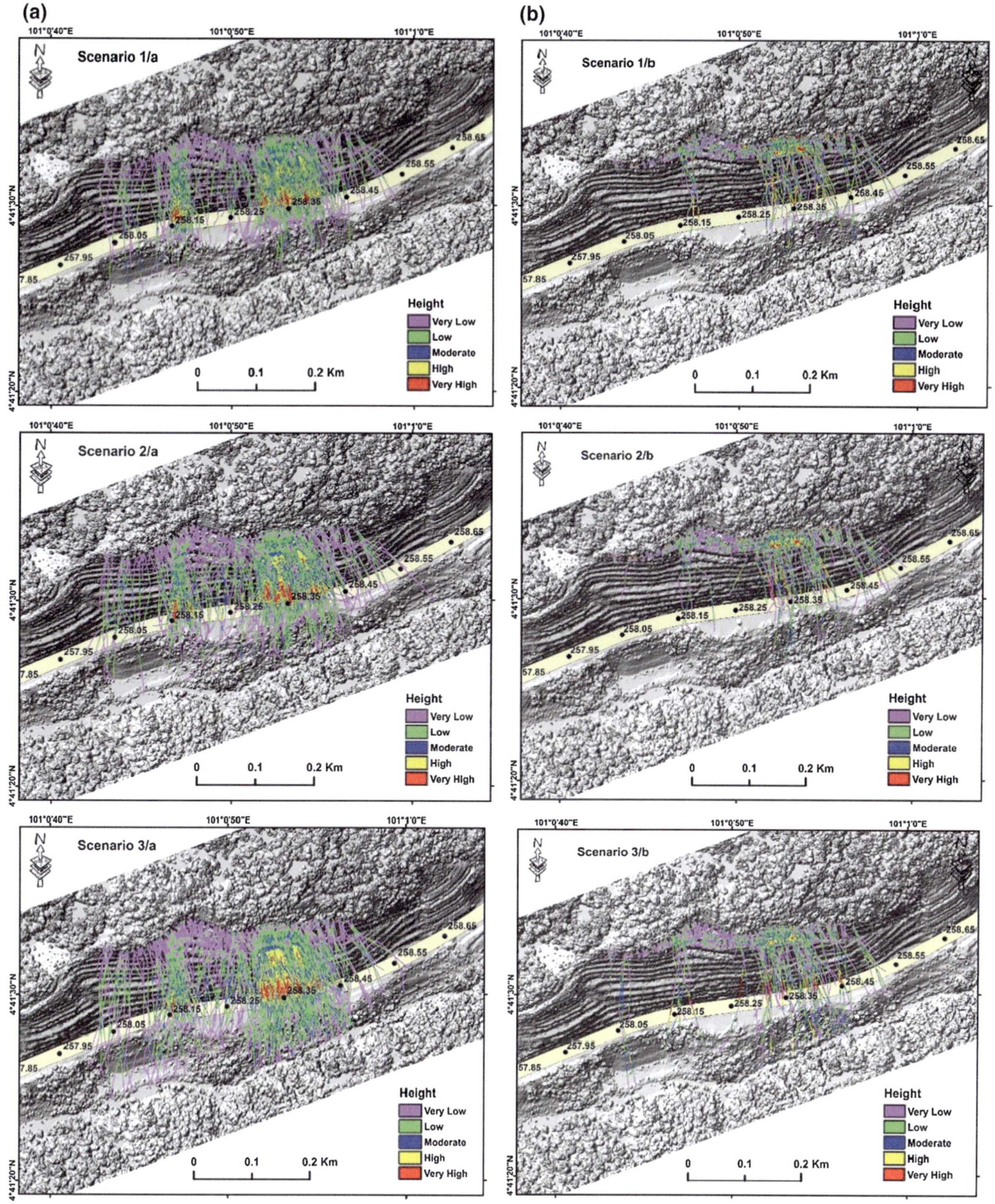

Fig. 16.17 Rockfall height in each scenario; **a** without; and **b** with barrier

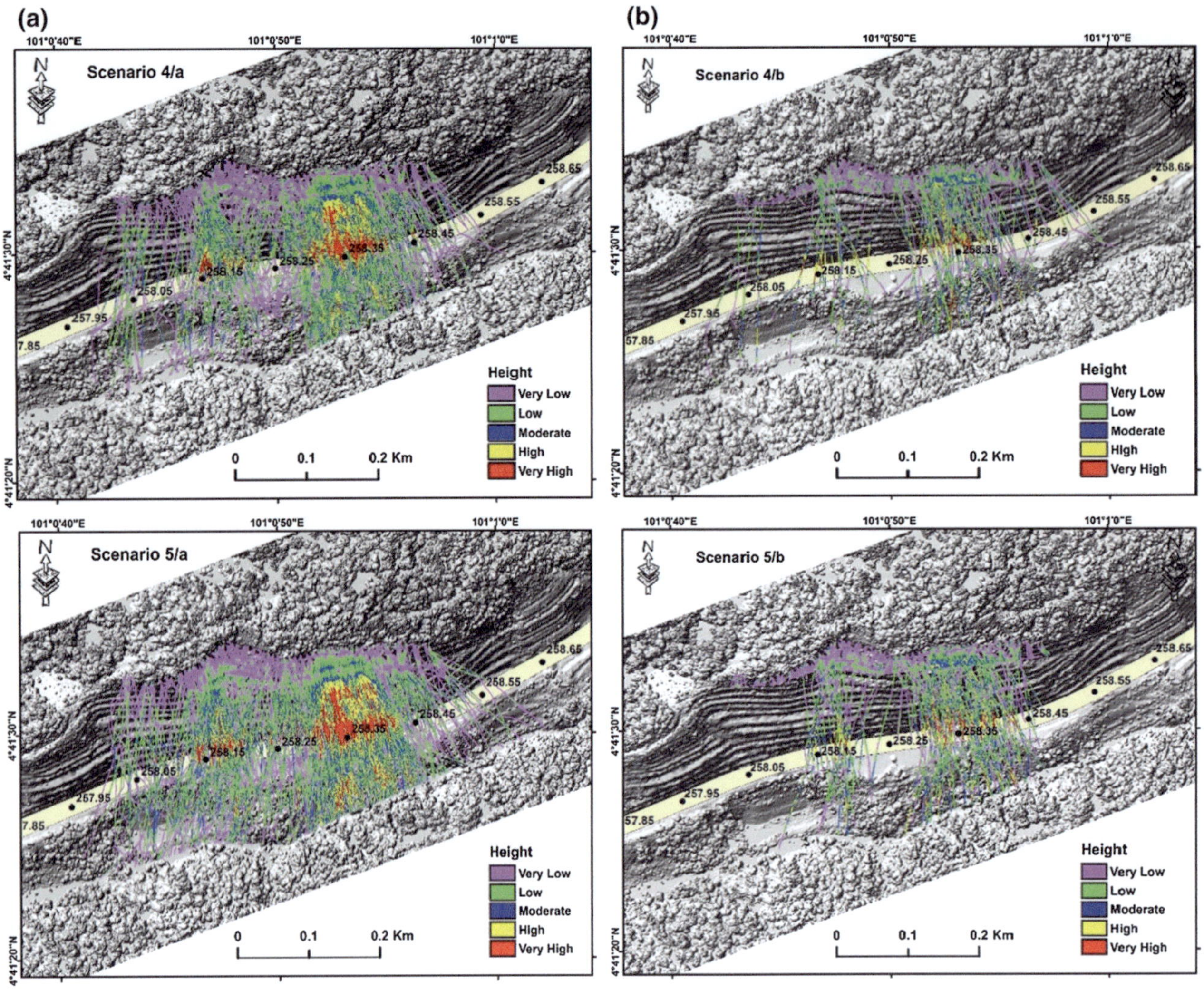

Fig. 16.17 (continued)

the expressway and within a distance before arriving at the expressway over the slope surface. However, this distance is different from scenario to another. The barrier efficiency eliminating rockfall hazard is clear in this figure. The barrier assists stopping most of the rockfall trajectories that have the highest height which support the suggested location of barrier in this research. This is because these trajectories had the lowest height before the barrier and the highest height after the barrier (over the expressway and within a distance before arriving at the expressway).

Figure 16.18 demonstrates the probability of rockfall height in five scenarios. Rockfall height is varied from scenario to another due to the difference in mechanical parameters. The maximum height was observed in fifth scenario (45 m) where the highest values of mechanical parameters are. However, most of the rockfall trajectories heights were less than 5 m in all scenarios that support the suggested barrier height in the current study.

16.9.4.3 Rockfall Velocity

Figure 16.19 illustrates rockfall velocity associated with rockfall trajectory without and with the barrier. Velocity raster was also classified into five classes ranging from very low to very high based on natural breaks classification algorithm. The highest velocity in all scenarios was observed over the expressway and within a distance before the expressway. The areas where the rockfall trajectories have the highest velocity were almost same in all scenarios.

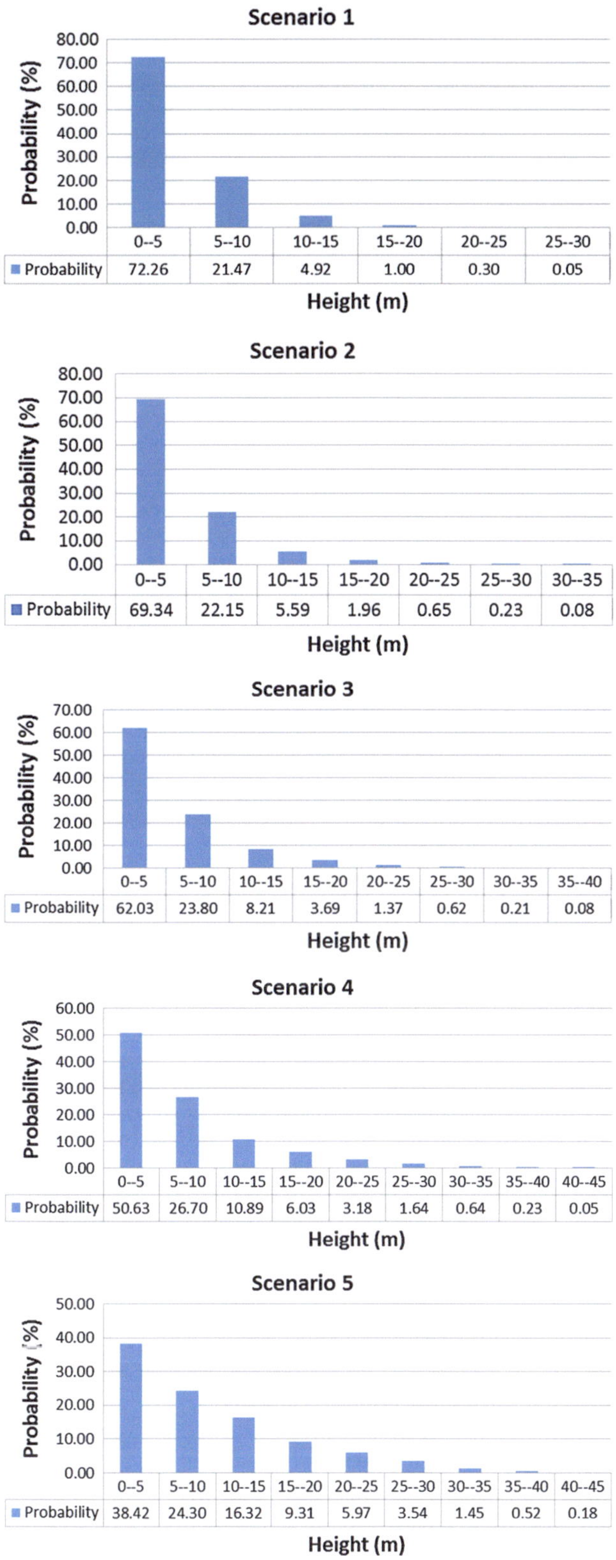

Fig. 16.18 Rockfall height probability in five scenarios

The maximum velocity values over the expressway were recorded from station 258.15 to 258.18 and from 258.30 to 258.40. The figure also illustrates the barrier efficiency eliminating rockfall hazard because of the barrier that stopped the rockfall trajectories that have the highest velocity.

Figure 16.20 shows the probability of rockfall velocity in five scenarios. The rockfall velocity was ranging from 0 to 40 m/s in the first and second scenarios, from 0 to 45 m/s in third and fourth scenario, and from 0 to 50 in the fifth scenario. However, the highest probability was recorded with a velocity range of 5–10 m/s in first, second, third and fourth scenarios and with a velocity range of 10–15 m/s in the fifth scenario.

16.9.4.4 Rockfall Energy

Figure 16.21 shows the rockfall energy in five scenarios without and with the barrier. The rockfall energy raster was also classified into five classes ranging from very low to very high based on natural breaks classification algorithm. The rockfall energy relies on rock velocity and mass. Because the rock mass is constant, rockfall energy is varied based on the difference in rockfall velocity. Therefore, the areas where the rockfall trajectories have the highest energy are almost same the areas where the rockfall trajectories have the highest velocity. Consequently, the highest rockfall energy was observed over the expressway from station 258.15 to 258.18 and from 258.30 to 258.40.

Figure 16.22 demonstrates the probability of rockfall kinetic energy in each scenario. The probability varies in each scenario due to the difference in the values of mechanical parameters. The highest kinetic energy is observed in the fifth scenario, whereas the lowest kinetic energy is recorded in the first scenario. The highest value in the first and second scenarios is 200 kj and in the third and fourth scenarios is 250 kj, whereas in the fifth scenario is 300 kj. However, in the first, second, third and fourth scenarios the highest probability of kinetic energy is between 0 and 10 kj, whereas in the fifth scenario it is between 10 and 50 kj.

16.9.5 Spatial Modelling-Based Rockfall Hazard Map

16.9.5.1 AHP

Rockfall characteristics (energy, frequency and height) are considered in the implementation of AHP. The values of intensity of importance are varied for the same criteria based on the difference in expert's opinion. Therefore, geometric

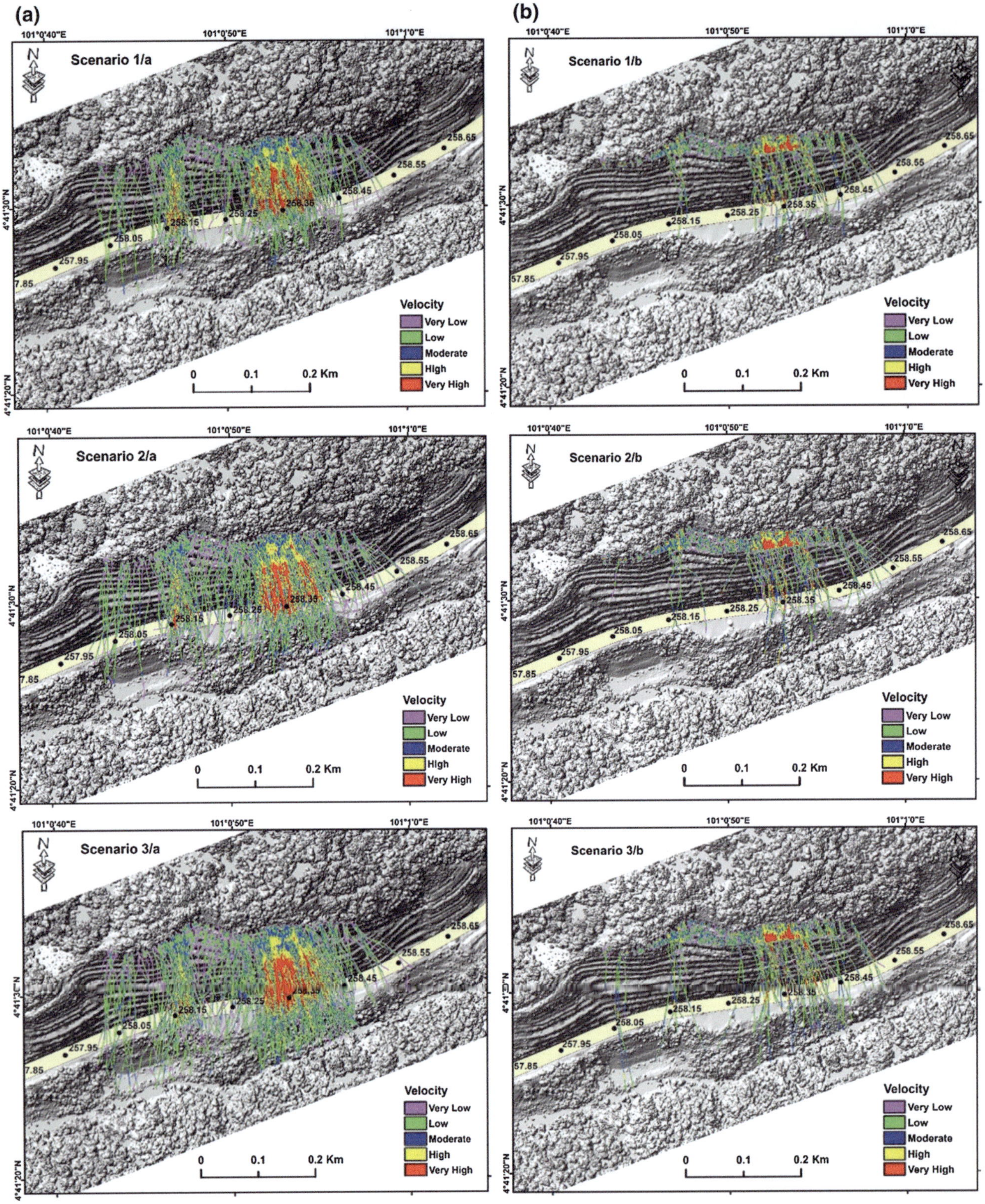

Fig. 16.19 Rockfall velocity **a** without; and **b** with barrier

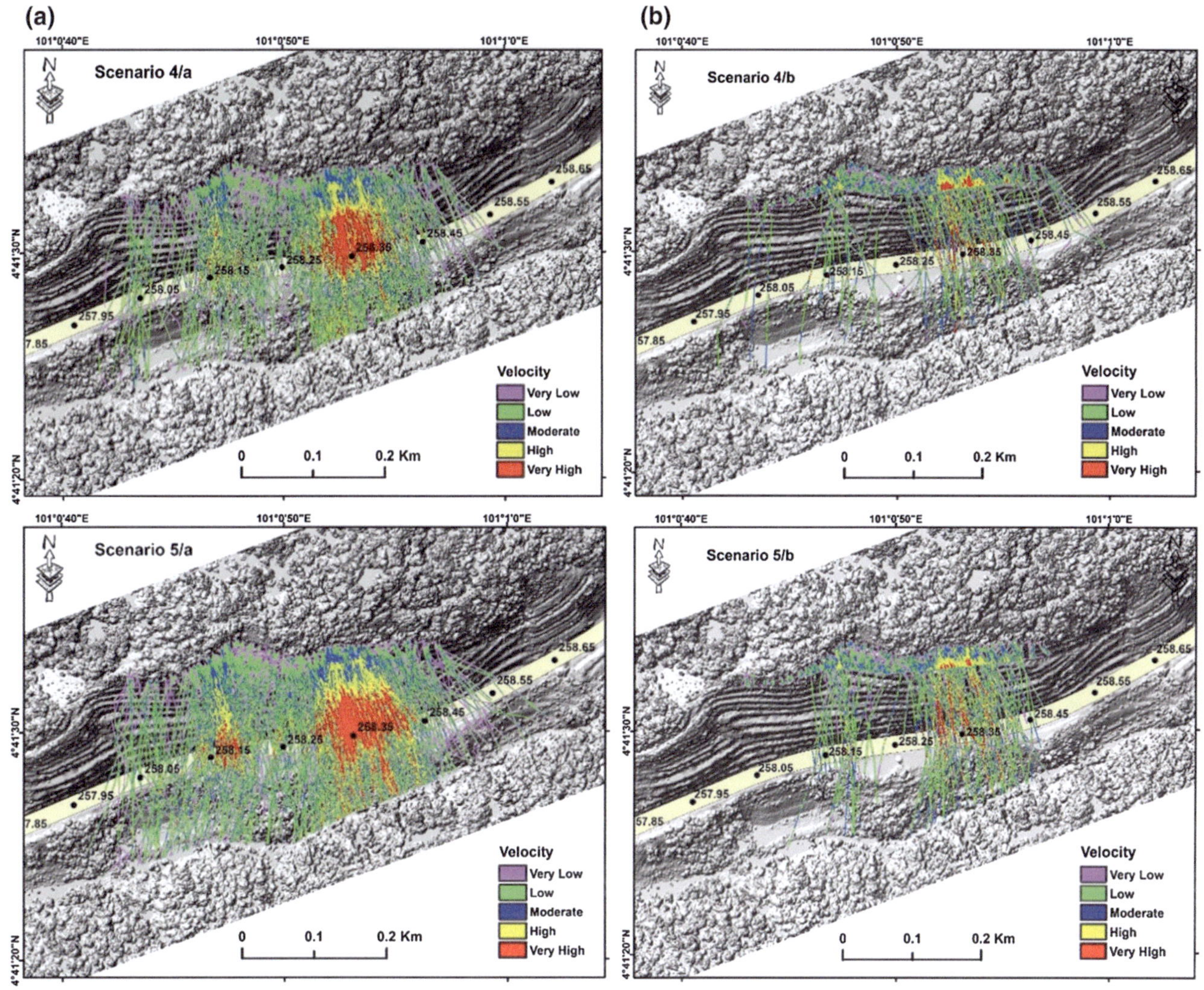

Fig. 16.19 (continued)

mean was applied to obtain the appropriate value of importance intensity for each criterion. After rounding the result to an integer, a judgment or pairwise comparisons matrix was obtained as shown in Table 16.8.

The principal eigenvalue and the corresponding normalized right eigenvector of the comparison matrix give relative importance of various criteria being compared. After normalizing process, the result matrix was as shown in Table 16.9.

After the matrix was normalized, the weight for each rockfall characteristics was derived as shown in Table 16.10, and the other values of AHP implementation were as shown in Table 16.11.

16.9.5.2 Rockfall Hazard Map

After the weights of rockfall characteristics were derived, spatial modelling was conducted based on rockfall energy, frequency and height. The result of the implementation of

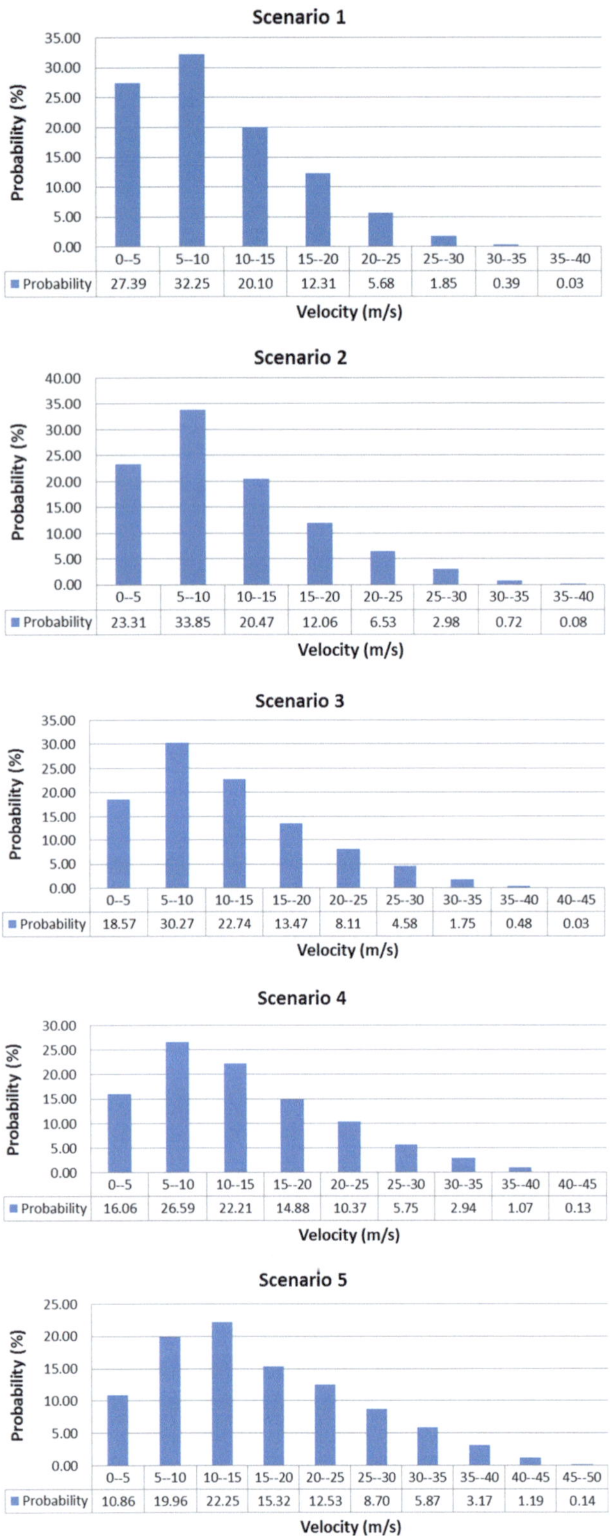

Fig. 16.20 Rockfall velocity probability in five scenarios

spatial modelling was rockfall hazard map of the study area in different scenarios. Figure 3.23 shows the rockfall hazard in multi-scenario without and with the barrier. The map was also classified into five classes ranging from very low to very high based on natural breaks classification algorithm. The rockfall hazard varied over the study area and also varied in each scenario for the same area. This is because of the difference in mechanical parameters (Rn, Rt and friction angle) from one scenario to another. The highest rockfall hazard was observed from station 258.15 to 258.18 and from 258.30 to 258.40 in the first, second, third and fourth scenarios and from 258.15 to 258.20 and 258.30 to 258.40 in the fifth scenario. Barrier efficiency eliminating rockfall hazard is also clear in Fig. 16.23. However, this efficiency was varied from one scenario to another due to the difference in mechanical parameters that affect rockfall behaviour in each scenario. In the first scenario, the barrier allowed to reduce the hazard rate from very high to low at station 258.15 and station 258.35. It also permitted to the area between stations 258.05 and 258.15 to be completely safe from the rockfall hazard after it was having a percentage of rockfall hazard. In the second scenario, the barrier allowed the area between stations 258.15 and 258.25 to be almost safe from rockfall hazard. It also assisted to reduce the rockfall hazard rate from very high to low for the area between stations 258.30 and 258.35. In the third scenario, the barrier aided the area between stations 257.95 and 258.05 to be completely safe and the area between stations 258.05 and 258.25 to be almost safe from the rockfall hazard. It also assisted to reduce the rate of rockfall hazard from very high to low at the area between stations 258.30 and 258.35. In the fourth scenario, the barrier effectively protected the area between stations 257.95 and 258.05. It also assisted to reduce the rockfall hazard from very high to low at station 258.15 and between stations 258.30 and 258.40. In the fifth scenario, the barrier effectively saved the area between stations 257.95 and 258.05. It also assisted to reduce the rate of the rockfall hazard from very high to low in the areas from stations 258.15 to 258.20 and from 258.30 to 258.42.

Figure 16.24 demonstrates rockfall hazard percentage along the expressway without and with the barrier. The highest hazard percentage was recorded between stations 258.35 and 258.45 and then from 258.35 to 258.25 and 258.25 to 258.15, respectively, in all scenarios. However, the rockfall hazard was varied from one scenario to another. For instance, the area between station 257.95 and 258.05 was free of hazard in the first scenario, whereas it

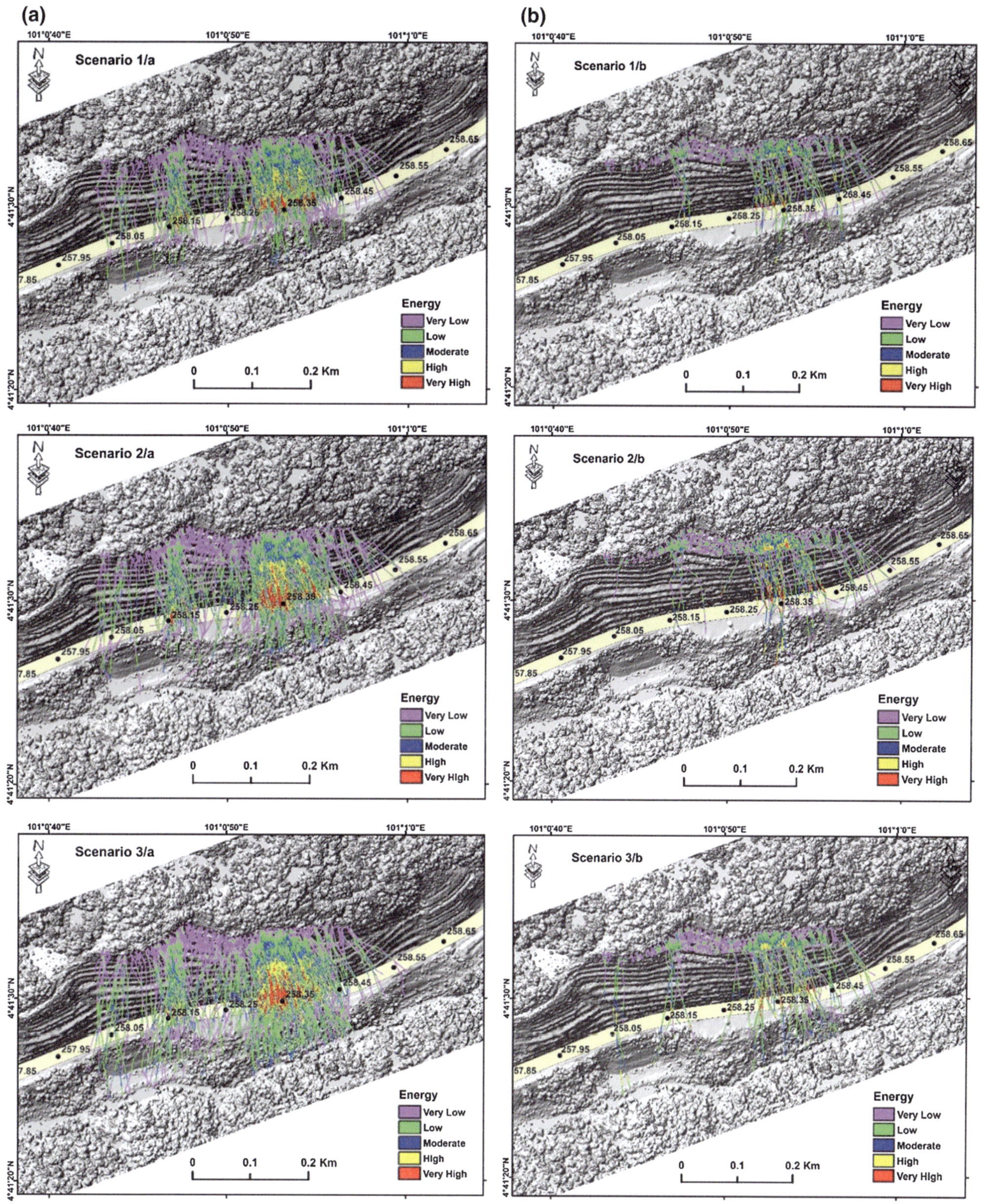

Fig. 16.21 Rockfall energy in five scenarios **a** without; and **b** with barrier

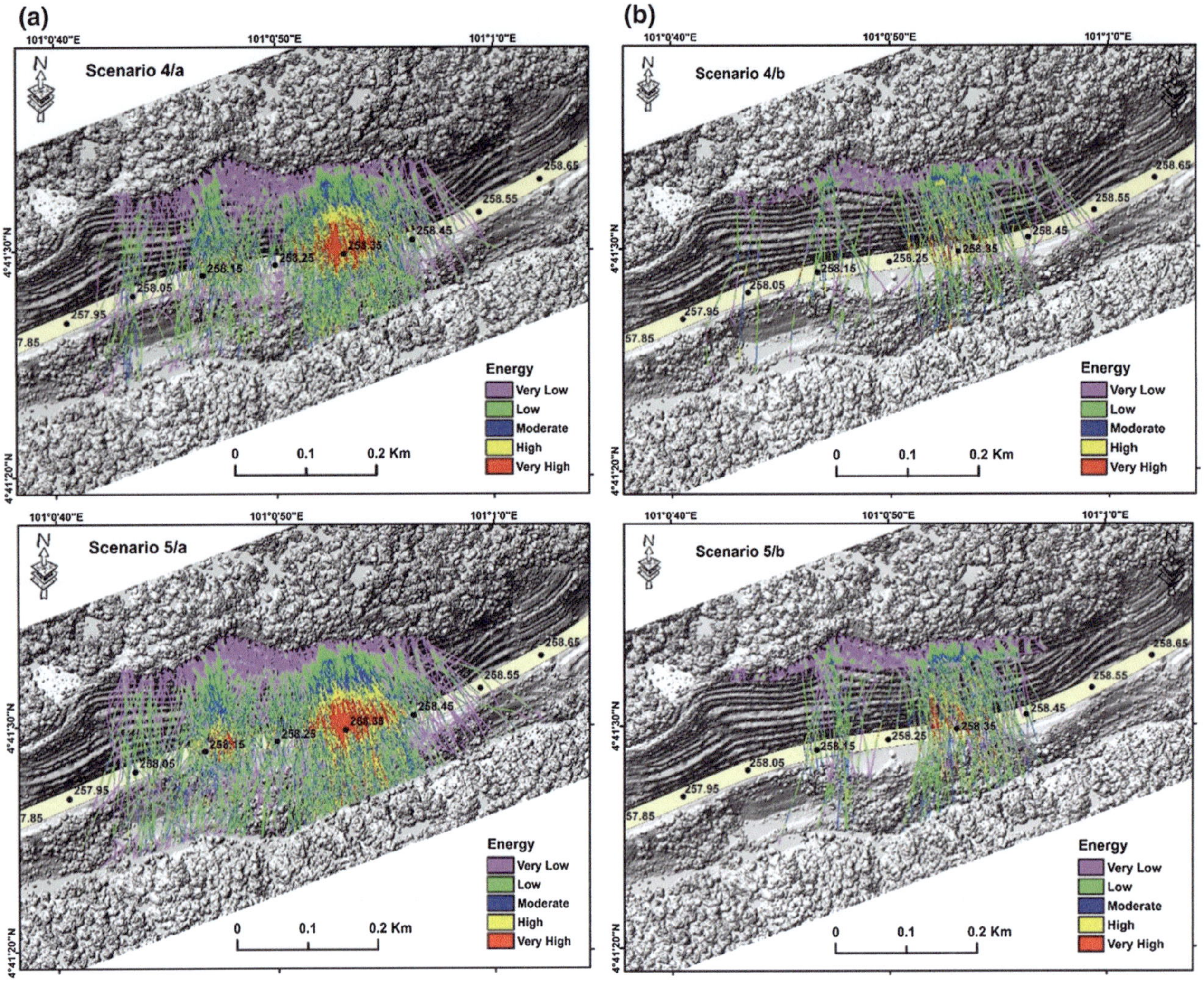

Fig. 16.21 (continued)

was having a percentage of hazard in the others scenarios. Also, the area between stations 258.55 and 258.65 was free of hazard in first, second, third and fourth scenarios while it was having a percentage of hazard in the fifth scenario. The barrier efficiency remedial rockfall hazard is also clear in this figure. The barrier efficiency was varied from one scenario to another, and the lowest efficiency was recorded in the fifth scenario due to the highest values of mechanical parameters. However, the barrier efficiency eliminating rockfall hazard is still acceptable. For example, in the first scenario, the area between station 258.05 and 258.15 was having a percentage of hazard, whereas in the simulation with suggested barrier it was free of rockfall hazard. In the first scenario, the barrier reduced the rockfall hazard percentage from 36.88 to 6.80% over the area between stations 258.35 and 258.45 and from 28.26 to 3.11% over the area between stations 258.25 and 258.35 and from 25.01 to 0.84% over the area between stations 258.15 and 258.25. In the second scenario, the barrier assisted to safe the areas between stations 257.95–258.05 and 258.05–258.15, from the rockfall hazard. It also reduced the rockfall hazard percentage from 37.26 to 5.83% over the area between stations 258.35 and 258.45 and from 28.98 to 3.43% over the area between stations 258.25 and 258.35 and from 24.64 to 1.32% over the areas between stations 258.15 and 258.25. In the third scenario, the barrier assisted to protect the area between stations 257.95 and 258.05 from the rockfall hazard. It also reduced the rockfall

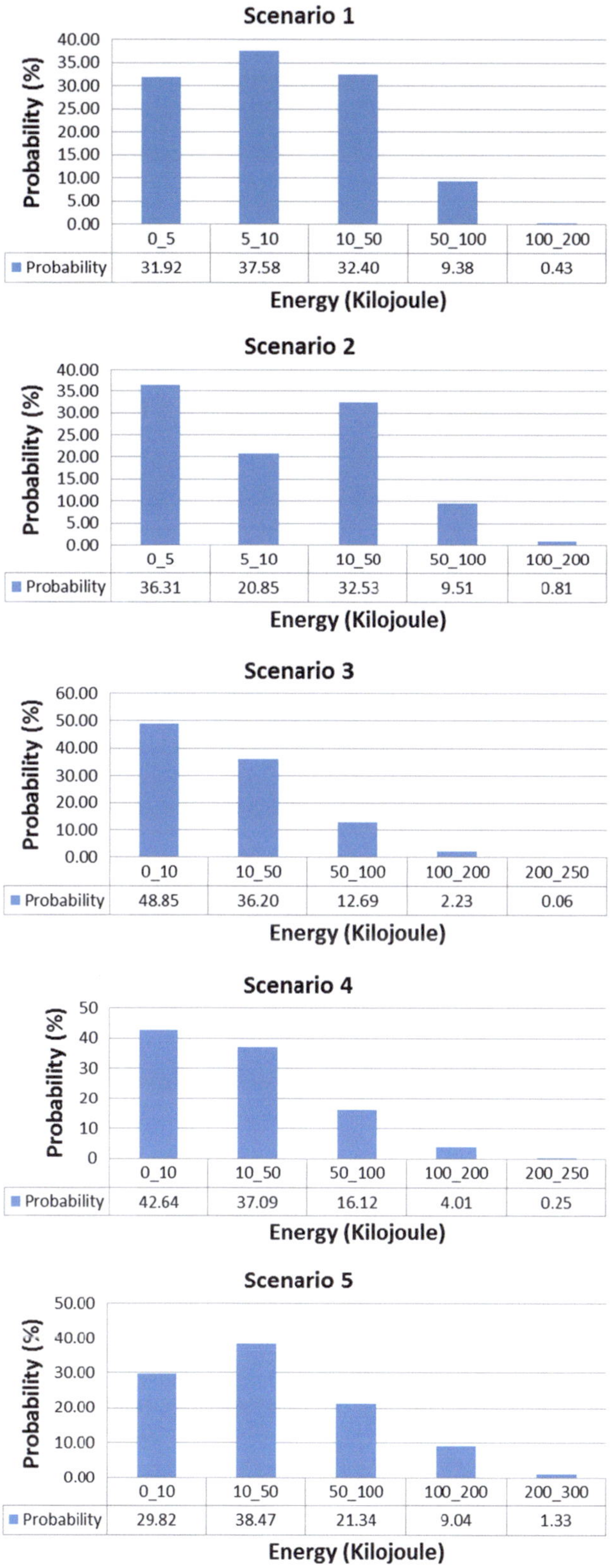

Fig. 16.22 Probability of rockfall kinetic energy in five scenarios

Table 16.8 Pairwise comparisons matrix among objectives/alternatives

Criteria/factors	Energy	Frequency	Height
Energy	1.00	3.00	5.00
Frequency	0.33	1.00	3.00
Height	0.20	0.33	1.00
Summation	1.53	4.33	9.00

Table 16.9 Normalized matrix

Criteria/factors	Energy	Frequency	Height
Energy	0.6522	0.6923	0.5556
Frequency	0.2174	0.2308	0.3333
Height	0.1304	0.0769	0.1111
Summation	1.0000	1.0000	1.0000

Table 16.10 Rockfall characteristics weight

Rockfall characteristic	Weight
Energy	0.6333
Frequency	0.2605
Height	0.1062

Table 16.11 Other values of AHP implementation

Consistency Ratio (CR)	0.03
Consistency Index (CI)	0.019357
Randomness Index (RI)	0.58

hazard percentage from 38.69 to 7.01% over the area between stations 258.35 and 258.45 and from 28.53 to 4.73% over the area between stations 258.25 and 258.35 and from 20.58 to 2.05% over the area between stations 258.15 and 258.25. In the fourth scenario, the barrier assisted to minimize the percentage of rockfall hazard from 32.66 to 7.35% over the area between stations 258.35 and 258.45 and from 26.35 to 6.14% over the area between stations 258.25 and 258.35 and from 24.00 to 4.01% over the area between stations 258.15 and 258.25. In the fifth scenario, the barrier assisted to safe the area between stations 258.55 and 258.65. It also aided to minimize the rockfall hazard percentage from 35.21 to 7.81% over the area between stations 258.35 and 258.45 and from 23.08 to 5.58% over the area between stations 258.25 and 258.35 and from 22.64 to 4.40% over the area between stations 258.15 and 258.25.

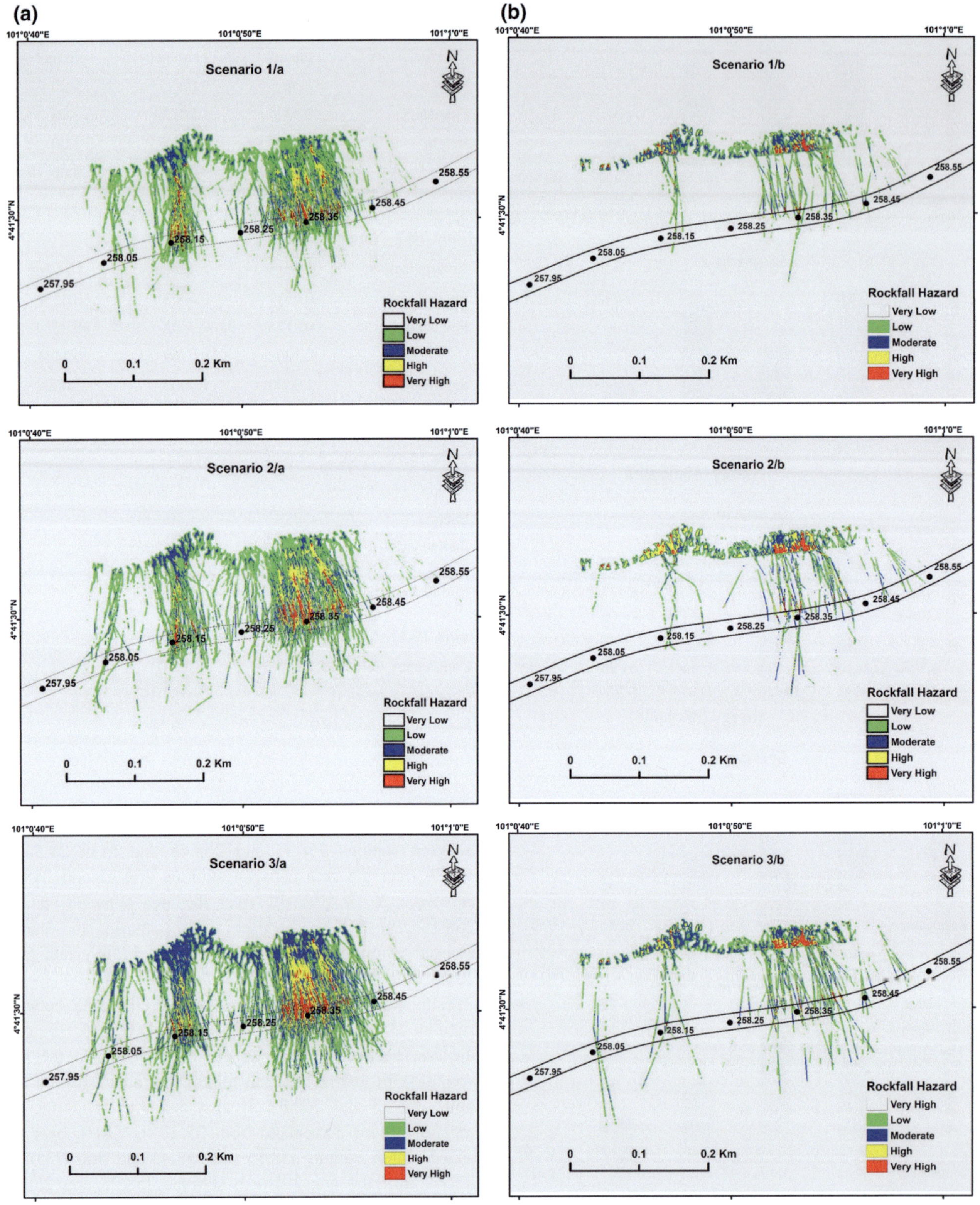

Fig. 16.23 Rockfall hazard map in five scenarios **a** without; and **b** with barrier

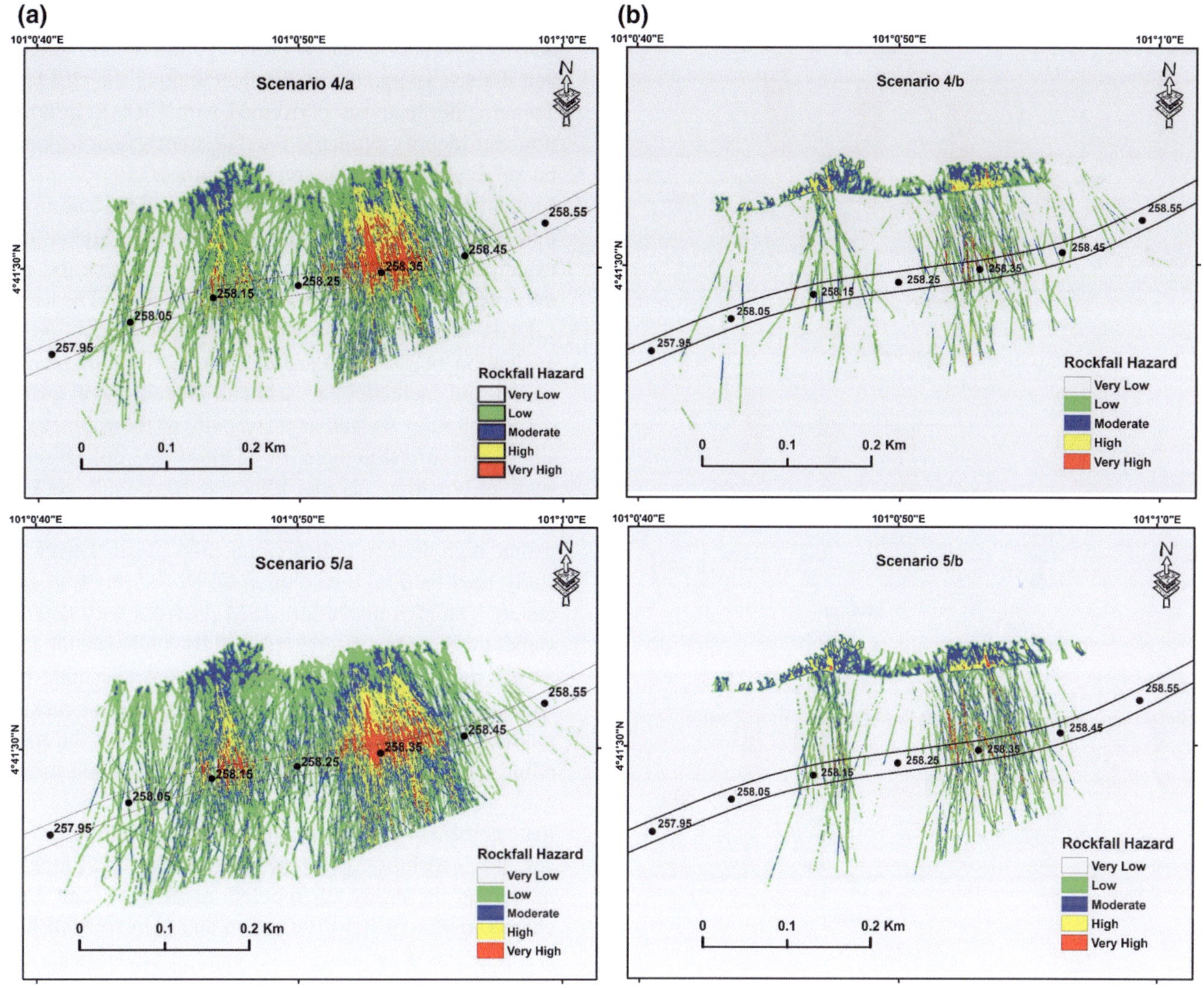

Fig. 16.23 (continued)

16.9.6 Validation

Historical data have been used in order to validate the final result of rockfall hazard along the expressway. The obtained results of rockfall hazard correspond to the historical records of rockfall incidents (Fig. 16.25), with only exception in one area. The areas where high hazard is observed are the same areas that recorded in historical data which are at station 258.15 and the area between stations 258.35 and 258.45. In the assessment, no high hazard observed in the area between stations 258.55 and 258.65, just small hazard percentage in the fifth scenario. This is because some mitigation measures have been taken in this area (Fig. 16.26).

16.10 Conclusion

1. The identification of rockfall source areas is one of the most significant elements in rockfall modelling. Rockfall source areas (seeder points) were identified based on multi-criteria method. After the DEM was generated, DEM derivates (slope, curvature and topographic contrast) were extracted. In addition to these criteria, the terrain type or LULC and the aerial photograph were considered in the implementation of multi-criteria method, and then the rockfall source areas were

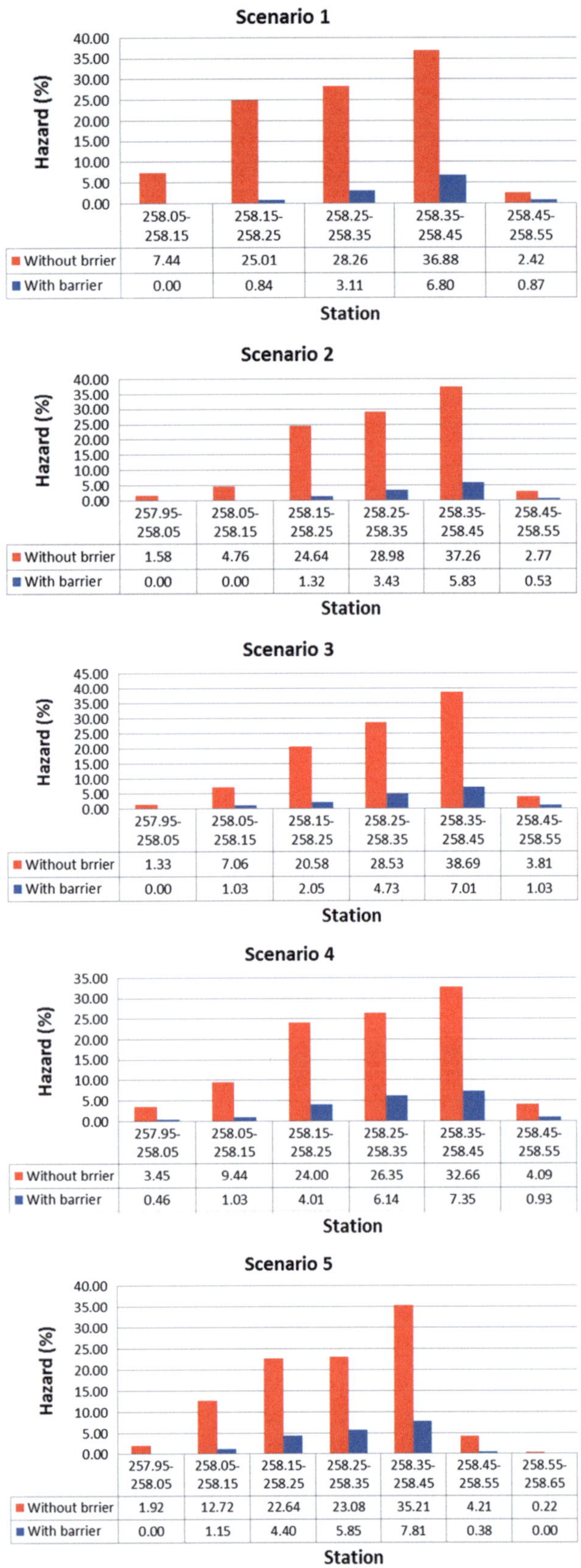

Fig. 16.24 Rockfall hazard percentage in five scenarios without and with barrier

identified along the study area. The result shows that the density of seeder points is different from area to another and the seeder points are distributed along the cliff face. Topographic analyses performed with LiDAR information can identify probable rockfall source regions based on slope angle and topographic contrast.

2. Modelling findings rely strongly on the DEM accuracy that affects the rockfall trajectories and the spatial distribution of rockfall frequency, height and energy, and thus risk and hazard assessment, in particular. The use of LiDAR illustrates the advantage of this process in the evaluation of rockfall hazards over the area of investigation and facilitates the accurate modelling of surface geomorphology as well as the capture of the geometry of significant infrastructures. In addition, the utilization of such information permits effective large-scale rockfall assessment. An airborne laser scanner (ALS) was used to gather high-density points cloud (3–4 pts./m^2) over the study area then high-resolution DEM (0.5 m) was generated. The high-resolution DEM provides very detailed slope topography that gives a fully understanding of terrain morphology. Thus, allow for more accurate and realistic rockfall modelling. The ArcGIS-based rockfall model provides powerful and unique tools for the modelling and analysis of 3D rockfall processes and hazard evaluation. The powerful functionality of GIS in data management, visualization and spatial modelling advances the capability of rockfall analysis. The rockfall modelling in terms of rockfall trajectories and their characteristics spatial distribution and then rockfall hazard map was done using a 3D rockfall model integrated into GIS environment. A kinematic algorithm based on lumped-mass approach was applied to derive rockfall trajectories and the velocity associated with them. The algorithm permits to simulate the rockfall motion in different modes (flying/bouncing, sliding/rolling and final deposition. The rockfall trajectories were derived in multi-scenario based on a range of mechanical parameters values (coefficients of restitution (Rn and Rt) and friction angle) related to geological setting and terrain type of the study area. The result shows that the final deposition or stopping distance of rockfall trajectories was various from scenario to another for the same rockfall source areas. This is due to the difference in the mechanical parameters in each scenario. The longest stopping distance was recorded in the fifth scenario due to the highest values of the mechanical parameters. After rockfall trajectories and velocity associated were obtained, raster modelling was implemented to derive rockfall characteristics spatial distribution in terms of rockfall frequency, height and energy. Since rockfall characteristics are related to rockfall trajectories, rockfall characteristics also varied for each scenario. The highest

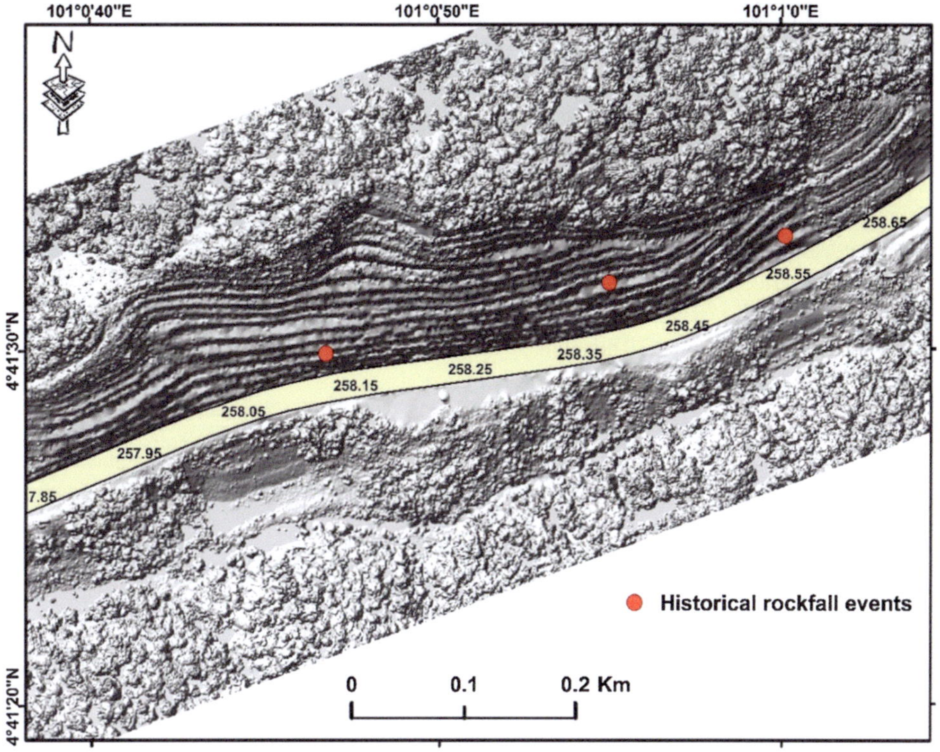

Fig. 16.25 Historical rockfall events along the study area

Fig. 16.26 Mitigation measures in the area between 258.55 and 258.65

values of rockfall characteristics along the expressway were illustrated in the result.

3. Spatial modelling was performed to derive rockfall hazard maps in multi-scenario. Expert's opinion was included in this step to obtain the proper weight of rockfall characteristics raster (energy, frequency and height) which were considered in deriving rockfall hazard map. After expert's opinion was collected, analytic

hierarchy process (AHP) was applied to derive appropriate weight for each characteristics raster. The weights were 0.6333, 0.2605 and 0.1062 for rockfall energy, frequency and height, respectively, which means the rockfall energy affects the rockfall hazard more than frequency and height. After the weights were derived, the spatial modelling was implemented considering rockfall energy, frequency and height and rockfall hazard map in multi-scenario was extracted. The percentage of rockfall hazard along the expressway was observed and demonstrated in five scenarios. The highest hazard percentage along the expressway was in the areas between stations 258.35 and 258.45, 258.25 and 258.35, and 258.15 and 258.25 descending. The result also shows the barrier efficiency remedial rockfall hazard along the expressway. The barrier efficiency was also varied from one scenario to another. However, the overall efficiency in all scenarios was acceptable, and the barrier eliminated the rockfall hazard effectively along the expressway.

4. A barrier location was suggested in order to mitigate rockfall hazard based on the result of rockfall characteristics. The barrier was suggested based on the lowest values of rockfall height and energy. The barrier was effectively stopping most of the rockfall trajectories. The barrier efficiency stopping the rockfall trajectories was 94.86, 93.75, 92.74, 89.63 and 84.86% in first, second, third, fourth and fifth scenarios, respectively, that supports the suggestion of barrier location in this study.
5. The result of this study suggests that the most crucial input parameters in rockfall modelling are the coefficients of restitution (Rn and Rt) and the friction angle. These parameters control the bouncing height of the falling block and the energy. The outcome of rockfall simulation demonstrates that when the values of the mechanical parameters (Rn, Rt and friction angle) are increased, the characteristics of rockfall trajectories (run-out distance, bouncing height, and energy) increase as well. Therefore, rockfall frequency, height, energy and, subsequently, the number of regions at risk are increased. The simulation of rockfall with a range of restitution coefficient and friction angle values is illustrated; this process comprehensively clarifies rockfall hazard and facilitates the identification of poor rockfall conditions. The procedure then supports the barrier design process and the barrier location to eliminate the rockfall hazard effect.
6. This research can aid in defining rockfall source areas and generates an extensive result related to rockfall hazard especially when the exact values of mechanical parameters are unknown. In addition, these findings can assist a barrier designer in developing barriers based on the worst rockfall hazard condition.

References

Ahmad, M., Umrao, R., Ansari, M., Singh, R., & Singh, T. (2013). Assessment of rockfall hazard along the road cut slopes of state highway-72, Maharashtra, India. *Geomaterials, 3*(1), 15–23.

Akin, M., Topal, T., & Akin, M. K. (2013). Evaluation of the rockfall potential of Kastamonu Castle Using 3-D analysis. In *Landslide science and practice* (pp. 335–340). Berlin: Springer.

Ansari, M., Ahmad, M., & Singh, T. (2014). Rockfall hazard analysis of Ellora Cave, Aurangabad, Maharashtra, India. *International Journal of Science and Research (IJSR), 3*(5), 427–431.

Assali, P., Grussenmeyer, P., Villemin, T., Pollet, N., & Viguier, F. (2014). Surveying and modeling of rock discontinuities by terrestrial laser scanning and photogrammetry: Semi-automatic approaches for linear outcrop inspection. *Journal of Structural Geology, 66,* 102–114.

Asteriou, P., Saroglou, H., & Tsiambaos, G. (2012). Geotechnical and kinematic parameters affecting the coefficients of restitution for rock fall analysis. *International Journal of Rock Mechanics and Mining Sciences, 54,* 103–113.

Ayalew, L., & Yamagishi, H. (2005). The application of GIS-based logistic regression for landslide susceptibility mapping in the Kakuda-Yahiko Mountains, Central Japan. *Geomorphology, 65*(1), 15–31.

Azzoni, A., La Barbera, G., & Zaninetti, A. (1995). Analysis and prediction of rockfalls using a mathematical model. International *Journal of Rock Mechanics and Mining Sciences & Geomechanics Abstracts*, 709.

Chai, B., Tang, Z., Zhang, A., Du, J., Su, H., & Yi, H. (2015). An uncertainty method for probabilistic analysis of buildings impacted by rockfall in a Limestone Quarry in Fengshan, Southwestern China. *Rock Mechanics and Rock Engineering, 48*(5), 1981–1996.

Chen, G., Zheng, L., Zhang, Y., & Wu, J. (2013). Numerical simulation in rockfall analysis: a close comparison of 2-D and 3-D DDA. *Rock Mechanics and Rock Engineering, 46*(3), 527–541.

Evans, I. S. (1977). The selection of class intervals. *Transactions of the Institute of British Geographers, 2*(1), 98–124.

Ferrari, F., Giani, G. P., & Apuani, T. (2013). Why can rockfall normal restitution coefficient be higher than one? *Rendiconti Online Società Geologica ItalianaSocietà Geologica Italiana*, 122.

Heckmann, T., & Schwanghart, W. (2013). Geomorphic coupling and sediment connectivity in an alpine catchment—Exploring sediment cascades using graph theory. *Geomorphology, 182,* 89–103.

Jaboyedoff, M., Dudt, J., & Labiouse, V. (2005). An attempt to refine rockfall hazard zoning based on the kinetic energy, frequency and fragmentation degree. *Natural Hazards and Earth System Science, 5*(5), 621–632.

Kenner, R., Bühler, Y., Delaloye, R., Ginzler, C., & Phillips, M. (2014). Monitoring of high alpine mass movements combining laser scanning with digital airborne photogrammetry. *Geomorphology, 206,* 492–504.

Keskin, İ. (2013). Evaluation of rock falls in an urban area: The case of Boğaziçi (Erzincan/Turkey). *Environmental Earth Sciences, 70*(4), 1619–1628.

Ku, C. (2012). Assessing rockfall hazards using a three-dimensional numerical model based on high resolution DEM. In *The Twenty-second International Offshore and Polar Engineering Conference* (p. 790). International Society of Offshore and Polar Engineers.

Lan, H., Derek Martin, C., & Lim, C. (2007). RockFall analyst: A GIS extension for three-dimensional and spatially distributed rockfall hazard modeling. *Computers & Geosciences, 33*(2), 262–279.

Lan, H., Martin, C. D., Zhou, C., & Lim, C. H. (2010). Rockfall hazard analysis using LiDAR and spatial modeling. *Geomorphology, 118* (1), 213–223.

Leine, R., Schweizer, A., Christen, M., Glover, J., Bartelt, P., & Gerber, W. (2013). Simulation of rockfall trajectories with consideration of rock shape. *Multibody System Dynamics, 32*(2), 1–31.

Loye, A., Jaboyedoff, M., & Pedrazzini, A. (2009). Identification of potential rockfall source areas at a regional scale using a DEM-based geomorphometric analysis. *Natural Hazards and Earth System Science, 9*(5), 1643–1653.

Ma, G., Matsuyama, H., Nishiyama, S., & Ohnishi, Y. (2011). Practical studies on rockfall simulation by DDA. *Journal of Rock Mechanics and Geotechnical Engineering, 3*(1), 57–63.

Macciotta, R., Martin, C. D., & Cruden, D. M. (2014). Probabilistic estimation of rockfall height and kinetic energy based on a three-dimensional trajectory model and Monte Carlo simulation. *Landslides, 12*(4), 1–16.

MacEachren, A. M. (1994). Some truth with maps: A primer on symbolization and design. *American Association of Geographers.*

Pradhan, B., Abokharima, M. H., Jebur, M. N., & Tehrany, M. S. (2014). Land subsidence susceptibility mapping at Kinta Valley (Malaysia) using the evidential belief function model in GIS. *Natural Hazards, 73*(2), 1019–1042.

Rammer, W., Brauner, M., Dorren, L., Berger, F., & Lexer, M. (2010). Evaluation of a 3-D rockfall module within a forest patch model. *Natural Hazards and Earth System Sciences, 10*(4), 699–711.

Saaty, T. (1980). *The analytic hierarchy process*. New York: McGraw-Hill.

Sabatakakis, N., Depountis, N., & Vagenas, N. (2015). Evaluation of rockfall restitution coefficients. *Engineering Geology for Society and Territory, 2*, 2023–2026.

Samodra, G., Chen, G., Sartohadi, J., Hadmoko, D., & Kasama, K. (2014). Automated landform classification in a rockfall-prone area, Gunung Kelir, Java. *Earth Surface Dynamics, 2*(1), 339–348.

Singh, P., Wasnik, A., Kainthola, A., Sazid, M., & Singh, T. (2013). The stability of road cut cliff face along SH-121: A case study. *Natural Hazards, 68*(2), 497–507.

Siqiao, Y., Hongmei, T., Hongkai, C., & Hui, Z. (2010, August). Stability evaluation of rockfall based on AHP-Fuzzy method. In *Seventh International Conference on Fuzzy Systems and Knowledge Discovery (FSKD)* (Vol. 3, pp. 1369-1373). IEEE.

Stephenne, N., Frippiat, C., Veschkens, M., Salmon, M., & Pacyna, D. (2014). Use of a Lidar high resolution digital elevation model for risk stability analysis. *EARSeL eProceedings, 13*(S1), 24–29.

Topal, T., Akin, M., & Ozden, U. A. (2007). Assessment of rockfall hazard around Afyon Castle, Turkey. *Environmental Geology, 53* (1), 191–200.

Vaidya, O. S., & Kumar, S. (2006). Analytic hierarchy process: An overview of applications. *European Journal of Operational Research, 169*(1), 1–29.

Wang, X., Frattini, P., Crosta, G., Zhang, L., Agliardi, F., Lari, S., et al. (2014). Uncertainty assessment in quantitative rockfall risk assessment. *Landslides, 11*(4), 711–722.

Wyllie, D. C. (2014). Calibration of rock fall modeling parameters. *International Journal of Rock Mechanics and Mining Sciences, 67*, 170–180.

Yusof, N. M., Pradhan, B., Shafri, H. Z. M., Jebur, M. N., & Yusoff, Z. (2015). Spatial landslide hazard assessment along the Jelapang Corridor of the North-South Expressway in Malaysia using high resolution airborne LiDAR data. *Arabian Journal of Geosciences, 8* (11), 1–12.

MIX
Papier aus verantwortungsvollen Quellen
Paper from responsible sources
FSC® C105338

If you have any concerns about our products,
you can contact us on
ProductSafety@springernature.com

In case Publisher is established outside the EU,
the EU authorized representative is:
Springer Nature Customer Service Center GmbH
Europaplatz 3, 69115 Heidelberg, Germany

Printed by Libri Plureos GmbH
in Hamburg, Germany